SOLUTIONS MANUAL TO ACCOMPANY STATISTICS AND PROBABILITY WITH APPLICATIONS FOR ENGINEERS AND SCIENTISTS

SOLUTIONS MANUAL TO ACCOMPANY STATISTICS AND PROBABILITY WITH APPLICATIONS FOR ENGINEERS AND SCIENTISTS

BHISHAM C. GUPTA

Professor of Statistics
University of Southern Maine
Portland, ME

IRWIN GUTTMAN

Professor Emeritus of Statistics
SUNY at Buffalo and
University of Toronto, Canada

Published by John Wiley & Sons, Inc., Hoboken, New Jersey
Published simultaneously in Canada

For general information on our other products and services or for technical support, please contact our Customer Care Department within the United States at (800) 762-2974, outside the United States at (317) 572-3993 or fax (317) 572-4002.

Wiley also publishes its books in a variety of electronic formats. Some content that appears in print may not be available in electronic formats. For more information about Wiley products, visit our web site at www.wiley.com.

Library of Congress Cataloging-in-Publication Data is available.

ISBN 9781118789698

10 9 8 7 6 5 4 3 2 1

CONTENTS

Chapter 19   345

2

DESCRIBING DATA GRAPHICALLY AND NUMERICALLY

PRACTICE PROBLEMS FOR SECTIONS 2.1 AND 2.2

1. See Section 2.1.2.
3. (a) All students of the graduation class.
   (b) All students in that professor's class.
   (c) GPA.
5. (a) Ratio (b) Ratio (c) Ratio (d) Nominal (e) Ratio (f) Ordinal (g) Ratio (h) Ratio (i) Interval (j) Ratio (k) Ratio (*l*) Ratio (*m*) Ordinal (*n*) Nominal.

PRACTICE PROBLEMS FOR SECTION 2.3

1. (a) Frequency Distribution Table:

Categories	Frequency	Percentages	Cumulative Frequency
1	12	24	12
2	12	24	24
3	8	16	32
4	11	22	43
5	7	14	50
Total	50	100	

   (b) See column 3 of the frequency distribution table in part (a).
   (c) Percentage of the customer in this sample survey was very satisfied or fairly satisfied is

$$24\% + 24\% = 48\%$$

Solutions Manual to Accompany Statistics and Probability with Applications for Engineers and Scientists, Bhisham C. Gupta and Irwin Guttman.
© 2014 John Wiley & Sons, Inc. Published 2014 by John Wiley & Sons, Inc.

3. (a) Frequency Distribution Table:

Categories	Frequency	Percentages	Cumulative Frequency
1	11	30.56	11
2	8	22.22	19
3	5	13.89	24
4	3	8.33	27
5	9	25	36
Total	36	100	

(b) See column 3 of the frequency distribution table in (a).

(c) Percentage of senior citizens who drive cars of category 1 or 3 is

$$30.56\% + 13.89\% = 44.45\%$$

5. Frequency Distribution Table:

Classes	Tally	Frequency	Relative Freq.	Percentages	Cumulative Freq.
[120–130)	///	3	3/30	10	3
[130–140)	/// //	5	5/30	16.67	8
[140–150)	///// ////	9	9/30	30	17
[150–160)	//	2	2/30	6.67	19
[160–170)	///// //	7	7/30	23.33	26
[170–180]	////	4	4/30	13.33	30
Total		30	1	100	

PRACTICE PROBLEMS FOR SECTION 2.4

1.

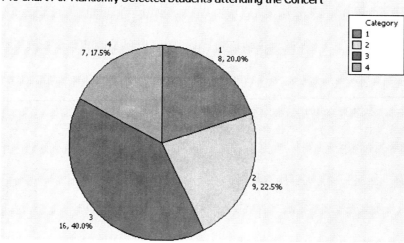

Pie Chart For Randomly Selected Students attending the Concert

3.

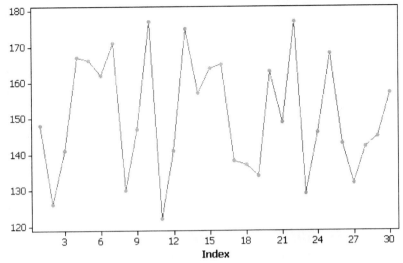

Line Graph Graduates Admitted Engineering Programs of a University

The line graph does not show any particular pattern. There are some dips and peaks which are occurring randomly.

5. Stem-and-leaf for the data in Problem 4 of Section 2.3
Leaf Unit = 1.0 (stems with increments 10)

```
15    1  555566667788999
(27)  2  011112234444555666677777888
3     3  000
```

Stem-and-leaf for the data in Problem 4 of Section 2.3
Leaf Unit = 1.0 (stems with increments 5)

```
15    1  555566667788999
(12)  2  011112234444
18    2  555666677777888
3     3  000
```

The stem-and-leaf diagram with increment 5 is more informative since the stems are not very large. Perhaps stems with increments 2 would be even better.

7. Using MINITAB The frequency distribution Table for the data in Problem 2.7

Tally for Discrete Variables: Classes

Classes	Count	Cumulative Count	Percent	Cumulative Percent
[12,15)	3	3	10.00	10.00
[15,18)	6	9	20.00	30.00
[18,21)	9	18	30.00	60.00
[21,24)	5	23	16.67	76.67
[24,27)	4	27	13.33	90.00
[27,30]	3	30	10.00	100.00
N=	30			

9. In this type of problem it is very helpful to first prepare a stem-and-leaf diagram for the given data and then prepare the frequency distribution table. This eliminates the chances of missing or counting any number more than once. The stem-

and-leaf diagram for these data is:

```
 11   2  00133567899
(12)  3  022355577889
  8   4  00011346
```

(a)

Class	Tally	Frequency or Count	Relative Frequency	Cumulative Frequency
[20–25)	/////	5	5/31	5
[25–30)	///// /	6	6/31	11
[30–35)	////	4	4/31	15
[35–40)	///// ///	8	8/31	23
[40–45)	///// //	7	7/31	30
[45–50)	/	1	1/31	31
Total		31	1	

(b) On 13 of 31 days 36 or more patients were treated.

11.

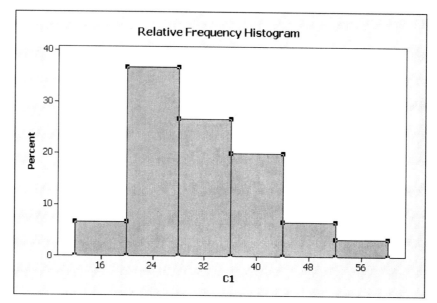

MINITAB printout of relative frequency histogram for the data in Problem 11

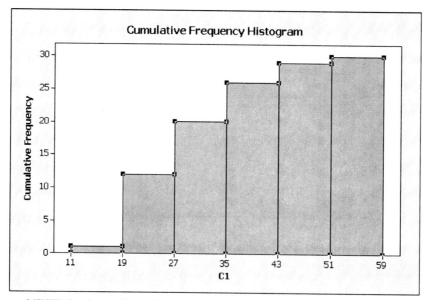

MINITAB printout of cumulative frequency histogram for the data in Problem 11

Ninety percent of the parts have life span between 20 and 48 months. Only one out of thirty parts has life span more than 50 months. Two out of thirty parts have life span less than 20 months.

13. (a) In this problem the maximum and minimum values of data points are 71.7 and 58.3 respectively. Thus, we have

$$\text{Range} = 71.7 - 58.3 = 13.4$$

Now we choose to have 7 classes $(1 + 3.3 \log 60 = 6.87)$, hence the class width is

$$\frac{13.4}{7} \cong 2$$

Thus the Frequency Distribution Table is as follows:

Class	Tally	Frequency	Relative Frequency	Cumulative Frequency
[58–60)	///// ///	8	8/60	8
[60–62)	///// ///// ////	14	14/60	22
[62–64)	///// ///// /	11	11/60	33
[64–66)	///// ////	9	9/60	42
[66–68)	///// ///	8	8/60	50
[68–70)	///// ////	9	9/60	59
[70–72]	/	1	1/60	60
Total		60	1	

(b)

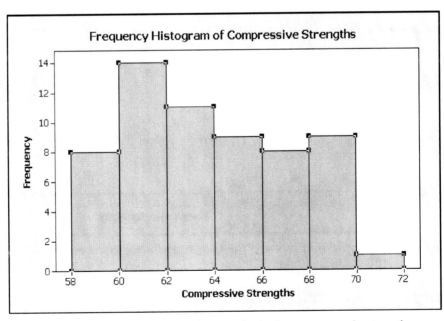

MINITAB printout of frequency histogram for the data on compressive strengths

(c)

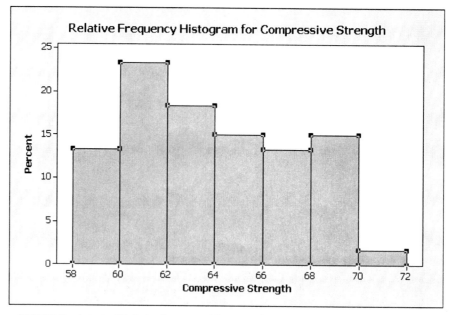

MINITAB printout of Relative frequency histogram for the data on compressive strengths

(d)

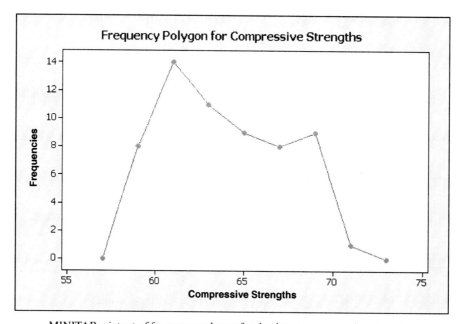

MINITAB printout of frequency polygon for the data on compressive strengths

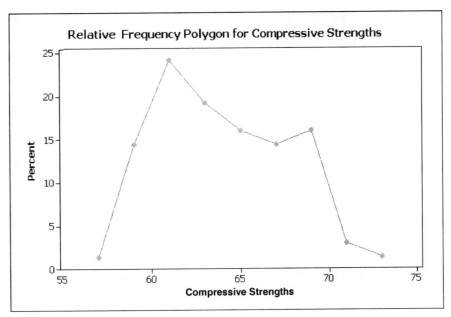

MINITAB printout of Relative frequency polygon for the data on compressive strengths

(e)

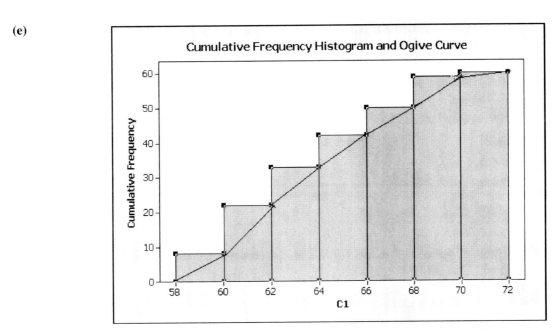

MINITAB printout of the cumulative frequency histogram and the Ogive
curve for the data on compressive strengths

15. (a) Using MINITAB the stem-and-leaf diagram for the data on consumption of electricity in kilowatt-hours (Leaf unit = 1):

4	20	6679
6	21	37
9	22	566
11	23	03
15	24	0177

15	25	00
13	26	0024458
6	27	8
5	28	06
3	29	0
2	30	5
1	31	0

(b) Median consumption of electricity in kilowatt-hours is 248.5 kilowatt-hours, maximum consumption is 310 kilowatt-hours, and minimum consumption is 206 kilowatt-hours.

PRACTICE PROBLEMS FOR SECTION 2.5

1. Using MINITAB, we have

   (a) Mean $= 120.02$, Median $= 120.10$, Mode 120.1.

   (b) Standard Deviation 1.84.

   (c) Since the mean median, and mode are almost equal the data is symmetric.

3. Using MINITAB, we have

   (a) Mean $= 22.356$, Median $= 23.00$

   (b) Standard Deviation $= 4.73$

   (c) $(\overline{X} - 2.5S, \overline{X} + 2.5S) = (10.531, 34.181)$. Thus, we can easily see that all the data points in the data set fall within 2.5 standard deviations of the mean.

5. Using MINITAB, we have

   (a) Mean $= 108.47$, Median $= 107.50$, and Mode $= 100$

   (b) Range $= 20$, Variance $= 50.10$, Standard Deviation $= 7.08$, and CV $= 6.53\%$.

7. John's GPA is the weighted mean of his grade points, that is

$$\text{GPA} = \frac{\sum_{i=1}^{6} w_i x_i}{\sum_{i=1}^{6} w_i} = \frac{5 \times 3.7 + 4 \times 4 + 4 \times 3.3 + 3 \times 4 + 3 \times 3.7 + 2 \times 4}{5 + 4 + 4 + 3 + 3 + 2} = \frac{78.8}{21} = 3.75238$$

9. Approximately 68% of the salaries should fall within one standard deviation of the mean, i.e., $(55,600 - 4,500, 55,600 + 4,500) = (51,100, 60,100)$.

   Approximately 95% of the salaries should fall within two standard deviations of the mean, i.e., $(55,600 - 9,000, 55,600 + 9,000) = (46,600, 64,600)$

11. Using MINITAB, we have

    (a) Mean $= 22.650$, Standard Deviation $= 1.461$

    (b) $\begin{aligned} (\overline{X} - S, \overline{X} + S) &= (21.189, 24.111) \\ (\overline{X} - 2S, \overline{X} + 2S) &= (19.728, 25.572) \\ (\overline{X} - 3S, \overline{X} + 3S) &= (18.267, 27.033) \end{aligned}$

    (c) The number of parts whose lengths fall within the intervals $(\overline{X} - 2S, \overline{X} + 2S)$, and $(\overline{X} - 3S, \overline{X} + 3S)$ is 100%. Thus, the Chebychev's Inequality is valid.

PRACTICE PROBLEMS FOR SECTION 2.6

1. (a) From Problem 4 of Section 2.3, we have the following frequency distribution table.

Classes	Tally	Frequency	Relative Freq.	Percentages	Cumulative Freq.
[15–17.5)	///// /////	10	10/45	22.22	10
[17.5–20)	/////	5	5/45	11.11	15
[20–22.5)	///// //	7	7/45	15.56	22
[22.5–25)	/////	5	5/45	11.11	27
[25-27.5)	///// ///// //	12	12/45	26.67	39
[27.5–30]	///// /	6	6/45	13.33	45
Total		45	1	100	

In order to find the desired numerical measures we first find the mid points for each class. That is,

$$\text{Mid Point of class } 1 = (15 + 17.5)/2 = 16.25$$

Mid points for other classes can be found just by adding the class width (2.5) to the mid points found above. Thus, we have

$$\text{Mid Point of class } 2 = 16.25 + 2.5 = 18.75$$
$$\text{Mid Point of class } 3 = 18.75 + 2.5 = 21.25$$
$$\text{Mid Point of class } 4 = 21.25 + 2.5 = 23.75$$
$$\text{Mid Point of class } 5 = 23.75 + 2.5 = 26.25$$
$$\text{Mid Point of class } 6 = 26.25 + 2.5 = 28.75$$

Now using the results of this section, we obtain

$$\overline{X}_G = \frac{\sum f_i x_i}{\sum f_i} = \frac{10 \times 16.25 + \cdots + 6 \times 28.75}{10 + 5 + \cdots + 6} = 22.47$$

To find the median we first find its position (rank) which is given by $(n+1)/2 = |(45+1)/2 = 23$. Thus, the median in this case falls in class 4. Now substituting the values of L, c, f and w, we obtain:

$$M_G = L + (c/f)w = 22.5 + (1/5) \times 2.5 = 23$$

The mode is the midpoint of a class or classes with highest frequency. In this problem class 5 has the highest frequency (12). Thus,

$$\text{Mode} = 26.25$$

(b) $$S^2 = \frac{1}{n-1}\left(\sum f_i x_i^2 - \frac{(\sum f_i x_i)^2}{n}\right) = \frac{1}{44}(23607.8125 - 22725.0347)$$
$$= \frac{1}{44}(882.778) = 20.063.$$
$$S = \sqrt{20.063} = 4.479.$$

3. (a) From Problem 6 of Section 2.3, we have the following frequency distribution table.

Classes	Tally	Frequency	Relative Freq.	Percentages	Cumulative Freq.
[98–102)	///// ///// /	11	11/40	27.5	11
[102–106)	/////	5	5/40	12.5	16
[106–110)	///// //	7	7/40	17.5	23
[110–114)	///// //	7	7/40	17.5	30
[114–118)	//	2	2/40	5	32
[118–122]	///// ///	8	8/40	20	40
Total		40	1	100	

In order to find the desired numerical measures we first find the mid points for each class. That is,

$$\text{Mid Point of class } 1 = (98 + 102)/2 = 100$$

Mid points for other classes can be found just by adding the class width (4) to the mid points found above. Thus, we have

$$\text{Mid Point of class } 2 = 100 + 4 = 104$$
$$\text{Mid Point of class } 3 = 104 + 4 = 108$$
$$\text{Mid Point of class } 4 = 108 + 4 = 112$$
$$\text{Mid Point of class } 5 = 112 + 4 = 116$$
$$\text{Mid Point of class } 6 = 116 + 4 = 120$$

Now using the results of this section, we obtain

$$\overline{X}_G = \frac{\sum f_i x_i}{\sum f_i} = \frac{11 \times 100 + \cdots + 8 \times 120}{11 + 5 + \cdots + 8} = \frac{4352}{40} = 108.8$$

To find the median we first find its position (rank) which is given by $(n + 1)/2 = (40 + 1)/2 = 21.5$. Thus, the median in this case falls in class 3. Now substituting the values of L, c, f and w, we obtain:

$$M_G = L + (c/f)w = 106 + (5.5/7) \times 4 = 109.14$$

The mode is the midpoint of a class or classes with highest frequency. In this problem class 1 has the highest frequency (11). Thus,

$$\text{Mode} = 100$$

(b)

$$S^2 = \frac{1}{n-1}\left(\sum f_i x_i^2 - \frac{(\sum f_i x_i)^2}{n}\right) = \frac{1}{39}(475648 - 473497.6)$$

$$= \frac{1}{39}(2150.4) = 55.138.$$

$$S = \sqrt{55.138} = 7.425$$

5. A frequency distribution table for the data in this problem is as shown below.

Classes	Tally	Frequency	Relative Freq.	Percentages	Cumulative Freq.
[40–44)	///// ///// /	11	11/36	30.56	11
[44–48)	/////	5	5/36	13.89	16
[48–52)	///// /	6	6/36	16.67	22
[52–56)	//	2	2/36	5.56	24
[56–60]	///// ///// //	12	12/36	33.33	36
Total		36	1	100.01	

$$\overline{X}_G = \frac{\sum f_i x_i}{\sum f_i} = \frac{1796}{36} = 49.89,$$

$$S_G^2 = \frac{1}{n-1}\left(\sum f_i x_i^2 - \frac{(\sum f_i x_i)^2}{n}\right) = \frac{1}{35}(91184 - 89600.4) = 45.246$$

$$S_G = \sqrt{45.246} = 6.727$$

The mean and the standard deviation obtained by using the grouped data are only the approximate values of the actual mean and the standard deviation of the data.

In this problem the actual mean and standard deviation, which are computed by using the given data, are 49.56 and 7.0. Clearly, in this problem the approximate values of the mean and the standard deviation are quite close to the actual values.

PRACTICE PROBLEMS FOR SECTIONS 2.7 AND 2.8

1. (a) Mean $= 12.026$, Variance $= 0.289$, Standard Deviation $= 0.537$.

(b) First Quartile $= 11.753$, Second Quartile $= 12.070$, Third Quartile $= 12.320$, and the Inter-quartile Range $= 12.320 - 11.753 = 0.567$.

(c)

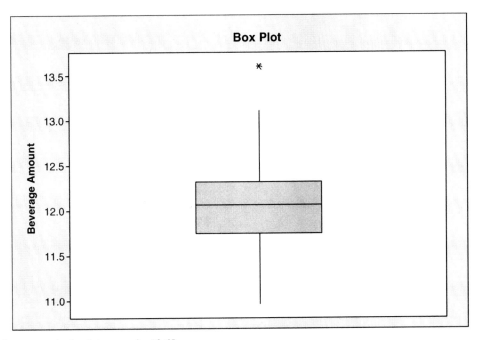

There is one outlier present in the data, namely, 13.60.

3.

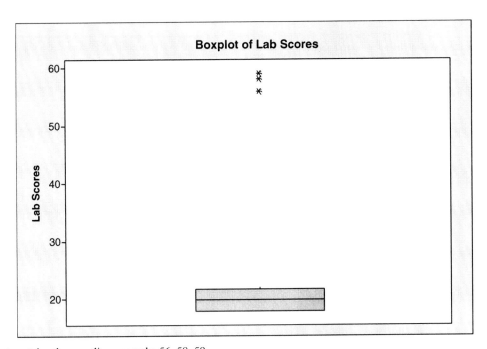

The data set contains three outliers, namely, 56, 58, 59.

5. (a) Data set I : Mean = 25.633, Standard deviation = 2.798.
Data set II : Mean = 51.194, Standard deviation = 5.966

(b) The coefficient of variations in these data sets are CV = 10.91% and CV = 11.65%, respectively.
Since the CV for data set II is slightly higher than data set I, data set II has more variability than the data set I.

7. (a) Reconsider the data in Problem 5 of Section 2.6. Using MINITAB, we obtain

$$\text{Mean} = 49.56, \text{Variance} = 49.00, \text{and Standard Deviation} = 7.00$$

(b) Quartile 1 = 43.00, Quartile 2 = 48.00, Quartile 3 = 57.75, and inter quartile range = 14.75.

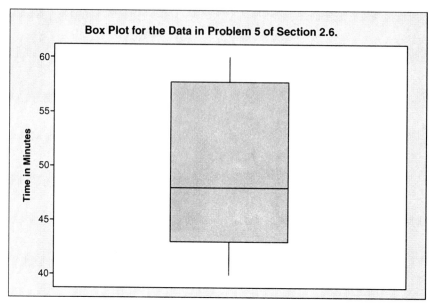

The data set does not contain any outlier.

PRACTICE PROBLEMS FOR SECTION 2.9

1. (a)

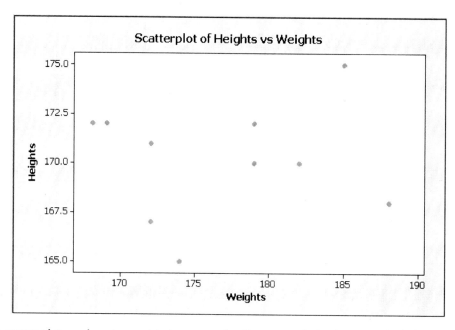

By observing the scatter plot we do not expect to have any significant correlation between the heights and weights.

(b) Correlation coefficient $= 0.075$, which is insignificant and it confirms our conclusion in part (a).

3. (a)

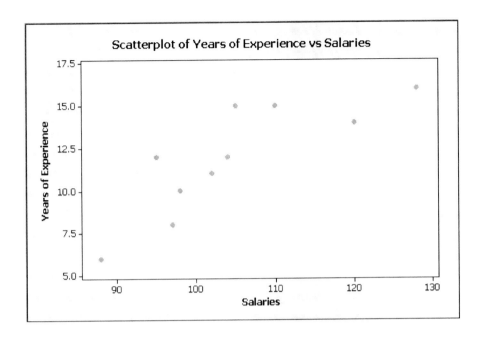

By observing the scatter plot we conclude that there is a strong positive correlation between the years of experience and the salaries.

(b) The computed correlation coefficient between the years of experience and the salaries is 0.822. Our conclusion in part (a) is consistent with the actual correlation coefficient.

REVIEW PRACTICE PROBLEMS

1.

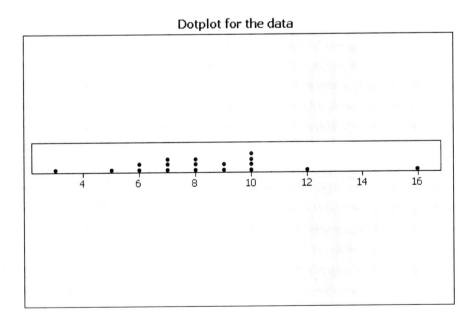

On the majority of days, six to ten workers did not come to work. On about eleven percent of the days, more than ten workers did not come to work.

3.

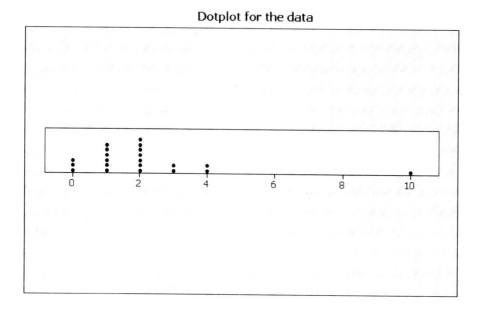

Most shifts had 1 or 2 machine breakdowns. In about 15% of the shifts, no machine had any breakdown. Very rarely, more than 4 machines had any breakdown.

5. (a)

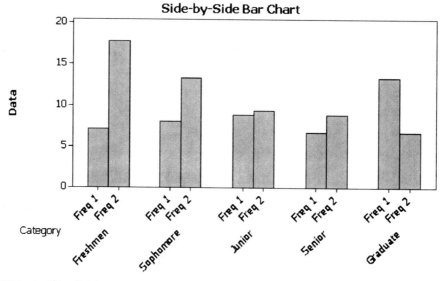

(b)

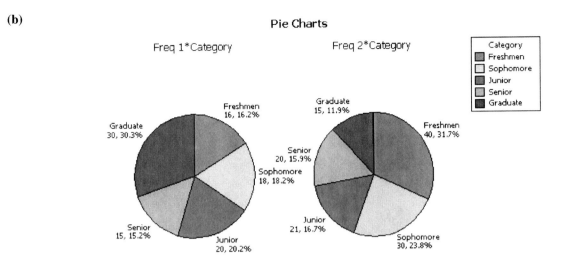

Pie Charts

Freq 1*Category   Freq 2*Category

Here we have the data from Concert 1 and Concert 2 in terms of frequencies of different categories. Side-by-Side bar charts are certainly more informative than two separate pie charts, since in side-by-side bar charts we can see the change in each category simultaneously which is not true in case of pie charts. For instance, from the side-by-side bar charts we can immediately see that there were more students from each class except the graduate class in concert 2 than in concert 1. Of course, we can also get similar information from pie charts but we will have to compare each category separately.

7. (a)

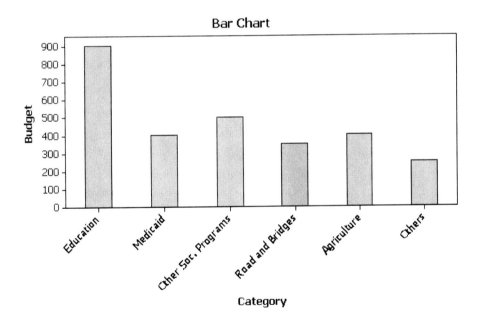

(b)

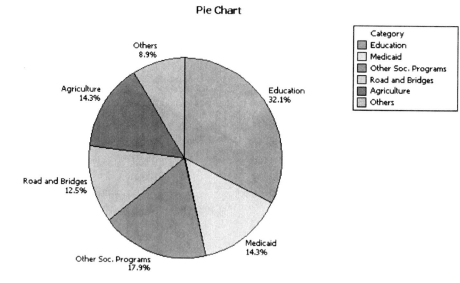

Pie Chart

(c) Percent of budget used for social programs $= (17.9 + 14.3)\% = 32.2\%$.

9. The reproduced data are the following:

32	35	37	40	43	46	48	49	51	52	52	57
58	63	65	66	66	69	69	71	75	75	77	78

11.

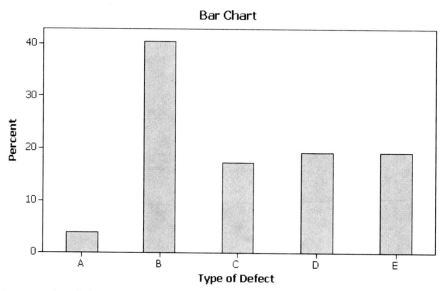

Bar Chart

Percent within all data.

About 40% of the defects are of type B whereas type A defects are only about 4%. Defects of type C, D, and E occurred with almost the same frequency.

13. (a)–(b)

Frequency Distribution Table

Number of Cars	Count	Relative Frequency	Percent	Cumulative Percent
1	9	9/50	18.00	18.00
2	18	18/50	36.00	54.00
3	13	13/50	26.00	80.00
4	7	7/50	14.00	94.00
5	3	3/50	6.00	100.00
Total	50	1	100	

(c)

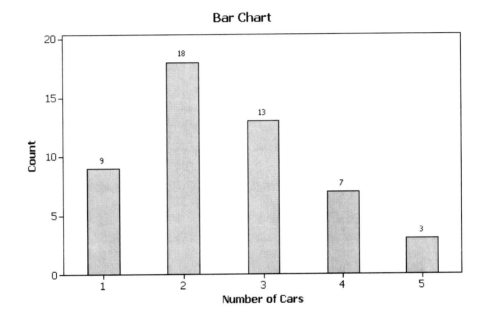

(d) Percentage of families who own at least three cars $= (26 + 14 + 6) = 46$.

(e) Percentage of families who own at most two cars $= (18 + 36) = 54$.

15. (a) The hypothetical data using the frequency distribution table in Problem 14 is

159	153	150	159	156	150	156	153	152	159
169	163	166	165	169	166	164	164	162	160
178	170	179	171	177	174	178	179	176	170
189	181	182	186	184	188	181	188	181	188
197	200	199	195	190	197	192	194	195	198
153	151	155	152	158	155	158	156	156	164
161	166	166	165	178	171	173	175	179	178
175	173	176	172	175	170	179	171	170	177
180	188	180	183	180	186	186	191	198	196
199	192	190	194	196	197	198	200	192	190

Note that these data are not the same as in Problem 14. However, since we used the frequency distribution table to generate these data for each class separately, the frequency distribution table should be exactly the same as in Problem 14. That is, the following frequency distribution table:

Frequency Distribution Table of Cholesterol Levels

Classes	Frequency	Relative Frequency	Cumulative Frequency
[150,160)	19	19/100	19
[160,170)	15	15/100	34
[170,180)	26	26/100	60
[180,190)	17	17/100	77
[190,200]	23	23/100	100
Total	100	1	

is the same as the frequency distribution table in Problem 14. Had we randomly generated the 100 data points falling between the lower limit 150 and the upper limit 200 rather than using each class separately, then the new frequency distribution table will almost certainly be different.

17. Using MINITAB we obtain

Variable	Mean	Median	Mode
Credit Hours	11.00	11.50	7

19. **(a)** Descriptive Statistics: Daily sales of Gas

Variable	Mean	Median	Mode
Daily sales of Gas	420.90	416.00	380, 398, 416, 430, 450

The data contain five mode values.

Mean is slightly greater than the median. Therefore, for all practical purposes we may consider the data is slightly right skewed.

(b) More Descriptive Statistics: Daily sales of Gas

Variable	St. Dev	Variance	Coefficient of Variation	Range
Daily sales of Gas	29.48	869.27	7.00%	110.00

23. The frequency distribution table for the data in Problem 9 of Section 2.4 is:

Class	Tally	Frequency or Count	Relative Frequency	Cumulative Frequency
[20–25)	/////	5	5/31	5
[25–30)	///// /	6	6/31	11
[30–35)	////	4	4/31	15
[35–40)	///// ///	8	8/31	23
[40–45)	///// //	7	7/31	30
[45–50)	/	1	1/31	31
Total		31	1	

Using the formulae for the grouped mean and standard deviation we obtain the following:
Descriptive Statistics for the Grouped Data:

Mean	Standard Deviation	Variance
33.95	7.55	56.99

Descriptive Statistics for the Ungrouped Data:

Mean	Standard Deviation	Variance
33.13	7.51	56.45

By comparing the two sets of results we find that the mean and the variance of the grouped data are approximately equal to the mean and standard deviation of the actual data.

25. (a) Descriptive Statistics: Length of Rods

Mean	Standard Deviation	Variance
125.00	5.40	29.18

The interval $\overline{X} \pm S = 125 \pm 5.40 = (119.60, 130.40)$ contains 26 data points or 65% of the data points.
The interval $\overline{X} \pm 2S = 125.0 \pm 2 \times 5.40 = (114.20, 135.80)$ contains all or 100% of the data points.
The interval $\overline{X} \pm 3S = 125.0 \pm 3 \times 5.40 = (108.80, 141.20)$ also contains all or 100% of the data points.

(b) For all practical purposes we can say that empirical rule holds for these data.

27. (a) Descriptive Statistics: Gas Mileage

Variable	Q1	Q2	Q3
Gas Mileage	32.0	33.0	35.0

(b) Interquartile Range: $35.0 - 32.0 = 3.0$

(c) Boxplot

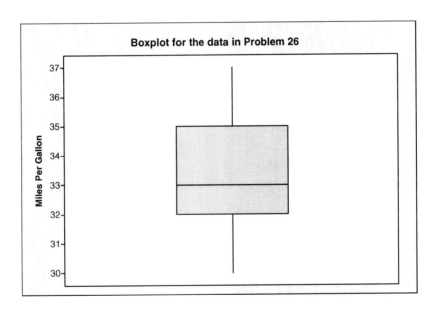

The data does not contain any outliers.

29. (a) Descriptive Statistics: Overtime Wages

Q1	Q2	Q3	IQR
30.50	36.00	41.50	11.00

(b) In order to find the percentage of the data that fall between quartiles 1 and 3 we first prepare a stem-and-leaf diagram.

Stem-and-leaf of Overtime Wages N $= 40$
Leaf Unit $= 1.0$

5	2	45688
(19)	3	0000024455566666888
16	4	00000022225566
2	5	00

(c) The interval $(Q_1, Q_2) = (30.50, 41.50)$ contains 20 data points or 50% of the data.

(d) Yes, because we were expecting the interval $(Q_1, Q_2) = (30.50, 41.50)$ to contain 50% of the data.

31. (a) Descriptive Statistics: Book Budget in dollars

Mean	Standard Deviation
779.0	85.6

(b) The interval $(\overline{X} - S, \overline{X} + S) = (779 - 85.60, 779 + 85.60) = (693.40, 864.60)$ contains 11 data points, or 55% of the data.

33. (a) The interval $(82, 98)$ contains the data points that fall within one standard deviation of the mean. Thus, by empirical rule it should contain approximately 68% of the days.

(b) The interval $(66, 114)$ contains the data points that fall within three standard deviations of the mean. Thus, by empirical rule it should contain approximately 100% of the days.

(c)

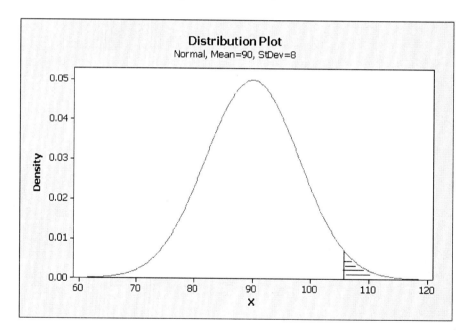

Number of days when more than 106 trees were cut is represented by the shaded area under the right tail, which is approximately 2.5%.

(d)

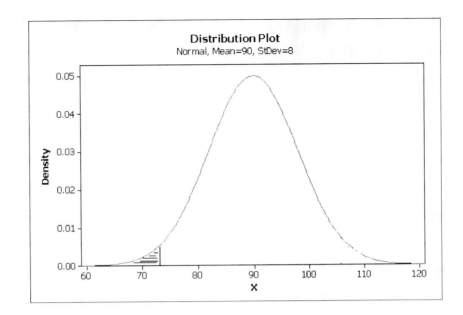

Distribution Plot
Normal, Mean=90, StDev=8

Number of days when less than 74 trees were cut is represented by the shaded area under the left tail, which is approximately 2.5%.

35. (a) Descriptive Statistics: GRE Scores

Mean	Standard Deviation
288.25	26.96

Since the data was coded by subtracting 2000, therefore, the actual mean is 2288.25. However, the standard deviation does not change.

(b) The desired interval $\overline{X} \pm 2S = 2288.25 \pm 2 \times 26.96 = (2234.33, 2342.17)$ contains all the data points or 100% of the data.

(c) Since the lower and upper quartiles are Q1 = 268.25, Q3 = 314.50, respectively. The range that contains the middle 50% of the observations is the interquartile range which is:

$$\text{Interquartile range} = 314.50 - 268.25 = 46.25.$$

37. (a) Mean = 12.733   **(b)** Standard Deviation = 2.840   **(c)** Coefficient of variation = 22.31%.

39.

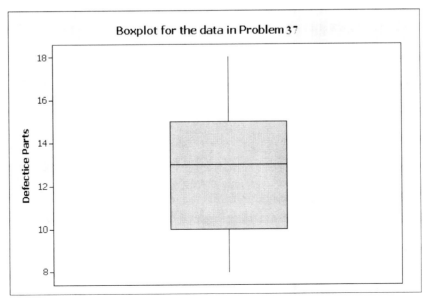

Boxplot for the data in Problem 37

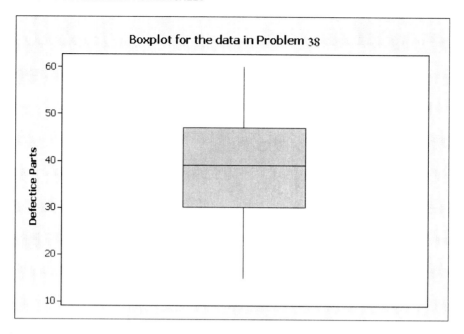

The data in Problem 37 is slightly left skewed whereas in problem 38 it is almost symmetric.

41. (a) Quartiles: No. of defective ball bearings in shipment 1

Q1	Q2	Q3
60.00	67.50	71.75

Quartiles: No. of defective ball bearings in shipment 2

Q1	Q2	Q3
42.25	49.50	55.00

(b)

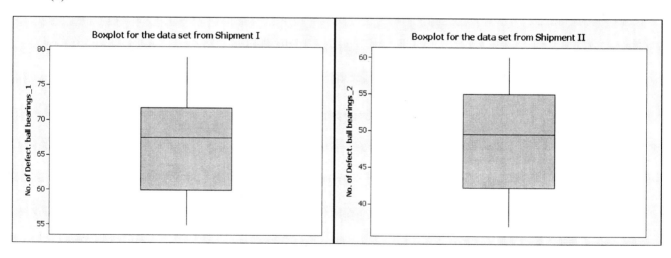

(c) The average number of defective ball bearings per shipment is higher in the first shipment than in the second shipment. However, there is more variability in the second shipment.

43. Pearson correlation between inflation rates and interest rates $= 0.314$. Based on these data we can say that there is a weak correlation between inflation rates and interest rates.

45. Pearson correlation between hours of sleep and test scores $= 0.252$. In this problem, the correlation between hours of sleep and test scores is quite weak.

3

ELEMENTS OF PROBABILITY

PRACTICE PROBLEMS FOR SECTIONS 3.2 AND 3.3

1. (a) $A \cup B$ (b) $A \cap B$ (c) $\overline{A} \cap \overline{B}$ or $\overline{(A \cup B)}$ (d) $(A \cap \overline{B}) \cup (\overline{A} \cap B)$ (e) $\overline{(A \cap B)}$.

3. (a) S = {HHH, HHT, HTH, THH, HTT, THT, TTH, TTT}.

   (b) S = {(H,1) (H,2) (H,3) (H,4) (H,5) (H,6) (T,1) (T,2) (T,3) (T,4) (T,5) (T,6)}

   (c) S = {(1,1)(1,2) ⋯ (1,6)(2,1)(2,2) ⋯ (2,6) ⋯ (6,6)}

   (d) S = {BBB, BBG, BGB, GBB, BGG, GBG, GGB, GGG}

   (e) S = {(HH,1)(HT,1)(TH,1)(TT,1)(HH,2)(HT,2)(TH,2)(TT,2) ⋯ (HH,6)(HT,6)(TH,6)(TT,6)}

5. Let C, E, and M be the events that are defined as follows:

   C: The selected student is a Chemical Engineering Major

   E: The selected student is a Electrical Engineering Major

   M: The selected student is a Mechanical Engineering Major.

   Selecting of three students can be considered a three step process, which is, first selecting one student and recording his/her major, then selecting the second and recording his/her major, and then selecting the third student and recording his/her major. Since each step has three possible outcomes, the total number sample points in the sample space are $3 \times 3 \times 3 = 27$, that is

   $$S = \{MMM, MMC, MME, MCM, MCC, MCE, MEM, MCE, MEE, CMM, CMC,$$
   $$CME, CCM, CCC, CCE, CEM, CCE, CEE, EMM, EMC, EME, ECM, ECC,$$
   $$ECE, EEM, ECE, EEE\}$$

   Now suppose that A is the event that at the most one of the three selected students is an Electrical Engineering major. Then, we have

   $$A = \{MMM, MMC, MME, MCM, MCC, MCE, MEM, MCE, CMM, CMC,$$
   $$CME, CCM, CCC, CCE, CEM, CCE, EMM, EMC, ECM, ECC\}$$

Solutions Manual to Accompany Statistics and Probability with Applications for Engineers and Scientists, Bhisham C. Gupta and Irwin Guttman.
© 2014 John Wiley & Sons, Inc. Published 2014 by John Wiley & Sons, Inc.

Hence

$$P(A) = 20/27.$$

7. (a) $A \cap B \cap C = \{4\}$.

   (b) $(A \cap B) \cup (C \cap D) = \{1, 4, 5, 7\}$.

   (c) $A \cap (B \cup C \cup D) = \{1, 3, 4, 7\}$.

   (d) $\overline{A} \cap \overline{B} = \overline{(A \cup B)} = \overline{\{1, 2, 3, 4, 6, 7, 8, 9\}} = \{5\}$.

   (e) $\varnothing$.

   (f) $\overline{A} \cap \overline{B} \cap \overline{C} \cap \overline{D} = \overline{(A \cup B \cup C \cup D)} = \overline{\{\}} = \varnothing$.

9. A, B and C are events shown in the following Venn diagram.

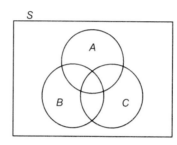

(a) $A \cap B \cap C$: The patient is diagnosed with liver cancer who needs a liver transplant and the hospital finds a matching liver on time.

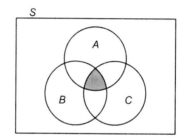

(b) $A \cap (B \cup C)$:

The patient is diagnosed with liver cancer and the patient needs a liver transplant but the hospital does not find the matching liver on time

or:

the patient is diagnosed with liver cancer and does need a liver transplant but the hospital finds a matching liver

or:

the patient is diagnosed with liver cancer who needs a liver transplant and the hospital finds a matching liver on time.

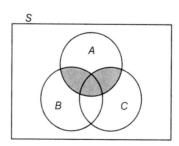

(c) $\overline{A} \cap \overline{B} = (\overline{A \cup B})$: The patient is not diagnosed with liver cancer and does not need a liver transplant.

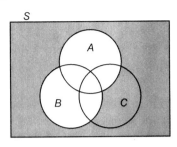

(d) $(\overline{A} \cap \overline{B} \cap \overline{C}) = (\overline{A \cup B \cup C})$: The patient is not diagnosed with liver cancer and does not need a liver transplant and the hospital does not find a matching liver.

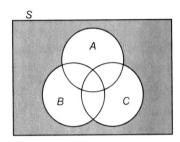

11. In This problem the sample space has 32 sample points, that is

$$\{CCCCC, CCCCN, CCCNC, CCNCC, CNCCC, NCCCC, CCCNN, \cdots, NNNNC, NNNNN\}$$

13. (a) {HHT, HTH, THH, HHH} **(b)** {HTT, THT, TTH, TTT} **(c)** {HHT, HTH, THH} **(d)** {TTT}

Since each possible outcome is equally likely, the probabilities for the events in parts (a) - (d) are given by 1/4, 1/4, 3/16, and 1/16 respectively.

PRACTICE PROBLEMS FOR SECTION 3.4

1. $\binom{4}{1} \times \binom{5}{1} = 20$

3. (a) In case no student can serve in multiple roles, the number of possible committees is

$$P(30, 4) = \frac{30!}{26!} = 657,720.$$

(b) In case any student can serve in multiple roles, the number of possible committees is

$$30 \times 30 \times 30 \times 30 = 810,000.$$

5. A customer can choose the four items, that is, (i) a soup (ii) a salad (iii) an entrée, and (iv) a dessert 4, 3, 10, and 4 ways, respectively. Thus, the total number of ways of choosing the four items simultaneously is $4 \times 3 \times 10 \times 4 = 480$ ways.

7. A hand of 13 cards from a well shuffled deck of 52 cards can be dealt in $\binom{52}{13}$ ways. Now a hand of 13 cards consisting of 5 spades, 4 diamond, and the remaining 4 of either club, heart or mixed can be dealt in

$$\binom{13}{5} \times \binom{13}{4} \times \binom{26}{4}$$

ways. Thus, the desired probability is

$$\frac{\binom{13}{5} \times \binom{13}{4} \times \binom{26}{5}}{\binom{52}{13}} = 0.02166.$$

9. An experiment in which a physician prescribes a drug to a patient can be carried out in three steps, where the first step consists of selecting a manufacturer, the second step is selecting the strength, and the third step is choosing the form of the medication, that is, either a capsule or a tablet. These three steps can respectively be carried out in 4, 5, and 2 ways. Thus, the total number of ways a drug can be prescribed to a patient is equal to

$$4 \times 5 \times 2 = 40.$$

11. There are 6 possible outcomes when a die is rolled, 2 possible outcomes when a coin is tossed and there are 4 possible outcomes when a card is drawn from a well shuffled regular deck of playing cards and its suit is noted. Thus, using the multiplication rule (or, a tree diagram), we obtain the total number of sample points in the sample space equal to

$$6 \times 2 \times 4 = 48.$$

PRACTICE PROBLEMS FOR SECTIONS 3.5 AND 3.6

1. Let A be the event that an odd number shows up and B be the event that 3 or 5 show up.

Then, we have
$$S = \{1, 2, 3, 4, 5, 6\}$$
$$A = \{1, 3, 5\}$$
$$B = \{3, 5\}$$

We are now interested in finding the conditional probability $(B|A)$.

$$P(B|A) = \frac{P(A \cap B)}{P(A)} = \frac{2/6}{3/6} = 2/3.$$

3. Given that A_1, A_2, A_3, A_4 and A_5 are mutually exclusive and exhaustive events in a sample space S. Let E be any other event in S. Then we are given that $P(A_1) = .2$, $P(A_2) = .1$, $P(A_3) = .15$ $P(A_4) = .3$, and $P(A_5) = .25$. Further, we are given that $P(E|A_1) = .2$, $P(E|A_2) = .1$, $P(E|A_3) = .35$, $P(E|A_4) = .3$, and $P(E|A_5) = .25$. We now wish to determine the probabilities $P(A_1|E)$, $P(A_2|E)$, $P(A_3|E)$, $P(A_4|E)$ and $P(A_5|E)$

Using the given information and the Venn diagram shown below, we have that

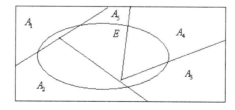

$$P(E) = P(E \cap A_1) + P(E \cap A_2) + P(E \cap A_3) + P(E \cap A_4) + P(E \cap A_5)$$
$$= P(E|A_1)P(A_1) + P(E|A_2)P(A_2) + P(E|A_3)P(A_3) + P(E|A_4)P(A_4) + P(E|A_5)P(A_5)$$
$$= .2 \times .2 + .1 \times .1 + .35 \times .15 + .3 \times .3 + .25 \times .25 = .255$$

Further, we have that

$$P(E \cap A_1) = P(E|A_1) \times P(A_1) = .2 \times .2 = .04$$

Thus, we have

$$P(A_1|E) = \frac{P(E \cap A_1)}{P(E)} = .04/.255 = .1569.$$

Similarly, we can determine that

$$P(A_2|E) = \frac{P(E \cap A_2)}{P(E)} = .01/.255 = .0392, P(A_3|E) = \frac{P(E \cap A_3)}{P(E)} = .0525/.255 = .2059,$$

$$P(A_4|E) = \frac{P(E \cap A_4)}{P(E)} = .09/.255 = .3529, \text{ and } P(A_5|E) = \frac{P(E \cap A_5)}{P(E)} = .0625/.255 = .2451.$$

Note that the sum of all of these conditional probabilities is equal to one. This is always true when the given information is the same and the sum of conditional probabilities is taken over all the events which are mutually exclusive and exhaustive.

5. **(a)** From the given information, we have

$$P(C_1) = P(C_2) = P(C_3) = P(C_4) = .25$$

Now suppose that B is an event that when a coin is tossed a head appears. Then we have

$$P(B|C_1) = 0.9, P(B|C_2) = 0.75, P(B|C_3) = 0.6, \text{ and } P(B|C_4) = 0.5$$

Now we are interested in determining the probability $P(C_2|B)$. In order to determining the probability $P(C_2|B)$ we first determine the probability $P(B)$, which is given by

$$\begin{aligned} P(B) &= P(B \cap C_1) + P(B \cap C_2) + P(B \cap C_3) + P(B \cap C_4) \\ &= P(B|C_1)P(C_1) + P(B|C_2)P(C_2) + P(B|C_3)P(C_3) + P(B|C_4)P(C_4) \\ &= .9 \times .25 + .75 \times .25 + .6 \times .25 + .5 \times .25 = .6875. \end{aligned}$$

Hence, $P(C_2|B) = \dfrac{P(B \cap C_2)}{P(B)} = \dfrac{.75 \times .25}{.6875} = 0.2727.$

(b) In this part we are interested to find the probability that if the outcome of the experiment was a tail, find the probability that coin C_4 was tossed. In other words, we are interested in determining the probability $P(C_4|\overline{B})$. From part (a), we have $P(\overline{B}) = 1 - P(B) = 1 - 0.6875 = 0.3125$.

$$P(C_4|\overline{B}) = \frac{P(\overline{B} \cap C_4)}{P(\overline{B})} = \frac{.50 \times .25}{.3125} = 0.4.$$

7. **(a)** Let C_1, C_2, C_3 and C_4 be the events that person chooses a Road I, Road II, Road III or Road IV, respectively. Then from the given information, we have

$$P(C_1) = P(C_2) = P(C_3) = P(C_4) = .25$$

Now suppose that B is an event that the person arrives late because he/she gets stuck in the traffic. Then we have

$$P(B|C_1) = 0.3, P(B|C_2) = 0.2, P(B|C_3) = 0.6, \text{ and } P(B|C_4) = 0.35$$

Now we are interested in determining the probability $P(C_3|B)$. In order to determining the probability $P(C_3|B)$ we first determine the probability $P(B)$, which is given by

$$P(B) = P(B \cap C_1) + P(B \cap C_2) + P(B \cap C_3) + P(B \cap C_4)$$
$$= P(B|C_1)P(C_1) + P(B|C_2)P(C_2) + P(B|C_3)P(C_3) + P(B|C_4)P(C_4)$$
$$= .3 \times .25 + .2 \times .25 + .6 \times .25 + .35 \times .25 = .3625.$$

Hence,

$$P(C_3|B) = \frac{P(B \cap C_3)}{P(B)} = \frac{.6 \times .25}{.3625} = 0.4138.$$

9. Let C_1, C_2, C_3, C_4 and C_5 be the events that the first fair coin, second fair coin, third fair coin, a two headed coin, or a two tailed coin is selected, respectively. Then from the given information, we have

$$P(C_1) = P(C_2) = P(C_3) = P(C_4) = P(C_5) = .20$$

Now suppose that B is an event a head appears both time. Then we have

$$P(B|C_1) = 0.25, P(B|C_2) = 0.25, P(B|C_3) = 0.25, P(B|C_4) = 1, P(B|C_5) = 0.00$$

Now we are interested in determining the probability $P(C_4|B)$. In order to determining the probability $P(C_4|B)$ we first determine the probability $P(B)$, which is given by

$$P(B) = P(B \cap C_1) + P(B \cap C_2) + P(B \cap C_3) + P(B \cap C_4) + P(B \cap C_5)$$
$$= P(B|C_1)P(C_1) + P(B|C_2)P(C_2) + P(B|C_3)P(C_3) + P(B|C_4)P(C_4) + P(B|C_5)P(C_5)$$
$$= .25 \times .20 + .25 \times .20 + .25 \times .20 + 1 \times .20 + 0 \times .20 = .3500.$$

Hence,

$$P(C_4|B) = \frac{P(B \cap C_4)}{P(B)} = \frac{.20}{.35} = 0.5714.$$

REVIEW PRACTICE PROBLEMS

1. The given information can be expressed using a Venn diagram as follows:

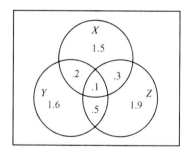

(a) Percentage of no defect $= 100 - 1.5 - 1.6 - 1.9 - 0.2 - 0.3 - 0.5 - 0.1 = 93.9\%.$

(b) Percentage with at least one defect $= 100 - 93.9 = 6.1\%$

(c) Percentage with no type X or type Y defect $= 93.9 + 1.9 = 95.8\%$

(d) Percentage with no more than one defect $= 100 - 0.2 - 0.3 - 0.5 - 0.1 = 98.9\%$

3. (a) $E_A \cap E_B$ contains 2 elements, $E_A \cap \overline{E}_B$ contains 5 elements, $\overline{E}_A \cap E_B$ contains 3 elements, $\overline{E}_A \cap \overline{E}_B$ contains 90 elements, and $E_A \cup E_B$ contains $5 + 2 + 3 = 10$ elements.

(Note that $\overline{(E_A \cup E_B)} = \overline{E}_A \cap \overline{E}_B$.)

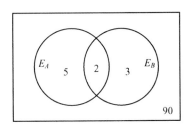

(b) Let G be the event corresponding to a pair of non-defective items. Let G_A be the event corresponding to a pair containing at least one type-A defective, and let G_B be the event corresponding to a pair containing at least one type-B defective.

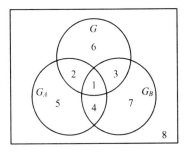

where

1:	$G \cap G_A \cap G_B$	5:	$\overline{G} \cap G_A \cap \overline{G}_B$
2:	$G \cap G_A \cap \overline{G}_B$	6:	$G \cap \overline{G}_A \cap \overline{G}_B$
3:	$G \cap \overline{G}_A \cap G_B$	7:	$\overline{G} \cap \overline{G}_A \cap G_B$
4:	$\overline{G} \cap G_A \cap G_B$	8:	$\overline{G} \cap \overline{G}_A \cap \overline{G}_B$

Now

$$\overline{G} \cap G_A \cap G_B = (E_A \cap E_B, E_A \cap \overline{E}_B) \cup (E_A \cap \overline{E}_B, \overline{E}_A \cap E_B) \cup (E_A \cap E_B, E_A \cap E_B)$$
$$\cup (E_A \cap E_B, \overline{E}_A \cap E_B) \cup (E_A \cap E_B, \overline{E}_A \cap \overline{E}_B)$$

so that $\overline{G} \cap G_A \cap G_B$ contains $10 + 15 + 1 + 6 + 180 = 212$ elements.

Similarly, $\overline{G} \cap G_A \cap \overline{G}_B$ contains 460 elements, $\overline{G} \cap \overline{G}_A \cap G_B$ contains 273 elements, and $G \cap \overline{G}_A \cap \overline{G}_B$ contains $\binom{90}{2} = 4005$ elements. The other four sets are empty. Note that $4005 + 273 + 460 + 212 = 4950 = \binom{100}{2} =$ number of ways of choosing a pair of items.

5. (a) P(10th defective item found on 50th item tested) is given by

$$\text{P(9 defective items in first 49)} \times \text{P(50th item is defective)} = \frac{\binom{10}{9}\binom{90}{40}}{\binom{100}{49}} \times \frac{1}{51} = 0.000119.$$

(b) $\dfrac{\binom{10}{10}\binom{90}{40}}{\binom{100}{50}} = 0.0005934.$ **(c)** $\dfrac{\binom{10}{10}\binom{70}{20}}{\binom{80}{30}} = 0.0000182.$

7. (a)

$$P(\text{2-motor plane forced down}) = (1 - .99)(1 - .99) = .0001$$

$$P(\text{4-motor plane force down}) = P(\text{all motors fail}) + P(\text{3 motors fail})$$

$$= (.01)(.01)(.01)(.01) + \binom{4}{3}(.01)^3(.99)$$

$$= .00000397 \approx .000004$$

Since

$$\frac{.0001}{.000004} = 25$$

Thus, 2-motor plane is more than 25 times more likely to be forced down by motor failures than the 4-motor plane.

(b)

$$P(\text{2-motor plane not forced down}) = 1 - .0001 = .9999$$
$$P(\text{4-motor plane not forced down}) = 1 - .00000397 = .99999603$$

(c)

$$P(\text{2-motor plane not forced down}) = 1 - p^2$$
$$P(\text{4-motor plane not forced down}) = 1 - \left[p^4 + \binom{4}{3}p^3(1 - p) \right]$$

9. (a) $P(\text{no failures}) = (1 - .001)^{1000} = (.999)^{1000}$

(b) $P(\text{1 failures}) = \binom{1000}{1}(.001)(.999)^{999} = (.999)^{999}$

(c) $P(k \text{ failures}) = \binom{1000}{k}(.001)^k(.999)^{1000-k}$

These probabilities may also be found using Poisson distribution (*to be studied in Chapter 4*) with $\lambda = 1000 \times .001 = 1$. Thus, we obtain

(a) P(no failure) = 0.368

(b) P(1 failure) = 0.368

(c) $P(k \text{ failures}) = \dfrac{e^{-1} \times 1^k}{k!} = \dfrac{e^{-1}}{k!}$.

11.

(a) $P(\text{oldest and youngest are chosen}) = \dfrac{\binom{2}{2}\binom{9}{3}}{\binom{11}{5}} \cong 0.182$

(b) $P(\text{third youngest chosen is sixth youngest overall}) = \dfrac{\binom{5}{2}\binom{5}{3}}{\binom{11}{5}} \cong 0.2165$

(c) $P(\text{at least three of the four youngest will be chosen}) = \dfrac{\binom{4}{3}\binom{7}{2} + \binom{4}{4}\binom{7}{1}}{\binom{11}{5}} \cong 0.1970$

13. (a) P(all faces alike) $= \dfrac{6}{6^6} = 1/6^5$.

(b) P(no two faces alike) $= \dfrac{6!}{6^6} = 5/324$.

(c) P(only five different faces) $= \dfrac{\dbinom{6}{5}\dbinom{5}{1} \times \dfrac{6!}{2!}}{6^6} = \dfrac{450}{6^6} = 25/2592$.

15. Let A be the event that letter 1 is in the correct envelope. Similarly, let B, C, D be the events that letters 2, 3, 4 are in their correct envelopes, respectively. Then

(a) P(at least one match) $= P(A \cup B \cup C \cup D) = 4p_1 - 6p_2 + 4p_4 - p_4$ (from 14(b))

where $\qquad\qquad p_1 = \dfrac{1}{4}, p_2 = \dfrac{2!}{4!} = \dfrac{1}{12}, p_3 = p_4 = \dfrac{1}{4!}$.

$\therefore P(\text{at least one match}) \qquad = 4\left(\dfrac{1}{4}\right) - 6\left(\dfrac{1}{12}\right) + 4\left(\dfrac{1}{4!}\right) - \dfrac{1}{4!}$

$$= 1 - \dfrac{1}{2!} + \dfrac{1}{3!} - \dfrac{1}{4!}$$

$$= \dfrac{5}{8}$$

(b) $p(\text{no matches}) = 1 - \dfrac{5}{8} = \dfrac{3}{8}$

(c) $P(\text{at least one match} \mid n \text{ letters}, n \text{ envelopes}) = 1 - \dfrac{1}{2!} + \dfrac{1}{3!} - \dfrac{1}{4!} + \cdots + \dfrac{1}{n!}$

$$\rightarrow 1 - e^{-1} = .632$$

P(no matches) $= 0.368$.

17. $P(A|D) = \dfrac{P(A \cap D)}{P(D)} = \dfrac{P(D|A) \cdot P(A)}{P(D)} = \dfrac{(.03)(.4)}{.0525} = .2285$

19. Define the following events:

C: individual has cancer

N: individual does not have cancer

Y: test is positive

$$P(C|Y) = \dfrac{P(Y \mid C) \cdot P(C)}{P(Y \mid C) \cdot P(C) + P(Y \mid N) \cdot P(N)}$$

$$= \dfrac{(.95)(.005)}{(.95)(.005) + (.10)(.995)} = 0.0456$$

20. Define the events:

W_1: first chip drawn is white

W_2: second chip drawn is white

B_1: first chip drawn is blue

$$P(W_2) = P(W_2 \cap W_1) + P(W_2 \cap B_1)$$

$$= P(W_2 \mid W_1) \cdot P(W_1) + P(W_2 \mid B_1) \cdot P(B_1)$$

$$= \left(\dfrac{9}{12}\right)\left(\dfrac{7}{10}\right) + \left(\dfrac{7}{9}\right)\left(\dfrac{3}{10}\right) = \dfrac{91}{120}$$

21. $P(B_1|W_2) = \dfrac{P(W_2|B_1) \times P(B_1)}{P(W_2)} = \dfrac{(7/9)(3/10)}{(91/120)} = \dfrac{4}{13}$.

23. Define the following events:

H: result is four heads, N: coin used is a nickel

Q: coin used is a quarter, D: coin used is a fake dime.

Then $P(H|D) = 1 \times 1 \times 1 \times 1 = 1$

$$P(D \mid H) = \frac{P(H \mid D) \cdot P(D)}{P(H \mid D) \cdot P(D) + P(H \mid N) \cdot P(N) + P(H \mid Q) \cdot P(Q)}$$

$$= \frac{(1)^4 \left(\frac{1}{3}\right)}{(1)^4 \left(\frac{1}{3}\right) + \left(\frac{1}{2}\right)^4 \left(\frac{1}{3}\right) + \left(\frac{1}{2}\right)^4 \left(\frac{1}{3}\right)}$$

$$= \frac{1}{1 + \frac{1}{16} + \frac{1}{16}} = \frac{8}{9}.$$

25. Given that A_1, A_2, A_3, and A_4 are mutually exclusive and exhaustive events in a sample space S. Let B be any other event in S. Then we are given that $P(A_1) = .2$, $P(A_2) = .1$, $P(A_3) = .4$ and $P(A_4) = .3$. Further, we are given that $P(B|A_1) = .4$, $P(B|A_2) = .1$, $P(B|A_3) = .6$, and $P(B|A_4) = .2$. We now wish to determine the probabilities $P(A_1|B)$, $P(A_2|B)$, $P(A_3|B)$, and $P(A_4|B)$.

Using the given information and the Venn diagram shown below, we have that

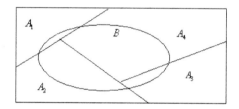

$$P(B) = P(B \cap A_1) + P(B \cap A_2) + P(B \cap A_3) + P(B \cap A_4)$$
$$= P(B|A_1)P(A_1) + P(B|A_2)P(A_2) + P(B|A_3)P(A_3) + P(B|A_4)P(A_4)$$
$$= .4 \times .2 + .1 \times .1 + .6 \times .4 + .2 \times .3$$
$$= .08 + .01 + .24 + .06 = .39$$

Further, we have that

$$P(B \cap A_1) = P(B|A_1) \times P(A_1) = .4 \times .2 = .08$$

Thus, we have

$$P(A_1|B) = \frac{P(B \cap A_1)}{P(B)} = .08/.39 = 8/39.$$

Similarly, we can determine that

$$P(A_2|B) = \frac{P(B \cap A_2)}{P(B)} = .01/.39 = 1/39, P(A_3|B) = \frac{P(B \cap A_3)}{P(B)} = .24/.39 = 24/39, \text{ and}$$

$$P(A_4|B) = \frac{P(B \cap A_4)}{P(B)} = .06/.39 = 6/39.$$

Note that the sum of all these conditional probabilities is equal to one. This is always true when the given information is the same and the sum of conditional probabilities is taken over all the events which are mutually exclusive and exhaustive.

27. The events A_1, A_2, A_3, and B are defined as follows:

A_1: Insurer had no accident in the past three years

A_2: Insurer had one accident in the past three years

A_3: Insurer had two or more accident in the past three years

B: Insurer will have an accident in the future.

Then, we are given that

$$P(A_1) = .60, \quad P(A_2) = .25, \quad P(A_3) = .15$$

Furthermore, we have

$$P(B|A_1) = .01, \quad P(B|A_2) = .03, \quad P(B|A_3) = .10$$

We now wish to determine the probabilities $P(A_1|B)$. Since A_1, A_2, A_3 events are mutually exclusive and exhaustive and B is another event in the same sample space, we have

$$\begin{aligned} P(B) &= P(B \cap A_1) + P(B \cap A_2) + P(B \cap A_3) \\ &= P(B|A_1)P(A_1) + P(B|A_2)P(A_2) + P(B|A_3)P(A_3) \\ &= .01 \times .60 + .03 \times .25 + .10 \times .15 = .0285 \end{aligned}$$

Further, we know that

$$P(B \cap A_1) = P(B|A_1) \times P(A_1) = .01 \times .60 = .0060$$

Thus, we have

(a) $P(A_1|B) = \dfrac{P(B \cap A_1)}{P(B)} = .0060/.0285 = .2105.$

Similarly, we can determine that

(b) $P(A_2|B) = \dfrac{P(B \cap A_2)}{P(B)} = .0075/.0285 = .2632,$

(c) $P(A_3|B) = \dfrac{P(B \cap A_3)}{P(B)} = .0150/.0285 = .5263.$

Note: These probabilities are frequently used to determine the future premiums.

29. Let M be an event that a voter is a male and F an event that the voter is a female. Then we have

$$P(M) = .55, \quad P(F) = .45$$

Further, suppose that C, D, and E are the events that are defined as follows:

C: A registered voter favors the Casino,

D: A registered voter does not favor the Casino,

E: A registered voter has no opinion for the Casino.

Then, we are given that

$$P(C|M) = .75, \quad P(D|M) = .20, \quad P(E|M) = .05,$$

and

$$P(C|F) = .40, \quad P(D|F) = .50, \quad P(E|F) = .10.$$

We now wish to determine the probabilities **(a)** $P(M|C)$ **(b)** $P(F|E)$ **(c)** $P(F|D)$.

From the given information we can easily see that the events A, B, and C are mutually exclusive and exhaustive. Therefore, we have

$$P(C) = P(C \cap M) + P(C \cap F)$$
$$= P(C|M)P(M) + P(C|F)P(F)$$
$$= .75 \times .55 + .40 \times .45 = .5925$$

Similarly, we have

$$P(D) = P(D \cap M) + P(D \cap F)$$
$$= P(D|M)P(M) + P(D|F)P(F)$$
$$= .20 \times .55 + .50 \times .45 = .3350$$

$$P(E) = P(E \cap M) + P(E \cap F)$$
$$= P(E|M)P(M) + P(E|F)P(F)$$
$$= .05 \times .55 + .10 \times .45 = .0725$$

Thus, we obtain

$$P(M|C) = \frac{P(M \cap C)}{P(C)} = \frac{P(C|M)P(M)}{P(C)} = \frac{.75 \times .55}{.5925} = 165/237$$

Similarly, we can determine that

$$P(F|E) = \frac{P(F \cap E)}{P(E)} = \frac{P(E|F)P(F)}{P(E)} = \frac{.10 \times .45}{.0725} = 18/29$$

$$P(F|D) = \frac{P(F \cap D)}{P(D)} = \frac{P(D|F)P(F)}{P(D)} = \frac{.50 \times .45}{.3350} = 45/67$$

31. (a) Since $P(X = 0) + P(X = 2) + \cdots + P(X = 5) = .10 + .09 + .20 + .15 + .16 + .20 = .90 < 1$, the given distribution does not represent a probability distribution. Since the sum of probabilities is not equal to one.

(b) Since $P(X = 0) = -.10$, which is negative. Hence, the given distribution does not represent a probability distribution.

(c) The sum of the probabilities is $P(X = 0) + P(X = 2) + \cdots + P(X = 5) = .20 + .09 + .20 + .15 + .16 + .20 = 1$, and each of the probabilities is > 0. Hence, the given distribution represents a probability distribution.

33. The sample space S is consist of twelve elements, that is

$$S = \{e_i = i | i = 1, 2, \cdots, 12\}$$

(a) From the Venn diagram the events A, B, and C in terms of the elements in S are:

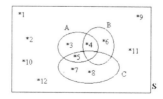

$$A = \{3, 4, 5\}, B = \{4, 6\}, \quad \text{and} \quad C = \{5, 7, 8\}$$

(b) Since the sample points in the sample space are equally likely, we have

(i) $P(A) = 3/12 = 1/4$

(ii) $P(B) = 2/12 = 1/6$

(iii) $P(C) = 3/12 = 1/4$

(iv) $P(A \cap B) = P\{4\} = 1/12$

(v) $P(A \cap B \cap C) = P(\varnothing) = 0$

(vi) $P(\overline{A \cap B \cap C}) = 1 - P(A \cap B \cap C) = 1 - P(\varnothing) = 1 - 0 = 1$

(vii) $P(A \cup B \cup C) = P\{3, 4, 5, 6, 7, 8\} = 6/12 = 1/2$

(viii) $P(\overline{A \cup B \cup C}) = 1 - P(A \cup B \cup C) = 1 - 1/2 = 1/2.$

4

DISCRETE RANDOM VARIABLES AND SOME IMPORTANT DISCRETE PROBABILITY DISTRIBUTIONS

PRACTICE PROBLEMS FOR SECTIONS 4.1 AND 4.2

1. **(a)** $P(X \leq 2) = P(X = 0) + P(X = 1) + P(X = 2) = 0.02 + 0.05 + 0.10 = 0.17$.

   (b) $P(2 < X < 5) = P(X = 3) + P(X = 4) = 0.15 + 0.18 = 0.33$.

   (c) $P(X \geq 5) = P(X = 5) = 0.50$.

   (d) $P(1 \leq X \leq 4) = P(X = 1) + P(X = 2) + P(X = 3) + P(X = 4) = 0.05 + 0.10 + 0.15 + 0.18 = 0.48$.

3. **(a)** Yes, since all probabilities are positive and sum of all probabilities is equal to 1.

   (b) No, since sum of the probabilities is greater than 1.

   (c) No, since the probability of $P(X = 2)$ is negative.

5.
$$\mu = \sum_{i=0}^{5} x_i p(x_i) = 0 \times 0.0010 + \cdots + 4 \times 0.3995 + 5 \times 0.2373 = 3.75$$

$$\sigma^2 = \sum_{i=0}^{5} x_i^2 p(x_i) - \mu^2 = 0^2 \times 0.0010 + \cdots + 4^2 \times 0.3995 + 5^2 \times 0.2373 - (3.75)^2 = 0.9375.$$

7. Let a random variable X denotes the sum of the points turn up on two dice. Then, it can easily be shown that

$$P(X = 2) = P\{(1, 1)\} = \frac{1}{36}$$
$$P(X = 3) = P\{(1, 2)(2, 1)\} = \frac{2}{36}$$
$$P(X = 4) = P\{(1, 3)(2, 2)(3, 1)\} = \frac{3}{36}$$
$$P(X = 5) = P\{(1, 4)(2, 3)(3, 2)(4, 1)\} = \frac{4}{36}$$
$$P(X = 6) = P\{(1, 5)(2, 4)(3, 3)(4, 2)(5, 1)\} = \frac{5}{36}$$
$$P(X = 7) = P\{(1, 6)(2, 5)(3, 4)(4, 3)(5, 2)(6, 1)\} = \frac{6}{36}$$

Solutions Manual to Accompany Statistics and Probability with Applications for Engineers and Scientists, Bhisham C. Gupta and Irwin Guttman.
© 2014 John Wiley & Sons, Inc. Published 2014 by John Wiley & Sons, Inc.

$$P(X = 8) \quad = P\{(2,6)(3,5)(4,4)(5,3)(6,2)\} = \frac{5}{36}$$

$$P(X = 9) \quad = P\{(3,6)(4,5)(5,4)(6,3)\} = \frac{4}{36}$$

$$P(X = 10) = P\{(4,6)(5,5)(6,4)\} = \frac{3}{36}$$

$$P(X = 11) = P\{(5,6)(6,5)\} = \frac{2}{36}$$

$$P(X = 12) = P\{(6,6)\} = \frac{1}{36}$$

Hence, the desired probability distribution is the following:

$X = x$	2	3	4	5	6	7	8	9	10	11	12
$P(X = x)$	1/36	2/36	3/36	4/36	5/36	6/36	5/36	4/36	3/36	2/36	1/36

$$\mu \; = E(X) = \sum xP(x) = [2 \times 1/36 + 3 \times 2/36 + \cdots + 12 \times 1/36] = 7$$

$$\sigma^2 = \sum x^2 P(x) - \mu^2 = [2^2 \times 1/36 + 3^2 \times 2/36 + \cdots + 12^2 \times 1/36] - (7)^2 = 5.83$$

9. (a) $E(Y) = E(3X) = 3 \times E(X) = 3 \times 7 = 21,$
$V(Y) = V(3X) = 9 \times V(X) = 9 \times 5.83 = 52.47$

(b) $E(Y) = E(2X + 5) = 2 \times E(X) + 5 = 2 \times 7 + 5 = 19,$
$V(Y) = V(2X + 5) = V(2X) = 4 \times V(X) = 4 \times 5.83 = 23.32.$

PRACTICE PROBLEMS FOR SECTIONS 4.3 AND 4.4

1. Since the sampling is done with replacement the probability distribution of the random variable X is uniform. That is

X	1	2	3	4	5	6	7	8	9	10
$p(x)$	.1	.1	.1	.1	.1	.1	.1	.1	.1	.1

Using Equation (4.3.2), we obtain

$$\mu = (n+1)/2 = (10+1)/2 = 5.5, \; \sigma^2 = (n^2 - 1)/12 = (10^2 - 1)/12 = 8.25.$$

3.

$$P(X = 0) = \frac{\binom{6}{0} \times \binom{6}{4}}{\binom{12}{4}} = 0.0303, P(X = 1) = \frac{\binom{6}{1} \times \binom{6}{3}}{\binom{12}{4}} = 0.2424,$$

$$P(X = 2) = \frac{\binom{6}{2} \times \binom{6}{2}}{\binom{12}{4}} = 0.4545, P(X = 3) = \frac{\binom{6}{3} \times \binom{6}{1}}{\binom{12}{4}} = 0.2424,$$

$$P(X = 4) = \frac{\binom{6}{4} \times \binom{6}{0}}{\binom{12}{4}} = 0.0303.$$

Thus the probability distribution of the random variable X is given by

X	0	1	2	3	4
$p(x)$	0.0303	0.2424	0.4545	0.2424	0.0303

Using equation (4.4.3) the mean and the variance of the random Variable X are given by

$$\mu = n\theta = 4(1/2) = 2, \theta = N_1/N = 6/12 = 1/2$$

$$\sigma^2 = \frac{N-n}{N-1}n\theta(1-\theta) = \frac{12-4}{12-1} \times 4 \times (1/2) \times (1 - 1/2) = 8/11 = 0.7273.$$

5. The sample space S consists of twenty elements, that is

$$S = \{e_i = i | i = 1, 2, \cdots, 20\}$$

and all of the sample points are equally likely. Thus, we can proceed to determine the desired probabilities as follows:

(a) $P(X > 15) = P(X = 16, 17, 18, 19, 20) = 5/20 = 0.25$

(b) $P(10 \leq X \leq 18) + P(X = 10, 11, 12, 13, 14, 15, 16, 17, 18) = 9/20 = 0.45$

(c) $P(X < 10) = P(X = 1, 2, 3, 4, 5, 6, 7, 8, 9) = 9/20 = 0.45.$

7. In this problem we have $N = 20$, $n = 6$, $N_1 = 8$. Thus, the desired probabilities are found as follows:

(a) $P(X = 2) = \dfrac{\binom{8}{2} \times \binom{12}{4}}{\binom{20}{6}} = 0.3576$

(b) $P(X \geq 3) = P(X = 3, 4, 5, 6) = \dfrac{\binom{8}{3} \times \binom{12}{3}}{\binom{20}{6}} + \cdots + \dfrac{\binom{8}{6} \times \binom{12}{0}}{\binom{20}{6}} = 0.4551.$

9. In this problem we have $N = 12$, $n = 4$, $N_1 = 5$. Thus, the desired probabilities are found as follows:

(a) $P(X = 2) = \dfrac{\binom{5}{2} \times \binom{7}{2}}{\binom{12}{4}} = 0.4242$

(b) $P(2 \leq X \leq 4) = 1 - P(X = 0) - P(X = 1) = 1 - \dfrac{\binom{5}{0} \times \binom{7}{4}}{\binom{12}{4}} - \dfrac{\binom{5}{1} \times \binom{7}{3}}{\binom{12}{4}} = 1 - .0707 - .3535 = .5758.$

(c) $P(X \leq 2) = P(X = 0) + P(X = 1) + P(X = 2) = .0707 + .3535 + .4242 = .8484.$

PRACTICE PROBLEMS FOR SECTIONS 4.5 AND 4.6

1. Let a random variable X denotes the number of PCV valves that are defective. Then the random variable X is having binomial distribution with $n = 16$ and $\theta = 0.02$. We are interested in finding the following probabilities:

(a) $P(X = 0) = 0.7238$ (b) $P(X \leq 1) = P(X = 0) + P(X = 1) = 0.7238 + 0.2363 = 0.9601$ (c) $P(X \geq 1) = 1 - P(X = 0) = 1 - 0.7238 = 0.2762.$

3. A Bernoulli experiment is a random experiment that has two possible outcomes, usually called "success" and "failure". We say that an experiment has Bernoulli trials when it consists of a number of independent Bernoulli experiments such that the probability of success is equal to θ that remains the same throughout the experiment. Let a random variable X denote the number of successes in n independent Bernoulli trials. Then the probability distribution of the random

variable X is binomial, since we have n independent trials, each trial having two possible outcomes, "success" and "failure", and that the probability of success in each trial θ remains the same throughout the experiment. The probability distribution of the random variable X in this problem is given by

$$P(X = x) = \binom{n}{x}(\theta)^x(1 - \theta)^{n-x}, \quad x = 0, 1, \cdots, n.$$

5. In this problem we are given that

$$n\theta = 12 \ and \ n\theta(1 - \theta) = 3$$

Solving these two equations for n and θ, we get $n = 16$ and $\theta = 0.75$. Having found the values of n and θ, we can now easily determine the desired probabilities:

(a) $P(X < 4) = P(X \leq 3) = P(X = 0) + P(X = 1) + P(X = 2) + P(X = 3)$

$$= \sum_{x=0}^{3}\binom{16}{x}(0.75)^x(1 - 0.75)^{16-x} = 0.000004.$$

(b) $P(4 \leq X < 11) = P(X \leq 10) - P(X \leq 3) = \sum_{x=0}^{10}\binom{16}{x}(0.75)^x(1 - 0.75)^{16-x} -$

$$\sum_{x=0}^{3}\binom{16}{x}(0.75)^x(1 - 0.75)^{16-x} = 0.189655 - 0.000004 = 0.189651.$$

(c) $P(X \geq 7) = 1 - P(X \leq 6) = 1 - \sum_{x=0}^{6}\binom{16}{x}(0.75)^x(1 - 0.75)^{16-x} = 1 - 0.001644 = 0.998356.$

7. Referring to Problem 5, we have

$$P(X = x) = \binom{16}{x}(0.75)^x(1 - 0.75)^{16-x}; \quad x = 0, 1, \cdots, 16$$

$$[\mu - 3\sigma, \mu + 3\sigma] = [12 - 3 \times 1.732, 12 + 3 \times 1.732] = [6.80, 17.20].$$

We are now interested in determining the probability

$$P(6.80 \leq X \leq 17.20) = P(6 < X \leq 17) = P(X \leq 17) - P(X \leq 6) = 1 - 0.00164 = 0.99836.$$

9. Let a random variable X denotes the number of components that function at a given time. Then the random variable X has a binomial distribution with $n = 8$ and $\theta = 0.60$. We are interested in finding the following probability:

$$P(X \geq 5) = 1 - P(X \leq 4) = 1 - \sum_{x=0}^{4}\binom{8}{x}(0.60)^x(1 - 0.60)^{8-x} = 1 - 0.405914 = 0.594086.$$

10. Let a random variable X denotes the number of personal bankruptcies due to medical bills. Then the random variable X has a binomial distribution with $n = 10$ and $\theta = 0.40$. We are interested in finding the following probability:

$$P(X \geq 4) = 1 - P(X \leq 3) = 1 - \sum_{x=0}^{3}\binom{10}{x}(0.40)^x(1 - 0.40)^{10-x} = 1 - 0.382281 = 0.617719.$$

PRACTICE PROBLEMS FOR SECTION 4.7

1. The desired probability in this problem is found using multinomial probability distribution with $n = 19$, $X_1 = 4, X_2 = 5, X_3 = 10, \theta_1 = 0.25, \theta_2 = 0.30, \theta_3 = 0.45$. Thus, we obtain

$$p = \frac{19!}{4! \times 5! \times 10!}(.25)^4(.30)^5(.45)^{10} = 0.03762.$$

3. The desired probability in this problem is found using multinomial probability distribution with $n = 10$, $X_1 = 3, X_2 = 7, X_3 = 0, X_4 = 0, \theta_1 = 0.55, \theta_2 = 0.20, \theta_3 = 0.15, \theta_4 = 0.10$. Thus, we obtain

$$p = \frac{10!}{3! \times 7! \times 0! \times 0!}(.55)^3(.20)^7(.15)^0(.10)^0 = 0.00025.$$

5. The desired probability in this problem is found using multinomial probability distribution with $n = 20$, $X_1 = 6, X_2 = 4, X_3 = 7, X_4 = 3, \theta_1 = 0.30, \theta_2 = 0.25, \theta_3 = 0.25, \theta_4 = 0.20$. Thus, we obtain

$$p = \frac{20!}{6! \times 4! \times 7! \times 3!}(.30)^6(.25)^4(.25)^7(.20)^3 = 0.00647.$$

7. The desired probability in this problem is found using multinomial probability distribution with $n = 60$, $X_1 = 10, X_2 = 15, X_3 = 10, X_4 = 25, \theta_1 = .15, \theta_2 = .18, \theta_3 = .27, \theta_4 = .40$. Thus, we obtain

$$p = \frac{60!}{10! \times 15! \times 10! \times 25!}(.15)^{10}(.18)^{15}(.27)^{10}(.40)^{25} = .00028.$$

PRACTICE PROBLEMS FOR SECTION 4.8

1. (a) $P(X < 5) = P(X = 0) + P(X = 1) + P(X = 2) + P(X = 3) + P(X = 4)$

$$= \sum_{x=0}^{4} \frac{e^{-3.5}(3.5)^x}{x!} = .0302 + .1057 + .1850 + .2158 + .1888 = .7255.$$

(b) $P(2 \le X \le 6) = P(X = 2) + P(X = 3) + P(X = 4) + P(X = 5) + P(X = 6) = \sum_{x=2}^{6} \frac{e^{-3.5}(3.5)^x}{x!}$

$$= .1850 + .2158 + .1888 + .1322 + .0771$$
$$= .7989.$$

(c) $P(X > 7) = P(X = 8) + P(X = 9) + P(X = 10) + P(X = 11) + P(X = 12) + P(X = 13)$

$$+ P(X = 14) + \ldots = \sum_{x=8}^{\infty} \frac{e^{-3.5}(3.5)^x}{x!}$$

$$= .0169 + .0066 + .0023 + .0007 + .0002 + .0001 + 0 + 0 + \ldots$$

$$= .0268.$$

Note that for $X > 13$, all probabilities become zero.

(d) $P(X \ge 5) = 1 - P(X \le 4) = 1 - \sum_{x=0}^{4} \frac{e^{-3.5}(3.5)^x}{x!} = 1 - .7255 = .2745.$

3. Let a random variable X denote the number of defective parts produced per shift. Then the random variable X has a Poisson distribution with $\lambda = 3$. We wish to find the following probabilities:

(a) $P(X = 4) = \frac{e^{-3}3^4}{4!} = 0.1680$

(b) Since we are interested in determining the probability that there are 7 defective parts in the next two shifts, the value of $\lambda = 2 \times 3 = 6$. Hence, we have

$$P(X > 7) = 1 - P(X \le 7) = 1 - \sum_{x=0}^{7} \frac{e^{-6}(6)^x}{x!} = 1 - 0.7440 = 0.2560.$$

(c) We now wish to the probability $P(X \leq 8)$ in three shifts, the new value of $\lambda = 3 \times 3 = 9$. Thus, we have

$$P(X \leq 8) = P(X = 0) + P(X = 1) + \cdots + P(X = 8) = \sum_{x=0}^{8} \frac{e^{-9}(9)^x}{x!} = 0.4556.$$

5. From the given information we know that on the average 2.5 customers come to the store every five minutes. Thus, if X is a random variable denoting the average number of customers coming to the store every five minutes, then the random variable X has a Poisson distribution with $\lambda = 2.5$. We now wish to determine the following probabilities:

(a) $P(X \geq 2) = 1 - P(X \leq 1) = 1 - P(X = 0) - P(X = 1)$

$$= 1 - .0821 - .2052 = 0.7127.$$

(b) $P(X = 4) = 0.1336.$

(c) $P(X \leq 6) = \sum_{x=0}^{6} \frac{e^{-2.5}(2.5)^x}{x!} = P(X = 0) + P(X = 1) + \cdots + P(X = 6) = .0821 + .2050 + \cdots + 0.278$

$$= 0.9858.$$

7. In this problem we approximate the binomial distribution with Poisson distribution with $\lambda = n\theta = 15,000 \times .0003 = 4.5$. Thus, the desired probability during a given year is given by (note that $\lambda = 2 \times 4.5 = 9$)

$$P(X \geq 10) = 1 - P(x \leq 9) = 1 - \sum_{x=0}^{9} \frac{e^{-9}(9)^x}{x!} = 1 - .5874 = 0.4126.$$

PRACTICE PROBLEMS FOR SECTION 4.9

1. In this problem, unlike the binomial probability distribution, the number of successes is fixed and the number of trials is a random variable. Thus, the desired probability is found by using the negative binomial distribution. That is,

$$P(X = 100) = \binom{99}{2}(.03)^3(.97)^{97} = 0.0068.$$

3. In this problem, unlike the binomial probability distribution, the number of successes is fixed and the number of trials is a random variable. Thus, the desired probability is found by using the negative binomial distribution. That is,

$$P(X = 20) = \binom{19}{8}(.6)^9(.4)^{11} = 0.03195$$

5. In this problem we want to find the probability that the 5th success in the 31st trial or after. It means the quality inspector did not find 5 defectives by inspecting 30 parts. Using the negative binomial distribution, we obtain

$$P(X \geq 31) = 1 - P(X \leq 30) = 1 - .1755 = .8245.$$

7. In this problem we want to find the probability that the quality engineer 3rd nonconforming part is the 40th part produced. Using the negative binomial distribution, we obtain

$$P(X = 40) = \binom{39}{2}(.05)^3(.95)^{37} = 0.01388.$$

REVIEW PRACTICE PROBLEMS

1. The number of defectives x has a hypergeometric distribution

$$h(x) = \frac{\binom{7}{x}\binom{43}{10-x}}{\binom{50}{10}}, \quad 0 \leq x \leq 7.$$

Using this result we can easily see that the desired probabilities are as follows:

$$P(X = 0) = 0.1867, P(X = 1) = 0.3843, P(X = 2) = 0.2964, P(X = 3) = 0.1098.$$
$$P(X = 4) = 0.0208, P(X = 5) = 0.0020, P(X = 6) = 0.0001, P(X = 7) = 0000.$$

3. Let y be the number of aces obtained after rolling 5 dice.

$$P(y \geq 1) = 1 - P(y = 0) = 1 - \left(\frac{5}{6}\right)^5 = .598$$

$$P(y = 1) = \binom{5}{1}\left(\frac{1}{6}\right)\left(\frac{5}{6}\right)^4 = \left(\frac{5}{6}\right)^5 = .402$$

$$P(y = 2) = \binom{5}{2}\left(\frac{1}{6}\right)^2\left(\frac{5}{6}\right)^3 = .161$$

5. $P(\text{no aces, kings, queens, or jacks in a 13-card hand}) = \dfrac{\binom{16}{0}\binom{36}{13}}{\binom{52}{13}} = 0.0036.$

7.

(a) $P(x \text{ spades}) = \dfrac{\binom{13}{x}\binom{39}{13-x}}{\binom{52}{13}}, \; 0 \leq x \leq 13$

(The number of spades in a thirteen-card hand is hypergeometric.)

(b) $P(y \text{ hearts}) = \dfrac{\binom{13}{y}\binom{39}{13-y}}{\binom{52}{13}}, \; 0 \leq y \leq 13$

(c) $P(x \text{ spades}, y \text{ hearts}) = \dfrac{\binom{13}{x}\binom{13}{y}\binom{26}{13-x-y}}{\binom{52}{13}}, \qquad 0 \leq x \leq 13, 0 \leq y \leq 13,$

$$0 \leq x + y \leq 13$$

9. Let y be the number of insured males that die from a certain kind of accident each year. To find $P(y > 3)$ we can use either the binomial distribution with $n = 10,000$ and $\theta = .000005$,

$$P(y > 3) = 1 - P(y \leq 3)$$

$$= 1 - \sum_{y=0}^{3} \binom{10000}{y}(.000005)^y(.999995)^{10000-y}$$

$$= 0.000.$$

This problem can also be done by using the Poisson approximation of the binomial distribution with $\lambda = .000005 \times 10000 = .05$. Thus, the desired probability is equal to

$$P(X > 3) = 1 - \sum_{x=0}^{3} \frac{e^{-.05}(.05)^x}{x!} = 0.000.$$

11. Let X be the number of seeds. Than X has a Poisson distribution with $\lambda = 4$.

(a) $\lambda = (10 \times 5)(4/100) = 2$

$$P(X > 2) = 1 - P(X \leq 2) = 1 - \sum_{x=0}^{2} \frac{e^{-2}2^x}{x!} = 1 - 0.6767 = 0.3233.$$

$\lambda = (5 \times 5)(4/100) = 1$

(b) $P(6 \text{ pieces will all be free of seeds}) = [P(x = 0)]^6 = \left[\frac{e^{-\mu}\mu^0}{0!}\right]^6 = \left[\frac{e^{-1}1^0}{0!}\right]^6$

$$= e^{-6} = .0025.$$

13. This is a multinomial experiment with various probabilities defined as follows:

$$P(\text{two heads}) = P(\text{two tails}) = \frac{1}{4}$$
$$P(\text{one head, one tail}) = \frac{1}{2}.$$

(a) $P(3 \text{ get 2H, 3 get 2T, 4 get 1H and 1T}) = \dfrac{10!}{3!3!4!}\left(\dfrac{1}{4}\right)^3\left(\dfrac{1}{4}\right)^3\left(\dfrac{1}{2}\right)^4 = 0.0641.$

(b) $P(\text{no head and tail combination}) = (.5)^{10} = 0.00098.$

15. $P(\text{total drawn is } x) = P(4 \text{ whites in first } x - 1 \text{ draws}) \cdot P(\text{white on } x^{th} \text{ draw})$

$$= \frac{\binom{10}{4}\binom{20}{x-5}}{\binom{30}{x-1}} \cdot \frac{6}{30 - x + 1}, \quad x = 5, 6, 7, \ldots, 25.$$

17. (a) $P(x \text{ defectives per box}) = \binom{m}{x}\theta^x(1-\theta)^{m-x}.$

(b) $P(\text{no defectives per box}) = (1 - \theta)^m = \varphi, \text{ say}$

$$P(y \text{ boxes are free of defectives}) = \binom{n}{y}\varphi^y(1-\varphi)^{n-y}$$

$$= \binom{n}{y}(1-\theta)^{my}(1 - [1-\theta]^m)^{n-y}$$

19. Let X be a random variable denoting the number of engineers who pass the Six Sigma black belt in the first attempt. Then X has a binomial distribution with $n = 10$ and $\theta = 0.6$. Thus, we have

$$P(X = 6) = \binom{10}{6}(.6)^6(.4)^4 = 0.2508$$

21. (a) $P(x \text{ trials to obtain 1}^{st} \text{ failure}) = (1-\theta)^{x-1}\theta.$ In this problem the random variable X has geometric distribution.

(b) $P(x \text{ trials to obtain } k^{\text{th}} \text{ failure}) = \begin{pmatrix} x - 1 \\ k - 1 \end{pmatrix} (1 - \theta)^{x-k} \theta^k$. In this problem the random variable X has a negative binomial distribution.

23.

$$P_A(6) = \begin{pmatrix} 5 \\ 3 \end{pmatrix} (.85)^4 (.15)^2$$

$$= 0.1175.$$

$$P_A(x) = \begin{pmatrix} x - 1 \\ 3 \end{pmatrix} (.85)^4 (.15)^{x-4}, \quad x = 4, 5, \dots.$$

25. If y is the number of trials needed to obtain one failure, then Y has a geometric (θ) distribution, so that, we have

$$p(y) = \begin{pmatrix} y - 1 \\ 0 \end{pmatrix} \theta^1 (1 - \theta)^{y-1}$$

$$= \theta(1 - \theta)^{y-1}, \quad y = 1, 2, \dots.$$

$$\sum_{y=1}^{\infty} p(y) = \sum_{y=1}^{\infty} \theta(1 - \theta)^{y-1} = \theta \sum_{y=1}^{\infty} (1 - \theta)^{y-1} = \theta \cdot \frac{1}{1 - 1 + \theta} = 1.$$

Note that

$$\sum_{y=1}^{\infty} (1 - \theta)^{y-1} = \frac{1}{\theta}.$$

$$E(y) = \sum_{y=1}^{\infty} y\theta(1 - \theta)^{y-1} = -\theta \sum_{y=1}^{\infty} \frac{d}{d\theta} [(1 - \theta)^y]$$

$$= -\theta \cdot \frac{d}{d\theta} \left[\sum_{y=1}^{\infty} (1 - \theta)^y \right] = -\theta \cdot \frac{d}{d\theta} \left(\frac{1 - \theta}{\theta} \right)$$

$$= (-\theta) \left(-\frac{1}{\theta^2} \right) = \frac{1}{\theta}.$$

$$E(y(y - 1)) = \sum_{y=1}^{\infty} y(y - 1)(1 - \theta)^{y-1} \cdot \theta$$

$$= \theta(1 - \theta) \sum_{y=1}^{\infty} y(y - 1)(1 - \theta)^{y-2}$$

$$= \theta(1 - \theta) \sum_{y=2}^{\infty} \frac{d^2}{d\theta^2} (1 - \theta)^y = \theta(1 - \theta) \frac{d^2}{d\theta^2} \left[\sum_{y=2}^{\infty} (1 - \theta)^y \right]$$

$$= \theta(1 - \theta) \frac{d^2}{d\theta^2} \left(\frac{(1 - \theta)^2}{\theta} \right) = \theta(1 - \theta) \cdot \frac{2}{\theta^3} = \frac{2(1 - \theta)}{\theta^2}.$$

$$\text{Var}(y) = \frac{2(1 - \theta)}{\theta^2} + \frac{1}{\theta} - \frac{1}{\theta^2} = \frac{1 - \theta}{\theta^2}$$

OR: from Problem 24, with $k = 1$,

$$E(y) = \frac{1}{\theta}, \quad \mathrm{Var}(y) = \frac{1 - \theta}{\theta^2}.$$

27. Let X be the number of articles which must be drawn until k defectives are found. From $N\theta$ defectives, choose $k - 1$, and from the remainder choose $x - k$. Then the last choice must be the kth defective. Therefore, for $X = k, k+1, \cdots, (N - N\theta + k)$,

$$p(x) = \frac{\binom{N\theta}{k-1}\binom{N-N\theta}{x-k}}{\binom{N}{x-1}} \cdot \frac{N\theta - k + 1}{N - x + 1}$$

$$= \frac{(N\theta)!}{(k-1)!(N\theta - k + 1)!} \cdot \frac{(N - N\theta)!}{(x-k)!(N - N\theta - x + k)!} \cdot \frac{(x-1)!(N - x + 1)!}{N!} \cdot \frac{N\theta - k + 1}{N - x + 1}$$

$$= \frac{(x-1)!}{(k-1)!(x-k)!} \cdot \frac{(N\theta)!(N - N\theta)!}{N!} \cdot \frac{(N - x)!}{(N\theta - k)!(N - N\theta - x + k)!}$$

$$= \frac{\binom{x-1}{k-1}\binom{N-x}{N\theta - k}}{\binom{N}{N\theta}}.$$

Note that since $\sum_{x=k}^{N-N\theta+k} p(x) = 1$, then $\binom{N}{N\theta} = \sum_{x=k}^{N-N\theta+k} \binom{x-1}{k-1}\binom{N-x}{N\theta - k}$.

Now, $E(x) = \sum_{x=k}^{N-N\theta+k} \frac{x \cdot \frac{(x-1)!}{(k-1)!(x-k)!}\binom{N-x}{N\theta - k}}{\binom{N}{N\theta}}$

$$= k \sum_{x=k}^{N-N\theta+k} \frac{\binom{x}{k}\binom{N-x}{N\theta - k}}{\binom{N}{N\theta}}$$

$$= k \sum_{x=(k+1)-1}^{N-N\theta-1+(k+1)} \frac{\binom{(x+1)-1}{(k+1)-1}\binom{(N+1)-(x+1)}{(N\theta+1)-(k+1)}}{\binom{N}{N\theta}}$$

$$= k \frac{\binom{N+1}{N\theta + 1}}{\binom{N}{N\theta}} = \frac{k(N+1)}{N\theta + 1}$$

29. Let a random variable X denote the number of bulbs that burn out before the warranty period. Then the random variable X has a binomial distribution with $n = 12$ and $\theta = 0.40$. We are interested in finding the following probabilities:

(a) $P(4 \leq X \leq 6) = P(X \leq 6) - P(X \leq 3) = \sum_{x=0}^{6} \binom{12}{x}(0.40)^x(1 - 0.40)^{12-x} - \sum_{x=0}^{3} \binom{12}{x}(0.40)^x(1 - 0.40)^{12-x}$.

$= 0.841788 - 0.225337 = 0.6165.$

(b) $P(X > 5) = 1 - P(X \leq 5) = 1 - \sum_{x=0}^{5} \binom{12}{x}(0.40)^x(1 - 0.40)^{12-x} = 1 - 0.665209 = 0.3348.$

(c) $P(X < 8) = P(X \leq 7) = \sum_{x=0}^{7} \binom{12}{x}(0.40)^x(1 - 0.40)^{12-x} = 0.9427.$

(d) $P(X = 0) = .0022.$

31. Let a random variable X denotes the number persons who get addicted with the drug Xanax. Then the random variable X has a binomial distribution with $n = 15$ and $\theta = 0.70$. We are interested in finding the following probabilities:

(a) $P(X > 10) = 1 - P(X \leq 10) = 1 - \sum_{x=0}^{10} \binom{15}{x}(0.70)^x(1 - 0.70)^{15-x} = 1 - 0.4845 = 0.5155.$

(b) $P(X < 8) = P(X \leq 7) = \sum_{x=0}^{7} \binom{12}{x}(0.70)^x(1 - 0.70)^{15-x} = 0.0500.$

(c) $P(10 \leq X \leq 12) = P(X \leq 12) - P(X \leq 9)$

$= \sum_{x=0}^{12} \binom{15}{x}(0.70)^x(1 - 0.70)^{15-x} - \sum_{x=0}^{9} \binom{15}{x}(0.70)^x(1 - 0.70)^{15-x} = 0.873172 - 0.278379 = 0.5948.$

33. From Section 4.10.2, we have

$$\mu = n\theta, \sigma^2 = \frac{N - n}{N - 1}n\theta(1 - \theta)$$

Where N is the population size, n the sample size, and θ is the fraction of the total population that is of interest. Thus, for example, in Problem 32, $N = 100$, $n = 10$, and $\theta = .08$. Thus, the mean, variance, and the standard deviation are:

$$\mu = n\theta = 10 \times .08 = .8$$
$$\sigma^2 = \frac{N - n}{N - 1}n\theta(1 - \theta) = \frac{100 - 10}{100 - 1} \times 10 \times .08(1 - .08) = 0.6691,$$
$$\sigma = \sqrt{0.6691} = 0.8180$$

35. In this problem the probability that a policy-holder will file a claim against a particular kind of accident is 0.003 and the number of drivers who are insured against such an accident is 2000. Since we can consider this problem as a rare event problem, we can model it using the Poisson distribution. Thus, the random variable X denoting the number of policy-holder filing a claim against that kind of accident has a Poisson distribution with $\lambda = 2000 \times .003 = 6$. We are now interested in determining the following probabilities:

(a) $P(X \geq 4) = 1 - P(X \leq 3) = 1 - \sum_{x=0}^{3} \frac{e^{-6}6^x}{x!} = 1 - 0.1512 = 0.8488.$

(b) $P(X > 10) = 1 - P(X \leq 10) = 1 - \sum_{x=0}^{10} \frac{e^{-6}6^x}{x!} = 1 - 0.9574 = 0.0426.$

(c) $P(5 \leq X \leq 8) = \sum_{x=0}^{8} \frac{e^{-6}6^x}{x!} - \sum_{x=0}^{4} \frac{e^{-6}6^x}{x!} = 0.8473 - 0.2851 = 0.5622.$

(d) $P(X < 2) = P(X \leq 1) = 0.0174.$

(e) $P(X > 2) = 1 - P(X \leq 2) = 1 - \sum_{x=0}^{2} \frac{e^{-6}6^x}{x!} = 1 - 0.0620 = 0.9380.$

37. We suppose that a random variable X denotes the number the number of customers arriving at a bank teller's window in 10 minutes then the random variable X has a Poisson distribution with $\lambda = 4$. Thus, we can determine the desired probabilities as follows:

(a) $P(X \geq 5) = 1 - P(X \leq 4) = 1 - \sum_{x=0}^{4} \frac{e^{-4}(4)^x}{x!} = 1 - 0.6288 = 0.3712$

(b) $P(X \leq 2) = \sum_{x=0}^{2} \frac{e^{-4}(4)^x}{x!} = 0.2381$

(c) $P(2 \leq X \leq 6) = P(X \leq 6) - P(X \leq 1) = \sum_{x=0}^{6} \frac{e^{-4}(4)^x}{x!} - \sum_{x=0}^{1} \frac{e^{-4}(4)^x}{x!} = .8893 - .0916 = .7977$

(d) $P(X < 6) = P(X \leq 5) = \sum_{x=0}^{5} \frac{e^{-4}(4)^x}{x!} = 0.7851$

39. (a) $P(X \geq 5) = 1 - P(X \leq 4) = 1 - 0.0003 = 0.9997$.
  (b) $P(X \leq 7) = 0.0210$.
  (c) $P(5 < X < 10) = P(X \leq 9) - P(X \leq 5) = 0.1275 - 0.0016 = 0.1259$.
  (d) $P(X = 8) = 0.0355$.
  (e) $P(X \leq 9) = 0.1275$.

41. Given a batch of 500 batteries that is known to have 60 defective batteries. The manufacturer will ship them only if a random sample of 20 batteries from that batch contains two or fewer defectives. Thus the probability that the batch will be shipped can be found by using the hypergeometric distribution. That is,

$$p(x) = \sum_{x=0}^{2} \frac{\binom{60}{x} \times \binom{440}{20-x}}{\binom{500}{20}} = 0.5615.$$

Note: Since the population size 500 is quite large and the sample size $n < .05N$, we can find the above probability by using a binomial distribution with $n = 20$ and $\theta = 60/500 = .12$. Using the binomial approximation, we obtain

$$p(x) = \sum_{x=0}^{2} \binom{20}{x}(.12)^x(.88)^{20-x} = 0.5631.$$

43. In order for a function to represent a probability function it must satisfy the two conditions, that is, (i) $p(x) \geq 0$ and (ii) $\sum_{x} p(x) = 1$. Thus, we have

(a) (i) $p(x) \geq 0$, for $x = 1, 2, \cdots, 6$.

  (ii) $\sum_{x=1}^{6} p(x) = 1/20 + 2/20 + 3/20 + 4/20 + 5/20 + 6/20 = 21/20 > 1$

  Hence, the given function does not represent a probability function.

(b) (i) $p(x) \geq 0$, for $x = 1, 2, \cdots, 7$.

  (ii) $\sum_{x=1}^{6} p(x) = 1/140 + 4/140 + 9/140 + 16/140 + 25/140 + 36/140 + 49/140 = 1$.

  Hence, the given function represents a probability function.

(c) (i) Since $p(x) < 0, for\ x = 2$.

  Hence, the given function does not represent a probability function.

45. (a) $\sum_{x=1}^{6} cx/20 = c(1/20 + 2/20 + \cdots + 6/20) = c(21/20) \Rightarrow c = 20/21$, will satisfy both the properties of a probability function.

(b) $\sum_{x=1}^{5} c(x^2 + 1) = c(60) \Rightarrow c = 1/60$, will satisfy both the properties of a probability function.

(c) $\sum_{x=1}^{6} c(x - 1) = c(15) \Rightarrow c = 1/15$, will satisfy both the properties of a probability function.

47. (a) In this part the mean and variance are given by

$$\mu = \sum xp(x) = 1 \times 1/21 + 2 \times 2/21 + \cdots + 6 \times 6/21 = 91/21 = 4.33,$$

$$\sigma^2 = V(X) = \sum_{x=1}^{6} x^2 p(x) - \mu^2 = \sum_{x=1}^{6} x^3/21 - \mu^2 = 1/21 \times (1 + 8 + \cdots + 216) - (4.33)^2$$
$$= 2.251$$

(b) In this Problem the probability function is given by

$$p(x) = (x^2 - 1)/50, x = 1, 2, \cdots, 5.$$

Thus, the mean and the variance are given by

$$\mu = \sum xp(x) = \sum_{x=1}^{5} x(x^2 - 1)/50 = 4.2,$$

$$\sigma^2 = V(X) = \sum_{x=1}^{5} x^2 p(x) - \mu^2 = \sum_{x=1}^{5} x^2(x^2 - 1)/50 - (4.2)^2 = 0.84.$$

49. In this problem the probability function is Bernoulli, that is

$$p(x) = \theta^x(1 - \theta)^{1-x}, \ x = 0, 1.$$

Thus, the moment generating function of Bernoulli distribution is given by

$$M_X(t) = E(e^{tX}) = e^{t \times 0} \times (1 - \theta) + e^{t \times 1} \times \theta = (1 - \theta) + \theta e^t.$$

Now the mean and variance using the above moment generating function are given by

$$\mu = \mu' = \left. \frac{\partial M_X(t)}{\partial t} \right|_{t=0} = \theta e^t |_{t=0} = \theta,$$

$$\mu'_2 = \left. \frac{\partial^2 M_X(t)}{\partial^2 t} \right|_{t=0} = \theta e^t |_{t=0} = \theta, \mu = \mu' = \left. \frac{\partial M_X(t)}{\partial t} \right|_{t=0} = \theta e^t |_{t=0} = \theta,$$

Hence, $\sigma^2 = \mu'_2 - \mu^2 = \theta - \theta^2 = \theta(1 - \theta)$.

51. Referring to Problem 48, we have

$$\mu = E(X + c) = E(X) + c = \frac{n + 1}{2} + c.$$

$$\sigma^2 = V(X + c) = V(X) = \frac{n^2 - 1}{12}.$$

Note that the mean has changed but the variance remains the same. Thus, by adding or subtracting a constant from a random variable does not change its variance.

53. Again, this is a rare event problem. Thus, it can be shown the number of joints becoming loose are distributed as a Poisson distribution with $\lambda = .001 \times 500 = .5$.
(a) $P(X = 3) = 0.01264$.
(b) $P(X \geq 2) = 1 - P(X \leq 1) = 1 - 0.9098 = 0.0902$.

(c) The circuit board becomes dysfunctional when one or more joints become loose in a two year period. The probability that the circuit board becomes dysfunction is equal to the probability $P(X \geq 1)$. Now using Poisson distribution with $\lambda = 2 \times .5 = 1$. Thus, we have

$$P(X \geq 1) = 1 - P(X = 0) = 1 - 0.3679 = 0.6321.$$

55. It is possible for the Xth part to be the first defective part only if the first $(X - 1)$ parts are nondefective. Suppose the probability that a randomly selected part is defective is θ, so that the probability that a randomly selected part is not defective is equal to $1 - \theta$. Thus, the probability function that the Xth part is the first defective is given by

$$p(x) = (1 - \theta) \times (1 - \theta) \times \cdots \times (1 - \theta) \times \theta = (1 - \theta)^{x-1}\theta, \quad x = 1, 2, \cdots$$

Hence,

$$\mu = \sum xp(x) = \sum_{x=1}^{\infty} x(1 - \theta)^{x-1}\theta = \theta \sum_{x=1}^{\infty} x(1 - \theta)^{x-1} = \theta\varphi_1, \tag{1}$$

so that

$$\varphi_1 = \sum_{x=1}^{\infty} x(1 - \theta)^{x-1} = 1 + 2(1 - \theta) + 3(1 - \theta)^2 + 4(1 - \theta)^3 + \cdots \tag{2}$$

If we now multiply both sides of (2) by $(1 - \theta)$ then we obtain

$$(1 - \theta)\varphi_1 = (1 - \theta) + 2(1 - \theta)^2 + 3(1 - \theta)^3 + 4(1 - \theta)^4 + \cdots \tag{3}$$

Now $(2) - (3)$, leads to

$$\varphi_1 - (1 - \theta)\varphi_1 = 1 + (1 - \theta) + (1 - \theta)^2 + (1 - \theta)^3 + (1 - \theta)^4 + \cdots = (1 - (1 - \theta))^{-1} = \frac{1}{\theta}.$$

Or $\varphi_1 = \dfrac{1}{\theta^2}$

Thus, from (1), we obtain

$$\mu = E(X) = \theta \times \frac{1}{\theta^2} = \frac{1}{\theta},$$

$$\sigma^2 = V(X) = E(X^2) - \mu^2 = \sum_{x=1}^{\infty} x^2 p(x) - \mu^2 = \sum_{x=1}^{\infty} x^2 (1 - \theta)^{x-1}\theta$$

$$= \theta(1 + 4(1 - \theta) + 9(1 - \theta)^2 + \cdots + n^2(1 - \theta)^{n-1} + \cdots)$$

$$= \theta\varphi_2, \text{say}$$

so that

$$\varphi_2 = (1 + 4(1 - \theta) + 9(1 - \theta)^2 + \cdots + n^2(1 - \theta)^{n-1} + \cdots) \tag{4}$$

If we now multiply both sides of (4) by $(1 - \theta)$ then we obtain

$$(1 - \theta)\varphi_2 = ((1 - \theta) + 4(1 - \theta)^2 + 9(1 - \theta)^3 - + \cdots + n^2(1 - \theta)^n + \cdots) \tag{5}$$

Subtracting (5) from (4), we obtain

$$
\begin{aligned}
\theta\varphi_2 &= (1 + 3(1-\theta) + 5(1-\theta)^2 + 7(1-\theta)^3 - + \cdots + (2n-1)(1-\theta)^{n-1} + \cdots) \\
&= 2\varphi_1 - (1 + (1-\theta) + (1-\theta)^2 + \cdots + (1-\theta)^n + \cdots) \\
&= 2\varphi_1 - (1 - (1-\theta))^{-1} \\
&= \frac{2}{\theta^2} - \frac{1}{\theta} = \frac{2-\theta}{\theta^2},
\end{aligned}
$$

or $\qquad \varphi_2 = \dfrac{2-\theta}{\theta^3}$

or $\qquad E(X^2) = \theta \times \dfrac{2-\theta}{\theta^3} = \dfrac{2-\theta}{\theta^2}.$

Hence, $\sigma^2 \qquad = V(X) = E(X^2) - \mu^2 = \dfrac{2-\theta}{\theta^2} - \dfrac{1}{\theta^2} = \dfrac{1-\theta}{\theta^2}.$

(c) $P(X = 20) = (1 - .07)^{19}(.07) = (.93)^{19}(.07) = .01763.$

57. In this problem the desired probabilities are found using the hypergeometric distribution with $N = 15, N_1 = 5, n = 8$.

(a) $P(X = 3) = \dfrac{\dbinom{5}{3} \times \dbinom{10}{5}}{\dbinom{15}{8}} = 0.3916.$

(b) $P(X \geq 2) = 1 - P(X \leq 1) = 1 - \displaystyle\sum_{x=0}^{1} \dfrac{\dbinom{5}{x} \times \dbinom{10}{8-x}}{\dbinom{15}{8}} = 1 - 0.1002 = 0.8998.$

(c) $P(X \leq 2) = \displaystyle\sum_{x=0}^{2} \dfrac{\dbinom{5}{x} \times \dbinom{10}{8-x}}{\dbinom{15}{8}} = 0.4266.$

5

CONTINUOUS RANDOM VARIABLES AND SOME IMPORTANT CONTINUOUS PROBABILITY DISTRIBUTIONS

PRACTICE PROBLEMS FOR SECTIONS 5.1 AND 5.2

1. Given that the random variable X has exponential distribution with $\lambda = 2$. Hence, we have

$$P(X \geq a) = \int_a^\infty \lambda e^{-\lambda x} dx = e^{-\lambda a} = e^{-2a}$$

(a) Using the above result, we obtain

$$P(X > 4) = e^{-2(4)} = e^{-8} = .0003.$$

(b) Using the above result, we obtain

$$P(X < 5) = 1 - P(X \geq 5) = 1 - e^{-10} = 1.0000.$$

(c) Again, using the above result, we obtain

$$P(2 < X < 7) = P(X < 7) - P(X < 2) = (1 - e^{-14}) - (1 - e^{-4}) = e^{-4} - e^{-14} = 0.0183.$$

(d) Since in case of continuous probability distribution the probability at any particular point is zero. Hence,

$$P(X = 5) = 0.$$

3. Since $\int_1^\infty f(x)dx = 1$. Thus, we obtain

$$\int_{11.5}^{12.5} 4cx\,dx = \left|2cx^2\right|_{11.5}^{12.5} = 48c \Rightarrow c = 1/48$$

That is,

$$f(x) = \begin{cases} (1/12)x, & 11.5 < x < 12.5 \\ 0, & elsewhere \end{cases}$$

Solutions Manual to Accompany Statistics and Probability with Applications for Engineers and Scientists, Bhisham C. Gupta and Irwin Guttman.
© 2014 John Wiley & Sons, Inc. Published 2014 by John Wiley & Sons, Inc.

(a) $P(X > 11.5) = 1 - P(X < 11.5) = 1 - 0 = 1$

(b) $P(X < 12.25) = \int_{11.5}^{12.25} (x/12)dx = |x^2/24|_{11.5}^{12.25} = 0.7422.$

(c) $P(11.75 < X < 12.25) = \int_{11.75}^{12.25} (x/12)dx = |x^2/24|_{11.75}^{12.25} = 0.5000.$

5. In this problem we are given that

$$f(x) = \begin{cases} 0, & x \leq 0 \\ cx^2(1-x), & 0 < x < 1 \end{cases}$$

We first find the value of c such that the given function represents a probability density function. That is, c is such that

$$\int_1^{\infty} f(x)dx = 1.$$

Thus, we obtain

$$\int_0^1 cx^2(1-x)dx = |c(x^3/3 - x^4/4)|_0^1 = 1 \Rightarrow c(1/3 - 1/4) = 1 \Rightarrow c = 12.$$

Now we are interested in determining the probability that the life of such a component is more than 8 (or 0.8 units) years. That is,

$$P(X > 0.8) = 1 - \int_0^{0.8} 12x^2(1-x)dx = 1 - 12 \times |x^3/3 - x^4/4|_0^{0.8} = 1 - 0.8192 = .1808.$$

7. The moment generating function is given by

$$M_x(t) = E(e^{tX}) = \lambda \int_0^{\infty} e^{-x(\lambda-t)}dx = (1 - t/\lambda)^{-1}.$$

$$\mu = \frac{\partial M_x(t)}{\partial t} = \left| \frac{\partial(1-t/\lambda)^{-1}}{\partial t} \right|_{t=0} = |(1/\lambda)(1-t/\lambda)^{-2}|_{t=0} = 1/\lambda.$$

$$\sigma^2 = \frac{\partial^2 M_x(t)}{\partial t^2} - (1/\lambda)^2 = \left| \frac{\partial^2(1-t/\lambda)^{-1}}{\partial t^2} \right|_{t=0} - (1/\lambda)^2 = 2/\lambda^2 - 1/\lambda^2 = 1/\lambda^2.$$

9. Using the results of Problem4, we obtain

$$(\mu - \sigma < X < \mu + \sigma) = (11.7164 < X < 12.2974)$$
$$(\mu - 2\sigma < X < \mu + 2\sigma) = (11.4259 < X < 12.5879)$$
$$(\mu - 3\sigma < X < \mu + 3\sigma) = (11.1354 < X < 12.8784)$$

Now from Problem 3, we obtain

(a) $P(\mu - \sigma < X < \mu + \sigma) = \int_{11.7164}^{12.2974} (x/12)dx = |x^2/24|_{11.7164}^{12.2974} = 0.5813.$

(b) $P(\mu - 2\sigma < X < \mu + 2\sigma) = \int_{11.4259}^{12.5879} (x/12)dx = \int_{11.5}^{12.5} (x/12)dx = 1,$ since $f(x) = 0$ outside the interval (11.5, 12.5).

(c) Using the same argument as in (b), we obtain $P(\mu - 3\sigma < X < \mu + 3\sigma) = 1.$
These probabilities, using the Empirical rule, are 0.6826, 0.9544, and 0.9973, respectively.

PRACTICE PROBLEMS FOR SECTION 5.3

1. Using MINITAB the mean and the standard deviation for the given set of data, we obtain

$$\overline{X} = 17.550, \quad S = 1.468$$

To find the percentage of data points that fall in intervals $(\overline{X} \pm 1.5S)$, $(\overline{X} \pm 2S)$, and $(\overline{X} \pm 3S)$ we first determine these intervals, that is

$$(\overline{X} \pm 1.5S) = (15.348, 19.752)$$
$$(\overline{X} \pm 2S) = (14.614, 20.486)$$
$$(\overline{X} \pm 3S) = (13.146, 21.954)$$

Now we can easily determine that the intervals $(\overline{X} \pm 1.5S)$, $(\overline{X} \pm 2S)$, and $(\overline{X} \pm 3S)$ contain 80%, 100% and 100% of the data points, respectively. According to Chebychev's Inequality these intervals contain at least 56%, 75%, 89% of the data points.

3. Using MINITAB the mean and the standard deviation for the given set of data, we obtain

$$\overline{X} = 10.900, \quad S = 3.463$$

To find the percentage of data points that fall in intervals $(\overline{X} \pm 1.5S)$, $(\overline{X} \pm 2S)$, and $(\overline{X} \pm 3S)$ we first determine these intervals, that is

$$(\overline{X} \pm 1.5S) = (5.7055, 16.0945)$$
$$(\overline{X} \pm 2S) = (3.974, 17.826)$$
$$(\overline{X} \pm 3S) = (0.511, 21.289)$$

Now we can easily determine that the intervals $(\overline{X} \pm 1.5S)$, $(\overline{X} \pm 2S)$, and $(\overline{X} \pm 3S)$ contain 90%, 95% and 100% of the data points, respectively. According to Chebychev's Inequality these intervals contain at least 56%, 75%, 89% of the data points.

5. Let X be the number of days that the heart surgery patients spend time in a hospital. We want to find the lower limit of the probability $P(3 \le X \le 9)$. Using Chebychev's Inequality, we obtain $P(3 \le X \le 9) = P(|X - 6| \le 3) = P(|X - 6| \le 2\sigma) \ge (1 - 1/2^2) = 0.75$. Thus, at least 75% of the heart surgery patients stay between 3 and 9 days (inclusive) in the hospital.

PRACTICE PROBLEMS FOR SECTION 5.4

1. Using the result of equation (5.4.6), we obtain

   (a) $P(X > 6) = 1 - P(X \le 6) = 1 - \dfrac{6 - 4}{10 - 4} = 1 - \dfrac{1}{3} = \dfrac{2}{3}.$

   (b) $P(5 < X < 8) = \dfrac{8 - 5}{10 - 4} = \dfrac{3}{6} = \dfrac{1}{2}.$

   (c) $P(X < 7) = \dfrac{7 - 4}{10 - 4} = \dfrac{3}{6} = \dfrac{1}{2}.$

3. Here, we have $a = 0$ and $b = 15$. Thus, using the results of equations (5.4.3) and (5.4.4), we obtain

$$\mu = E(X) = \frac{0 + 15}{2} = 7.5, \quad \sigma^2 = \frac{(15 - 0)^2}{12} = \frac{75}{4} = 18.75, \quad \sigma = \sqrt{18.75} = 4.33.$$

5. Let the random variable X denote the hourly wages, which is distributed uniformly over the interval $(20, 32)$. We are now interested in finding the probability $P(X > 25)$, which is given by

$$P(X > 25) = 1 - P(X \le 25) = 1 - \frac{25 - 20}{32 - 20} = 1 - \frac{5}{12} = \frac{7}{12}.$$

7. The random variable X has uniform distribution over the interval [2, 4]. Hence

$$\mu = \frac{2+4}{2} = 3, \quad \sigma^2 = \frac{(4-2)^2}{12} = 0.333, \quad \sigma = \sqrt{0.333} = 0.577.$$

Thus, we obtain

$$P(\mu - 2\sigma < X < \mu + 2\sigma) = P(1.846 < X < 4.154) = P(2 < X < 4) = 1,$$

Since the probability density function for $X < 2$ and $X > 4$ is zero.

PRACTICE PROBLEMS FOR SECTION 5.5

1. (a) $P(8 < X < 12) = P\left(\dfrac{8-10}{1.5} < \dfrac{X-10}{1.5} < \dfrac{12-10}{1.5}\right)$

$\qquad = P(-1.33 < Z < 1.33)$

$\qquad = P(Z < 1.33) - P(Z \leq -1.33)$

$\qquad = 0.9082 - 0.0918 = 0.8164.$

(b) $P(X \leq 12) = P\left(\dfrac{X-10}{1.5} \leq \dfrac{12-10}{1.5}\right)$

$\qquad = P(Z \leq 1.33) = 0.9082.$

(c) $P(X \geq 8.5) = P\left(\dfrac{X-10}{1.5} \geq \dfrac{8.5-10}{1.5}\right)$

$\qquad = P(Z \geq -1) = 1 - P(Z \leq -1) = 1 - .1587 = .8413.$

3. From the given information, we have

$$\mu = 1.8, \sigma = 0.25$$

(a) $P(1.5 < X < 2) = P\left(\dfrac{1.5-1.8}{.25} < \dfrac{X-1.8}{.25} < \dfrac{2-1.8}{.25}\right)$

$\qquad = P(-1.2 < Z < .8)$

$\qquad = P(Z < .8) - P(Z \leq -1.2)$

$\qquad = 0.7881 - 0.1151 = 0.6730.$

(b) $P(X \geq 1.55) = P\left(\dfrac{X-1.8}{.25} \geq \dfrac{1.55-1.8}{.25}\right)$

$\qquad = P(Z \geq -1) = 1 - P(Z \leq -1) = 1 - .1587 = .8413.$

(c) $P(X \leq 2.2) = P\left(\dfrac{X-1.8}{.25} \leq \dfrac{2.2-1.8}{.25}\right)$

$\qquad = P(Z \leq 1.6) = .9452.$

5. From the given information, we have

$$\mu = 16.2, \quad \sigma = 0.1$$

(a) $P(15.5 < X < 16.2) = P\left(\dfrac{15.5 - 16.2}{0.1} < \dfrac{X - 16.2}{0.1} < \dfrac{16.2 - 16.2}{0.1}\right)$

$$= P(-7 < Z < 0)$$

$$= P(Z < 0) - P(Z \leq -7)$$

$$= 0.5 - 0.0 = 0.5.$$

(b) $P(X \geq 16.4) = P\left(\dfrac{X - 16.2}{0.1} \geq \dfrac{16.4 - 16.2}{0.1}\right)$

$$= P(Z \geq 2) = .0228.$$

(c) $P(X \leq 16.1) = P\left(\dfrac{X - 16.2}{0.1} \leq \dfrac{16.1 - 16.2}{0.1}\right)$

$$= P(Z \leq -1) = .1587.$$

7. From the given information, we have

$$\mu = 155, \quad \sigma = 7$$

(a) $P(145 < X < 165) = P\left(\dfrac{145 - 155}{7} < \dfrac{X - 155}{7} < \dfrac{165 - 155}{7}\right)$

$$= P(-1.43 < Z < 1.43)$$

$$= P(Z < 1.43) - P(Z \leq -1.43)$$

$$= 0.9236 - .0764 = 0.8472.$$

(b) $P(X \geq 150) = P\left(\dfrac{X - 155}{7} \geq \dfrac{150 - 155}{7}\right)$

$$= P(Z \geq -0.71) = 1 - P(Z \leq -0.71) = 1 - .2389 = .7611.$$

(c) $P(X \leq 169) = P\left(\dfrac{X - 155}{7} \leq \dfrac{169 - 155}{7}\right)$

$$= P(Z \leq 2) = .9772.$$

PRACTICE PROBLEMS FOR SECTION 5.6

1. Since the random variables X and Y are independent

(a) $\mu_{X+Y} = \mu_X + \mu_Y = 10 + 15 = 25$

$\sigma_{X+Y}^2 = \sigma_X^2 + \sigma_Y^2 = 9 + 16 = 25$

$\sigma_{X+Y} = \sqrt{\sigma_X^2 + \sigma_Y^2} = \sqrt{25} = 5$

$$P(X + Y \geq 33) = P\left(\dfrac{X + Y - 25}{5} \geq \dfrac{33 - 25}{5}\right) = P(Z \geq 1.6) = 0.0548.$$

(b) $\mu_{X-Y} = \mu_X - \mu_Y = 10 - 15 = -5$

$\sigma_{X-Y}^2 = \sigma_X^2 + \sigma_Y^2 = 9 + 16 = 25$

$\sigma_{X-Y} = \sqrt{\sigma_X^2 + \sigma_Y^2} = \sqrt{25} = 5$

$$P(-8 \leq X - Y \leq 6) = P\left(\frac{-8+5}{5} \leq \frac{X-Y+5}{5} \leq \frac{6+5}{5}\right)$$
$$= P(-.67 \leq Z \leq 2.2)$$
$$= P(Z \leq 2.2) - P(Z \leq -.67)$$
$$= .9861 - .2514 = .7347.$$

(c) $P(20 \leq X + Y \leq 28) = P\left(\dfrac{20-25}{5} \leq \dfrac{X+Y-25}{5} \leq \dfrac{28-25}{5}\right)$
$$= P(-1 \leq Z \leq 0.6)$$
$$= P(Z \leq 0.6) - P(Z \leq -1)$$
$$= 0.7257 - 0.1587 = 0.5670.$$

(d) $\mu_{X-2Y} = \mu_X - 2\mu_Y = 10 - 2 \times 15 = -20$
$\sigma_{X-2Y}^2 = \sigma_X^2 + 4\sigma_Y^2 = 9 + 4 \times 16 = 73$
$\sigma_{X-2Y} = \sqrt{\sigma_X^2 + 4\sigma_Y^2} = \sqrt{73} = 8.54$

$$P(X - 2Y \leq -10) = P\left(\frac{X-2Y+20}{8.54} \leq \frac{-10+20}{8.54}\right) = P(Z \leq 1.17) = 0.8790.$$

3. (a) $\mu_{X+Y} = \mu_X + \mu_Y = 70 + 75 = 145$
$\sigma_{X+Y}^2 = \sigma_X^2 + \sigma_Y^2 = 64 + 100 = 164$
$\sigma_{X+Y} = \sqrt{\sigma_X^2 + \sigma_Y^2} = \sqrt{164} = 12.806$

$$P(X + Y \geq 145) = P\left(\frac{X+Y-145}{12.806} \geq \frac{145-145}{12.806}\right) = P(Z \geq 0) = 0.5.$$

(b) $\mu_{X-Y} = \mu_X - \mu_Y = 70 - 75 = -5$
$\sigma_{X-Y}^2 = \sigma_X^2 + \sigma_Y^2 = 64 + 100 = 164$
$\sigma_{X-Y} = \sqrt{\sigma_X^2 + \sigma_Y^2} = \sqrt{164} = 12.806$

$$P(-18 \leq X - Y \leq 16) = P\left(\frac{-18+5}{12.806} \leq \frac{X-Y+5}{5} \leq \frac{16+5}{12.806}\right) \cong P(-1.01 < Z < 1.64)$$
$$= P(Z \leq 1.64) - P(Z \leq -1.01) = 0.9495 - 0.1562 = 0.7933$$

(c) $P(122 \leq X + Y \leq 168) = P\left(\dfrac{122-145}{12.806} \leq \dfrac{X+Y-145}{12.806} \leq \dfrac{168-145}{12.806}\right)$
$$\cong P(-1.80 \leq Z \leq 1.80)$$
$$= P(Z \leq 1.80) - P(Z \leq -1.80)$$
$$= 0.9641 - 0.0359 = 0.9282.$$

5. All three sections of MCAT test carry the same weight and we wish to determine the distribution of $U = X + Y + Z$. Since the random variables X, Y, and Z are independently normally distributed the random variable U is also normally distributed. The mean and the standard deviation of the random variable U are given by

$$\mu_u = \mu_{X+Y+Z} = \mu_X + \mu_Y + \mu_Z = 10 + 12 + 10 = 32$$
$$\sigma_u^2 = \sigma_{X+Y+Z}^2 = \sigma_X^2 + \sigma_Y^2 + \sigma_Z^2 = (2.6)^2 + (1.2)^2 + (1.3)^2 = 9.89$$
$$\sigma_U = \sqrt{9.89} = 3.145$$

PRACTICE PROBLEMS FOR SECTIONS 5.7 AND 5.8

1. We use MINITAB to verify whether the given set of data is drawn from a normal population.

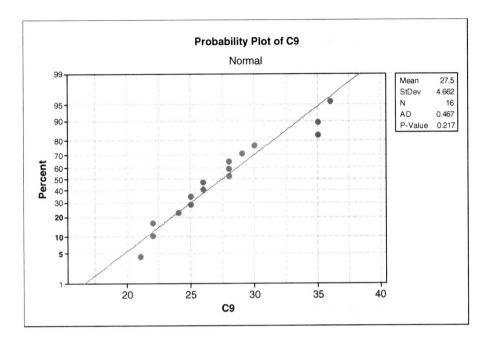

The above graph shows that all of the data points either fall on a straight line or are very close to the straight line. Moreover, the p-value is 0.217, which is greater 0.05 the level that is usually very widely accepted level of significance. Hence, we can conclude that these data have been drawn from a normal population. The mean and the standard deviations are 27.5 and 4.662 respectively.

3. The normality test can be carried out by using one of the statistical packages discussed in this book. Thus, for example, using MINITAB, the normality plot is as shown below.

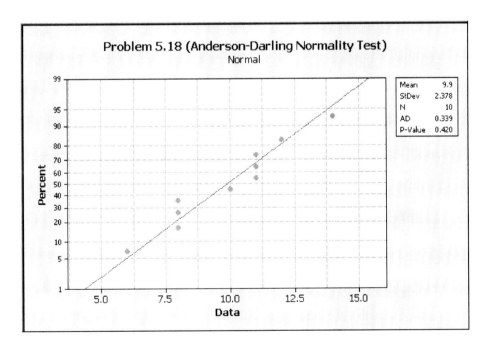

Clearly all of the points either fall on a straight line or are very close to the straight line. Moreover, the p-value is 0.420, which is greater 0.05 the level that is usually very widely accepted level of significance. Hence we conclude that the data come from a normal population.

5. (a) Using binomial tables, we have

$$P(4 \leq X \leq 12) = P(X \leq 12) - P(X \leq 3) = 0.9790 - 0.0160 = 0.9630.$$

(b) The mean and the standard deviation of the binomial distribution in (a) are

$$\mu = n\theta = 8, \quad \text{and} \quad \sigma = \sqrt{n\theta(1 - \theta)} = 2.19.$$

Now using the normal approximation, we have

$$P(4 \leq X \leq 12) \cong P\left(\frac{3.5 - 8}{2.19} \leq \frac{X - 8}{2.19} \leq \frac{12.5 - 8}{2.19}\right)$$
$$= P(-2.05 \leq Z \leq 2.05) = P(Z \leq 2.05) - (Z \leq -2.05) = 0.9798 - 0.0202 = 0.9596,$$

which is approximately equal to the exact probability obtained in (a).

7. In order to find the desired probabilities using the normal approximation of the binomial distribution we first determine its mean and the standard deviation, which are given by

$$\mu = 500 \times .03 = 15, \quad \sigma = \sqrt{500(.03)(.97)} = 3.814$$

(a) $P(X \geq 10) = P\left(Z \geq \frac{9.5 - 15}{3.814}\right) = P(Z \geq -1.44) = .9251.$

(b) $P(15 \leq X \leq 20) = P\left(\frac{14.5 - 15}{3.814} \leq Z \leq \frac{20.5 - 15}{3.814}\right) = P(-.13 \leq Z \leq 1.44) = .4251 + .0517 = .4768.$

9. In this problem the phenomena of passengers arriving on time can be modeled by the binomial distribution with $n = 212$ *and* $\theta = .92$. The seats in the plane will be empty only if less than 200 passengers show up on time. Thus, finding the probability that some seats in the plane will be empty is equivalent to finding the probability that less than 200 passengers show up on time. That is,

$$P(X < 200)$$

where X is the number of passengers showing up on time. To find this probability we use the normal approximation to the binomial distribution with mean and the standard deviation given by

$$\mu = 212 \times .92 = 195.04, \quad \sigma = \sqrt{212(.08)(.92)} = 3.95$$

Thus, the desired probability is given by

$$PX < 200) = P\left(Z \leq \frac{199.5 - 195.04}{3.95}\right) = P(Z \leq 1.13) = .8708.$$

PRACTICE PROBLEMS FOR SECTION 5.9.1

1. To use MINITAB we first find the location parameter μ and the scale parameter σ. Using equations (5.9.2) and (5.9.3), we have

$$\text{Mean} = e^{\mu + \sigma^2/2},$$

or

$$e^{\mu + \sigma^2/2} = 6,000 \tag{1}$$

Variance $= (5{,}000)^2 = e^{2\mu+2\sigma^2} - e^{2\mu+\sigma^2} = e^{2\mu+2\sigma^2} - (6{,}000)^2$, that is

$$e^{2\mu+2\sigma^2} = (5{,}000)^2 + (6{,}000)^2 \tag{2}$$

Taking the natural log on both sides of equation (1) and (2) and solving them for μ and σ, we obtain the location and the scale parameters equal to

$$\mu = 8.44, \quad \sigma = 0.723.$$

Now using MINITAB we obtain the

$$P(X \geq 8{,}000) = 1 - P(X \leq 8{,}000) = 1 - 0.7754 = 0.2246.$$

3. Using the given information we obtain

$$E(X) = e^{\mu+\sigma^2/2} = e^{2+(0.5)^2/2} = 8.373.$$

$$\begin{aligned} Var(X) &= e^{2\mu+2\sigma^2} - e^{2\mu+\sigma^2} \\ &= e^{2(2)+2(.5)^2} - e^{2(2)+(.5)^2} = 19.91. \end{aligned}$$

5. Using the information given we know that the random variable Y has lognormal distribution with parameters $\mu = 2, \sigma^2 = 4$. Thus, the mean and the standard deviation of the random variable Y are given by

(a) $\mu = e^{\mu+\sigma^2/2} = e^4 = 54.6, \sigma^2 = e^{2\mu+2\sigma^2} - e^{2\mu+\sigma^2} = e^{12} - e^8 = 159773.83, \sigma = 399.72.$

(b) Since $\ln Y$ is distributed as normal with mean 2 and standard deviation 2, we have

$$P(Y > 250) = P(\ln Y > \ln 250) = P(X > 5.52) = P\left(\frac{X-2}{2} > \frac{5.52-2}{2}\right)$$
$$= P(Z > 1.76) = 1 - .9608 = .0392.$$

(c) $P(100 < Y < 200) = P(\ln 100 < \ln Y < \ln 200) = P(4.605 < X < 5.298)$

$$= P\left(\frac{4.605-2}{2} < Z < \frac{5.298-2}{2}\right) = P(1.30 < Z < 1.65) = .9505 - .9032 = .0473.$$

PRACTICE PROBLEMS FOR SECTION 5.9.2

1. Using MINITAB, we obtain
   (a) $P(T \geq 5) = 1 - P(T \leq 5) = 1 - 0.7135 = 0.2865.$
   (b) $P(3 \leq T \leq 6) = P(T \leq 6) - P(T \leq 3) = 0.7769 - 0.5276 = 0.2493.$
   (c) $P(T \leq 4) = 0.6321$
   (d) $P(T < 5) = 0.7135.$

3. In this problem the lapse time between two accidents has the exponential distribution with mean 15 days. Using equation (5.9.7) the probability that time between two successive accidents is more than 20 days is given by

$$P(X \geq 20) = e^{-20/15} = 0.2635.$$

5. In this problem we first find the probability that a computer will function for more than five years, that is, its failure time is more than five years. This probability is given by

$$P(X > 5) = e^{-5/3} = 0.1889.$$

Now the probability that 15 of the 20 computers will be functioning after five years can be found by using the binomial distribution with $\theta = .1889$. Thus, the desired probability is given by

$$P(X = 15) = 0.0000001.$$

7. The time T, in minutes, between the arrivals of two successive patients in an emergency room has an exponential distribution with mean 20 minutes, that is

$$\mu = 1/\lambda = 20 \Rightarrow \lambda = 1/20$$

(a) $P(T > 30) = e^{-(1/20)30} = e^{-1.5} = .2231$.

(b) $P(12 < T < 18) = P(T < 18) - P(T < 12) = (1 - e^{-(1/20)18}) - (1 - e^{-(1/20)12}) = e^{-0.6} - e^{-0.9} = .1422$.

(c) $P(T < 25) = 1 - e^{(1/20)25} = 1 - e^{-1.25} = .7135$.

PRACTICE PROBLEMS FOR SECTIONS 5.9.3 AND 5.9.4

1. In this problem we first find the probability that a computer will function for more than five years, that is, its failure time is more than five years. Using MINITAB, we have

$$P(X > 5) = 1 - P(X \le 5) = 1 - 0.9329 = 0.0671$$

Now the probability that 12 of the 15 computers will be functioning after five years can be found by using binomial distribution. Thus, the desired probability is given by

$$P(X = 12) = 0.0000.$$

3. In this problem we first determine values of the parameters of the gamma distribution. We are given

$$\text{Mean} = \gamma/\lambda = 8 \quad \text{and} \quad \text{Variance} = \gamma/\lambda^2 = 4$$

Solving these equations for γ and λ, we obtain

$$\lambda = 2, \quad \gamma = 16.$$

Now using MINITAB, we obtain

(a) $P(X \le 12) = 0.9656$.

(b) $P(X \ge 6) = 1 - P(X \le 6) = 1 - 0.1556 = 0.8444$.

(c) $P(6 \le X \le 10) = P(X \le 10) - P(X \le 6) = 0.8435 - 0.1556 = 0.6879$.

5. $\text{Mean} = \mu = \alpha\Gamma\left(1 + \frac{1}{\beta}\right) = 6 \times \Gamma(1 + 1/0.5) = 6 \times 2 = 12$.

$$\sigma^2 = \alpha^2\left[\Gamma\left(1 + \frac{2}{\beta}\right) - \left(\Gamma\left(1 + \frac{1}{\beta}\right)\right)^2\right] = (6)^2 \times \{\Gamma(1 + 2/0.5) - [\Gamma(1 + 1/0.5)]^2\} = 720.$$

The desired probabilities can easily be found by using one of the statistical packages. Here we find these probabilities by using MINITAB (**Calc** > **Probability Distributions** > **Weibull**). Thus, we have

(b) $P(X > 10) = 1 - P(X \le 10) = 1 - 0.7250 = 0.2750$.

(c) $P(X \le 15) = 0.7943$.

7. Using MINITAB with shape parameter $\gamma = 2$ and scale parameter $1/\lambda = 1/2 = 0.5$, we obtain

(a) $P(T \ge 1) = 1 - P(T \le 1) = 1 - .5940 = .4060$.

(b) $P(T \le 2) = .9084$.

(c) $P(1 \le T \le 2) = P(T \le 2) - P(T \le 1) = .9084 - .5940 = .3144$.

9. (a) The expected life of the battery is: $\mu = \alpha\Gamma\left(1 + \dfrac{1}{\beta}\right) = 2\Gamma\left(1 + \dfrac{1}{0.5}\right) = 2\Gamma(3) = 4$ years.

   (b) Using MINITAB with shape parameter $\beta = 0.5$, and scale parameter $\alpha = 2$, we obtain

$$P(T > 5) = 1 - P(T \le 5) = 1 - .7943 = .2057.$$

REVIEW PRACTICE PROBLEMS

1. $X \sim N(1500, (200)^2)$

Standardize X by using $\dfrac{X - \mu}{\sigma} = \dfrac{X - 1500}{200}$

(a) $P(X < 1400) = P\left(\dfrac{X - 1500}{200} < \dfrac{1400 - 1500}{200}\right) = P\left(Z < -\dfrac{1}{2}\right)$

$$= 1 - P\left(Z < \dfrac{1}{2}\right) = 1 - 0.6915 = 0.3085.$$

(b) $P(X > 1700) = P\left(\dfrac{X - 1500}{200} > \dfrac{1700 - 1500}{200}\right) = P(Z > 1)$

$$= 1 - P(Z < 1) = 1 - 0.8413 = 0.1587.$$

(c) Find A so that $P(X > A) = 0.05$

$$\Rightarrow P\left(Z > \dfrac{A - 1500}{200}\right) = 0.05 \quad \text{or} \quad P\left(Z < \dfrac{A - 1500}{200}\right) = 0.95$$

Thus, $\dfrac{A - 1500}{200} = 1.645,$ so $A = 1829.$

(d) Find B so that $P(1500 - B < X < 1500 + B) = 0.95$

$$\Rightarrow P\left(\dfrac{-B}{200} < Z < \dfrac{B}{200}\right) = 0.95 \quad \Rightarrow \quad P\left(Z < \dfrac{B}{200}\right) = 0.975$$

Thus, $\dfrac{B}{200} = 1.96,$ so $B = 392.$

3. Let X be the weight of fillings in ounces. Then, $X \sim N(16.30, (0.15)^2)$.
   (a) The percentage of fillings under weight is estimated by $P(x < 16.00)$.

$$P(X < 16.00) = P\left(Z < \dfrac{16.00 - 16.30}{0.15}\right) = P(Z < -2) = 1 - P(Z < 2)$$
$$= 1 - 0.97725 = 0.02275.$$

Thus 2.275% are under weight.

(b) The percentage within 16.3 ± 0.2 ounces is estimated by

$$P(16.1 < X < 16.5) = P\left(\frac{16.10 - 16.30}{0.15} < Z < \frac{16.50 - 16.30}{0.15}\right)$$
$$= P(-1.333 < Z < 1.333)$$
$$= P(Z < 1.333) - P(Z < -1.333) = 0.8164.$$

Thus, the percentage between 16.3 ± 0.2 is 81.64%.

5. Let X be the diameter of the ball bearings. Then, $X \sim N(0.2497, (0.0002)^2)$.
Further, we are given the specification limits on the diameters are between 0.2497 and 0.2503.

(a)
$$P(\text{defective bearing}) = P(X > 0.2503) + P(X < 0.2497)$$
$$= P\left(Z > \frac{0.2503 - 0.2497}{0.0002}\right) + P\left(Z < \frac{0.2497 - 0.2497}{0.0002}\right)$$
$$= P(Z > 3) + P(Z < 0) = 0.00135 + 0.5 = 0.50135$$

50.135% are defective.

(b) By applying the result of Problem 4, the interval (0.2500 ± 0.0003) will contain the maximum percentage of good ball bearings if $\mu = 0.2500$ (and hence the minimum percentage of defectives).

(c)
$$P(\text{defective bearing if } \mu = 0.2500) = 1 - P(0.2497 < X < 0.2503)$$
$$= 1 - P\left(\frac{0.2497 - 0.2500}{0.0002} < Z < \frac{0.2503 - 0.2500}{0.0002}\right)$$
$$= 1 - P(-1.5 < Z < 1.5) = 1 - 0.86638 = 0.13362$$

13.362% are defective.

7. We use MINITAB (**Stat. > Basic Statistics > Normality Test**) to verify if these data come from a normal population. Since all of the data points either fall on a straight line or close to the straight line and the p-value is 0.174 which is greater than 0.05 the usual acceptable level of significance, we conclude that the these data come from a normal population.

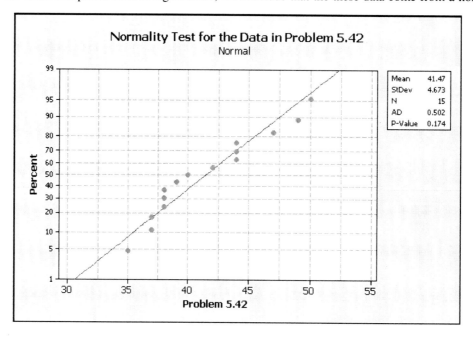

9. Let X be the number of defective articles. Then, $X \sim \text{Bin}(400, 0.2)$. To use the normal approximation, we need

$$\mu_x = n\theta = 400(0.2) = 80 \quad \text{and}$$

$$\sigma_x^2 = n\theta(1 - \theta) = 400(0.2)(0.8) = 64 \quad \text{so}$$

$$\sigma_x = \sqrt{n\theta(1 - \theta)} = 8.$$

(a)
$$P(X > 100) \cong P(X \geq 100.5) = P\left(Z \geq \frac{100.5 - 80}{8}\right)$$
$$= P(Z \geq 2.5625) \quad \text{(interpolate linearly)}$$
$$= 1 - 0.9948 = 0.0052.$$

(b)
$$P(80 - K < X < 80 + K) \cong P(80.5 - K \leq X \leq 79.5 + K)$$
$$= P\left(\frac{80.5 - K - 80}{8} \leq Z \leq \frac{79.5 + K - 80}{8}\right)$$
$$= P\left(\frac{0.5 - K}{8} \leq Z \leq \frac{-0.5 + K}{8}\right) = 0.95$$

Thus, $\dfrac{K - 0.5}{8} = 1.96$, so $K = 16.18$, i.e. $K = 17$.

11. The number of defectives X has a binomial distribution with $n = 400, \theta = 0.1$. So that

$$\mu_x = n\theta = 40$$

$$\sigma_x^2 = n\theta(1 - \theta) = 36$$

$$\sigma_x = 6.$$

(a) 15% of a sample of 400 is 60 items.

$$P(X > 60) \cong P(X \geq 60.5) = P\left(Z \geq \frac{60.5 - 40}{6}\right) = P(Z \geq 3.417)$$
$$= 0.0004$$

(b)
$$P(40 - K < X < 40 + K) \cong P\left(\frac{40.5 - K - 40}{6} \leq Z \leq \frac{39.5 + K - 40}{6}\right)$$
$$= P\left(\frac{0.5 - K}{6} \leq Z \leq \frac{-0.5 + K}{6}\right) = 0.90$$

Thus, $\dfrac{K - 0.5}{6} = 1.645$, so $K = 10.37$, i.e. $K = 11$.

13. The number of "sixes", X, is $\text{Bin}\left(720, \dfrac{1}{6}\right)$. Therefore

$$\mu_x = 720\left(\frac{1}{6}\right) = 120$$

$$\sigma_x^2 = 720\left(\frac{1}{6}\right)\left(\frac{5}{6}\right) = 100$$

$$\sigma_x = 10.$$

(a)
$$P(X > 130) \cong P(X \geq 130.5) = P\left(Z \geq \frac{130.5 - 120}{10}\right)$$
$$= P(Z \geq 1.05) = 1 - \Phi(1.05) = 0.1469.$$

(b)
$$P(100 \leq X \leq 140) \cong P(99.5 \leq X \leq 140.5)$$
$$= P\left(\frac{99.5 - 120}{10} \leq Z \leq \frac{140.5 - 120}{10}\right)$$
$$= P(-2.05 \leq Z \leq 2.05) = 2\Phi(2.05) - 1 = 0.9596.$$

15. Let X be total weight of an article in oz. Then, $E(X) = 2.05 + 3.10 + 10.5 = 15.65$, $\mathrm{Var}(x) = (0.03)^2 + (0.04)^2 + (0.12)^2 = 0.0169$, and $\sigma(X) = 0.13$.

(a)
$$P(X > 16) = P\left(Z > \frac{16 - 15.65}{0.13}\right) = P(Z > 2.692) = 1 - \Phi(2.692) = 0.0036.$$

(b) Let T be the total weight of four randomly picked articles. Then

$$E(T) = 4(15.65) = 62.6, \mathrm{Var}(T) = 4(0.0169) = 0.0676, \text{ and } \sigma(T) = 0.26.$$

$$P(T < K) = 0.95 \quad \Rightarrow \quad P\left(Z < \frac{K - 62.6}{0.26}\right) = 0.95$$

Thus,
$$\frac{K - 62.6}{0.26} = 1.645, \quad \text{so} \quad K = 63.03.$$

17. Let X be the total resistance of the resistor. Then

$$E(X) = 3(200) + 4(150) + 1(250) = 1450,$$

$$\mathrm{Var}(X) = 3(2)^2 + 4(3)^2 + 1(1)^2 = 49, \quad \text{and} \quad \sigma(x) = 7.$$

(a)
$$P(X < K) = 0.05 \quad \Rightarrow \quad P\left(Z < \frac{K - 1450}{7}\right) = 0.05$$

Thus,
$$\frac{K - 1450}{7} = -1.645, \quad \text{so} \quad K = 1438.49.$$

(b) Let $\overline{X}$ be the mean resistance of four such resistors. Then

$$E(\overline{X}) = 1450; \quad \mathrm{Var}(\overline{X}) = \frac{49}{4} = 12.25; \quad \sigma(\overline{X}) = 3.5.$$

$$P(\overline{X} > 1443) = P\left(Z > \frac{1443 - 1450}{3.5}\right) = P(Z > -2.00) = 0.97725.$$

19. In this problem the mean and the standard deviation are $\mu = 0$, $\sigma = 1$. Thus, using the normal distribution tables, we have

(a) $P(Z \leq 2.11) = 0.9826$
(b) $P(Z \geq -1.2) = 1 - P(Z \leq -1.2) = 0.8849.$
(c) $P(-1.58 \leq Z \leq 2.40) = P(Z \leq 2.40) - P(Z \leq -1.58) = 0.9918 - 0.0571 = 0.9347.$
(d) $P(Z \geq 1.96) = 1 - P(Z \leq 1.96) = 1 - 0.9750 = 0.0250.$
(e) $P(Z \leq -1.96) = 0.0250.$

21. Since $\lambda = 2$, the mean of the exponential distribution is $1/2 = 0.5$. Thus, we obtain

(a) $P(X > 1) = e^{-2(1)} = 0.1353.$
(b) $P(X > 2) = e^{-2(2)} = 0.0183.$
(c) $P(1 \leq X \leq 2) = P(X \leq 2) - P(X \leq 1) = e^{-2} - e^{-4} = 0.1170.$
(d) $P(X > 0) = 1 - P(X \leq 0) = 1 - 0 = 1.$

23. The desired probability can easily be found by using one of the statistical packages. Here we find this probability using MINITAB. Thus, we have

(a) $P(X \geq 450) = 1 - P(X \leq 450) = 1 - 0.8801 = 0.1199$.

(b) $P(X \geq 750) = 1 - P(X \leq 750) = 1 - 0.9290 = 0.0710$.

25. Using equation (5.9.19), we obtain

(a) $P(X < 800) = 1 - e^{-(x/\alpha)^\beta} = 1 - e^{-((0.001)(800))^2} = 0.4727$.

(b) $P(X > 1000) = e^{-(x/\alpha)^\beta} = e^{-((0.001)(1000))^2} = 0.3679$.

$P(1000 < X < 1500) = P(X \leq 1500) - P(X \leq 1000)$

(c)
$$= \left(1 - e^{-((0.001)(1500))^2}\right) - \left(1 - e^{-((0.001)(1000))^2}\right)$$

$$= 0.8946 - 0.6321 = 0.2625.$$

27. By using the memory less property of the exponential distribution, we obtain

$$P(X \geq 5 + 9|X > 5) = P(X \geq 9) = 0.4066.$$

29. Using equation (5.9.17), we obtain

(a) $\mu = E(X) = \dfrac{\gamma}{\lambda} = 3/1 = 3$ and $\sigma^2 = V(X) = \dfrac{\gamma}{\lambda^2} = 3/1 = 3$.

(b) $\mu = E(X) = \dfrac{\gamma}{\lambda} = 6/1.5 = 4$ and $\sigma^2 = V(X) = \dfrac{\gamma}{\lambda^2} = 6/(1.5)^2 = 6/2.25 = 2.67$.

31.
$$F(x) = \begin{cases} 0, & x \leq 0 \\ x^n, & 0 < x \leq 1, n \geq 1 \\ 1, & x > 1. \end{cases}$$

(a)
$$f(x) = \begin{cases} nx^{n-1}, & 0 < x \leq 1 \\ 0, & \text{elsewhere.} \end{cases}$$

(b) Let x_{50} be the median.

$$\text{Then } \int_0^{x_{50}} nx^{n-1}dx = \frac{1}{2}, \text{ by the definition of the } p^{\text{th}} \text{ percentile.}$$

$$\text{So } x^n|_0^{x_{50}} = \frac{1}{2}, \quad \text{and} \quad x_{50}^n = \frac{1}{2}, \quad \therefore x_{50} = \left(\frac{1}{2}\right)^{1/n} = \frac{1}{\sqrt[n]{2}}.$$

(c)
$$\mu = E(X) = \int_0^1 x \cdot nx^{n-1}dx = \frac{n}{n+1}x^{n+1}|_0^1 = \frac{n}{n+1}$$

$$E(X^2) = \int_0^1 x^2 \cdot nx^{n-1}dx = \frac{n}{n+2}x^{n+2}|_0^1 = \frac{n}{n+2}$$

Thus,
$$\sigma^2 = E(X^2) - \mu^2 = \frac{n}{n+2} - \left(\frac{n}{n+1}\right)^2 = \frac{n(n+1)^2 - n^2(n+2)}{(n+1)^2(n+2)}$$

$$= \frac{n}{(n+1)^2(n+2)}.$$

33.
$$f(x) = \begin{cases} 3x^2, & 0 < x < 1 \\ 0, & \text{otherwise.} \end{cases}$$

$$F(x) = \int_0^x 3t^2 dt = t^3|_0^x = x^3$$

(a)

$$\textit{Thus, the c.d.f. is } F(x) = \begin{cases} 0, & x \leq 0 \\ x^3, & 0 < x < 1 \\ 1, & x \geq 1. \end{cases}$$

(b)

$$F\left(\frac{1}{3}\right) = \left(\frac{1}{3}\right)^3 = \frac{1}{27} = 0.037$$

$$F\left(\frac{9}{10}\right) = \left(\frac{9}{10}\right)^3 = 0.729$$

$$P\left(\frac{1}{3} < x \le \frac{1}{2}\right) = F\left(\frac{1}{2}\right) - F\left(\frac{1}{3}\right) = \left(\frac{1}{2}\right)^3 - \left(\frac{1}{3}\right)^3 = \frac{19}{216} = 0.088.$$

(c) $P(x \le a) = \frac{1}{4} \Rightarrow F(a) = \frac{1}{4} \Rightarrow a^3 = \frac{1}{4} \Rightarrow a = \frac{1}{\sqrt[3]{4}}.$

(d)

$$\mu = \int_0^1 x \cdot 3x^2 dx = \frac{3}{4}x^4 \Big|_0^1 = \frac{3}{4} = 0.75.$$

$$E(x^2) = \int_0^1 x^2 \cdot 3x^2 dx = \frac{3}{5}x^5 \Big|_0^1 = \frac{3}{5} = 0.60.$$

Thus,
$$\sigma_x^2 = E(x^2) - \mu^2 = \frac{3}{5} - \left(\frac{3}{4}\right)^2 = \frac{3}{80} = 0.0375.$$

35. (a) Clearly $f(x) \ge 0$ for all positive values of c.

$$\int_0^3 cx(3 - x)dx = c\left(\frac{3x^2}{2} - \frac{x^3}{3}\right)\Big|_0^3 = (9/2)c \Rightarrow (9/2)c = 1, \quad for\ c = 2/9.$$

(b) Again, $f(x) \ge 0$ for all positive values of c.

$$\int_0^3 cx^2(3 - x)dx = c\left(\frac{3x^3}{3} - \frac{x^4}{4}\right)\Big|_0^3 = (27 - 81/4)c \Rightarrow (27/4)c = 1, \quad for\ c = 4/27.$$

(c) Again, in this case $f(x) \ge 0$ for all positive values of c.

$$\int_0^1 cx^3(1 - x)dx = c\left(\frac{x^4}{4} - \frac{x^5}{5}\right)\Big|_0^1 = (1/4 - 1/5)c \Rightarrow (1/20)c = 1, \quad for\ c = 20.$$

37. (a) From Problem 36, we know that $\mu = 2$ and $\sigma = 0.577$. Thus, we have

$$P(|X - \mu| \le 2\sigma) = P(|X - 2| \le 1.154) = P(-1.154 \le X - 2 \le 1.154) = P(0.846 \le X \le 3.154)$$

$$= \int_{0.846}^{3.154} 1/2dx = \int_1^3 1/2dx = 1$$

Whereas the lower bound of this probability using Chebychev's Inequality is 3/4 or 75%. Hence, the Chebychev's Inequality holds.

(b) From Problem 36(c), we know that $\mu = 3/5$ and $\sigma = 1/5$. Thus, we obtain

$$P(|X - \mu| \le 2\sigma) = P(|X - 3/5| \le 2/5) = P(-2/5 \le X - 3/5 \le 2/5) = P(0.2 \le X \le 1)$$

$$= \int_{0.2}^1 12x^2(1 - x)dx = \left(12\left(\frac{x^3}{3} - \frac{x^4}{4}\right)\right)\Big|_{0.2}^1 = 1 - 0.0272 = 0.9728.$$

Whereas the lower bound of this probability using Chebychev's Inequality is 3/4 or 75%. Hence, the Chebychev's Inequality holds.

39. The desired probabilities in this problem can be determined by using one of the statistical packages discussed in this book. Here, we determine these probabilities using MINITAB.

   (a) $P(40 < X < 120) = P(X < 120) - P(X < 40) = 0.8488 - 0.1429 = 0.7059.$

   (b) $P(X > 80) = 1 - P(X \leq 80) = 1 - 0.5665 = 0.4335.$

   (c) $P(X < 100) = 0.7350.$

41. Using equation (5.11.8), we obtain

$$\mu = \gamma/\lambda = 4/0.05 = 80, \quad \sigma^2 = \gamma/\lambda^2 = 4/(.05)^2 = 1600, \quad \text{and} \quad \sigma = 40.$$

43. The probabilities in this problem can be determined by using one of the statistical packages discussed in this book. Here, we determine these probabilities using MINITAB.

   (a) $P(X > 10) = 1 - P(X \leq 10) = 1 - 0.6321 = 0.3679.$

   (b) $P(X < 15) = 0.7769.$

   (c) $P(10 < X < 15) = P(X < 15) - PX < 10) = 0.7769 - 0.6321 = 0.1448.$

45. **(a)** From Chebychev's Inequality we know that

$$P(|X - \mu| \leq 3\sigma) \geq 88.8\%$$

Thus, using the information provided in Problem 44, we obtain

or $\qquad\qquad\qquad\qquad\qquad P(|X - 2| \leq 5.20) \geq 88.8\%$

or $\qquad\qquad\qquad\qquad\qquad P(-3.20 \leq X \leq 7.20) \geq 88.8\%$

Thus, the interval that contains at least 88.8% of the waiting period is (0, 7.20). Note that the lower limit is zero since the time cannot be negative.

   (b) The exact probability is given by

$$P(T \leq 7.2) = \int_0^{7.2} \frac{(2/3)^{4/3}}{\Gamma(4/3)} t^{1/3} e^{-(2/3)t} dt = 0.9835.$$

47. The average number of accidents taking place in three months is 1.5, that is, $\lambda = 1.5$. Let the random variable X denote the number of accidents taking place in three months. Then we want to find the probability

   (a) $P(X \geq 2) = 1 - P(X < 2) = 1 - P(X \leq 1) = 1 - 0.5578 = 0.4422.$

   (b) $P(X < 2) = P(X \leq 1) = 0.5578.$

49. If a random variable X has a lognormal distribution then we know that the random variable (ln X) is normally distributed. Thus, we have

   (a) $P(3,500 \leq X \leq 9,500) = P(\ln 3,500 \leq \ln X \leq \ln 9,500) = P(8.16 \leq \ln X \leq 9.16)$

$$= \left(\frac{8.16 - 5}{3} \leq \frac{\ln X - 5}{3} \leq \frac{9.16 - 5}{3}\right) = P(1.05 \leq Z \leq 1.39)$$

$$= P(Z \leq 1.39) - P(Z \leq 1.05) \cong 0.0646.$$

   (b) $P(1,500 \leq X \leq 2,500) = P(\ln 1,500 \leq \ln X \leq \ln 2,500) = P(7.31 \leq \ln X \leq 7.82)$

$$= \left(\frac{7.31 - 5}{3} \leq \frac{\ln X - 5}{3} \leq \frac{7.82 - 5}{3}\right) = P(0.77 \leq Z \leq 0.94)$$

$$= P(Z \leq 0.94) - P(Z \leq 0.77) = 0.8264 - 0.7794 \cong 0.0470.$$

51. Using Equation (5.9.2) – (5.9.3), we obtain

$$Mean = e^{\mu + \sigma^2/2} = e^{2+4/2} = e^4 = 54.60.$$

$$Variance = e^{2\mu + 2\sigma^2} - e^{2\mu + \sigma^2} = e^{12} - e^8 = e^8(e^4 - 1) = 1.59773 \times 10^5.$$

53. Using the given information and using equations (5.9.22) and (5.9.23), we obtain

(a) $\mu = \alpha \Gamma\left(1 + \dfrac{1}{\beta}\right) = 300 \times \Gamma\left(1 + \dfrac{1}{.25}\right) = 300 \times \Gamma(5) = 300 \times 24 = 7,200.$

(b)

$$\sigma^2 = \alpha^2\left(\Gamma\left(1 + \frac{2}{\beta}\right) - \left(\Gamma\left(1 + \frac{1}{\beta}\right)\right)^2\right) = (300)^2 \times \left[\Gamma(9) - (\Gamma(5))^2\right]$$

$$= (300)^2 \times [8! - (4!)^2] = (300)^2 \times (40320 - 576) = 3.57696 \times 10^9.$$

55. In this problem we determine the probabilities by using MINITAB. Thus, we obtain.

(a) $P(X \leq 14,000) = 0.8867.$

(b) $P(12,000 \leq X \leq 16,000) = P(X \leq 16,000) - P(X \leq 12,000) = 0.8995 - 0.8710 = 0.0285.$

(c) $P(X \geq 16,000) = 1 - P(X \leq 16,000) = 1 - 0.8995 = 0.1005.$

57. In this problem we determine the probabilities by using MINITAB. Thus, we obtain.

(a) $P(X \geq 25) = 1 - P(X \leq 25) = 1 - 0.7135 = 0.2835.$

(b) $P(15 < X < 25) = P(X \leq 25) - P(X \leq 15) = 0.7135 - 0.5276 = 0.1859.$

(c) $P(X \geq 20) = 1 - P(X \leq 20) = 1 - 0.6321 = 0.3679.$

59. In this problem we are interested in finding the probability of a certain number of breakdowns in a given time period. To determine these probabilities we use the Poisson distribution. From Problem 58, we have a mean $= 1/\lambda = 1.2 \Rightarrow \lambda = 1/1.2 = 0.83$. Now we are interested in finding the probability of a certain number of breakdowns in two units of time, that is, 1000 hours. Thus, in this problem the value of $\lambda = 2(.83) = 1.66$.

(a) $P(X > 2) = 1 - P(X \leq 2) = 1 - 0.7677 = 0.2323.$

(b) $P(X \geq 2) = 1 - P(X \leq 1) = 1 - 0.5058 = 0.4942.$

(c) $P(X < 2) = P(X \leq 1) = 0.5058.$

61. Using the given information the mean and the standard deviation of the binomial distribution are given by

$$\mu = n\theta = 225 \times 0.2 = 45, \quad \sigma = \sqrt{n\theta(1 - \theta)} = \sqrt{225(.2)(.8)} = 6$$

Thus, using the normal approximation to the binomial distribution, we obtain

(a) $P(X \leq 60) = P\left(\dfrac{X - 45}{6} \leq \dfrac{60.5 - 45}{6}\right) = P(Z \leq 2.58) = 0.9951.$

(b) $P(X \geq 57) = P\left(\dfrac{X - 45}{6} \geq \dfrac{56.5 - 45}{6}\right) = P(Z \geq 1.92) = 1 - P(Z \leq 1.92) = 1 - 0.9726 = 0.0274.$

(c) $P(80 \leq X \leq 100) = P\left(\dfrac{79.5 - 45}{6} \leq \dfrac{X - 45}{6} \leq \dfrac{100.5 - 45}{6}\right) = P(5.75 \leq Z \leq 9.25) = 0.$

63. The random variable is uniformly distributed over the interval [0, 20]. Thus, its density function is given by

$$f(x) = \frac{1}{20 - 0} = \frac{1}{20}$$

(a) $P(X < 5) = \displaystyle\int_0^5 (1/20)dx = 1/4.$

(b) $P(3 < X < 16) = \displaystyle\int_3^{16} (1/20)dx = 13/20.$

(c) $P(X > 12) = \int\limits_{12}^{20} (1/20)dx = 2/5.$

65. Since the random variable is uniformly distributed over the interval [19, 20], its mean and variance are given by

(a) $\mu = \dfrac{a+b}{2} = \dfrac{19+20}{2} = 19.5, \quad \sigma^2 = \dfrac{(b-a)^2}{12} = \dfrac{(20-19)^2}{12} = 1/12.$

(b) $P(|X - \mu| \le 2\sigma) = P(|X - 19.5| \le 0.577) = P(18.923 \le X \le 20.077) = P(19 \le X \le 20) = 100\%$

67. Given that the failure time T has the lognormal distribution with parameters μ and σ^2.

Thus, the random variable $\ln T$ is normally distributed with mean and variance μ and σ^2, respectively. The estimates μ and σ^2 are the sample mean and sample variance of the data obtained by taking the natural log of the given data. The natural log of the given data are

5.59471	5.33272	5.36598	5.53733	6.60530	6.36303
5.58725	6.58617	5.03695	5.72359	6.08450	5.37064

Thus, the desired estimates are the sample mean and the sample variance of these data, that is

$$\hat{\mu} = \overline{X} = 5.766, \quad \hat{\sigma}^2 = S^2 = 0.271$$

69. Proceeding in the same manner as in Problem 67, the natural logs of the given data are

1.60944	1.94591	2.19722	2.56495	3.17805	3.46574
3.55535	3.63759	3.73767	3.85015	3.89182	3.95124

Thus, the desired estimates are the sample mean and the sample variance of these data, that is

$$\hat{\mu} = \overline{X} = 3.132, \quad \hat{\sigma}^2 = S^2 = 0.691.$$

6

DISTRIBUTION OF FUNCTIONS OF RANDOM VARIABLES

PRACTICE PROBLEMS FOR SECTION 6.2

1. (a)

$$E(U) = E(5X + 3Y + 8Z)$$
$$= 5E(X) + 3E(Y) + 8E(Z)$$
$$= 5 \times 2 + 3 \times 7 + 8 \times 9 = 103$$

$$V(U) = V(5X + 3Y + 8Z)$$
$$= 5^2 V(X) + 3^2 V(Y) + 8^2 V(Z)$$
$$= 25 \times 2 + 9 \times 7 + 64 \times 9 = 689.$$

Note that random variable U is not distributed as Poison.

(b)

$$E(U) = \sum_{i=1}^{5} E(X_i) + \sum_{j=1}^{3} E(Y_j) + \sum_{k=1}^{8} E(Z_k).$$
$$= 5 \times 2 + 3 \times 7 + 8 \times 9 = 103$$

$$V(U) = \sum_{i=1}^{5} V(X_i) + \sum_{j=1}^{3} V(Y_j) + \sum_{k=1}^{8} V(Z_k)$$
$$= 5 \times 2 + 3 \times 7 + 8 \times 9 = 103.$$

Note that in this case U is also has a Poisson random variable with mean 103. That is, the sum of independent Poisson random variables is also Poisson random variable. However, as noted in part (a) not every linear combination of Poisson random variables is a Poisson random variable.

3. Let *X, Y,* and *Z* be the thickness of individual layers of material I, II, and III, respectively. Let *T* be the thickness of the sole. Then, we have

$$T = X + Y + Z_1 + Z_2,$$

Solutions Manual to Accompany Statistics and Probability with Applications for Engineers and Scientists, Bhisham C. Gupta and Irwin Guttman.
© 2014 John Wiley & Sons, Inc. Published 2014 by John Wiley & Sons, Inc.

where Z_1 *and* Z_2 are the thicknesses of layer 1 and 2 of material III. Thus, the mean and the variance of T are given by

$$E(T) = E(X) + E(Y) + E(Z_1) + E(Z_2) = 0.20 + 0.30 + 0.15 + 0.15 = 0.80,$$
$$V(T) = V(X) + V(Y) + 2V(Z_i) = (0.02)^2 + (0.01)^2 + 2(0.03)^2 = 0.0023.$$

5.

$$f_1(x_1) = \int_0^{1-x_1} 2dx_2 = 2(1 - x_1), \quad 0 \le x_1 \le 1.$$

$$f_2(x_2) = \int_0^{1-x_2} 2dx_1 = 2(1 - x_2), \quad 0 \le x_2 \le 1.$$

$$E(X_1) = \int_0^1 2x_1(1 - x_1)dx_1 = \left| x_1^2 - (2/3)x_1^3 \right|_0^1 = 1 - 2/3 = 1/3.$$

$$E(X_2) = \int_0^1 2x_2(1 - x_2)dx_2 = \left| x_2^2 - (2/3)x_2^3 \right|_0^1 = 1 - 2/3 = 1/3.$$

$$E(X_1X_2) = \int_0^1 \int_0^{1-x_2} 2x_1x_2dx_1dx_2 = \int_0^1 x_1^2 \big|_0^{1-x_2} x_2dx_2 = \int_0^1 (1 - x_2)^2 x_2dx_2$$

$$= \left| \left(\frac{x_2^2}{2} + \frac{x_2^4}{4} - \frac{2x_2^3}{3} \right) \right|_0^1 = 1/2 + 1/4 - 2/3 = 1/12.$$

$$Cov(X_1, X_2) = E(X_1X_2) - E(X_1)E(X_2) = \frac{1}{12} - \frac{1}{3} \times \frac{1}{3} = -1/36.$$

$$Var(X_1) = E(X_1^2) - (E(X_1))^2$$

$$E(X_1^2) = \int_0^1 2x_1^2(1 - x_1)dx_1 = 2\int_0^1 (x_1^2 - x_1^3)dx_1 = 2\left[\frac{x_1^3}{3} - \frac{x_1^4}{4} \right]_0^1 = 1/6.$$

Hence,

$$Var(X_1) = 1/6 - (1/3)^2 = 1/6 - 1/9 = 1/18.$$

Similarly,

$$Var(X_2) = 1/18.$$

Thus, we have

$$\rho = \frac{Cov(X_1, X_2)}{\sqrt{Var(X_1)} \times \sqrt{Var(X_2)}} = \frac{-1/36}{\sqrt{1/18} \times \sqrt{1/18}} = -1/2.$$

7. From Example 6.2.7 the joint probability function of the random variables X_1 and X_2 is as shown below.

X_2	X_1 0	1	2
0	1/3	0	1/3
1	0	1/3	0

From the joint probability function we can easily see that the marginal probability functions of the random variables X_1 and X_2 are

$$p(x_1) = 1/3, \quad x_1 = 0, 1, 2$$

$$p(x_2) = \begin{cases} 2/3, & x_2 = 0 \\ 1/3, & x_2 = 1 \end{cases}$$

respectively. The conditional probability functions are given by

$$p(x_1 | x_2 = 0) = \begin{cases} 1/2, & x_1 = 0, 2 \\ 0, & x_1 = 1 \end{cases}$$

$$p(x_1 | x_2 = 1) = \begin{cases} 0, & x_1 = 0, 2 \\ 1, & x_1 = 1 \end{cases}$$

$$p(x_2 | x_1 = 0) = \begin{cases} 1, & x_2 = 0 \\ 0, & x_2 = 1 \end{cases}$$

$$p(x_2 | x_1 = 1) = \begin{cases} 0, & x_2 = 0 \\ 1, & x_2 = 1 \end{cases}$$

$$p(x_2 | x_1 = 2) = \begin{cases} 1, & x_2 = 0 \\ 0, & x_2 = 1 \end{cases}$$

$$E(X_1 | X_2 = 0) = 0 \times 1/2 + 2 \times 1/2 + 1 \times 0 = 1$$

$$E(X_1^2 | X_2 = 0) = 0^2 \times 1/2 + 2^2 \times 1/2 + 1^2 \times 0 = 2$$

$$Var(X_1 | X_2 = 0) = E(X_1^2 | X_2 = 0) - [E(X_1 | X_2 = 0)]^2 = 2 - 1 = 1$$

$$E(X_1 | X_2 = 1) = 0 \times 0 + 2 \times 0 + 1 \times 1 = 1$$

$$E(X_1^2 | X_2 = 1) = 0^2 \times 0 + 2^2 \times 0 + 1^2 \times 1 = 1$$

$$Var(X_1 | X_2 = 1) = E(X_1^2 | X_2 = 1) - [E(X_1 | X_2 = 1)]^2 = 1 - 1 = 0$$

Note that $Var(X_1 | X_2 = 1) = 0$, because the whole probability is assigned to only one point.
Similarly, we can easily find the conditional means and variances for $(X_2 | X_1)$.

9. In this problem we given the joint probability distribution of random variables (X, Y) and we are interested in finding the distribution of the random variable $U = X - Y$. We now define another random variable $V = X + Y$, so that the inverse transformation is given by

$$X = (U + V)/2 \text{ and } Y = (V - U)/2$$

and the Jacobian of transformation is given by

$$J = \det \begin{pmatrix} \dfrac{\partial x}{\partial u} & \dfrac{\partial x}{\partial v} \\ \dfrac{\partial y}{\partial u} & \dfrac{\partial y}{\partial v} \end{pmatrix} = \det \begin{pmatrix} 1/2 & 1/2 \\ -1/2 & 1/2 \end{pmatrix} = 1/2$$

so that $|J| = 1/2$. Hence, the joint probability density function of the random variables (U, V) is given by

$$h(u, v) = f((u+v)/2, (v-u)/2)|J|$$

$$= \frac{1}{2}e^{-(u+v)/2}, \qquad for\ 0 \le u \le v < \infty,$$

$$= 0, \qquad\qquad elsewhere$$

Thus, the marginal density function of the random variable U is given by

$$h_1(u) = \frac{1}{2}\int_u^\infty e^{-(u+v)/2}dv = \frac{1}{2}e^{-(u/2)}\int_u^\infty e^{-(v/2)}dv$$

$$= e^{-(u/2)}\left|-e^{-(v/2)}\right|_u^\infty = e^{-(u/2)}(0 + e^{-(u/2)})$$

$$= e^{-u}, \qquad for\ u \ge 0,$$

$$= 0, \qquad\qquad elsewhere.$$

PRACTICE PROBLEMS FOR SECTIONS 6.3 AND 6.4

1. Suppose $X_1, \cdots, X_n$ is a random sample from a population having standard normal distribution. Then we know that the m.g.f. of a random variable $Y = X_1 + \cdots + X_n$ is the product of the moment generating functions of X_i's($M_{x_i}(t) = e^{(1/2)t^2}$). Thus, we have

$$M_u(t) \quad = e^{(1/2)t^2} \times e^{(1/2)(-t)^2} = e^{t^2}.$$

$$Mean \quad = \frac{\partial(e^{t^2})}{\partial t}\bigg|_{t=0} = 0$$

$$Var(V) \quad = \frac{\partial^2(e^{t^2})}{\partial t^2}\bigg|_{t=0} = 2$$

3. Proceeding in the same manner as in Problem 1 above, we obtain

$$M_v(t) \quad = e^{(1/2)(2t)^2} \times e^{(1/2)(3t)^2} = e^{(13/2)t^2}.$$

$$Mean \quad = \frac{\partial(e^{(13/2)t^2})}{\partial t}\bigg|_{t=0} = 0$$

$$Var(V) \quad = \frac{\partial^2(e^{(13/2)t^2})}{\partial t^2}\bigg|_{t=0} = 13.$$

5. Here we are interested in determining the moment generating function of the random variable U, where its probability distribution is given by

$$f(u) = e^{-u} \quad u > 0$$

Hence, we have

$$M_u(t) = \int_0^\infty e^{tu}e^{-u}du = \int_0^\infty e^{-u(1-t)}du$$

$$= \frac{1}{(1-t)} = (1-t)^{-1}$$

This is the moment generating function of an exponential random variable with $\lambda = 1$.

REVIEW PRACTICE PROBLEMS

1. From equation (5.9.16) of Chapter 5, we know that the m.g.f. of the gamma distribution is given by

$$M_X(t) = (1 - t/\lambda)^{-\gamma}$$

Furthermore, from Chapter 5 we know that if random variables $X_1, X_2, \cdots, X_n$ are independent random variables then the m.g.f. of the random variable $Y = \sum_{i=1}^{n} X_i$ is given by

$$M_Y(t) = \prod_{i=1}^{n} M_{X_i}(t)$$

Thus, in this problem the m.g.f. of the random variable $Y = \sum_{i=1}^{n} X_i$ is given by

$$M_Y(t) = \prod_{i=1}^{n} M_{X_i}(t) = \prod_{i=1}^{n} (1 - t/\lambda)^{-\gamma_i} = (1 - t/\lambda)^{-\Sigma \gamma_i}.$$

This is the moment generating function of the gamma random variable with parameter (λ, γ), where $\gamma = \sum \gamma_i$.

3. Given that the random variables $X_1, X_2, \cdots, X_n$ are independent and having Poisson distribution with parameters $\lambda_1, \lambda_2, \cdots, \lambda_n$, respectively. From Chapter 4 we know that m.g.f. of a Poisson random variable X_i, is given by

$$M_{X_i} = e^{\lambda_i(e^t - 1)}$$

Thus, the m.g.f. of the random variable $Y = \sum_{i=1}^{n} X_i$ is given by

$$M_Y(t) = \prod_{i=1}^{n} M_{X_i} = \prod_{i=1}^{n} e^{\lambda_i(e^t - 1)} = e^{(\Sigma \lambda_i)(e^t - 1)}$$

This is the moment generating function of a Poisson random variable with $\lambda = \sum \lambda_i$.

5. Let X and Y be the resistances of the first and second components, respectively, and let T be the total resistance. Given $\mu_x = 200$, $\sigma_x = 2$, $\mu_y = 150$, $\sigma_y = 3$, we have

$$\mu_T = E(T) = \mu_x + \mu_y = 200 + 150 = 350$$
$$\sigma_T^2 = \text{Var}(T) = \sigma_x^2 + \sigma_y^2 = 2^2 + 3^2 = 13$$
$$\sigma_T = \sqrt{13} = 3.606.$$

7. $\mu_A = 0.0100 \qquad \mu_B = 0.0050 \qquad \mu_C = 0.0025$
$\sigma_A = 0.0005 \qquad \sigma_B = 0.0003 \qquad \sigma_C = 0.0001.$

Let T be the thickness of a laminated strip, with A_i, B_j, C_k the thicknesses of the layers. Then

$$T = A_1 + A_2 + B_1 + B_2 + B_3 + C_1 + C_2 + C_3 + C_4.$$
$$\mu_T = E(T) = 2(0.0100) + 3(0.0050) + 4(0.0025) = 0.0450$$
$$\sigma_T^2 = \text{Var}(T) = 2(0.0005)^2 + 3(0.0003)^2 + 4(0.0001)^2 = 81 \times 10^{-8}$$
or
$$\sigma_T = 0.0009.$$

9. $100\theta_1\%$ have diameters less than 0.4950

$100\theta_2\%$ have diameters between 0.4950 and 0.5050

$100\theta_3\%$ have diameters greater than 0.5050

So that $\theta_3 = 1 - \theta_1 - \theta_2$. Then

$$p(x_1, x_2) = \frac{n!}{x_1! x_2! (n - x_1 - x_2)!} \theta_1^{x_1} \theta_2^{x_2} (1 - \theta_1 - \theta_2)^{n - x_1 - x_2}.$$

(a) $p_2(x_2) = \sum_{x_1} \frac{n!}{x_1! x_2! (n - x_1 - x_2)!} \theta_1^{x_1} \theta_2^{x_2} (1 - \theta_1 - \theta_2)^{n - x_1 - x_2}$

$$= \frac{n!}{x_2!} \theta_2^{x_2} \sum_{x_1} \frac{1}{x_1! (n - x_1 - x_2)!} \theta_1^{x_1} (1 - \theta_1 - \theta_2)^{n - x_1 - x_2}$$

$$= \frac{n!}{x_2!} \cdot \frac{\theta_2^{x_2}}{(n - x_2)!} \sum_{x_1=0}^{n - x_2} \frac{(n - x_2)!}{x_1! ((n - x_2) - x_1)!} \theta_1^{x_1} ((1 - \theta_2) - \theta_1)^{(n - x_2) - x_1}$$

$$= \frac{n!}{x_2!} \cdot \frac{\theta_2^{x_2}}{(n - x_2)!} (\theta_1 + 1 - \theta_2 - \theta_1)^{n - x_2} \quad \text{[by the binomial theorem]}$$

$$= \frac{n!}{x_2! (n - x_2)!} \theta_2^{x_2} (1 - \theta_2)^{n - x_2}$$

$$= \binom{n}{x_2} \theta_2^{x_2} (1 - \theta_2)^{n - x_2}$$

That is, $X_2 \sim \text{Bin}(n, \theta_2)$.

To find the marginal of X_2, we are considering only two categories: 'X_2' and 'not X_2'. Thus the distribution simplifies from a multinomial to a binomial.

(b) Since X_1 and X_2 are binomial with parameters θ_1 and θ_2 respectively,

$$E(X_1) = n\theta_1 \qquad\qquad E(X_2) = n\theta_2$$

$$\text{Var}(X_1) = n\theta_1(1 - \theta_1) \qquad \text{Var}(X_2) = n\theta_2(1 - \theta_2).$$

(c) $\text{Cov}(X_1, X_2) = E(X_1 X_2) - E(X_1) \cdot E(X_2)$

$$E(X_1 X_2) = \sum_{0 \le x_1 \le n,\, 0 \le x_2 \le n,\, 0 \le x_1 + x_2 \le n} x_1 x_2 \frac{n!}{x_1! x_2! (n - x_1 - x_2)!} \theta_1^{x_1} \theta_2^{x_2} (1 - \theta_1 - \theta_2)^{n - x_1 - x_2}.$$

Thus,

$$E(X_1 X_2) = \sum_{x_1=1}^{n-1} \sum_{x_2=1}^{n - x_1} \frac{n!}{(x_1 - 1)! (x_2 - 1)! (n - x_1 - x_2)!}$$

$$\cdot \theta_1^{x_1} \theta_2^{x_2} (1 - \theta_1 - \theta_2)^{n - x_1 - x_2}$$

$$= \sum_{k=0}^{n-2} \sum_{j=0}^{n-k-2} \frac{n!}{k! j! (n - k - j - 2)!} \theta_1^{k+1} \theta_2^{j+1} (1 - \theta_1 - \theta_2)^{n - k - j - 2}$$

$$= n(n - 1)\theta_1 \theta_2 \sum_{k=0}^{n-2} \sum_{j=0}^{n-k-2} \frac{(n - 2)!}{k! j! (n - k - j - 2)!} \times \theta_1^k \theta_2^j (1 - \theta_1 - \theta_2)^{n - k - j - 2}$$

The double sum sums to 1 since the expression to the right is a multinomial density. That is,

$$\sum_{k=0}^{n-2} \frac{(n-2)!}{k!} \cdot \frac{\theta_1^k}{(n-2-k)!} \cdot \sum_{j=0}^{n-2-k} \frac{\theta_2^j (1-\theta_1-\theta_2)^{n-2-j-k}(n-2-k)!}{j!(n-2-j-k)!}$$

$$= \sum_{k=0}^{n-2} \binom{n-2}{k} \theta_1^k \cdot \sum_{j=0}^{n-2-k} \binom{n-2-k}{j} \theta_2^j (1-\theta_1-\theta_2)^{n-2-j-k}$$

$$= \sum_{k=0}^{n-2} \binom{n-2}{k} \theta_1^k (1-\theta_1)^{n-2-k}$$

$$= (\theta_1 + 1 - \theta_1)^{n-2} = 1.$$

Hence, $E(X_1 X_2)$ $\qquad = n(n-1)\theta_1\theta_2,$

and $\text{Cov}(x_1, x_2) = n(n-1)\theta_1\theta_2 - n\theta_1 \times n\theta_2$
$$= n\theta_1\theta_2(n-1-n)$$
$$= -n\theta_1\theta_2.$$

11. Let X be a randomly-chosen digit. Then

$$\mu_x = \frac{1}{10}(0+1+2+\cdots+9) = \frac{45}{10} = 4.5,$$

$$\sigma_x^2 = \frac{1}{10}(0^2+1^2+2^2+\cdots+9^2) - \left(\frac{45}{10}\right)^2 = \frac{33}{4} = 8.25.$$

Let T be the sum of n digits. Then

$$T = \sum_{i=1}^{n} X_i, \quad \text{and} \quad \mu_T = 4.5n, \quad \sigma_T^2 = 8.25n.$$

13.

$$\theta = \frac{100}{10,000} = 0.01; \quad X = \# \text{ defectives}; \quad E(X) = n\theta = 0.01n$$

$$\text{Var}(X) = \frac{N-n}{N-1} n\theta(1-\theta) = \frac{10,000-n}{9999} n(0.01)(0.99).$$

The percentage of defectives in the sample is given by $100\dfrac{X}{n}$.

Now we wish to determine the probability

$$P\left(0.1 \le 100\frac{x}{n} \le 1.9\right) \ge 0.96 \Rightarrow P(0.001n \le x \le 0.019n) \ge 0.96.$$

Now $\qquad P(0.001n \le x \le 0.019n)$

$$= P(0.001n - 0.01n \le x - 0.01n \le 0.019n - 0.01n)$$
$$= P(|x - 0.01n| \le 0.009n) \ge 0.96.$$

By Chebyshev inequality, we obtain

$$P(|x - 0.01n| \le k\sigma) \ge 1 - \frac{1}{k^2} = 0.96, \Rightarrow 1 - \frac{1}{k^2} = 0.96 \Rightarrow \frac{1}{k^2} = 0.04$$

$$\Rightarrow k^2 = 25 \Rightarrow k = 5.$$

Thus, $\qquad\qquad\qquad k\sigma = 0.009n.$

In other words, we have

$$0.009n = 5\sqrt{\frac{10{,}000 - n}{9999}\, n(0.01)(0.99)} = 5\sqrt{\frac{10{,}000 - n}{9999}}\,(99)\left(10^{-4}\right)n$$

$$= 5 \times 10^{-2}\sqrt{\frac{10{,}000 - n}{101}}\,\sqrt{n}$$

$$\Rightarrow 0.9\sqrt{n} = 5\sqrt{\frac{10{,}000 - n}{101}} \Rightarrow 0.81n = 25\left(\frac{10{,}000 - n}{101}\right)$$

$$\Rightarrow 81.81n = 250{,}000 - 25n \Rightarrow 106.81n = 250{,}000 \Rightarrow n = 2340.6 \Rightarrow n = 2341.$$

15. Let X be the number of claims. Then

$$\mu_x = E(X) = (50{,}000)(0.01) = 500; \quad \sigma_x^2 = (50{,}000)(0.01)(0.99) = 495;$$
$$\sigma_x = \sqrt{495}.$$

Let T be the total value of claims. Then

$$T = 5000X; \quad \mu_T = E(T) = 5000\mu_x = 2{,}500{,}000; \quad \sigma_T = 5000\sigma_x = 5000\sqrt{495}.$$
$$P(2{,}500{,}000 - k \le T \le 2{,}500{,}000 + k) = P(|T - 2{,}500{,}000| \le k) \ge 0.99.$$

Now, by the Chebyshev inequality,

$$P(|T - 2{,}500{,}000| \le k\sigma_T) \ge 1 - \frac{1}{k^2};$$

Thus, we have

$$P(|T - 2{,}500{,}000| \le k) \ge 1 - \frac{\sigma_T^2}{k^2} = 0.99;$$

$$\frac{\sigma_T^2}{k^2} = 0.01, \quad \text{or} \quad k^2 = 100(5000)^2(495) \Rightarrow k = 10(5000)\sqrt{495} = 1{,}112{,}430.$$

17. The probability function $p(x)$ is determined as follows:

$$p(1) = \frac{\#\,\text{ways of assigning correct guesses to exactly one}}{24} = \frac{8}{24}$$

$$p(2) = \frac{\#\,\text{ways of assigning correct guesses to exactly two}}{24} = \frac{6}{24}$$

$$p(3) = \frac{\#\,\text{ways of assigning correct guesses to exactly three}}{24} = 0$$

$$p(4) = \frac{\#\,\text{ways of assigning all correct guesses}}{24} = \frac{1}{24}$$

$$p(0) = 1 - p(1) - p(2) - p(3) - p(4) = \frac{9}{24}.$$

Thus, the mean and the variance are given by

$$\mu_x = E(x) = 1 \cdot \frac{8}{24} + 2 \cdot \frac{6}{24} + 4 \cdot \frac{1}{24} = 1$$

$$\sigma_x^2 = \text{Var}(x) = 1^2 \cdot \frac{8}{24} + 2^2 \cdot \frac{6}{24} + 4^2 \cdot \frac{1}{24} - 1^2 = 2 - 1 = 1.$$

19. The binomial distribution has probability function

$$b(x) = \binom{n}{x}\theta^x(1-\theta)^{n-x}, x = 0, 1, \ldots, n.$$

The moment-generating function is

$$
\begin{aligned}
M_x(t) &= E(e^{tX}) \\
&= \sum_{x=0}^{n}\binom{n}{x}e^{tX}\theta^x(1-\theta)^{n-x} \\
&= \sum_{x=0}^{n}\binom{n}{x}(e^t\theta)^x(1-\theta)^{n-x} \\
&= (1-\theta+\theta e^t)^n \qquad \text{by the Binomial Theorem}
\end{aligned}
$$

Now
$$\frac{dM_x(t)}{dt} = n\theta e^t(1-\theta+\theta e^t)^{n-1},$$

and
$$\frac{d^2M_x(t)}{dt^2} = n(n-1)(1-\theta+\theta e^t)^{n-2}\theta^2 e^{2t} + n\theta e^t(1-\theta+\theta e^t)^{n-1}.$$

Thus, we obtain

$$\mu_1' = \frac{dM_x(0)}{dt} = n\theta e^0\left(1-\theta+\theta e^0\right)^{n-1} = n\theta$$

and
$$\mu_2' = \frac{d^2M_x(0)}{dt^2} = n(n-1)\theta^2 + n\theta.$$

Therefore
$$E(X) = \mu = \mu_1' = n\theta$$

and
$$\text{Var}(X) = \sigma^2 = \mu_2' - \left(\mu_1'\right)^2 = n(n-1)\theta^2 + n\theta - n^2\theta^2 = n\theta(1-\theta).$$

21. The joint probability function of X_1 and X_2 is given by

$$p(x_1, x_2) = p_2(x_2) \cdot p(x_1|x_2) = \frac{\mu^{x_2}e^{-\mu}}{x_2!} \cdot \binom{x_2}{x_1}\theta^{x_1}(1-\theta)^{x_2-x_1}, \quad x_1 = 0, 1, \ldots, x_2$$
$$x_2 = 0, 1, 2, \ldots.$$

Therefore the marginal distribution of X_1 is

$$
\begin{aligned}
p_1(x_1) &= \sum_{x_2=x_1}^{\infty}\frac{\mu^{x_2}e^{-\mu}}{x_2!} \cdot \binom{x_2}{x_1}\theta^{x_1}(1-\theta)^{x_2-x_1} \\
&= e^{-\mu}\theta^{x_1}\sum_{x_2=x_1}^{\infty}\frac{\mu^{x_2}}{x_2!} \cdot \frac{x_2!}{x_1!(x_2-x_1)!}(1-\theta)^{x_2-x_1} \\
&= \frac{e^{-\mu}(\mu\theta)^{x_1}}{x_1!}\sum_{x_2=x_1}^{\infty}\frac{\mu^{x_2-x_1}(1-\theta)^{x_2-x_1}}{(x_2-x_1)!} = \frac{e^{-\mu}(\mu\theta)^{x_1}}{x_1!}\sum_{x_2=x_1}^{\infty}\frac{[\mu(1-\theta)]^{x_2-x_1}}{(x_2-x_1)!} \\
&= \frac{e^{-\mu}(\mu\theta)^{x_1}}{x_1!} \times e^{\mu(1-\theta)} \\
&= \frac{e^{-\mu\theta}(\mu\theta)^{x_1}}{x_1!},
\end{aligned}
$$

Thus, the marginal distribution of X_1 is a Poisson$(\mu\theta)$ distribution.

23.

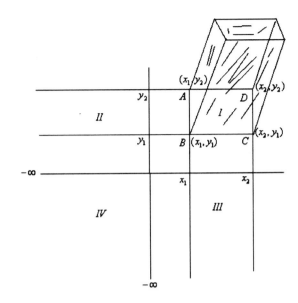

Referring to the above diagram we have $P(x_1 < X < x_2, y_1 < Y < y_2)$ = Volume between the surface representing the joint density function $f(x, y)$ and the area (I) covered by the rectangle ABCD in the x-y plane = Volume between the surface representing the joint density function $f(x, y)$ and the area (I + II + III + IV) in the x-y plane - Volume between the surface representing the joint density function $f(x, y)$ and the area (II + IV) in the x-y plane - Volume between the surface representing the joint density function $f(x, y)$ and the area (III + IV) in the x-y plane + Volume between the surface representing the joint density function $f(x, y)$ and the area (IV) in the x-y plane. That is,

$$P(x_1 < X < x_2, y_1 < Y < y_2) = F(x_2, y_2) - F(x_1, y_2) - F(x_2, y_1) + F(x_1, y_1).$$

25. **(a)** In this problem we are given that $f(x, y)$ is a p.d.f. of a bivariate normal. Thus, we first need to determine the suitable values of the parameters $\mu_1, \mu_2, \sigma_1, \sigma_2$, and ρ so that the expression

$$-\frac{25}{18}(x^2 + y^2 + (4/5)x - (14/5)y - (8/5)xy + 17/5)$$

represents the exponent of the bivariate normal, that is

$$-\frac{1}{2(1-\rho^2)}\left(\left(\frac{x-\mu_1}{\sigma_1}\right)^2 - 2\rho\left(\frac{x-\mu_1}{\sigma_1}\right)\left(\frac{y-\mu_2}{\sigma_2}\right) + \left(\frac{y-\mu_2}{\sigma_2}\right)^2\right)$$

By equating $-\dfrac{1}{2(1-\rho^2)} = -\dfrac{25}{18}$ and solving for ρ, we obtain $\rho = 4/5$. Now by comparing the coefficients of x^2, y^2, and xy, we can easily see that $\sigma_1 = \sigma_2 = 1$. Substituting the values of σ_1, σ_2, and ρ and again comparing the coefficients of x, y, xy, and the constant term, we obtain the following set of equations

$$\mu_1^2 + \mu_2^2 - (8/5)\mu_1\mu_2 = 17/5$$
$$(8/5)\mu_1 - 2\mu_2 = -14/5$$
$$(8/5)\mu_2 - 2\mu_1 = 4/5$$

Solving these equations for μ_1, and μ_2, we obtain $\mu_1 = 2, \mu_2 = 3$. Thus, $f(x, y)$ is a p.d.f. of a bivariate normal distribution with $\mu_1 = 2, \mu_2 = 3, \sigma_1 = \sigma_2 = 1$, and $\rho = 4/5$.

(b) The value of c is given by

$$c = \frac{1}{2\pi\sigma_1\sigma_2\sqrt{1-\rho^2}} = \frac{1}{2\pi\sqrt{1-(4/5)^2}} = \frac{5}{6\pi}$$

(c) Using equations (6.2.31) and (6.2.32), we obtain

$$f_1(x) = \frac{1}{\sqrt{2\pi}}e^{-(1/2)(x-2)^2}, \quad -\infty < x < \infty$$

$$f_2(y) = \frac{1}{\sqrt{2\pi}}e^{-(1/2)(y-3)^2}, \quad -\infty < y < \infty$$

7

SAMPLING DISTRIBUTIONS

PRACTICE PROBLEMS FOR SECTION 7.1

1. (a) All employees of the manufacturing company
   (b) All the chips manufactured in that batch
   (c) All the voters in that metropolitan area

3. Estimate of Mean$(\hat{T}) = N\overline{X} = (10,000) \times 39.60 = \$396,000$
   Estimate of Var$(\hat{T}) = 39.4384 \times 10^8$

5. Using MINITAB the mean and the standard deviation of the given data set are given by

$$\overline{X} = 32.20, \quad S = 5.25$$

PRACTICE PROBLEMS FOR SECTION 7.2

1. Because the sample size is large (>30). Using the Central Limit Theorem we can state that the sampling distribution of $\overline{X}$ is approximately normal with mean $\mu_{\overline{X}} = 28$ and standard deviation $\sigma_{\overline{X}} = \dfrac{9}{\sqrt{36}} = 1.5$.

3. (a) Standard Error will decrease from $\dfrac{\sigma}{6}$ to $\dfrac{\sigma}{8}$

   (b) Standard Error will decrease from $\dfrac{\sigma}{10}$ to $\dfrac{\sigma}{20}$

   (c) Standard Error will decrease from $\dfrac{\sigma}{9}$ to $\dfrac{\sigma}{18}$

   (d) Standard Error will decrease from $\dfrac{\sigma}{16}$ to $\dfrac{\sigma}{24}$

5. Since the sample size is large, using the Central Limit Theorem we can state that $\overline{X}$ is approximately normally distributed with mean $\mu_{\overline{X}} = 140$, and standard deviation $\sigma_{\overline{X}} = \dfrac{35}{\sqrt{49}} = 5$. Thus, we obtain

   (a) $P(\overline{X} > 145) = P\left(\dfrac{\overline{X} - 140}{5} > \dfrac{145 - 140}{5}\right) = P(Z > 1) = 1 - P(Z \leq 1) = 1 - 0.8413 = 0.1587$

Solutions Manual to Accompany Statistics and Probability with Applications for Engineers and Scientists, Bhisham C. Gupta and Irwin Guttman.
© 2014 John Wiley & Sons, Inc. Published 2014 by John Wiley & Sons, Inc.

(b) $P(\overline{X} < 140) = P\left(\dfrac{\overline{X} - 140}{5} < \dfrac{140 - 140}{5}\right) = P(Z < 0) = 0.5000$

(c)

$$P(132 < \overline{X} < 148) = P\left(\dfrac{132 - 140}{5} < \dfrac{\overline{X} - 140}{5} < \dfrac{148 - 140}{5}\right)$$
$$= P(-1.6 < Z < 1.6) = 0.9452 - 0.0548 = 0.8904$$

7. (a) In this problem we are given $n = 100$ and $\theta = 0.5$. Thus, we have $n\theta > 5$ and $n(1 - \theta) > 5$, therefore, the sample proportion $\hat{\theta}$ is approximately normal with mean $= \mu_{\hat{\theta}} = \theta = 0.5$ and standard error $\sigma_{\hat{\theta}} = \sqrt{\dfrac{\theta(1 - \theta)}{n}} = \sqrt{\dfrac{(0.5)(0.5)}{100}} = 0.05$

(b) $P(\hat{\theta} > 0.60) = P\left(\dfrac{\hat{\theta} - 0.5}{0.05} > \dfrac{0.6 - 0.5}{0.05}\right) = P(Z > 2.00) = 0.0228$

PRACTICE PROBLEMS FOR SECTION 7.3

1. Using the Chi-Square Tables given in Appendix A, we obtain
 (a) $P(\chi^2_{15} \geq 24.996) = 0.05$
 (b) $P(\chi^2_{15} \geq 6.262) = 0.975$
 (c) $P(\chi^2_{15} \leq 6.262) = 0.025$
 (d) $P(\chi^2_{15} \geq 7.261) = 0.95$
 (e) $P(\chi^2_{15} \leq 7.261) = 0.05$

3. Using the F-distribution Tables given in Appendix A, we obtain
 (a) $F_{6,8;0.05} = 3.58$
 (b) $F_{8,10,0.01} = 5.06$
 (c) $F_{6,10,0.05} = 3.22$
 (d) $F_{10,11,0.025} = 3.53$

5. Using MINITAB, for example, we obtain the value of x (i.e., we are determining the inverse probabilities, i.e., given the probabilities and we need to find the value of the random variable)
 Problem (1): 24.9958, 6.2621, 6.2621, 7.2609, 7.2609 [**Calc** > **Probability Distributions** > **Chi-Square**].
 Problem (2): 2.1009, 1.7247, 2.6025, 1.3722, 3.0545 [**Calc** > **Probability Distributions** > ***t*** . . .].
 Problem (3): 3.5806, 5.0567, 3.2172, 3.5257 [**Calc** > **Probability Distributions** > ***F*** . . .].
 Problem (4): 0.3433, 0.2328, 0.4296, 0.3238 [**Calc** > **Probability Distributions** > ***F*** . . .].
 [*Note: For MINITAB instructions see Review Practice Problems 12, 14, and 16.*]

7. (a) $P(3.247 < \chi^2_{10} < x) = P(\chi^2_{10} < x) - P(\chi^2_{10} < 3.247) = P(\chi^2_{10} < x) - 0.025 = 0.95 \Rightarrow P(\chi^2_{10} < x) = 0.975$.
 Now using the Chi-Square tables, we obtain $x = 20.483$
 (b) $P(8.260 < \chi^2_{20} < x) = P(\chi^2_{20} < x) - P(\chi^2_{20} < 8.260) = P(\chi^2_{20} < x) - 0.01 = 0.965 \Rightarrow P(\chi^2_{20} < x) = 0.975$.
 Now using the Chi-Square tables, we obtain $x = 34.170$
 (c) $P(13.120 < \chi^2_{25} < x) = P(\chi^2_{25} < x) - P(\chi^2_{25} < 13.120) = P(\chi^2_{25} < x) - 0.025 = 0.95 \Rightarrow P(\chi^2_{25} < x) = 0.975$.
 Now using the Chi-Square tables, we obtain $x = 40.646$

PRACTICE PROBLEMS FOR SECTION 7.4

1.

$$f(x) = \begin{cases} 0, & x \leq 0 \\ 1, & 0 < x \leq 1, \\ 0, & x > 1 \end{cases} \qquad F(x) = \begin{cases} 0, & x \leq 0 \\ x, & 0 < x \leq 1. \\ 1, & x > 1 \end{cases}$$

(a) $g\left(x_{(n)}\right) = nF^{n-1}\left(x_{(n)}\right)f\left(x_{(n)}\right) = \begin{cases} nx_{(n)}^{n-1}, & 0 < x_{(n)} < 1 \\ 0, & \text{otherwise.} \end{cases}$

(b) $g\left(x_{(1)}\right) = n\left[1 - F\left(x_{(1)}\right)\right]^{n-1}f\left(x_{(1)}\right) = \begin{cases} n\left(1 - x_{(1)}\right)^{n-1}, & 0 < x_{(1)} < 1 \\ 0, & \text{otherwise.} \end{cases}$

(c)

$$g\left(x_{(r)}\right) = \frac{n!}{(r-1)!(n-r)!}F^{r-1}\left(x_{(r)}\right)\left[1 - F(x_{(r)})\right]^{n-r}f\left(x_{(r)}\right)$$

$$= \begin{cases} \dfrac{n!}{(r-1)!(n-r)!}x_{(r)}^{r-1}\left(1 - x_{(r)}\right)^{n-r}, & 0 < x_{(r)} < 1 \\ 0, & \text{otherwise.} \end{cases}$$

3. In this problem the breaking strength of the chain is the same as the breaking strength of the weakest link. From the solution of Problem 1 above we know that

$$g(x_{(1)}) = 100[1 - F(x_{(1)})]^{99}f(x_{(1)}) = 100(e^{-\lambda x_{(1)}})^{99}\lambda e^{-\lambda x_{(1)}} = 100\lambda e^{-100\lambda x_{(1)}}$$

$$P(\text{breaking strength of the chain} > y) = \int_y^\infty 100\lambda e^{-100\lambda x_{(1)}}dx_{(1)} = -e^{-100\lambda x_{(1)}}\Big|_y^\infty = e^{-100\lambda y}$$

5. (a) Let a random variable X denote the time taken by a manager of the company to drive from one plant to another. Then we have

$$f(x) = \begin{cases} \dfrac{1}{30 - 15} = \dfrac{1}{15}, & 15 \le x \le 30 \\ 0, & \textit{elsewhere} \end{cases}$$

Hence the distribution function $F(x)$ is given by

$$F(x) = \int_{15}^x (1/15)dx = [(x - 15)/15]$$

Thus, using (7.4.3) the distribution function of the random variable

$$Y = X_{(n)} = Max(X_1, X_2, \cdots, X_n)$$

is given by

$$G(y) = \frac{1}{(15)^n}(y - 15)^n \Rightarrow g(y) = \frac{n}{(15)^n}[(y - 15)]^{n-1}, \ 15 \le y \le 30$$

and 0 elsewhere

(b)

$$\mu = E(Y) = \frac{n}{(15)^n}\int_{15}^{30} y(y - 15)^{n-1}dy = \frac{n}{(15)^n}\left(\frac{y(y - 15)^n}{n}\Big|_{15}^{30} - \frac{1}{n}\int_{15}^{30}(y - 15)^n dy\right)$$

$$= \frac{n}{(15)^n}\left(\frac{30(15)^n}{n} - \frac{1}{n}\int_{15}^{30}(y - 15)^n dy\right) = 30 - \frac{1}{(15)^n}\left(\frac{(y - 15)^{n+1}}{n + 1}\Big|_{15}^{30}\right) = 30 - \frac{15}{n + 1}$$

$$= \frac{15(2n + 1)}{n + 1}$$

7. (a) From (7.4.23), we have

$$g_{X_{(m+1)}}(x)dx = \frac{(2m+1)!}{(m!)^2} F^m(x)[1 - F(x)]^m f(x)dx. \tag{1}$$

In Problem 5, we determined that

$$F(x) = \int_{15}^{x} (1/15)dx = [(x - 15)/15]$$

Now substituting the value of $F(x)$ and $m = 10$ in (1), we obtain

$$g_{X_{(11)}}(x) = \frac{(21)!}{(10!)^2} \left(\frac{x-15}{15}\right)^{10} \left(\frac{30-x}{15}\right)^{10} \frac{1}{15} = (2.586584 \times 10^5) \left(\frac{x-15}{15}\right)^{10} \left(\frac{30-x}{15}\right)^{10}.$$

(b) From the solution of Problem 5 above, we obtain

$$\mu = 30 - \frac{15}{n+1} = 30 - 15/22 = 29.32.$$

REVIEW PRACTICE PROBLEMS

1. From the given information we see that the sample size is greater than 30. Thus, using the Central Limit Theorem we know that $\overline{X}$ is approximately normally distributed with mean $\mu_{\overline{X}} = 120$, and standard error $\sigma_{\overline{X}} = \frac{10}{\sqrt{36}} = 1.6667$. Hence, we obtain

(a) $P(\overline{X} > 122) = P\left(\frac{\overline{X} - 120}{1.6667} > \frac{122 - 120}{1.6667}\right)$

$$= P(Z > 1.20) = 1 - P(Z \le 1.20) = 1 - 0.8849 = 0.1151.$$

(b) $P(\overline{X} < 115) = P(Z < -3.00) = 0.0013.$

(c) $P(116 < \overline{X} < 123) = P(-2.40 < Z < 1.80) = 0.9641 - 0.0082 = 0.9559.$

3. (a) In this problem we are given $n = 500$ and $\theta = 0.8$. Thus, we have $n\theta > 5$ and $n(1 - \theta) > 5$, therefore, the sample proportion $\hat{\theta}$ is approximately normal with mean $\mu_{\hat{\theta}} = \theta = 0.8$ and standard deviation $\sigma_{\hat{\theta}} = \sqrt{\frac{\theta(1-\theta)}{n}} = \sqrt{\frac{(0.8)(0.2)}{500}} = 0.0179$

(b) $P(\hat{\theta} \ge 0.75) = P\left(\frac{\hat{\theta} - 0.8}{0.0179} \ge \frac{0.75 - 0.8}{0.0179}\right) = P(Z \ge -2.80) = 0.9974.$

5. (a) $P(\text{given brick has crushing strength} \le x \text{ psi}) = F(x);$
$P(\text{given brick has crushing strength} > x \text{ psi}) = 1 - F(x);$
$P(\text{all 100 bricks have crushing strength} > x \text{ psi}) = [1 - F(x)]^{100}.$

(b) Using (7.4.12), the probability that the smallest element is in the interval $(x, x + dx)$ is $g(x)dx = 100[1 - F(x)]^{99} f(x)dx.$

7. In this problem we are given that the diameters of ball bearings are normally distributed with mean $\mu = 1.20$ cm and standard deviation $\sigma = 0.05$ and the sample size is 25. Thus, the sample average or sample mean $\overline{X}$ is also normally distributed with mean $\mu = 1.20$ cm and standard deviation $\sigma/\sqrt{n} = .05/\sqrt{25} = .01$. Thus, we have

(a)

$$P(1.18 < \overline{X} < 1.22) = P\left(\frac{1.18 - 1.20}{.01} < \frac{\overline{X} - 1.20}{.01} < \frac{1.22 - 1.20}{.01}\right)$$

$$= P(-2 < Z < 2) = 0.9772 - 0.0228 = 0.9544.$$

(b) $P\left(1.19 < \bar{X} < 1.215\right) = P\left(\dfrac{1.19 - 1.20}{.01} < \dfrac{\bar{X} - 1.20}{.01} < \dfrac{1.215 - 1.20}{.01}\right)$

$$= P(-1 < Z < 1.5) = 0.9332 - 0.1587 = 0.7745.$$

9. $g\left(x_{(n)}\right) = n\left[F\left(x_{(n)}\right)\right]^{n-1} f\left(x_{(n)}\right)$

Let $\quad t = F\left(x_{(n)}\right), \qquad \dfrac{dx_{(n)}}{dt} = \dfrac{1}{f\left(x_{(n)}\right)}.$

Then $\quad h(t) = n t^{n-1} f\left(x_{(n)}\right) \dfrac{1}{f\left(x_{(n)}\right)} = n t^{n-1}, \qquad 0 \le t \le 1.$

Thus $\quad E\left(F\left(x_{(n)}\right)\right) = E(t) = \displaystyle\int_0^1 t \cdot n t^{n-1} \, dt = n \int_0^1 t^n \, dt = \dfrac{n t^{n+1}}{n+1}\bigg|_0^1 = \dfrac{n}{n+1},$

and $\quad E\left(F^2\left(x_{(n)}\right)\right) = \displaystyle\int_0^1 t^2 n t^{n-1} \, dt = n \int_0^1 t^{n+1} \, dt = \dfrac{n}{n+2},$

so $\mathrm{Var}\left(F\left(x_{(n)}\right)\right) = \dfrac{n}{n+2} - \dfrac{n^2}{(n+1)^2} = \dfrac{n\left((n+1)^2 - n(n+2)\right)}{(n+1)^2(n+2)}$

$$= \dfrac{n}{(n+1)^2(n+2)}.$$

11.
$$f(x) = \begin{cases} \dfrac{1}{\theta}, & 0 < x \le \theta \\ 0, & \text{elsewhere} \end{cases}, \qquad F(x) = \begin{cases} 0, & x \le 0 \\ \dfrac{x}{\theta}, & 0 < x \le \theta \\ 1, & x \ge \theta. \end{cases}$$

The p.d.f. of $X_{(n)}$ is given by

$$g\left(x_{(n)}\right) = n\left(\dfrac{x_{(n)}}{\theta}\right)^{n-1} \dfrac{1}{\theta} = \begin{cases} \dfrac{n}{\theta^n} x_{(n)}^{n-1}, & 0 < x_{(n)} < \theta \\ 0, & \text{otherwise.} \end{cases}$$

Now, $\quad P\left(X_{(n)} < \theta < \dfrac{X_{(n)}}{\sqrt[n]{1-\gamma}}\right) = P\left(\theta\sqrt[n]{1-\gamma} < X_{(n)} < \theta\right)$

$$= \int_{\theta\sqrt[n]{1-\gamma}}^{\theta} \dfrac{n}{\theta^n} x_{(n)}^{n-1} \, dx_{(n)} = \dfrac{n}{\theta^n} \cdot \dfrac{x_{(n)}^n}{n}\bigg|_{\theta\sqrt[n]{1-\gamma}}^{\theta}$$

$$= \dfrac{\theta^n - \theta^n(1-\gamma)}{\theta^n} = \gamma$$

Thus, $\quad \left(x_{(n)}, \dfrac{x_{(n)}}{\sqrt[n]{1-\gamma}}\right) \quad$ is a $100\gamma\%$ confidence interval for θ.

13. From Theorem 7.3.11, it follows that if S_1^2 and S_2^2 are sample variances of two independent samples taken from two normal populations with variances σ_1^2 and σ_2^2, respectively then the statistic

$$\dfrac{S_1^2/\sigma_1^2}{S_2^2/\sigma_2^2}$$

is distributed as an F distribution with $n_1 - 1$ and $n_2 - 1$ degrees of freedom, where n_1 and n_2 are sample sizes. In this problem we are given that $\sigma_1^2 = \sigma_2^2$, $n_1 = 16$ and $n_2 = 13$. Thus, it follows immediately that

$$\frac{S_x^2}{S_y^2}$$

is distributed as an F distribution with $n_1 - 1 = 15$ and $n_2 - 1 = 12$ degrees of freedom.

15. From Theorem 7.3.5, we have that

$$\frac{(n-1)S^2}{\sigma^2}$$

is distributed as a Chi-Square distribution with $(n-1)$ degrees of freedom. Thus, in the present problem

$$\frac{20S^2}{144} = \frac{S^2}{7.2}$$

is distributed as Chi-Square with 20 degrees of freedom or $S^2 \sim 7.2\chi_{20}^2$.

$$E\left(\frac{S^2}{7.2}\right) = 20 \Rightarrow E(S^2) = 7.2 \times 20 = 144,$$

$$Var\left(\frac{S^2}{7.2}\right) = 2 \times 20 \Rightarrow var(S^2) = (7.2)^2 \times 40 = 2073.6.$$

17. In this problem the life (T) of the system is the same as that of the weakest component. That is,

$$T = Min.(X_1, X_2, X_3)$$

From equation (7.4.11) we know that the c.d.f. $G(t)$ of the smallest element in the sample is given by

$$G(t) = 1 - [1 - F(t)]^n$$

Now from equation (5.10.3) of Chapter 5, we have

$$F(t) = 1 - e^{-\lambda t}$$

Thus, substituting the value of $F(x)$, we obtain

$$G(t) = 1 - [1 - (1 - e^{-\lambda t})]^n = 1 - e^{-n\lambda t}, g(t) = n\lambda e^{-n\lambda t}$$

Hence, $P(T > t) = 1 - G(t)$.

In this problem we are given that

$$n = 3, \lambda = 0.1, \text{and } t = 10$$

Thus, we have $P(T > 10) = 1 - G(10) = 1 - (1 - e^{-3 \times (0.1) \times 10}) = e^{-3} = 0.0498$.

19. From equation (5.9.19) of Chapter 5, we have

$$F(t) = P(T \le t) = 1 - e^{-[(t-\tau)/\alpha]^\beta}$$

In this problem we are given that $\alpha = 2$, $\beta = 0.5$, and $\tau = 0$. Thus, we have

$$F(t) = P(T \le t) = 1 - e^{-[(t/2)]^{0.5}}$$

For Problem 17, we have

$$G(t) = 1 - [1 - F(t)]^n = 1 - [1 - (1 - e^{-(t/2)^{0.5}})]^n = 1 - e^{-n(t/2)^{0.5}} = 1 - e^{-3(t/2)^{0.5}}, \qquad \text{since } n = 3$$

Thus, we have

$$P(T > 10) = 1 - G(10) = 1 - (1 - e^{-3(10/2)^{0.5}}) = e^{-3(10/2)^{0.5}} = e^{-6.71} = 0.0012.$$

For problem 18, we have

$$P(T > 15) = 1 - P(T \leq 15) = 1 - G(15) = 1 - (1 - e^{-(15/2)^{0.5}})^3 = 1 - (1 - 0.0646)^3 = 1 - 0.8184 = 0.1816.$$

8

ESTIMATION OF POPULATION PARAMETERS

PRACTICE PROBLEMS FOR SECTION 8.2

1. Let $(X_1, X_2, \cdots, X_n)$ be a random sample from a Bernoulli population with mean (parameter) θ.

The sample mean is $\overline{X} = \sum X_i/n = X/n$, where X is the number of successes. Since θ is the mean of a Bernoulli population. Hence, we obtain

$$\hat{\theta} = \overline{X}.$$

3. (a) $E(\hat{\mu}_1) = E(2X_1 - 2X_2 + X_3) = 2E(X_1) - 2E(X_2) + E(X_3) = 2\mu - 2\mu + \mu = \mu$

$E(\hat{\mu}_2) = E(2X_1 - 3X_2 + 2X_3) = 2E(X_1) - 3E(X_2) + 2E(X_3) = 2\mu - 3\mu + 2\mu = \mu$

Hence both $\hat{\mu}_1$ and $\hat{\mu}_2$ are unbiased estimators of μ.

(b) $V(\hat{\mu}_1) = V(2X_1 - 2X_2 + X_3) = 4V(X_1) + 4V(X_2) + V(X_3) = 4\sigma^2 + 4\sigma^2 + \sigma^2 = 9\sigma^2$

$V(\hat{\mu}_2) = V(2X_1 - 3X_2 + 2X_3) = 4V(X_1) + 9V(X_2) + 4V(X_3) = 4\sigma^2 + 9\sigma^2 + 4\sigma^2 = 17\sigma^2$

Since $V(\hat{\mu}_1) < V(\hat{\mu}_2) \Rightarrow \hat{\mu}_1$ is a better estimator.

5. Using MINITAB we obtain

$$\overline{X} = 9.850, S^2 = 1.818$$

Now from (8.2.20) we know that the maximum likelihood estimator of a population mean and population variance are given by

$$\hat{\mu} = \overline{X}, \ \hat{\sigma}^2 = \frac{1}{n}\sum_{i=1}^{n}\left(X_i - \overline{X}\right)^2.$$

respectively.

(a) Thus, the maximum likelihood estimator of a population mean is given by $\overline{X} = 9.850$.

Solutions Manual to Accompany Statistics and Probability with Applications for Engineers and Scientists, Bhisham C. Gupta and Irwin Guttman.
© 2014 John Wiley & Sons, Inc. Published 2014 by John Wiley & Sons, Inc.

(b) The maximum likelihood estimator of a population variance is given by

$$\frac{(n-1)}{n}S^2 = \frac{19}{20} \times 1.818 = 1.7271.$$

7. The probability function of each X_i is given by

$$p(x_i) = \frac{e^{-\lambda}\lambda^{x_i}}{x_i!}.$$

Thus, the likelihood function is given by

$$l(\theta|x_1, \ldots, x_n) = l(\lambda|x_1, \ldots, x_n) = \prod p(x_i) = \frac{e^{-n\lambda}\lambda^{\sum x_i}}{\prod(x_i!)}$$

Then the log likelihood function is given by

$$L = \ln l(\lambda|x_1, \ldots, x_n) = \ln\frac{e^{-n\lambda}\lambda^{\sum x_i}}{\prod(x_i!)} = -n\lambda + \sum x_i\ln\lambda + \ln\prod(x_i!)$$

Now taking the partial derivative with respect to λ and setting it equal to zero, we obtain the normal equation

$$-n + \frac{\sum x_i}{\hat{\lambda}} = 0 \Rightarrow \hat{\lambda} = \frac{\sum x_i}{n} = \overline{X}$$

9. The probability function of each X_i is given by

$$p(x_i) = 1/\theta$$

Thus, the likelihood function is given by

$$l(\theta|x_1, \ldots, x_n) = 1/\theta^n$$

Since l is strictly decreasing in θ, the largest value of l is obtained by selecting θ as small as possible. However, $\theta > X_i$ for every i, so that $\theta \geq Max(X_i)$. Thus, the maximum likelihood estimator $\hat{\theta} = Max(X_1, X_2, \cdots, X_n) = X_{(n)}$.

To show that $\hat{\theta} = Max(X_1, X_2, \cdots, X_n) = X_{(n)}$ is not an unbiased estimator of θ consider the p.d.f of the largest element $Y = X_{(n)} = Max(X_1, X_2, \cdots, X_n)$. From Chapter 7, we have

$$g(y) = \frac{ny^{n-1}}{\theta^n}, 0 < y < \theta$$

so that

$$E(Y) = \frac{n}{\theta^n}\int_0^\theta y^n dy = \frac{n}{n+1}\theta.$$

Hence $Y = X_{(n)}$ is not an unbiased estimator of θ.

10. **(a)** From Problem 9, we have

$$E(X_{(n)}) = \frac{n}{n+1}\theta \Rightarrow E\left(\frac{n+1}{n}X_{(n)}\right) = \frac{n+1}{n} \times \frac{n}{n+1}\theta = \theta$$

$$\Rightarrow U = \frac{n+1}{n}X_{(n)} \text{ is an unbiased estimator of } \theta.$$

(b) Let $U = \dfrac{n+1}{n} X_{(n)}$. Then we want to find the variance of U, that is

$$V(U) = V\left(\frac{n+1}{n} X_{(n)}\right) = \left(\frac{n+1}{n}\right)^2 V(X_{(n)})$$

$$E(X_{(n)}^2) = \frac{n}{\theta^n} \int_0^\theta (X_{(n)})^{n+1} dX_{(n)} = \frac{n}{\theta^n} \frac{(X_{(n)})^{n+2}}{n+2}\bigg|_0^\theta = \frac{n}{n+2}\theta^2$$

Hence

$$V(U) = \left(\frac{n+1}{n}\right)^2 V(X_{(n)}) = \left(\frac{n+1}{n}\right)^2 \left[E(X_{(n)}^2) - (E(X_{(n)})^2)\right] = \left(\frac{n+1}{n}\right)^2 \left(\frac{n}{n+2}\theta^2 - \left(\frac{n}{n+1}\theta\right)^2\right)$$

$$= \left(\frac{n+1}{n}\right)^2 \times \frac{n\theta^2}{(n+1)^2(n+2)} = \frac{\theta^2}{n(n+2)}$$

(c) Let X be a random variable distributed uniformly over the interval $(0, \theta)$. Then we know that

$$E(X) = \frac{\theta}{2} \text{ and } V(X) = \frac{\theta^2}{12}$$

Thus, we have

$$E(2\overline{X}) = 2E((X_1 + X_2 + \cdots X_n)/n) = (2/n)E(X_1 + X_2 + \cdots X_n)$$

$$= (2/n)(\theta/2 + \theta/2 + \cdots + \theta/2) = \theta$$

$$V(2\overline{X}) = 4V(\overline{X}) = 4 \times \frac{\theta^2}{12n} = \theta^2/3n.$$

(d) $\dfrac{V(U)}{V(2\overline{X})} = \dfrac{\theta^2}{n(n+2)} \times \dfrac{3n}{\theta^2} = 3/(n+2)$

Thus, for $n \geq 1, V(U) \leq V(2\overline{X}) \Rightarrow U$ is a better estimator of θ,

PRACTICE PROBLEMS FOR SECTION 8.3

1. (a) Since in this problem sample size is 36 (>30), applying Central Limit Theorem and using one sample Z statistics in MINITAB we can find the desired confidence interval. Thus, by taking the following steps:

1. Enter the data in column C1 of the **Worksheet** window.

2. Since the population standard deviation is not known, calculate the sample standard deviation of these data using one of the MINITAB procedures discussed earlier in Chapter 2. We will find S = 3.808.

3. In the pull-down menu Select **Stat** > **Basic Statistics** > **1-Sample Z**. This will prompt a dialog box **1-Sample Z** to appear on the screen. Note that since the sample size is greater than 30, we select the command **1-Sample Z** instead of **1-Sample t.**

    Enter C1 in the box below **Samples in columns,** and the value of the standard deviation in a box next to **Standard deviation.**

4. Check **Options**, which will prompt another dialog box to appear. Enter the desired confidence level in the box next to **Confidence level** and under **alternative** option select **not equal** by using the down arrow. In each dialog box Click **OK.** The MINITAB output appears in the session window as given below.

    We obtain a 95% confidence interval for the mean of all care givers in the United States. That is

$$(45.562, 48.049).$$

(b) To find lower and upper one-sided confidence intervals in step 4, select under the **alternative** option greater than/less than by using the down arrow. Thus, in this problem 95% lower and upper one-sided confidence intervals are given by $(45.762, \infty)$ and $(0, 47.849)$, respectively. Note that if we use **1-Sample t** instead of **1-Sample Z,** the confidence intervals are slightly wider.

3. Again proceeding exactly in the same manner as in Problem 1 above, we obtain 90% confidence intervals as given below.
(a) $(8.164, 9.036)$ (b) $(8.260, \infty)$, $(0, 8.940)$.

5. From the given data, we obtain

$$n = 10 \quad \overline{X} = 8.993 \quad S = 0.538$$

Since in this problem the sample size is small, the population is normal, and the standard deviation is unknown. Thus, using (8.3.8), we obtain a 95% confidence interval

$$(8.608, 9.378).$$

7. From the given data, we obtain

$$n = 16 \quad \overline{X} = 18.5 \quad S = 3.246$$

Since in this problem the sample size is small, the population is normal, and the standard deviation is unknown. Thus, using (8.3.8), we obtain a 99% confidence interval

$$(16.109, 20.891).$$

9. Using the summary statistic in Problem 8, we obtain one sided upper and lower confidence intervals $(0, 87.605)$ and $(85.130, \infty)$ respectively.

In Problem 11 the population is normal with known variance. We determine the confidence intervals for the population mean using MINITAB as follows:

1. From the pull-down menu Select <u>S</u>tat > <u>B</u>asic Statistics > **1-Sample Z.** This will prompt a dialog box **1-Sample Z** to appear on the screen. Note that we use **1-Sample Z** since the populations are normal with known variance. In this dialog box select summarized data and enter the sample size, mean, and standard deviation.

2. Check **Options,** which will prompt another dialog box to appear. Enter the desired confidence level in the box next to **Confidence level** and under **alternative** option select **not equal** by using the down arrow. In each dialog box Click **OK.** The MINITAB output for Problem 11 is given below.

11. $(1301.7, 1513.3)$. We may conclude with 99% confidence that $\mu = 1400$, since it lies within the 99% C.I.
The confidence intervals in Problem 11, through application of the formulas derived in this chapter is obtained as follows:

$$\text{The } 100(1 - \alpha)\% \text{ C.I. for } \mu \text{ is } \overline{x} \pm z_{\alpha/2} \frac{\sigma}{\sqrt{n}}$$

Given $n = 30$, $\overline{x} = 1407.5$, $\sigma = 225$, $\alpha = 0.01$, $z_{0.005} = 2.576$, then

$$\overline{x} \pm z_{\alpha/2} \frac{\sigma}{\sqrt{n}} = 1407.5 \pm (2.576)\left(\frac{225}{\sqrt{30}}\right) = 1407.5 \pm 105.82.$$

Thus a 99% C.I. for μ is 1407.5 ± 105.82, or $(1301.68, 1513.32)$.
Thus, we can conclude with 99% confidence that $\mu = 1400$, since it lies within the 99% C.I.

PRACTICE PROBLEMS FOR SECTION 8.4

1.
The $100(1-\alpha)\%$ C.I. for $\mu_1 - \mu_2$ is $(\overline{x}_1 - \overline{x}_2) \pm z_{\alpha/2}\sqrt{\dfrac{\sigma_1^2}{n_1} + \dfrac{\sigma_2^2}{n_2}}$.

Given $n_1 = 10$, $\bar{x}_1 = 170.2$, $\sigma_1^2 = 225$, $n_2 = 12$, $\bar{x}_2 = 176.7$, $\sigma_2^2 = 256$, $\alpha = 0.05$, $z_{0.025} = 1.96$, then

$$(\bar{x}_1 - \bar{x}_2) \pm z_{\alpha/2}\sqrt{\frac{\sigma_1^2}{n_1} + \frac{\sigma_2^2}{n_2}} = (170.2 - 176.7) \pm 1.96\sqrt{\frac{225}{10} + \frac{256}{12}} = -6.5 \pm 12.98.$$

Thus a 95% C.I. for $\mu_1 - \mu_2$ is -6.5 ± 12.98, or $(-19.48, 6.48)$.

3. Calculating the sample means, variances, and standard deviations, we get

$$\bar{X}_A = 1362.5, \quad s_A^2 = 40990.7222, \quad s_A = 202.46, \quad n_A = 10$$
$$\bar{X}_B = 1225.8, \quad s_B^2 = 65790.6222, \quad s_B = 256.50, \quad n_B = 10.$$

Here, we have $n_A = n_B (= n)$, so, for s_p^2 we get

$$s_p^2 = \frac{(n_A - 1)s_A^2 + (n_B - 1)s_B^2}{n_A + n_B - 2} = \frac{s_A^2 + s_B^2}{2} = 53390.6722,$$

$\Rightarrow s_p = 231.06$. Also, $t_{18;0.025} = 2.101$.

Assuming normality and equality of variances, the 95% C.I. for $\mu_A - \mu_B$ is

$$(\bar{X}_A - \bar{X}_B) \pm t_{n_A+n_B-2;\alpha/2} \cdot s_p\sqrt{\frac{1}{n_A} + \frac{1}{n_B}}$$

$$= (1362.5 - 1225.8) \pm (2.101)(231.06)\sqrt{\frac{1}{10} + \frac{1}{10}} = 136.7 \pm 217.107$$

or $(-80.4, 353.8)$. The sample evidence supports the assumption of equal means at the 5% level of significance, since 0 $(= \mu_A - \mu_B)$ is in the 95% C.I.

By using MINITAB the confidence interval is $(-80, 354)$. Note that this confidence interval is slightly different from the one obtained earlier. This is due to some rounding errors.

5. The sample means and variances are

$$\bar{x}_1 = 7.5, \quad s_1^2 = 0.1356, \quad n_1 = 10$$
$$\bar{x}_2 = 7.18, \quad s_2^2 = 0.0284, \quad n_2 = 10.$$

The d.f., m, is calculated as follows:

$$c = \frac{s_1^2/n_1}{s_1^2/n_1 + s_2^2/n_2} = 0.8268;$$

$$\frac{1}{m} = \frac{c^2}{n_1 - 1} + \frac{(1 - c)^2}{n_2 - 1} = \frac{1}{9}\left((0.8268)^2 + (0.1732)^2\right) = 0.0793,$$

$\Rightarrow m = 12.61$, so we use $m = 12$ d.f. Also, $t_{12;0.005} = 3.055$.

Thus the 99% C.I. for $\mu_1 - \mu_2$ is

$$(\bar{x}_1 - \bar{x}_2) \pm t_{m;\alpha/2}\sqrt{\frac{s_1^2}{n_1} + \frac{s_2^2}{n_2}} = (7.5 - 7.18) \pm (3.055)\sqrt{\frac{0.1356}{10} + \frac{0.0284}{10}}$$

$$= 0.32 \pm 0.391$$

or $(-0.071, 0.711)$.

The data does not support the claim of increased average batch yield since 0 is included in the interval. That is, $\mu_1 - \mu_2 = 0$ $(\mu_1 = \mu_2)$ is supported by the data.

*Problems 6 thru 10 of this section are done by using MINITAB. The procedure of using MINITAB is approximately the same as discussed in Section 8.3 (**Stat** > **Basic Statistics** > **2 - Sample** t). Note that MINITAB does not provide the 2-Sample Z test.*

7. The MINITAB output is the following (Variances are assumed not equal):

Two-sample T for T1 vs T2

	N	Mean	StDev	SE Mean
T1	10	9.468	0.333	0.11
T2	10	10.045	0.241	0.076

Difference = mu (T1) − mu (T2)
Estimate for difference: −0.577
95% CI for difference: (−0.852, −0.302).

9. The $100(1-\alpha)\%$ C.I. for $\mu_1 - \mu_2$ is $(\bar{x}_1 - \bar{x}_2) \pm z_{\alpha/2}\sqrt{\dfrac{\sigma_1^2}{n_1} + \dfrac{\sigma_2^2}{n_2}}$

Sample	N	Mean	StDev	SE Mean
1	20	80.85	4.00	0.89
2	20	99.90	5.00	1.1

Difference = mu (1) − mu (2)
Estimate for difference: −19.05
Hence, 99% CI for difference of the two means is given by

$$(\bar{x}_1 - \bar{x}_2) \pm z_{\alpha/2}\sqrt{\frac{\sigma_1^2}{n_1} + \frac{\sigma_2^2}{n_2}} = (80.85 - 99.90) \pm 2.575\sqrt{((16/20 + 25/20)} = (-22.74, -15.36)$$

10. The MINITAB output is the following (Variances are assumed equal):

Two-sample T for Drying Time 1 vs Drying Time 2

	N	Mean	StDev	SE Mean
Drying Time 1	16	10.839	0.712	0.18
Drying Time 2	16	12.983	0.704	0.18

Difference = mu (Drying Time 1) − mu (Drying Time 2)
Estimate for difference: −2.143
95% CI for difference: (−2.654, −1.632).

PRACTICE PROBLEMS FOR SECTIONS 8.5 AND 8.6

1. Given $n = 12$ and $s^2 = 86.2$, a 95% C.I. for σ^2 is

$$\left(\frac{(n-1)s^2}{\chi^2_{n-1;\alpha/2}}, \frac{(n-1)s^2}{\chi^2_{n-1;1-\alpha/2}}\right) = \left(\frac{(11)(86.2)}{21.9200}, \frac{(11)(86.2)}{3.81575}\right) = (43.257, 248.496).$$

3. (a) From the given information, we have

$$n_A = 220, \quad \overline{X}_A = 2.57 \quad S_A = 0.57$$
$$n_B = 220, \quad \overline{X}_B = 2.66 \quad S_B = 0.48$$

Using MINITAB **(Stat > Basic Statistics > 2 variance)**, we obtain a 95% confidence interval for the ratio of two variance σ_1^2/σ_2^2, which is

$$(1.081, 1.839)$$

Since the interval does not contain '1', we reject the null hypothesis that variances are equal

(b) A 95% C.I. for $\mu_A - \mu_B$ under this assumption is

$$(-0.1888, 0.0088)$$

Since the interval contains '0', so we do not reject the null hypothesis that means are equal

5. Using the given information we obtain the following confidence intervals.

95% confidence interval for the population variance:

$$\left(\frac{(n_1 - 1)s^2}{\chi_{n-1,\alpha/2}^2} \leq \sigma^2 \leq \frac{(n_1 - 1)s^2}{\chi_{n-1,1-\alpha/2}^2} \right) = \left(\frac{(25 - 1)(5^2)}{39.3641} \leq \sigma^2 \leq \frac{(25 - 1)(5^2)}{12.4011} \right)$$

$$= (15.2423 \leq \sigma^2 \leq 48.3828)$$

95% confidence interval for the population standard deviation:

$$\left(\sqrt{\frac{(n - 1)s^2}{\chi_{n-1,\alpha/2}^2}} \leq \sigma \leq \sqrt{\frac{(n - 1)s^2}{\chi_{n-1,1-\alpha/2}^2}} \right) = \left(\sqrt{15.2423} \leq \sigma \leq \sqrt{48.3828} \right) = (3.9041 \leq \sigma \leq 6.9557)$$

95% lower confidence limit for the population standard deviation:

$$\sqrt{\frac{(n - 1)s^2}{\chi_{n-1,\alpha}^2}} = \sqrt{\frac{(25 - 1)(5^2)}{36.4151}} = 4.0591$$

95% upper confidence limit for the population standard deviation:

$$\sqrt{\frac{(n - 1)s^2}{\chi_{n-1,1-\alpha}^2}} = \sqrt{\frac{(25 - 1)(5^2)}{13.8484}} = 6.5823$$

The lower one-sided confidence limit, 4.0591, is greater than the lower two-sided confidence limit, 3.9043. Also the upper one-sided confidence limit, 6.5823, is less than the upper two-sided confidence limit, 6.9557. The difference in these confidence limits has to do with the way the alpha value was used in each of the equations. For the two sided confidence interval we used $\alpha/2$ on each side and for the one-sided confidence limits we used α on each side.

7. The sample mean and sample variance for the given set of data with $n_2 = 15$ are given by

$$\overline{X}_2 = 31.20 \quad S_2^2 = 4.89$$

95% confidence interval for the population variance:

$$\left(\frac{(n_1 - 1)S_1^2}{\chi_{n-1,\alpha/2}^2} \leq \sigma_2^2 \leq \frac{(n_1 - 1)S_1^2}{\chi_{n-1,1-\alpha/2}^2} \right) = \left(\frac{(15 - 1)(4.89)}{26.1190} \leq \sigma_2^2 \leq \frac{(15 - 1)(4.89)}{5.6287} \right) = (2.62, 12.16)$$

99% lower confidence interval for the population variance

$$\left(\frac{(n_1 - 1)S_1^2}{\chi_{n-1,\alpha}^2} \le \sigma_2^2 \le \infty \right) = \left(\frac{(15-1)(4.89)}{29.1413} \le \sigma_2^2 \le \infty \right) = \left(2.35 \le \sigma_1^2 \le \infty \right)$$

99% upper confidence interval for the population variance:

$$\left(0 \le \sigma_2^2 \le \frac{(n_1 - 1)S_1^2}{\chi_{n-1,1-\alpha}^2} \right) = \left(0 \le \sigma_2^2 \le \frac{(15-1)(4.89)}{4.6604} \right) = \left(0 \le \sigma_2^2 \le 14.69 \right).$$

PRACTICE PROBLEMS FOR SECTION 8.7

1. A point estimate of $\theta = \hat{\theta} = \dfrac{200}{500} = 0.4$.

   95% confidence interval for the population proportion is given by

   $$\hat{\theta} \pm z_{\frac{\alpha}{2}} \sqrt{\frac{\hat{\theta}(1-\hat{\theta})}{n}} = 0.4 \pm 1.96 \sqrt{\frac{(0.4)(0.6)}{500}} = 0.4 \pm 0.0429 = (0.3571, 0.4429).$$

3. Let θ_1 and θ_2 be the proportions of the populations in favoring Brand X before and after the advertising campaign. Then point estimates of θ_1 and θ_2 are given by

   $$\hat{\theta}_1 = 16/100 = 0.16, \hat{\theta}_2 = 50/200 = 0.25.$$

   95% confidence interval for the difference in proportions of the population favoring Brand X before and after the advertising campaign, is given by

   $$\left(\hat{\theta}_1 - \hat{\theta}_2 \right) \pm z_{\frac{\alpha}{2}} \sqrt{\frac{\hat{\theta}_1\left(1-\hat{\theta}_1\right)}{n_1} + \frac{\hat{\theta}_2\left(1-\hat{\theta}_2\right)}{n_2}} = (0.16 - 0.25) \pm 1.96 \sqrt{\frac{(0.16)(0.84)}{100} + \frac{(0.25)(0.75)}{200}}$$

   or
   $$(0.09 - 0.0936, 0.09 + 0.0936) = (-0.0036, 0.1836)$$

   Since the confidence interval contains 0, we can say at the 5% level of significance, that the proportions of population favoring Brand X before and after the advertising campaign are the same.

5. Point estimates of θ_1 and θ_2 are $\hat{\theta}_1 = \dfrac{40}{800} = 0.05$, $\hat{\theta}_2 = \dfrac{50}{600} = 0.083$, respectively.

   Thus, a 95% confidence interval for the difference between two population proportions is given by:

   $$\left(\hat{\theta}_1 - \hat{\theta}_2 \right) \pm z_{\frac{\alpha}{2}} \sqrt{\frac{\hat{\theta}_1\left(1-\hat{\theta}_1\right)}{n_1} + \frac{\hat{\theta}_2\left(1-\hat{\theta}_2\right)}{n_2}} = (0.05 - 0.083) \pm 1.96 \sqrt{\frac{(0.05)(0.95)}{800} + \frac{(0.083)(0.917)}{600}}$$

   $$= -0.033 \pm 0.0267 = (-0.0597, -0.0063).$$

   This confidence interval does not include zero, so we can conclude at the 5% level of significance, that the percentages of persons favoring the nuclear plant in two different states are different.

   Problems 6 thru 9 of this section are done by using MINITAB. Procedure of using MINITAB is exactly the same as discussed earlier. Given below are the steps that are needed to construct confidence intervals for one population and two population proportions using MINITAB.

1. In MINITAB, select **Stat > Basic Statistics > 1 Proportion,** this will prompt a dialog box titled **1 Proportion.** Since we have summarized data, click the circle next to **Summarized Data** and enter the number of trials and the number of events (successes) in the appropriate boxes. Check **Options,** which will prompt another dialog box to appear. Enter the desired confidence level in the box next to **Confidence level** and from the **Alternative** option select **not equal, less than, or greater than** depending upon whether two-sided, one-sided upper or one sided lower confidence intervals are sought. Click **OK** on dialog boxes. The MINITAB output will show up in the session window.

1. Select **Stat > Basic Statistics > 2 Proportions.** This will prompt a dialog box **2- Proportions.**

2. Since we have summarized data, check the circle next to **Summarized Data** and enter the number of trials and the number of events (successes) in the appropriate boxes. [Here "success" means a bulb failed before 5,000 hours].

3. Check **Options,** which prompts another dialog box to appear. Enter the desired confidence level in the box next to **Confidence level** and from the **Alternative** option select **not equal, less than, or greater than** depending upon whether two-sided, one-sided upper or one sided lower confidence intervals are sought. Click **OK** on dialog boxes. The MINITAB output will show up in the session window as given below.

7. Again, using the steps discussed above we obtain the following MINITAB output.

Machine	X	N	Sample p
1	12	225	0.053333
2	16	400	0.040000

Difference $= p_1 - p_2$
Estimate for difference: 0.0133333
95% CI for difference: $(-0.0217, 0.0484)$.

9. In this problem the MINITAB output is as given blow (using normal approximation).

Sample	X	N	Sample p	90% CI
1	200	500	0.4000	(0.3640, 0.4360)

PRACTICE PROBLEMS FOR SECTION 8.8

1. $E = \dfrac{width\ of\ confidence\ interval}{2} = \dfrac{20}{2} = 10$, $z_{\alpha/2} = z_{0.02/2} = 2.33$.

   Now, Using Equation (8.8.3), we obtain

   $$n = \frac{z_{\alpha/2}{}^2 \sigma^2}{E^2} = \frac{(2.33^2)(30^2)}{10^2} = 48.86 \cong 49.$$

3.
   (a) $n = \dfrac{z_{\alpha/2}^2[\theta_1(1-\theta_1) + \theta_2(1-\theta_2)]}{E^2} = \dfrac{(1.96^2)((0.5)(0.5) + (0.5)(0.5))}{(0.04^2)} = 1200.5 \cong 1201$

   (b) $n = \dfrac{z_{\alpha/2}^2[\theta_1(1-\theta_1) + \theta_2(1-\theta_2)]}{E^2} = \dfrac{(1.96^2)((0.5)(0.5) + (0.5)(0.5))}{(0.035^2)} = 1568$

   (c) $n = \dfrac{z_{\alpha/2}^2[\theta_1(1-\theta_1) + \theta_2(1-\theta_2)]}{E^2} = \dfrac{(1.96^2)((0.5)(0.5) + (0.5)(0.5))}{(0.03^2)} = 2134.22 \cong 2135.$

5.
   (a) $n = \dfrac{z_{\alpha/2}^2[\theta_1(1-\theta_1) + \theta_2(1-\theta_2)]}{E^2} = \dfrac{(1.96^2)((0.5)(0.5) + (0.5)(0.5))}{(0.04^2)} = 1200.5 \cong 1201.$

   (b) $n = \dfrac{z_{\alpha/2}^2[\theta_1(1-\theta_1) + \theta_2(1-\theta_2)]}{E^2} = \dfrac{(1.96^2)((0.35)(0.65) + (0.47)(0.53))}{(0.04^2)} = 1144.32 \cong 1145.$

7. In this problem we do not have any information about θ the proportion of residents who favor the construction of a new high school. So we take $\theta = 0.5$. Hence the required sample size is

$$n = \frac{z_{\alpha/2}^2 \theta(1-\theta)}{E^2} = \frac{(2.33^2)(.5)(.5)}{(.03)^2} = 1508.027 \cong 1509.$$

9. In this problem the width of the confidence interval is equal to 6 years and standard deviation is known to be 8 years. Hence using (8.8.3) the required sample size is

$$n = \frac{z_{\alpha/2}^2 \sigma^2}{E^2} = \frac{(2.33^2)(8^2)}{3^2} = 38.605 \cong 39.$$

REVIEW PRACTICE PROBLEMS

1.
The C.I. for μ, $\quad \overline{X} \pm z_{\alpha/2} \dfrac{\sigma}{\sqrt{n}}$, must have length less than or equal to L, so

$$\overline{X} + z_{\alpha/2} \frac{\sigma}{\sqrt{n}} - \left(\overline{X} - z_{\alpha/2} \frac{\sigma}{\sqrt{n}} \right) \le L,$$

$$\text{i.e.,} \quad 2z_{\alpha/2} \frac{\sigma}{\sqrt{n}} \le L.$$

For a 99% C.I., $\alpha = 0.01$, and $z_{0.005} = 2.575$, so that

$$\sqrt{n} = \frac{2z_{\alpha/2}\sigma}{L} = \frac{2(2.575)\sigma}{L} = \frac{(5.15)\sigma}{L} \Rightarrow n = \left(\frac{5.15\sigma}{L} \right)^2$$

3. The sample mean and standard deviation are $\overline{x} = 7.92$ and $s = 0.0183$ with $n = 4$.

    (a) Since σ is unknown, population is assumed normal, and n is small, the $100(1-\alpha)\%$ C.I. for μ is $\overline{X} \pm t_{n-1;\alpha/2}s\sqrt{n}$. For $\alpha = 0.01$, $t_{3;0.005} = 5.841$, so

$$\overline{X} \pm t_{n-1;\alpha/2} \times \frac{s}{\sqrt{n}} = 7.92 \pm (5.841)\left(\frac{0.0183}{\sqrt{4}} \right) = 7.92 \pm 0.053.$$

Thus a 99% C.I. for μ is 7.92 ± 0.053, or $(7.867, 7.973)$.

    (b)

$$\text{The 95\% C.I. for } \sigma \text{ is} \quad \left(\frac{\sqrt{n-1}(s)}{\chi_{n-1;\alpha/2}}, \frac{\sqrt{n-1}(s)}{\chi_{n-1;1-\alpha/2}} \right).$$

where $n = 4$, $\alpha = 0.05$, we have $\chi_{3;0.025}^2 = 9.34840$, $\chi_{3;0.975}^2 = 0.215795$. Thus a 95% C.I. for σ is given by

$$\left(\frac{\sqrt{n-1}(s)}{\sqrt{\chi_{n-1;\alpha/2}^2}}, \frac{\sqrt{n-1}(s)}{\sqrt{\chi_{n-1;1-\alpha/2}^2}} \right) = \left(\sqrt{\frac{(n-1)s^2}{\chi_{3;0.025}^2}}, \sqrt{\frac{(n-1)s^2}{\chi_{3;0.975}^2}} \right)$$

$$\cong \left(\sqrt{\frac{0.001}{9.34840}}, \sqrt{\frac{0.001}{0.215795}} \right)$$

$$= (0.0103, 0.0681).$$

5. The sample mean, variance, and standard deviation are $\bar{x} = \dfrac{2}{8} = 0.25$,

$$s^2 = \frac{1}{7}\left(256 - \frac{4}{8}\right) = 36.50, \quad \text{and} \quad s = 6.0415, \ \text{with} \ n = 8.$$

(a) The $100(1-\alpha)\%$ C.I. for μ is $\bar{x} \pm t_{n-1:\alpha/2} \cdot s/\sqrt{n}$.
For $\alpha = 0.05$, $t_{7:0.025} = 2.365$, so the C.I. is

$$0.25 \pm (2.365)\left(\frac{6.0415}{\sqrt{8}}\right) = 0.25 \pm 5.05 \,.$$

Thus a 95% C.I. for μ is 0.25 ± 5.05, or $(-4.80, 5.30)$.

(b)

$$\text{The } 100(1-\alpha)\% \text{ C.I. for } \sigma \text{ is} \quad \left(\sqrt{\frac{(n-1)s^2}{\chi^2_{n-1:\alpha/2}}}, \ \sqrt{\frac{(n-1)s^2}{\chi^2_{n-1:1-\alpha/2}}}\right).$$

For $\alpha = 0.05$, we have $\chi^2_{7:0.025} = 16.0128$ and $\chi^2_{7:0.975} = 1.68987$, so the C.I. is

$$\left(\sqrt{\frac{255.5}{16.0128}}, \ \sqrt{\frac{255.5}{1.68987}}\right) = (3.994, 12.296).$$

7. Given $n = 9$ and $s = 38$, a 95% C.I. for σ is

$$\left(\frac{\sqrt{n-1}(s)}{\sqrt{\chi^2_{n-1:\alpha/2}}}, \ \frac{\sqrt{n-1}(s)}{\sqrt{\chi^2_{n-1:1-\alpha/2}}}\right) = \left(\frac{\sqrt{8}(38)}{\sqrt{17.5346}}, \ \frac{\sqrt{8}(38)}{\sqrt{2.17973}}\right) = (25.67, 72.80).$$

9. The sample means and variances are

$$\bar{x}_A = 64.16, \quad s_A^2 = 4.4338, \quad s_A = 2.106, \quad n_A = 10$$
$$\bar{x}_B = 63.49, \quad s_B^2 = 8.2788, \quad s_B = 2.877, \quad n_B = 10 \,.$$

As in earlier problems, since $n_A = n_B \, (= n)$, we get

$$s_p^2 = \frac{s_A^2 + s_B^2}{2} = 6.3563,$$

$\Rightarrow s_p = 2.521$. Also, $t_{18:0.05} = 1.734$.
Assuming normality and equality of variances, the 90% C.I. for $\mu_A - \mu_B$ is

$$(\bar{x}_A - \bar{x}_B) \pm t_{18:0.05} \cdot s_p \sqrt{\frac{1}{n_A} + \frac{1}{n_B}}$$

$$= (64.16 - 63.49) \pm (1.734)(2.521)\sqrt{\frac{1}{10} + \frac{1}{10}} = 0.67 \pm 1.955$$

or $(-1.285, 2.625)$.

11. The sample means and variances are

$$\bar{x}_A = 87.4, \quad s_A^2 = 63.3, \quad n_A = 5$$
$$\bar{x}_B = 93.78, \quad s_B^2 = 17.69, \quad n_B = 9 \,.$$

Then $s_p^2 = \dfrac{(n_A - 1)s_A^2 + (n_B - 1)s_B^2}{n_A + n_B - 2} = \dfrac{4(63.3) + 8(17.69)}{12} = 32.89$,

$\Rightarrow s_p = 5.735$. Also, for $\alpha = 0.01$, we have $t_{12;0.005} = 3.055$.

Thus, under the assumption of normality and equal variances, the 99% C.I. for $\mu_A - \mu_B$ is

$$(\overline{x}_A - \overline{x}_B) \pm t_{n_A + n_B - 2; \alpha/2} \times s_p \sqrt{\frac{1}{n_A} + \frac{1}{n_B}}$$

$$= (87.4 - 93.78) \pm (3.055)(5.735)\sqrt{\frac{1}{5} + \frac{1}{9}} = -6.38 \pm 9.77$$

or $(-16.15, 3.39)$.

13. We are given the following:

$$\overline{x}_1 = 35.84, \quad s_1^2 = 130.4576, \quad n_1 = 25$$

$$\overline{x}_2 = 30.60, \quad s_2^2 = 53.0604, \quad n_2 = 25,$$

We can calculate the degrees of freedom m as follows:

$$c = \frac{130.4576/25}{130.4576/25 + 53.0604/25} = 0.711;$$

$$\frac{1}{m} = \frac{(0.711)^2}{24} + \frac{(0.289)^2}{24} = 0.02454,$$

$\Rightarrow \quad m = 40.74$, so we use $m = 40$ degrees of freedom. Also, we have $t_{40;0.025} = 2.021$.

Thus the 95% C.I. for $\mu_1 - \mu_2$ is

$$(\overline{x}_1 - \overline{x}_2) \pm t_{m;\alpha 2}\sqrt{\frac{s_1^2}{n_1} + \frac{s_2^2}{n_2}} = (35.84 - 30.60) \pm (2.021)\sqrt{\frac{130.4576}{25} + \frac{53.0604}{25}}$$

$$= 5.24 \pm 5.475$$

or $(-0.235, 10.715)$.

15. From Problem 8, we have

$$s_A = 1.3, \quad s_A^2 = 1.69, \quad n_A = 4$$

$$s_B = 1.5, \quad s_B^2 = 2.25, \quad n_B = 4.$$

For $\alpha = 0.05$, $F_{3,3;0.025} = 15.439$, so a 95% C.I. for σ_A^2/σ_B^2 is

$$\left(\frac{s_A^2}{s_B^2} \cdot \frac{1}{F_{n_A - 1, n_B - 1;\alpha/2}}, \frac{s_A^2}{s_B^2} \cdot \frac{1}{F_{n_A - 1, n_B - 1;1 - \alpha/2}}\right) = \left(\frac{1.69}{2.25} \cdot \frac{1}{15.439}, \frac{1.69}{2.25} \cdot 15.439\right)$$

$$= (0.0486, 11.5964)$$

Since the interval includes the value '1', the data bears out the statement "$\sigma_A^2/\sigma_B^2 = 1$" at the 95% level.

17. Since σ is unknown, but n is large, the 95% C.I. for μ is $\overline{X} \pm z_{0.025} \cdot \dfrac{s}{\sqrt{n}}$ where $n = 100$, $\overline{X} = 515.87$, $s = 14.34$. Thus the C.I. is

$$515.87 \pm (1.96)\left(\frac{14.34}{\sqrt{100}}\right) = 515.87 \pm 2.811$$

or $(513.059, 518.681)$.

19. Given $n = 105$, $\bar{x} = 68.05$, and $s = 2.79$, the 95% C.I. for μ is

$$\bar{x} \pm z_{0.025} \cdot \frac{s}{\sqrt{n}} = 68.05 \pm (1.96)\left(\frac{2.79}{\sqrt{105}}\right) = 68.05 \pm 0.534$$

or $(67.516, 68.584)$.

21. We are given

$$n_1 = 121, \quad \bar{x}_1 = 3.6, \quad s_1^2 = 1.96$$
$$n_2 = 121, \quad \bar{x}_2 = 3.2, \quad s_2^2 = 3.24.$$

(a) $F_{120,120;0.025} = 1.4327$, so a 95% C.I. for σ_1^2/σ_2^2 is

$$\left(\frac{1.96}{3.24} \cdot \frac{1}{1.4327}, \frac{1.96}{3.24} \cdot 1.4327\right) = (0.422, 0.867).$$

(b) σ_1^2 and σ_2^2 are unknown and cannot be assumed equal; but, n_1 and n_2 are large. Thus, the 95% C.I. for $\mu_1 - \mu_2$ is

$$(\bar{x}_1 - \bar{x}_2) \pm z_{0.025}\sqrt{\frac{s_1^2}{n_1} + \frac{s_2^2}{n_2}} = (3.6 - 3.2) \pm (1.96)\sqrt{\frac{1.96}{121} + \frac{3.24}{121}}$$

$$= 0.4 \pm 0.406$$

or $(-0.006, 0.806)$.

23. We are given the following:

$$n_1 = 100 \quad \bar{X}_1 = 23 \quad S_1 = 4$$
$$n_2 = 100 \quad \bar{X}_2 = 25 \quad S_2 = 6$$

(a) Using MINITAB, we have $F_{99,99;0.005} = 1.6854$.
So a 99% C.I. for σ_1^2/σ_2^2 is

$$\left(\frac{4^2}{6^2} \cdot \frac{1}{1.6854}, \frac{4^2}{6^2} \cdot 1.6854\right) = (0.2637, 0.7491).$$

Since the interval does not contain '1', we cannot assume $\sigma_1^2 = \sigma_2^2$.

(b) A 99% C.I. for $\mu_1 - \mu_2$ under this assumption is

$$(\bar{X}_1 - \bar{X}_2) \pm z_{0.005}\sqrt{\frac{s_1^2}{n_1} + \frac{s_2^2}{n_2}} = (23 - 25) \pm (2.575)\sqrt{\frac{16}{100} + \frac{36}{100}}$$

$$= -2 \pm 1.8568$$

or $(-3.8568, -0.1432)$.

25. We are given the following:

$$n_1 = 27, \quad \bar{x}_1 = 13.540, \quad s_1 = 0.476$$
$$n_2 = 45, \quad \bar{x}_2 = 11.691, \quad s_2 = 0.519.$$

(a) The 95% C.I. for σ_1^2/σ_2^2 is

$$\left(\frac{s_1^2}{s_2^2} \cdot \frac{1}{F_{n_1-1,n_2-1;\alpha/2}}, \frac{s_1^2}{s_2^2} \cdot \frac{1}{F_{n_1-1,n_2-1;1-\alpha/2}}\right)$$

We need $F_{26,44;0.025}$ and $F_{44,26;0.025}$. Using MINITAB, we find $F_{26,44;0.025} = 1.9479$ and $F_{44,26;0.025} = 2.0748$

Thus, we have

$$\left(\frac{(0.476)^2}{(0.519)^2} \cdot \frac{1}{1.9479}, \frac{(0.476)^2}{(0.519)^2} \cdot 2.0748\right) = (0.4318, 1.7452).$$

Since the interval contains '1', we may assume $\sigma_1^2 = \sigma_2^2$.

(b) Under the assumption in part (a), a 95% C.I. for $\mu_1 - \mu_2$ is

$$(\bar{x}_1 - \bar{x}_2) \pm z_{0.025} \cdot s_p \sqrt{\frac{1}{n_1} + \frac{1}{n_2}}.$$

$$s_p^2 = \frac{26(0.476)^2 + 44(0.519)^2}{70} = 0.2535.$$

Thus the C.I. is

$$(13.540 - 11.691) \pm 1.96\sqrt{0.2535}\sqrt{\frac{1}{27} + \frac{1}{45}} = 1.849 \pm 0.2402$$

or $(1.6088, 2.0892)$.

27. In MINITAB, select **Stat > Basic Statistics > 1 Proportion,** which will prompt a dialog box titled **1 Proportion.** Since we have summarized data, click the circle next to **Summarized Data** and enter the number of trials and the number of events (successes) in the appropriate boxes. Check **Options,** which will prompt another dialog box to appear. Enter 0.95 the desired confidence level in the box next to **Confidence level** and select **Use test and interval based on normal distribution.** Click **OK** on dialog boxes. The MINITAB output will show up in the session window as given below.

Sample	X	N	Sample θ	95% CI
1	9	15	0.60	$(0.3521, 0.8479)$.

29. Each x_i has p.d.f. $f(x_i; \mu) = \dfrac{1}{\sigma_0\sqrt{2\pi}} \exp\left(-\dfrac{(x_i - \mu)^2}{2\sigma_0^2}\right)$,

$$\text{so} \quad \ell(\mu|x_1, \ldots, x_n) = \left(\frac{1}{\sigma_0\sqrt{2\pi}}\right)^n \exp\left(-\frac{1}{2\sigma_0^2}\sum (x_i - \mu)^2\right)$$

$$\text{and} \quad L(\mu|x_1, \ldots, x_n) = -n\ln\left(\sigma_0\sqrt{2\pi}\right) - \frac{1}{2\sigma_0^2}\sum (x_i - \mu)^2.$$

Taking the partial derivative with respect to μ gives

$$\frac{\partial L}{\partial \mu} = \frac{2}{2\sigma_0^2}\sum (x_i - \mu) = \frac{\Sigma(x_i - \mu)}{\sigma_0^2}.$$

Setting the partial derivative equal to zero and denoting the MLE of μ by $\hat{\mu}$ gives

$$\sum (x_i - \hat{\mu}) = 0.$$

Thus

$$\hat{\mu} = \frac{\Sigma x_i}{n} = \bar{X} \text{ is the MLE of } \mu.$$

Note that $E(\hat{\mu}) = E(\bar{X}) = E(X) = \mu$, so $\hat{\mu} = \bar{X}$ is an unbiased estimator of μ.

$$\text{Also, } \bar{X} \sim N\left(\mu, \frac{\sigma_0^2}{n}\right).$$

31. The margin of error with probability 95% is given by

$$E = \pm z_{0.025} \frac{\sigma}{\sqrt{n}} = \pm 1.96 \frac{1.5}{\sqrt{36}} = \pm 0.49$$

33. From the given information in this problem, we have

$$\overline{X} = 12, S = 0.6, \quad n = 64$$

Since the sample size is large, a 99% confidence interval for the population mean is given by

$$\left(\overline{X} - z_{\alpha/2} \frac{S}{\sqrt{n}} \le \mu \le \overline{X} + z_{\alpha/2} \frac{S}{\sqrt{n}} \right) = \left(12 - 2.575 \frac{0.6}{\sqrt{64}} \le \mu \le 12 + 2.575 \frac{0.6}{\sqrt{64}} \right)$$
$$= (11.8069, 12.1931).$$

35. From the information given, we have

$$\overline{X}_1 = 295{,}000, \quad \overline{X}_2 = 305{,}000, \quad S_1 = 10{,}600, \quad S_2 = 12{,}800, \quad n_1 = 100, \quad n_2 = 121$$

A 98% confidence interval for the difference of two population means is given by

$$\left(\overline{X}_1 - \overline{X}_2 \right) \pm z_{\alpha/2} \sqrt{\frac{S_1^2}{n_1} + \frac{S_2^2}{n_2}} = (295{,}000 - 305{,}000) \pm 2.33 \sqrt{\frac{10{,}600^2}{100} + \frac{12{,}800^2}{121}}$$
$$= -10{,}000 \pm 3667.5485$$
$$= (-13{,}667.5485, -6{,}332.4515).$$

37. **(a)** In this problem we are given that

$$n_1 = 49, \quad \overline{X}_1 = 79, \quad S_1^2 = 30$$
$$n_2 = 36, \quad \overline{X}_2 = 86, \quad S_2^2 = 40$$

Hence, a 95% confidence interval for the difference between two population means is given by

$$\left(\overline{X}_1 - \overline{X}_2 \right) \pm z_{\alpha/2} \sqrt{\frac{S_1^2}{n_1} + \frac{S_2^2}{n_2}} = (79 - 86) \pm 1.96 \sqrt{\frac{30}{49} + \frac{40}{36}} = -7 \pm 2.573$$
$$= (-9.573, -4.427)$$

(b) A 98% lower confidence limit for the difference between two population means is given by

$$\left(\overline{X}_1 - \overline{X}_2 \right) - z_{\alpha} \sqrt{\frac{S_1^2}{n_1} + \frac{S_2^2}{n_2}} = (79 - 86) - 2.06 \sqrt{\frac{30}{49} + \frac{40}{36}} = -7 - 2.704 = -9.704$$

Hence, a 98% one-sided lower confidence interval is $(-9.704, \infty)$
A 98% upper confidence limit for the difference between two population means is given by

$$\left(\overline{X}_1 - \overline{X}_2 \right) + z_{\alpha} \sqrt{\frac{S_1^2}{n_1} + \frac{S_2^2}{n_2}} = (79 - 86) + 2.06 \sqrt{\frac{30}{49} + \frac{40}{36}} = -7 + 2.704 = -4.296$$

Hence, a 98% one-sided upper confidence interval is $(-\infty, -4.4296)$.

39. From the given information, we have

$$\hat{\theta}_1 = \frac{72}{120} = 0.6, (1 - \hat{\theta}_1) = 0.40, \hat{\theta}_2 = \frac{110}{150} = 0.73, (1 - \hat{\theta}_2) = 0.27$$

A 95% confidence interval for the difference between two population proportions is given by

$$
\begin{aligned}
\left(\hat{\theta}_1 - \hat{\theta}_2\right) \pm z_{\frac{\alpha}{2}}\sqrt{\frac{\hat{\theta}_1(1 - \hat{\theta}_1)}{n_1} + \frac{\hat{\theta}_2(1 - \hat{\theta}_2)}{n_2}} &= (0.6 - 0.73) \pm 1.96\sqrt{\frac{(0.6)(0.4)}{120} + \frac{(0.73)(0.27)}{150}} \\
&= -0.13 \pm 0.1128 \\
&= (-0.2428, -0.0172).
\end{aligned}
$$

41. Referring to Problems 38 and 40, we have

$$n = \frac{\left(z_{\alpha/2}\right)^2 \hat{\theta}(1 - \hat{\theta})}{E^2} = \frac{\left(1.96^2\right)(0.02)(0.98)}{(0.05)^2} = 30.12 \cong 31$$

Thus, the sample size decreases by 90.

43. Using MINITAB $F_{n_2-1,n_1-1,\alpha} = F_{14,9,0.05} = 3.02547$. Thus a 95% upper confidence interval for the ratio of two variances is given by

$$\left(0, F_{n_2-1,n_1-1,\alpha}\frac{S_1^2}{S_2^2}\right) = \left(0, F_{14,9,0.05}\frac{30^2}{38^2}\right) = \left(0, 3.02547 \times \frac{30^2}{38^2}\right) = (0, 1.8857).$$

45. Referring to Problem 1 of Section 8.7, we have a point estimate of $\theta = \hat{\theta} = \frac{200}{500} = 0.4$.

99% confidence interval for the population proportion is given by

$$\hat{\theta} \pm z_{\frac{\alpha}{2}}\sqrt{\frac{\hat{\theta}(1 - \hat{\theta})}{n}} = 0.4 \pm 2.575\sqrt{\frac{(0.4)(0.6)}{500}} = 0.4 \pm 0.0564 = (0.3436, 0.4564).$$

47. Referring to Problem 3 of Section 8.7, let θ_1 and θ_2 be the proportions of the populations favoring Brand X before and after the advertising campaign. Then point estimates of θ_1 and θ_2 are given by

$$\hat{\theta}_1 = 16/100 = 0.16, \quad \hat{\theta}_2 = 50/200 = 0.25.$$

99% confidence interval for the difference in proportions of the population in favoring Brand X before and after the advertising campaign, is given by

$$\left(\hat{\theta}_1 - \hat{\theta}_2\right) \pm z_{\frac{\alpha}{2}}\sqrt{\frac{\hat{\theta}_1\left(1 - \hat{\theta}_1\right)}{n_1} + \frac{\hat{\theta}_2\left(1 - \hat{\theta}_2\right)}{n_2}} = (0.16 - 0.25) \pm 2.575\sqrt{\frac{(0.16)(0.84)}{100} + \frac{(0.25)(0.75)}{200}}$$

or

$$(-0.09 - 0.1230, -0.09 + 0.1230) = (-0.2130, 0.0330)$$

Since the confidence interval contains 0, we can say that at the 1% level of significance that the proportion of the population favoring Brand X before and after the advertising campaign is the same.

49. Referring to Problem 5 of Section 8.7, the point estimates of θ_1 and θ_2 are $\hat{\theta}_1 = \frac{40}{800} = 0.05$, $\hat{\theta}_2 = \frac{50}{600} = 0.083$, respectively.

Thus, a 99% confidence interval for the difference between two population proportions is given by:

$$\left(\hat{\theta}_1 - \hat{\theta}_2\right) \pm z_{\frac{\alpha}{2}}\sqrt{\frac{\hat{\theta}_1\left(1 - \hat{\theta}_1\right)}{n_1} + \frac{\hat{\theta}_2\left(1 - \hat{\theta}_2\right)}{n_2}} = (0.05 - 0.083) \pm 2.575\sqrt{\frac{(0.05)(0.95)}{800} + \frac{(0.083)(0.917)}{600}}$$

$$= -0.033 \pm 0.0351 = (-0.0681, 0.0021).$$

By increasing the confidence coefficient the size (or width) of the confidence interval has increased. This confidence interval includes zero, so we can conclude at the 1% level of significance that that the percentages of persons favoring the nuclear plant in two different states are not different. In Problem 5 of Section 8.7, we noted that at the 5% level of significance that these percentages are different.

9

HYPOTHESIS TESTING

PRACTICE PROBLEMS FOR SECTION 9.2

1. Using the given information the null hypothesis and the• alternative hypothesis may be described as follows:

$$H_0 : \mu = 4 \quad versus \quad H_1 : \mu > 4.$$

Since the sample size is large, the test statistic is

$$Z \sim \frac{\overline{X} - 4}{S/\sqrt{n}}$$

Hence, at $\alpha = .05$, we reject H_0 if observed $Z > 1.645$. The observed Z is

$$Z = \frac{\overline{X} - 4}{S/\sqrt{n}} = \frac{4.2 - 4}{0.5/\sqrt{36}} = 2.4,$$

which is greater than 1.645. Thus, we reject H_0.

3. The Type II error will be smaller. The actual value of Type II error is

$$\beta = P\left(Z < \frac{\mu_0 - \mu_1}{\sigma/\sqrt{n}} + z_\alpha \right) = P\left(Z < \frac{4 - 5}{0.5/\sqrt{36}} + 1.645 \right) = P(Z < -10.355) = 0.$$

So our conjecture was valid.

5. Using (9.2.10), we obtain

$$n \geq \frac{(z_\alpha + z_\beta)^2 \sigma^2}{(\mu_1 - \mu_0)^2} = \frac{(1.645 + 1.28)^2 \times (0.5)^2}{(4.2 - 4)^2} = 53.47 \cong 54.$$

Solutions Manual to Accompany Statistics and Probability with Applications for Engineers and Scientists, Bhisham C. Gupta and Irwin Guttman.
© 2014 John Wiley & Sons, Inc. Published 2014 by John Wiley & Sons, Inc.

PRACTICE PROBLEMS FOR SECTION 9.3

1. We wish to test $H_0 : \mu = 800$ vs. $H_1 : \mu = \mu_1 > 800$. At $\alpha = 0.01$, $z_{0.01} = 2.33$, and we reject H_0 if

$$\overline{X} > \mu_0 + \frac{\sigma}{\sqrt{n}} z_\alpha = 800 + \frac{30}{\sqrt{100}}(2.33) = 806.99.$$

Since the observed $\overline{X} = 812 > 806.99$, we reject H_0 at the 1% level.
<u>OR</u>

$$Z = \frac{\overline{X} - \mu_0}{\sigma/\sqrt{n}} = \frac{812 - 800}{30/\sqrt{100}} = 4. \text{We reject } H_0 \text{ since } Z = 4 > z_{0.01} = 2.33.$$

The power function is

$$\gamma(\mu_1) = P(\overline{X} > 806.99 | \mu = \mu_1) = P\left(z > \frac{806.99 - \mu_1}{30/\sqrt{100}}\right)$$

$$\gamma(\mu_1) = 1 - \Phi\left(\frac{806.99 - \mu_1}{3}\right).$$

Thus the power at $\mu = 810$ is

$$\gamma(810) = 1 - \Phi\left(\frac{806.99 - 810}{3}\right) = 1 - \Phi(-1.007) = 0.8431.$$

Similarly, we obtain

μ_1 :	800	802	804	806	808	810	812	814
$\gamma(\mu_1)$:	0.01	0.05	0.16	0.37	0.63	0.84	0.95	0.99

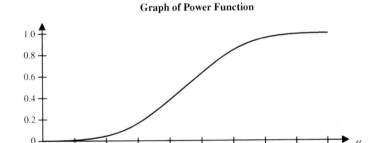

Graph of Power Function

3. From the given information, we formulate the hypothesis as

$$H_0 : \mu = 800 \quad \text{versus} \quad H_1 : \mu = \mu_1 < 800$$

We test this hypothesis using MINITAB, taking the following steps:
(i) Select **Stat** > **Basic Statistics** > **1 sample Z.** This will prompt a dialog box **1-sample Z** to appear on the screen.
(ii) In this dialog box enter sample size (100) and mean (785).

(iii) Check the **Perform hypothesis test** and Select **Options** and make the necessary entries in the new dialog box that appears. Click **OK** on each of the dialog box. The MINITAB output will appear in the session window as given below.

Test of $H_0 : \mu = 800$ Vs $H_1 : \mu < 800$
The assumed standard deviation $= 30$

N	Mean	SE Mean	99% Upper Bound	Z	P
100	785.00	3.00	791.98	−5.00	0.000

Since the p-value is less than 1% level of significance we reject the null hypothesis.
To determine power of the test we first find the Type II error (β). That is,

$$\beta = P\left(Z > \frac{\mu_0 - \mu_1}{\sigma/\sqrt{n}} - z_\alpha\right) = P\left(Z > \frac{800 - 790}{30/\sqrt{100}} - 2.575\right) = P(Z > 0.7583) = 0.2241$$

Power of the test is equal to $1 - \beta = 1 - 0.2241 = 0.7759$

5. In this problem the given test is a two tail test. Thus, the rejection region at the α level of significance is given by

$$\left(\left|\frac{\overline{X} - \mu}{\sigma/\sqrt{n}}\right| > z_{\alpha/2} \Big| \mu = \mu_0\right) = \alpha,$$

From this it can easily be seen that the critical points in terms of $\overline{X}$ are $\mu_0 \pm z_{\alpha/2} \times \sigma/\sqrt{n} = .25 \pm 2.575(.0015/\sqrt{10}) \Rightarrow$ the critical region in terms $\overline{X}$ is ($\overline{X} < 0.2488$ or $\overline{X} > 0.2512$). To determine the power of the test we first find the Type II error (β). That is,

$$\beta = P\left(\frac{.2500 - .2490}{.0015/\sqrt{10}} - 2.575 < Z < \frac{.2500 - .2490}{.0015/\sqrt{10}} + 2.575\right)$$

$$= P(-0.467 < Z < 4.683) = 0.6797.$$

Hence, power of the test at $\mu = .2490$ is equal to $1 - \beta = 1 - 0.6797 = 0.3203$.

PRACTICE PROBLEMS FOR SECTION 9.4

We perform testing of hypotheses in Problems 1 to 6, using MINITAB. To do this we take the following steps: (Note that in all these problems for test statistic we will be using the t- statistics)

(i) Select **Stat** > **Basic Statistics** > **1-sample** t. This will prompt a dialog box **1-sample** t to appear on the screen.

(ii) In this dialog box select either Samples in Column (if raw data is available) or Summarized data and enter the available information.

(iii) Check **Perform hypothesis test** and Select **Options** and make the necessary entries in the new dialog box that appears. Click **OK** on each of the dialog box. The MINITAB output will appear in the session window

1. The MINITAB printout is shown below.
Test of mu $= .5$ versus less than $.5$ (Test: $H_0 : \mu = .5$ Vs $H_1 : \mu < .5$)

N	Mean	StDev	SE Mean	95% Upper Bound	T	P
10	0.453	0.370	0.117	0.667	−0.40	0.349

Since the p-value (0.349) is greater than the level of significance (.05) we do not reject the null hypothesis.

3. The MINITAB printout is shown below.
   Test of mu $= 110$ versus not $= 110$ (Test: $H_0 : \mu = 110$ Vs $H_1 : \mu \neq 110$)

Variable	N	Mean	StDev	SE Mean	99% CI	T	P
Cholesterol	25	104.52	10.53	2.11	(98.63, 110.41)	-2.60	0.016

   Since the p-value (0.016) is greater than the level of significance (.01) we do not reject the null hypothesis. Also, note that the 99% confidence interval contains 110 the value of μ *under* H_0. Hence, we do not reject the null hypothesis.

5. The MINITAB printout is shown below.
   Test of mu $= 13.5$ versus not $= 13.5$ (Test: $H_0 : \mu = 13.5$ Vs $H_1 : \mu \neq 13.5$)

Variable	N	Mean	StDev	SE Mean	95% CI	T	P
Voltage	16	13.520	0.706	0.176	(13.144, 13.896)	0.11	0.911

   Since the p-value (0.911) is greater than the level of significance (.05) we do not reject the null hypothesis. Also, note that the 99% confidence interval contains 13.5, the value of μ *under* H_0. Hence, we do not reject the null hypothesis.

7. Test $H_0 : \mu = 7.13$ versus $H_1 : \mu = \mu_1 > 7.13$ at the $\alpha = 0.05$ level.

   Reject $\qquad H_0$ if $\dfrac{\overline{X} - \mu_0}{\sigma / \sqrt{n}} > z_\alpha, or$

   $$\overline{X} > \mu_0 + \frac{\sigma}{\sqrt{n}} z_\alpha = 7.13 + \frac{0.0216}{\sqrt{10}}(1.645) = 7.141.$$

   Since the observed value of $\overline{X} = \dfrac{50.05}{7} = 7.15, \quad S = 0.0216, \quad p - value = .025,$ we reject H_0 at the 5% level.

PRACTICE PROBLEMS FOR SECTION 9.5

1. To test $H_0 : \mu = 4.90$ versus $H_1 : \mu < 4.90$, calculate the value of the test statistic

   $$Z = \frac{(\overline{X} - \mu_0)\sqrt{n}}{S} = \frac{(4.34 - 4.90)\sqrt{50}}{1.93} = -2.052.$$

   At the $\alpha = 0.05$ level of significance, reject H_0 if $Z < -1.645$. Since the observed Z is in the rejection region, we reject H_0; i.e. the sample evidence supports the claim that desert conditions do decrease the life of a canteen.

3. We wish to test $H_0 : \mu = 90$ versus $H_1 : \mu = \mu_1 > 90$.

   $$z = \frac{(\overline{X} - \mu_0)\sqrt{n}}{S} = \frac{(94.2 - 90)\sqrt{30}}{8.5} = 2.706.$$

   At the $\alpha = 0.01$ level of significance, reject H_0 if $z > 2.33$. Since the observed $z = 2.706$, we reject H_0; i.e. the sample evidence supports the claim that remedial reading has helped.

5. We test this hypothesis using MINITAB, taking the steps given in Problem 4 above.
   The MINITAB printout is shown below.
   Test of mu $= 0.25$ Vs > 0.25
   The assumed standard deviation $= 0.00554$

Variable	N	Mean	StDev	SE Mean	95% Lower Bound	Z	P
Diameter:	49	0.249388	0.005537	0.000791	0.248086	-0.77	0.780

   Since the p-value (.780) is greater than the significance level (.05), so we do not reject the null hypothesis H_0, i.e., the sample evidence supports the claim that mean diameter of rivets is 0.25 inch.

PRACTICE PROBLEMS FOR SECTION 9.6

1. We wish to test $H_0 : \mu_s = \mu_c$ versus $H_1 : \mu_s < \mu_c$.

Since σ_s and σ_c are known, the test statistic is $Z = \dfrac{\overline{X}_s - \overline{X}_c}{\sqrt{\dfrac{\sigma_s^2}{n_s} + \dfrac{\sigma_c^2}{n_c}}}$, and we reject H_0 if $z < -z_\alpha$.

From the data, $\overline{X}_s = 8.7867$ and $\overline{X}_c = 8.9533$, so

$$Z = \frac{8.7867 - 8.9533}{\sqrt{\dfrac{(0.12)^2}{3} + \dfrac{(0.12)^2}{3}}} = -1.7003.$$

At the 5% level of significance, reject H_0 if $z_{obs} < -1.645$; thus we reject H_0. That is, the sample evidence supports the claim that the state lab has a downward bias relative to the company lab.

3. We wish to test $H_0 : \mu_A - \mu_B = 120$ versus $H_1 : \mu_A - \mu_B \neq 120$.

$$\text{Given }\; \sigma_A = 200, \quad n_A = 25, \quad \overline{X}_A = 1610,$$
$$\sigma_B = 200, \quad n_B = 25, \quad \overline{X}_B = 1455,$$

$$|Z| = \frac{\left|(\overline{X}_A - \overline{X}_B) - (\mu_A - \mu_B)\right|}{\sqrt{\dfrac{\sigma_A^2}{n_A} + \dfrac{\sigma_B^2}{n_B}}} = \frac{\left|(1610 - 1455) - 120\right|}{\sqrt{\dfrac{(200)^2}{25} + \dfrac{(200)^2}{25}}} = 0.6187.$$

At the 5% level of significance, reject H_0 if $|z| > z_{0.025} = 1.96$.
Since the observed $|z| = 0.6187$, we do not reject H_0.
The power function is

$$\gamma(\delta) = 1 - \left[\Phi\left(1.96 - \frac{\delta - 120}{40\sqrt{2}} \right) - \Phi\left(-1.96 - \frac{\delta - 120}{40\sqrt{2}} \right) \right].$$

$$\gamma(100) = 1 - [\Phi(2.314) - \Phi(-1.606)] = 0.0645.$$

δ:	−120	−80	−40	0	40	80	120	160
$\gamma(\delta)$:	0.989	0.942	0.807	0.564	0.293	0.109	0.050	0.109

δ:	200	240	280	320	360
$\gamma(\delta)$:	0.293	0.564	0.807	0.942	0.989

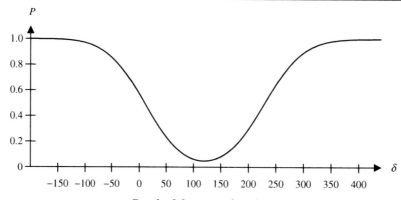

Graph of the power function:

5. From the given information and data sets I and II we obtain

$$\sigma_I^2 = 0.5 \quad n_I = 16 \quad \overline{X}_I = 10.839$$
$$\sigma_{II}^2 = 0.5 \quad n_{II} = 16 \quad \overline{X}_{II} = 12.983$$

We wish to test $H_0 : \mu_I - \mu_{II} = 0$ Versus $H_1 : \mu_I - \mu_{II} \neq 0$
   Observed value of test statistic is

$$Z_{obs} = \frac{\overline{X}_I - \overline{X}_{II}}{\sqrt{\left(\dfrac{\sigma_I^2}{n_I} + \dfrac{\sigma_{II}^2}{n_{II}}\right)}} = \frac{10.839 - 12.983}{\sqrt{\left(\dfrac{0.5}{16} + \dfrac{0.5}{16}\right)}} = -8.576$$

Since $|Z_{obs}| = |-8.576| > z_{0.025} = 1.96$, we reject H_0 at the 5% level of significance. The p-value is 0.00.

PRACTICE PROBLEMS FOR SECTION 9.7

1. We wish to test $H_0 : \mu_1 = \mu_2$ versus $H_1 : \mu_1 > \mu_2$ (where μ_1 is the mean before modification).
   Given $\sigma_1 = \sigma_2 = 0.05$, $n_1 = n_2 = 50$, $\overline{X}_1 = 4.091$, $\overline{X}_2 = 4.075$, we have

$$z_{obs} = \frac{\overline{X}_1 - \overline{X}_2}{\sqrt{\dfrac{\sigma_1^2}{n_1} + \dfrac{\sigma_2^2}{n_2}}} = \frac{4.091 - 4.075}{\sqrt{\dfrac{(0.05)^2}{50} + \dfrac{(0.05)^2}{50}}} = 1.6.$$

At the 5% level of significance, reject H_0 if $z_{obs} > 1.645$. Since $z_{obs} = 1.6$, we do not reject H_0, and conclude that the sample evidence supports the claim that modification has not changed the amount of filling.

3. Here we wish to test the hypothesis $H_0 : \mu_A = \mu_B$ Versus $H_1 : \mu_A \neq \mu_B$.
   In this problem we perform the testing of hypothesis using MINITAB (**Stat > Basic Statistics > 2- Sample t**). The MINITAB output is shown below. Further, in this problem we assume that the two populations variances are equal.
   Two-Sample T-Test and CI: Brand A Vis., Brand B Vis.
   Two-sample T for Brand A Vis. vs. Brand B Vis.

	N	Mean	StDev	SE Mean
Brand A Vis.	8	10.2850	0.0177	0.0063
Brand B Vis.	8	10.2888	0.0223	0.0079

Difference = mu (Brand A Vis.) − mu (Brand B Vis.)
Estimate for difference: −0.0038
95% CI for difference: (−0.0254, 0.0179)
T-Test of difference = 0 (vs. not = 0): T-Value = −0.37 P-Value = 0.715 DF = 14
Both use Pooled StDev = 0.0202
Since the p-value (0.715) is greater that the level of significance $\alpha = .05$, we do not reject the null hypothesis.

5. In this problem we wish to test the hypothesis

$$H_0 : \mu_A = \mu_B \quad Versus \quad H_1 : \mu_A < \mu_B$$

Here we use the paired t-test and in order to carry out this test we assume the differences are normally distributed with mean μ_d and standard deviation σ_d. Using MINITAB (**Stat > Basic Statistics > Paired t**). The MINITAB output is shown below.

Paired T for Before - After

	N	Mean	StDev	SE Mean
Before	10	152.00	8.21	2.59
After	10	137.00	6.51	2.06
Difference	10	15.00	11.97	3.79

95% upper bound for mean difference: 21.94

T-Test of mean difference $= 0$ (vs < 0): T-Value $= 3.96$ p-value $= 0.000$.

Since the p-value (0.000) is less than the level of significance (0.05), we reject the null hypothesis, and conclude the sample evidence supports the claim that the drug therapy is effective.

7. Here we wish to test the hypothesis $H_0 : \mu_A = \mu_B$ Versus $H_1 : \mu_A \neq \mu_B$.

In this problem we perform the testing of hypothesis using MINITAB (**Stat > Basic Statistics > 2- Sample t**). The MINITAB output is shown below. In this problem we do not assume the two population variances are equal.

Two-sample T for Brand A vs Brand B

	N	Mean	StDev	SE Mean
Brand A	8	10.2850	0.0177	0.0063
Brand B	8	10.2888	0.0223	0.0079

Difference $=$ mu (Brand A) $-$ mu (Brand B)

Estimate for difference: -0.0038

95% CI for difference: $(-0.0255, 0.0180)$

T-Test of difference $= 0$ (vs. not $= 0$): T-Value $= -0.37$ p-value $= 0.716$ DF $= 13$

Since the p-value (0.716) is greater than the level of significance (0.05) we do not reject H_0. That is, the sample evidence supports the claim that the brand viscosities are not significantly different.

PRACTICE PROBLEMS FOR SECTION 9.8

1. We know that for large n, we have

$$ Z \cong \frac{n\hat{\theta} - n\theta}{\sqrt{n\theta(1 - \theta)}} \sim N(0, 1) \quad \text{or} \quad Z \cong \frac{\hat{\theta} - \theta}{\sqrt{\frac{\theta(1 - \theta)}{n}}} \sim N(0, 1) $$

(a) To test $H_0 : \theta = \theta_0$ versus $H_1 : \theta = \theta_1 > \theta_0$ at level α, reject H_0 if

observed

$$ Z = \frac{\hat{\theta} - \theta_0}{\sqrt{\frac{\theta_0(1 - \theta_0)}{n}}} > z_\alpha. $$

(b) To test $H_0 : \theta = \theta_0$ versus $H_1 : \theta \neq \theta_0$ at level α, reject H_0 if

observed

$$ |Z| = \left| \frac{\hat{\theta} - \theta_0}{\sqrt{\frac{\theta_0(1 - \theta_0)}{n}}} \right| > z_{\alpha/2}. $$

3. From the given information, we formulate the hypothesis as

$$H_0 : \theta_1 = \theta_2 \text{ versus } H_1 : \theta_1 > \theta_2$$

We test this hypothesis using MINITAB, and proceed as follows:

1. Select **Stat** > **Basic Statistics** > **2 Proportion.** This will prompt a dialog box **2 Proportion** to appear on the screen.
2. In this dialog box enter Number of events and Number of Trials.
3. Check the **Perform hypothesis test** and Select **Options** and make the necessary entries in the new dialog box that appears. In this dialog box do not forget to check the box next to use pooled estimate of p (we use θ instead) for test. Click **OK** on each of the dialog boxes. The Minitab output will show up in the session window as given below.

Test and CI for Two Proportions

Sample	X	N	Sample p
1	146	1350	0.108148
2	113	1300	0.086923

Difference = p (1) − p (2)
Estimate for difference: 0.0212251
99% CI for difference: (−0.00842472, 0.0508749)
Test for difference = 0 (vs. not = 0): Z = 1.84, P-Value = 0.066
Since the p-value (0.066) is greater than the level of significance (0.01), we do not reject the null hypothesis, and conclude the sample evidence does not supports the claim that the production process has improved from January to March.

5. From the given information, we formulate the hypothesis as

$$H_0 : \theta = 0.20 \quad \text{versus} \quad H_1 : \theta = \theta_1 \neq 0.20$$

Using MINITAB we have the following results
Test of $\theta = 0.2$ versus $\theta \neq 0.2$ (Using exact distribution)

Sample	X	N	Sample θ	99% CI	Exact P-Value
1	136	1000	0.136000	(0.109373, 0.166166)	0.000

Test for One Proportion (Using the normal approximation)
Test of $\theta = 0.2$ vs. θ not $= 0.2$

Sample	X	N	Sample θ	95% CI	Z-Value	P- Value
1	136	1000	0.136000	(0.108078, 0.163922)	−5.06	0.000

The p-value whether we use the exact distribution or the normal approximation is 0.000. Hence, in both cases we reject the null hypothesis. In other words, the oil company's claim that 20% of the homes are heated by oil is not valid.

7. From the given information the point estimates of θ_1 and θ_2 are given by

$$\hat{\theta}_1 = \frac{20}{500} = 0.04, \quad \hat{\theta}_2 = \frac{30}{600} = 0.05$$

(a) We are now interested in testing at the 5% level of significance the hypothesis

$$H_0 : \theta_1 - \theta_2 = 0 \quad against \quad H_1 : \theta_1 - \theta_2 < 0.$$

From Equation (9.8.5) the pivotal quantity (see Chapter 8) for $\theta_1 - \theta_2$ is the test statistic

$$Z = \frac{(\hat{\theta}_1 - \hat{\theta}_2) - (\theta_1 - \theta_2)_0}{\sqrt{\theta_1(1 - \theta_1)/n_1 + \theta_2(1 - \theta_2)/n_2}}$$

Since under the null hypothesis $\theta_1 - \theta_2 = 0$, we substitute the values of $\hat{\theta}_1$, $\hat{\theta}_2$ and $\theta_1 - \theta_2 = 0$ in the numerator of above test statistic and the values of θ_1 and θ_2 in the denominator. However, here θ_1 and θ_2 are unknown but under the null hypothesis $\theta_1 = \theta_2 = \theta$ (say), we can estimate θ by pooling the two samples, that is

$$\hat{\theta} = (X_1 + X_2)/(n_1 + n_2).$$

We then replace θ_1 and θ_2 in the denominator with $\hat{\theta}$. In this example, we have

$$\hat{\theta} = (20 + 30)/(500 + 600) = 0.0455.$$

Thus, the value of the test statistic computed under the null hypothesis is given by

$$Z = \frac{(0.04 - 0.05) - 0}{\sqrt{(0.0455)(0.9545)/500 + (0.0455)(0.9545)/600}} = -0.7924$$

(a) The p-value is 0.214, which is larger than the significance level (.01) and therefore we do not reject the null hypothesis. In parts (b) and (c) the observed value of test statistic does not change, so it can be shown the p-value in (b) is 0.786 and in (c) is 0.428. Thus, we do not reject H_0 in either case.

PRACTICE PROBLEMS FOR SECTION 9.9

1. To test $H_0 : \sigma = 0.05$ versus $H_1 : \sigma = \sigma_1 > 0.05$, reject H_0 if the observed

$$\chi^2 > \chi^2_{n-1;\alpha} \quad \text{where} \quad \chi^2 = \frac{(n-1)S^2}{\sigma_0^2}.$$

$$\chi^2 = \frac{(n-1)S^2}{\sigma_0^2} = \frac{8(0.07)^2}{(0.05)^2} = 15.68.$$

At the $\alpha = 0.05$ level of significance, reject H_0 if $\chi^2 > \chi^2_{8;0.05} = 15.5073$.

Since the observed $\chi^2 = 15.68$, we reject H_0. That is, the sample value of S is significantly larger than the claimed value of σ.

3. From the given set of data we obtain, $S = 0.013038$, $n = 5$.

To test $H_0 : \sigma = 0.01$ versus $H_1 : \sigma > 0.01$ at the 5% significance level, reject H_0 if $\chi^2 > \chi^2_{4;0.05} = 9.48773$

Observed

$$\chi^2_4 = \frac{4(0.013038)^2}{(0.01)^2} = 6.800,$$

so we do not reject H_0. p-value = 0.147.

5. From the given set of data we obtain, $S^2 = 0.00083$, $n = 5$.

To test $H_0 : \sigma^2 = 0.9$ versus $H_1 : \sigma^2 \neq 0.9$ for $\alpha = 0.05$,

$$\chi^2 = \frac{4(0.00083)}{0.9} = 0.0037.$$

Reject H_0 if $\chi^2 < \chi^2_{4;0.975} = 0.484419$ or if $\chi^2 > \chi^2_{4;0.025} = 11.1433$.

Since the observed χ^2 falls in the rejection region, we reject H_0.

PRACTICE PROBLEMS FOR SECTION 9.10

1. From the given data we obtained the following:

$$\overline{X}_1 = 0.14067, \qquad \overline{X}_2 = 0.1385,$$
$$S_1^2 = 0.000007867, \qquad S_2^2 = 0.000007100,$$
$$n_1 = 6, \qquad n_2 = 6.$$

(a) Test $H_0 : \dfrac{\sigma_1^2}{\sigma_2^2} = 1$ against $H_1 : \dfrac{\sigma_1^2}{\sigma_2^2} \neq 1$ at the $\alpha = 0.01$ level

$$\text{Reject } H_0 \text{ if observed } F, \text{i.e.,} \quad F = \frac{S_1^2}{S_2^2} < F_{5,5;0.995} = \frac{1}{14.94} = 0.067,$$

or if $F > F_{5,5;0.005} = 14.94$.
Now, we note that the observed value of F is:

$$F = \frac{S_1^2}{S_2^2} = 1.108,$$

Hence, we do not reject H_0.

(b) From part (a), it follows that we may assume $\sigma_1^2 = \sigma_2^2$.
We wish to test $H_0 : \mu_1 - \mu_2 = 0$ versus $H_1 : \mu_1 - \mu_2 \neq 0$.

$$S_p^2 = \frac{S_1^2 + S_2^2}{2} = 0.000007483 \Rightarrow S_p = 0.0027355$$

$$|t_{10}| = \frac{|\overline{X}_1 - \overline{X}_2|}{\sqrt{S_p^2 \left(\frac{1}{n_A} + \frac{1}{n_B}\right)}} = \frac{|0.14067 - 0.1385|}{\sqrt{(0.000007483)\left(\frac{1}{6} + \frac{1}{6}\right)}} = 1.372.$$

Reject H_0 at the 5% level if $|t_{10}| > t_{10;0.025} = 2.228$.
Thus, we do not reject H_0 and conclude from the sample evidence that there is no significant difference between the mean resistance of the two lots.

3. In this problem we wish to test at 5% level of significance the hypothesis

$$H_0 : \sigma_1^2/\sigma_2^2 = 1 \quad \text{versus} \quad H_1 : \sigma_1^2/\sigma_2^2 \neq 1.$$

From the data in Problem 8 of Section 8.4, we obtain

$$n_1 = 20 \quad S_1^2 = 40.45 \quad n_2 = 20 \quad S_2^2 = 28.95$$

Reject H_0 if the observed value $F = \dfrac{S_1^2}{S_2^2} < F_{19,19;0.975} \cong \dfrac{1}{2.5310} = 0.3951$ or if $F = \dfrac{S_1^2}{S_2^2} > F_{19,19;0.025} \cong 2.5310$

But observed $F = \dfrac{S_1^2}{S_2^2} = \dfrac{40.45}{28.95} = 1.3972$, which does not fall in the rejection region, so we do not reject H_0.

5. In this problem we wish to test at 5% level of significance the hypothesis

$$H_0 : \sigma_1^2/\sigma_2^2 = 1 \quad \text{versus} \quad H_1 : \sigma_1^2/\sigma_2^2 \neq 1.$$

From the given data, we obtain

$$n_1 = 20 \quad S_1^2 = 12.842 \quad n_2 = 20 \quad S_2^2 = 26.38$$

Reject H_0 if the observed value $F = \dfrac{S_1^2}{S_2^2} < F_{19,19;0.975} \cong \dfrac{1}{2.5310} = 0.3951$ or if $F = \dfrac{S_1^2}{S_2^2} > F_{19,19;0.025} \cong 2.5310$

But observed $F = \dfrac{S_1^2}{S_2^2} = \dfrac{12.842}{26.38} = 0.4868$, which does not fall in the rejection region, so we do not reject H_0.

PRACTICE PROBLEMS FOR SECTIONS 9.11 AND 9.12

1. (a) We are given that the yields are normally distributed with unknown mean μ and known standard deviation 20. In part (a) we want to determine a 95% confidence interval for the population mean μ. From the formula given in Table 9.11.1, for the confidence interval of the mean with confidence coefficient $(1 - \alpha)$ when σ is known, we have

$$\left(\overline{X} - z_{\alpha/2}\sigma/\sqrt{n}, \overline{X} + z_{\alpha/2}\sigma/\sqrt{n} \right)$$

Now substituting the values of $\overline{X}, \sigma, n$, and $z_{\alpha/2}$, we have a desired 95% confidence interval

$$(630 - 1.96(20/\sqrt{50}), 630 + 1.96(20/\sqrt{50}) = (624.46, 635.54).$$

(b) In this part we want to test the hypothesis

$$H_0 : \mu = 650 \quad \text{versus} \quad H_1 : \mu \neq 650.$$

Since the confidence interval does not contain 650 the value of μ under the null hypothesis, we reject the null hypothesis.

3. (a) We are given that the yields are normally distributed with unknown mean μ and unknown variance σ^2. In part (a) we want to determine a 95% confidence interval for the population mean μ. From the formula given in Table 9.11.1, for the confidence interval for the mean with confidence coefficient $(1 - \alpha)$ when σ is unknown, we have

$$\left(\overline{X} - t_{n-1,\alpha/2} \, S/\sqrt{n}, \ \overline{X} + t_{n-1,\alpha/2} \, S/\sqrt{n} \right)$$

Now substituting the values of $\overline{X}, S, n$, and $t_{\alpha/2}$, we have a desired 95% confidence interval

$$(19.4 - 3.18(1.5/\sqrt{4}), 19.4 - 3.18(1.5/\sqrt{4})) = (17.015, 21.785).$$

(b) Since the confidence interval contains 18 the value of μ under the null hypothesis, we do not reject the null hypothesis.

5. (a) In this problem we are given that the percentage of chlorine in batch of polymer obtained by two analysts are normally distributed with unknown means μ_1 and μ_2 and unknown but equal standard deviations. In part (a) we want to determine a 95% confidence interval for the difference of two population means. From the formula given in Table 9.11.1, for the confidence interval for the difference of two means with confidence coefficient $(1 - \alpha)$ when σ's are unknown and sample sizes are small, we have

$$\left((\overline{X}_1 - \overline{X}_2) - t_{n_1+n_2-2,\frac{\alpha}{2}} S_p \sqrt{1/n_1 + 1/n_2}, \ (\overline{X}_1 - \overline{X}_2) + t_{n_1+n_2-2,\frac{\alpha}{2}} S_p \sqrt{1/n_1 + 1/n_2} \right)$$

Now substituting the values of $\overline{X}_1, S_p, n_1, \overline{X}_2, n_2$, and $t_{n_1+n_2-2,\alpha/2}$, we have a desired 95% confidence interval as

$$\left((12.21 - 11.83) - 2.101 \times 0.7577 \sqrt{1/10 + 1/10}, (12.21 - 11.83) + 2.101 \times 0.7577 \sqrt{1/10 + 1/10} \right)$$

or

$$(-0.332, 1.092).$$

(b) Since the confidence interval contains 0, the value of $\mu_1 - \mu_2$ under the null hypothesis, we do not reject the null hypothesis.

7. In part (a) we want to determine a 95% confidence interval for the population proportion θ. From the formula given in Table 9.11.1, confidence interval for the proportion of one population with confidence coefficient $(1 - \alpha)$, we have

$$\left(\hat{\theta} - z_{\alpha/2}\sqrt{\hat{\theta}(1 - \hat{\theta})/n}, \ \hat{\theta} + z_{\alpha/2}\sqrt{\hat{\theta}(1 - \hat{\theta})/n} \right)$$

Now substituting the values of $\hat{\theta}$ and $z_{\alpha/2}$ we have a desired 95% confidence interval as

$$\left(0.6 - 1.96\sqrt{0.6(1 - 0.6)/45}, \quad 0.6 + 1.96\sqrt{0.6(1 - 0.6)/45} \right)$$

or

$$(0.4569, 0.7431)$$

(b) Since the confidence interval contains "0.5" the value of θ under the null hypothesis, we do not reject the null hypothesis.

9. We wish to test $H_0 : \mu = \mu_0 = 20$ versus $H_1 : \mu = \mu_1 > 20$ sequentially with

$$\mu_1 = 21, \alpha = 0.05 \quad \text{and} \quad \beta = 0.05.$$

Now, $f_0(x) = \dfrac{1}{\sqrt{2\pi\sigma^2}} e^{-\dfrac{(x - \mu_0)^2}{2\sigma^2}}$ and $\dfrac{1}{\sqrt{2\pi\sigma_1^2}} e^{-\dfrac{(x - \mu_1)^2}{2\sigma^2}}$, so

$$\frac{f_1(x_1)\cdots f_1(x_m)}{f_0(x_1)\cdots f_0(x_m)} = \frac{\left(\dfrac{1}{\sqrt{2\pi\sigma^2}}\right)^m \exp\left(-\dfrac{\sum(x_i - \mu_1)^2}{2\sigma^2} \right)}{\left(\dfrac{1}{\sqrt{2\pi\sigma^2}}\right)^m \exp\left(-\dfrac{\sum(x_i - \mu_0)^2}{2\sigma^2} \right)}$$

$$\frac{f_1(x_1)\cdots f_1(x_m)}{f_0(x_1)\cdots f_0(x_m)} = \exp\left[\frac{1}{8}\left(\sum(x_i - 20)^2 - \sum(x_i - 21)^2 \right) \right].$$

Also, $A \cong \dfrac{\beta}{1 - \alpha} = \dfrac{0.05}{0.95}$ and $B \cong \dfrac{1 - \beta}{\alpha} = \dfrac{0.95}{0.05}$.

So continue sampling as long as

$$\ln\left(\frac{0.05}{0.95}\right) < \left[\frac{1}{8}\left(\sum(x_i - 20)^2 - \sum(x_i - 21)^2 \right) \right] < \ln\left(\frac{0.95}{0.05}\right),$$

$$-2.9444 < \left[\frac{1}{8}\left(\sum(x_i - 20)^2 - \sum(x_i - 21)^2 \right) \right] < 2.9444$$

If $\left[\dfrac{1}{8}\left(\sum (x_i - 20)^2 - \sum (x_i - 21)^2\right)\right] < -2.9444,$ stop sampling and do not reject H_0.

If $\left[\dfrac{1}{8}\left(\sum (x_i - 20)^2 - \sum (x_i - 21)^2\right)\right] > 2.9444,$ stop sampling and reject H_0.

REVIEW PRACTICE PROBLEMS

1.

(a) Given $\mu = 1500$, $\sigma = 200$, $n = 25$, $\alpha = 0.01$, $\overline{X} = 1380$, we wish to test the hypothesis

$$H_0 : \mu = 1500 \quad \text{versus} \quad H_1 : \mu = \mu_1 < 1500. \text{ We reject } H_0 \text{ if}$$

$$\overline{X} < \mu_0 - \frac{\sigma}{\sqrt{n}} z_\alpha = 1500 - \frac{200}{\sqrt{25}}(2.326) = 1406.96.$$

Since the observed $\overline{X} = 1380$, we reject H_0 at the 1% significance level.

OR

$$\text{Reject } H_0 \text{ if } Z < -z_\alpha \quad \text{where} \quad Z = \frac{\overline{X} - \mu_0}{\sigma/\sqrt{n}} = \frac{1380 - 1500}{200/\sqrt{25}} = -3.$$

that is, if $Z < -2.326$. Since the observed value falls in the rejection region we reject H_0.

(b) The power function is given by

$$\gamma(\mu_1) = P\left(\overline{X} < 1406.96 \,|\, \mu = \mu_1 < 1500\right) = P\left(Z < \frac{1406.96 - \mu_1}{200/\sqrt{25}}\right)$$

$$\gamma(\mu_1) = \Phi\left(\frac{1406.96 - \mu_1}{40}\right).$$

Thus the power test at $\mu = 1400$ is $\gamma(1400) = \Phi\left(\dfrac{6.96}{40}\right) = \Phi(0.174) \cong 0.5675$

(c)

μ_1:	1300	1325	1350	1375	1400	1425	1450	1475
$\gamma(\mu_1)$:	0.996	0.98	0.92	0.79	0.57	0.33	0.14	0.04

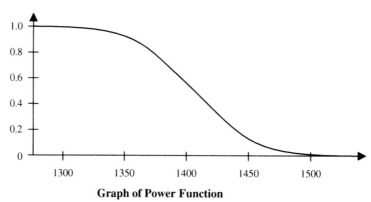

Graph of Power Function

3. Here we wish to test $H_0 : \mu = \mu_0$ versus $H_1 : \mu = \mu_1 < \mu_0$ where the population variance is σ_0^2, the significance level is α, and the power at μ_1 is $1 - \beta$.

$$\gamma(\mu_1) = 1 - \beta = P\left(\overline{X} < \mu_0 - \frac{\sigma_0}{\sqrt{n}} z_\alpha \Big| \mu = \mu_1\right) = P\left(Z < \frac{\mu_0 - \mu_1}{\sigma_0/\sqrt{n}} - z_\alpha\right)$$

Thus,
$$\beta = P\left(Z > \frac{\mu_0 - \mu_1}{\sigma_0/\sqrt{n}} - z_\alpha\right), \quad \text{so} \quad z_\beta = \frac{\mu_0 - \mu_1}{\sigma_0/\sqrt{n}} - z_\alpha.$$

Solving for n,
$$\frac{\sqrt{n}(\mu_0 - \mu_1)}{\sigma_0} = z_\alpha + z_\beta, \quad \text{or} \quad \sqrt{n} = \frac{(z_\alpha + z_\beta)\sigma_0}{\mu_0 - \mu_1}$$

That is,
$$n = \frac{(z_\alpha + z_\beta)^2 \sigma_0^2}{(\mu_0 - \mu_1)^2}$$

Note that it can easily be shown that if the test s a two tail test then

$$n = \frac{(z_{\alpha/2} + z_\beta)^2 \sigma_0^2}{(\mu_0 - \mu_1)^2}$$

5. Tes $H_0 : \mu = 40$ versus $H_1 : \mu \neq 40$ at $\alpha = 0.01$.

Reject H_0 if
$$\left|\overline{X} - 40\right| > \frac{1.2}{\sqrt{10}}(2.575) = 0.9771;$$

i.e. reject H_0 if $\overline{X} < 39.0229$ or $\overline{X} > 40.9771$.
  The power function is

$$\gamma(\mu_1) = P\left(\overline{X} < 39.0229 \big| \mu = \mu_1\right) + P\left(\overline{X} > 40.9771 \big| \mu = \mu_1\right)$$

Graph for Problem 5

$$\gamma(\mu_1) = P\left(z < (39.0225 - \mu_1)/(1.2/\sqrt{10})\right) + P\left(z > (40.9775 - \mu_1)/(1.2/\sqrt{10})\right)$$

μ_1	38	38.2	38.4	38.6	38.8	39	39.2	39.4	39.6	39.8	40	40.2
$\gamma(\mu_1)$	.996	.985	0.95	0.87	0.72	0.52	0.32	0.16	0.06	0.01	0.02	0.06

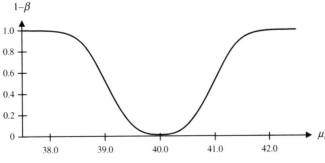

Graph of the power function:

7. We wish to test $H_0 : \mu = 20$ versus $H_1 : \mu \neq 20$.
From the data, the observed value of $|t_{13}|$ is

$$|t_{13}| = \left| \frac{(\overline{X} - \mu_0)\sqrt{n}}{S} \right| = \left| \frac{(20.82 - 20)\sqrt{14}}{2.20} \right| = 1.395.$$

At the $\alpha = 0.05$ level of significance, reject H_0 if $|t_{13}| > t_{13;0.025} = 2.160$.
Thus we do not reject H_0; i.e. the sample evidence supports the claim that the mean weight is 20 ounces.

9. We wish to test $H_0 : \mu = 1260$ versus $H_1 : \mu \neq 1260$.

$$\overline{X} = 1263.56, \quad S = 5.98143, \quad n = 9$$

$$|t_8| = \left| \frac{(\overline{X} - \mu_0)\sqrt{n}}{S} \right| = \left| \frac{(1263.56 - 1260)\sqrt{9}}{5.98143} \right| = 1.786.$$

At the $\alpha = 0.05$ significance level, reject H_0 if observed $|t_8| > t_{8;0.025} = 2.306$.
Since the observed $|t_8| = 1.786$, we do not reject H_0; i.e. the sample evidence supports the claim that the results are consistent with the published value.

11. To test $H_0 : \sigma = 0.05$ versus $H_1 : \sigma = \sigma_1 > 0.05$, reject H_0 if the observed

$$\chi^2 > \chi^2_{n-1;\alpha} \quad \text{where} \quad \chi^2 = \frac{(n-1)S^2}{\sigma_0^2}.$$

$$\chi^2 = \frac{(n-1)S^2}{\sigma_0^2} = \frac{15(0.07)^2}{(0.05)^2} = 29.4.$$

At the $\alpha = 0.05$ level of significance, reject H_0 if $\chi^2 > \chi^2_{15;0.05} = 24.9958$.
Since the observed $\chi^2 = 29.4$ falls in the rejection region, we reject H_0.

13. Test $H_0 : \dfrac{\sigma_1^2}{\sigma_2^2} = 1$ versus $H_1 : \dfrac{\sigma_1^2}{\sigma_2^2} \neq 1$ at the $\alpha = 0.05$ level

Reject H_0 if observed $F = \dfrac{S_1^2}{S_2^2} < F_{5,5;0.975} = \dfrac{1}{7.15} = 0.1399$, or if $F > F_{5,5;0.025} = 7.15$

But it can be verified that of $S_1^2 = S_2^2 = .00001$, so that the observed value $F = 1$, which does not fall in the rejection region, so we do not reject H_0.
(b) From part (a), we may assume $\sigma_1^2 = \sigma_2^2$.
We wish to test $H_0 : \mu_1 - \mu_2 = 0$ versus $H_1 : \mu_1 - \mu_2 \neq 0$.

$$S_p^2 = \frac{S_1^2 + S_2^2}{2} = 0.00001 \Rightarrow S_p = 0.003162$$

$$|t_{10}| = \frac{|\overline{X}_1 - \overline{X}_2|}{\sqrt{S_p^2\left(\frac{1}{n_A} + \frac{1}{n_B}\right)}} = \frac{|0.24067 - 0.1885|}{\sqrt{(0.00001)\left(\frac{1}{6} + \frac{1}{6}\right)}} = 28.57$$

Reject H_0 at the 5% level if $|t_{10}| > t_{10;0.025} = 2.228$.
Thus we reject H_0 and conclude from the sample evidence that there is difference in the means of two batches.

15. From the given data we obtain the following:

$$\overline{X}_A = 4.0775, \qquad \overline{X}_B = 3.2800,$$
$$S_A^2 = 0.289584, \qquad S_B^2 = 0.434418,$$
$$S_A = 0.53813, \qquad S_B = 0.659104,$$
$$n_A = 12, \qquad n_B = 12.$$

(a) To test $H_0 : \dfrac{\sigma_A^2}{\sigma_B^2} = 1$ versus $H_1 : \dfrac{\sigma_A^2}{\sigma_B^2} \neq 1$ at the $\alpha = 0.01$ level,

reject H_0 if $F_{11,11} = \dfrac{S_A^2}{S_B^2} < F_{11,11;0.995} = 0.1877$, or if $F_{11,11} > F_{11,11;0.005} = 5.3273$.

But observed $F = \dfrac{S_A^2}{S_B^2} = \dfrac{.289584}{.434418} = 0.666,$

so we do not reject H_0 at 1% level of significance.

(b) In view of (a), we may assume $\sigma_A^2 = \sigma_B^2$.
We wish to test $H_0 : \mu_A - \mu_B = 0$ versus $H_1 : \mu_A - \mu_B > 0$.

$$S_p^2 = \frac{S_A^2 + S_B^2}{2} = 0.36200.$$

Reject H_0 at the 1% level if $\qquad t_{22} = \dfrac{\overline{X}_A - \overline{X}_B}{S_p \sqrt{\frac{1}{n_A} + \frac{1}{n_B}}} > t_{22;0.01} = 2.508.$

Observed $\qquad t_{22} = \dfrac{4.0775 - 3.2800}{\sqrt{(0.36200)\left(\dfrac{1}{12} + \dfrac{1}{12}\right)}} = 3.247.$

Thus we reject H_0; i.e., the sample evidence supports the claim that method A is better than method B.

17. We are given the following:

$$\overline{X}_1 = 128.4, \qquad \overline{X}_2 = 96.5,$$
$$S_1^2 = 117.1, \qquad S_2^2 = 227.7,$$
$$n_1 = 8, \qquad n_2 = 8.$$

(a) Test $H_0 : \sigma_1^2 = \sigma_2^2$ versus $H_1 : \sigma_1^2 \neq \sigma_2^2$ at $\alpha = 0.01$.

Reject H_0 if $F_{7,7} = \dfrac{S_1^2}{S_2^2} < F_{7,7;0.995} = 0.1125$, or if $F_{7,7} > F_{7,7;0.005} = 8.8854.$

The observed value of F is: $\quad F = \dfrac{S_1^2}{S_2^2} = \dfrac{117.1}{227.7} = 0.514,$

so we do not reject H_0 at the 1% level of significance.

(b) In view of (a), we may assume $\sigma_1^2 = \sigma_2^2$.

To test $H_0 : \mu_1 - \mu_2 = 0$ versus $H_1 : \mu_1 - \mu_2 > 0$ at $\alpha = 0.01$, reject H_0 if

$$t_{14} = \frac{\overline{X}_1 - \overline{X}_2}{\sqrt{S_w^2\left(\dfrac{1}{n_1} + \dfrac{1}{n_2}\right)}} > t_{14;0.01} = 2.624.$$

$$S_p^2 = \frac{117.1 + 227.7}{2} = 172.4.$$

Observed $t_{14} = \dfrac{128.4 - 96.5}{\sqrt{(172.4)\left(\dfrac{1}{8} + \dfrac{1}{8}\right)}} = 4.859$, so we reject H_0; i.e., the sample

evidence supports the claim that there is a difference between treated and untreated plots.

19. (a) We are given the following:

$$\begin{array}{ll} \overline{X}_1 = 22.9 & \overline{X}_2 = 21.5 \\ S_1^2 = 6.75 & S_2^2 = 7.25 \\ n_1 = 5 & n_2 = 7 \end{array}$$

We wish to test $H_0 : \sigma_1^2 = \sigma_2^2$ versus $H_1 : \sigma_1^2 \neq \sigma_2^2$.

Reject H_0 at $\alpha = 0.01$ if $F = \dfrac{S_1^2}{S_2^2} < F_{4,6;0.995}$ or   if $F > F_{4,6;0.005}$.

Since, $F_{4,6;0.995} = 1/F_{6,4;0.005} = 1/21.975 = .0455$ and $F_{4,6;0.005} = 12.028$,

and

$$F_{obs} = \frac{S_1^2}{S_2^2} = \frac{6.75}{7.26} = 0.930.$$

Hence we do not reject H_0.

(b) In view of (a), we assume $\sigma_1^2 = \sigma_2^2$ and note that both n_1 and n_2 are not large. So to test $H_0 : \mu_1 - \mu_2 = 0$ versus $H_1 : \mu_1 - \mu_2 > 0$ at $\alpha = 0.01$, reject H_0 if

$$t_{10} = \frac{\overline{X}_1 - \overline{X}_2}{\sqrt{S_p^2\left(\dfrac{1}{n_1} + \dfrac{1}{n_2}\right)}} > t_{10;0.01} = 2.764.$$

$$S_p^2 = \frac{6.75 + 7.25}{2} = 7.00.$$

The observed value of test statistic is: $t_{10} = \dfrac{22.9 - 21.5}{\sqrt{7\left(\dfrac{1}{5} + \dfrac{1}{7}\right)}} = 0.9036,$

Therefore we do not reject H_0. So the sample evidence does not support the claim that the morning shift is more efficient. That is, there is no difference between the efficiency of two shifts.

21. We are given the following:

$$\overline{X}_1 = 21.78, \qquad \overline{X}_2 = 20.71,$$
$$S_1^2 = 3.11, \qquad S_2^2 = 2.40,$$
$$n_1 = 40, \qquad n_2 = 40.$$

(a) Test $H_0 : \sigma_1^2 = \sigma_2^2$ versus $H_1 : \sigma_1^2 \neq \sigma_2^2$ at $\alpha = 0.05$.

Reject H_0 at the 5% level if observed $F_{39,39} = \dfrac{S_1^2}{S_2^2} < F_{39,39;0.975}$, or if $F_{39,39} > F_{39,39;0.025}$.

Since $F_{39,39;0.025} \cong 1.8907$. Therefore, we reject H_0 if

$$\text{observed } F < \frac{1}{1.8907} = 0.5289 \quad \text{or} \quad \text{observed } F > 1.8907.$$

But observed $F = \dfrac{S_1^2}{S_2^2} = \dfrac{3.11}{2.40} = 1.2958$, so we do not reject H_0.

(b) In view of (a), we may assume $\sigma_1^2 = \sigma_2^2$.

Then we will reject $H_0 : \mu_1 = \mu_2$ in favor of $H_1 : \mu_1 > \mu_2$ at the $\alpha = 0.05$ level of significance if the observed

$$Z = \frac{\overline{X}_1 - \overline{X}_2}{S_p \sqrt{\dfrac{1}{n_1} + \dfrac{1}{n_2}}} > z_{0.05} = 1.645.$$

$$S_p^2 = \frac{3.11 + 2.40}{2} = 2.755.$$

Thus, the observed value of $Z = \dfrac{21.78 - 20.71}{\sqrt{(2.755)\left(\dfrac{1}{40} + \dfrac{1}{40}\right)}} = 2.883$,

Hence, we reject H_0 at 5% level of significance.

23. We are given the following:

$$\overline{X}_1 = 268.8, \qquad \overline{X}_2 = 255.4,$$
$$S_1 = 20.2, \qquad S_2 = 26.8,$$
$$n_1 = 70, \qquad n_2 = 70.$$

(a) Test $H_0 : \sigma_1^2 = \sigma_2^2$ versus $H_1 : \sigma_1^2 < \sigma_2^2$ at the $\alpha = 0.01$ level.

Reject H_0 if observed $F_{69,69} = \dfrac{S_2^2}{S_1^2} > F_{69,69;0.01} \cong 1.76094$.

But Observed $F = \dfrac{(26.8)^2}{(20.2)^2} = 1.7602$,

so we do not reject H_0 at the 1% level of significance. However, at the 5% level of significance we would reject the null hypothesis.

(b) We wish to test $H_0 : \mu_1 = \mu_2$ versus $H_1 : \mu_1 > \mu_2$ at the 5% level. In view of part (a), and since both n_1 and n_2 are large, we reject H_0 if

$$\text{observed } Z = \frac{\overline{X}_1 - \overline{X}_2}{\sqrt{\dfrac{S_1^2}{n_1} + \dfrac{S_2^2}{n_2}}} > z_{0.05} = 1.645.$$

Since observed $Z = \dfrac{268.8 - 255.4}{\sqrt{\dfrac{(20.2)^2}{70} + \dfrac{(26.8)^2}{70}}} = 3.341 > z_{0.05},$ we reject H_0 at the 5% level of significance.

25. The sample of differences is $\{1, 0.5, 1, 0.5, 0.5, 0.5, 2.5, 1.5\}$, so $\overline{X}_d = 1.0$, $S_d^2 = 0.50$, $n = 8$. To test $H_0 : \mu_d = 0$ versus $H_1 : \mu_d \neq 0$ at $\alpha = 0.05$, reject H_0 if

$$\text{observed } |t_7| = \left| \frac{\overline{X}_d \sqrt{n}}{S_d} \right| > t_{7;0.025} = 2.365.$$

Since observed $|t_7| = \left| \dfrac{1.0}{\sqrt{\dfrac{0.50}{8}}} \right| = 4.00 > t_{7;0.025},$

we reject H_0 and conclude from the samples that the methods differ significantly at the 5% level.

27. Given $\overline{X} = -2.430$, $S_d = 0.925$, $n = 200$, we wish to test $H_0 : \mu_d = -1$ versus $H_1 : \mu_d \neq -1$ at the 5% level.

$$\text{Thus reject } H_0 \text{ if observed } |Z| = \left| \frac{\overline{X} - \mu_0}{S_d/\sqrt{n}} \right| > z_{0.025} = 1.96.$$

Since observed $|Z| = \left| \dfrac{-2.430 - (-1)}{0.925/\sqrt{200}} \right| = 21.86$, we reject H_0 and conclude from the sample evidence that $\mu_d \neq -1$.

29. Test $H_0 : \sigma_A^2 = \sigma_B^2$ versus $H_1 : \sigma_A^2 \neq \sigma_B^2$ at the 1% level. Reject H_0 if observed $F < F_{5,5;0.995} = 0.0669$, or if observed $F > F_{5,5;0.005} = 14.940$.

But observed $F = \dfrac{S_A^2}{S_B^2} = \dfrac{8.17}{13.07} = 0.6251$, so we do not reject H_0 and conclude that $\sigma_A^2 = \sigma_B^2$ at the 1% level of significance.

31. Test $H_0 : \sigma_1^2 = \sigma_2^2$ versus $H_1 : \sigma_1^2 \neq \sigma_2^2$ at the 1% level. Reject H_0 if observed $F < F_{219,204;0.995}$, or if observed $F > F_{219,204;0.005}$.

Since, $F_{219,204;0.995} = 1/1.4260 = 0.7013$ and $F_{219,204;0.005} = 1.4287$; and the observed

$$F = \frac{S_1^2}{S_2^2} = \frac{(0.57)^2}{(0.48)^2} = 1.410,$$

so we do not reject H_0 at the 1% level of significance.

33. Test $H_0 : \sigma_{\text{I}}^2 = \sigma_{\text{II}}^2$ versus $H_1 : \sigma_{\text{I}}^2 \neq \sigma_{\text{II}}^2$ at $\alpha = 0.01$.

From the data, $S_{\text{I}}^2 = 0.62500$, $S_{\text{II}}^2 = 0.69643$, $n_{\text{I}} = n_{\text{II}} = 8$.

Reject H_0 if bserved $F < F_{7,7;0.995} = \dfrac{1}{8.8854} = 0.1125$ or if observed $F > F_{7,7;0.005} = 8.8854$. But observed $F =$

$\dfrac{S_{\text{I}}^2}{S_{\text{II}}^2} = \dfrac{0.62500}{0.69643} = 0.897$, so we do not reject H_0 and conclude that $\sigma_{\text{I}}^2 = \sigma_{\text{II}}^2$ at the 1% level.

35. Test $H_0 : \theta = 0.10$ versus $H_1 : \theta = 0.20$ sequentially with $\alpha = 0.10$ and $\beta = 0.05$.

$$f_0(x) = \binom{n}{x}\theta_0^x(1-\theta_0)^{n-x} \text{ and } f_1(x) = \binom{n}{x}\theta_1^x(1-\theta_1)^{n-x}.$$

Thus $\dfrac{f_1(x_1)\cdots f_1(x_m)}{f_0(x_1)\cdots f_0(x_m)} = \dfrac{\left[\prod\binom{n}{x_i}\right]\theta_1^{\sum x_i}(1-\theta_1)^{mn-\sum x_i}}{\left[\prod\binom{n}{x_i}\right]\theta_0^{\sum x_i}(1-\theta_0)^{mn-\sum x_i}} = \left(\dfrac{\theta_1}{\theta_0}\right)^{\sum x_i}\left(\dfrac{1-\theta_1}{1-\theta_0}\right)^{mn-\sum x_i}.$

Also, $A \cong \dfrac{\beta}{1-\alpha} = \dfrac{0.05}{0.90}$ and $B \cong \dfrac{1-\beta}{\alpha} = \dfrac{0.95}{0.10}$,   so

$$A < \frac{f_1(x_1)\cdots f_1(x_m)}{f_0(x_1)\cdots f_0(x_m)} < B \text{ becomes } \frac{0.05}{0.90} < \left(\frac{\theta_1}{\theta_0}\right)^{\sum x_i}\left(\frac{1-\theta_1}{1-\theta_0}\right)^{mn-\sum x_i} < \frac{0.95}{0.10}.$$

Taking natural logarithms, we have

$$\ln\left(\frac{0.05}{0.90}\right) < \left(\sum x_i\right)\ln\left(\frac{\theta_1}{\theta_0}\right) + \left(mn - \sum x_i\right)\ln\left(\frac{1-\theta_1}{1-\theta_0}\right) < \ln\left(\frac{0.95}{0.10}\right)$$

$$\Rightarrow \quad \ln\left(\frac{0.05}{0.90}\right) < \left(\sum x_i\right)\left(\ln\left(\frac{0.20}{0.10}\right) - \ln\left(\frac{0.80}{0.90}\right)\right) + mn\ln\left(\frac{0.80}{0.90}\right) < \ln\left(\frac{0.95}{0.10}\right)$$

$$\Rightarrow \quad -2.8904 < 0.811\sum x_i - 0.1178\,mn < 2.2513.$$

Thus sampling continues as long as $-2.8904 < 0.693\sum x_i - 0.1178\,mn < 2.2513$.

If $0.693\sum x_i - 0.1178\,mn < -2.8904$,   stop sampling and do not reject H_0.

If $0.693\sum x_i - 0.1178\,mn > 2.2513$,   stop sampling and reject H_0.

37. From the given information, we wish to test the following hypothesis

$$H_0 : \mu = \mu_0 = 7 \quad \text{versus} \quad H_1 : \mu = \mu_1 > 7$$

Test Statistic:

$$Z = \frac{\overline{X} - \mu}{s/\sqrt{n}} = \frac{7.5 - 7.0}{1.4/\sqrt{49}} = 2.5$$

$$\text{Rejection Region}: RR = \{Z \geq z_\alpha\} = \{Z \geq z_{0.01}\} = \{Z \geq 2.33\}$$

Since $2.5 > 2.33$, we reject the null hypothesis H_0.

39. (a) $p - value = P(Z \leq z) = P(Z \leq -6.0) = 0.0000$.

$$H_0 : \mu = \mu_0 = 500 \quad \text{versus} \quad H_1 : \mu = \mu_1 < 500$$

Since the $p - value = 0.0000 < 0.05 = \alpha$, we reject the null hypothesis H_0.

(b) $H_0 : \mu = \mu_0 = 500$ versus $H_1 : \mu = \mu_1 \neq 500$
$p - value = 2P(Z \geq |z|) = 2P(Z \geq 6.0) = 2(0.0000) = 0.0000$, Note that we multiply by 2 because the test is a two-tail test. Since the $p - value = 0.0000 < 0.05 = \alpha$, we reject the null hypothesis H_0.

41. In this problem the hypothesis that we wish to test is:

$$H_0 : \mu = \mu_0 = 18 \qquad \text{versus} \qquad H_1 : \mu = \mu_1 \neq 18$$

Test Statistic:

$$Z = \frac{\overline{X} - \mu}{s/\sqrt{n}} = \frac{18.2 - 18}{1.2/\sqrt{64}} = 1.33$$

Rejection Region: $RR = \left\{ |Z| \geq z_{\alpha/2} \right\} = \left\{ |Z| \geq z_{0.025} \right\} = \left\{ |Z| \geq 1.96 \right\}$
Since $|1.33| < 1.96$, we do not reject the null hypothesis H_0.

$$p - value = 2P(Z \geq |z|) = 2P(Z \geq 1.33) = 2(0.0918) = 0.1836$$

Since the $p - value = 0.1836 > 0.05 = \alpha$, we do not reject the null hypothesis H_0.
The power of the test of the test at $\mu_1 = 18.5$ is given by

$$
\begin{aligned}
1 - \beta &= 1 - P\left(\frac{\mu_0 - \mu_1}{S/\sqrt{n}} - z_{\alpha/2} < Z < \frac{\mu_0 - \mu_1}{S/\sqrt{n}} + z_{\alpha/2} \right) \\
&= 1 - P\left(\frac{18 - 18.5}{1.2/\sqrt{64}} - 1.96 < Z < \frac{18 - 18.5}{1.2/\sqrt{64}} + 1.96 \right) \\
&= 1 - P(-5.29 < Z < -1.37) = 0.9147
\end{aligned}
$$

43. (a) In this problem we wish to test the following hypothesis

$$H_0 : \mu_1 - \mu_2 = 0 \quad \text{versus} \quad H_1 : \mu_1 - \mu_2 < 0$$

In this problem the sample sizes are large, therefore, the test statistic is

$$Z = \frac{(\overline{X}_1 - \overline{X}_2) - (\mu_1 - \mu_2)}{\sqrt{\dfrac{S_1^2}{n_1} + \dfrac{S_2^2}{n_2}}} = \frac{(68.8 - 81.5)}{\sqrt{\dfrac{5.1^2}{49} + \dfrac{7.4^2}{49}}} = -9.8918$$

Rejection Region: $RR = \{ Z \leq z_\alpha \} = \{ Z \leq z_{0.05} \} = \{ Z \leq -1.645 \}$
Since $-9.8918 < -1.645$, we reject the null hypothesis H_0.

(b) Type II Error at $\mu_1 - \mu_2 = -5$ is given by

$$\beta = P\left(Z > \frac{(\mu_1 - \mu_2)_0 - (\mu_1 - \mu_2)_1}{\sqrt{\dfrac{S_1^2}{n_1} + \dfrac{S_2^2}{n_2}}} - z_\alpha \right) = P\left(Z > \frac{0 + 5}{\sqrt{\dfrac{5.1^2}{49} + \dfrac{7.4^2}{49}}} - 1.645 \right)$$

$$= P(Z > 2.25) = 0.0122.$$

Power of the test at $\mu_1 - \mu_2 = -5$ is given by

$$1 - \beta = 1 - 0.0122 = 0.9878.$$

45. (a) In this problem we want to test the following hypothesis

$$H_0 : \mu = \mu_0 = 15 \quad \text{versus} \quad H_1 : \mu = \mu_1 > 15$$

Rejection Region: $RR = \{ Z \geq z_\alpha \} = \{ Z \geq z_{0.01} \} = \{ Z \geq 2.326 \}$

Since the sample size is large, and σ is known the test statistic is given by

$$Z = \frac{\overline{X} - \mu}{\sigma/\sqrt{n}} = \frac{15.5 - 15}{1.4/\sqrt{64}} = 2.86$$

Since $2.86 > 2.33$, we reject the null hypothesis H_0.

(b) Type II Error at $\mu = 16$ is given by

$$\beta = P\left(Z < \frac{\mu_0 - \mu_1}{\sigma/\sqrt{n}} + z_\alpha\right) = P\left(Z < \frac{15 - 16}{1.4/\sqrt{64}} + 2.33\right) = P(Z < -3.38) = 0.0004$$

Thus, the power of the test at $\mu = 16$ is

$$1 - \beta = 1 - P(Z < -3.38) = 0.9996.$$

47. In this problem we want to test the following hypothesis

$$H_0 : \mu = \mu_0 = 5{,}000 \quad \text{versus} \quad H_1 : \mu = \mu_1 < 5{,}000$$

Assuming that the sample comes from a normal population, the test statistics is given by

$$T = \frac{\overline{x} - \mu}{s/\sqrt{n}} = \frac{4{,}858 - 5{,}000}{575/\sqrt{16}} = -0.99$$

Rejection Region is given by

$$RR = \left\{T \leq -t_{n-1,\alpha}\right\} = \left\{T \leq -t_{15,0.05}\right\} = \left\{T \leq -1.753\right\}$$

Since $-0.99 > -1.753$, we do not reject the null hypothesis H_0.

$$p - value = P(T \leq t) = P(T \leq -0.99) = P(T \geq 0.99) > P(T \geq 1.341) = 0.100.$$

$$p - value > 0.100$$

Using MINITAB we can find that the p-*value* $= .1689$.

49. Again, in this problem we want to test the following hypothesis

$$H_0 : \mu_1 - \mu_2 = 0 \quad \text{versus} \quad H_1 : \mu_1 - \mu_2 \neq 0$$

Since we are assuming that the populations are normal with unequal variance, the degree of freedom for the t-distribution is given by

$$m = \frac{\left(\frac{s_1^2}{n_1} + \frac{s_2^2}{n_2}\right)^2}{\frac{\left(s_1^2/n_1\right)^2}{n_1 - 1} + \frac{\left(s_2^2/n_s\right)^2}{n_2 - 1}} = \frac{\left(\frac{9^2}{12} + \frac{10^2}{14}\right)^2}{\frac{\left(9^2/12\right)^2}{12 - 1} + \frac{\left(10^2/14\right)^2}{14 - 1}} = 23.93 = 24$$

and the test Statistic is given by

$$T = \frac{\left(\overline{X}_1 - \overline{X}_2\right) - (\mu_1 - \mu_2)}{\sqrt{\frac{s_1^2}{n_1} + \frac{s_2^2}{n_2}}} = \frac{(110 - 115) - 0}{\sqrt{\frac{9^2}{12} + \frac{10^2}{14}}} = -1.34$$

The rejection region is given by

$$RR = \{|T| \geq t_{m,\alpha/2}\} = \{|T| \geq t_{24,0.005}\} = \{|T| \geq 2.797\}$$

Since $|-1.34| < 2.797$, we do not reject the null hypothesis H_0. Using t-distribution table

$$p - value = 2P(T \geq t) = 2P(T \geq 1.34) > 2P(T > 1.383) = 0.20 \Rightarrow p - value > 0.20$$

Note that with the assumption of unequal variances our conclusion did not change.

51. From the given data we have

$$\overline{X} = 16.05, S = 2.198$$

(a) The hypothesis that we wish to test in this problem is

$$H_0 : \mu \geq 16 \quad \text{vs.} \quad H_1 : \mu < 16$$

(b) Given the population is normal, the test statistic is given by

$$T = \frac{\overline{X} - \mu}{S/\sqrt{n}} = \frac{16.05 - 16}{2.198/\sqrt{19}} = 0.0992$$

The rejection region is given by

$$RR = \{T \leq -t_{n-1,\alpha}\} = \{T \leq -t_{18,0.05}\} = \{T \leq -1.734\}$$

Since $0.0992 > -1.734$, we do not reject the null hypothesis H_0

(c) $p - value = P(t_{18} < 0.0992) > 0.50$. Using MINITAB we find that the p-value =.539.

53. From the given data the mean and the variances of the differences are given by

$$\overline{X}_d = \frac{-36}{10} = -3.6$$

$$S_d^2 = \frac{1}{n-1}\left(\sum d_i^2 - \frac{(\sum d_i)^2}{n}\right) = 5.156$$

$$S_d = \sqrt{S_d^2} = \sqrt{5.156} = 2.271.$$

(a) The hypothesis we wish to test is

$$H_0 : \mu_d = 0 \quad \text{vs.} \quad H_1 : \mu_d < 0$$

Assuming that the differences are normally distributed, the test statistic is given by

$$T = \frac{\overline{X}_d - \mu_d}{S_d/\sqrt{n}} = \frac{-3.6 - 0}{2.271/\sqrt{10}} = -5.01$$

and the rejection region is given by

$$RR = \{T < -t_{n-1,\alpha}\} = \{T < -t_{9,0.01}\} = \{T < -2.821\}$$

Since $-5.01 < -2.821$, we reject the null hypothesis H_0.

$$p - value = P(t_9 \leq t) = P(t_9 \leq -5.01) = P(t_9 \geq 5.01) < 0.001.$$

55. (a) In this problem the hypothesis we wish to test is

$$H_0 : \theta = .5 \quad versus \quad H_1 : \theta \neq .5$$

(b) From the given information the test statistic is

$$Z = \frac{\hat{\theta} - \theta_0}{\sqrt{\frac{\theta_0(1 - \theta_0)}{n}}} = \frac{.3 - .5}{\sqrt{\frac{.5(1 - .5)}{100}}} = -4.0$$

To find the level of significance at which the null hypothesis is rejected is equivalent to find the p-value. The p-value is $=$ $2P(Z \geq |-4|) = 2P(Z \geq 4) = 0.00$. Thus, the null hypothesis will be rejected at any level.

57. In this problem the hypothesis we wish to test is

$$H_0 : \theta = .1 \quad versus \quad H_1 : \theta > .1$$

From the given data, we obtain

$$\hat{\theta} = \frac{18}{140} = 0.129$$

The test statistic in this problem is given by

$$Z = \frac{\hat{\theta} - \theta_0}{\sqrt{\frac{\theta_0(1 - \theta_0)}{n}}} = \frac{.129 - .1}{\sqrt{\frac{.1(.9)}{140}}} = 1.14$$

and the rejection region is given by

$$RR = \{Z \geq z_\alpha\} = \{Z \geq z_{0.05}\} = \{Z \geq 1.645\}$$

Since $1.14 < 1.645$, we do not reject the null hypothesis H_0.

$$p - value = P(Z \geq z) = P(Z \geq 1.14) = 0.1271.$$

59. In this problem the hypothesis we wish to test is

$$H_0 : \sigma^2 = 0.2 \quad vs. \quad H_1 : \sigma^2 \neq 0.2$$

From the given data set, we have

$$n = 12, \ \overline{X} = 15.8583, S = 0.4757$$

Assuming normality the test statistic in this problem is given by

$$\chi^2 = \frac{(n-1)s^2}{\sigma^2} = \frac{(12-1)(0.4757^2)}{0.2} = 12.446$$

and the rejection region is given by

$$RR = \left\{ \chi^2 \leq \chi^2_{n-1,1-\alpha/2} \right\} \text{ or } \left\{ \chi^2 \geq \chi^2_{n-1,\alpha/2} \right\} = \left\{ \chi^2 \leq \chi^2_{11,0.995} \right\} \text{ or } \left\{ \chi^2 \geq \chi^2_{11,0.005} \right\}$$

That is,

$$\left\{ \chi^2 \leq 2.6032 \right\} \text{ or } \left\{ \chi^2 \geq 26.7569 \right\}$$

Since $2.6032 < 12.446 < 26.7569$, we do not reject the null hypothesis H_0.

61. In this problem the hypothesis we wish to test is

$$H_0 : \sigma_1^2 = \sigma_2^2 \quad \text{versus} \quad H_1 : \sigma_1^2 < \sigma_2^2$$

From the given information the test statistic is given by

$$F = \frac{S_1^2}{S_2^2} = \frac{(4,930)^2}{(5,400)^2} = 0.8335.$$

and the rejection region is given by

$$RR = \left\{ F \leq F_{n_1-1,n_2-1,1-\alpha} \right\} = \left\{ F \leq 1/F_{59,54,0.05} \right\} = \left\{ F \leq 1/1.5594 = 0.6412 \right\}$$

Since $0.8335 > .6412$, we do not reject the null hypothesis H_0.

63. In this problem the hypothesis we wish to test is

$$H_0 : \sigma_1^2 = \sigma_2^2 \quad \text{vs.} \quad H_1 : \sigma_1^2 \neq \sigma_2^2$$

Further, from the given data, we have
Portfolio I: $n_1 = 12$, $\overline{X}_1 = 1.125$, $S_1 = 1.5621$
Portfolio II: $n_2 = 12$, $\overline{X}_2 = 0.733$, $S_2 = 2.7050$
Moreover, the populations are given to be normal. Thus, the statistic is given by

$$F = \frac{S_1^2}{S_2^2} = \frac{1.5621^2}{2.7050^2} = 0.3335$$

and the rejection region is given by

$$\begin{aligned} RR &= \left\{ F \leq \frac{1}{F_{n_2-1,n_1-1,\alpha/2}} \right\} \quad \text{or} \quad \left\{ F \geq F_{n_1-1,n_2-1,\alpha/2} \right\} \\ &= \left\{ F \leq \frac{1}{F_{11,11,0.025}} \right\} \quad \text{or} \quad \left\{ F \geq F_{11,11,0.025} \right\} \\ &= \left\{ F \leq 0.2882 \right\} \quad \text{or} \quad \left\{ F \geq 3.47 \right\} \end{aligned}$$

Since $0.2882 < 0.3335 < 3.47$, we do not reject the null hypothesis H_0.

65. Referring to Problem 1 of Section 8.7, a 95% confidence interval for the population proportion is given by

$$(0.3571 \leq \theta \leq 0.4429)$$

which does not contains the value of θ under the null hypothesis, i.e., 0.45. Thus, we reject the null hypothesis.

67. Referring to Problem 3 of Section 8.7, point estimates of θ_1 and θ_2 are given by

$$\hat{\theta}_1 = 16/100 = 0.16, \hat{\theta}_2 = 50/200 = 0.25.$$

We are now interested in testing at the 5% level of significance the hypothesis

$$H_0 : \theta_1 - \theta_2 = 0 \quad against \quad H_1 : \theta_1 - \theta_2 \neq 0.$$

From Equation (9.8.5) the pivotal quantity (see Chapter 8) for $\theta_1 - \theta_2$ is the test statistic

$$Z = \frac{(\hat{\theta}_1 - \hat{\theta}_2) - (\theta_1 - \theta_2)_0}{\sqrt{\theta_1(1 - \theta_1)/n_1 + \theta_2(1 - \theta_2)/n_2}}$$

Since under the null hypothesis $\theta_1 - \theta_2 = 0$, we substitute the values of $\hat{\theta}_1, \hat{\theta}_2$ and $\theta_1 - \theta_2 = 0$ in the numerator of above test statistic and the values of $\hat{\theta}_1$ and $\hat{\theta}_2$ in the denominator. However, here θ_1 and θ_2 are unknown but under the null hypothesis $\theta_1 = \theta_2 = \theta$ (say), we can estimate θ by pooling the two samples, that is

$$\hat{\theta} = (X_1 + X_2)/(n_1 + n_2).$$

We then replace $\hat{\theta}_1$ and $\hat{\theta}_2$ in the denominator with $\hat{\theta}$. In this example, we have

$$\hat{\theta} = (16 + 50)/(100 + 200) = 0.22.$$

Thus, the value of the test statistic computed under the null hypothesis is given by

$$Z = \frac{(0.16 - 0.25) - 0}{\sqrt{(0.22)(0.78)/100 + (0.22)(0.78)/200}} = \frac{-0.09}{0.0507} = -1.775$$

The rejection regions are as shown in the following diagram, that is

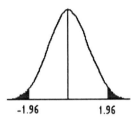

-1.96   1.96

Clearly, the value of the test statistic does not fall in the rejection region. Hence, we do not reject the null hypothesis, p-value $= 0.076$.

69. Referring to Problem 5 of Section 8.7, point estimates of θ_1 and θ_2 are given by

$$\hat{\theta}_1 = \frac{40}{800} = 0.05, \hat{\theta}_2 = \frac{50}{600} = 0.083$$

We are now interested in testing at the 5% level of significance the hypothesis

$$H_0 : \theta_1 - \theta_2 = 0 \quad against \quad H_1 : \theta_1 - \theta_2 \neq 0.$$

From Equation (9.8.5) the pivotal quantity (see Chapter 8) for $\theta_1 - \theta_2$ is the test statistic

$$Z = \frac{(\hat{\theta}_1 - \hat{\theta}_2) - (\theta_1 - \theta_2)_0}{\sqrt{\theta_1(1 - \theta_1)/n_1 + \theta_2(1 - \theta_2)/n_2}}$$

Since under the null hypothesis $\theta_1 - \theta_2 = 0$, we substitute the values of $\hat{\theta}_1, \hat{\theta}_2$ and $\theta_1 - \theta_2 = 0$ in the numerator of above test statistic and the values of $\hat{\theta}_1$ and $\hat{\theta}_2$ in the denominator. However, here $\hat{\theta}_1$ and $\hat{\theta}_2$ are unknown but under the null hypothesis $\hat{\theta}_1 = \hat{\theta}_2 = \theta$ (say), we can estimate θ by pooling the two samples, that is

$$\hat{\theta} = (X_1 + X_2)/(n_1 + n_2).$$

We then replace $\hat{\theta}_1$ and $\hat{\theta}_2$ in the denominator with $\hat{\theta}$. In this example, we have

$$\hat{\theta} = (40 + 50)/(800 + 600) = 0.0643.$$

Thus, the value of the test statistic computed under the null hypothesis is given by

$$Z = \frac{(0.05 - 0.083) - 0}{\sqrt{(0.0643)(0.9357)/800 + (0.0643)(0.9357)/600}} = \frac{-0.033}{0.01325} = -2.49$$

The rejection regions are as shown in the following diagram, that is

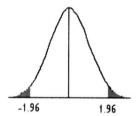

Clearly, the value of the test statistic falls in the rejection region. Hence, we reject the null hypothesis, p-value $= 0.0128$.

10

ELEMENTS OF RELIABILITY THEORY

PRACTICE PROBLEMS FOR SECTION 10.1

1. We first find the cumulative probability that the fuel pump works at time t = 10 years. To obtain this probability using MINITAB we proceed as follows:

   (i) In Column C1 enter the data, which in this case is 10.

   (ii) From the bar-menu select **Calc** > **Probability Distributions** > **Weibull.**

   (iii) In the dialog box titled "Weibull Distribution" that appears check the circle for **Cumulative probability** and then enter the value of shape parameter 1, scale parameter 30, threshold parameter 0, and enter C1 in the input column and then click **OK**.

   (iv) In the session window we obtain

$$F(10) = P(X \leq 10) = 0.2835.$$

   Then from Equation (10.1.1), we have

$$R(10) = 1 - F(10) = 1 - 0.2835 = 0.7165.$$

   Thus, the fuel pump in the aircraft gas turbine engine will be working at 10 years with probability 0.7165. Now using Equation (10.1.3) the hazard rate function at $t = 10$ is given by

$$h(10) = \frac{f(10)}{R(10)} = \frac{0.0239}{0.7165} = 0.0333.$$

   That is, the instantaneous failure rate of the fuel pump at $t = 10$ years is approximately .0333. (for finding the value of $f(10)$, check **Probability density** instead of **Cumulative probability**).

3. In this problem, we first the cumulative probability that the hard drive is still functioning at time t = 3,000 hours. To obtain this probability using MINITAB we proceed as follows:

   (i) In Column C1 enter the data, which in this case is 3,000.

   (ii) From the bar-menu select **Calc** > **Probability Distributions** > **Exponential**.

Solutions Manual to Accompany Statistics and Probability with Applications for Engineers and Scientists, Bhisham C. Gupta and Irwin Guttman.
© 2014 John Wiley & Sons, Inc. Published 2014 by John Wiley & Sons, Inc.

(iii) In the dialog box titled "Exponential Distribution" that appears check the circle for **Cumulative probability** and then enter the value of scale parameter 5000 (mean of the distribution) threshold parameter 0, and enter C1 in the input and then click **OK**.

In the session window we obtain

$$F(3,000) = P(X \leq 3,000) = 0.4519.$$

Equation (10.1.1), we have

$$R(3,000) = 1 - F(3,000) = 1 - 0.4519 = 0.4481.$$

Thus, a hard drive will still be function at time $t = 3,000$ hours with probability 0.4519.

Now using Equation (10.1.3) the hazard rate function at $t = 3,000$ is given by

$$h(3,000) = \frac{f(3,000)}{R(3,000)} = \frac{0.0001}{0.4519} = 0.0002.$$

That is, the instantaneous failure rate of a hard drive at $t = 3,000$ hours is approximately .0002. (To determine f(3,000), see part 4 of Problem 1)

5. In this problem we first find the cumulative hazard function of the life of the transmission, which is given by

$$H(t) = \int_0^t \frac{\alpha \beta u^{\beta-1}}{1 + \alpha u^\beta} du = \ln(1 + \alpha t^\beta)$$

Now using Equation (10.26), we obtain

$$f(t) = h(t) \times \exp(-\ln(1 + \alpha t^\beta)) = \frac{\alpha \beta t^{\beta-1}}{1 + \alpha t^\beta} \times \frac{1}{1 + \alpha t^\beta} = \frac{\alpha \beta t^{\beta-1}}{(1 + \alpha t^\beta)^2}.$$

Thus, we have

$$F(t) = \int_0^t f(u) du = \int_0^t \frac{\alpha \beta u^{\beta-1}}{(1 + \alpha u^\beta)^2} du = 1 - \frac{1}{(1 + \alpha t^\beta)}.$$

Hence, the survival function is given by

$$R(t) = 1 - F(t) = \frac{1}{1 + \alpha t^\beta}.$$

PRACTICE PROBLEMS FOR SECTION 10.2

1. Using the result (10.2.3) we obtain the following:

$$\hat{\mu} = \frac{964 + 1002 + 1067 + 1099 + 1168 + 1260}{6} + \frac{(12 - 6) \times (1260)}{6} = 2353.33.$$

3. The 95% confidence interval for the mean was determined in Problem 2, which is

$$(1210.11, 6412.63).$$

Now, using the results (10.2.7) and (10.2.9) the $100(1 - \alpha)\%$ confidence interval for the reliability at any time t is given by

$$[\exp(-\lambda_U t), \quad \exp(-\lambda_L t)]$$

where

$$\begin{cases} \lambda_L = \dfrac{\hat{\lambda}}{2k} \chi^2_{2k, 1-\alpha/2}, \\ \lambda_U = \dfrac{\hat{\lambda}}{2k} \chi^2_{2k, \alpha/2} \end{cases}$$

Substituting the values of $k, \hat{\lambda}$, and χ^2, we obtain, since $k = 6$ and $\alpha/2 = .025$, that

$$\lambda_L = \frac{1}{12(2353.33)} \times 4.4038 = .00016, \lambda_U = \frac{1}{12(2353.33)} \times 23.3367 = .00083.$$

Hence, a 95% confidence interval for the reliability of the part at t $= 1180$ hours is given by

$$\left(e^{-.00083 \times 1180}, e^{-.00016 \times 1180}\right) = \left(e^{-.9794}, e^{-.1888}\right) = (0.3755, 0.8280)$$

5. The hazard rate for the exponential is the constant $\lambda = 1/\mu$. Hence, an estimate of the hazard rate is given by

$$\hat{\lambda} = \frac{1}{145.8} = 0.0069.$$

Using the result (10.2.6)), the 99% confidence interval for μ is given by

$$\left(\frac{2k\hat{\mu}}{\chi^2_{2k:.005}}, \frac{2k\hat{\mu}}{\chi^2_{2k:.995}}\right) = \left(\frac{2(5)(145.8)}{25.1882}, \frac{2(5)(145.8)}{2.1559}\right) = (57.8842, 676.2837)$$

Hence, 99% confidence interval for the hazard rate is given by (1/676.2837, 1/57.8842), that is

$$(.0015, .0173)$$

PRACTICE PROBLEMS FOR SECTIONS 10.3 AND 10.4

1. (a) In this problem we are dealing with the case $\mu_0 = 120, \mu_1 = 72, \alpha = .05$, and $\beta = .05$. Then referring to (10.3.1a)-(10.3.1d), we find

$$a_0 = -530 \quad a_1 = 530 \quad b = 91.95.$$

Thus, when plotting T_m versus m, as long as T_m falls between the two parallel lines $a_j + bm, j = 0, 1$, that is, if T_m is such that

$$-530 + 91.95m \leq T_m \leq 530 + 91.95m,$$

then sampling continues, and if T_m falls below $-530 + 91.95m$, the sampling stops and we reject H_0 in favor of $H_1 : \mu = \mu_1 = 72$, while if T_m falls above $530 + 91.95m$, sampling stops and we do not reject $H_0 : \mu_0 = 120$.

(b) In this part we are dealing with the case $\mu_0 = 120, \mu_1 = 72, \alpha = .05$, and $\beta = .10$. Then referring to (10.3.1a)-(10.3.1d), we find

$$a_0 = -520 \quad a_1 = 405 \quad b = 91.95.$$

Thus, when plotting T_m versus m, as long as T_m falls between the two parallel lines $a_j + bm, j = 0, 1$, that is, if T_m is such that

$$-520 + 91.95m \leq T_m \leq 405 + 91.95m,$$

then sampling continues and if T_m falls below $-520 + 91.95m$, sampling stops and we reject H_0 in favor of $H_1 : \mu = \mu_1 = 72$, but if T_m falls above $405 + 91.95m$, sampling stops and we do not reject $H_0 : \mu_0 = 120$.

3. In this problem the test was concluded after the seventh hard drive failed. Thus, as in Example 10.3, the point estimate (MLE) of the mean time between failures is given by

$$\hat{\mu} = \frac{[3056 + 3245 + 3375 + 3450 + 3524 + 3875 + 4380]}{7} + \frac{(15 - 7)(4380)}{7} = 8,563.57, \text{hours}.$$

Now using Equation (10.2.6), we obtain, say, 95% confidence interval for the mean μ

$$\left(\frac{2(7)(8563.57)}{26.1190}, \frac{2(7)(8563.57)}{5.6287} \right) = (4590.14, 21299.76).$$

5. In this problem we first determine a point estimate of the mean time between failures, which is given by

(a) $\hat{\mu} = \dfrac{[3056 + 3245 + 3375 + 3450 + 3524 + 3875 + 4380]}{7} + \dfrac{8(4500)}{7} = 8,700.71.$

(b) Using the Equation (10.2.8), we obtain, $\hat{R}(t) = \exp(-\hat{\lambda}t)$ at t = 8,000 that

$$\hat{R}(8,000) = \exp(-(1/8,700.71) \times 8,000) = \exp(-0.9195) = 0.3987.$$

(c) The estimate of the hazard rate, is given by

$$\hat{\lambda} = 1/8700.71 = 1.1493 \times 10^{-4}.$$

(d) In order to find a 95% confidence interval for the hazard rate we have to first find a 95% confidence interval for the mean time between failures. From the previous problems of this section, the 95% confidence interval is

$$\left(\frac{2(7)(8700.71)}{26.1190}, \frac{2(7)(8700,71)}{5.62872} \right) = (4663.65, 21640.79).$$

(e) A 95% confidence interval for the hazard rate is given by

$$(4.6209 \times 10^{-5}, 2.14424 \times 10^{-4}).$$

(f) In this part the estimate of the mean time to failure with 95% confidence is given by

$$\frac{2k\hat{\mu}}{\chi^2_{2k,.05}} = \frac{2(7)8700.71}{23.6848} = 5,142.96.$$

6. To fit Weibull distribution to the **non-censored or censored** data we proceed as follows:

   (i) Enter the data in **Column C1** and create a censoring **Column C2** by entering 1for non-censored observations and 0 for censoring observations. In this case all values in column C2 are 1 since we are using only non-censored values.

   (ii) From the bar-menu select **Stat > Reliability/Survival > Distribution Analysis (Right Censoring) > Parametric Distribution Analysis.**

   (iii) A dialog box **Parametric Distribution Analysis- Right Censoring** appears. In this dialog box enter the data column under variables. Then select **Censor** option and enter censoring column in the box under the censoring column click **OK** and then select the assumed distribution. Now select other options such as, for example,

Estimate . . . and select the estimation method- least square or maximum likelihood method; **Graphs . . .** under this select any desired plots such as Probability plot, Survival plot, Cumulative Failure plot, Confidence interval, Hazard plot and then click **OK.**

(iv) Finally select one of the Assumed Distributions.

(v) Click **OK.** The final results will appear in the **Session Window.**

Non-censored Data

Estimation Method: **Least Squares**

Distribution: **Weibull**

Parameter Estimates

Parameter	Estimate	Standard Error	95.0% Normal CI Lower	95.0% Normal CI Upper
Shape	9.72880	2.17827	6.27301	15.0884
Scale	3725.15	156.487	3430.73	4044.84

Characteristics of Distribution

	Estimate	Standard Error	95.0% Normal CI Lower	95.0% Normal CI Upper
Mean (MTTF)	3539.76	168.871	3223.78	3886.71
Standard Deviation	437.062	80.1670	305.080	626.142

Estimation Method: **Maximum Likelihood**

Parameter Estimates

Parameter	Estimate	Standard Error	95.0% Normal CI Lower	95.0% Normal CI Upper
Shape	8.43125	2.30473	4.93414	14.4070
Scale	3749.62	178.930	3414.83	4117.25

Characteristics of Distribution

	Estimate	Standard Error	95.0% Normal CI Lower	95.0% Normal CI Upper
Mean(MTTF)	3540.06	189.473	3187.52	3931.60
Standard Deviation	500.054	115.638	317.816	786.789

Censored Data

Estimation Method: **Least Squares**

Distribution: **Weibull**

Censoring Information Count

Uncensored value	7
Right censored value	8

Parameter Estimates

Parameter	Estimate	Standard Error	95.0% Normal CI Lower	95.0% Normal CI Upper
Shape	8.32064	2.04860	5.13553	13.4812
Scale	4233.61	158.890	3933.37	4556.77

Log-Likelihood $= -66.005$

Goodness-of-Fit

Anderson-Darling (adjusted) $= 57.872$

Characteristics of Distribution

	Estimate	Standard Error	95.0% Normal CI Lower	95.0% Normal CI Upper
Mean (MTTF)	3994.50	176.712	3662.74	4356.31
Standard Deviation	571.269	116.503	383.044	851.985

Estimation Method: **Maximum Likelihood**

Distribution: **Weibull**

Censoring Information Count

Uncensored value	7
Right censored value	8

Parameter Estimates

Parameter	Estimate	Standard Error	95.0% Normal CI Lower	95.0% Normal CI Upper
Shape	5.54095	1.89738	2.83210	10.8408
Scale	4723.42	367.416	4055.51	5501.34

Log-Likelihood $= -63.444$

Goodness-of-Fit
Anderson-Darling (adjusted) $= 57.690$

Characteristics of Distribution

	Estimate	Standard Error	95.0% Normal CI Lower	95.0% Normal CI Upper
Mean (MTTF)	4362.51	307.738	3799.19	5009.35
Standard Deviation	909.541	308.537	467.816	1768.36

7. (See Problem 6 above)

Non-censored Data
Estimation Method: **Least Squares**

Distribution: **Lognormal**
Censoring Information Count

Uncensored value	7

Parameter Estimates

Parameter	Estimate	Standard Error	95.0% Normal CI Lower	95.0% Normal CI Upper
Location	8.17064	0.0489177	8.07476	8.26652
Scale	0.129424	0.0448173	0.0656536	0.255136

Log-Likelihood $= -51.871$

Goodness-of-Fit
Anderson-Darling (adjusted) $= 1.871$

Characteristics of Distribution

	Estimate	Standard Error	95.0% Normal CI Lower	95.0% Normal CI Upper
Mean (MTTF)	3565.34	175.630	3237.20	3926.73
Standard Deviation	463.379	166.050	229.568	935.323

Estimation Method: **Maximum Likelihood**

Distribution: **Lognormal**

Censoring Information Count

Uncensored value   7

Parameter Estimates

Parameter	Estimate	Standard Error	95.0% Normal CI Lower	Upper
Location	8.17064	0.0418082	8.08870	8.25258
Scale	0.110614	0.0295629	0.0655115	0.186769

Log-Likelihood $= -51.715$

Goodness-of-Fit
Anderson-Darling (adjusted) $= 1.988$

Characteristics of Distribution

	Estimate	Standard Error	95.0% Normal CI Lower	Upper
Mean (MTTF)	3557.30	149.178	3276.61	3862.03
Standard Deviation	394.694	108.684	230.076	677.094

Censored Data
Estimation Method: **Least Squares**

Distribution: **Lognormal**

Censoring Information Count

Uncensored value   7
Right censored value   8

Parameter Estimates

Parameter	Estimate	Standard Error	95.0% Normal CI Lower	Upper
Location	8.33254	0.0581836	8.21850	8.44658
Scale	0.203895	0.0502785	0.125750	0.330601

Log-Likelihood $= -63.086$

Goodness-of-Fit
Anderson-Darling (adjusted) $= 57.717$

Characteristics of Distribution

	Estimate	Standard Error	95.0% Normal CI Lower	Upper
Mean (MTTF)	4244.29	256.492	3770.20	4777.98
Standard Deviation	874.460	241.328	509.132	1501.93

Estimation Method: **Maximum Likelihood**

Distribution: **Lognormal**

Censoring Information Count

Uncensored value   7
Right censored value   8

Parameter Estimates

			95.0% Normal CI	
Parameter	Estimate	Standard Error	Lower	Upper
Location	8.38907	0.0801996	8.23188	8.54626
Scale	0.242931	0.0741466	0.133561	0.441859

Log-Likelihood $= -62.711$

Goodness-of-Fit
Anderson-Darling (adjusted) $= 57.677$

Characteristics of Distribution

			95.0% Normal CI	
	Estimate	Standard Error	Lower	Upper
Mean (MTTF)	4530.44	409.495	3794.92	5408.52
Standard Deviation	1117.02	422.331	532.395	2343.63

9. Suppose $((X_1, X_2, \cdots, X_n)$ is a random sample from a lognormal population with parameters μ *and* σ^2. Now define a new random variable $Y_i = \ln X_i$, then $(Y_1, Y_2, \cdots, Y_n)$ is a random sample from the normal population with mean μ and variance σ^2. Thus, in this problem to estimate the parameters of the lognormal distribution we estimate the mean and the variance of the normal population using log to the base e (ln) of times to failure, we obtain

$$3.83 \quad 3.89 \quad 3.99 \quad 4.06 \quad 4.09 \quad 4.16 \quad 4.19 \quad 4.23 \quad 4.28 \quad 4.36$$

The (MLE) of the mean μ and the variance σ^2 are given by

$$\hat{\mu} = \overline{X} = 4.1080 \text{ and } \hat{\sigma}^2 = \frac{1}{n}\sum_{i=1}^{n}(X_i - \overline{X})^2 = 0.02592.$$

Estimate of mean of time to failure $= e^{4.1080 + .02592/2} = e^{4.12096} = 61.618$
    Estimate of variance of time to failure $= e^{2(4.1080) + 2(.02592)} - e^{2(4.1080) + (.02592)} = e^{8.26784} - e^{8.24192} = 99.7.$

REVIEW PRACTICE PROBLEMS

1. Since failure times have an exponential distribution, $f(t) = \lambda e^{-\lambda t}$, $F(t) = 1 - e^{-\lambda t}$, $\mu = \frac{1}{\lambda}$, $R(t) = e^{-\lambda t}$.
    The reliability must be at least 0.90 at 1000 days of service, so

$$0.90 = e^{-1000\lambda} \quad \Rightarrow \quad \lambda = \frac{-\ln(0.90)}{1000} = 1.054 \times 10^{-4}.$$

$$\text{Thus MTBF} = \mu = \frac{1}{\lambda} = \frac{1}{1.054 \times 10^{-4}} = 9491 \text{ hrs.}$$

$$\text{The hazard rate is } h(t) = \frac{f(t)}{R(t)} = \frac{\lambda e^{-\lambda t}}{e^{-\lambda t}} = \lambda = 1.054 \times 10^{-4}.$$

3. This results follows from the result (10.2.3). Substituting $f(t) = \lambda e^{-\lambda t}$ and $F(t) = 1 - e^{-\lambda t}$ in (10.2.2), the joint density function of $t_{(1)}, \ldots, t_{(k)}$ is

$$\frac{n!}{(n-k)!}\lambda^k \exp\left\{-\lambda\left[\sum_{i=1}^{k} t_{(i)} + (n-k)t_{(k)}\right]\right\}.$$

So the likelihood function is

$$\ell\left(\lambda|t_{(1)},\ldots,t_{(k)}\right) = \frac{n!}{(n-k)!}\lambda^k \exp\left\{-\lambda\left[\sum_{i=1}^{k} t_{(i)} + (n-k)t_{(k)}\right]\right\},$$

and the log likelihood is

$$L = \ln\frac{n!}{(n-k)!} + k\,\ln\,\lambda - \lambda\left[\sum_{i=1}^{k} t_{(i)} + (n-k)t_{(k)}\right].$$

Differentiating with respect to λ, we have $\dfrac{\partial L}{\partial \lambda} = \dfrac{k}{\lambda} - \left[\sum_{i=1}^{k} t_{(i)} + (n-k)t_{(k)}\right].$

Setting the derivative equal to zero and denoting the solution by $\hat{\lambda}$, we have the maximum likelihood estimator for λ:

$$\hat{\lambda} = \frac{k}{T_k}, T_k = \sum_{i=1}^{k} t_{(i)} + (n-k)t_{(k)}.$$

5. The hazard rate is constant, so the life distribution is exponential with

$$f(t) = \lambda e^{-\lambda t}, R(t) = e^{-\lambda t}, \quad \text{and} \quad \mu = \frac{1}{\lambda}.$$

Then $\hat{\mu} = \dfrac{\sum_{i=1}^{k} t_{(i)} + (n-k)t_0}{k} = \dfrac{8929 + 8(3000)}{4} = 8232.25$, and

$$\hat{\lambda} = \frac{1}{\hat{\mu}} = 1.2147 \times 10^{-4}.$$

Since $R(t) = e^{-\lambda t}$, the reliability at $t = 4000$ is

$$\hat{R}(4000) = \exp\left(-\left(1.2147 \times 10^{-4}\right)(4000)\right) = 0.615.$$

7. The estimated MTBF is $\hat{\mu} = \dfrac{\sum_{i=1}^{k} t_{(i)} + (n-k)t_{(k)}}{k} = \dfrac{7036 + 7(1530)}{5} = 3549.2.$

(a) The 95% confidence interval for μ (MBTF), $k = 5$, is given by

$$\left(\frac{2k\hat{\mu}}{\chi^2_{2k;\alpha/2}}, \frac{2k\hat{\mu}}{\chi^2_{2k;1-\alpha/2}}\right) = \left(\frac{2(5)(3549.2)}{20.4831}, \frac{2(5)(3549.2)}{3.24697}\right) = (1732.7, 10930.8)$$
$$\cong (1733, 10931).$$

The estimate of μ with 95% confidence is

$$\frac{2k\hat{\mu}}{\chi^2_{2k;\alpha}} = \frac{2(5)(3549.2)}{18.3070} = 1938.7 \cong 1939.$$

(b) The estimated reliability is $\hat{R}(t) = \exp\left(-\hat{\lambda}t\right) = \exp(-t/\hat{\mu})$.

At time $t = 2000$, $\hat{R}(2000) = \exp(-2000/3549.2) = 0.569$.

An estimate of reliability at $t = 2000$, with 95% confidence, is $\exp(-2000/1938.7) = 0.356$.

Thus the estimated reliability at $t = 2000$ is 0.569, but the estimated reliability at $t = 2000$ with 95% confidence is 0.356.

(c) To find t such that $R(t) = 0.5$, note that $\hat{R}(t) = e^{-t/\hat{\mu}}$, so

$$t = -\hat{\mu}\ln\left(\hat{R}\right) = \hat{\mu}\ln\left(\frac{1}{\hat{R}}\right).$$

Thus $t = (3549.2)\ln\left(\dfrac{1}{0.5}\right) = (3549.2)\ln 2 = 2460$; i.e., at $t = 2460$ hrs, the reliability will be 0.5.

9. We wish to test $H_0 : \mu = \mu_0 = 3000$ against $H_1 : \mu = \mu_1 = 2500$ using a sequential sampling plan. The α and β risks are 0.05 and 0.10, respectively.

(a) Measure the time to failure when items are replaced on failure.

$$T_m = \sum_{i=1}^{m} t_i, \text{ where } t_i\text{'s are the failure times.}$$

We will continue testing as long as $a_1 + bm \leq T_m \leq a_0 + bm$.

$$A \cong \frac{\beta}{1 - \alpha} = \frac{0.10}{0.95}, \quad B \cong \frac{1 - \beta}{\alpha} = \frac{0.90}{0.05},$$

$$\frac{\mu_1\mu_0}{\mu_1 - \mu_0} = \frac{(2500)(3000)}{-500} = -15000,$$

$$\text{so } b = \frac{\mu_1\mu_0}{\mu_1 - \mu_0}\ln\frac{\mu_1}{\mu_0} = (-15000)\ln\frac{2500}{3000} = 2735,$$

$$a_0 = \frac{\mu_1\mu_0}{\mu_1 - \mu_0}\ln A = (-15000)\ln\frac{0.10}{0.95} = 33769,$$

$$a_1 = \frac{\mu_1\mu_0}{\mu_1 - \mu_0}\ln B = (-15000)\ln\frac{0.90}{0.05} = -43356.$$

Thus, sampling continues if

$$-43356 + 2735m \leq T_m \leq 33769 + 2735m.$$

If $T_m > 33769 + 2735m$, sampling stops and we do not reject $H_0 : \mu = 3000$.

If $T_m < -43356 + 2735m$, sampling stops and we reject H_0 in favor of $H_1 : \mu = 2500$.

(b) When items are *not* replaced on failure, the same procedure is used, except that

$$T_m = \sum_{i=1}^{m} t_i + (n-m)(t-t_m) \quad \text{instead of} \quad T_m = \sum_{i=1}^{m} t_i.$$

11. Weibull

Distribution Analysis: Time to Failure

Variable: C1

Censoring Information Count

Uncensored values   7

Right censored values   3

Censoring value: C2 = 0

Estimation Method: Maximum Likelihood

Distribution: Weibull

Parameter Estimates

Parameter	Estimate	Standard Error	95.0% Normal CI	
			Lower	Upper
Shape	4.14380	1.35491	2.18313	7.86536
Scale	27.4021	2.50973	22.8993	32.7902

Log-Likelihood = −26.093

Goodness-of-Fit

Anderson-Darling (adjusted) = 22.421

Characteristics of Distribution

	Estimate	Standard Error	95.0% Normal CI	
			Lower	Upper
Mean (MTTF)	24.8875	2.28494	20.7889	29.7941
Standard Deviation	6.76240	2.02787	3.75705	12.1718

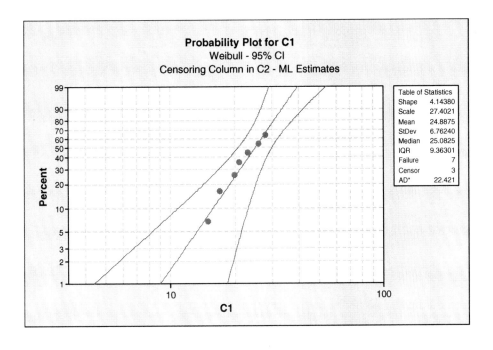

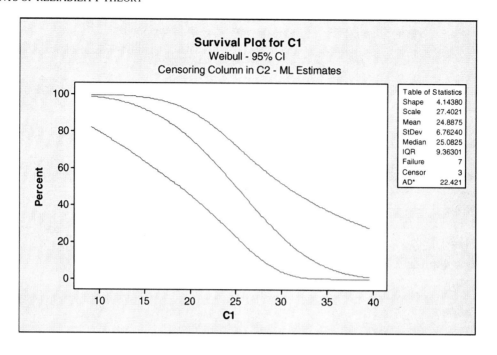

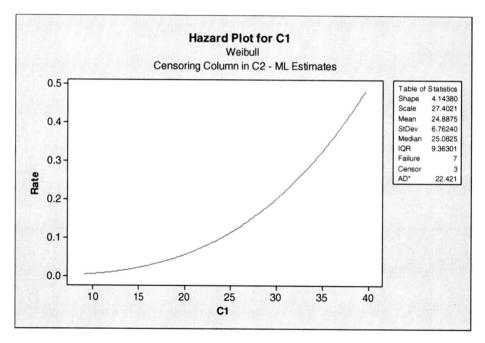

Lognormal
Variable: C1
Censoring Information Count
Uncensored values   7
Right censored values   3

Censoring value: $C2 = 0$

Estimation Method: Maximum Likelihood

Distribution: Lognormal

Parameter Estimates

			95.0% Normal CI	
Parameter	Estimate	Standard Error	Lower	Upper
Location	3.19794	0.102364	2.99731	3.39857
Scale	0.303692	0.0871909	0.173002	0.533109

Log-Likelihood $= -25.636$

Goodness-of-Fit
Anderson-Darling (adjusted) $= 22.420$

Characteristics of Distribution

			95.0% Normal CI	
	Estimate	Standard Error	Lower	Upper
Mean (MTTF)	25.6375	2.84758	20.6220	31.8727
Standard Deviation	7.96891	2.89214	3.91268	16.2302

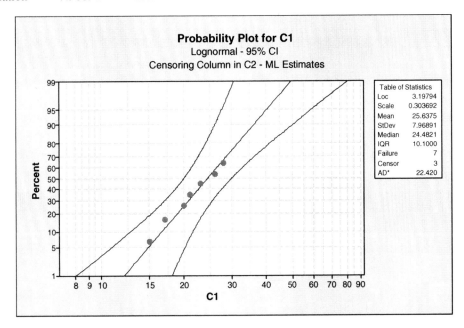

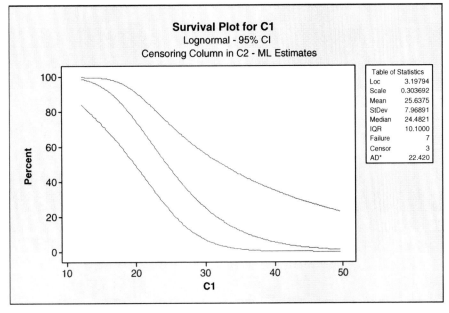

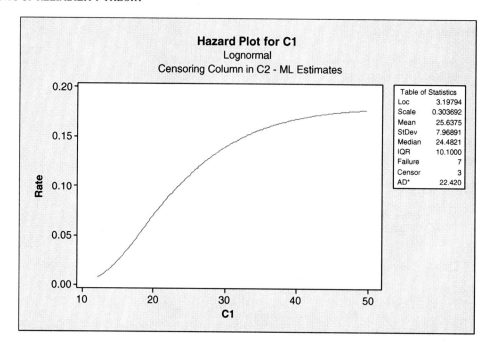

Hazard Plot for C1
Lognormal
Censoring Column in C2 - ML Estimates

Table of Statistics	
Loc	3.19794
Scale	0.303692
Mean	25.6375
StDev	7.96891
Median	24.4821
IQR	10.1000
Failure	7
Censor	3
AD*	22.420

13. From the solution of Problem 12, we have a 95% confidence interval for the mean is $(13.6201, 31.7795)$.

11

STATISTICAL QUALITY CONTROL PHASE I CONTROL CHARTS

1. Solutions for this chapter have been prepared primarily using MINITAB.

3. MINITAB by default uses a pooled estimate of the population standard deviation σ, but we may change it to $\overline{R}/d_2$ in the dialog box by selecting: **Xbar-R options** > **Estimate** > **Rbar.**

PRACTICE PROBLEMS FOR SECTIONS 11.3 AND 11.4

1. *Common Causes or Random Causes*
   Common causes or random causes refer to the various sources of variation within a process that is in statistical control (possess statistical stability). These type of causes behave like a constant system of chance causes. While individual measured values may all be different, as a group they tend to have an identity that can be explained by a statistical distribution which can generally be characterized by:

   1. A location parameter, the mean μ

   2. A dispersion parameter, the variance σ^2 or standard deviation σ, and the shape (the pattern of variation - symmetrical, right skewed, or left skewed).

   Assignable Causes or Special Causes
   Assignable causes or special causes refer to any source of variation that cannot be adequately explained by a single distribution of the process, given that the process was in statistical control. Unless all the assignable causes of variation are identified and corrected, they will continue to affect the process output in an unpredictable way. Any process with assignable causes is considered unstable and hence not in statistical control. However, any process free of assignable causes is considered stable and therefore in statistical control. Assignable causes can be corrected by local or particular/ specific actions, while the common causes or random causes can be corrected only by the action on the system.

3. *A control chart may be defined as:*
   1. A graphic device for describing in concrete terms the state of statistical control of a monitored process.

   2. A graphic device for judging whether or not control has been attained and thus detecting whether or not assignable causes are present, and

   3. A device for attaining a stable process.

5. A *run* occurs when a number of successive items possessing the same characteristics are observed. An *average run length* (ARL) is the average number of points plotted or the number of subgroups inspected before a point falls outside the control limits, where these thresholds indicate the process is out of control.

Solutions Manual to Accompany Statistics and Probability with Applications for Engineers and Scientists, Bhisham C. Gupta and Irwin Guttman.
© 2014 John Wiley & Sons, Inc. Published 2014 by John Wiley & Sons, Inc.

In Shewhart control charts, the ARL is determined by using the following formula:

$$ARL = \frac{1}{p}$$

where p is the probability that any point falls outside the control limits when the process is assumed to be in statistical control.

There are advantages and disadvantages of having a large ARL value. An advantage of having a large ARL is that if the process is in statistical control then we do not encounter many false alarms. However, the disadvantage is that if there is any shift in the process, we do not find it soon enough in order to avoid producing nonconforming products.

The advantage of having a small ARL is that if the process is not in statistical control then we find it quicker and consequently we avoid producing nonconforming products. Conversely, the disadvantage is that we can have too many false alarms.

7. In this problem the upper and lower control limits are

$$\mu_0 + 3\sigma_{\bar{x}} = \mu_0 + 3\frac{\sigma}{\sqrt{9}} = \mu_0 + 1\sigma,$$

$$\mu_0 - 3\sigma_{\bar{x}} = \mu_0 - 3\frac{\sigma}{\sqrt{9}} = \mu_0 - 1\sigma,$$

The Type II error (β) is the probability of not detecting a shift. Thus, the Type II for given shifts is determined as follows:

(i) Shift $= 0.5$ standard deviations: In this case $\overline{X} \sim N(\mu_0 + 0.5\sigma, (\sigma/3)^2)$. Thus, we obtain

$$\beta = P\left(\frac{\mu_0 - \sigma - (\mu_0 + 0.5\sigma)}{\sigma/3} \leq Z \leq \frac{\mu_0 + \sigma - (\mu_0 + 0.5\sigma)}{\sigma/3}\right)$$
$$= P(Z \leq 1.5) - P(Z \leq -4.5) = 0.9332.$$

(ii) Shift $= 0.75$ standard deviations: In this case $\overline{X} \sim N(\mu_0 + 0.75\sigma, (\sigma/3)^2)$. Thus, we obtain

$$\beta = P\left(\frac{\mu_0 - \sigma - (\mu_0 + 0.75\sigma)}{\sigma/3} \leq Z \leq \frac{\mu_0 + \sigma - (\mu_0 + 0.75\sigma)}{\sigma/3}\right)$$
$$= P(Z \leq .75) - P(Z \leq -5.25) = 0.7734.$$

(iii) Shift $= 1.0$ standard deviation: In this case $\overline{X} \sim N(\mu_0 + \sigma, (\sigma/3)^2)$. Thus, we obtain

$$\beta = P\left(\frac{\mu_0 - \sigma - (\mu_0 + \sigma)}{\sigma/3} \leq Z \leq \frac{\mu_0 + \sigma - (\mu_0 + \sigma)}{\sigma/3}\right)$$
$$= P(Z \leq 0) - P(Z \leq -6) = 0.5000.$$

(iv) Shift $= 1.5$ standard deviations: In this case $\overline{X} \sim N(\mu_0 + 1.5\sigma, (\sigma/3)^2)$. Thus, we obtain

$$\beta = P\left(\frac{\mu_0 - \sigma - (\mu_0 + 1.5\sigma)}{\sigma/3} \leq Z \leq \frac{\mu_0 + \sigma - (\mu_0 + 1.5\sigma)}{\sigma/3}\right)$$
$$= P(Z \leq -1.5) - P(Z \leq -7.5) = 0.0668.$$

(v) Shift $= 2.0$ standard deviations: In this case $\overline{X} \sim N(\mu_0 + 2\sigma, (\sigma/3)^2)$. Thus, we obtain

$$\beta = P\left(\frac{\mu_0 - \sigma - (\mu_0 + 2\sigma)}{\sigma/3} \leq Z \leq \frac{\mu_0 + \sigma - (\mu_0 + 2\sigma)}{\sigma/3}\right)$$
$$= P(Z \leq -3.0) - P(Z \leq -9.0) = 0.0013.$$

If the sample size is increased, then the value of β decreases. The ARL in each of the cases is given by $[1/(1 - \beta)]$. Thus ARLs in parts (i) - (v) are 15, 5, 2, 1, 1, respectively.

9. The average time to detect each shift is 44, 15, 6, 2, 1 hour, respectively.

PRACTICE PROBLEMS FOR SECTION 11.5

1. (a) The process standard deviation is given by

$$\hat{\sigma} = \overline{R}/d_2$$

where the values of d_2 for different sample sizes are tabulated in Table VIII of Appendix I. Thus substituting the values of d_2 and $\overline{R}$, we obtain

$$\hat{\sigma} = \overline{R}/d_2 = 12.60/2.326 = 5.417.$$

(b) The control limits for the $\overline{X}$ are given by

$$\text{LCL} = \overline{\overline{x}} - A_2\overline{R}, \quad \text{UCL} = \overline{\overline{x}} + A_2\overline{R}.$$

where the values of A_2 for different sample sizes are also tabulated in Table VIII of Appendix I. Thus, substituting the values of A_2, $\overline{\overline{X}}$, and $\overline{R}$, we obtain

$$\text{LCL} = 40.50 - 0.577(12.60) = 33.2298$$
$$\text{UCL} = 40.50 + 0.577(12.60) = 47.7702$$

(c) The control limits for the R are given by

$$\text{LCL} = D_3\overline{R}, \quad CL = \overline{R}, \quad \text{UCL} = D_4\overline{R}.$$

where the values of D_3 and D_4 for different sample sizes are tabulated in Table VIII of Appendix I. Thus, substituting the values of D_3, D_4, and $\overline{R}$, we obtain

$$\text{LCL} = 0 \times (12.6) = 0$$
$$\text{UCL} = 2.114 \times (12.6) = 26.6364.$$

3. (a) The process standard deviation is given by

$$\hat{\sigma} = \overline{R}/d_2$$

where the values of d_2 for different sample sizes are tabulated in Table VIII of Appendix I. Thus, substituting the values of d_2 and $\overline{R}$, we obtain

$$\hat{\sigma} = \overline{R}/d_2 = 14.30/2.059 = 6.945.$$

(b) The control limits for the $\overline{X}$ are given by

$$\text{LCL} = \overline{\overline{x}} - A_2\overline{R}, \quad \text{UCL} = \overline{\overline{x}} + A_2\overline{R}.$$

where the values of A_2 for different sample sizes are also tabulated in Table VIII of Appendix I. Thus, substituting the values of A_2, $\overline{\overline{X}}$, and $\overline{R}$, we obtain

$$\text{LCL} = 60.25 - 0.729(14.30) = 49.8253$$
$$\text{UCL} = 60.25 + 0.729(14.30) = 70.6747$$

(c) The control limits for the R are given by

$$\text{LCL} = D_3\overline{R}. \quad \text{UCL} = D_4\overline{R}.$$

where the values of D_3 and D_4 for different sample sizes are tabulated in Table VIII of Appendix I. Thus, substituting the values of D_3, D_4, and $\overline{R}$, we obtain

$$\text{LCL} = 0 \times (14.3) = 0$$
$$\text{UCL} = 2.282 \times (14.30) = 32.6326.$$

5. **(a)** $\hat{\mu} = \overline{\overline{X}} = 20.066$, $\hat{\sigma} = \overline{R}/d_2 = 1.6332/2.326 = 0.7021$.

(b) The percentage of nonconforming rods is equal to $100p\%$, where

$$p = P(X < LSL) + P(X > USL) = P(X < 19.05) + P(X > 20.95)$$

$$= P\left(\frac{X - 20.066}{0.7021} < \frac{19.05 - 20.066}{0.7021}\right) + P\left(\frac{X - 20.066}{0.7021} > \frac{20.95 - 20.066}{0.7021}\right)$$

$$= P(Z < -1.4471) + P(Z > 1.2590) = 0.0735 + 0.1038 = 0.1773.$$

Thus, the percentage of nonconforming rods is equal to 17.73%.

7. $n = 5, A_3 = 1.427, D_4 = 2.114, D_3 = 0$

$$\overline{\overline{X}} = \frac{\overline{X}_1 + \overline{X}_2 + \cdots + \overline{X}_m}{m} = \frac{20.040 + 20.242 + \cdots + 20.290}{25} = 20.2274$$

$$\overline{S} = \frac{S_1 + S_2 + \cdots + S_m}{m} = \frac{0.557360 + 0.371645 + \cdots + 0.526260}{25} = 0.5036$$

$$UCL_{\overline{X}} = \overline{\overline{X}} + A_3\overline{S} = 20.2274 + 1.427(0.5036) = 20.9460$$

$$CL\overline{X} = \overline{\overline{X}} = 20.2274$$

$$LCL_{\overline{X}} = \overline{\overline{X}} - A_3\overline{S} = 20.2274 - 1.427(0.5036) = 19.5087$$

$$UCL_{\overline{S}} = B_4\overline{S} = 2.089(0.5036) = 1.0520$$

$$CL_{\overline{S}} = \overline{S} = 0.5036$$

$$LCL_{\overline{S}} = B_3\overline{S} = 0(0.5036) = 0.00.$$

The control limits are as shown in the following control charts.

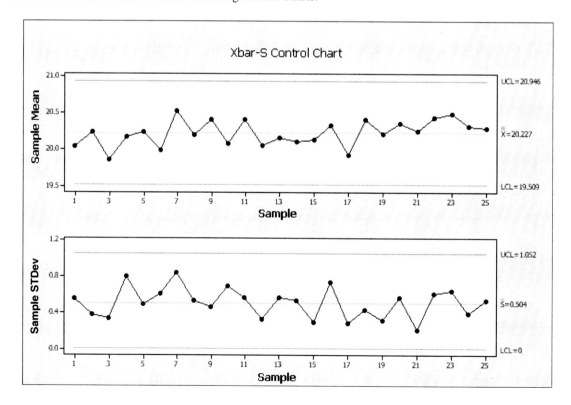

Since no points fall outside the control limits and there are no specific patterns, the conclusions are the same as in problem 4.

9. The subgroup means and standard deviations for the new data are given below.

$\overline{X}$	S
20.306	0.428987
20.182	0.387195
20.116	0.278532
20.898	0.252230
19.928	0.311480
20.770	0.294364
20.244	0.375606
20.272	0.480801
19.820	0.750233
19.934	0.453244
20.532	0.608457
20.130	0.289223
20.138	0.409475
20.762	0.374727
20.346	0.478571
20.530	0.437950
20.136	0.297456
20.532	0.446004
20.164	0.341950
20.884	0.177426

In Problem 7, the control limits for $\overline{X}$ and S Control Charts are (19.509, 20.946) and (0, 1.052), respectively. Clearly, all the values of $\overline{X}$ and S fall within their control limits.

However, there are eight or more consecutive values of S which are less than $\overline{S} = 0.504$. Hence, in S Control Chart there are runs of length eight points or more that fall below the center line. Thus, the process is out of control. However, note that this is a positive development in the sense that the process variation is smaller. So an investigation should be launched to determine what caused the standard deviation to become smaller so that these conditions could be perpetuated here and implemented elsewhere.

11. As noted in earlier problems, MINITAB does not provide dialog commands to construct $\overline{X}$-R Control Charts when we are given the means and the ranges, rather than the raw data. Thus, the following control charts have been created using macro from MINITAB. On the website of the book you can find a macro in Excel to construct these charts

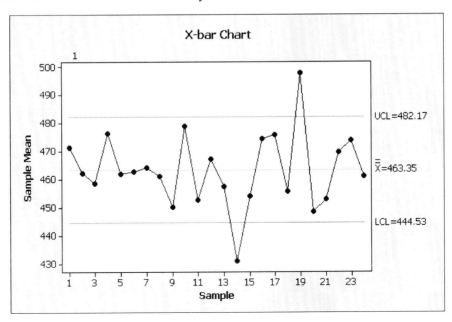

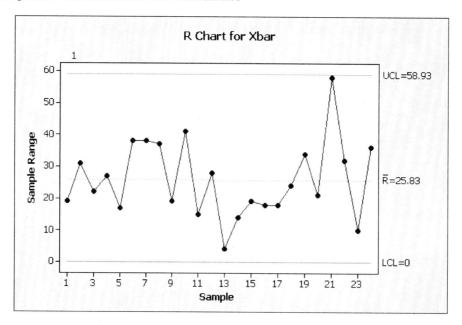

In $\overline{X}$ Control Chart there are two points that fall outside the control limits, so that the process is not in statistical control.

In order to check whether or not the process is behaving according to the specifications one must first bring the process under statistical control.

13. As noted in earlier problems, MINITAB does not provide dialog commands to construct $\overline{X}$-R Control Charts when we are given the means and the ranges, rather than the raw data. Thus, the following control charts have been created using macro. In $\overline{X}$-R Control Charts we note that no points fall outside the control limits nor are there specific patterns indicating any abnormalities. Thus the process is in statistical control.

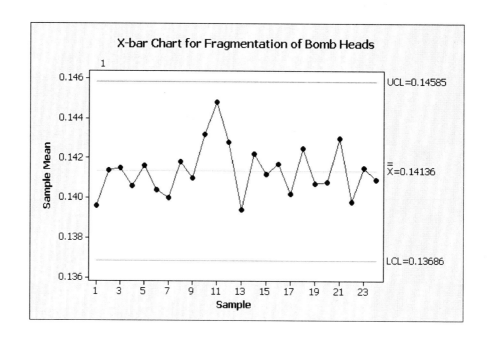

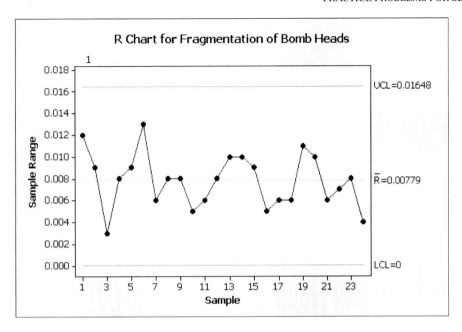

PRACTICE PROBLEMS FOR SECTION 11.6

1. A *p* Control Chart and an *np* Control Chart may be described as follows:
   p Control Chart - It plots percent or fraction of nonconforming in a sample (subgroup) where the sample size can be constant or variable.
   np Control Chart - It plots number of nonconforming units in a sample (subgroup) where sample size is constant.

3. The *p* control chart for the data in this problem is as shown below.

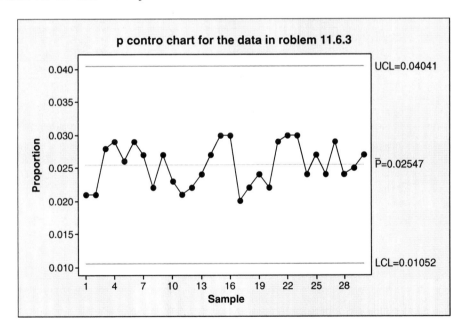

The above control chart indicates that the process is in statistical control.

5. (a) $UCL = \bar{c} + 3\sqrt{\bar{c}} = 52.93 + 3\sqrt{52.93} = 74.76$
   $CL = \bar{c} = 52.93$
   $LCL = \bar{c} - 3\sqrt{\bar{c}} = 52.93 - 3\sqrt{52.93} = 31.11$

(b) The control chart for the given set of data is as shown below.

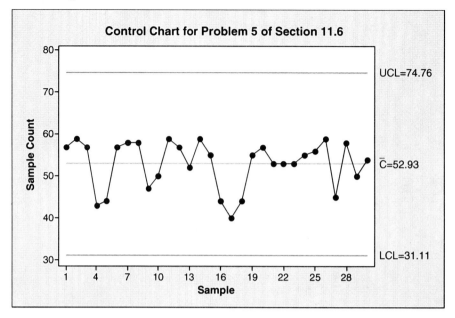

The above control chart indicates that the process is in statistical control.

7.

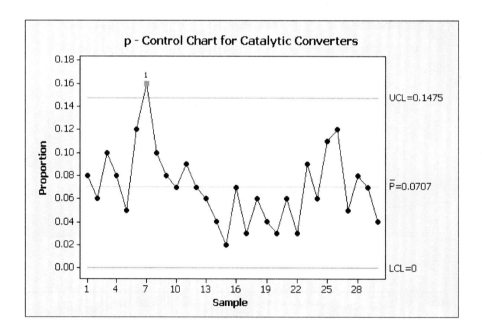

Since the point corresponding to the sample point 7 falls outside the upper control limit, the process is not in statistical control. So we may now proceed to construct the revised control chart by ignoring sample 7. In order to do so we assume that the reasons for the sample point seven to fall outside the control limit have been determined and corrective actions have been taken.

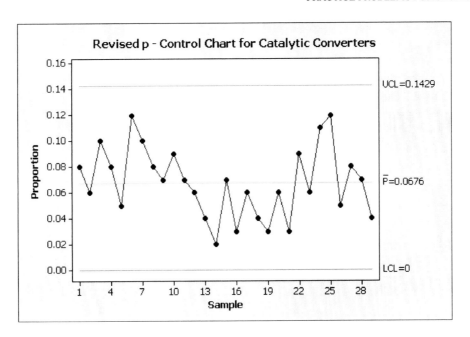

9. We are given in this problem the number of conformities per square yard; we assume that they follow Poisson distribution. The c Control Chart is an appropriate control chart for this process. Thus, using the given data, we obtain

$$\hat{\lambda} = \bar{c} = \frac{c_1 + c_2 + c_3 + \dots + c_m}{m} = \frac{3 + 7 + \dots + 8}{24} = 8.46$$

and the three-sigma upper and lower control limits are

$$UCL = \bar{c} + 3\sqrt{\bar{c}} = 8.46 + 3\sqrt{8.46} = 17.18$$

and

$$LCL = \bar{c} - 3\sqrt{\bar{c}} = 8.46 - 3\sqrt{8.46} = 0$$

respectively.

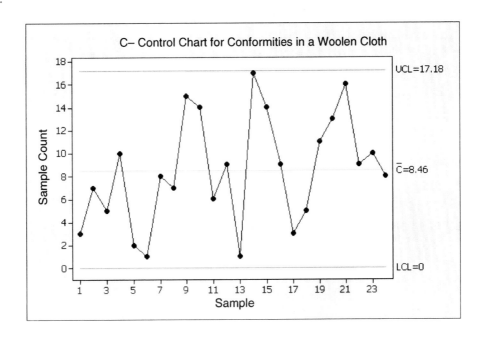

Since no points fall outside the control limits and nor are there any specific pattern indicating abnormalities in the process. However, note that sample point 14 falls very close to the upper control limit. Thus, if some preventive measures are not taken, the process may go out of statistical control. Thus, it is important we take some steps to improve the process so that it does not go out of control in the near future.

11. Using the data in Problem 10, we obtain

$$\bar{u} = \frac{c_1 + \cdots + c_m}{n_1 + \cdots + n_m} = 9.95$$

$$UCL = \bar{u} + 3\sqrt{\bar{u}/n} = 9.95 + 3\sqrt{9.95/1.1725} = 18.69$$

$$LCL = \bar{u} - 3\sqrt{\bar{u}/n} = 9.95 - 3\sqrt{9.95/1.1725} = 1.21$$

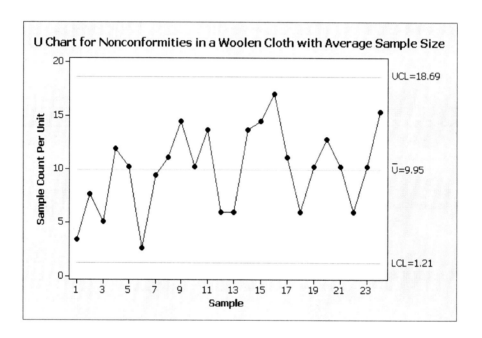

Again, the control chart shows that the process is under statistical control. Thus, the conclusion in this problem is the same as in Problem 10.

13. Referring to Problem 12, the ARL when the process is under control is given by

$$ARL = [1/(1 - \beta)] = [1/(1 - .9506)] = 20.24,$$

or

$$ARL = 21.$$

(a) In this problem we first determine the probability of Type I error. That is,

$$\alpha = \Pr(\hat{p} > UCL|p) = \Pr(\hat{p} > .105|p = .065) = \Pr(Z > 1.9334) = 1 - 0.9734 = 0.0266$$

Thus, ARL is given by

$$ARL = 1/0.0266 = 38.$$

(b) Again, we first determine the probability of Type I error. That is,

$$\alpha = \Pr(\hat{p} > UCL|p) = \Pr(\hat{p} > .105|p = .070) = \Pr(Z > 1.6346) = 1 - 0.9489 = 0.0511.$$

Thus, ARL is given by

$$ARL = 1/0.0511 = 20.$$

(c) Again, we first determine the probability of Type I error. That is,

$$\alpha = \Pr(\hat{p} > UCL|p) = \Pr(\hat{p} > .105|p = .075) = \Pr(Z > 1.3573) = 1 - 0.9127 = 0.0873.$$

Thus, ARL is given by

$$ARL = 1/0.0873 = 12.$$

PRACTICE PROBLEMS FOR SECTION 11.7

1. In this problem we first need to estimate the process standard deviation. That is

$$\hat{\sigma} = \overline{R}/d_2 = 5.6/2.326 = 2.408$$

Further the lower and upper specification limits are

$$LSL = 14, \quad USL = 26$$

Thus, we obtain

$$\hat{C}_p = \frac{26 - 14}{6 \times 2.408} = 0.8306$$

Since $\hat{C}_p < 1$, we conclude that the process is not capable.

3. In this problem we first need to estimate the process standard deviation. That is

$$\hat{\sigma} = \overline{R}/d_2 = 8.6/2.059 = 4.177$$

Further the lower and upper specification limits are

$$LSL = 23, \quad USL = 49$$

Thus, we obtain

$$\hat{C}_p = \frac{49 - 23}{6 \times 4.177} = 1.0374.$$

Since $\hat{C}_p > 1$, we conclude that the process is capable.

5. Since the value of $\hat{C}_p$ is not sensitive to the location of center of the process the value of $\hat{C}_p$ does not change no matter how much the process mean has shifted. For this reason $\hat{C}_p$ is not a very efficient coefficient of process capability (also, see solution of Problem 4).

REVIEW PRACTICE PROBLEMS

1. **(a)** The *X*-bar and *R* charts for the given data are as shown below

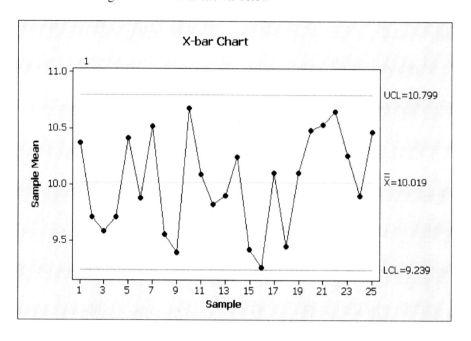

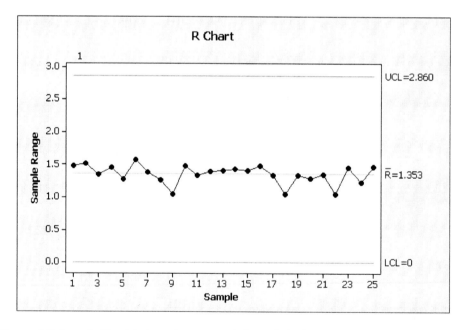

(b) The *X*-bar and *R* charts in Part (a) show that process is in statistical control. Thus, the trial control limits can be used for future control. However, note that in *X*-bar Control Chart point 16 came very close to the lower control limit. Thus, one should remain very vigilant for the process since the process is still somewhat vulnerable.

3. In Example 11.5.4 after ignoring sample 22 the process parameter estimates are

$$\bar{\bar{x}} = 15.162 \text{ and } \hat{\sigma}_x = \overline{R}/d_2 = 0.0355/2.059 = .017.$$

Hence, the percentage of nonconforming diameters by using new spec limits is estimated to be:

$$= 100 \times [P(x < LSL) + P(x > USL)]$$

$$= 100 \times [P(x < 15.10) + P(x > 15.20)]$$

$$= 100 \times \left[P\left(z \leq \frac{15.10 - 15.162}{0.017} \right) + P\left(z \geq \frac{15.20 - 15.162}{0.017} \right) \right]$$

$$= 100 \times [P(z < -3.647) + P(z > 2.235)]$$

$$= 100 \times [0.0001 + 0.0128] = 1.29\%.$$

5. (a) Using $\overline{\overline{X}}$ and $\overline{S}$ values obtained in Problem 4 the estimates of the process mean μ and standard deviation estimate the process mean μ and standard deviation σ are given by

$$\hat{\mu} = \overline{\overline{X}} = 10.019, \hat{\sigma} = \overline{S}/c_4 = 0.501/0.9400 = 0.5330.$$

(b) The spec limits so that no more than 5% of the bolts are nonconforming are clearly given by

$$\hat{\mu} \pm 1.96\hat{\sigma} = 10.019 \pm 1.96 \times 0.5330$$
$$= 10.019 \pm 1.045$$

Thus, the spec limits are (8.974, 11.064).

7. Since the process in Problem 6, is in statistical control, we do not need to find the new control limits and to find the ARL we find the probability of Type I error. That is,

$$\alpha = \Pr(\overline{X} < LCL) + \Pr(\overline{X} > UCL) = \Pr(\overline{X} < LCL) + 1 - \Pr(\overline{X} < UCL)$$

$$= \Pr(\overline{X} < 9.304) + 1 - \Pr(\overline{X} < 10.734)$$

$$= \Pr\left(\frac{\overline{X} - 10.019}{0.5330/\sqrt{5}} < \frac{9.304 - 10.019}{0.5330/\sqrt{5}} \right) + 1 - \Pr\left(\frac{\overline{X} - 10.019}{0.5330/\sqrt{5}} < \frac{10.734 - 10.019}{0.5330/\sqrt{5}} \right)$$

$$= 0.00135 + 1 - 0.99865 = 0.0027.$$

Thus, the average run length (ARL) is given by

$$ARL = 1/0.0027 = 370.$$

9. Since the process has an upward shift of one standard deviation, that is, the mean has shifted to $1.9930 + 0.1790 = 2.1720$. Thus, the probability of detecting in the first subsequent sample is equal to the probability of $\overline{X}$ falling outside the control limits, which is given by

$$\Pr(\overline{X} < LCL) + \Pr(\overline{X} > UCL) = \Pr(\overline{X} < LCL) + 1 - \Pr(\overline{X} < UCL)$$

$$= \Pr(\overline{X} < 1.7529) + 1 - \Pr(\overline{X} < 2.2332)$$

$$= \Pr\left(\frac{\overline{X} - 2.1720}{0.1790/\sqrt{5}} < \frac{1.7529 - 2.1720}{0.1790/\sqrt{5}} \right) + 1 - \Pr\left(\frac{\overline{X} - 2.1720}{0.1790/\sqrt{5}} < \frac{2.2332 - 2.1720}{0.1790/\sqrt{5}} \right)$$

$$= 0.0000 + 1 - 0.7777 = 0.2223.$$

11. (a) From the given information we have

$$\overline{\overline{X}} = 450/30 = 15, \overline{R} = 80/30 = 8/3 = 2.667$$

Thus, we have

$$UCL = \overline{\overline{X}} + A_2\overline{R} = 15 + .577 \times 8/3 = 16.539$$

$$CL \ \ = \overline{\overline{X}} = 15$$

$$LCL \ = \overline{\overline{X}} - A_2\overline{R} = 15 - .577 \times 8/3 = 13.461$$

Thus, the X-bar Control Chart is as shown below.

The control limits for R Control Chart are given by

$$UCL_{\overline{R}} = D_4 \times \overline{R} = 2.114 \times 2.667 = 5.638$$
$$CL_{\overline{R}} \ \ = \overline{R} = 2.667$$
$$LCL_{\overline{R}} \ = D_3\overline{R} = 0(2.667) = 0.00.$$

(b) Using the results of Part (a), we obtain

$$\hat{\mu} = \overline{\overline{X}} = 15, \quad \hat{\sigma} = \overline{R}/d_2 = (8/3)/2.326 = 1.1465$$

(c) In this part we determine the estimated value of the process capability index C_p

$$\hat{C}_p = \frac{USL - LSL}{6\hat{\sigma}} = \frac{18.5 - 11.5}{6 \times 1.1465} = \frac{7}{6.879} = 1.0176 > 1$$

Since the estimated value of the process capability index C_p is greater than 1, the process is capable.

13. From the given information we have

$$\overline{\overline{X}} = 450/30 = 15, \overline{S} = 35.5/30 = 1.1833$$

Thus, we have

$$UCL = \overline{\overline{X}} + A_3\overline{S} = 15 + 1.427 \times 1.1833 = 16.6886$$

$$CL \ \ = \overline{\overline{X}} = 15$$

$$UCL = \overline{\overline{X}} - A_3\overline{S} = 15 - 1.427 \times 1.1833 = 13.3114$$

Thus, the X-bar Control Chart is as shown below.

$$UCL = 16.6886$$

$$CL = 15$$

$$LCL = 13.3114$$

The control limits for the S Control Chart are given by

$$UCL_{\overline{S}} = B_4\overline{S} = 2.089(1.1833) = 2.4719$$

$$CL_{\overline{S}} = \overline{S} = 1.1833$$

$$LCL_{\overline{S}} = B_3\overline{S} = 0(1.1833) = 0.00.$$

15. The 99% X-bar and S Control Chart for the data in Problem 8, using MINITAB is as shown below. Note that the upper and lower control limits shown in the charts as if the calculations are done using the z-values ± 2.6. But the actual calculations are done using the z-value ± 2.575. The X-bar and S Control Charts below show that the process is in statistical control.

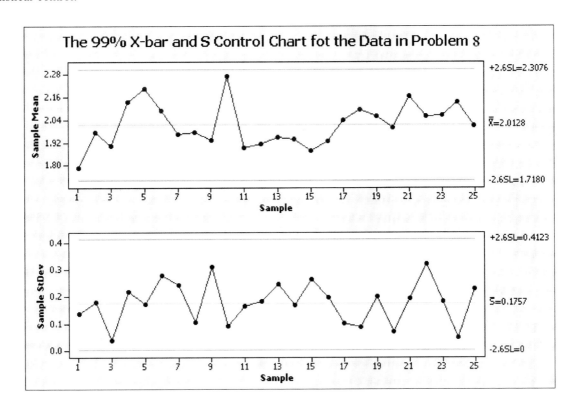

17. Since the process has a downward shift of one standard deviation, that is, the mean has shifted to $1.9930 - 0.1790 = 1.8140$. Thus, the probability of detecting the shift in the first subsequent sample is equal to the probability of $\overline{X}$ falling outside the control limits, which is given by

$$\Pr(\overline{X} < LCL) + \Pr(\overline{X} > UCL) = \Pr(\overline{X} < LCL) + 1 - \Pr(\overline{X} < UCL)$$

$$= \Pr(\overline{X} < 1.7529) + 1 - \Pr(\overline{X} < 2.2332)$$

$$= \Pr\left(\frac{\overline{X} - 1.8140}{0.1790/\sqrt{5}} < \frac{1.7529 - 1.8140}{0.1790/\sqrt{5}}\right) + 1 - \Pr\left(\frac{\overline{X} - 1.8140}{0.1790/\sqrt{5}} < \frac{2.2332 - 1.8140}{0.1790/\sqrt{5}}\right)$$

$$= 0.223 + 1 - 1.0000 = 0.223.$$

19. (a) We obtain the p Control Chart for the given data by using MINITAB.

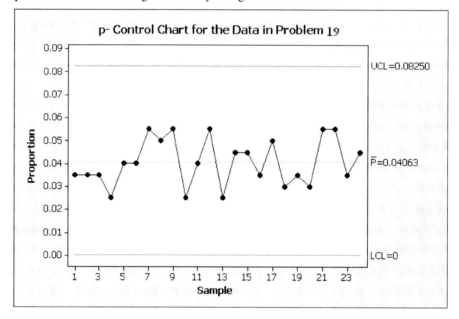

(b) Since the data are in statistical control no data point needs to be ignored. Hence, the control chart remains the same.

21. The p Control Chart for the given data is obtained by using MINITAB.

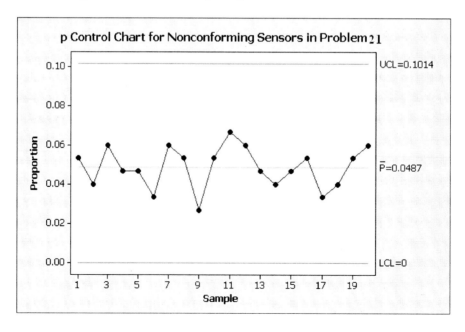

The *p* Control Chart clearly indicates that the process is in statistical control.

(b) Since the data are in statistical control no data point needs to be ignored. Hence, the control chart remains the same.

23. In Problem 21, (i) no one point falls outside the lower or upper control limits, (ii) no consecutive seven points fall on either side of the center line, and from the two control chart with 2-sigma and 1-sigma warning limits given below we note that (iii) no two out of three consecutive points fall beyond two-sigma warning limits, or (iv) no four out of five consecutive points fall beyond one-sigma warning limits. Hence, the process is in statistical control.

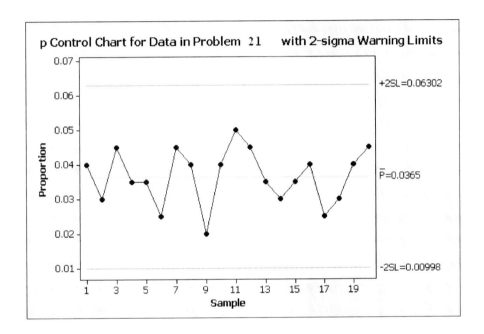

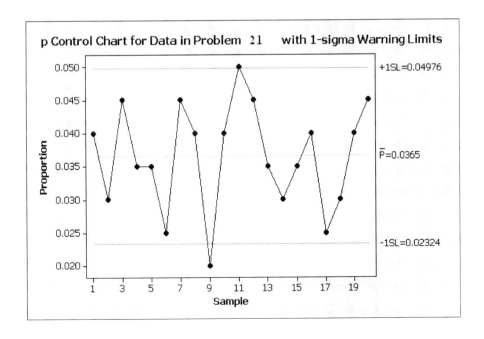

25. (a) In this problem clearly an appropriate control chart for the number of nonconforming belts is *np* Control Chart. We construct the *np* Control Chart using MINITAB.

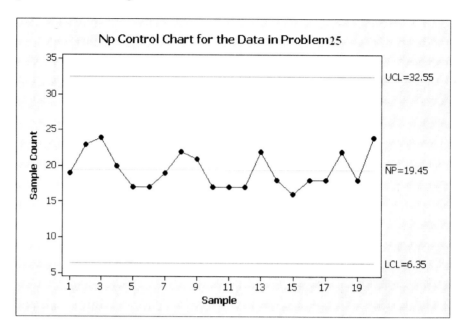

(b) Since in the control chart in part (a) we note that no sample points fall beyond the control limits and there are no unusual patterns, the process is in statistical control. Hence, the control chart does not change.

27. (a) From the given information we obtain

$$\hat{\theta} = \hat{p} = 50/(25 \times 60) = 0.0333$$

(b) The control limits for the *np* control chart are

$$UCL = 2 + 3\sqrt{n\hat{p}(1 - \hat{p})} = 2 + 3\sqrt{60 \times 0.0333(1 - 0.0333)} = 6.17$$

$$CL = 2$$

$$LCL = 2 - 3\sqrt{n\hat{p}(1 - \hat{p})} = 2 - 3\sqrt{60 \times 0.0333(1 - 0.0333)} = 0$$

(c) From Part (a), we have

$$\hat{\theta} = \hat{p} = 50/(25 \times 60) = 0.0333$$

Now we are interested in determining the probability of finding 48 nonconforming cylinder heads in a total sample of 1200. Using the Poisson approximation to the binomial ($\lambda = 1200 \times 0.0333 \cong 40$), we obtain

$$\Pr(X = 48|\lambda = 40) = 0.027.$$

29. After removing the two days production and the nonconforming chips the total production on the remainder of days (23) is 21,620 and 370 respectively. Now proceeding in the same manner as in Problem 28, we obtain

Average number of chips inspected per day $= 21,620/23 = 940$

Estimated fraction of nonconforming chips $\hat{\theta} = \hat{p} = 370/21620 = 0.0171$.

The control limits for a *p* Control Chart with $n = 940, \hat{p} = 0.0171$ are given by

$$UCL = \hat{p} + 3\sqrt{\hat{p}(1 - \hat{p})/n} = 0.0171 + 3\sqrt{0.0171(1 - 0.0171)/940} = 0.0298$$

$$CL = 0.0171$$

$$LCL = \hat{p} - 3\sqrt{\hat{p}(1 - \hat{p})/n} = 0.0171 - 3\sqrt{0.0171(1 - 0.0171)/940} = 0.0044.$$

31. From the given data it can easily be seen that

$$\bar{c} = \frac{\sum_{i=1}^{n} c_i}{n} = \frac{370}{20} = 18.5$$

The c Control Chart for the given data is given below. We constructed the control chart using MINITAB. From the control chart we observe that no points fall outside the control limits and there are no special patterns showing any abnormalities that would indicate the process to be out control. Hence, the process is in statistical control and this control chart can be used for future use at least until the next 25–30 sample period and then one should re-evaluate the control chart again and decide whether to continue with the same control chart or it should be revised.

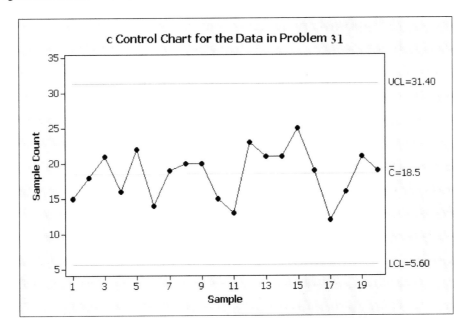

33. We construct the u Control Chart for the given data using MINITAB. The control chart is shown below. From the control chart we observe that no point falls outside the control limits and there are no special patterns showing any abnormalities that would indicate the process to be out control. Hence, the process is in statistical control and this control chart can be used for future use at least until the next 25–30 sample period and then one should re-evaluate the control chart again and decide whether to continue with the same control chart or it should be revised.

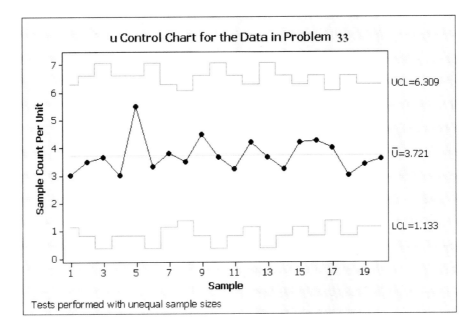

35. In this problem we are given that $\bar{u} = 10$. Thus, using the given information the desired control limits are given by

$$UCL = \bar{u} + 3\sqrt{\bar{u}/n} = 10 + 3\sqrt{10/8} = 13.3541$$
$$CL \;\;= \bar{u} = 10$$
$$LCL \;= \bar{u} - 3\sqrt{\bar{u}/n} = 10 - 3\sqrt{10/8} = 6.6459.$$

12

STATISTICAL QUALITY CONTROL PHASE II CONTROL CHARTS

Solutions of this Chapter have been prepared primarily using MINITAB.

PRACTICE PROBLEMS FOR SECTION 12.2

1. (i) The Shewhart $\overline{X}$-R Control Charts are very effective in detecting large shifts (i.e. magnitude 1.5σ or larger), but they are not effective in detecting smaller shifts. On the other hand, CUSUM Control Charts are very effecting in detecting smaller shifts. (ii) In $\overline{X}$-R Control Charts the rational sample size is usually 4 or 5, whereas in CUSUM Control Charts samples of size one are often used. (iii) The $\overline{X}$-R Control Charts use information from current sample, whereas CUSUM Control Charts use information not only from the current sample but also from all preceding samples. (iv) The Shewhart $\overline{X}$-R Control Charts are used for initial stage processes (since new processes can usually experience larger shifts), whereas CUSUM Control Charts are used for more mature processes because they usually experience smaller shifts.

3. The CUSUM Control Charts plot the cumulative sum of the deviations of sample values (or sample means, if the sample size is greater than one) from the target value.

5. Suppose the sample values of the first 10 sample are $X_1, X_2, X_3, \cdots, X_{10}$ and the target value of the process mean is μ_0. Then the first ten points plotted on a CUSUM Control Chart are sample numbers (i) versus $S_i^+ = Max(0, Z_i - k + S_{i-1}^+)$ or $S_i^- = Min(0, Z_i + k + S_{i-1}^-)$, $i = 1, 2, 3, \cdots, 10$, where $k = \frac{|\mu_1 - \mu_0|}{2\sigma}$ depending upon whether we wish to detect an upward or downward shift. Where Z_i is the standardized value of X_i, S_0 is assigned the value zero, and k is equal to half the size of the shift that we would like to detect. For example, if we would like to detect a shift of one standard deviation then $k = 0.5$.

PRACTICE PROBLEMS FOR SECTION 12.3

1. In this problem we are given the target value $= 40$, $\sigma = 2$, $n = 4$, $h = 5$, and $k = 0.5$, where h is the decision interval and k is the reference value for the CUSUM chart. Note that standardized reference value k is 0.5 since we want to detect a shift of one standard deviation and after standardizing the sample mean, $\sigma_{\overline{Z}} = 1$. Here we develop the two one-sided Tabular CUSUM Control Charts. The details of all the calculations are self explanatory from the following table.

Solutions Manual to Accompany Statistics and Probability with Applications for Engineers and Scientists, Bhisham C. Gupta and Irwin Guttman.
© 2014 John Wiley & Sons, Inc. Published 2014 by John Wiley & Sons, Inc.

Tabular CUSUM Control Chart for the Data in Problem 1, Using $k = 0.5$ and $h = 5.0$

Observations	$\overline{X}_i$	$\overline{Z}_i = \dfrac{\overline{X}_i - 40}{2/\sqrt{4}}$	$S_i^+ = Max$ $(0, \overline{Z}_i - k + S_{i-1}^+)$	$S_i^- = Min$ $(0, \overline{Z}_i + k + S_{i-1}^-)$
34 36 38 38	36.50	−3.50	0.00	−3.00
36 40 36 40	38.00	−2.00	0.00	−4.50
34 37 43 38	38.00	−2.00	0.00	−6.00
34 45 42 46	41.75	1.75	1.25	−3.75
44 40 35 34	38.25	−1.75	0.00	−5.00
35 43 41 38	39.25	−0.75	0.00	−5.25
34 44 40 36	38.50	−1.50	0.00	−6.25
36 41 45 41	40.75	0.75	0.25	−5.00
46 43 34 36	39.75	−0.25	0.00	−4.75
41 34 42 36	38.25	−1.75	0.00	−6.00
43 34 35 40	38.00	−2.00	0.00	−7.50
41 39 38 36	38.50	−1.50	0.00	−8.50
45 35 46 45	42.75	2.75	2.25	−5.25
41 39 40 38	39.50	−0.50	1.25	−5.25
34 35 46 44	39.75	−0.25	0.50	−5.00
34 38 42 39	38.25	−1.75	0.00	−6.25
35 43 34 38	37.50	−2.50	0.00	−8.25
42 37 39 37	38.75	−1.25	0.00	−9.00
35 44 34 38	37.75	−2.25	0.00	−10.75
42 34 42 40	39.50	−0.50	0.00	−10.75

The Tabular CUSUM Control Chart clearly shows that there is a downward shift of more than one standard deviation. This shift was detected immediately after the third sample was plotted. To clarify it further we present below two one-sided Graphical CUSUM Control Charts. Our conclusion using the Tabular CUSUM Control Chart is confirmed by the Graphical CUSUM Control Charts.

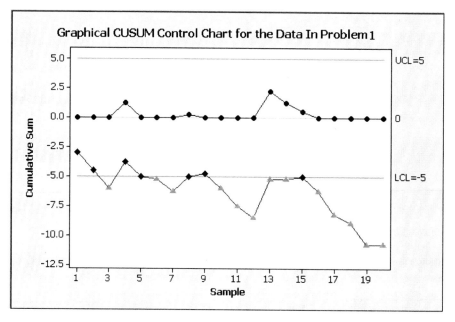

3. In this problem we reconsider the data from Problem 1 with sample size one; that is, by taking the first observation from each sample with standardized reference value $k = 0.5$ and the decision interval $h = 4.77$. The Tabular CUSUM Control Charts and two one-sided Graphical CUSUM Control Charts are given below.

Tabular CUSUM Control Chart for the Data in Problem 1, Using $k = 0.5$ and $h = 4.77$

Observations	$Z_i = \dfrac{X_i - 40}{2}$	$S_i^+ = Max$ $(0, Z_i - k + S_{i-1}^+)$	$S_i^- = Min$ $(0, Z_i + k + S_{i-1}^-)$
34	−3.0	0.0	−2.5
36	−2.0	0.0	−4.0
34	−3.0	0.0	−6.5
34	−3.0	0.0	−9.0
44	2.0	1.5	−6.5
35	−2.5	0.0	−8.5
34	−3.0	0.0	−11.0
36	−2.0	0.0	−12.5
46	3.0	2.5	−9.0
41	0.5	2.5	−8.0
43	1.5	3.5	−6.0
41	0.5	3.5	−5.0
45	2.5	5.5	−2.0
41	0.5	5.5	−1.0
34	−3.0	2.0	−3.5
34	−3.0	0.0	−6.0
35	−2.5	0.0	−8.0
42	1.0	0.5	−6.5
35	−2.5	0.0	−8.5
42	1.0	0.5	−7.0

Again, the downward shift is detected immediately after the third sample is plotted. Note that by using samples of size one we were able to detect an upward shift, which we did not detect when we used samples of size 4. Further, the upward shift was detected after the hirteenth sample was plotted.

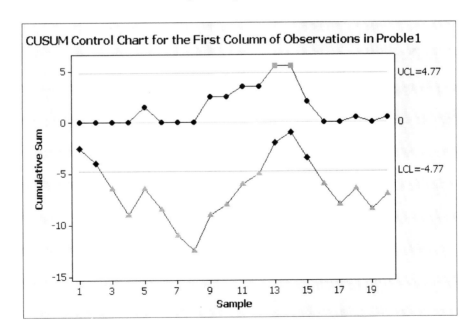

5. Note that by using the FIR feature we were able to detect the downward shift immediately after plotting the first sample, whereas without using the FIR feature ($(S_0^+ = h/2 = 2.5, S_0^- = -h/2 = -2.5)$ the shift was detected after plotting the third sample point (see Solution of Problem 1).

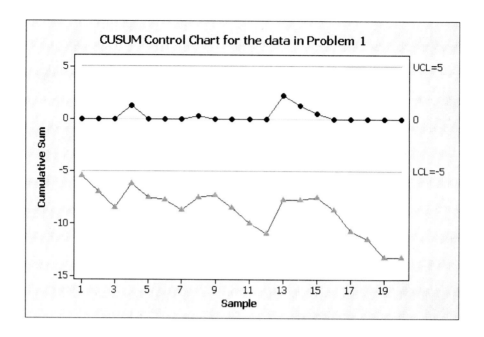

Tabular CUSUM Control Chart for the Data in Problem 1, Using FIR Feature

Observations	$\overline{X}_i$	$\overline{Z}_i = \dfrac{\overline{X}_i - 40}{2/\sqrt{4}}$	$S_i^+ = Max$ $(0, \overline{Z}_i - k + S_{i-1}^+)$	$S_i^- = Min$ $(0, \overline{Z}_i + k + S_{i-1}^-)$
34 36 38 38	36.50	−3.50	0.00	−5.50
36 40 36 40	38.00	−2.00	0.00	−7.00
34 37 43 38	38.00	−2.00	0.00	−8.50
34 45 42 46	41.75	1.75	1.25	−6.25
44 40 35 34	38.25	−1.75	0.00	−7.50
35 43 41 38	39.25	−0.75	0.00	−7.75
34 44 40 36	38.50	−1.50	0.00	−8.75
36 41 45 41	40.75	0.75	0.25	−7.50
46 43 34 36	39.75	−0.25	0.00	−7.25
41 34 42 36	38.25	−1.75	0.00	−8.50
43 34 35 40	38.00	−2.00	0.00	−10.00
41 39 38 36	38.50	−1.50	0.00	−11.00
45 35 46 45	42.75	2.75	2.25	−7.75
41 39 40 38	39.50	−0.50	1.25	−7.75
34 35 46 44	39.75	−0.25	0.50	−7.50
34 38 42 39	38.25	−1.75	0.00	−8.75
35 43 34 38	37.50	−2.50	0.00	−10.75
42 37 39 37	38.75	−1.25	0.00	−11.50
35 44 34 38	37.75	−2.25	0.00	−13.25
42 34 42 40	39.50	−0.50	0.00	−13.25

7. The standardized means $\overline{Z}_i$ are

1.417	−1.000	−1.583	2.333	−0.333	0.417	−0.333	0.167	−1.333	−0.750
1.250	0.500	−0.250	0.333	−0.500	−0.167	1.750	−0.583	0.250	−0.667

The two one-sided CUSUM Control Charts for the standardized means $\overline{Z}_i$ prepared by using MINITAB are shown below. The control charts clearly indicate the process is in statistical control.

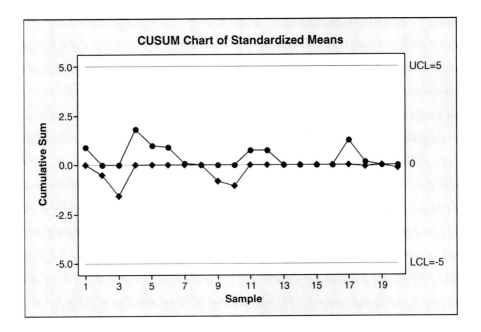

9. In this problem we present both the Graphical CUSUM and the Tabular Control Charts for controlling process variability. The following CUSUM Control Chart for process variability is prepared by using MINITAB. Note that MINITAB does not provide a direct procedure to prepare Control Charts for controlling process variability. Thus, to prepare the Control Charts for controlling process variability, we first determine the values of $V_i's$ (see Section 12.3.4) and then use the same procedure as used for constructing CUSUM Charts for the process mean. Both the two one-sided Graphical and Tabular Control Charts indicate that the process variability is in statistical control.

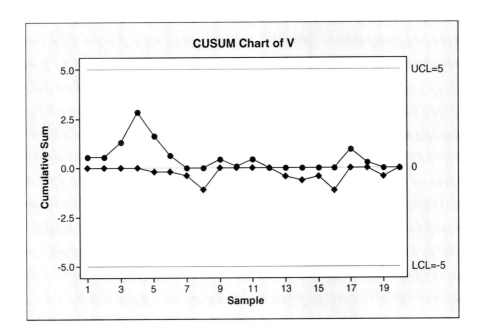

Tabular CUSUM Control Chart for the Data in Problem 7, for Controlling Variability

Sample #	V_i	$V_i^+ = \mathrm{Max}$ $[0, v_i - k + V_{i-1}^+]$	$V_i^- = \mathrm{Min}$ $[0, v_i + k + V_{i-1}^-]$
1	1.05512	0.55512	0.00000
2	0.51003	0.56515	0.00000
3	1.25016	1.31531	0.00000
4	2.02156	2.83687	0.00000
5	−0.70100	1.63587	−0.20100
6	−0.50574	0.63013	−0.20674
7	−0.70100	0.00000	−0.40774
8	−1.18553	0.00000	−1.09328
9	0.95330	0.45330	0.00000
10	0.12615	0.07944	0.00000
11	0.84823	0.42768	0.00000
12	−0.32921	0.00000	0.00000
13	−0.92264	0.00000	−0.42264
14	−0.70100	0.00000	−0.62364
15	−0.32921	0.00000	−0.45285
16	−1.18553	0.00000	−1.13838
17	1.43517	0.93517	0.00000
18	−0.16687	0.26830	0.00000
19	−0.92264	0.00000	−0.42264
20	−0.01577	0.00000	0.00000

PRACTICE PROBLEMS FOR SECTION 12.4

1. The Moving Average Control Chart for the data in Problem 1 is shown below, from which we conclude that the process is in statistical control. Here we estimate the standard deviation by using the moving range of length 2 and the mean as the average of all the 25 observations. Thus, for the given data we obtain

$$\overline{\overline{X}} = \overline{X} = 11.44, \text{ since sample size is one}$$

$$\overline{R} = 4.875$$

$$\hat{\sigma} = \overline{R}/d_2 = 4.875/1.128 \cong 4.32$$

Note that for the value of d_2 we use $d_2 = 2$, since we are using moving range of length 2. To determine the control limits we proceed as follows:

For time period $i = 1$, these limits are

$$\mathrm{UCL} = 11.44 + 3\frac{4.32}{\sqrt{1}} = 24.40$$

$$\mathrm{LCL} = 11.44 - 3\frac{4.32}{\sqrt{1}} = -1.52$$

For time period $i = 2$, these limits are

$$\mathrm{UCL} = 11.44 + 3\frac{4.32}{\sqrt{2}} = 20.60$$

$$\mathrm{LCL} = 11.44 - 3\frac{4.32}{\sqrt{2}} = 2.28$$

For time period $i = 3$, these limits are

$$UCL = 11.44 + 3\frac{4.32}{\sqrt{3}} = 18.92$$

$$LCL = 11.44 - 3\frac{4.32}{\sqrt{3}} = 3.96$$

Similarly, for time period i, with $4 \leq i \leq 25$, these limits are

$$UCL = 11.44 + 3\frac{4.32}{\sqrt{4}} = 17.92$$

$$LCL = 11.44 - 3\frac{4.32}{\sqrt{4}} = 4.96$$

The values of $M'_i s$ are calculated as follows:

$M_1 = 15, M_2 = (15 + 11)/2 = 13, M_3 = (15 + 11 + 8)/3 = 11.33, M_4 = (15 + 11 + 8 + 15)/4 = 12.25$
$M_5 = (11 + 8 + 15 + 6)/4 = 10, \ldots$

Moving Average Control Chart for the Data in Problem 1 with m = 4

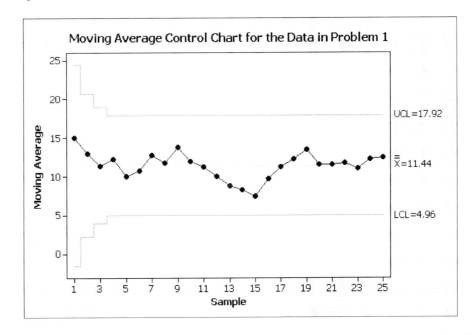

3. The Moving Average Control Chart for the data in this Problem with $m = 3$ is shown below. By observing this control chart we conclude that the process is in statistical control. However, there is a run of seven points above the center line; that is, the process seems to demonstrate a trend that may be investigated further. Further, since the plotted points are inherently dependent, the usual run rules (see Section 11.5) are not applicable in Phase II Control Charts discussed in this chapter.

Here we estimate the standard deviation by using the moving range of length 2, and the mean as the average of all the 20 observations. Thus, for the given data we obtain

$$\overline{\overline{X}} = \overline{X} = 35.93, \text{ since sample size is one}$$

$$\overline{R} = 0.7632$$

$$\hat{\sigma} = \overline{R}/d_2 = 0.7632/1.128 = 0.6766$$

Note that for the value of d_2 we use sample size 2 since we are using moving range of length 2. To determine the control limits we proceed as follows:

For time period $i = 1$, these limits are

$$UCL = 35.93 + 3\frac{0.6766}{\sqrt{1}} = 37.9597$$

$$LCL = 35.93 - 3\frac{0.6766}{\sqrt{1}} = 33.9003$$

For time period $i = 2$, these limits are

$$UCL = 35.93 + 3\frac{0.6766}{\sqrt{2}} = 37.3652$$

$$LCL = 35.93 - 3\frac{0.6766}{\sqrt{2}} = 34.4948$$

For time period $i = 3$, these limits are

$$UCL = 35.93 + 3\frac{0.6766}{\sqrt{3}} = 37.1018$$

$$LCL = 35.93 - 3\frac{0.6766}{\sqrt{3}} = 34.7582$$

The values of $M_i's$ are calculated as follows:

$$M_1 = 35, M_2 = (35 + 36.9)/2 = 35.95, M_3 = (35 + 36.9 + 36.6)/3 = 36.1667,$$
$$M_4 = (36.9 + 36.6 + 35.9)/3 = 36.4667$$

Similarly, we can calculate other values of control limits and plotted points. These values calculated by MINITAB are shown in the following table.

Sample #	M_i	*LCL*	*UCL*
1	35.0000	33.9003	37.9597
2	35.9500	34.4948	37.3652
3	36.1667	34.7582	37.1018
4	36.4667	34.7582	37.1018
5	36.1333	34.7582	37.1018
6	36.1000	34.7582	37.1018
7	35.9667	34.7582	37.1018
8	36.2000	34.7582	37.1018
9	35.8667	34.7582	37.1018
10	36.3000	34.7582	37.1018
11	36.0667	34.7582	37.1018
12	36.1333	34.7582	37.1018
13	35.8000	34.7582	37.1018
14	35.6667	34.7582	37.1018
15	35.8667	34.7582	37.1018
16	35.9667	34.7582	37.1018
17	35.8000	34.7582	37.1018
18	35.7333	34.7582	37.1018
19	35.4000	34.7582	37.1018
20	35.7000	34.7582	37.1018

Moving Average Control Chart for the Data in Problem 3 with $m = 3$

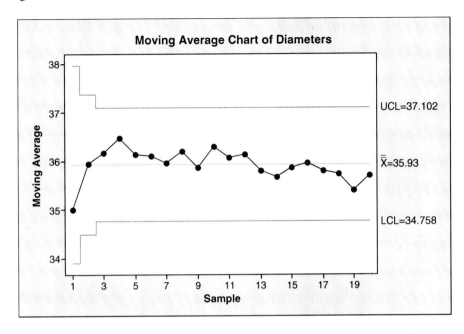

5. The Moving Average Control Chart for the data in this Problem with $m = 5$ is shown below. By observing this control chart we conclude that the process is in statistical control.

Again, in this problem, we estimate the standard deviation by using the moving range of length 2, and the mean as the average of all the 20 observations. Thus, for the given data we obtain

$$\overline{\overline{X}} = \overline{X} = 35.93, \text{ since sample size is one}$$
$$\overline{R} = 0.7632$$
$$\hat{\sigma} = \overline{R}/d_2 = 0.7632/1.128 = 0.6766$$

Note that for the value of d_2 we use sample size 2 since we are using moving range of length 2. The control limits and the plotted points are determined in usual manner. However, these values, determined by MINITAB, are given in the following table. All the points fall within the control limits; so the process is in statistical control. However, there is a run of 13 points above the center line; that is, the process seems to demonstrate a trend that may be investigated further. Further, since the plotted points are inherently dependent, the usual run rules (see Section 11.5) are not applicable in Phase II Control Charts discussed in this chapter.

Comparing the results in Problems 3, 4, and 5, we notice that as the span value is increasing, the control limits are becoming narrower. Moreover, as the span value is increasing, the pattern of the plotted points is changing. For example, in Problem 3 there is a run of seven points above the center line, in Problem 4 there is a run of 11 points above the center line, and in Problem 5 there is a run of 13 points above the center line. Thus, as the span value is increasing the sensitivity of the plotted points is increasing.

Sample #	M_i	LCL	UCL
1	35.0000	33.9003	37.9597
2	35.9500	34.4948	37.3652
3	36.1667	34.7582	37.1018
4	36.1000	34.9152	36.9448
5	36.0600	35.0223	36.8377
6	36.3600	35.0223	36.8377
7	36.0800	35.0223	36.8377
8	36.0800	35.0223	36.8377

(continued)

(*Continued*)

Sample #	M_i	LCL	UCL
9	36.0000	35.0223	36.8377
10	36.1800	35.0223	36.8377
11	36.0600	35.0223	36.8377
12	36.1000	35.0223	36.8377
13	35.9400	35.0223	36.8377
14	35.9400	35.0223	36.8377
15	35.8400	35.0223	36.8377
16	35.8800	35.0223	36.8377
17	35.7400	35.0223	36.8377
18	35.8000	35.0223	36.8377
19	35.7200	35.0223	36.8377
20	35.6400	35.0223	36.8377

Moving Average Control Chart for the Data in Problem 3 with m = 5

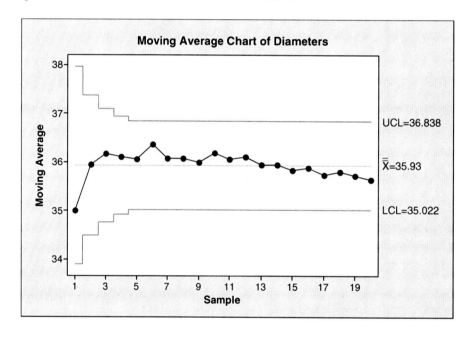

PRACTICE PROBLEMS FOR SECTION 12.5

1.

$$z_1 = \lambda x_1 + (1 - \lambda)z_0$$
$$= 0.20 \times 15 + (1 - 0.20) \times 12$$
$$= 12.60$$

$$z_2 = \lambda x_2 + (1 - \lambda)z_1$$
$$= 0.20 \times 11 + (1 - 0.20) \times 12.60$$
$$= 11.48$$

$$\ldots\ldots$$

The control limits *with* $\lambda = 0.20, L = 2.962, \mu_0 = 12, \sigma = 3.80$ are determined using equations (12.5.12) − (12.5.17) as below.

For time period $i = 1$, these limits are

$$\text{UCL} = 12 + 2.962 \times 3.8 \sqrt{\left(\frac{0.2}{2 - 0.2}\right)\left[1 - (1 - 0.2)^2\right]} = 14.2511$$

$$\text{LCL} = 12 - 2.962 \times 3.8 \sqrt{\left(\frac{0.2}{2 - 0.2}\right)\left[1 - (1 - 0.2)^2\right]} = 9.7489$$

For time period $i = 2$, these limits are

$$\text{UCL} = 12 + 2.962 \times 3.8 \sqrt{\left(\frac{0.2}{2 - 0.2}\right)\left[1 - (1 - 0.2)^4\right]} = 14.8828$$

$$\text{LCL} = 12 - 2.962 \times 3.8 \sqrt{\left(\frac{0.2}{2 - 0.2}\right)\left[1 - (1 - 0.2)^4\right]} = 9.0172$$

Similarly, for period $i = 25$ (i.e., when i is fairly large), good approximation of these limits are given by:

$$\text{UCL} = 12 + 2.962 \times 3.8 \sqrt{\left(\frac{0.2}{2 - 0.2}\right)} = 15.752$$

$$\text{LCL} = 12 - 2.962 \times 3.8 \sqrt{\left(\frac{0.2}{2 - 0.2}\right)} = 8.248.$$

The EMWA Control Chart for the given data is shown below. All the plotted points fall inside the control limits. Thus, the process is in statistical control. However, there is a run of 15 points below the center line; that is, the process seems to demonstrate a trend that may be investigated further. However, since the plotted points are inherently dependent, the usual run rules (see Section 11.5) are not applicable in Phase II Control Charts discussed in this chapter.

Further, note that the control chart shows the limits as $\pm 3\text{SL}$ but the actual calculations are made by using $\pm 2.962\text{SL}$ (see above).

EMWA Control Chart for the Data in Problem 1 with L = 2.962 and $\lambda = 0.20$

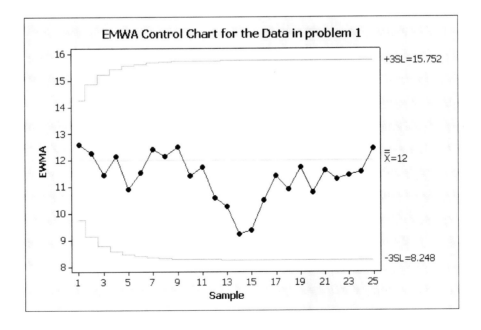

3. The EMWA Control Charts using $\lambda = 0.2$, 0.25 and 0.3 are shown below. The plotted points and the control limits are determined in the same manner as in solution of Problem 1. Note that by increasing the weight of EWMA (λ), the control limits have become wider. Also, the patterns of plotted points have changed since the value of λ effects the calculations of the plotted points (see solution of Problem 1). All three control charts indicate that the process is in statistical control.

EMWA Control Chart for the Data in Problem 1 with $L = 3$ and $\lambda = 0.2$

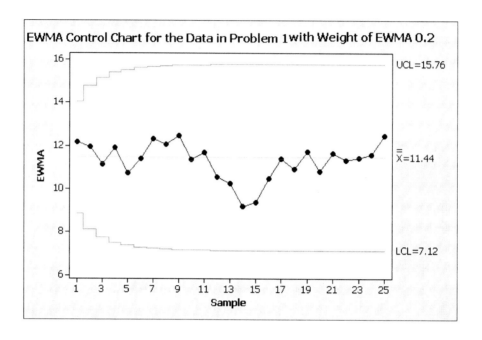

EMWA Control Chart for the Data in Problem 1 with $L = 3$ and $\lambda = 0.25$

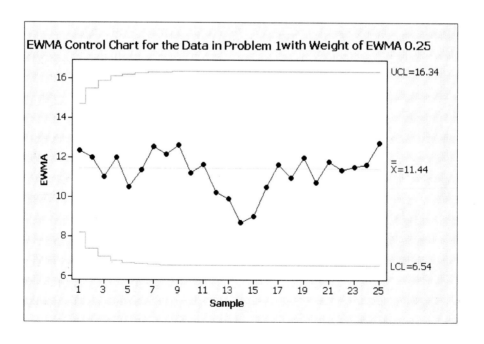

EMWA Control Chart for the Data in Problem 1 with $L = 3$ and $\lambda = 0.3$

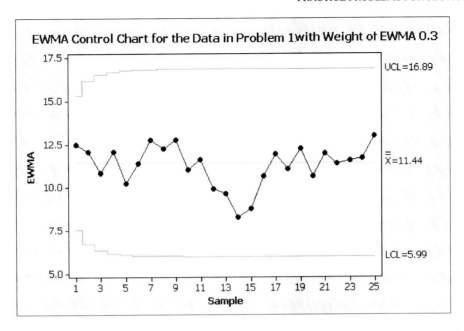

5. The EWMA Control Chart for the data in Problem 4 above with $\lambda = 0.1, L = 2.8$ is shown below. Also, given below are the values of Plotted points Z_i's. The control chart shows that the process is in statistical control.

$$35.900, \; 35.900, \; 36.000, \; 35.970, \; 36.033, \; 36.010, \; 35.999$$
$$35.949, \; 35.944, \; 35.980, \; 36.032, \; 36.038, \; 35.985, \; 35.886$$
$$35.958, \; 35.972, \; 35.925, \; 35.842, \; 35.938, \; 35.934$$

Minitab printout EWMA Control Chart for the data in Problem 4 with $\lambda = 0.1, L = 2.8$.

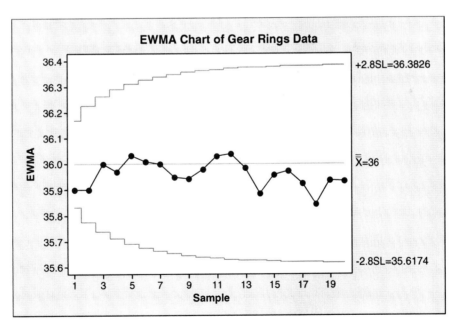

7. The EWMA Control Charts for the data in Problem 4 with $L = 2.6; \lambda = 0.05$ is show below. Note that the control charts show that the process is in statistical control. Note that by changing the values of L and λ control limits have become narrower. All the points fall within the control limits, so the process is in statistical control. There is a run of 8 points below the center line; that is, the process seems to demonstrate a trend that may be investigated further. However, since the plotted points are inherently dependent, the usual run rules (see Section 11.5) are not applicable in Phase II Control Charts discussed in this chapter.

Minitab printout EWMA Control Chart for the data in Problem 4 with $\lambda = 0.05, L = 2.6$.

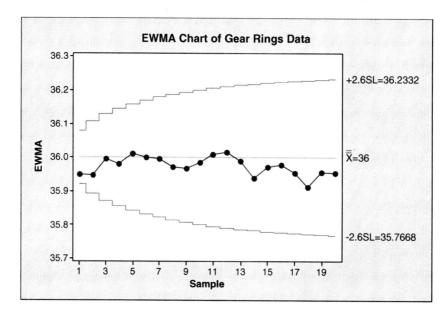

REVIEW PRACTICE PROBLEMS

1. The $\overline{X} - R$ Control Chart and both Tabular and Graphical CUSUM Control Charts are shown below. The $\overline{X} - R$ Control Chart indicates that process has gone out of control at the eleventh plotted point. The CUSUM Control Chart also indicates that the process has gone out of statistical control. The 11th point indicated that a shift in the process mean has taken place while the 12th plotted point has fallen above the decision interval; the process is declared to be out of control. We plot in CUSUM charts the standardized Z-values and $k = 0.5$, which is one half of the shift that we wanted to detect. We may comment here that it is not very common when the $\overline{X} - R$ Control Chart is very efficient to detect one-sigma shift. However, in this case we find that the observations in sample 11 have moved upward significantly and consequently the corresponding $\overline{X}$ is significantly large (i.e. shift seems larger than one-sigma). The $\overline{X} - R$ Control Charts are usually more efficient to detect large shifts.

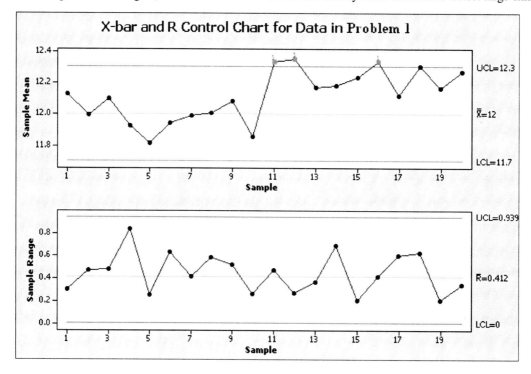

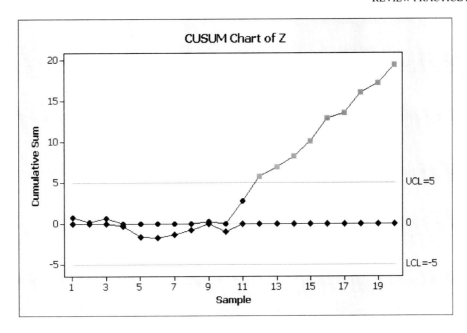

$\overline{X}_i$	$\overline{Z}_i = \dfrac{\overline{X}_i - 12}{0.2/\sqrt{4}}$	$S_i^+ = Max$ $(0, \overline{Z}_i - k + S_{i-1}^+)$	$S_i^- = Min$ $(0, \overline{Z}_i + k + S_{i-1}^-)$
12.1248	1.24750	0.7475	0.00000
11.9944	−0.05600	0.1915	0.00000
12.0996	0.99600	0.6875	0.00000
11.9207	−0.79275	0.0000	−0.29275
11.8150	−1.85050	0.0000	−1.64325
11.9384	−0.61625	0.0000	−1.75950
11.9897	−0.10325	0.0000	−1.36275
12.0054	0.05425	0.0000	−0.80850
12.0805	0.80475	0.3047	0.00000
11.8512	−1.48800	0.0000	−0.98800
12.3293	3.29325	2.7933	0.00000
12.3482	3.48175	5.7750	0.00000
12.1661	1.66150	6.9365	0.00000
12.1827	1.82725	8.2637	0.00000
12.2341	2.34125	10.1050	0.00000
12.3300	3.30025	12.9052	0.00000
12.1171	1.17125	13.5765	0.00000
12.2996	2.99625	16.0727	0.00000
12.1621	1.62125	17.1940	0.00000
12.2687	2.68675	19.3807	0.00000

3. The MA Control Chart for the data in Problem 2 is shown below. The plotted points can easily be calculated, which are

$$M_1 = 12.2271, M_2 = 12.2155, M_3 = 12.1161, M_4 = 12.1557, \cdots, M_{20} = 12.1581.$$

The control limits at the fifth point become constant at UCL = 12.268, LCL = 11.732. Of course CL remains fixed having a constant value of 12. The 15th sample point falls outside the control limits. Hence the process is out of control.

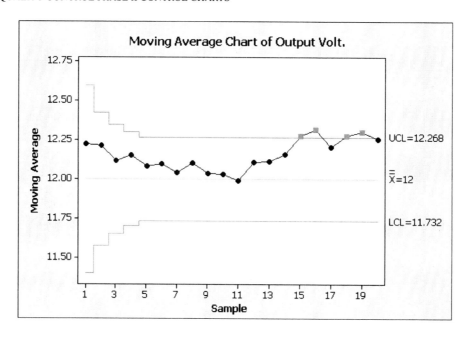

5. The EWMA Control Chart for the data in this problem is shown below. The plotted points can easily be calculated, which are

$$z_1 = 21.8064, z_2 = 23.5567, z_3 = 24.0692, z_4 = 24.6456, \cdots, z_{10} = 25.4310, \cdots, z_{20} = 24.5339$$

The control limits at the seventh point become constant at UCL = 29.5, LCL = 20.4. Of course CL always had a constant value of 25. All the points fall within the control limits; so the process is in statistical control. There is a run of 9 points below the center line; that is, the process seems to demonstrate a trend that may be investigated further. However, since the plotted points are inherently dependent, the usual run rules (see Section 11.5) are not applicable in Phase II Control Charts discussed in this chapter.

In this problem the value of $\lambda = 0.4$ is quite large. It will be useful if the quality control team also considers a EWMA Control Chart for a smaller value of λ.

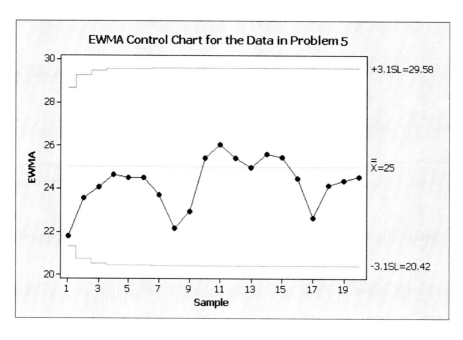

7. The Tabular CUSUM Control Chart for the data in Problem 5 with the standardized reference value k = 0.5 and the decision interval h = 4.77 is shown below. The Tabular CUSUM Control Chart does not indicate the process is out of control since no value in column 3 is greater than the reference value 4.77, and no value in column 4 is less than the reference value - 4.77. It may be worth, however, to redoing this problem by using the FIR feature.

Observations	$Z_i = \dfrac{X_i - 24}{2.70}$	$S_i^+ = Max$ $(0, Z_i - k + S_{i-1}^+)$	$S_i^- = Min$ $(0, Z_i + k + S_{i-1}^-)$
17.0161	−2.58663	0.00000	−2.08663
26.1820	0.80815	0.30815	−0.77848
24.8380	0.31037	0.11852	0.00000
25.5103	0.55937	0.17789	0.00000
24.2961	0.10967	0.00000	0.00000
24.4898	0.18141	0.00000	0.00000
22.5242	−0.54659	0.00000	−0.04659
19.8843	−1.52433	0.00000	−1.07093
24.0591	0.02189	0.00000	−0.54904
29.1806	1.91874	1.41874	0.00000
26.9380	1.08815	2.00689	0.00000
24.4623	0.17122	1.67811	0.00000
24.3371	0.12485	1.30296	0.00000
26.5040	0.92741	1.73037	0.00000
25.2357	0.45767	1.68804	0.00000
22.9902	−0.37400	0.81404	0.00000
19.8725	−1.52870	0.00000	−1.02870
26.4289	0.89959	0.39959	0.00000
24.6567	0.24322	0.14281	0.00000
24.8073	0.29900	0.00000	0.00000

9. The MA Control Chart for the data in Problem 5 using 99% probability control limits and m = 5 is shown below. The center line is fixed having a constant value of 24.21, which is the same as in Problem 6. The control limits at the fifth point become constant at UCL = 26.82, LCL = 21.60. The plotted points can easily be calculated, which are

$$M_1 = 17.0161, M_2 = 21.5990, M_3 = 22.6787, M_4 = 23.3866, \cdots, M_{20} = 23.7511$$

The very first point falls outside the control limits. Hence the process is out of control. Thus, our conclusion is the same as in Problem 6. Note that for 99% control limits we used L = 2.575 even though the control chart shows L = 2.6.

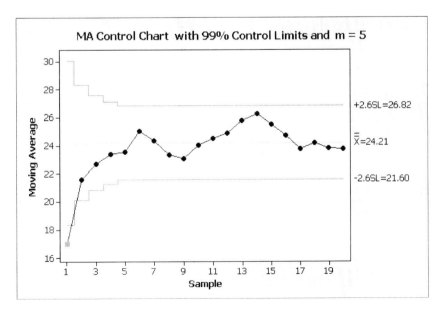

11. The Tabular CUSUM Control Chart for the data on Ball Bearings with the FIR feature, which is set at $h/2$, is shown below. Again, no plotted points fall outside the decision interval. Hence, the process is clearly in statistical control. Note that if the process is very much in statistical control, then the FIR feature does not have any significant effect on the CUSUM Control Chart.

Observations	$Z_i = \dfrac{X_i - 46}{8.5}$	$S_i^+ = Max$ $(0, Z_i - k + S_{i-1}^+)$	$S_i^- = Min$ $(0, Z_i + k + S_{i-1}^-)$
45	−0.11765	1.88235	−2.11765
31	−1.76471	0.00000	−3.38235
49	0.35294	0.00000	−2.52941
44	−0.23529	0.00000	−2.26471
52	0.70588	0.20588	−1.05882
53	0.82353	0.52941	0.00000
51	0.58824	0.61765	0.00000
60	1.64706	1.76471	0.00000
51	0.58824	1.85294	0.00000
54	0.94118	2.29412	0.00000
34	−1.41176	0.38235	−0.91176
53	0.82353	0.70588	0.00000
52	0.70588	0.91176	0.00000
44	−0.23529	0.17647	0.00000
39	−0.82353	0.00000	−0.32353
55	1.05882	0.55882	0.00000
52	0.70588	0.76471	0.00000
53	0.82353	1.08824	0.00000
32	−1.64706	0.00000	−1.14706
31	−1.76471	0.00000	−2.41176

13. The MA Control Chart for the data in Problem 10 is shown below. The plotted points can easily be calculated, which are

$$m = 3 : M_1 = 45.0000, M_2 = 38.0000, M_3 = 41.6667, M_4 = 41.3333, \cdots, M_{20} = 38.6667$$

$$m = 4 : M_1 = 45.0000, M_2 = 38.0000, M_3 = 41.6667, M_4 = 42.2500, \cdots, M_{20} = 42.0000$$

$$m = 5 : M_1 = 45.0000, M_2 = 38.0000, M_3 = 41.6667, M_4 = 42.2500, \cdots, M_{20} = 44.6000$$

The control limits for m = 3, 4, and 5 become constant at the 3rd, 4th, and 5th point with UCL = 58.64, LCL = 33.36; UCL = 56.94, LCL = 35.06; UCL = 55.79, LCL = 36.21, respectively. Of course, in all cases CL is fixed having a constant value of 46.00. The patterns of the plotted points have changed with the value of m. In all three cases all the points fall within the control limits, so the process is in statistical control. However, there is a run of 7 or more points above the center line; that is, the process seems to demonstrate a trend that may be investigated further. However, since the plotted points are inherently dependent, the usual run rules (see Section 11.5) are not applicable in Phase II Control Charts discussed in this chapter.

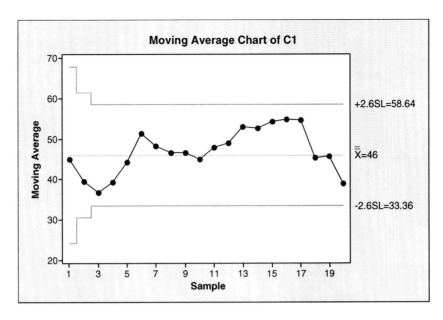

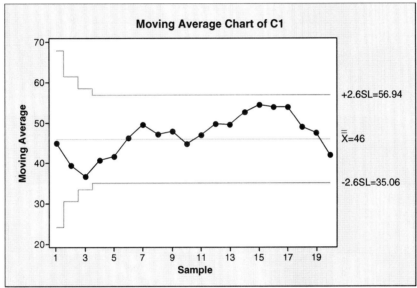

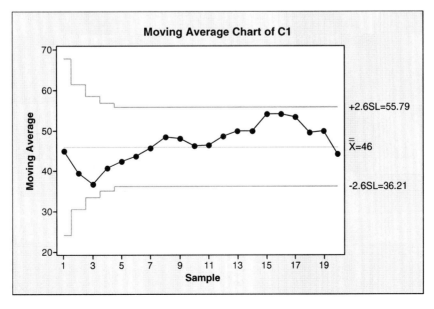

15. The EMWA Control Charts, using $L = 2.5$, 3.0, and 3.5, are shown below.

For calculations of the plotted points and the control limits, see solution of Problem 1 of Section 12.5. Note that by changing the value of L, the control limits have changed (i.e. as the value of L increases the control limits become wider). Our conclusion in this problem remained the same as in Problem 14.

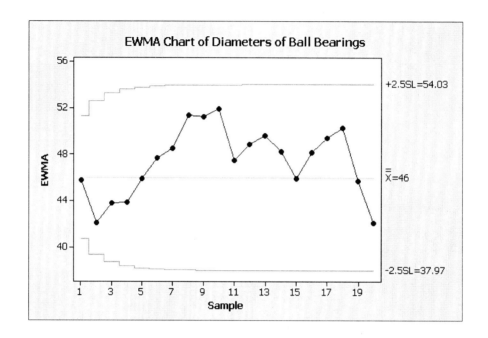

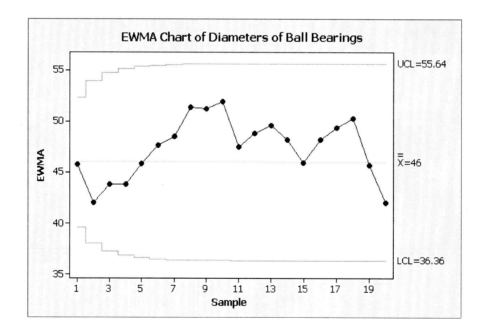

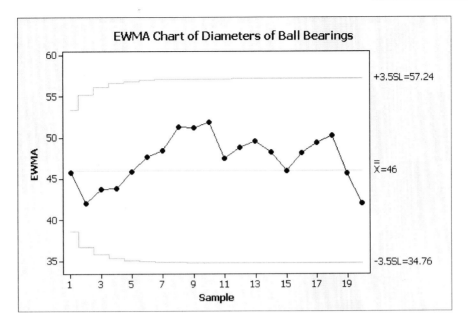

EWMA Chart of Diameters of Ball Bearings

17. The Tabular CUSUM Control Chart for the data on Ball Bearings with the FIR feature, which is set at $3h/4$, is shown below. Again, no plotted points fall outside the decision interval. However, the second plotted point came close to falling outside the decision interval. This indicates the process had a slight downward shift, but not enough to be out of control. Hence, the process is in statistical control. Our final conclusion is the same as in Problem 16; that is, the process is in statistical control.

Observations	$Z_i = \dfrac{X_i - 46}{8.5}$	$S_i^+ = Max$ $(0, Z_i - k + S_{i-1}^+)$	$S_i^- = Min$ $(0, Z_i + k + S_{i-1}^-)$
45	−0.11765	3.13235	−3.36765
31	−1.76471	0.86765	−4.63235
49	0.35294	0.72059	−3.77941
44	−0.23529	0.00000	−3.51471
52	0.70588	0.20588	−2.30882
53	0.82353	0.52941	−0.98529
51	0.58824	0.61765	0.00000
60	1.64706	1.76471	0.00000
51	0.58824	1.85294	0.00000
54	0.94118	2.29412	0.00000
34	−1.41176	0.38235	−0.91176
53	0.82353	0.70588	0.00000
52	0.70588	0.91176	0.00000
44	−0.23529	0.17647	0.00000
39	−0.82353	0.00000	−0.32353
55	1.05882	0.55882	0.00000
52	0.70588	0.88235	0.00000
53	0.82353	1.20588	0.00000
32	−1.64706	0.00000	−1.14706
31	−1.76471	0.00000	−2.41176

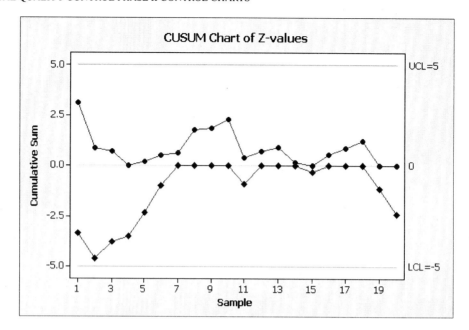

19. The Tabular CUSUM Control Chart for the data on tread depth in Problem 18, with the FIR feature set at $h/2$, is shown below. Again, no plotted points fall outside the decision interval. Hence, the process is in statistical control.

Observations	$Z_i = \dfrac{X_i - 6.7685}{0.2454}$	$S_i^+ = Max$ $(0, Z_i - k + S_{i-1}^+)$	$S_i^- = Min$ $(0, Z_i + k + S_{i-1}^-)$
6.28	−1.99063	0.00937	−3.99063
7.06	1.18786	0.69723	−2.30277
6.50	−1.09413	0.00000	−2.89690
6.76	−0.03464	0.00000	−2.43154
6.82	0.20986	0.00000	−1.72168
6.92	0.61736	0.11736	−0.60432
6.86	0.37286	0.00000	0.00000
7.15	1.55460	1.05460	0.00000
6.57	−0.80888	0.00000	−0.30888
6.48	−1.17563	0.00000	−0.98452
6.64	−0.52363	0.00000	−1.00815
6.94	0.69886	0.19886	0.00000
6.49	−1.13488	0.00000	−0.63488
7.14	1.51385	1.01385	0.00000
7.16	1.59535	2.10921	0.00000
7.10	1.35086	2.96007	0.00000
7.08	1.26936	3.72942	0.00000
6.48	−1.17563	2.05379	−0.67563
6.40	−1.50163	0.05216	−1.67726
6.54	−0.93113	0.00000	−2.10839

21. The EMWA Control Charts, using $L = 2.5$, 3.0, and 3.5, are shown below.

For calculations of the plotted points and the control limits see solution of Problem 1 of Section 12.5.

Note that by changing the value of L, the control limits have changed - the value of L is increasing and the control limits are becoming wider. For $L = 2.5$, the 1st and 17th plotted points came very close to the control limits, but the process for all three values of L is in statistical control. The value of L does not affect the values of the plotted points, and for $L = 2.5$ the control limits are narrower and the process is in control. Therefore for larger values of L, the conclusion does not change. If, however, the process for $L = 2.5$ were out of control, then our conclusion could have changed, that is, the control chart with larger values of L *could* show that the process is in statistical control.

EWMA control charts for the data in Problem 18 using $L = 2.5$ and $\lambda = 0.25$.

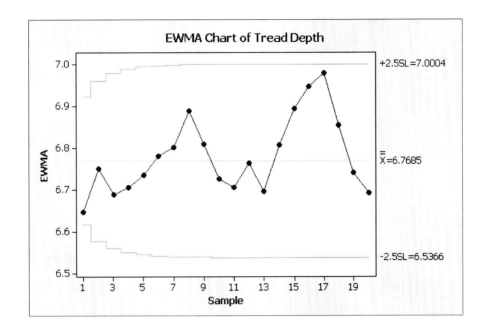

EWMA control charts for the data in Problem 18 using $L = 3$ and $\lambda = 0.25$.

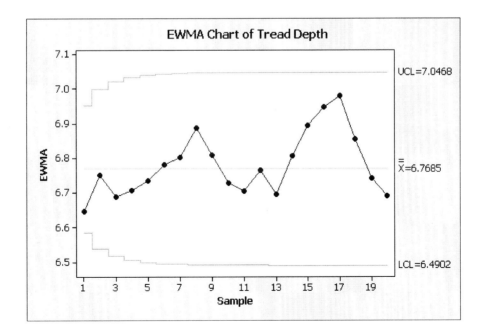

EWMA control charts for the data in Problem 18 using $L = 3.5$ and $\lambda = 0.25$.

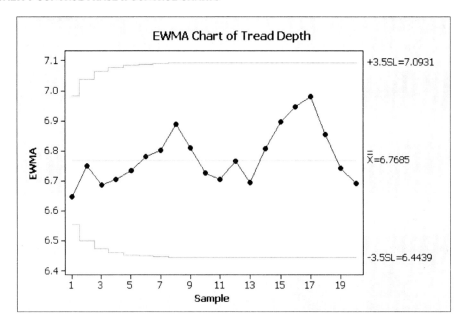

23. The Tabular CUSUM Control Chart for the data on tread depth is shown below. No plotted points fall outside the decision interval. Thus, based on this control chart we can say that the process is in statistical control. Thus, in this problem, our conclusion is the same as in Problem 18 (i.e. the new values of k and h had no effect on the final outcome).

Observations	$Z_i = \dfrac{X_i - 6.7685}{0.2454}$	$S_i^+ = Max$ $(0, Z_i - k + S_{i-1}^+)$	$S_i^- = Min$ $(0, Z_i + k + S_{i-1}^-)$
6.28	-1.99063	0.00000	-1.49063
7.06	1.18786	0.68786	0.00000
6.50	-1.09413	0.00000	-0.59413
6.76	-0.03464	0.00000	-0.12877
6.82	0.20986	0.00000	0.00000
6.92	0.61736	0.11736	0.00000
6.86	0.37286	0.00000	0.00000
7.15	1.55460	1.05460	0.00000
6.57	-0.80888	0.00000	-0.30888
6.48	-1.17563	0.00000	-0.98452
6.64	-0.52363	0.00000	-1.00815
6.94	0.69886	0.19886	0.00000
6.49	-1.13488	0.00000	-0.63488
7.14	1.51385	1.01385	0.00000
7.16	1.59535	2.10921	0.00000
7.10	1.35086	2.96007	0.00000
7.08	1.26936	3.72942	0.00000
6.48	-1.17563	2.05379	-0.67563
6.40	-1.50163	0.05216	-1.67726
6.54	-0.93113	0.00000	-2.10839

25. The MA Control Chart for the data in Problem 24 is shown below. The plotted points can easily be calculated, which are

$$m = 3 : M_1 = 21.5000, M_2 = 21.0500, M_3 = 20.2667, M_4 = 20.5333, \cdots, M_{20} = 21.7000$$
$$m = 5 : M_1 = 21.5000, M_2 = 21.0500, M_3 = 20.2667, M_4 = 20.7750, \cdots, M_{20} = 22.2400$$

The control limits for m = 3 and 5 become constant at the 3rd and 5th point with UCL = 23.910, LCL = 20.090 and UCL = 23.480, LCL = 20.520, respectively. Of course CL is fixed having a constant value of 22. The patterns of the plotted points have changed with the value of m. In control chart for m = 3, the process is out of control. In the case of m = 5, all the points fall inside the control limits, and so the control chart indicates the process is in statistical control. However, when m = 5, there are runs of more than seven points below and above the center line. This means that the process seems to have a trend that may be investigated further. However, since the plotted points are inherently dependent the usual run rules (see Section 11.5) are not applicable in Phase II Control Charts discussed in this chapter.

MA Control Chart for the data in Problem 25 with $m = 3$

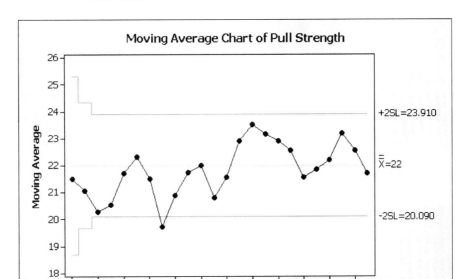

MA Control Chart for the data in Problem 25 with $m = 5$

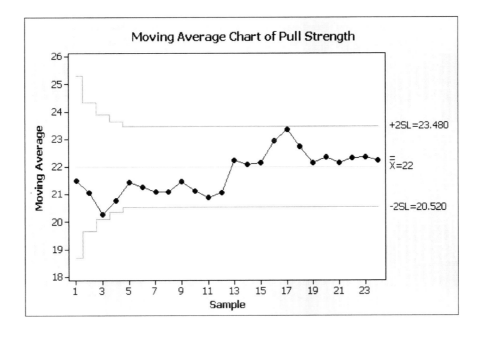

13

ANALYSIS OF CATEGORICAL DATA

PRACTICE PROBLEMS FOR SECTION 13.2

1. In this problem we want to test the null hypothesis that an artist has the same chance of being born in any month of the year, that is

$$H_0: \theta_i = 1/12, i = 1, 2, \cdots, 12$$

against

$$H_1: \text{ All } \theta_i\text{'s are not the same}$$

From the calculation given below we note that $n\theta_i \geq 5$ for each i and thus we can apply the χ^2 test without merging any categories. In this problem, the number of categories $k = 12$. Using the data in this problem, we obtain

f_i	$n\theta_i$	$(f_i - n\theta_i)^2$	$\dfrac{(f_i - n\theta_i)^2}{n\theta_i}$
68	63	25	0.39683
78	63	225	3.57143
67	63	16	0.25397
60	63	9	0.14286
61	63	4	0.06349
51	63	144	2.28571
50	63	169	2.68254
60	63	9	0.14286
67	63	16	0.25397
61	63	4	0.06349
73	63	100	1.58730
60	63	9	0.14286
756	756	730	11.59

The test statistic is given by

$$\chi^2 = \sum_{i=1}^{k} \frac{(f_i - n\theta_i)^2}{n\theta_i}$$

It can be seen from the above table that the observed value of the test statistic is 11.59. Further, the distribution of the test statistic is approximately Chi-Square with 11 ($=12-1$) degrees of freedom. From the Chi-Square table we have $\chi^2_{11;.05} = 19.68 > 11.59$, that is the tabulated value of the test statistic is greater than the observed value. Thus, we do not reject the null hypothesis. In other words, based on these data we can conclude that an artist has the same chance of being born in any month of the year.

This problem can be done by using one of the statistical packages. For instance using MINITAB we can proceed as follows:

1. Enter the data in columns C1 (Frequency).

2. Select **Stat > Tables > Goodness-of –fit Test (One Variable)**.

3. Enter C1 in the box next to **Observed counts**.

4. Specify under Test your null hypothesis. For example, in the present problem check equal proportions and then click **OK**. The output appears in the Session Window as

Category	Observed	Proportion	Expected	Test Contribution to Chi-Sq
1	68	0.0833333	63	0.39683
2	78	0.0833333	63	3.57143
3	67	0.0833333	63	0.25397
4	60	0.0833333	63	0.14286
5	61	0.0833333	63	0.06349
6	51	0.0833333	63	2.28571
7	50	0.0833333	63	2.68254
8	60	0.0833333	63	0.14286
9	67	0.0833333	63	0.25397
10	61	0.0833333	63	0.06349
11	73	0.0833333	63	1.58730
12	60	0.0833333	63	0.14286

N	DF	Chi-Sq	P-Value
756	11	11.5873	0.395

Since the p-value is greater than $\alpha = .05$. Hence, we do not reject the null hypothesis.

3. In this problem we wish test the null hypothesis that the sample comes from a Poisson population. We do this problem using MINITAB as follows:

1. Enter the variable (violations) in columns C1 and Frequency variable (weeks) in column C2.

2. Select **Stat > Basic Statistics > Goodness-of –fit Test for Poisson**.

3. Enter C1 in the box next to **Variable** and C2 next to **Frequency variable**.

4. Click **OK**.

The final result appears in the Session Window as follows:

Goodness-of-Fit Test for Poisson Distribution for Problem 3 of Section 13.2

Data column: Violations

Frequency column: Weeks

Poisson mean for Violations = 1.95

Violations	Observed	Poisson Probability	Expected	Contribution to Chi-Sq
0	25	0.142274	11.3819	16.2935
1	15	0.277434	22.1948	2.3323
2	8	0.270499	21.6399	8.5974
3	14	0.175824	14.0659	0.0003
4	7	0.085714	6.8571	0.0030
>=5	11	0.048255	3.8604	13.2045

N	DF	Chi-Sq	P-Value
80	4	40.4310	0.000

Note that the expected frequency (3.8604) is less than 5. Thus, we merge the last two cells and reenter the data in the MINITAB again so that the number of violations will be entered as 4 with frequency 18 weeks. The MINITAB automatically considers the last entry as 4 or more violations. Thus the new result will appear in the Session Window as follows:

Goodness-of-Fit Test for Poisson Distribution for Problem 3 of Section 13.2

Data column: Violations

Frequency column: Weeks

Poisson mean for Violations = 1.8125

Violations	Observed	Poisson Probability	Expected	Contribution to Chi-Sq
0	25	0.163246	13.0596	10.9170
1	15	0.295882	23.6706	3.1761
2	8	0.268144	21.4515	8.4350
3	14	0.162003	12.9603	0.0834
>=4	18	0.110725	8.8580	9.4351

N	DF	Chi-Sq	P-Value
80	3	32.0465	0.000

Note that the degrees of freedom is given by
Number of Categories (after merging) $- 1 - 1 = 5 - 1 - 1 = 3$
Since distribution mean was not known so we lose one degree of freedom for estimating it.

Since the p-value is zero which is less than 1% the level of significance, we reject the null hypothesis that the data come from a Poisson population.

5. In this problem we wish test the null hypothesis that the sample comes from a Poisson population. We do this problem using MINITAB as follows:

   1. Enter the variable (Gamma Particles) in columns C1 (Frequency) and Frequency variable (time intervals) in column C2.

   2. Select **Stat > Basic Statistics > Goodness-of –fit Test for Poisson**.

3. Enter C1 in the box next to **Variable** and C2 next to **Frequency variable**.
4. Click **OK**.

The final result appears in the Session Window as follows:

Goodness-of-Fit Test for Poisson Distribution for Problem 5 of Section 13.2

Data column: Gamma Particles

Frequency column: Intervals

Poisson mean for Gamma Particles $= 3.28$

Gamma Particles	Observed	Poisson Probability	Expected	Contribution to Chi-Sq
$<= 1$	5	0.161049	8.0524	1.15709
2	10	0.202410	10.1205	0.00143
3	12	0.221302	11.0651	0.07899
4	14	0.181467	9.0734	2.67506
5	5	0.119043	5.9521	0.15231
$>= 6$	4	0.114730	5.7365	0.52565

N	DF	Chi-Sq	P-Value
50	4	4.59054	0.332

Since the p-value is 0.332, which is greater than 5% the level of significance, we do not reject the null hypothesis that the data come from a Poisson population.

7. The solution of this problem is obtained using MINITAB (follow the instruction given in Problem 5 above).

Goodness-of-Fit Test for Poisson Distribution

```
Data column: No. of Cars
Frequency column: Frequency
Poisson mean for No. of Cars = 5.28
```

No. of Cars	Observed	Poisson Probability	Expected	Contribution to Chi-Sq
$<=2$	20	0.102964	10.2964	9.1469
3	10	0.124933	12.4933	0.4976
4	12	0.164911	16.4911	1.2231
5	10	0.174146	17.4146	3.1569
6	8	0.153248	15.3248	3.5011
7	11	0.115593	11.5593	0.0271
8	14	0.076291	7.6291	5.3201
$>=9$	15	0.087913	8.7913	4.3848

N	N*	DF	Chi-Sq	P-Value
100	0	6	27.2576	0.000

Since the p-value is 0.000, which is less than 5% the level of significance, we reject the null hypothesis that the data come from a Poisson population. In other words, the data do not come from a Poisson population.

9. From the given data we first estimate the population mean and standard deviation. That is,

$$\hat{\mu} = \overline{X} = 31.3, \hat{\sigma} = S = 4.146$$

Here, we first create a frequency distribution table, find the corresponding probabilities and then find the expected frequencies. Then we perform the following hypothesis:

H_0: The sample comes from a normal population

H_1: The sample does not come from a normal population

Class	Observed Frequencies (f_i)	Probabilities	Expected Frequency
<26	7	0.100	4
[26-28)	4	0.112	4.48
[28-30)	1	0.163	6.52
[30-32)	6	0.191	7.64
[32-34)	6	0.176	7.04
>34	16	0.258	10.32
	40	1.000	40

Since the expected frequencies of the first two classes are less than 5, so we merge them before calculating the observed value of Chi-Square. The observed value of Chi-Square is $\chi_2^2 = \dfrac{(11-8.48)^2}{8.48} + \dfrac{(1-6.52)^2}{6.52} + \dfrac{(6-7.64)^2}{7.64}$ $+ \dfrac{(6-7.04)^2}{7.04} + \dfrac{(16-10.32)^2}{10.32} = 9.054.$

Note that the degrees of freedom is 2, since after merging the first two classes we are left with five classes. Further, we lose two degrees of freedom for estimating the mean and the variance. The Table value of $\chi_{2;.01}^2 = 9.2103$, which is greater than the observed value. Hence, we do not reject the null hypothesis that the data come from a normal population. It can easily be seen that the p-value is 0.0108.

PRACTICE PROBLEMS FOR SECTION 13.3

1. In this problem we have a group of engineers who are classified with respect to two characteristics of interest, that is, their starting salary and their school of graduation public or private. We want to test a hypothesis that the two characteristics of interest are independent. That is,

H_0: Starting Salaries are independent of School of Graduation

against

H_1: Starting Salaries are not independent of School of Graduation

To carry out this test we employ

$$\chi^2 = \sum_{j=1}^{s} \sum_{i=1}^{r} \left[\frac{(f_{ij} - f_i f_{.j}/n)^2}{f_i f_{.j}/n} \right]$$

The test statistics for large n is approximately distributed as Chi-Square with $(r-1)(s-1)$ degrees of freedom. Note that r and s are the number of categories in which each characteristic is classified.

The observed value of χ^2 is given by

$$\chi^2 = \left[\frac{(12 - (20 \times 44)/106)^2}{(20 \times 44)/106}\right] + \left[\frac{(18 - (42 \times 44)/106)^2}{(42 \times 44)/106}\right] + \cdots + \left[\frac{(16 - (22 \times 62)/106)^2}{(22 \times 62)/106}\right] = 4.924$$

The test statistics is distributed as Chi-Square with $(2 - 1) \times (4 - 1) = 3$ degrees of freedom.

The tabulated value of the test statistic is $\chi^2_{3;.05} = 7.815$, which is greater than the observed value of 4.924. Hence, we do not reject the null hypothesis that the starting salaries are independent of the school of graduation. This problem can also be done by using one of the statistical packages discussed in this book. For example, using MINITAB we may proceed as follows:

Enter the data as shown below:

C1 Row	C2 Column	C3 Frequency
Public	$40,000–$49,999	12
Private	$40,000–$49,999	8
Public	$50,000–$59,999	18
Private	$50,000–$59,999	24
Public	$60,000–$69,999	8
Private	$60,000–$69,999	14
Public	Over $70,000	6
Private	Over $70,000	16

- Select **Stat** > **Tables** > **Cross Tabulation and Chi –Square**
- In the dialog box that appears, enter Rows, Columns, and Frequencies in boxes next to them and click **OK**. The MINITAB output that appears in the session window is shown here.

```
            $40k-$49.999k  $50k-$59.999k  $60k-$69.999k  over$70k  All
Private     8              24             14             16        62
            11.70          24.57          12.87          12.87     62.00
Public      12             18             8              6         44
            8.30           17.43          9.13           9.13      44.00
All         20             42             22             22        106
            20.00          42.00          22.00          22.00     106.00

Cell Contents: Count
               Expected count
Pearson Chi-Square = 4.924, DF = 3, P-Value = 0.177
```

Since the p-value is greater than the level of significance, we do not reject the null hypothesis.

Note: Sometimes the MINITAB gives output with columns or rows in a different order than in the given data, but the cell entries are then correctly done since MINITAB usually lists categories in alphabetic order.

3. Here we are given a group of women who are classified based on two characteristics of interest, that is, their education and their marriage life and we wish to test the hypothesis

H_0: Education and Marriage Life are independent

against

$$H_1: \text{Education and Marriage Life are not independent}$$

This problem can also be done by using one of the statistical packages discussed in this book. For example, to use MINITAB we proceed as follows:

Enter the two categories in columns C1, C2 and Frequencies in column C3. Then

- Select **Stat** > **Tables** > **Cross Tabulation and Chi –Square**
- In the dialog box that appears, enter Rows, Columns, and Frequencies in boxes next to them and click **OK**. The MINITAB output that appears in the session window is shown here.

Note that the expected frequencies (second entry in each cell) are computed by the MINITAB.

Tabulated statistics: Education, Marriage Life

```
Using frequencies in Frequency

Rows: Education Columns: Marriage Life

                      <10     10-<20    20-<30     ≥30      All
Grad.Degree            20        25        35       20      100
                      (23)      (26)      (25)     (26)     100
High School            37        31        20       12      100
                      (23)      (26)      (25)     (26)     100
Professional Degree    10         8        30       52      100
                      (23)      (26)      (25)     (26)     100
Undergrad. Degree      25        40        15       20      100
                      (23)      (26)      (25)     (26)     100
All                    92       104       100      100      400
                       92       104       100      100      400

Cell Contents:      Count
                    Expected count
```

Pearson Chi-Square $= 83.742$, DF $= 9$, P-Value $= 0.000$

Since the *p*-value is less than 5% the level of significance, we reject the null hypothesis that "education" and "marriage life" are independent. That is, "education" and "marriage life" are not independent.

Note: Sometimes the MINITAB give output with columns or row in different order than in the given data but the cell entries are then correctly done. Since MINITAB usually lists categories in alphabetic order.

5. In this problem we wish to test that the level of education among immigrants is independent of their region of migration. We perform this test with the help of MINITAB using the following instructions.

   1. Enter in the MINITAB worksheet as shown in Example 13.3.4.
   2. Select **Stat** > **Tables** > **Cross Tabulation and Chi –Square**
   3. In the dialog box that appears, enter Rows, Columns, and Frequencies in boxes next to them and click **OK**. When entering the data we denote Asian Countries, European Countries, Middle-eastern countries and African Countries by AC, EC, MC and AFC, respectively. The MINITAB output that appears in the session window is shown here.

Tabulated statistics: Level of Edu., Countries of Origin

Using frequencies in Frequency
Rows: Level of Edu. Columns: Countries of Origin

	AC	AFC	EU	MC	All
Highly edu	90.00	30.00	80.00	40.00	240.00
	82.45	39.75	73.62	44.17	240.00
Not well edu	70.00	70.00	50.00	60.00	250.00
	85.89	41.41	76.69	46.01	250.00
Well edu	120.00	35.00	120.00	50.00	325.00
	111.66	53.83	99.69	59.82	325.00
All	280.00	135.00	250.00	150.00	815.00
	280.00	135.00	250.00	150.00	815.00

Cell Contents: Count
                Expected count
Pearson Chi-Square = 53.207, DF = 6, P-Value = 0.000

Since the *p*-value is less than 5% the level of significance, we reject the null hypothesis that the level of education among immigrants is independent of their region of migration.

PRACTICE PROBLEMS FOR SECTION 13.4

1. Here we have three groups of patients and each group is treated with different medication. We wish to test the null hypothesis that these groups are homogeneous with respect to the relief they get. That is

H_0: Three groups of patients are homogeneous with respect to the relief they get

against

H_1: Three groups of patients are not homogeneous with respect to the relief they get

We now proceed to do this problem using MINITAB. Enter the two categories in columns C1, C2 and Frequencies in column C3. Then

- Select **Stat** > **Tables** > **Cross Tabulation and Chi –Square**
- In the dialog box that appears, enter Rows, Columns, and Frequencies in boxes next to them and click **OK**. The MINITAB output that appears in the session window is shown here.

Tabulated statistics: Level of Relief, Medication

Using frequencies in Frequency
Rows: Level of Relief Columns: Medication

	Acetaminophen	Ibuprofen	Motrin	All
Complete Relief	36.00	40.00	38.00	114.00
	38.00	38.00	38.00	114.00
Little or No Relief	18.00	22.00	12.00	52.00
	17.33	17.33	17.33	52.00
Some Relief	26.00	18.00	30.00	74.00
	24.67	24.67	24.67	74.00
All	80.00	80.00	80.00	240.00
	80.00	80.00	80.00	240.00

Cell Contents: Count
                Expected count

Pearson Chi-Square = 6.161, DF = 4, P-Value = 0.187. Since the p-value is greater than 1% the level of significance, we do not reject the null hypothesis. That is, the patients get the same kind of relief with either medication.

Note: Sometimes the MINITAB give output with columns or row in different order than in the given data but the cell entries are then correctly done. Since MINITAB usually lists categories in alphabetic order.

3. Here we are given five types of industries. We wish to test the hypothesis that the hiring in all sectors is equally affected by higher interest rates. In other words, all sectors are equally prone to higher interest rates for hiring new employees or they are homogeneous.

H_0: Five sectors are equally affected by higher interest for hiring new employees

against

H_1: Five machines are not equally affected by higher interest for hiring new employees

We proceed to do this problem using MINITAB. Enter the two categories in columns C1, C2 and Frequencies in column C3. Then

- Select **Stat** > **Tables** > **Cross Tabulation and Chi –Square**
- In the dialog box that appears, enter Rows, Columns, and Frequencies in boxes next to them and click **OK**. The MINITAB output that appears in the session window is shown here.

Tabulated statistics: Affect, Industry Type

```
Using frequencies in Frequency
Rows: Affect Columns: Industry Type

        Food   General  Insurance  Manufacturing  Pharmaceutical  All
No       35      52       45           25             45          202
        33.67   40.40    40.40        40.40          47.13        202
Yes      15       8       15           35             25           98
        16.33   19.60    19.60        19.60          22.87         98
All      50      60       60           60             70          300
         50      60       60           60             70          300


Cell Contents: Count
               Expected count
```

Pearson Chi-Square = 30.227, DF = 4, P-Value = 0.000.

Since the p-value is less than 10% the level of significance, we reject the null hypothesis that the five sectors are equally affected by higher interest rates. That is, five sectors are not equally prone to higher interest rates.

Note: Sometimes the MINITAB gives output with columns or rows in a different order than in the given data, but the cell entries are then correctly done, since MINITAB usually lists categories in alphabetic order.

5. Here we are given three groups of people with different income levels. We wish to test the hypothesis that the three groups of people are homogeneous with respect to the marital status.

H_0: Three groups of people are homogeneous with respect to the marital status

H_1: Three groups of people are not homogeneous with respect to the marital status

We proceed to do this problem using MINITAB. Enter the two categories in columns C1, C2 and Frequencies in column C3. Then

- Select **Stat** > **Tables** > **Cross Tabulation and Chi –Square**
- In the dialog box that appears, enter Rows, Columns, and Frequencies in boxes next to them and click **OK**. The MINITAB output that appears in the session window is shown here.

Tabulated statistics: Income Level, Marital Status

```
Using frequencies in Frequency
Rows: Income Level Columns: Marital Status

               Divorced    Married  Separated     Single        All
$100K-$250K          11         44         12         43        110
                   7.42      35.75      13.20      53.63     110.00
Less than $100K       7         20         16         22         65
                   4.39      21.13       7.80      31.69      65.00
Over $250K            9         66         20        130        225
                  15.19      73.13      27.00     109.69     225.00
All                  27        130         48        195        400
                  27.00     130.00      48.00     195.00     400.00


Cell Contents: Count
              Expected count
Pearson Chi-Square = 27.829, DF = 6, P-Value = 0.000
```

Since the p-value is less than 5% the level of significance, we reject the null hypothesis that the three groups of people are homogeneous with respect to marital status.

REVIEW PRACTICE PROBLEMS

1. Here we wish to test the hypothesis

$$H_0: P(H) = P(T) = \frac{1}{2}. \quad \text{Versus} \quad H_1: P(H) \neq P(T)$$

Given $n = 1000$, $\quad \theta_1 = \theta_2 = \frac{1}{2}$, $\quad f_1 = \#\text{heads} = 462$, $\quad f_2 = \#\text{tails} = 538$,

$$\chi^2 = \sum_{i=1}^{2} \frac{(f_i - n\theta_i)^2}{n\theta_i} = \frac{(462 - 500)^2}{500} + \frac{(538 - 500)^2}{500} = 5.776.$$

We reject the hypothesis of a 'true' coin at the 5% level if $\chi^2 \geq \chi^2_{1;0.05} = 3.84146$.

Since observed $\chi^2 = 5.776$, so we reject the null hypothesis and conclude that the coin is not 'true'. The p-value in this problem is 0.0162.

3. From the given data we first estimate the population mean and standard deviation. That is,

$$\hat{\mu} = \overline{X} = 175.43, \hat{\sigma} = S = 14.98$$

Here, we first create a frequency distribution table, find the corresponding probabilities and then find the expected frequencies. Then we perform the following hypothesis.

$$H_0: \text{The sample comes from a normal population}$$

$$H_1: \text{The sample does not come from a normal population}$$

Frequency Distribution Table of Cholesterol Levels

Classes	Frequency	Probabilities	Expected Frequency
[150,160)	19	0.151	15.1
[160,170)	15	0.207	20.7
[170,180)	26	0.262	26.2
[180,190)	17	0.215	21.5
[190,200]	23	0.165	16.5
Total	100	1	

Here the observed value of the test statistic is

$$\chi^2 = \frac{(19-15.1)^2}{15.1} + \frac{(15-20.7)^2}{20.7} + \frac{(26-26.2)^2}{26.2} + \frac{(17-21.5)^2}{21.5} + \frac{(23-16.5)^2}{16.5} = 6.0808.$$

Note that the degrees of freedom $= k - c - 1$, where k is the number of classes (5), and c is the number of parameters estimated (2). Thus, degrees of freedom $= k - c - 1 = 2$.

Since $\chi^2_{2}; 0.05 = 5.99147$, which is less than the observed value. Hence, we reject the hypothesis that the data came from a normal population. p-value $= 0.0478$.

5. To test whether the roulette wheel is true, test the hypothesis H_0: $\theta_i = \dfrac{1}{19}$, $i = 1, 2, \ldots, 19$; i.e., each pair is equally likely.

$$n\theta_i = 380\left(\frac{1}{19}\right) = 20$$

$$\chi^2 = \sum_{i=1}^{19} \frac{(f_i - n\theta_i)^2}{n\theta_i} = \frac{(24-20)^2}{20} + \cdots + \frac{(16-20)^2}{20} = 19.30.$$

Reject H_0 at the 1% level if $\chi^2 > \chi^2_{18;0.01} = 34.8053$.

Since observed $\chi^2 = 19.30$, we do not reject the hypothesis that the roulette wheel is fair.

7. Estimate the probability of a hit.

$$\widehat{\theta}_p = \frac{181 + 160}{216 + 216} = \frac{341}{432} = .7894.$$

We want to test the hypothesis that $\theta_A = \theta_B = \hat{\theta}_p$.

f_i	$n\widehat{\theta}_p$	f_i	$n\widehat{\theta}_q$
181	170.5	35	45.5
160	170.5	56	45.5

$$\chi^2 = \frac{(181-170.5)^2}{170.5} + \frac{(160-170.5)^2}{170.5} + \frac{(35-45.5)^2}{45.5} + \frac{(56-45.5)^2}{45.5} = 6.1393$$

Reject the hypothesis if $\chi^2 > \chi^2_{1;0.05} = 3.84146$.

Since observed $\chi^2 = 6.1393$, so we reject the hypothesis that the probabilities of type A and type B each registering a hit are equal.

9.

		Defective		Nondefective		$f_{i.}$
Shift	1	52	(63.28)	921	(909.72)	973
	2	61	(62.63)	902	(900.37)	963
	3	73	(60.09)	851	(863.91)	924
	$f_{.j}$	186		2674		2860

The expected frequencies are calculated from $\dfrac{f_{i.}f_{.j}}{n}$, are in brackets in the above table.

$$\chi^2 = \sum\sum \frac{(f_{ij} - f_{i.}f_{.j}/n)^2}{f_{i.}f_{.j}/n} = \frac{(52 - 63.28)^2}{63.28} + \cdots + \frac{(851 - 863.91)^2}{863.91} = 5.1625.$$

Reject the hypothesis of independence at the 5% level if $\chi^2 > \chi^2_{2;0.05} = 5.99147$.

Since the observed value $\chi^2 = 5.1625$, we do not reject the hypothesis that the time of work does not significantly affect the quality of results at the 5% level.

11.

	Webbed $\left(\dfrac{3}{4}\right)$	Non-webbed $\left(\dfrac{1}{4}\right)$
White $\left(\dfrac{3}{4}\right)$	94	33
	(90)	(30)
Colored $\left(\dfrac{1}{4}\right)$	28	5
	(30)	(10)

$$\chi^2 = \sum_i \sum_j \frac{(f_{ij} - n\theta_i\tau_j)^2}{n\theta_i\tau_j} = \frac{(94 - 90)^2}{90} + \cdots + \frac{(5 - 10)^2}{10} = 3.111.$$

The expected frequencies are given by $n\theta_i\tau_j$, and are in brackets of the above table.

Reject the null hypothesis of independence at the 5% level if $\chi^2 > \chi^2_{3;0.05} = 7.81473$.

Since the observed value $\chi^2 = 3.111$, we do not reject the hypothesis that the modes of classification are independent.

13. Here we are interested in testing that the quality of bulbs manufactured in the United States, Canada, and Mexico is the same. In other words, the quality of bulbs is independent of where they are manufactured. That is,

H_0: The populations of bulbs manufactured in the United States, Canada, and Mexico are homogeneous

H_1: The populations of bulbs manufactured in the United States, Canada, and Mexico are not homogeneous

We now proceed to do this problem using MINITAB. Enter the two categories in columns C1, C2 and Frequencies in column C3. Then

- Select **Stat > Tables > Cross Tabulation and Chi –Square**
- In the dialog box that appears, enter Rows, Columns, and Frequencies in boxes next to them and click **OK**. The MINITAB output that appears in the session window is shown here.

Tabulated statistics: Quality, Country of Production

```
Using frequencies in Frequency
Rows: Quality Columns: Country of Production

                                  United
                  Canada  Mexico  States  All
Defective             14      17      15   46
                    14.4    15.3    16.4   46
Non-defective        280     295     320  895
                   279.6   296.7   318.6  895
All                  294     312     335  941
                     294     312     335  941
Cell Contents: Count
               Expected count
```

Pearson Chi-Square $= 0.342$, DF $= 2$, p-Value $= 0.843$. Since the p-value is greater than 1% the level of significance, we do not reject the null hypothesis that the quality of bulbs manufactured in the United States, Canada, and Mexico is the same. In other words three populations of bulbs are homogeneous.

Note: Sometimes MINITAB gives output with columns or rows in a different order (alphabetic order) than in the given data, but the cell entries are then correctly done since MINITAB usually lists categories in alphabetic order.

15. In this problem first we need to construct a frequency distribution table and then calculate the probabilities of a randomly selected observation falling in a given interval under the assumption that the data come from a normal distribution, using (MLE) estimate of the mean and the standard deviation ($\hat{\mu} = 85.36$, $\hat{\sigma} = 5.932$). Having done all this then we use the Chi-Square Goodness-of-Fit test to verify whether our assumption of normality is valid or not. The frequency distribution table for these data is shown below.

Classes	Tally	Frequencies	Probabilities Assuming Normality
≤ 74	/	1	0.027744
[74,80)	//// ///	9	0.183111
[80,86)	///// ///// ///	13	0.183111
[86,92)	///// ///// ///// ////	19	0.868505
≥ 92	///// ///	8	0.131495
Total		50	1.00

The hypothesis we want to test is the following:

H_0: The data come from a normal population
H_1: The data do not come from a normal population

Now using the Chi-Square Goodness-of –Fit test we get the following calculations

Chi-Square Goodness-of-Fit Test for Observed Counts in Variable: C3

```
                        Test              Contribution
Category  Observed  Proportion  Expected   to Chi-Sq
1                1    0.027744    1.3872     0.10807
2                9    0.155367    7.7683     0.19528
3               13    0.359848   17.9924     1.38524
4               19    0.325547   16.2773     0.45541
5                8    0.131495    6.5747     0.30896


N    DF   Chi-Sq    P-Value
50   2    2.45297   0.293322
```

Note that the frequency in the first cell is less than 5. So we merge the first two classes and recalculate the expected frequencies. The new results obtained are as shown below.

Chi-Square Goodness-of-Fit Test for Observed Counts in Variable: C3

Category	Observed	Proportion	Expected	Test Contribution to Chi-Sq
1	10	0.183111	9.1555	0.07789
2	13	0.359848	17.9924	1.38524
3	19	0.325547	16.2773	0.45541
4	8	0.131495	6.5747	0.30896

N	DF	Chi-Sq	P-Value
50	1	2.22751	0.135572.

Since the p-value is greater than 1% the level of significance we do not reject the null hypothesis that the data come from a normal population.

Note: In this case the computer does not take into consideration whether or not we estimated the mean and the standard deviation. Since in this problem we estimated the mean and the standard deviations without using MINITAB, the MINITAB output will give the degrees of freedom equal to # categories $- 1 = 4 - 1 = 3$, whereas the actual degrees of freedom is equal to # categories - # of parameters estimated $- 1 = 4 - 2 - 1 = 1$. So we have to enter manually the correct degrees of freedom and find the correct p-value.

Note: Using the Anderson-Darling test we arrive at the same conclusion that the data come from a normal population. However, the Anderson-Darling test gives the p-value of 0.514.

17. In this problem we wish to test the hypothesis that the given data come from a discrete uniform population with $\theta_i = 0.125, i = 1, 2, \cdots, 8$. That is,

H_0: The data come from a discrete uniform population with $\theta_i = 0.125, i = 1, 2, \cdots, 8$

H_1: The data do not come from a discrete uniform population with $\theta_i = 0.125, i = 1, 2, \cdots, 8$

The testing of hypothesis in this problem is performed using MINITAB as follows:

1. Enter the variable (frequencies) in columns C1 and the corresponding probabilities in column C2.
2. Select **Stat > Tables > Goodness-of –fit Test (One Variable)**.
3. Enter C1 in the box next to **Observed counts**.
4. Check an appropriate circle under **Test**. For example, in the present problem check specific proportions, enter the column containing these probabilities and then click **OK**.
   The final result appears in the Session Window as follows:

Chi-Square Goodness-of-Fit Test for Observed Counts in Variable: Violation Tick

Category	Observed	Proportion	Expected	Test Contribution to Chi-Sq
1	19	0.125	20.875	0.16841
2	25	0.125	20.875	0.81512
3	21	0.125	20.875	0.00075
4	24	0.125	20.875	0.46781
5	15	0.125	20.875	1.65344
6	28	0.125	20.875	2.43189
7	20	0.125	20.875	0.03668
8	15	0.125	20.875	1.65344

N	DF	Chi-Sq	P-Value
167	7	7.22754	0.406

Since the p-value is greater than 1% the level of significance we do not reject the null hypothesis that the data come from a discrete population with probability $\theta_i = 0.125, i = 1, 2, \cdots, 8$.

19. In this problem, first we need to construct a frequency distribution table and then calculate the probabilities of a randomly selected observation falling in a given interval under the assumption that the data come from a exponential distribution. Using an estimate of the mean we obtain ($\hat{\lambda} = \overline{X} = 85.36$). Note that MINITAB uses the scale parameter equal to the mean, whereas in the text we use mean $= 1/\lambda$. Having done all this then we use the Chi-Square Goodness-of-Fit test to verify whether our assumption of exponential distribution is valid or not. The frequency distribution table for these data is shown below.

Classes	Tally	Frequencies	Probabilities Assuming Exponential Distribution
≤ 10	//	2	0.232234
[10,20)	//// ///// //	12	0.178302
[20,30)	///// ///	8	0.136894
[30,40)	/////	5	0.105102
[40,50)	///// //	7	0.080694
[50,60)	///// //	7	0.061954
[60,70)	///// /	6	0.047566
≥ 70	///	3	0.157254

The hypothesis we want to test is the following:

$$H_0: \text{The data come from an exponential population}$$

$$H_1: \text{The data do not come from an exponential population}$$

Now using the Chi-Square Goodness-of –Fit test we get the following calculations

Chi-Square Goodness-of-Fit Test for Observed Counts in Variable: Frequency_1

Category	Observed	Proportion	Expected	Test Contribution to Chi-Sq
1	2	0.232234	11.6117	7.95618
2	12	0.178302	8.9151	1.06749
3	8	0.136894	6.8447	0.19500
4	5	0.105102	5.2551	0.01239
5	7	0.080694	4.0347	2.17935
6	7	0.061954	3.0977	4.91587
7	6	0.047566	2.3783	5.51510
8	3	0.157254	7.8627	3.00734

Note that 3 categories are with expected value frequencies less than 5. Thus, we merge category 5 with category 6 and category 7 with category 8 and recalculate the expected frequencies. The new results obtained are as shown below.

Chi-Square Goodness-of-Fit Test for Observed Counts in Variable: Frequency_1

Category	Observed	Proportion	Expected	Test Contribution to Chi-Sq
1	2	0.232234	11.6117	7.95618
2	12	0.178302	8.9151	1.06749
3	8	0.136894	6.8447	0.19500
4	5	0.105102	5.2551	0.01239
5	14	0.142648	7.1324	6.61263
6	9	0.204820	10.2410	0.15038

N	DF	Chi-Sq	P-Value
50	4	15.9941	0.003

Since the p-value is less than 5% the level of significance, we reject the null hypothesis that the data come from an exponential distribution.

Note: In this case the computer does take into consideration whether or not we have estimated the mean and the standard deviation. Since we estimated the parameter of the exponential distribution without using MINITAB, the MINITAB output will give the degrees of freedom equal to # categories $- 1 = 6 - 1 = 5$, whereas the actual degrees of freedom is equal to # categories - # of parameters estimated $- 1 = 6 - 1 - 1 = 4$. So we have to enter manually the correct degrees of freedom and find the correct p-value.

21. Here we are interested in testing whether the quality of memory cards purchased from three manufacturers (M_1, M_2, M_3) have the same quality. In other words, the memory cards from the three manufacturers (M_1, M_2, M_3) are homogeneous with respect to defectives and non-defectives.

H_0: The populations of memory cards manufactured by manufacturers (M_1, M_2, M_3) are homogeneous

H_1: The populations of memory cards manufactured by manufacturers (M_1, M_2, M_3) are not homogeneous

We now proceed to do this problem using MINITAB. Enter the two categories in columns C1, C2 and Frequencies in column C3. Then

- Select **Stat** > **Tables** > **Cross Tabulation and Chi –Square**
- In the dialog box that appears, enter Rows, Columns, and Frequencies in boxes next to them and click **OK**. The MINITAB output that appears in the session window is shown here.

Tabulated statistics: Quality, Manufacturer

```
Using frequencies in Frequency
Rows: Quality Columns: Manufacturer
```

	M1	M2	M3	All
Defective	32	52	28	112
	37.3	37.3	37.3	112.9
Non-defective	468	448	472	1388
	462.7	462.7	462.7	1388.1
All	500	500	500	1500
	500	500	500	1500

```
Cell Contents: Count
               Expected count
```

Pearson Chi-Square $= 9.572$, DF $= 2$, p-Value $= 0.008$. Since the p-value is less than 1%, the level of significance, we reject the null hypothesis that the quality of memory cards manufactured by the three manufacturers (M_1, M_2, M_3) is the same.

23. In this problem we want to test the hypothesis that the effectiveness of a medication is independent of its form, namely, tablet, suspension, or injection. That is,

H_0: The effectiveness of a medication is independent of its form, namely, tablet, suspension, or injection

H_1: The effectiveness of a medication is not independent of its form, namely, tablet, suspension, or injection

We now proceed to do this problem using MINITAB. Enter the two categories in columns C1, C2 and Frequencies in column C3. Then

- Select **Stat** > **Tables** > **Cross Tabulation and Chi –Square**
- In the dialog box that appears, enter Rows, Columns, and Frequencies in boxes next to them and click **OK**. The MINITAB output that appears in the session window is shown here.

Tabulated statistics: Effectiveness, Medication

```
Using frequencies in Frequency
Rows: Effectiveness Columns: Medication

            Injection   Suspension    Tablet       All
Average          22          20          32         74
                24.15       22.59       27.26      74.00
High             28          23          20         71
                23.17       21.67       26.16      71.00
Low              12          15          18         45
                14.68       13.74       16.58      45.00
All              62          58          70        190
                62.00       58.00       70.00     190.00


Cell Contents: Count
               Expected count
```

Pearson Chi-Square = 4.578, DF = 4, p-Value = 0.333. Since the p-value is greater than 1%, the level of significance, we do not reject the null hypothesis that the effectiveness of a medication is independent of its from.

Note: Sometimes MINITAB gives output with columns or rows in a different order (alphabetic order) than in the given data, but the cell entries are then correctly done since MINITAB usually lists categories in alphabetic order.

14

NONPARAMETRIC TESTS

PRACTICE PROBLEMS FOR SECTION 14.2

1. In this problem two tests are applied to determine the hardness of a metal which is used in SUV bumpers. We wish to test a hypothesis that the two tests produce equivalent results.

   This problem can be done by using one of the statistical packages discussed in this book. We test the proposed hypothesis in this problem using MINITAB by taking the following steps.

   1. Enter the paired data in two columns C1 and C2, that is, Test I values in column C1 and Test II values in column C2. Then create another column of differences, that is, $C3 = C1 - C2$.

   2. Select **Stat > Nonparametric > 1-sample sign.**

   3. Enter C3 in the box below Variables, and select **Test median** and enter 0, the specified value of the median under the null hypothesis (in this case it is zero). Then select the appropriate **Alternative** hypothesis (in this problem not equal) and click **OK**. The hypothesis we wish to test is the following:

   $$H_0 : M = 0 \quad Vs \quad H_1 : M \neq 0$$

   The output that appears in the session window is given below.

   Sign Test for Median: Differences

   ```
   Sign test of median = 0.00000 versus not = 0.00000

                   N     Below    Equal    Above       P      Median
   Differences:   10       4        0        6     0.7539     2.000
   ```

   Since the p-value is greater than 5% the level of significance, we do not reject the null hypothesis. That is, the median hardness of the metal by both tests is the same.

3. In this problem we wish to test a hypothesis that the median placement score of candidates who have applied for admission at a public university is 115 versus the hypothesis that the median score is greater than 115. That is,

   $$H_0 : M = 115 \quad Vs \quad H_1 : M > 115$$

Solutions Manual to Accompany Statistics and Probability with Applications for Engineers and Scientists, Bhisham C. Gupta and Irwin Guttman.
© 2014 John Wiley & Sons, Inc. Published 2014 by John Wiley & Sons, Inc.

This problem can be done by using one of the statistical packages discussed in this book. We test the proposed hypothesis in this problem using MINITAB by taking the following steps.

1. Enter the data in Column C1.

2. Select **Stat > Nonparametric > 1-sample sign.**

3. Enter C1 in the box below Variables, and select **Test median** and enter 115, the specified value of the median under the null hypothesis. Then select the appropriate **Alternative** hypothesis (in this problem greater than) and click **OK.** The output that appears in the session window is given below.

Sign Test for Median: Scores of Placement Test

```
Sign test of median = 115.0 versus > 115.0

                               N     Below    Equal    Above       P      Median
Scores of Placement Test:     20      11        2        7     0.8811     112.5
```

Since the p-value is greater than 5% the level of significance, we do not reject the null hypothesis. That is, the median score of the placement test for all the applicants is equal to 115.

5. In this problem we wish to test a hypothesis that the median placement score of candidates who have applied for admission at a public university is 115 versus the hypothesis that the median score is greater than 115. That is,

$$H_0 : M = 115 \quad Vs \quad H_1 : M > 115$$

Again, this problem can be done by using one of the statistical packages discussed in this book. We test the proposed hypothesis in this problem using MINITAB by taking the following steps.

1. Enter the data in Column C1.

2. Select **Stat > Nonparametric > 1-sample Wilcoxon.**

3. Enter C1 in the box below Variables, and select **Test median** and enter 115, the specified value of the median under the null hypothesis. Then select the appropriate **Alternative** hypothesis (in this problem greater than) and click **OK.** The output that appears in the session window is given below.

Wilcoxon Signed Rank Test: Placement Scores

```
Test of median = 115.0 versus median > 115.0

             N for      Wilcoxon                 Estimated
     N       Test       Statistic        P        Median
    20        18          51.0         0.936       112.3
```

Since the p-value is greater than 5% the level of significance, we do not reject the null hypothesis. That is, the median score of the placement test for all the applicants is equal to 115.

7. In this problem we wish to test a hypothesis that the median cholesterol level of American males between the ages of 30 and 40 years is 150 mg/dl versus the median cholesterol level is not equal to 150 mg/dl.

$$H_0 : M = 150 \quad Vs \quad H_1 : M \neq 150$$

Again, this problem can be done by using one of the statistical packages discussed in this book. We test the proposed hypothesis in this problem using MINITAB by taking the following steps.

1. Enter the data in Column C1.

2. Select **Stat > Nonparametric > 1-sample sign.**

3. Enter C1 in the box below Variables, and select **Test median** and enter 150, the specified value of the median under the null hypothesis. Then select the appropriate **Alternative** hypothesis (in this problem not equal) and click **OK.** The output that appears in the session window is given below.

Sign Test for Median: C1

```
Sign test of median = 150.0 versus not = 150.0

     N     Below    Equal    Above      P       Median
    20      12        0        8     0.5034     144.5
```

Since the p-value is greater than 5% the level of significance, we do not reject the null hypothesis. That is, the median cholesterol level of American males in that age group is 150 mg/dL.

9. In this problem we wish to test a hypothesis that an asthma treatment drug is ineffective versus effective. In other words, we wish to test a hypothesis that the measured forced vital capacity is the same after the administration of the asthma treatment drug versus the measured forced vital capacity is higher after the administration of the asthma treatment drug.

This problem can be done by using one of the statistical packages discussed in this book. We test the proposed hypothesis in this problem using MINITAB by taking the following steps.

1. Enter the paired data in two columns C1 and C2, that is, X-values in column C1 and X' – values in column C2. Then create another column of differences, i.e., C3 = C1 − C2.

2. Select **Stat** > **Nonparametric** > **1-sample Wilcoxon.**

3. Enter C3 in the box below Variables, and select **Test median** and enter 0, the specified value of the median under the null hypothesis. Then select the appropriate **Alternative** hypothesis (in this problem less than) and click **OK**. The output that appears in the session window is given below.

$$H_0 : M = 0 \quad Vs \quad H_1 : M < 0$$

Wilcoxon Signed Rank Test: Before-After

```
Test of median = 0.000000 versus median < 0.000000

                      N for    Wilcoxon                Estimated
               N      Test     Statistic       P        Median
Before-After   8        8         5.0        0.040      -406.0
```

Since the p-value is less than 5% the level of significance, we reject the null hypothesis. That is, the measured forced vital capacity is higher after the administration of the asthma treatment drug.

PRACTICE PROBLEMS FOR SECTION 14.3

1. In this problem we have two independent samples of measurements on thickness of coated film (coded data) on chips provided by two suppliers. We assume that the populations of measurements on thickness of coated film (coded data) on all chips provided by the two suppliers have continuous c.d.f $F_1(x)$ and $F_2(x)$, respectively. We wish to test a hypothesis that

$$H_0 : F_1(x) \equiv F_2(x)$$

$$H_1 : \text{The two c.d.f.'s are not identical}$$

This problem can be done by using one of the statistical packages discussed in this book. We test the proposed hypothesis in this problem using MINITAB by taking the following steps.

1. Enter the data from two samples in columns C1 and C2, respectively.

2. Select **Stat** > **Nonparametric** > **Mann-Whitney.**

3. Enter C1 and C2 in the boxes next to **First sample** and **Second Sample**, respectively. Then select the **confidence level**, that is, $1 - \alpha$, the appropriate **Alternative** hypothesis (in this problem not equal), and click **OK**. The output that appears in the session window is given below.

Mann-Whitney Test and CI: Supplier I, Supplier II

```
               N      Median
Supplier I     8       27.50
Supplier II    8       22.50

Point estimate for ETA1-ETA2 is 5.00
95.9 Percent CI for ETA1-ETA2 is (-1.00,12.00)
W = 83.0
```

```
Test of ETA1 = ETA2 versus ETA1 not = ETA2 is significant at 0.1278
The test is significant at 0.1250 (adjusted for ties)
```

Since the *p*-value is 0.1250 which is greater than 5% the level of significance, we do not reject the null hypothesis that the two c.d.f. are identical.

3. Here we have two sets of measurements of time taken to complete a project by two independent groups of technicians from two different shifts. We assume that the populations of measurements of time taken to complete a project by technicians from two different shifts have continuous c.d.f $F_1(x)$ and $F_2(x)$, respectively. We wish to test a hypothesis that

$$H_0 : F_1(x) \equiv F_2(x)$$

Against

$$H_1 : \text{The two c.d.f.'s are not identical}$$

This problem can be done by using one of the statistical packages discussed in this book. We test the proposed hypothesis in this problem using MINITAB by taking the following steps.

1. Enter the data from two samples in columns C1 and C2, respectively.
2. Select **Stat > Nonparametric > Mann-Whitney.**
3. Enter C1 and C2 in the boxes next to **First sample** and **Second Sample**, respectively. Then select the **confidence level**, that is, $1 - \alpha$, the appropriate **Alternative** hypothesis (in this problem not equal), and click **OK**. The output that appears in the session window is given below.

Mann-Whitney Test and CI: Technicians - Shift I, Technicians-shift II

```
                           N        Median
Technicians - Shift I     10        25.000
Technicians - Shift II    10        21.000

Point estimate for ETA1-ETA2 is 3.500
95.5 Percent CI for ETA1-ETA2 is (1.000,5.999)
W = 135.5
Test of ETA1 = ETA2 versus ETA1 not = ETA2 is significant at 0.0233
The test is significant at 0.0223 (adjusted for ties)
```

Since the *p*-value is 0.0223 which is less than 5% the level of significance, we reject the null hypothesis that the two c.d.f.'s are identical.

5. Here we have two sets of measurements of drying times of two brands of oil-based paint. We assume that the populations of measurements of drying times have continuous c.d.f $F_1(x)$ and $F_2(x)$, respectively. We wish to test a hypothesis that

$$H_0 : F_1(x) \equiv F_2(x)$$

$$H_1 : \text{The two c.d.f.'s are not identical}$$

This problem can be done by using one of the statistical packages discussed in this book. We test the proposed hypothesis in this problem using MINITAB by taking the following steps.

1. Enter the data from two samples in columns C1 and column C2, respectively.
2. Select **Stat > Nonparametric > Mann-Whitney.**
3. Enter C1 and C2 in the boxes next to **First sample** and **Second Sample**, respectively. Then select the **confidence level**, that is, $1 - \alpha$, the appropriate **Alternative** hypothesis (in this problem not equal), and click **OK**. The output that appears in the session window is given below.

Mann-Whitney Test and CI: Brand 1, Brand 2

```
                N        Median
Brand   1       8         9.000
Brand   2       8         9.100
```

```
Point estimate for ETA1-ETA2 is -0.400
95.9 Percent CI for ETA1-ETA2 is (-1.200, 0.500)
W = 59.5
Test of ETA1 = ETA2 vs ETA1 not = ETA2 is significant at 0.4008
The test is significant at 0.3984 (adjusted for ties)
```

Since the p-value is 0.3984 which is greater than 5% the level of significance, we do not reject the null hypothesis that the two c.d.f.'s are identical.

PRACTICE PROBLEMS FOR SECTION 14.4

1. In this problem we have $n_1 = $ number of wins $= 21$, $n_2 = $ number of losses $= 19$, and the number of runs $= 19$. The number of games is even ($40 = 2n$, or $n = 20$). Since n is large, it follows that U, the number of runs, is approximately normally distributed with mean, variance, and standard deviation equal to

$$\mu_u = n + 1 = 20 + 1 = 21$$

$$\sigma_u^2 = \frac{n(n-1)}{2n-1} = \frac{20(20-1)}{2 \times 20 - 1} = 9.7436$$

$$\sigma_u \cong 3.1215$$

respectively.

   A two-tailed test for the hypothesis of randomness at the 5% level of significance is given by the following rule: Reject the hypothesis if $|U - 21| \geq (3.1215)(1.96) = 6.1180$; otherwise do not reject the hypothesis.

   In this case we have $|U - 21| = |19 - 21| = 2$, so we do not reject the hypothesis and conclude that the sequence is random, i.e., does not exhibit significant non-randomness.

3. In this problem we have $n_1 = $ number of nonconforming $= 11$, $n_2 = $ number of conforming $= 34$, and the number of runs $= 17$. The number of games is odd ($45 = 2n + 1$, or $n = 22$). Since n is large, it follows that U, the number of runs, is approximately normally distributed with mean, variance and standard deviation equal to

$$\mu_u = n + 1 = 22 + 1 = 23$$

$$\sigma_u^2 = \frac{n(n-1)}{2n-1} = \frac{22(22-1)}{2 \times 22 - 1} = 10.74$$

$$\sigma_u \cong 3.277$$

respectively.

   A two-tailed test for the hypothesis of randomness at the 5% level of significance is given by the following rule: Reject the hypothesis if $|U - 23| \geq (3.277)(1.96) \cong 6.423$; otherwise do not reject the hypothesis.

   Since in this case we have $|U - 23| = |17 - 23| = 6$, we do not reject the hypothesis and conclude that the sequence is random, i.e., does not exhibit significant non-randomness.

5. In this problem we have $n_1 = $ number of males $= 13$, $n_2 = $ number of females $= 12$, and the number of runs $= 13$. Since the number of games is even ($25 = 2n + 1$, or $n = 12$). Since n is large, it follows that U, the number of runs, is approximately normally distributed with mean, variance, and standard deviation equal to

$$\mu_u = n + 1 = 12 + 1 = 13$$

$$\sigma_u^2 = \frac{n(n-1)}{2n-1} = \frac{12(12-1)}{2 \times 12 - 1} = \frac{132}{23} = 5.739$$

$$\sigma_u \cong 2.3956$$

respectively.

   A two-tailed test for the hypothesis of randomness at the 5% level of significance is given by the following rule: Reject the hypothesis if $|U - 13| \geq (2.3956)(1.96) = 4.6954$; otherwise do not reject the hypothesis.

Since in this case we have $|U - 13| = |13 - 13| = 0$, we do not reject the hypothesis and conclude that the sequence is random, i.e., does not exhibit significant non-randomness.

7. In this problem we wish to test the following hypothesis:

H_0 : Two samples come from populations having identical c.d.f.'s.

H_1 : Two samples do not come from populations having identical c.d.f.'s

Here, we first pool the two samples and arrange the observations in order of magnitude (descending or ascending) as shown below. The underlined observations are residency interviews granted to medical students from school A. Then we replace each observation by a 0 or 1, based upon the observation from school A or school B. The sequence for the given data, for example, is

7	7	7	7	8	8	8	8	9	9	9	9	9
9	9	9	9	9	9	9	10	10	10			

The sequence above after replacing the observations with 0 and 1 is

1 1 1 1 1 1 0 0 1 1 1 1 1 1 0 0 0 0 0 0 0 0 0

which gives $U = 4$. Since $n_1 = 11$ and $n_2 = 12$ are reasonably large, the distribution of U is approximately normal with mean and variance equal to

$$\mu_u = \frac{2n_1 n_2}{n_1 + n_2} + 1$$

$$\sigma_u^2 = \frac{2n_1 n_2(2n_1 n_2 - n_1 - n_2)}{(n_1 + n_2)^2(n_1 + n_2 - 1)}$$

Now substituting the vales of n_1 and n_2 we obtain

$$\mu_u = \frac{2(11)(12)}{11 + 12} + 1 = 12.478$$

$$\sigma_u^2 = \frac{2(11)(12)(2(11)(12) - 11 - 12)}{(11 + 12)^2(11 + 12 - 1)} = 5.467$$

$$\sigma_u = 2.338$$

Now, we reject the null hypothesis using the normal approximation at the 5% level of significance if

$$|U - 12.478| \geq 2.338(1.96) \cong 4.58248.$$

Otherwise, do not reject the null hypothesis. Here, the observed value of U is 4 so that

$$|4 - 12.478| = 8.478 > 4.58248.$$

Hence, we reject the null hypothesis that the two samples come from populations having identical c.d.f.'s.

PRACTICE PROBLEMS FOR SECTION 14.5

1. Let $(X_1, X_1'), (X_2, X_2'), \cdots, (X_{10}, X_{10}')$ be the scores of ten randomly selected students in their final exam of DE and QM. Each pair $(X_i, X_i'), i = 1, 2, \cdots, 10$, represents a pair of scores in the final exam of each student. Now, by arranging the scores in each subject in ascending order and ranking them, we obtain the results shown below. Note that if more than one student had scored the same points, then each score was assigned an average rank, that is, ties were broken in the usual manner. In this example, the hypothesis that we wish to test is

$H_0 : X$ and X' are not correlated

$H_1 : X$ and X' are positively or negatively correlated

X_i	X_i'	$R(X_i)$	$R(X_i')$	d_i	d_i^2
87	82	2	3.5	−1.5	2.25
92	91	4	9.5	−5.5	30.25
93	81	5	1.5	3.5	12.25
94	88	6	7.5	−1.5	2.25
97	83	9	5.0	4.0	16.00
86	86	1	6.0	−5.0	25.00
98	82	10	3.5	6.5	42.25
95	88	7	7.5	−0.5	0.25
88	91	3	9.5	−6.5	42.25
96	81	8	1.5	6.5	42.25

$$\sum_{i=1}^{10} d_i^2 = 2.25 + 30.25 + \cdots + 42.25 = 215$$

Thus, the observed value of the test statistic is given by

$$|r_s| = \left| 1 - \frac{6(215)}{10(100 - 1)} \right| = |-0.3030| = 0.3030$$

From Table XVI of appendix A we find the critical value of the test statistics r_s for $n = 10$ and a two-sided test at the 5% level of significance is 0.6364 which is greater than the observed value of r_s. Hence, we do not reject the null hypothesis that the scores obtained by students in DE and QM are independent.

3. Let $(X_1, X_1'), (X_2, X_2'), \cdots, (X_{12}, X_{12}')$ be the heights and scores of 12 randomly selected basketball players. Each pair $(X_i, X_i'), i = 1, 2, \cdots, 12$, represents the height and scores in a game of the same player. Now, by arranging the heights and scores in ascending order and ranking them, we obtain the results shown below. Note that if more than one player had the same height or had the same score, then each height or each score was assigned an average rank, that is, ties were broken in the usual manner. In this example, the hypothesis that we wish to test is

$H_0 : X$ and X' are not correlated

$H_1 : X$ and X' are positively or negatively correlated

X_i	X_i'	$R(X_i)$	$R(X_i')$	d_i	d_i^2
201	16	9.5	2	7.5	56.25
199	22	8.0	7	1.0	1.00
194	25	3.0	12	−9.0	81.00
188	22	1.0	7	−6.0	36.00
198	21	6.5	5	1.5	2.25
201	23	9.5	10	−0.5	0.25
202	23	11.0	10	1.0	1.00
197	22	5.0	7	−2.0	4.00
192	19	2.0	4	−2.0	4.00
195	23	4.0	10	−6.0	36.00
205	17	12.0	3	9.0	81.00
198	15	6.5	1	5.5	30.25

$$\sum_{i=1}^{10} d_i^2 = 56.25 + 1 + \cdots + 30.25 = 333$$

Thus, the observed value of the test statistic is given by

$$|r_s| = \left|1 - \frac{6(333)}{12(144 - 1)}\right| = |-0.1643| = 0.1643$$

From Table XVI of appendix A we find the critical value of the test statistics r_s for $n = 12$ and a two-sided test at the 5% level of significance is 0.5804 which is greater than the observed value of r_s. Hence, we do not reject the null hypothesis that the scores of players are independent of their heights.

5. Let $(X_1, X_1'), (X_2, X_2'), \cdots, (X_{12}, X_{12}')$ be the IQ and test scores of 12 randomly selected candidates. Each pair $(X_i, X_i'), i = 1, 2, \cdots, 12$, represents the IQ and test scores of the same candidate. Now, by arranging the IQs and scores in ascending order and ranking them, we obtain the results shown below. Note that if more than one candidate had the same IQ or had the same test score, then each IQ or each test score was assigned an average rank, that is, ties were broken in the usual manner. In this example, the hypothesis that we wish to test is

$$H_0 : X \text{ and } X' \text{ are not correlated}$$

$$H_1 : X \text{ and } X' \text{ are positively or negatively correlated}$$

X_i	X_i'	$R(X_i)$	$R(X_i')$	d_i	d_i^2
92	79	2	2.5	−0.5	0.25
117	88	7	9	−2	4
100	79	3	2.5	0.5	0.25
90	86	1	7.5	−6.5	42.25
130	86	11.5	7.5	4	16
108	84	5	5.5	−0.5	0.25
121	77	8	1	7	49
130	95	11.5	10	1.5	2.25
105	82	4	4	0	0
123	95	9	10	−1	1
129	84	10	5.5	4.5	20.25
114	95	6	10	−4	16

$$\sum_{i=1}^{10} d_i^2 = 0.25 + 4 + \cdots + 16 = 151.5$$

Thus, the observed value of the test statistic is given by

$$|r_s| = \left|1 - \frac{6(151.5)}{12(144 - 1)}\right| = 0.4703$$

From Table XVI of appendix A we find the critical value of the test statistics r_s for $n = 12$ and a two-sided test at the 5% level of significance is 0.5804 which is greater than the observed value of r_s. Hence, we do not reject the null hypothesis that the IQs of the candidates are independent of their test scores.

7. Let $(X_1, X_1'), (X_2, X_2'), \cdots, (X_{10}, X_{10}')$ be the systolic blood pressures taken by a doctor and a nurse of 10 randomly selected patients. Each pair $(X_i, X_i'), i = 1, 2, \cdots, 10$, represents the systolic blood pressures taken by a doctor and her nurse. Now, by arranging the systolic blood pressures taken by a doctor and her nurse in ascending order and ranking them, we obtain the results shown below. Note that if more than one patient had the same systolic blood pressure, then each observation was assigned an average rank, that is, ties were broken in the usual manner. In this example, the hypothesis that we wish to test is

$$H_0 : X \text{ and } X' \text{ are not correlated}$$
$$H_1 : X \text{ and } X' \text{ are positively or negatively correlated}$$

X_i	X'_i	$R(X_i)$	$R(X'_i)$	d_i	d_i^2
122	135	2	6.5	−4.5	20.25
130	130	9	4.5	4.5	20.25
115	140	1	10	9	81
130	135	9	6.5	2.5	6.25
124	128	3.5	3	0.5	0.25
125	137	5	9	4	16
130	130	9	4.5	4.5	20.25
129	125	7	2	5	25
126	123	6	1	5	25
124	136	3.5	8	4.5	20.25

$$\sum_{i=1}^{10} d_i^2 = 20.25 + 20.25 + \cdots + 20.25 = 234.5$$

Thus, the observed value of the test statistic is given by

$$|r_s| = \left| 1 - \frac{6(234.5)}{10(100 - 1)} \right| = 0.4212$$

From Table XVI of appendix A we find the critical value of the test statistics r_s for $n = 10$ and a two-sided test at the 5% level of significance is 0.6364 which is greater than the observed value of r_s. Hence, we do not reject the null hypothesis that the systolic blood pressure taken by a doctor is independent of the systolic blood pressure taken by her nurse.

REVIEW PRACTICE PROBLEMS

1. Here we wish to test a hypothesis that median sugar level of patients who take 1000 mg of Metformin (500 mg/twice a day) is 140 mg/DL. That is,

$$H_0 : M = 140 \quad Vs \quad H_1 : M \neq 140$$

This problem can be done by using one of the statistical packages discussed in this book. We test the proposed hypothesis in this problem using MINITAB by taking the following steps.

1. Enter the data in Column C1.
2. Select **Stat > Nonparametric > 1-sample sign.**
3. Enter C1 in the box below Variables, and select **Test median** and enter 140, the specified value of the median under the null hypothesis. Then select the appropriate **Alternative** hypothesis and click **OK** (in this problem not equal). The output that appears in the session window is given below.

Sign Test for Median: Sugar Level

```
Sign test of median = 140.0 versus not = 140.0

                N    Below   Equal   Above    P    Median
Sugar Level    15      8       0       7     1.0    135.0
```

Since the p-value is greater than 5% the level of significance, we do not reject the null hypothesis. That is, the median sugar level of all patients who take 1000 mg of Metformin a day is 140 mg/DL.

3. In this problem we wish to test the following hypothesis

$$H_0 : \text{Coating has no significant effect for corrosion}$$

Versus

$$H_1 : \text{Coating has significant effect for corrosion}$$

$$\text{Let } Y_i = \begin{cases} 1, & \text{if } x_i - x_i' > 0 \\ 0, & \text{otherwise.} \end{cases}$$

Then the values of Y_i are $\{1,0,1,1,1,0,1,1,0,1,1,1,1,1,0,1,1,1,0,1\}$. Note that i^{th} observation is discarded if $x_i = x_i'$.

$$r = \sum_{i=1}^{n} Y_i = 15, n = 20.$$

Using the values from Table 14.2.2 of this chapter, reject H_0 if $r \geq n - r_{\alpha/2}$ or $r \leq r_{\alpha/2}$.

In this problem, at $\alpha = 0.05$ level of significance, $r_{\alpha/2} = 6, n - r_{\alpha/2} = 14$. Since $r = 15$, we reject the null hypothesis and conclude that coating has significant effect for corrosion. In other words, the coated pipes have significantly less corrosion than those uncoated pipes.

5. $m = n = 30, \quad T = 1085, \quad W = mn + \dfrac{m(m+1)}{2} - T = 280.$

Since both m and n are large, we have

$$\frac{W - \frac{mn}{2}}{\sqrt{\dfrac{mn(m+n+1)}{12}}} \sim N(0,1).$$

Hence,

$$Z = \frac{W - \frac{mn}{2}}{\sqrt{\dfrac{mn(m+n+1)}{12}}} = \frac{280 - \frac{(30)(30)}{2}}{\sqrt{\dfrac{(30)(30)(61)}{12}}} = \frac{280 - 450}{\sqrt{4575}} = -2.513.$$

Since $Z = -2.513 < -z_{0.01} = -2.33$, we conclude that W is significantly small at the $\alpha = 0.01$ level.

7. In this example we first determine the median of the given data, which in this case is 462. Then by using the median test, we determine the number of runs above and below the median and determine whether the number of runs are either unusually high or unusually low.

This test can be done by using one of the statistical packages discussed in this book. We test the proposed hypothesis in this problem using MINITAB by taking the following steps.

1. Enter the data in Column C1.

2. Select **Stat** > **Nonparametric** > **Run test**.

3. Enter C1 in the box below Variables, and select **Above and Below** and enter the value of the median that we determined earlier (462). Then click **OK**. The output that appears in the session window is given below.

Runs Test: Sample Means

Runs test for Sample Means

Runs above and below the median (462)

The observed number of runs = 16

The expected number of runs = 13

12 observations above the median and 12 below the median

The p-value is 0.210. Since the p-value is greater than 5% the level of significance, we do not reject the null hypothesis that the sample of sample means is random.

9. $U = 39; 2n = 100, n = 50$;

$$E(U) = n + 1 = 51, \sigma_u^2 = \frac{n(n-1)}{2n-1} = \frac{50(49)}{99} = 24.747, \sigma_u = 4.975.$$

The value of u is significantly low at the 5% level if

$$Z = \frac{u - 51}{4.975} < -1.645$$

Here, we have

$$\frac{U - 51}{4.975} = \frac{39 - 51}{4.975} = -2.412,$$

Thus, we reject the one-tailed null hypothesis of randomness and conclude that the value of U is significantly low at the 5% level.

11. Since, we have $58 = 2n \Rightarrow n = 29$; $x_{(29)} = 4.779, x_{(30)} = 4.781, Median(x) = 4.780$.

If $x_i < 4.780$, write 'b'; if $x_i > 4.780$, write 'a'. Thus we obtain the sequence

$$\underline{aa}\,\underline{b}\,\underline{a}\,\underline{bbb}\,\underline{aa}\,\underline{bbbb}\,\underline{a}\,\underline{b}\,\underline{a}\,\underline{bbb}\,\underline{a}\,\underline{bb}\,\underline{aa}\,\underline{b}\,\underline{aaaa}\,\underline{b}\,\underline{aa}\,\underline{b}\,\underline{a}\,\underline{b}\,\underline{aaa}\,\underline{bbb}\,\underline{a}\,\underline{bb}\,\underline{aa}\,\underline{b}\,\underline{aa}\,\underline{bb}\,\underline{a}\,\underline{b}\,\underline{aa}\,\underline{bb}\,\underline{a}.$$

Here, $U = 33, n = 29, E(U) = 30, \sigma_u^2 = \dfrac{29(28)}{57} = 14.2456,$

and $\sigma_u = 3.774.$

Reject the hypothesis of randomness at the $\alpha = 0.05$ level if

$$|U - 30| \geq (3.774)(1.96) = 7.397.$$

Otherwise, we do not reject the null hypothesis. Here we have

$$|U - 30| = |33 - 30| = 3 < 7.397.$$

Hence, we do not reject the null hypothesis of randomness.

13. Here we let

$$Y = \begin{cases} 1, & \text{if } x > x' \quad (\text{type } A > \text{type } B) \\ 0, & \text{otherwise.} \end{cases}$$

Then the values of Y are $\{0,0,0,0,0,0,0,0,0,1,0,0,1,0\}$. Note that i^{th} observation is discarded if $x_i = x_i'$.

$$r = \sum Y_i = 2, n = 14.$$

Using the values from Table 14.2.2 of this chapter, reject H_0 if $r \geq n - r_{\alpha/2}$ or $r \leq r_{\alpha/2}$.

From Table 14.2.2 of this chapter we obtain at $\alpha = 0.05$, $r_{\alpha/2} = 3$, $n - r_{\alpha/2} = 11$.

Reject the hypothesis of equivalent gauges if $r > 11$ or $r < 3$. Thus, $r = 2$ is significantly small at the $\alpha = 0.05$ level. That is, at the 5% level of significance we reject the null hypothesis that the two gauges are giving similar results.

15. Let $(X_1, X_1'), (X_2, X_2'), \cdots, (X_{12}, X_{12}')$ be the scores of 12 randomly selected students in their final exam of Calculus and Physics. Each pair $(X_i, X_i'), i = 1, 2, \cdots, 12$, represents a pair of scores in the final exam of the same student. Now by

arranging the scores in each subject in ascending order and ranking them, we obtain the results shown below. Note that if more than one student had scored the same point, then each score was assigned an average rank, that is, ties were broken in the usual manner. In this example, the hypothesis that we wish to test is

$$H_0 : X \text{ and } X' \text{ are not correlated}$$

$$H_1 : X \text{ and } X' \text{ are positively or negatively correlated}$$

X_i	X'_i	$R(X_i)$	$R(X'_i)$	d_i	d_i^2
95	91	10.5	8.5	2.0	4.00
96	91	12.0	8.5	3.5	12.25
94	95	8.5	11.5	−3.0	9.00
94	90	8.5	7.0	1.5	2.25
87	82	1.0	1.0	0.0	0.00
88	95	2.0	11.5	−9.5	90.25
89	85	3.0	4.0	−1.0	1.00
90	84	4.0	3.0	1.0	1.00
95	83	10.5	2.0	8.5	72.25
93	87	6.5	5.5	1.0	1.00
92	87	5.0	5.5	−0.5	0.25
93	92	6.5	10.0	−3.5	12.25

$$\sum_{i=1}^{10} d_i^2 = 4 + 12.25 + \cdots + 12.25 = 205.5$$

Thus, the observed value of the test statistic is given by

$$|r_s| = \left| 1 - \frac{6(205.5)}{12(144 - 1)} \right| = |0.2815| = 0.2815$$

From Table XVI of appendix A we find the critical value of the test statistics r_s for $n = 10$ and a two-sided test at the 5% level of significance is 0.5804 which is greater than the observed value of r_s. Hence, we do not reject the null hypothesis that the scores of the students in Calculus and Physics are independent.

15

SIMPLE LINEAR REGRESSION ANALYSIS

Solutions of this Chapter have been prepared using MINITAB.

PRACTICE PROBLEMS FOR SECTION 15.2

1. (a) The scatter plot for the data in this problem is shown below. It does not indicate that a least squares line would provide a good fit.

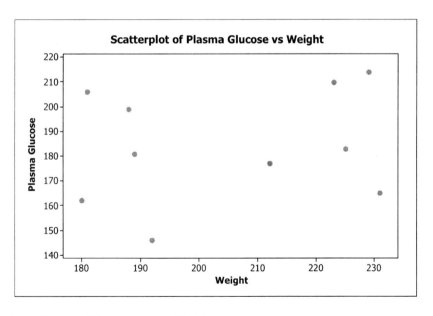

(b) Regression Analysis: Plasma Glucose versus Weight

Solutions Manual to Accompany Statistics and Probability with Applications for Engineers and Scientists, Bhisham C. Gupta and Irwin Guttman.
© 2014 John Wiley & Sons, Inc. Published 2014 by John Wiley & Sons, Inc.

```
The regression equation is
Plasma Glucose = 136 + 0.237 Weight

Predictor      Coef      SE Coef       T         P
Constant     135.70       77.04      1.76     0.116
Weight        0.2371      0.3741     0.63     0.544

S = 23.4806   R-Sq = 4.8%   R-Sq(adj) = 0.0%

Analysis of Variance
Source          DF        SS        MS       F       P
Regression       1      221.4     221.4    0.40   0.544
Residual Error   8     4410.7     551.3
Total            9     4632.1
```

Clearly the above analysis shows that the least squares line does not provide a good fit. In fact, the R-square value is less than 5%.

(c) The least squares estimates of the regression coefficients are: $\hat{\beta}_0 = 135.70 \cong 136, \hat{\beta}_1 = 0.2371$.

3. The scatter plot for the data in this problem is shown below. It does not indicate that a least squares line would provide a good fit.

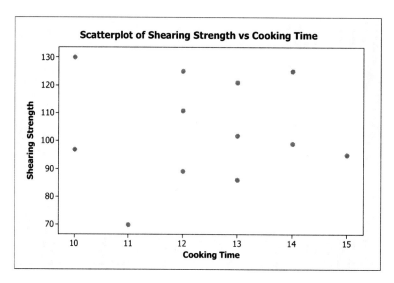

(b) Regression Analysis: Shearing Strength versus Cooking Time

```
The regression equation is
Shearing Strength = 105 - 0.11 Cooking Time

Predictor        Coef     SE Coef       T        P
Constant       105.47      46.74      2.26    0.048
Cooking Time    -0.105      3.737     -0.03    0.978

S = 19.3891    R-Sq = 0.0%   R-Sq(adj) = 0.0%

Analysis of Variance
Source          DF        SS        MS       F       P
Regression       1        0.3       0.3    0.00   0.978
Residual Error  10     3759.4     375.9
Total           11     3759.7
```

Clearly the above analysis shows that the least squares line does not provide a good fit. The R-square value is 0%, and p-value is 0.978. In other words, the above analysis shows that there is relationship between the shearing strength and the cooking time of the pulp.

(c) The least squares estimates of the regression coefficients are: $\hat{\beta}_0 = 105.47, \hat{\beta}_1 = -0.105$.

5. The scatter plot for the data in this problem is shown below. It indicates that a least squares line would provide a good fit.

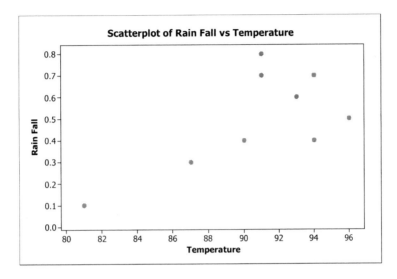

(b) Regression Analysis: Rain Fall versus Temperature

```
The regression equation is
Rain Fall = - 2.53 + 0.0333 Temperature


Predictor        Coef    SE Coef        T       P
Constant       -2.533      1.193    -2.12   0.066
Temperature   0.03333    0.01314     2.54   0.035


S = 0.168325   R-Sq = 44.6%   R-Sq(adj) = 37.7%


Analysis of Variance


Source             DF       SS       MS      F      P
Regression          1  0.18233  0.18233   6.44  0.035
Residual Error      8  0.22667  0.02833
Total               9  0.40900


Unusual Observations


Obs.  Temperature  Rain Fall    Fit    SE Fit   Residual  St Resid
  8          81.0     0.1000  0.1667   0.1381   -0.0667     -0.69 X


X denotes an observation whose X value gives it large leverage.
```

Clearly, the above analysis shows that the least squares line is a good fit. The p-value is .035, which is less than 5%. The R-square value is about 45% which indicates that there are factors other than temperature which may cause rain fall. Also, note that observation 8 is unusually low.

(c) The least squares estimates of the regression coefficients are: $\hat{\beta}_0 = -2.533, \hat{\beta}_1 = 0.0333$.

7. (a) The scatter plot for the data in this problem is shown below. It does not indicate that a least squares line would provide a good fit.

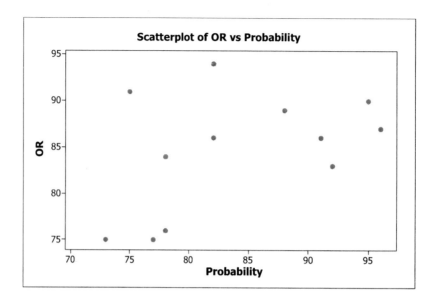

(b) Regression Analysis: OR versus Probability

```
The regression equation is
OR = 55.5 + 0.347 Probability

Predictor      Coef     SE Coef       T       P
Constant      55.53       18.81    2.95   0.014
Probability  0.3472      0.2232    1.56   0.151

S = 6.01079   R-Sq = 19.5%   R-Sq(adj) = 11.4%

Analysis of Variance

Source           DF      SS       MS      F       P
Regression        1    87.37    87.37   2.42   0.151
Residual Error   10   361.30    36.13
Total            11   448.67
```

(b) The above analysis shows that the least squares line is not a good fit. The p-value of .151 is greater than 5%, so that we do not reject the null hypothesis $H_0 : \beta_1 = 0$. The R-square value is under 20%.

(c) The least square estimates of the regression coefficients are: $\hat{\beta}_0 = 55.53, \hat{\beta}_1 = 0.3472$.

9. (a) The scatter plot for the data in this problem is shown below. It indicates that a least square line would provide a good fit.

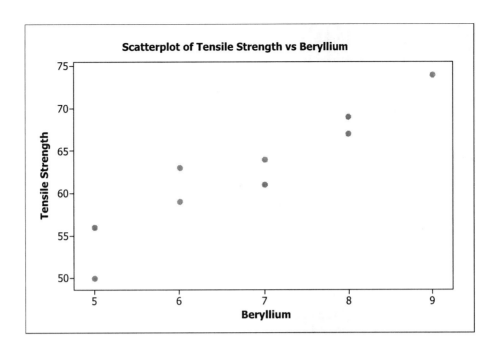

Scatterplot of Tensile Strength vs Beryllium

(b) Regression Analysis: Tensile Strength versus Beryllium

```
The regression equation is
Tensile Strength = 29.8 + 4.83 Beryllium

Predictor        Coef     SE Coef        T        P
Constant       29.829      4.562      6.54    0.000
Beryllium      4.8286     0.6608      7.31    0.000

S = 2.60612    R-Sq = 88.4%    R-Sq(adj) = 86.8%

Analysis of Variance
Source            DF          SS        MS       F       P
Regression         1      362.68    362.68   53.40   0.000
Residual Error     7       47.54      6.79
Total              8      410.22
```

Clearly the above analysis shows that the least squares line provides a very good fit. The p-value is 0, so that we reject the null hypothesis $H_0 : \beta_1 = 0$ at any level of significance. The R-square value is over 88% which indicates that Beryllium is a very important factor for increasing the tensile strength of copper wire.

(c) The least square estimates of the regression coefficients are: $\hat{\beta}_0 = 29.829$, $\hat{\beta}_1 = 4.8286$.

11. (a) The scatter plot for the data in this problem is shown below

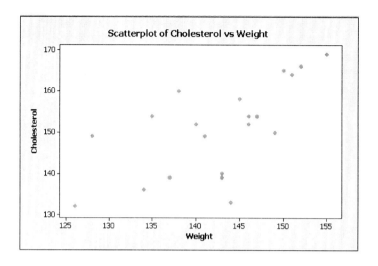

Using MINITAB (**Stat** > **Regression** > **Regression**), we obtain

(b) Regression Analysis: Cholesterol versus Weight

```
The regression equation is
Cholesterol = 18.8 + 0.926 Weight


Predictor     Coef       SE Coef        T       P
Constant      18.80       37.54      0.50    0.623
Weight        0.9259      0.2631     3.52    0.002


S = 8.88661 R-Sq = 40.8% R-Sq(adj) = 37.5%


Analysis of Variance

Source          DF         SS         MS        F       P
Regression       1      978.26     978.26    12.39    0.002
Residual Error  18     1421.49      78.97
Total           19     2399.75


Unusual Observations


Obs.  Weight  Cholesterol   Fit    SE Fit  Residual  St Resid
 10      144     133.00    152.14   2.03    -19.14    -2.21R


R denotes an observation with a large standardized residual.


Predicted Values for New Observations at X = 144.5
New
Obs.    Fit     SE Fit       90% CI            90% PI
  1   152.60     2.06    (149.04, 156.17)  (136.79, 168.42)
```
(We use this result later in Section 11.3)

(b) The fitted regression model is: $\hat{Y} = 18.8 + 0.9259\,X$. The above analysis shows that the least squares line provides a good fit. The p-value of .002 is less than 5%. However, the R-square value is under 41% which indicates that there are factors other than weight which effect the Cholesterol level.

(c) The least square estimates of the regression coefficients are: $\hat{\beta}_0 = 18.80$, $\hat{\beta}_1 = 0.9259$.

PRACTICE PROBLEMS FOR SECTIONS 15.3 AND 15.4

1. The MINITAB output for Problem 12 of Section 15.2 is given here, augmented with information about Y when $X = 3.50$.

Regression Analysis: Y versus X

```
The regression equation is
Y = 9.27 + 1.44 X

Predictor       Coef     SE Coef      T        P
Constant      9.2727     0.4632    20.02    0.000
X             1.4364     0.1465     9.81    0.000

S = 1.53610   R-Sq = 91.4%   R-Sq(adj) = 90.5%

Analysis of Variance

Source           DF      SS       MS      F       P
Regression        1   226.95   226.95  96.18  0.000
Residual Error    9    21.24     2.36
Total            10   248.18

Predicted Values for New Observations

New Obs.    Fit      SE Fit        95% CI             95% PI
    1    14.300     0.691    (12.737, 15.863)   (10.490, 18.110)

Values of Predictors for New Observations

New Obs.        X
    1         3.50
```

From the above MINITAB output, we obtain

(a) $\hat{\sigma}^2 = \text{MSE} = 2.36$

(b) $\hat{\sigma}_{\hat{\beta}_0} = 0.4632$, $\hat{\sigma}_{\hat{\beta}_1} = 0.1465$. Hence the 95% confidence intervals for β_0 *and* β_1 are given by

$$\hat{\beta}_0 \pm t_{n-2;\alpha/2}\hat{\sigma}_{\hat{\beta}_0} = (9.2727 \pm 2.262 \times (0.4632)) = (8.2249, 10.3205), \text{and}$$

$$\hat{\beta}_1 \pm t_{n-2;\alpha/2}\hat{\sigma}_{\hat{\beta}_1} = (1.4364 \pm 2.262(0.1465)) = (1.1050, 1.7678)$$

(c) To find a 95% confidence interval for $E(Y|X = 3.5)$, we first have find $\hat{Y}$ at $X = 3.5$, which is given by

$$\hat{Y} = 9.27 + 1.44(3.5) = 14.31$$

Now from equation (15.4.14) of this chapter, the 95% confidence interval for $E(Y|X = 3.5)$ is given by

$$\left(\hat{Y} \pm t_{n-2;\alpha/2}S\sqrt{\left(\frac{1}{n} + \frac{(X - \overline{X})^2}{S(X^2)}\right)}\right), \text{where } S = \sqrt{MSE} = \sqrt{\hat{\sigma}^2},$$

that is

$$\left(14.31 \pm 2.262(1.5362)\sqrt{\left(\frac{1}{11} + \frac{(3.5 - 0)^2}{110}\right)}\right) = (12.737, 15.863).$$

(d) Now from equation (15.4.21a) of this chapter, the 95% prediction interval for Y is given by

$$\left(\hat{Y} \pm t_{n-2;\alpha/2} S \sqrt{\left(1 + \frac{1}{n} + \frac{(X - \overline{X})^2}{S(X^2)} \right)} \right)$$

that is

$$\left(14.31 \pm 2.262(1.5362) \sqrt{\left(1 + \frac{1}{11} + \frac{(3.5 - 0)^2}{110} \right)} \right) = (10.490, 18.110).$$

Note that these intervals are the same as found by using MINITAB (**Stat** > **Regression** > **Regression** > **Prediction intervals for new observation**).

(e) Comparing the confidence interval in part (c) with the prediction interval in part (d) we clearly see that the confidence interval is much smaller than the prediction interval. Indeed, it is always the case that a confidence interval for the expected value of a response variable is always narrower than that of the prediction interval with same confidence coefficient for an individual response.

3. The MINITAB output for Problem 11 of Section 15.2 is given here, augmented with the information about Y when X = 144.5:

Regression Analysis: Cholesterol Level versus Weight

```
The regression equation is
Cholesterol Level = 18.8 + 0.926 Weight
```

Predictor	Coef	SE Coef	T	P
Constant	18.80	37.54	0.50	0.623
Weight	0.9259	0.2631	3.52	0.002

```
S = 8.88661    R-Sq = 40.8%   R-Sq(adj) = 37.5%
```

Analysis of Variance

Source	DF	SS	MS	F	P
Regression	1	978.26	978.26	12.39	0.002
Residual Error	18	1421.49	78.97		
Total	19	2399.75			

Unusual Observations

Obs.	Weight	Cholesterol Level	Fit	SE Fit	Residual	St Resid
10	144	133.00	152.14	2.03	-19.14	-2.21R

R denotes an observation with a large standardized residual.

Predicted Values for New Observations

New Obs.	Fit	SE Fit	95% CI	95% PI
1	152.60	2.06	(148.28, 156.92)	(133.44, 171.76)

Values of Predictors for New Observations when X =144.5

New Obs.	Weight
1	144.5

From the above MINITAB output we obtain

(a) $\hat{\sigma}^2 = \text{MSE} = 78.97$

(b) $\hat{\sigma}_{\hat{\beta}_0} = 37.54$, $\hat{\sigma}_{\hat{\beta}_1} = 0.2631$. Hence the 95% confidence intervals for β_0 *and* β_1 are given by

$$\hat{\beta}_0 \pm t_{n-2;\alpha/2}\hat{\sigma}_{\hat{\beta}_0} = (18.80 \pm 2.101 \times 37.54) = (-60.0715, 97.6715), \text{ and}$$

$$\hat{\beta}_1 \pm t_{n-2;\alpha/2}\hat{\sigma}_{\hat{\beta}_1} = (0.9259 \pm 2.101 \times 0.2631) = (0.3731, 1.4787)$$

respectively.

(c) From the MINITAB output, we have that the 95% confidence interval for the true mean of Y when $X = 144.5$ is: (148.28, 156.92).

(d) The 95% prediction interval for Y when $X = 144.5$ is: (133.44, 171.76).

(e) Comparing the confidence interval in part (c) with the prediction interval in part (d) we clearly see that the confidence interval with the same confidence coefficient is much smaller than the prediction interval. This is always the case - the confidence interval for the expected value of response variable is always narrower than that of the prediction interval of an individual response.

5. The MINITAB output for Problem 2 of Section 15.2 is given here, augmented with the information about Y when $X = 35.0$:

```
The regression equation is
Yield = 85.6 - 0.079 Reaction Time


Predictor        Coef      SE Coef       T         P
Constant         85.62      11.66      7.34     0.000
Reaction Time   -0.0786      0.3145   -0.25     0.808

S = 5.35815 R-Sq = 0.7% R-Sq(adj) = 0.0%


Analysis of Variance
Source            DF        SS        MS        F        P
Regression         1       1.79      1.79     0.06     0.808
Residual Error     9     258.39     28.71
Total             10     260.18


Predicted Values for New Observations when X = 35.0


New Obs.    Fit      SE Fit        95% CI            95% PI
    1      82.86      1.70     (79.01, 86.72)    (70.14, 95.58)


Values of Predictors for New Observations
               Reaction
New Obs          Time
    1            35.0
```

From the above MINITAB output we obtain

(a) $\hat{\sigma}^2 = \text{MSE} = 28.71$

(b) $\hat{\sigma}_{\hat{\beta}_0} = 11.66$, $\hat{\sigma}_{\hat{\beta}_1} = 0.3145$. Hence, the 95% confidence intervals for β_0 *and* β_1 are given by

$$\hat{\beta}_0 \pm t_{n-2;\alpha/2}\hat{\sigma}_{\hat{\beta}_0} = (85.62 \pm 2.262 \times 11.66) = (59.2451, 111.9949), \text{ and}$$

$$\hat{\beta}_1 \pm t_{n-2;\alpha/2}\hat{\sigma}_{\hat{\beta}_1} = (-0.0786 \pm 2.262 \times 0.3145) = (-0.7899, 0.6328)$$

respectively.

(c) From the MINITAB output, we have that the 95% confidence interval for the true mean of the chemical yield when $X = 35$ is: (79.01, 86.72).

(d) The 95% prediction interval for the chemical yield when $X = 35$ is: (70.14, 95.58).

7. The MINITAB output for Problem 4 of Section 15.2 is given here, augmented with the information about Y when $X = 32.0$:

Regression Analysis: Hypertension versus Noise Level

```
The regression equation is
Hypertension = 45.1 + 1.06 Noise Level

Predictor       Coef     SE Coef        T         P
Constant       45.06       12.17     3.70     0.003
Noise Level   1.0612      0.4132     2.57     0.026

S = 6.31604    R-Sq = 37.5%    R-Sq(adj) = 31.8%

Analysis of Variance

Source           DF         SS       MS        F        P
Regression        1     263.18   263.18     6.60    0.026
Residual Error   11     438.82    39.89
Total            12     702.00

Unusual Observations
        Noise
Obs    Level    Hypertension    Fit    SE Fit    Residual    St Resid
  2     33.0           68.00   80.08      2.37      -12.08      -2.06R

R denotes an observation with a large standardized residual.
Predicted Values for New Observations when x =32.0

New Obs      Fit     SE Fit        95% CI             95% PI
      1    79.02       2.11   (74.38, 83.66)    (64.36, 93.68)

Values of Predictors for New Observations

               Noise
New Obs        Level
      1         32.0
```

From the above MINITAB output, we obtain

(a) $\hat{\sigma}^2 = \text{MSE} = 39.89$

(b) $\hat{\sigma}_{\hat{\beta}_0} = 12.17$, $\hat{\sigma}_{\hat{\beta}_1} = 0.4132$. Hence, the 99% confidence intervals for β_0 and β_1 are given by

$$\hat{\beta}_0 \pm t_{n-2;\alpha/2}\hat{\sigma}_{\hat{\beta}_0} = (45.06 \pm 3.106 \times 12.17) = (7.26, 82.86), \text{and}$$

$$\hat{\beta}_1 \pm t_{n-2;\alpha/2}\hat{\sigma}_{\hat{\beta}_1} = (1.0612 \pm 3.106 \times 0.4132) = (-0.2222, 2.3446)$$

respectively.

(c) From the MINITAB output, we have that the 95% confidence interval for the true mean of hypertension when $X = 32$ is: (74.38, 83.66).

(d) The 95% prediction interval for hypertension when $X = 32$ is: (64.36, 93.68).

9. The MINITAB output for Problem 6 of Section 15.2 is given here, augmented with the information about Y when $X = 9.50$:

Regression Analysis: Number of Streaks versus Moisture

```
The regression equation is
Number of Streaks = 0.75 + 1.25 Moisture

Predictor      Coef     SE Coef      T        P
Constant      0.750      8.808      0.09     0.935
Moisture      1.2500     0.8195     1.53     0.178

S = 5.31115   R-Sq = 27.9%   R-Sq(adj) = 15.9%

Analysis of Variance

Source           DF       SS        MS        F       P
Regression        1      65.63     65.63     2.33    0.178
Residual Error    6     169.25     28.21
Total             7     234.88

Predicted Values for New Observations when x = 9.50

New Obs     Fit      SE Fit        95% CI            95% PI
   1       12.62      2.05      (7.61, 17.64)     (-1.30, 26.55)

Values of Predictors for New Observations

New Obs       Moisture
   1            9.50
```

From the above MINITAB output, we obtain

(a) $\hat{\sigma}^2 = \text{MSE} = 28.21$

(b) $\hat{\sigma}_{\hat{\beta}_0} = 8.808$, $\hat{\sigma}_{\hat{\beta}_1} = 0.8195$. Hence, the 99% confidence intervals for β_0 and β_1 are given by

$$\hat{\beta}_0 \pm t_{n-2;\alpha/2}\hat{\sigma}_{\hat{\beta}_0} = (0.750 \pm 3.707 \times 8.808) = (-31.9013, 33.4013), \text{ and}$$
$$\hat{\beta}_1 \pm t_{n-2;\alpha/2}\hat{\sigma}_{\hat{\beta}_1} = (1.25 \pm 3.707 \times 0.8195) = (-1.7879, 4.2879)$$

respectively.

(c) From the MINITAB output, we have that the 95% confidence interval for the true mean of the number of streaks when $X = 9.5$ is: (7.61, 17.64).

(d) The 95% prediction interval for the the number of streaks when $X = 9.50$ is: (0, 26.55). Note that the lower prediction limit cannot be negative, so we replace it with zero.

11. The MINITAB output for Problem 8 of Section 15.2 is given here, augmented with the information about Y when $X = 16.0$:

Regression Analysis: Yield versus Distance

```
The regression equation is
Yield = 4.30 + 5.70 Distance

Predictor      Coef     SE Coef      T        P
Constant      4.300      8.613      0.50     0.635
Distance      5.7000     0.5761     9.89     0.000
```

```
S = 3.97073    R-Sq = 94.2%    R-Sq(adj) = 93.3%
```

Analysis of Variance

```
Source            DF      SS       MS       F        P
Regression         1    1543.3   1543.3   97.88    0.000
Residual Error     6      94.6     15.8
Total              7    1637.9
```

Predicted Values for New Observations

```
New Obs    Fit      SE Fit      95% CI           95% PI
   1     95.50      1.58    (91.64, 99.36)   (85.05, 105.95)
```

Values of Predictors for New Observations

```
New Obs    Distance
   1         16.0
```

From the above MINITAB output, we obtain

(a) $\hat{\sigma}^2 = \text{MSE} = 15.8$

(b) $\hat{\sigma}_{\hat{\beta}_0} = 8.613$, $\hat{\sigma}_{\hat{\beta}_1} = 0.5761$. Hence, the 95% confidence intervals for β_0 and β_1 are given by

$$\hat{\beta}_0 \pm t_{n-2;\alpha/2}\hat{\sigma}_{\hat{\beta}_0} = (4.3 \pm 2.447 \times 8.613) = (-16.7760, 25.3760), \text{and}$$

$$\hat{\beta}_1 \pm t_{n-2;\alpha/2}\hat{\sigma}_{\hat{\beta}_1} = (5.7 \pm 2.447 \times 0.5761) = (4.2903, 7.1097)$$

respectively.

(c) From the MINITAB output, we have that the 95% confidence interval for the true mean of yield when $X = 16$ is: (91.64, 99.36).

(d) The 95% prediction interval for yield when $X = 16$ is: (85.05, 105.95).

PRACTICE PROBLEMS FOR SECTION 15.5

In this section we perform the testing of hypotheses at the α level of significance by using the $100(1-\alpha)\%$ confidence intervals for the corresponding parameters that are determined in Sections 15.3 and 15.4.

1. The 95% confidence intervals for β_0 and β_1 in this problem are given by (see solution to the Problem 4 of the Practice Problems of Sections 15.3 and 15.4):

$$\hat{\beta}_0 \pm t_{n-2;\alpha/2}\hat{\sigma}_{\hat{\beta}_0} = (135.7 \pm 2.306 \times 77.04) = (-41.9542, 313.3542)$$

$$\hat{\beta}_1 \pm t_{n-2;\alpha/2}\hat{\sigma}_{\hat{\beta}_1} = (0.2371 \pm 2.306 \times 0.3741) = (-0.6256, 1.0998)$$

Since the 95% confidence intervals for both β_0 and β_1 contain "0", we do not reject at the 5% level of significance the null hypotheses given here, i.e.,

$$H_0: \beta_0 = 0 \quad Vs \quad H_1: \beta_0 \neq 0, \text{and } H_0: \beta_1 = 0 \quad Vs \quad H_1: \beta_1 \neq 0$$

3. The 99% confidence intervals for β_0 and β_1 in this problem are given by (see solution to the Problem 6 of the Practice Problems of Sections 15.3 and 15.4):

$$\hat{\beta}_0 \pm t_{n-2;\alpha/2}\hat{\sigma}_{\hat{\beta}_0} = (105.47 \pm 3.169 \times 46.74) = (-42.6491, 253.5891)$$

$$\hat{\beta}_1 \pm t_{n-2;\alpha/2}\hat{\sigma}_{\hat{\beta}_1} = (-0.105 \pm 3.169 \times 3.737) = (-11.9476, 11.7376)$$

The confidence interval for β_0 contains "0", so we do not reject the null hypothesis $H_0 : \beta_0 = 0$ at the 1% level of significance. Also, the confidence interval for β_1 contains "0", so we do not reject the null hypothesis $H_0 : \beta_1 = 0$ at the 1% level of significance.

5. The 99% confidence intervals for β_0 *and* β_1 in this problem are given by (see solution to the Problem 8 of the Practice Problems of Sections 15.3 and 15.4):

$$\hat{\beta}_0 \pm t_{n-2;\alpha/2}\hat{\sigma}_{\hat{\beta}_0} = (-2.533 \pm 3.355 \times 1.193) = (-6.5355, 1.4695)$$

$$\hat{\beta}_1 \pm t_{n-2;\alpha/2}\hat{\sigma}_{\hat{\beta}_1} = (0.0333 \pm 3.355 \times 0.01314) = (-0.0108, 0.0774)$$

The confidence interval for β_0 contains "0", so we do not reject the null hypothesis $H_0 : \beta_0 = 0$ at the 1% level of significance. Also, the confidence interval for β_1 contains "0", so we do not reject the null hypothesis $H_0 : \beta_1 = 0$ at the 1% level of significance.

7. The 99% confidence intervals for β_0 *and* β_1 in this problem are given by (see solution to the Problem 10 of the Practice Problems of Sections 15.3 and 15.4):

$$\hat{\beta}_0 \pm t_{n-2;\alpha/2}\hat{\sigma}_{\hat{\beta}_0} = 55.53 \pm 3.169 \times 18.81 = (-4.0789, 115.1389)$$

$$\hat{\beta}_1 \pm t_{n-2;\alpha/2}\hat{\sigma}_{\hat{\beta}_1} = 0.3472 \pm 3.169 \times 0.2232 = (-0.3601, 1.0545)$$

The confidence interval for β_0 contains "0", so we do not reject the null hypothesis $H_0 : \beta_0 = 0$ at the 1% level of significance. Also, the confidence interval for β_1 contains "0", so we do not reject the null hypothesis $H_0 : \beta_1 = 0$ at the 1% level of significance.

9. The 95% confidence intervals for β_0 *and* β_1 in this problem are given by (see solution to the Problem 12 of the Practice Problems of Sections 15.3 and 15.4):

$$\hat{\beta}_0 \pm t_{n-2;\alpha/2}\hat{\sigma}_{\hat{\beta}_0} = (29.829 \pm 2.365 \times 4.562) = (19.0399, 40.6181)$$
$$\hat{\beta}_1 \pm t_{n-2;\alpha/2}\hat{\sigma}_{\hat{\beta}_1} = (4.8286 \pm 2.365 \times 0.6608) = (3.2658, 6.3914)$$

The confidence interval for β_0 does not contain "0", so we reject the null hypothesis $H_0 : \beta_0 = 0$ at the 5% level of significance. Also, the confidence interval for β_1 does not contain "0", so, again, we reject the null hypothesis $H_0 : \beta_1 = 0$ at the 5% level of significance.

11. The 95% confidence intervals for β_0 *and* β_1 in this problem are given by (see solution to the Problem 3 of the Practice Problems of Sections 15.3 and 15.4):

$$\hat{\beta}_0 \pm t_{n-2;\alpha/2}\hat{\sigma}_{\hat{\beta}_0} = (18.80 \pm 2.101 \times 37.54) = (-60.0715, 97.6715)$$

$$\hat{\beta}_1 \pm t_{n-2;\alpha/2}\hat{\sigma}_{\hat{\beta}_1} = (0.9259 \pm 2.101 \times 0.2631) = (0.3731, 1.4787)$$

The confidence interval for β_0 contains "0", so we do not reject the null hypothesis $H_0 : \beta_0 = 0$ at the 5% level of significance. However, the confidence interval for β_1 does not contain "0", so we reject the null hypothesis $H_0 : \beta_1 = 0$ at the 5% level of significance.

PRACTICE PROBLEMS FOR SECTION 15.6

In this Section for all but problem # 4, ANOVA tables have been developed and are given in solutions of problems of Section 15.2. We will reproduce these tables here, as, they are needed for the solutions to the problems of this section.

1. From the MINITAB output for problem 12 of Section 15.2, the ANOVA table is

ANOVA

Source	DF	SS	MS	F	P
Regression	1	226.95	226.95	96.18	0.000
Residual Error	9	21.24	2.36		
Total	10	248.18			

Since the p-value is 0.000, which is less than 5%, we conclude that the model provides a good fit.

3. (a) From the MINITAB output for Problem 11 of Section 15.2, the ANOVA table is:

ANOVA

```
Source            DF      SS        MS        F       P
Regression         1     978.26    978.26   12.39   0.002
Residual Error    18    1421.49     78.97
Total             19    2399.75
```

(b) From the ANOVA table in part (a), the p-value is 0.002, which is less than 1%. Hence, we conclude that the model provides a good fit.

(c) The observed level of significance is 0.002.

(d) The coefficient of determination is given by

$$R^2 = \frac{SS_{Reg.}}{SS_{Total}} = \frac{978.26}{2399.75} = 40.8\%.$$

Since the R-square value is just over 40%, we conclude that weight affects the cholesterol level, but also that there are other factors too that are affecting the cholesterol level.

5. (a) From the MINITAB output for Problem 1 of Section 15.2, the ANOVA table is:

```
Analysis of Variance

Source            DF      SS        MS       F       P
Regression         1     221.4     221.4    0.40    0.544
Residual Error     8    4410.7     551.3
Total              9    4632.1
```

(b) From the ANOVA table in part (a), the p-value is 0.544, which is greater than 5%. Hence, we conclude that the model does not provide a good fit.

(c) The observed level of significance is 0.544.

(d) The coefficient of determination is given by

$$R^2 = \frac{SS_{Reg.}}{SS_{Total}} = \frac{221.4}{4632.1} = 4.78\%.$$

Since the R-square value is under 5%, and the p-value is very high, we conclude that the fitted model is not good for predicting plasma glucose, using the weight of a person.

7. (a) From the MINITAB output for Problem 3 of Section 15.2, the ANOVA table is:

ANOVA

```
Source            DF      SS        MS       F       P
Regression         1      0.3       0.3     0.00    0.978
Residual Error    10    3759.4     375.9
Total             11    3759.7
```

(b) From the ANOVA table in part (a), the p-value is 0.978, which is much greater than 5%. Hence, the model does not provide a good fit.

(c) The observed level of significance is 0.978.

(d) The coefficient of determination is given by

$$R^2 = \frac{SS_{Reg.}}{SS_{Total}} = \frac{0.3}{3759.7} = 0.008\%.$$

Since the R-square value is very small and the p-value is very high, we conclude that the fitted model is not good for predicting shearing strength using cooking time.

9. (a) From the MINITAB output for Problem 8 of Section 15.2, the ANOVA table is:

ANOVA

```
Source          DF      SS      MS      F       P
Regression      1       1543.3  1543.3  97.88   0.000
Residual Error  6       94.6    15.8
Total           7       1637.9
```

(b) From the ANOVA table in part (a), the p-value is 0.000, which is less than 1%. Hence, we conclude that the model provides a good fit.

(c) The observed level of significance is 0.000.

(d) The coefficient of determination is given by

$$R^2 = \frac{SS_{Reg.}}{SS_{Total}} = \frac{1543.3}{1637.9} = 94.22\%.$$

Since the R-square value is over 94%, and the p-value is 0.000, we conclude that the distance between the plants has a very significant effect on the yield of cotton.

PRACTICE PROBLEMS FOR SECTION 15.7

In this section the plots for all the problem have been prepared using MINITAB.

1.

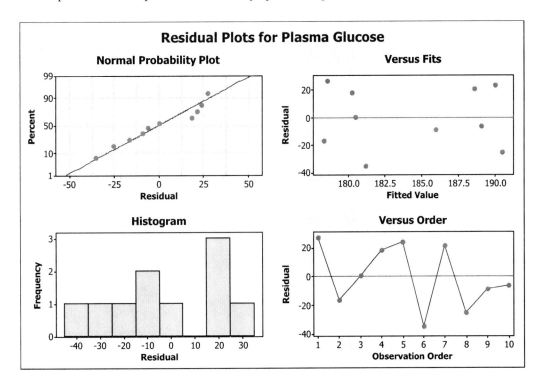

(a) The normal probability plot shows that we may conclude that the normality assumption is valid.

(b) Plots of the residuals verses fitted values and order of the observations do not show any significant departures from model assumptions.

3.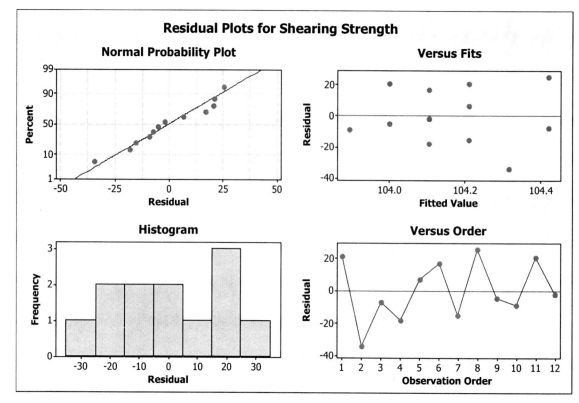

(a) The normal probability plot shows that we may conclude that the normality assumption is valid.

(b) Plots of the residuals verses fitted values and order of the observations do not show any significant departures from model assumptions.

5.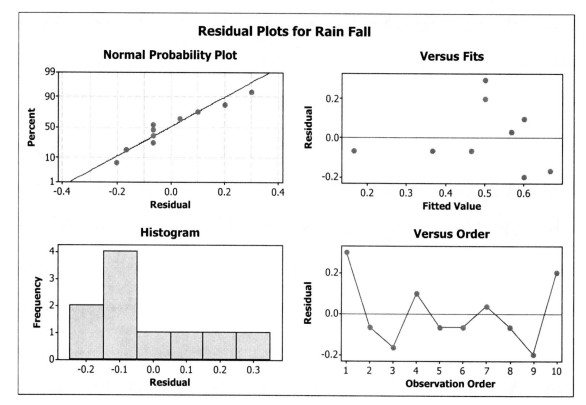

(a) The normal probability plot shows that we may conclude that the normality assumption is valid.

(b) Plots of the residuals verses fitted values and order of the observations do not show any significant departures from model assumptions.

7.

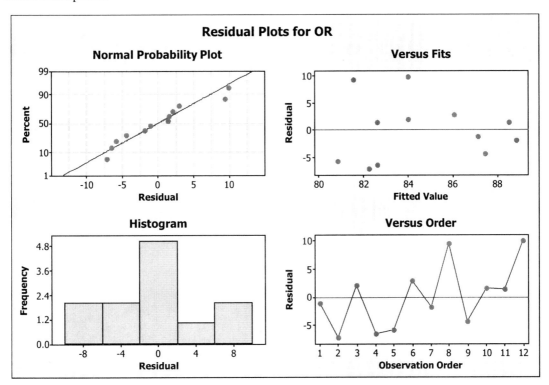

(a) The normal probability plot shows that we may conclude that the normality assumption is valid.

(b) Plots of the residuals verses fitted values and order of the observations do not show any departures from model assumptions.

9.

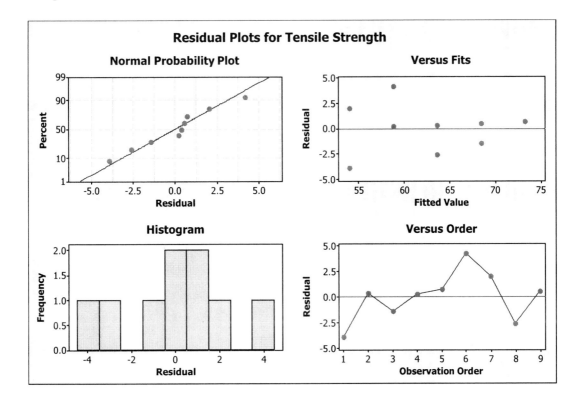

(a) The normal probability plot shows that we may conclude that the normality assumption is valid.

(b) Plots of the residuals verses fitted values and order of the observations do not show any departures from model assumptions.

11.

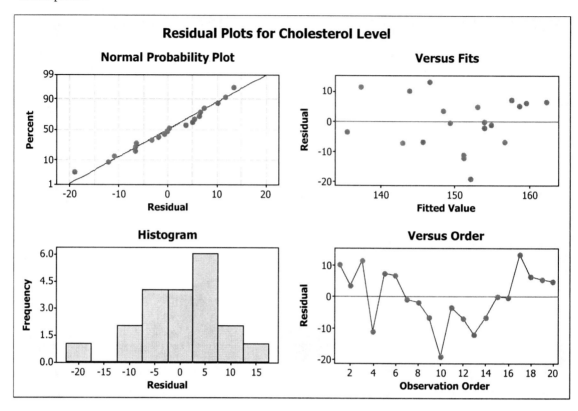

(a) The normal probability plot shows that we may conclude that the normality assumption is valid.

(b) Plots of the residuals verses fitted values and order of the observations do not show any departures from model assumptions.

PRACTICE PROBLEMS FOR SECTION 15.8

In this section to perform the necessary analysis we use the same MINITAB instructions as used in Section 15.2.

1. In this problem we first determine the new responses by using the log Y transformation and then entering the new data in MINITAB worksheet. Then by select **Stat > Regression > Regression** and making the necessary entries we obtain the following MINITAB output:

Regression Analysis: log(Hypertension) versus Noise Level

```
The regression equation is
log(Hypertension) = 1.71 + 0.00579  Noise Level

Predictor      Coef     SE Coef       T        P
Constant    1.71009     0.06847    24.97    0.000
Noise Level 0.005789    0.002324    2.49    0.030

S = 0.0355311    R-Sq = 36.1%    R-Sq(adj) = 30.2%
```

Analysis of Variance

Source	DF	SS	MS	F	P
Regression	1	0.007831	0.007831	6.20	0.030
Residual Error	11	0.013887	0.001262		
Total	12	0.021718			

Unusual Observations

Obs	Noise Level	log(Hypertension)	Fit	SE Fit	Residual	St Resid
2	33.0	1.83251	1.90112	0.01331	-0.06861	-2.08R

R denotes an observation with a large standardized residual.

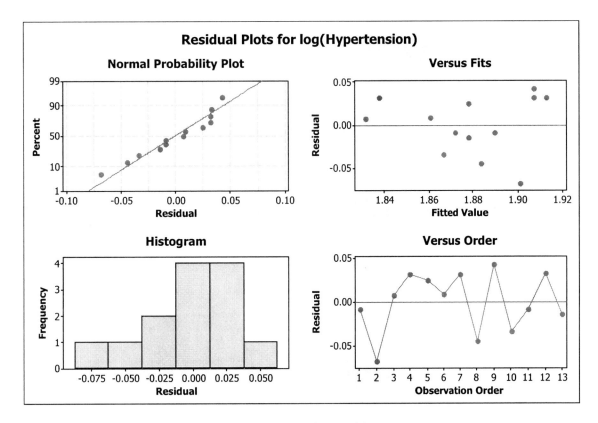

(a) From the above output we have the fitted linear regression model:

$$\log(\text{Hypertension}) = 1.71 + 0.00579 \text{ Noise Level}$$

(b) The least square regression coefficients are

$$\hat{\beta}_0 = 1.71, \text{ and } \hat{\beta}_1 = 0.00579.$$

(c) From the above MINITAB output we note the p-values for both regression coefficients are less than 5%. Hence, in both cases we reject the corresponding null hypothesis. Further, note that the results by using the log transformation have changed the magnitude of the regression coefficients. However, the p-values and the value of R-square value are almost the same. Also, the residual plots in this problem and the original problem 4 resemble with each other.

3. Again, taking the same steps as in Problem 1, we obtain the following MINITAB output:

Regression Analysis: Log(Hypertension) versus Log(Noise Level)

```
The regression equation is
Log (Hypertension) = 1.37 + 0.349 Log (Noise Level)


Predictor           Coef     SE Coef     T        P
Constant            1.3693   0.2219      6.17     0.000
Log (Noise Level)   0.3491   0.1519      2.30     0.042


S = 0.0365212    R-Sq = 32.4%    R-Sq(adj) = 26.3%

Analysis of Variance

Source           DF     SS         MS         F       P
Regression       1      0.007046   0.007046   5.28    0.042
Residual Error   11     0.014672   0.001334
Total            12     0.021718
```

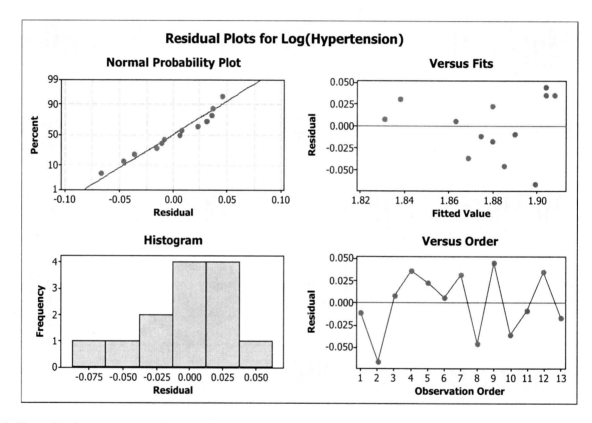

(a) From the above output we have the fitted linear regression model:

$$\text{Log(Hypertension)} = 1.37 + 0.349\,\text{Log(Noise Level)}$$

(b) The least square regression coefficients are

$$\hat{\beta}_0 = 1.37, and\ \hat{\beta}_1 = 0.349.$$

(c) From the above MINITAB output we note that the p-values for both regression coefficients are less than 5%. Hence, in both cases we reject the corresponding null hypotheses. Further, note that the results by using the log transformation on both the response and the predictor variables are practically the same. The p-value for the coefficient of Log (Noise level) has slightly increased, while the value of R-square has slightly decreased. The residual plots in this problem and the original problem 4 resemble with each other.

PRACTICE PROBLEMS FOR SECTION 15.9

1. In this problem we wish to test at the 5% level of significance the hypothesis

$$H_0 : \rho = 0 \quad versus \quad H_1 : \rho \neq 0.$$

In order to perform a test of this hypothesis, we first need to determine the value of the sample correlation coefficient r, where

$$r = \sum_{i=1}^{n} (x_i - \bar{x})(y_i - \bar{y}) \bigg/ \sqrt{\sum_{i=1}^{n} (x_i - \bar{x})^2 (y_i - \bar{y})^2}.$$

Using the data of Problem 1 of Section 15.2, we have that

$$r = 0.219.$$

Now from (15.9.3) the test statistic is

$$\frac{r}{\sqrt{1 - r^2}} \sqrt{n - 2}$$

Substituting the value of $r = 0.219$ and $n = 10$, we find that the observed value of the test statistic is

$$\frac{r}{\sqrt{1 - r^2}} \sqrt{n - 2} = 0.6348$$

Here, we reject the null hypothesis if the absolute value of the observed test statistic is such that

$$\left| \frac{r}{\sqrt{1 - r^2}} \sqrt{n - 2} \right| > t_{n-2;\alpha/2} = 2.306.$$

Otherwise we do not reject the hypothesis. Since in this case the absolute value of the observed test statistic is less than 2.306, we do not reject the null hypothesis.

3. In this problem we wish to test at 5% level of significance the hypothesis

$$H_0 : \rho = 0 \quad versus \quad H_1 : \rho \neq 0.$$

In order to perform a test of this hypothesis, we first need to determine the value of the sample correlation coefficient r, where

$$r = \sum_{i=1}^{n} (x_i - \bar{x})(y_i - \bar{y}) \bigg/ \sqrt{\sum_{i=1}^{n} (x_i - \bar{x})^2 (y_i - \bar{y})^2}.$$

Using the data of from Problem 4 of Section 15.2, we have that

$$r = 0.612.$$

Now from (15.9.3) the test statistic is

$$\frac{r}{\sqrt{1 - r^2}} \sqrt{n - 2}.$$

Substituting the value of $r = 0.612$ and $n = 13$, we find that the observed value of the test statistic

$$\frac{r}{\sqrt{1-r^2}}\sqrt{n-2} = 2.5665$$

We reject the null hypothesis if the absolute value of the observed test statistic is such that

$$\left|\frac{r}{\sqrt{1-r^2}}\sqrt{n-2}\right| > t_{n-2;\alpha/2} = 2.201,$$

Otherwise we do not reject the hypothesis. Since in this case the absolute value of the observed test statistic is greater than 2.201, we reject the null hypothesis.

5. In this problem we wish to test at 5% level of significance the hypothesis

$$H_0 : \rho = 0 \quad \textit{versus} \quad H_1 : \rho < 0.$$

In order to perform the test of this hypothesis we first need to determine the value of the sample correlation coefficient r, where

$$r = \sum_{i=1}^{n}(x_i - \bar{x})(y_i - \bar{y}) \Big/ \sqrt{\sum_{i=1}^{n}(x_i - \bar{x})^2(y_i - \bar{y})^2}.$$

Using the data from Problem 10 of Section 15.2, we obtain

$$r = 0.624.$$

Now from (15.9.3) the test statistic is

$$\frac{r}{\sqrt{1-r^2}}\sqrt{n-2}.$$

Substituting the value of $r = 0.624$ and $n = 20$, we obtain the observed value of the test statistics

$$\frac{r}{\sqrt{1-r^2}}\sqrt{n-2} = 3.3879.$$

We reject the null hypothesis if the value of the observed test statistic is such that

$$\frac{r}{\sqrt{1-r^2}}\sqrt{n-2} < -t_{n-2;\alpha} = -1.734,$$

Otherwise we do not reject the hypothesis. Since in this case the observed value of the test statistics is greater than -1.734, we do not reject the null hypothesis.

REVIEW PRACTICE PROBLEMS

1. Since $E(Y|x) = \eta = \beta_1 x$, the method of least squares reduces to minimizing

$$Q(\beta) = \sum_{i=1}^{n}(y_i - \beta_1 x_i)^2.$$

We have

$$\frac{\partial Q}{\partial \beta_1} = 2\sum_{i=1}^{n}(y_i - \beta_1 x_i)(-x_i), \text{ so that } -\sum x_i y_i + \sum b_1 x_i^2 = 0 \Rightarrow b_1 = \frac{\sum x_i y_i}{\sum x_i^2},$$

where we denote the solution to $\dfrac{\partial Q}{\partial \beta_1} = 0$ by b_1.

Hence, the estimated regression line of Y on X is $\hat{Y} = b_1 X$.

Assuming that the Y_i's are independent, with expectation $\beta_1 x_i$ and variance σ^2,

$$E(b_1) = \frac{\sum x_i E(Y_i)}{\sum x_i^2} = \frac{\sum x_i \beta_1 x_i}{\sum x_i^2} = \beta_1, \text{ so } b_1 \text{ is unbiased for } \beta_1;$$

$$\text{and Var}(b_1) = \frac{\sum \text{Var}(x_i Y_i)}{\sum x_i^2} = \frac{\sum x_i \cdot \text{Var}(Y_i)}{\left(\sum x_i^2\right)^2} = \frac{\sum x_i^2 \cdot \sigma^2}{\left(\sum x_i^2\right)^2} = \frac{\sigma^2}{\sum x_i^2}.$$

Note that $E(\hat{Y}) = x \cdot E(b_1) = x\beta_1 = \eta$.

Assuming that the Y_i's are normally distributed, $b_1 \sim N\left(\beta_1, \frac{\sigma^2}{\sum x_i^2}\right)$.

Thus a confidence interval for β_1 is $\left(b_1 \pm t_{n-1:\alpha/2}\sqrt{\frac{S^2}{\sum x_i^2}}\right)$, and a confidence interval for

$$\eta_0 = x_0\beta_1 \text{ when } x = x_0 \text{ is } \left(\hat{y}_{x_0} \pm t_{n-1:\alpha/2}\sqrt{\frac{x_0^2 S^2}{\sum x_i^2}}\right), y_{x_0} = x_0 b_1$$

Here, $S^2 = \dfrac{\sum (y_i - \hat{y}_i)^2}{n-1} = \dfrac{\sum (y_i - b_1 x_i)^2}{n-1} = \dfrac{\sum y_i^2 - b_1^2 \sum x_i^2}{n-1}$, as is easily shown.

3. From the data:

$\sum x_i = 720,$	$\sum y_i = 348,$	$\sum x_i y_i = 18980,$
$\sum x_i^2 = 49200,$	$\sum y_i^2 = 10884,$	$n = 12.$
$\bar{x} = 60,$	$\bar{y} = 29,$	

we find that

$$S_{xx} = \sum x_i^2 - n\bar{x}^2 = 49200 - 12(60)^2 = 6000,$$

$$S_{xy} = \sum x_i y_i - n\bar{x}\bar{y} = 18980 - 12(60)(29) = -1900,$$

$$S_{yy} = \sum y_i^2 - n\bar{y}^2 = 10884 - 12(29)^2 = 792;$$

Now using (15.2.13), we have

$$\hat{\beta}_1 = \frac{S_{xy}}{S_{xx}} = \frac{-1900}{6000} = -0.3167, \quad \hat{\beta}_0 = \bar{y} - \hat{\beta}_1\bar{x} = 29 - \left(\frac{-1900}{6000}\right)(60) = 48.00$$

Thus, the fitted regression line is

$$\hat{Y} = 48.00 - 0.3167X.$$

Also, we find that

$$SSE = S_{yy} - \hat{\beta}_1 S_{xy} = 792 - \left(\frac{-1900}{6000}\right)(-1900) = 190.33,$$

so that $S^2 = \dfrac{SSE}{n-2} = \dfrac{190.33}{10} = 19.033.$

Since $t_{10;0.025} = 2.228$, a 95% C.I. for β_0 is

$$\left(\hat{\beta}_0 \pm t_{n-2;0.025} \sqrt{\left(\frac{1}{n} + \frac{\bar{x}^2}{S_{xx}} \right) S^2} \right) = (48.00 \pm 2.228 \times 3.6064) = (48.00 \pm 8.04),$$

or $(39.96, 56.04)$

A 95% C.I. for β_1 is

$$\hat{\beta}_1 \pm t_{n-2;0.025} \sqrt{\frac{S^2}{S_{xx}}} = (-0.3167 \pm 2.228 \times 0.0563) = (-0.3167 \pm 0.1255),$$

or $(-0.442, -0.191)$

A 95% C.I. for $\eta_0 = \beta_0 + \beta_1 x_0$ is

$$\hat{y}_0 \pm t_{n-2;0.025} \sqrt{\left(\frac{1}{n} + \frac{(x_0 - \bar{x})^2}{S_{xx}} \right) S^2} = \hat{y}_0 \pm (2.228) \sqrt{\left(\frac{1}{12} + \frac{(x_0 - 60)^2}{6000} \right)(19.033)},$$

where $\hat{Y}_0 = \hat{\beta}_0 + \hat{\beta}_1 x_0 = 48.0 - 0.3167 x_0$.

5. From the data:

$$
\begin{array}{lll}
\sum x_i = 105, & \sum y_i = 56.92, & \sum x_i y_i = 1076.2, \\
\sum x_i^2 = 2275, & \sum y_i^2 = 554.6594, & n = 6. \\
\bar{x} = 17.5, & \bar{y} = 9.4867, &
\end{array}
$$

we find that $S_{xx} = 437.500$, $S_{xy} = 80.100$, $S_{yy} = 14.6783$.

Thus, we obtain $\hat{\beta}_1 = \dfrac{S_{xy}}{S_{xx}} = 0.183086$, $\hat{\beta}_0 = 6.28267$,

Hence, the fitted regression line is

$$\hat{Y} = 6.283 + 0.183X.$$

Also, $SSE = 0.0131676$ and $S^2 = 0.0032919$.

Since $t_{4;0.025} = 2.776$, a 95% C.I. for β_0 is

$$\left(6.283 \pm (2.776) \sqrt{\left(\frac{1}{6} + \frac{(17.5)^2}{437.50} \right)(0.0032919)} \right) = (6.283 \pm 2.776 \times 0.05341)$$

$$= (6.283 \pm 0.148)$$

or $(6.135, 6.431)$

A 95% C.I. for β_1 is

$$\left(0.183 \pm (2.776) \sqrt{\frac{(0.0032919)}{437.50}} \right) = (0.183 \pm 2.776 \times 0.002743) = (0.183 \pm 0.0076)$$

or $(0.175, 0.191)$

A 95% C.I. for η_0 is

$$\hat{Y}_0 \pm (2.776) \sqrt{\left(\frac{1}{6} + \frac{(x_0 - 17.5)^2}{437.50} \right)(0.0032919)},$$

where $\hat{Y}_0 = 6.283 + 0.183 x_0$.

7. From the data:

$\sum x_i = 42,$	$\sum y_i = 24.3,$	$\sum x_i y_i = 183.6,$
$\sum x_i^2 = 364,$	$\sum y_i^2 = 101.17,$	$n = 6.$
$\bar{x} = 7,$	$\bar{y} = 4.05,$	

we find that $S_{xx} = 70$, $S_{xy} = 13.5$, $S_{yy} = 2.755$.

Thus, we obtain $\hat{\beta}_1 = 0.192857$, $\hat{\beta}_0 = 2.7$, and thus $\hat{Y} = 2.7 + 0.193X$.

Also, $SSE = 0.15142857$ and $S^2 = 0.037857$.

Since $t_{4;0.025} = 2.776$, a 95% C.I. for β_0 is

$$\left(2.7 \pm (2.776)\sqrt{\left(\frac{1}{6} + \frac{(7)^2}{70}\right)(0.037857)}\right) = (2.7 \pm 2.776 \times 0.1811) = (2.7 \pm 0.503) = (2.197, 3.203)$$

A 95% C.I. for β_1 is

$$\left(0.193 \pm (2.776)\sqrt{\frac{0.037857}{70}}\right) = (0.193 \pm 0.0646) = (0.127, 0.258)$$

A 95% C.I. for η_0 is

$$\hat{Y}_0 \pm (2.776)\sqrt{\left(\frac{1}{6} + \frac{(x_0 - 7)^2}{70}\right)(0.037857)}.$$

$X = x_0$:	2	4	6	8	10	12
$\hat{Y}_0$:	3.0857	3.4714	3.8571	4.2429	4.6286	5.0143
95% confidence band:	±0.3909	±0.2934	±0.2298	±0.2298	±0.2934	±0.3909

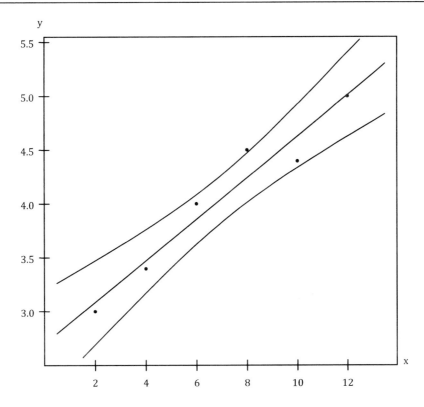

9. From the data:

$\sum x_i = 1660,$	$\sum y_i = 165.2,$	$\sum x_i y_i = 31685,$
$\sum x_i^2 = 300400,$	$\sum y_i^2 = 3563.48,$	$n = 10.$
$\bar{x} = 166.0,$	$\bar{y} = 16.52,$	

we find that $S_{xx} = 24840, S_{xy} = 4261.8, S_{yy} = 834.376.$

Hence $\hat{\beta}_1 = 0.17157, \hat{\beta}_0 = -11.96063, SSE = 103.179, S^2 = 12.897.$
We wish to test the hypothesis: $H_0 : \beta_1 = 0$ versus $H_1 : \beta_1 \neq 0.$
We reject H_0 at the 5% level if

$$\frac{|\hat{\beta}_1|}{\sqrt{S^2/S_{xx}}} > t_{n-2;0.025} = t_{8;0.025} = 2.306.$$

But observed value of the test statistic $\dfrac{|\hat{\beta}_1|}{\sqrt{S^2/S_{xx}}}$ is

$$\frac{0.17157}{\sqrt{12.897/24840}} = 7.53,$$

which is greater than 2.306, so that we reject H_0 and conclude that $\beta_1 \neq 0.$
Thus, the fitted regression line is

$$\hat{y} = -11.96063 + 0.17157x.$$

The estimate of $E(Y|X = 145)$ is

$$\hat{Y} = -11.96063 + (0.17157)(145) = 12.917.$$

The 95% C.I. for $E(Y|X = 145)$ is

$$\hat{Y} \pm t_{8;0.025} \sqrt{\left(\frac{1}{n} + \frac{(x_0 - \bar{x})^2}{S_{xx}}\right)S^2}$$

$$= \left(12.917 \pm (2.306)\sqrt{\left(\frac{1}{10} + \frac{(145 - 166)^2}{24840}\right)(12.897)}\right) = (12.917 \pm 2.842)$$

$$= (10.075, 15.759)$$

11. From the data:

$\sum x_i = 1556.3,$	$\sum y_i = 553.9$	$\sum x_i y_i = 57464.62,$
$\sum x_i^2 = 161520.87,$	$\sum y_i^2 = 20555.8,$	$n = 15.$

$$S_{xx} = 49.5573, \quad S_{xy} = 89.0267, \quad S_{yy} = 168.5333.$$

Then $\quad r = \dfrac{S_{xy}}{\sqrt{S_{xx}S_{yy}}} = \dfrac{89.0267}{\sqrt{(49.5573)(168.5333)}} = 0.9741.$

To test the hypothesis

$$H_0 : \rho = 0 \quad \text{against} \quad H_1 : \rho \neq 0$$

we reject the null hypothesis H_0 at the 5% level of significance if the observed value of the test statistic

$$\frac{|r|\sqrt{n-2}}{\sqrt{1-r^2}} > t_{n-2;\alpha/2} = t_{13;0.025} = 2.160.$$

Since the observed value of the test statistic is

$$\frac{|0.9741|\sqrt{13}}{\sqrt{1-(0.9741)^2}} = 15.546,$$

which is greater than 2.160. Hence we reject H_0 and conclude that $\rho \neq 0$.

To find a 95% C.I. for ρ, we first find an approximate $100(1-\alpha)\%$ C.I. for $\frac{1}{2}\ln\left(\frac{1+\rho}{1-\rho}\right)$. This is given by

$$\frac{1}{2}\ln\left(\frac{1+r}{1-r}\right) \pm z_{\alpha/2} \cdot \frac{1}{\sqrt{n-3}}.$$

Hence, with $r = 0.9741$ and $\alpha = 0.05$, we have

$$\left(2.1672 \pm (1.96)\frac{1}{\sqrt{12}}\right) = (2.1672 \pm 0.5658) = (1.6014, 2.7330).$$

From this we can easily determine a 95% C.I. for ρ to be $(0.9217, 0.9915)$.

13. From the data:

$\sum x_i = 60.22,$	$\sum y_i = 1127,$	$\sum x_i y_i = 8432.05,$
$\sum x_i^2 = 450.7948,$	$\sum y_i^2 = 157949,$	$n = 10.$

we find that $\quad S_{xx} = 88.14996, \quad S_{xy} = 1645.256, \quad S_{yy} = 30936.1.$

Hence $\quad r = \dfrac{S_{xy}}{\sqrt{S_{xx}S_{yy}}} = 0.9963.$

To test the hypothesis

$$H_0 : \rho = 0 \quad \text{against} \quad H_1 : \rho \neq 0$$

we reject the null hypothesis H_0 at the 5% level of significance if the observed value of the test statistic

$$\frac{|r|\sqrt{n-2}}{\sqrt{1-r^2}} > t_{n-2;\alpha/2} = t_{8;0.025} = 2.306.$$

Since the observed value of the test statistics is

$$\frac{|0.9963|\sqrt{8}}{\sqrt{1-(0.9963)^2}} = 32.7835,$$

which is greater than 2.306. Hence we reject H_0 and conclude that $\rho \neq 0$.

To find a 95% C.I. for ρ, we first find an approximate $100(1 - \alpha)\%$ C.I. for $\frac{1}{2} \ln\left(\frac{1+\rho}{1-\rho}\right)$. This is given by

$$\frac{1}{2} \ln\left(\frac{1+r}{1-r}\right) \pm z_{\alpha/2} \cdot \frac{1}{\sqrt{n-3}}.$$

Hence, with $r = 0.9963$ and $\alpha = .05$, we have

$$\left(3.1492 \pm (1.96)\frac{1}{\sqrt{7}}\right) = (3.1492 \pm 0.7408) = (2.4084, 3.8900).$$

From this we can easily determine a 95% C.I. for ρ to be $(0.9839, 0.9992)$.

15. (a) From the data:

$\sum x_i = 143,$	$\sum y_i = 141.4,$	$\sum x_i y_i = 1869.57,$
$\sum x_i^2 = 1943.44,$	$\sum y_i^2 = 1830.58,$	$n = 11.$

we find that $S_{xx} = 84.44$, $S_{xy} = 31.37$, $S_{yy} = 12.947273$.
Hence $\hat{\beta}_1 = 0.3715$ and $\hat{\beta}_0 = 8.0250$
Thus, the fitted regression line is

$$\hat{Y} = 8.0250 + 0.3715X.$$

(b) Reversing the roles of x and y:

$\sum x_i = 141.4,$	$\sum y_i = 143,$	$\sum x_i y_i = 1869.57,$
$\sum x_i^2 = 1830.58,$	$\sum y_i^2 = 1943.44,$	$n = 11.$

we find that $S_{xx} = 12.947273$, $S_{xy} = 31.37$, $S_{yy} = 84.44$.
Hence $\hat{\beta}_1 = 2.4229$ and $\hat{\beta}_0 = 18.1453$
Thus, the fitted regression line is

$$\hat{Y} = 18.1453 + 2.4229X.$$

(c) Assumptions are such that the independent variable is measured without error, while the dependent variable has a random error component.

17. (a) From the data:

$\sum x_i = 110.5,$	$\sum y_i = 249.4,$	$\sum x_i y_i = 4863.6,$
$\sum x_i^2 = 2136.25,$	$\sum y_i^2 = 11137.38,$	$n = 7.$

we find that $S_{xx} = 391.9286$, $S_{xy} = 926.6429$, $S_{yy} = 2251.6143$.
Hence $\hat{\beta}_1 = 2.3643$ and $\hat{\beta}_0 = -1.6938$
Thus, the fitted regression line is

$$\hat{Y} = -1.6938 + 2.3643X.$$

(b) The best estimate of Y at $x = 30$ is obtained by substituting $x = 30$ into the fitted regression line, i.e., $\hat{y}_{30} = 69.2356$.

(c) Solve the regression equation for the value of x when $\hat{Y} = 95$, i.e., $95 = \hat{\beta}_0 + \hat{\beta}_1 x_{95}$. This yields

$$\hat{x}_{95} = \frac{95 + 1.6938}{2.3643} = 40.897.$$

Thus, we require a tower of about 41 feet.

19. Let the estimate of the slope parameter be denoted by $\hat{\tau}$. Now, here the independent variable is t, dependent variable v and $\hat{\tau}$ is observed to be (see solution of Problem 18)

$$\hat{\tau} = \frac{6165.1}{204} = 30.221$$

But Again, from Problem 1, a 95% confidence interval for the slope parameter τ is given by

$$\hat{\tau} \pm t_{n-1;\alpha/2}\hat{\sigma}_{\hat{\tau}} = \hat{\tau} \pm t_{n-1;.025}\hat{\sigma}_{\hat{\tau}}$$

where

$$\hat{\sigma}_{\hat{\tau}}^2 = \frac{\hat{\sigma}^2}{\sum\limits_{i=1}^{n} t_i^2}$$

with ($n = 8$)

$$\hat{\sigma}^2 = \frac{\sum\limits_{i=1}^{n}(v_i - \hat{v}_i)^2}{n-1} = \frac{\sum\limits_{i=1}^{n}v_i^2 - b\sum\limits_{i=1}^{n}v_i t_i}{n-1} = \frac{186442 - 30.221 \times 6165.1}{7} = 17.9852$$

Thus, we have

$$\hat{\sigma}_{\hat{\tau}}^2 = \frac{17.9852}{204} = 0.088$$

Hence, a 95% confidence interval for slope parameter τ is given by

$$\hat{\tau} \pm t_{7;.025}\hat{\sigma}_{\hat{\tau}} = (30.2211 \pm 2.365 \times \sqrt{0.088}) = (30.2211 \pm 0.7016) = (29.5195, 30.9227).$$

21. No it is not reasonable to find a prediction interval for the observation Y when $X = 11$ or find a confidence interval for $E(Y|X)$ when $X = 11$. Since the value 11 of the predictor variable is outside the experimental range and outside the experimental range we do not know whether or not the fitted model is good. In other words, we do not know the shape of the model outside the experimental range.

23. Using MINITAB we obtain the following plots:

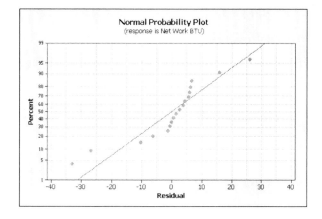

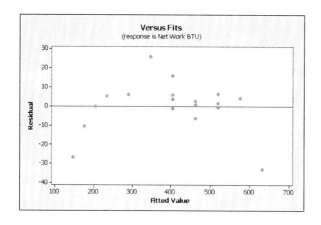

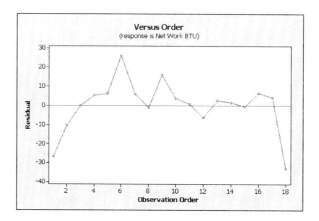

All three of the residual plots suggest some abnormalities (specifically departures from normality, lack of constancy of variance, and presence of outliers). One way to remedy these abnormalities is to try out some transformations or consider a quadratic model, and perform a residual analysis to check on the suitability of the transformations used or of the quadratic model. In fact, in this problem a quadratic model is a better fit. The fitted quadratic model is

$$Y = 18.38 + 764.7X - 175.6X^2$$

25. To the hypothesis

$$H_0 : \rho = 0 \quad Versus \quad H_1 : \rho > 0$$

we first find the value of r, which is observed to be

$$r = \frac{S_{xy}}{\sqrt{S_{xx}S_{yy}}} = 0.548$$

We reject H_0 if the observed value of test statistic is such that

$$\frac{r}{\sqrt{1 - r^2}} \sqrt{n - 2} > t_{n-2;.025} = t_{8;.05} = 1.860$$

Since the test statistic for testing the above hypothesis is

$$\frac{r}{\sqrt{1-r^2}}\sqrt{n-2} \sim t_{n-2}$$

Substituting the values of r and n, we obtain the observed value of the test statistic is equal to 1.853.

Since the observed value of the test statistic is less than the critical value, we do not reject the null hypothesis at 5% level of significance. The observed level of significance in this case is approximately 0.0505, which is slightly greater than the significance level.

27. (a) Using MINITAB we obtain the following scatter plot:

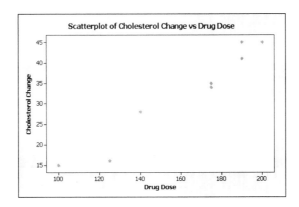

From the above scatter plot we suggest that a linear model would be appropriate.

(b) Using MINITAB the output of the regression analysis for the given data is given below.

Regression Analysis: Cholesterol Change versus Drug Dose

```
The regression equation is
Cholesterol Change = - 19.7 + 0.322 Drug Dose

Predictor      Coef     SE Coef      T        P
Constant     -19.682     5.322     -3.70    0.010
Drug Dose     0.32159    0.03219    9.99    0.000

S = 3.06192    R-Sq = 94.3%    R-Sq(adj) = 93.4%

Analysis of Variance

Source           DF      SS       MS       F        P
Regression        1    935.62   935.62   99.80    0.000
Residual Error    6     56.25     9.38
Total             7    991.87
```

Now, from the F table we obtain $F_{1,6;.05} = 5.99$, and we note that (see above) the observed value of F-statistic is 99.80, which is greater than 5.99, and the p-value is equal to 0.000. Hence, we can conclude that a linear regression model is an appropriate fit at a 5% significance level or at any significance level.

(c) Residual Analysis:

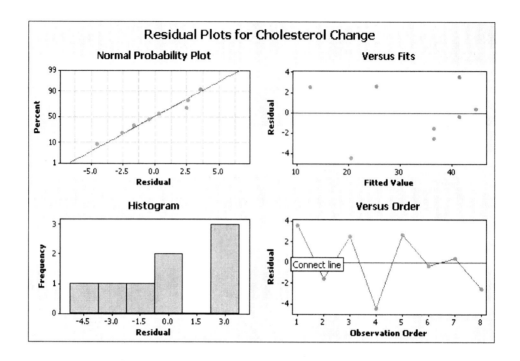

The above residual plots do not indicate any specific abnormalities with usual assumptions made for a linear regression model.

29. The MINITAB output along with the residual plots is given below.

Output for the model: $Y_i = \beta_0 + \beta_1 X_i + \varepsilon_i, \quad i = 1, \cdots, 10$

Regression Analysis: Y versus X

```
The regression equation is
Y = 31.4 + 0.017 X

Predictor      Coef     SE Coef       T       P
Constant     31.442     6.000      5.24    0.001
X            0.0174     0.2594     0.07    0.948

S = 8.78387    R-Sq = 0.1%    R-Sq(adj) = 0.0%

Analysis of Variance

Source          DF      SS       MS       F       P
Regression       1     0.35     0.35    0.00    0.948
Residual Error   8   617.25    77.16
Total            9   617.60
```

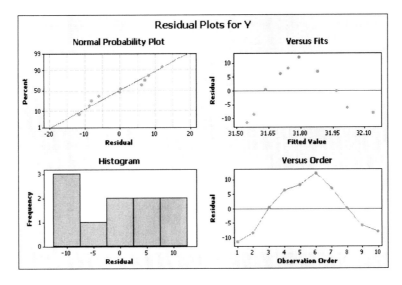

Output for the model: $Y_i = \tau_0 + \tau_1 X_i + \tau_2 X^2 + \varepsilon_i, \quad i = 1, \cdots, 10$ (We will discuss this kind of problems, again, in Chapter 16)

Regression Analysis: Y versus X, X-Square

```
The regression equation is
Y = 5.60 + 3.05 X - 0.0677 X-Square


Predictor      Coef      SE Coef      T        P
Constant       5.597     5.592        1.00     0.350
X              3.0456    0.5777       5.27     0.001
X Square      -0.06774   0.01263     -5.36     0.001


S = 4.15396    R-Sq = 80.4%    R-Sq(adj) = 74.9%


Analysis of Variance

Source          DF       SS        MS        F        P
Regression      2        496.81    248.41    14.40    0.003
Residual Error  7        120.79    17.26
Total           9        617.60
```

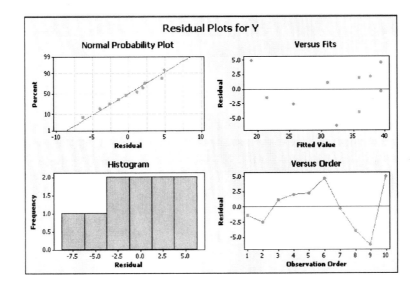

From the above regression analysis, ANOVA tables and residual plots we can clearly see that the quadratic model is a better fit for the given data.

31. The MINITAB output along with the residual plots is given below.

(a) Regression Analysis: Yield versus Temp.

The regression equation is

Yield = 58.3 + 0.433 Temp.

Predictor	Coef	SE Coef	T	P
Constant	58.327	3.126	18.66	0.000
Temp.	0.4329	0.1032	4.19	0.000

$$S = 7.25820 \quad R\text{-Sq} = 40.4\% \quad R\text{-Sq(adj)} = 38.1\%$$

Analysis of Variance

Source	DF	SS	MS	F	P
Regression	1	926.96	926.96	17.60	0.000
Residual Error	26	1369.72	52.68		
Total	27	2296.68			

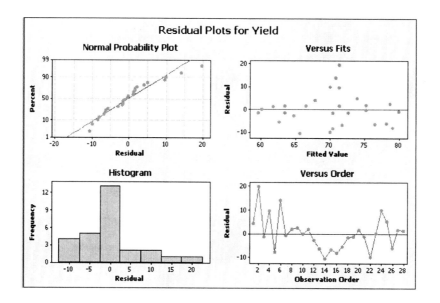

The normal probability indicates that the normality assumption is satisfactory. However, the plot of residuals versus fitted values does indicate non-constancy of variance. In fact, the plot of residuals versus observations order also presents a slight concern about the assumption of independence, since too many consecutive points show negative errors.

(b) Using the MINITAB output, a 95% Confidence Interval for β_0 is given by

$$\hat{\beta}_0 \pm t_{n-2;\alpha/2} \hat{\sigma}_{\hat{\beta}_0}$$
$$= \hat{\beta}_0 \pm t_{26;.025} \hat{\sigma}_{\hat{\beta}_0}$$
$$= (58.327 \pm 2.056 \times 3.126)$$
$$= (58.327 \pm 6.427)$$
$$= (51.900, 64.754).$$

Similarly, again using the MINITAB output, a 95% Confidence Interval for β_1 is given by

$$\hat{\beta}_1 \pm t_{n-2;\alpha/2} \times \hat{\sigma}_{\hat{\beta}_1}$$
$$= \hat{\beta}_1 \pm t_{n-2;\alpha/2} \times \hat{\sigma}_{\hat{\beta}_1}$$
$$= (0.433 \pm 2.056 \times 0.1032)$$
$$= (0.433 \pm 0.212)$$
$$= (0.221, 0.645).$$

(c) The 95% simultaneous confidence region is for β_0 and β_1 is provided by the set of values of β_0 and β_1 which satisfy the inequality

$$n(\hat{\beta}_0 - \beta_0)^2 + 2 \sum x_i(\hat{\beta}_0 - \beta_0)(\hat{\beta}_1 - \beta_1) + \sum x_i^2(\hat{\beta}_1 - \beta_1)^2 \leq 2S^2 F_{2,n-2;\alpha}$$

that is, the set of values of (β_0, β_1) contained by the ellipse

$$28(58.327 - \beta_0)^2 + 1524(58.327 - \beta_0)(0.433 - \beta_1) + 25684(0.433 - \beta_1)^2 = 2(52.68)(3.3690)$$

or

$$28(58.327 - \beta_0)^2 + 1524(58.327 - \beta_0)(0.433 - \beta_1) + 25684(0.433 - \beta_1)^2 = 354.958.$$

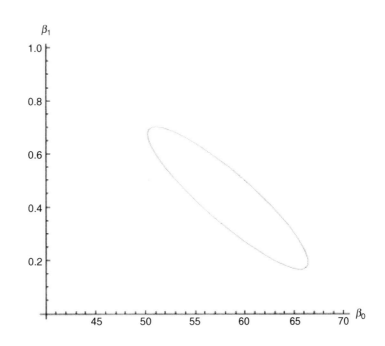

(d) A $100(1-\alpha)\%$ prediction interval for a new observation is given by

$$\hat{Y} \pm t_{n-2;\alpha/2}S\sqrt{\left(1+\frac{1}{n}+\frac{(x-\bar{x})^2}{S_{xx}}\right)}, \text{ with } \hat{Y} = 58.327 + 0.433X$$

Thus, predicted observations and prediction intervals at $X = 26, 34, 43$ are

X	Fit	SE Fit	95% PI
26	69.58	1.38	(54.40, 84.77)
34	73.05	1.54	(57.79, 88.30)
43	76.95	2.13	(61.39, 92.49)

respectively.

33. **(a)** In this problem we wish to test at 5% level of significance the hypothesis

$$H_0 : \rho = 0 \quad versus \quad H_1 : \rho > 0.$$

In order to perform this hypothesis we first need to determine the value of sample correlation coefficient r. That is,

$$r = \sum_{i=1}^{n}(x_i - \bar{x})(y_i - \bar{y}) \Big/ \sqrt{\sum_{i=1}^{n}(x_i - \bar{x})^2 \sum_{i=1}^{n}(y_i - \bar{y})^2}.$$

Using the data from Problem 3 of Section 15.2, we obtain

$$r = -0.872$$

Now from (15.9.3) the test statistic is

$$\frac{r}{\sqrt{1-r^2}}\sqrt{n-2}$$

Substituting the value of $r = -0.872$ and $n = 12$, we find that the observed value of the test statistics

$$\frac{r}{\sqrt{1-r^2}}\sqrt{n-2} = \frac{-.872}{\sqrt{1-(-.872)^2}} \times \sqrt{10} = -5.633.$$

We are dealing with the right-tail test of hypothesis, so that we reject the null hypothesis if the value of the observed test statistic is greater than

$$t_{n-2;\alpha} = t_{10;.05} = 1.812,$$

Otherwise we do not reject the hypothesis. Since in this case the observed value of the test statistics $-5.633l$ is less than 1.812, so we do not reject the null hypothesis.

(b) It can easily be seen that p-value is approximately 0.972. Since the p-value is much greater than the level of significance, we do not reject the null hypothesis, which supports our conclusion in part (a).

35. Using MINITAB (**Stat > Regression > Regression**) we obtain the following output
Regression Analysis: Loss in MDD versus Fe

```
The regression equation is
Loss in MDD = 129.787 - 24.02 Fe


Predictor      Coef      SE Coef      T       P
Constant     129.787     1.403      92.52    0.000
Fe           -24.020     1.280     -18.77    0.000


S = 3.05778     R-Sq = 97.0%     R-Sq(adj) = 96.7%
```

ANOVA

```
Source           DF      SS        MS       F       P
Regression        1     3293.8    3293.8   352.27  0.000
Residual Error   11      102.9       9.4
Total            12     3396.6


Predicted Values for New Observations when x = 1.5


New
Obs.      Fit      SE Fit       95% CI            95% PI
  1      93.757    1.167   (91.187, 96.326)  (86.553, 100.961)
```

(a) From the above MINITAB output, we have the fitted simple linear regression model as

$$\hat{Y} = 129.787 - 24.02X$$

(b) The ANOVA Table is shown in the MINITAB output.

(c) From the ANOVA Table given above we note that the p-value for regression is 0.000, which is less than 5% the level of significance. Hence, we reject the null hypothesis, and conclude that the fitted model is valuable.

37. Using MINITAB (**Stat** > **Regression** > **Regression**) we obtain the following output

Regression Analysis: Log(Loss in MDD) versus Fe

```
The regression equation is
Log(Loss in MDD) = 2.118 - 0.0984 Fe


Predictor      Coef        SE Coef      T        P
Constant     2.11765      0.00552    383.93    0.000
Fe          -0.098408     0.005032   -19.56    0.000


S = 0.0120236     R-Sq = 97.2%     R-Sq(adj) = 96.9%


Analysis of Variance


Source           DF      SS          MS         F       P
Regression        1     0.055285    0.055285   382.42  0.000
Residual Error   11     0.001590    0.000145
Total            12     0.056876
```

```
Unusual Observations

              Log(Loss
Obs.     Fe   in MDD)      Fit     SE Fit    Residual
  2     0.48   2.09342   2.07042   0.00388    0.02300


Predicted Values for New Observations (x = 1.50)

New
Obs.      Fit       SE Fit        95% CI              95% PI
  1     1.97004    0.00459   (1.95994,  1.98015)  (1.94172, 1.99837)
```

(a) The regression equation is

$$\text{Log(Loss in MDD)} = 2.118 - 0.0984(\text{Fe})$$

(b) From the above MINITAB the estimate of $\text{Log}(Y) = 1.97004$ for $X = 1.50$. Thus, estimated value of Y at $X = 1.50$ is 93.334. In Problem 35, we found the estimated value of Y at $X = 1.50$ to be equal to 93.757. Comparing the two values, we may conclude that they are practically the same.

16

MULTIPLE LINEAR REGRESSION ANALYSIS

PRACTICE PROBLEMS FOR SECTION 16.3

1. A first-order multiple linear regression model in three predictor variables can be set up as

$$Y_i = \beta_0 + \beta_1 X_{i1} + \beta_2 X_{i2} + \beta_3 X_{i3} + \varepsilon_i, \quad i = 1, 2, \cdots n$$

where X_1, X_2, X_3 are 3 independent variables, $\beta_0, \beta_1, \beta_2, \beta_3$ are the regression coefficients and ε is a random error incurred when observing Y at (X_1, X_2, X_3). We assume $E(\varepsilon) = 0, V(\varepsilon) = \sigma^2$, so that

$$E(Y_i) = \beta_0 + \beta_1 X_{i1} + \beta_2 X_{i2} + \beta_3 X_{i3}, \quad V(Y) = \sigma^2$$

In addition we assume that Y_i's are independent. This model is called *linear* because it is a linear function in regression parameters.

3. The least squares method proceeds by minimizing the sum of squared deviations of the observed Y_i $(i = 1, 2, \cdots, n)$ from true means $E(Y_i) = \beta_0 + \beta_1 X_{i1} + \beta_2 X_{i2} + \beta_3 X_{i3}$, that is, by minimizing

$$Q(\beta_0, \beta_1, \beta_2, \beta_3) = \sum_{i=1}^{n} (Y_i - \beta_0 - \beta_1 X_{i1} - \beta_2 X_{i2} - \beta_3 X_{i3})^2 \quad (16.3.1)$$

over choices of $(\beta_0, \beta_1, \beta_2, \beta_3)$. The minimization of Q in (16.3.1) is achieved by taking the partial derivatives of Q with respect to $\beta_0, \beta_1, \beta_2, and \beta_3$ and equating them to zero. When equating to zero, we denote the solution of the resulting equations by $\hat{\beta} = (\hat{\beta}_0, \hat{\beta}_1, \hat{\beta}_2, \hat{\beta}_3)$. Hence we have the $(3 + 1) = 4$ equations:

$$\frac{\partial Q}{\partial \beta_0} = -2\sum_{i=1}^{n} \left(Y_i - \hat{\beta}_0 - \hat{\beta}_1 X_{i1} - \hat{\beta}_2 X_{i2} - \hat{\beta}_3 X_{i3}\right) = 0$$

$$\frac{\partial Q}{\partial \beta_1} = -2\sum_{i=1}^{n} \left(Y_i X_{i1} - \hat{\beta}_0 X_{i1} - \hat{\beta}_1 X_{i1}^2 - \hat{\beta}_2 X_{i2} X_{i1} - \hat{\beta}_3 X_{i3} X_{i1}\right) = 0$$

$$\frac{\partial Q}{\partial \beta_2} = -2\sum_{i=1}^{n} \left(Y_i X_{i2} - \hat{\beta}_0 X_{i2} - \hat{\beta}_1 X_{i1} X_{i2} - \hat{\beta}_2 X_{i2}^2 - \hat{\beta}_3 X_{i3} X_{i2}\right) = 0$$

$$\frac{\partial Q}{\partial \beta_3} = -2\sum_{i=1}^{n} \left(Y_i X_{i3} - \hat{\beta}_0 X_{i3} - \hat{\beta}_1 X_{i1} X_{i3} - \hat{\beta}_2 X_{i2} X_{i3} - \hat{\beta}_3 X_{i3}^2\right) = 0$$

Solutions Manual to Accompany Statistics and Probability with Applications for Engineers and Scientists, Bhisham C. Gupta and Irwin Guttman.
© 2014 John Wiley & Sons, Inc. Published 2014 by John Wiley & Sons, Inc.

We have denoted the solutions to the above 4 equations by $\hat{\beta}_i$'s, which are called the least squares estimators of the regression coefficients β_i's. Also we may put the above equations in standard form, referred to as *normal equations* as follows: Expressions in $\hat{\beta}_i$'s $(i = 0, 1, 2, 3)$ appear on the left hand side, and expressions in $Y_i's$ appear on the right hand side of the equations. We then obtain the *normal equations*:

$$n\hat{\beta}_0 + \hat{\beta}_1\sum_{i=1}^{n}X_{i1} + \hat{\beta}_2\sum_{i=1}^{n}X_{i2} + \hat{\beta}_3\sum_{i=1}^{n}X_{i3} = \sum_{i=1}^{n}Y_i$$

$$\sum_{i=1}^{n}\hat{\beta}_0X_{i1} + \hat{\beta}_1\sum_{i=1}^{n}X_{i1}^2 + \hat{\beta}_2\sum_{i=1}^{n}X_{i2}X_{i1} + \hat{\beta}_3\sum_{i=1}^{n}X_{i3}X_{i1} = \sum_{i=1}^{n}Y_iX_{i1}$$

$$\sum_{i=1}^{n}\hat{\beta}_0X_{i2} + \hat{\beta}_1\sum_{i=1}^{n}X_{i1}X_{i2} + \hat{\beta}_2\sum_{i=1}^{n}X_{i2}^2 + \hat{\beta}_3\sum_{i=1}^{n}X_{i3}X_{i2} = \sum_{i=1}^{n}Y_iX_{i2}$$

$$\sum_{i=1}^{n}\hat{\beta}_0X_{i3} + \hat{\beta}_1\sum_{i=1}^{n}X_{i1}X_{i3} + \hat{\beta}_2\sum_{i=1}^{n}X_{i2}X_{i3} + \hat{\beta}_3\sum_{i=1}^{n}X_{i3}^2 = \sum_{i=1}^{n}Y_iX_{i3}$$

Now solving this system of 4 normal equations, and denoting the solutions of the unknown parameters by $\hat{\beta}_0, \hat{\beta}_1, \hat{\beta}_2,$ and $\hat{\beta}_3$, we obtain the least squares estimators for $\beta_0, \beta_1, \beta_2,$ and β_3.

5. From Problem 4, we have

$$\begin{pmatrix} \hat{\beta}_0 \\ \hat{\beta}_1 \\ \hat{\beta}_2 \end{pmatrix} = \begin{pmatrix} 20 & 1540 & 1502 \\ 1540 & 119797 & 115678.9 \\ 1502 & 115678.9 & 113104 \end{pmatrix}^{-1} \times \begin{pmatrix} 1908 \\ 148248.6 \\ 143626 \end{pmatrix}$$

Now on the right hand side, utilizing the indicated inverse matrix, we obtain

$$\begin{pmatrix} \hat{\beta}_0 \\ \hat{\beta}_1 \\ \hat{\beta}_2 \end{pmatrix} = \begin{pmatrix} 22.7459 & -0.0583103 & -0.242423 \\ -0.0583103 & 0.0008231 & -0.0000675 \\ -0.242423 & -0.0000675 & 0.003297 \end{pmatrix} \times \begin{pmatrix} 1908 \\ 148248.6 \\ 143626 \end{pmatrix}$$

$$= \begin{pmatrix} -63.5645 \\ 1.0742 \\ 1.0153 \end{pmatrix}$$

Thus, we have

$$\hat{\beta}_0 = -63.5645, \hat{\beta}_1 = 1.0742, \hat{\beta}_2 = 1.0153.$$

(b) From part (a) the fitted regression model (regression plane) is

$$\hat{Y} = -63.5645 + 1.0742X_1 + 1.0153X_2$$

Substituting the values of $X_1 = 75$ and $X_2 = 70$, in the fitted regression model, we obtain

$$\hat{Y} = 88.0716.$$

7. Using MINITAB (**Stat > Regression > Regression**) and using the predictor variables $X_2, X_3, X_4,$ and X_5, we obtain the MINITAB output as given below.

Regression Analysis: Y versus X2, X3, X4, X5

```
The regression equation is
Y = 7.46 - 0.030 X2 + 0.521 X3 - 0.102 X4 - 2.16 X5
```

```
Predictor     Coef      SE Coef         T          P
Constant     7.458       7.226       1.03      0.320
X2          -0.0297      0.2633      -0.11      0.912
X3           0.5205      0.1359       3.83      0.002
X4          -0.10180     0.05339     -1.91      0.077
X5          -2.161       2.395       -0.90      0.382
```

```
S = 0.882737    R-Sq = 67.2%    R-Sq(adj) = 57.8%
```

Analysis of Variance

```
Source           DF      SS       MS        F        P
Regression        4   22.3119   5.5780    7.16    0.002
Residual Error   14   10.9091   0.7792
Total            18   33.2211
```

```
Source           DF    Seq SS
X2                1    2.0711
X3                1   15.8899
X4                1    3.7165
X5                1    0.6343
```

Unusual Observations

```
Obs.    X2      Y       Fit    SE Fit   Residual   St Resid
 2     19.8   8.300   10.065   0.225    -1.765      -2.07R
```

```
R denotes an observation with a large standardized residual.
```

Predicted Values for New Observations

```
New Obs.     Fit     SE Fit       95% CI              95% PI
1           8.996    0.472    (7.982, 10.009)    (6.848, 11.143)
```

Values of Predictors for New Observations

```
New Obs.       X2       X3       X4       X5
1            20.0     30.0     90.0     2.00
```

From the above MINITAB output, we have the following:

(a) $\hat{Y} = 7.46 - 0.030\,X_2 + 0.521\,X_3 - 0.102\,X_4 - 2.16\,X_5$

(b) From the ANOVA table, we note that the p-value for the regression is 0.002, which is less than $\alpha = 0.05$, the level of significance. Hence, we conclude that the first order model for Y on (X_2, X_3, X_4, X_5) is significant.

(c) From the MINITAB output, we have that estimated value for the pull strength when $X_2 = 20$, $X_3 = 30$, $X_4 = 90$, and $X_5 = 2.0$. is 8.996 and the a 95% prediction interval is (6.848, 11.143).

9. In this problem, in order to test the hypothesis $H_0 : \beta_4 = \beta_5 = \beta_6 = 0$ versus H_1 : at least one of the parameters $\beta_4, \beta_5, \beta_6$ is not equal to zero, we first fit the reduced model, that is

$$Y = \beta_0 + \beta_1 X_1 + \beta_2 X_2 + \beta_3 X_3 + \varepsilon$$

Again using MINITAB, we obtain the following MINITAB output for the reduced model.

Regression Analysis: Y versus X1, X2, X3

```
The regression equation is
Y = - 4.31 + 0.521 X1 - 0.346 X2 + 0.563 X3

Predictor       Coef        SE Coef       T          P
Constant       -4.311       7.181        -0.60      0.557
X1              0.5207      0.5979        0.87       0.398
X2             -0.3458      0.2447       -1.41       0.178
X3              0.5627      0.1508        3.73       0.002


S = 0.984058    R-Sq = 56.3%    R-Sq(adj) = 47.5%


Analysis of Variance
Source           DF      SS       MS        F         P
Regression        3   18.6955   6.2318     6.44      0.005
Residual Error   15   14.5255   0.9684
Total            18   33.2211
```

(a) To test the hypothesis described in this problem we use the following test statistic (see (16.3.26) – here $SSR_1 =$ Regression sum of squares found when fitting Y on $(X_1, X_2, X_3, X_4, X_5, X_6)$ and $SSR_2 =$ Regression sum of squares found when fitting Y on (X_1, X_2, X_3).

$$F = \frac{(SSR_1 - SSR_2)/r}{SSE_1/(n - (k+1))}$$

Now from the MINITAB outputs in Problem 8 and 9 of this section, we have

$$SSR_1 = 23.6286, \quad SSR_2 = 18.6955, \quad SSE_1 = 9.5925, \quad n = 19, \quad k = 6, \quad r = 3$$

Substituting these values in the above test statistic, we obtain observed value of the test statistic, that is

$$F = \frac{(23.6286 - 18.6955)/3}{9.5925/12} = 2.057$$

From F-table, $F_{3,12;.05} = 3.4903$, which is greater than the observed value. Thus, we do not reject the null hypothesis $H_0 : \beta_4 = \beta_5 = \beta_6 = 0$.

(b) the desired p-value can easily be found to be equal to 0.1596, that is $P(F_{3,12} > 2.057) = .1596$.

11. To fit the $E(Y) = \beta_0 + \beta_1 X_1 + \beta_2 X_2$ using MINITAB (**Stat > Regression > Regression**) and using the predictor variables X_1, and X_2 we obtain the MINITAB output as given below.

Regression Analysis: Y versus X1, X2

```
The regression equation is
Y = 158 + 15.5 X1 - 0.911 X2

Predictor       Coef        SE Coef       T          P
Constant       158.21       18.01        8.78       0.000
X1              15.48        11.62        1.33       0.215
X2             -0.9106       0.1556      -5.85       0.000
S = 5.56253 R-Sq = 79.5% R-Sq(adj) = 74.9%
```

```
Analysis of Variance

Source            DF          SS        MS        F        P
Regression         2      1077.61    538.81    17.41    0.001
Residual Error     9       278.48     30.94
Total             11      1356.09
```

(a) Thus, using the above MINITAB output, we obtain the fitted model as

$$\hat{Y} = 158 + 15.5\,X1 - 0.911\,X2$$

(b) To test the hypothesis $H_0 : \beta_1 = \beta_2 = 0$ against the alternative that at least one of β_1, β_2 is not equal to zero, we use ANOVA table shown in the above MINITAB output. From the ANOVA table we find that p-value for the regression is 0.001. Hence, we reject the null hypothesis, that is the fitted model is significant. From the above MINITAB output, we note that the p-values for testing the hypotheses $H_0 : \beta_1 = 0$ *versus* $H_1 : \beta_1 \neq 0$, and $H_0 : \beta_2 = 0$ *versus* $H_1 : \beta_2 \neq 0$, are 0.215 and 0.000, respectively. Hence, we reject the hypothesis $H_0 : \beta_2 = 0$ while we fail to reject the null hypothesis $H_0 : \beta_1 = 0$.

PRACTICE PROBLEMS FOR SECTION 16.4

1. Using MINITAB (**Stat** > **Regression** > **General Regression**) we obtain the following output. Note that in the dialog box, you must include in the box under Model the following: all the indicator variable including the categorical variable; further in the box under categorical just enter the categorical variable (This is a new dialog box in version "16" of MINITAB. Using this version we just use one predictor variable irrespective of the number of categories and in the column of the predictor variables one can denote the levels by using any symbols or numbers).

The output for these data is:

General Regression Analysis: Y versus X1, X2
    Regression Equation

$$
\begin{array}{ll}
\text{X2} & \\
\text{A} & Y = 121.404 - 1.45977\,X1 \\
\text{B} & Y = 127.382 - 1.45977\,X1
\end{array}
$$

Coefficients

Term	Coef	SE Coef	T	P
Constant	124.393	28.7395	4.32830	0.003
X1	−1.45977	0.9543	−1.52968	0.170
X2				
A	−2.989	3.2781	−0.91165	0.392

Summary of Model

S = 7.96140   R-Sq = 54.56%   R-Sq(adj) = 41.58%

Note that for the MINITAB output, we have used indicator variables −1 for Lab A and 1 for Lab B, so that the coefficient of B is 2.989. Thus, the fitted model may be described as

$$\hat{Y} = 124.393 - 1.460X_1 + 2.989\,\text{Lab}$$

3. Using MINITAB (**Stat** > **Regression** > **General Regression**) we obtain the following output.
General Regression Analysis: Y versus X1, X2

```
Regression Equation
X2
-1    Y    = 21.8029 + 0.038751 X1
1     Y    = 28.1439 + 0.038751 X1
```

```
Coefficients

Term         Coef        SE Coef      T           P
Constant     24.9734     7.05457      3.54003     0.004
X1            0.0388     0.11911      0.32533     0.750
X2
 -1           3.1705     1.27284      2.49087     0.027
```

```
Summary of Model

S = 3.56698        R-Sq = 44.68%        R-Sq(adj) = 36.17%
PRESS = 243.092    R-Sq(pred) = 18.70%
```

```
Analysis of Variance

Source        DF    Seq SS     Adj SS     Adj MS      F          P
Regression     2    133.597    133.597    66.7983    5.25006    0.021315
X1             1     54.656      1.347     1.3466    0.10584    0.750111
X2             1     78.941     78.941    78.9409    6.20442    0.027051
Error         13    165.403    165.403    12.7233
Lack-of-Fit   10    104.403    104.403    10.4403    0.51346    0.814021
Pure Error     3     61.000     61.000    20.3333
Total         15    299.000
```

From the above MINITAB output, we note that the fitted model is

X2

Private : $\hat{Y} = 21.8029 + 0.038751\,X1$

Public : $\hat{Y} = 28.1439 + 0.038751\,X1$

5. Using MINITAB (**Stat > Regression > General Regression**) we obtain the following output. Note that in the column of X_5 the categorical variable, we enter the first seven entries as Wood, next seven Vinyl, next seven Stucco and the last seven entries as Brick.

General Regression Analysis: Y versus X1, X2, X3, X4, X5

```
Regression Equation
X5
Brick Y  = -18.3568 + 2.84465 X1 + 28.6835 X2 - 2.41353 X3 + 25.8673 X4
Stucco Y = -27.8076 + 2.84465 X1 + 28.6835 X2 - 2.41353 X3 + 25.8673 X4
Vinyl Y  = -5.33475 + 2.84465 X1 + 28.6835 X2 - 2.41353 X3 + 25.8673 X4
Wood Y   = -31.013  + 2.84465 X1 + 28.6835 X2 - 2.41353 X3 + 25.8673 X4
```

```
Coefficients

Term         Coef        SE Coef      T           P
Constant    -20.6280     16.0247     -1.28726     0.213
X1            2.8447      1.8701      1.52112     0.144
X2           28.6835      7.7671      3.69296     0.001
X3           -2.4135     11.5394     -0.20916     0.836
X4           25.8673      8.0309      3.22095     0.004
```

```
X5
  Brick        2.2712      5.6474      0.40217     0.692
  Stucco      -7.1796      5.6366     -1.27375     0.217
  Vinyl       15.2933      6.4561      2.36883     0.028
```

Note that X5 for wood is equal to $-2.2712 + 7.1796 - 15.2933 = -10.3851$.

Summary of Model

S = 16.8621   R-Sq = 92.14%   R-Sq(adj) = 89.39%
PRESS = 11748.6   R-Sq(pred) = 83.76%

Analysis of Variance

Source	DF	Seq SS	Adj SS	Adj MS	F	P
Regression	7	66662.4	66662.4	9523.19	33.4934	0.000000
X1	1	57219.2	657.9	657.88	2.3138	0.143884
X2	1	4096.3	3877.7	3877.68	13.6380	0.001441
X3	1	813.6	12.4	12.44	0.0437	0.836442
X4	1	2425.2	2949.8	2949.80	10.3745	0.004286
X5	3	2108.1	2108.1	702.69	2.4714	0.091380
Error	20	5686.6	5686.6	284.33		
Total	27	72349.0				

Fits and Diagnostics for Unusual Observations

Obs	Y	Fit	SE Fit	Residual	St Resid
4	116	83.944	8.74001	32.0563	2.22301 R
11	114	150.144	9.56958	-36.1436	-2.60334 R

R denotes an observation with a large standardized residual.

Predicted Values for New Observations at (X1=25, X2=4, X3=3, X4=2)

	Fit	SE Fit	95% CI	95% PI
Wood:	199.332	12.3104	(173.653, 225.011)	(155.782, 242.881)
Vinyl:	225.010	8.9277	(206.387, 243.633)	(185.210, 264.809)
Stucco:	202.537	11.9704	(177.567, 227.507)	(159.401, 245.672)
Brick:	211.988	11.8966	(187.172, 236.804)	(168.941, 255.034)

$$\text{Brick:} \quad \hat{Y} = -18.3568 + 2.84465\,X1 + 28.6835\,X2 - 2.41353\,X3 + 25.8673\,X4$$
$$\text{Stucco:} \quad \hat{Y} = -27.8076 + 2.84465\,X1 + 28.6835\,X2 - 2.41353\,X3 + 25.8673\,X4$$
$$\text{Vinyl:} \quad \hat{Y} = -5.33475 + 2.84465\,X1 + 28.6835\,X2 - 2.41353\,X3 + 25.8673\,X4$$
$$\text{Wood:} \quad \hat{Y} = -31.013 + 2.84465\,X1 + 28.6835\,X2 - 2.41353\,X3 + 25.8673\,X4$$

Further, note that the p-value for the overall model is "zero" and the R^2 value is quite high. Hence, we consider overall model is quite good but β_1, β_3, and β_5 are not significantly different from zero at the 5% level of significance, since their corresponding p-values are greater than 5%.

Note: *Since in this problem the categorical variable has four categories, in the design matrix the MINITAB uses (1, 0, 0), (0, 1, 0), (0, 0, 1) and (−1, −1, −1) for Brick, Stucco, Vinyl, and Wood, respectively)*

7. Using MINITAB (**Stat** > **Regression** > **Stepwise**). Under Methods, we use $\alpha = .15$, for alpha to enter a predictor variable.

Stepwise Regression: Y versus X1, X2, X3, X4

Alpha-to-Enter: 0.15 Alpha-to-Remove: 0.15 (MINITAB uses these values by default)
Response is Y on 4 predictors, with N = 28

Step	1	2	3
Constant	-28.38	-36.38	-27.84
X1	10.8	8.1	4.6
T-Value	9.92	6.70	2.74
P-Value	0.000	0.000	0.011
X4		28.9	24.1
T-Value		3.37	3.08
P-Value		0.002	0.005
X2			20.5
T-Value			2.75
P-Value			0.011
S	24.1	20.4	18.1
R-Sq	79.09	85.63	89.08
R-Sq(adj)	78.28	84.48	87.71
Mallows Cp	20.6	8.7	3.3

Thus, the fitted model is

$$\hat{Y} = -27.84 + 4.6X_1 + 20.5X_2 + 24.1X_4$$

This model is quite similar to the one obtained in Example 16.4.2, except it did not include the predictor variable X_3. R-sq and R-sq(adj) values are almost the same.

9. Using MINITAB (**Stat** > **Regression** > **General Regression**), we obtain the following output. Note that in this problem we have five predictor variables four of which are quantitative and one is qualitative (siding). The qualitative variable has four categories, which we denote here by 0,1, 2, and 3, for Wood, Vinyl, Stucco, and Brick, respectively. As mentioned earlier the version 16 of MINITAB requires to use only one predictor variable for each qualitative variable irrespective of how many categories it has.

General Regression Analysis: Y versus X1, X2, X3, X4, X5

Regression Equation

X5
0	Y = -31.013 + 2.84465 X1 + 28.6835 X2 - 2.41353 X3 + 25.8673 X4
1	Y = -5.33475 + 2.84465 X1 + 28.6835 X2 - 2.41353 X3 + 25.8673 X4
2	Y = -27.8076 + 2.84465 X1 + 28.6835 X2 - 2.41353 X3 + 25.8673 X4
3	Y = -18.3568 + 2.84465 X1 + 28.6835 X2 - 2.41353 X3 + 25.8673 X4

Coefficients

Term	Coef	SE Coef	T	P
Constant	-20.6280	16.0247	-1.28726	0.213
X1	2.8447	1.8701	1.52112	0.144
X2	28.6835	7.7671	3.69296	0.001
X3	-2.4135	11.5394	-0.20916	0.836
X4	25.8673	8.0309	3.22095	0.004
X5				
0	-10.3849	5.9488	-1.74572	0.096
1	15.2933	6.4561	2.36883	0.028
2	-7.1796	5.6366	-1.27375	0.217

```
Summary of Model
S = 16.8621      R-Sq = 92.14%   R-Sq(adj) = 89.39%
PRESS = 11748.6   R-Sq(pred) = 83.76%
```

Analysis of Variance

Source	DF	Seq SS	Adj SS	Adj MS	F	P
Regression	7	66662.4	66662.4	9523.19	33.4934	0.000000
X1	1	57219.2	657.9	657.88	2.3138	0.143884
X2	1	4096.3	3877.7	3877.68	13.6380	0.001441
X3	1	813.6	12.4	12.44	0.0437	0.836442
X4	1	2425.2	2949.8	2949.80	10.3745	0.004286
X5	3	2108.1	2108.1	702.69	2.4714	0.091380
Error	20	5686.6	5686.6	284.33		
Total	27	72349.0				

Fits and Diagnostics for Unusual Observations

Obs.	Y	Fit	SE Fit	Residual	St Resid
4	116	83.944	8.74001	32.0563	2.22301 R
11	114	150.144	9.56958	-36.1436	-2.60334 R

R denotes an observation with a large standardized residual.

The AVOVA table shows the p-value for regression is 0.000, so the fitted model is significant at any level of significance.

Note: *Since in this problem the categorical variable has four categories, in the design matrix the MINITAB uses (1, 0, 0), (0, 1, 0), (0, 0, 1) and (−1, −1, −1), respectively for Brick, Stucco, Vinyl, and Wood, irrespective of what we used, i. e., 0, 1, 2, and 3 to denote these categories.*

PRACTICE PROBLEMS FOR SECTIONS 16.6 AND 16.7

1. Using MINITAB (**Stat** > **Regression** > **Stepwise**) we obtain the following output.
   (a) **Stepwise Regression: Y versus X1, X2, X3, X4, X5, X6, X7**

   ```
   Alpha-to-Enter: 0.15 Alpha-to-Remove: 0.15 (MINITAB uses these values by default)

   Response is Y on 7 predictors, with N = 38
   ```

Step	1	2	3	4
Constant	4.941	6.202	8.318	9.214
X4	1.57	1.40	1.12	1.19
T-Value	7.73	8.60	6.42	6.73
P-Value	0.000	0.000	0.000	0.000
X6		-1.89	-2.76	-2.61
T-Value		-4.87	-6.13	-5.79
P-Value		0.000	0.000	0.000
X7			-1.22	-1.09
T-Value			-3.06	-2.73
P-Value			0.004	0.010
X5				-0.32
T-Value				-1.56
P-Value				0.128

S	1.27	0.996	0.895	0.876
R-Sq	62.42	77.59	82.42	83.63
R-Sq(adj)	61.37	76.31	80.87	81.64
Mallows Cp	35.4	9.4	2.5	2.2
PRESS	63.9565	40.4058	33.9935	33.0817
R-Sq(pred)	58.68	73.90	78.04	78.63

From the above MINITAB printout we can see that the predictor variables X_1, X_2, and X_3 have been excluded from the model since their p-value should be greater than 0.25. Also, p-value for X_5 is greater than .05. This indicates that the quality of the wine depends more upon the X_4 flavor and the region (predictor variables X_6 and X_7 are the indicator variables representing the region). With theses predictor variables in the R-sq(pred) value is also reasonable high. Thus the fitted model may be described as

$$\hat{Y} = 8.318 + 1.12X_4 - 2.76X_6 - 1.22X_7$$

(b) Using the PRESS statistics criterion we arrive at the same model as in part (a).

3. Using MINITAB (**Stat** > **Regression** > **Stepwise**) and forcing the predictor variables X_1, X_2 into the model we obtain the following output.

Stepwise Regression: Y versus X1, X2, X3, X4, X5, X6, X7

Alpha-to-Enter: 0.2 Alpha-to-Remove: 0.25

Response is Y on 7 predictors, with N = 38

Step	1	2	3	4	5
Constant	6.184	3.354	6.306	9.028	9.144
X1	-0.3	1.1	-0.5	-0.7	-0.2
T-Value	-0.13	0.63	-0.38	-0.51	-0.12
P-Value	0.898	0.532	0.709	0.615	0.903
X2	1.34	0.48	0.31	-0.00	0.09
T-Value	5.92	1.71	1.36	-0.00	0.38
P-Value	0.000	0.096	0.184	0.998	0.707
X4		1.21	1.16	1.10	1.14
T-Value		4.04	4.82	4.94	5.18
P-Value		0.000	0.000	0.000	0.000
X6			-1.82	-2.83	-2.52
T-Value			-4.42	-5.25	-4.43
P-Value			0.000	0.000	0.000
X7				-1.24	-1.00
T-Value				-2.63	-2.03
P-Value				0.013	0.051
X5					-0.33
T-Value					-1.48
P-Value					0.149
S	1.49	1.24	0.997	0.918	0.902
R-Sq	50.05	66.25	78.80	82.56	83.71
R-Sq(adj)	47.20	63.27	76.23	79.84	80.56
Mallows Cp	60.3	32.3	11.2	6.2	6.1
PRESS	92.8220	66.0752	47.3507	40.9682	40.0942
R-Sq(pred)	40.03	57.31	69.41	73.53	74.10

The fitted model is: $\hat{Y} = 9.144 - 0.2X1 + 0.09\,X2 + 1.14X4 - 0.33X5 - 2.52X6 - X7$

Note that by forcing the predictor variables X_1 and X_2 into the model, the model did not change since the regression coefficient for X_2 is almost zero and for X_1 the p-value is almost 90%. Moreover, by forcing the predictor variables X_1 and X_2 into the model, the value of PRESS statistic has rather edged up. Thus, we can conclude that the model in Problem 1 is better.

5. Using MINITAB we ran three different procedures (1) using all the predictor variables in the model $Y = \beta_0 + \beta_1 X_1 + \beta_2 X_2 + \beta_3 X_3 + \beta_{11} X_1^2 + \beta_{22} X_2^2 + \beta_{33} X_3^2 + \varepsilon$ (2) Only using the predictor variables that turned out to be significant in (1), that is, X_3 and X_1^2, and (3) using the stepwise regression method. All the three MINITAB outputs are shown below. Each time we arrived at the same model (except some minor changes in regression coefficients), that is

$$\hat{Y} = 0.148 - 0.00725\,X_3 + 1.59\,X_1^2$$

Regression Analysis: Y versus X1, X2, X3, X1*X1, X2*X2, X3*X3

The regression equation is
Y = - 0.143 + 0.063 X1 + 0.00443 X2 - 0.00666 X3 + 1.56 X1*X1 - 0.00342 X2*X2
    - 0.00079 X3*X3

Predictor	Coef	SE Coef	T	P
Constant	-0.1430	0.4881	-0.29	0.773
X1	0.0632	0.1135	0.56	0.584
X2	0.004427	0.006811	0.65	0.523
X3	-0.006665	0.002801	-2.38	0.028
X1*X1	1.5635	0.1462	10.69	0.000
X2*X2	-0.003416	0.006433	-0.53	0.602
X3*X3	-0.000785	0.001124	-0.70	0.493

S = 0.0233065   R-Sq = 98.6%   R-Sq(adj) = 98.1%
PRESS = 0.0255248   R-Sq(pred) = 96.43%

Analysis of Variance

Source	DF	SS	MS	F	P
Regression	6	0.70474	0.11746	216.23	0.000
Residual Error	19	0.01032	0.00054		
Total	25	0.71506			

Unusual Observations

Obs	X1	Y	Fit	SE Fit	Residual	St Resid
25	7.3	0.25100	0.20624	0.01143	0.04476	2.20R
26	7.6	0.00002	-0.01081	0.02165	0.01083	1.26 X

R denotes an observation with a large standardized residual.
X denotes an observation whose X value gives it large leverage.

Regression Analysis: Y versus X3, X1*X1

The regression equation is
Y = 0.148 - 0.00725 X3 + 1.59 X1*X1

Predictor	Coef	SE Coef	T	P
Constant	0.14772	0.01420	10.40	0.000
X3	-0.007246	0.001415	-5.12	0.000
X1*X1	1.58934	0.07201	22.07	0.000

```
S = 0.0218644        R-Sq = 98.5%    R-Sq(adj) = 98.3%
PRESS = 0.0129044    R-Sq(pred) = 98.20%
```

Analysis of Variance

Source	DF	SS	MS	F	P
Regression	2	0.70406	0.35203	736.39	0.000
Residual Error	23	0.01100	0.00048		
Total	25	0.71506			

Source	DF	Seq SS
X3	1	0.47119
X1*X1	1	0.23288

Unusual Observations

Obs	X3	Y	Fit	SE Fit	Residual	St Residual
15	6.6	0.23200	0.18544	0.00490	0.04656	2.19R
16	5.0	0.30600	0.26031	0.00434	0.04569	2.13R
26	20.7	0.00002	-0.00227	0.01794	0.00229	0.18 X

R denotes an observation with a large standardized residual.
X denotes an observation whose X value gives it large leverage.

Stepwise Regression: Y versus X1, X2, X3, X1*X1, X2*X2, X3*X3

```
Alpha-to-Enter: 0.15 Alpha-to-Remove: 0.15 (MINITAB uses these values by default)
Response is Y on 6 predictors, with N = 26
Step              1            2
Constant      0.08324      0.14772

X1*X1          1.860        1.589
T-Value        26.56        22.07
P-Value        0.000        0.000

X3                         -0.0072
T-Value                     -5.12
P-Value                     0.000

S             0.0313       0.0219
R-Sq          96.71        98.46
R-Sq(adj)     96.57        98.33
Mallows Cp    21.3          0.2
PRESS         0.027522     0.012904
R-Sq(pred)    96.15        98.20
```

(b) The value of C_p is very small which means total mean squared error is small. If the value of C_p is small and is close to p (but smaller than p), where $p =$ the number of terms in the model then the model has smaller bias due to sampling error.

7. (a) Using MINITAB and stepwise regression technique following output is obtained.

Stepwise Regression: Y versus X1, X2, X3, X4, X5, X6, X1*X5, X5*X6

```
Alpha-to-Enter: 0.15 Alpha-to-Remove: 0.15 (MINITAB uses these values by default)

Response is Y on 8 predictors, with N = 19
```

```
Step             1              2
Constant      -8.971          4.656

X3             0.59           0.51
T-Value        3.94           3.91
P-Value        0.001          0.001

X4                           -0.124
T-Value                      -2.78
P-Value                       0.013

S              1.01           0.855
R-Sq          47.72          64.76
R-Sq(adj)     44.65          60.35
Mallows Cp     6.3            1.3
PRESS         21.5665        15.3498
R-Sq(pred)    35.08          53.79
```

This gives us the same model as in Problem 6 above. The stepwise process did not include the interaction terms in the final model.

(b) Both models fit equally good, since the PRESS and/or R^2_{Pred} criteria are the same.

PRACTICE PROBLEMS FOR SECTION 16.8

1. Using MINITAB we obtain the following output:

Binary Logistic Regression: Y versus X

```
Link Function: Logit

Response Information

Variable      Value          Count
Y               1             4 (Event)
                0             6
              Total          10
```

Logistic Regression Table

Predictor	Coef	SE Coef	Z	P	Odds Ratio	95% CI Lower	95% CI Upper
Constant	-22.1696	14.0708	-1.58	0.115			
X	0.706755	0.452642	1.56	0.118	2.03	0.83	4.92

```
Log-Likelihood = -4.899
Test that all slopes are zero: G = 3.662, DF = 1, P-Value = 0.056
```

Goodness-of-Fit Tests

Method	Chi-Square	DF	P
Pearson	3.62166	6	0.728

The fitted logistic model is

$$\hat{Y} = \frac{e^{-22.1696+0.706755X_1}}{1 + e^{-22.1696+0.706755X_1}}$$

and using logit transformation given in (16.8.4), we obtain

$$\hat{\eta} = -22.1696 + 0.706755X_1$$

This can be interpreted (based on these data) as follows: the estimated increase in probability of a candidate getting into medical school with an increase of one point in MCAT scores is $(e^{0.706755} - 1)100\% \% = 102.7\%$ or in other words, the chances of getting into medical school with an increase of one point in MCAT scores becomes a little over double.

3. Using MINITAB we obtain the following output:

Binary Logistic Regression: Y versus X

```
Link Function: Logit

Response Information

Variable      Value         Count
Y                 1             8    (Event)
                  0             7
              Total            15
```

Logistic Regression Table

Predictor	Coef	SE Coef	Z	P	Odds Ratio	95% CI Lower	CI Upper
Constant	-1.59105	2.44992	-0.65	0.516			
X	0.117781	0.163469	0.72	0.471	1.12	0.82	1.55

```
Log-Likelihood = -10.093
Test that all slopes are zero: G = 0.542, DF = 1, P-Value = 0.462
Goodness-of-Fit Tests
```

Method	Chi-Square	DF	P
Pearson	10.9492	8	0.205

The fitted logistic model is

$$\hat{Y} = \frac{e^{-1.59105 + 0.117881X_1}}{1 + e^{-1.59105 + 0.117881X_1}}$$

and using logit transformation given in (16.8.4), we obtain

$$\hat{\eta} = -1.59105 + 0.117881X_1$$

This can be interpreted (based on these data) as follows: the estimated increase in probability for an engineer's performance to become satisfactory with an increase of one year service is $(e^{0.117881} - 1)100\%$, that is, 12.5%.

REVIEW PRACTICE PROBLEMS

1. The model

$$Y_i = \beta_0 + \beta_1 X_{i1} + \beta_2 X_{i2} + \beta_3 X_{i3} + \beta_4 X_{i4} + \varepsilon_i \qquad i = 1, 2, \cdots, 10,$$

using matrix notation can be written as:

$$\mathbf{Y} = \mathbf{X}\beta + \varepsilon,$$

where

$$Y[10 \times 1] = \begin{bmatrix} Y_1 \\ Y_2 \\ \vdots \\ Y_{10} \end{bmatrix}, \quad X[10 \times 5] = \begin{bmatrix} 1 & X_{11} & X_{12} & \cdots & X_{14} \\ 1 & X_{21} & X_{22} & \cdots & X_{2,4} \\ \vdots & \vdots & \vdots & & \vdots \\ 1 & X_{10,1} & X_{10,2} & \cdots & X_{10,4} \end{bmatrix}, \quad \beta[5 \times 1] = \begin{bmatrix} \beta_0 \\ \beta_1 \\ \vdots \\ \beta_4 \end{bmatrix}, \quad \varepsilon[10 \times 1] = \begin{bmatrix} \varepsilon_1 \\ \varepsilon_2 \\ \vdots \\ \varepsilon_n \end{bmatrix}$$

The least squares normal equations are given by

$$(X'X)\hat{\beta} = X'Y$$

and assuming that X is a full-rank matrix, the least squares estimators for regression coefficient are given by

$$\hat{\beta} = (X'X)^{-1}X'Y.$$

3. We know from equation (16.4.15) that

$$H = X(X'X)^{-1}X'$$

Thus, we have

$$H^2 = H \times H = (X(X'X)^{-1}X') \times (X(X'X)^{-1}X') = X(X'X)^{-1}X'X(X'X)^{-1}X' = X(X'X)^{-1}X' = H$$

which shows that the *HAT* matrix H is an idempotent matrix.

Now using the distributive law and the fact that H is an idempotent matrix, we have

$$(I - H)^2 = (I - H)(I - H) = (I - H)I - (I - H)H$$
$$= I - H - H + H^2 = I - H - H + H = I - H.$$

that is,

$$(I - H)^2 = I - H.$$

Thus, (I-H) is also an idempotent matrix.

5. Using the data of Problem 16.7 and the predictor variables X_1, X_3, X_4, and X_6, the MINITAB output for the fitted model is as given below.

Regression Analysis: Y versus X1, X3, X4, X6

```
The regression equation is
Y = 0.41 + 0.715 X1 + 0.485 X3 - 0.137 X4 + 1.07 X6

Predictor       Coef        SE Coef         T          P
Constant        0.405       7.285           0.06       0.956
X1              0.7153      0.5649          1.27       0.226
X3              0.4853      0.1357          3.57       0.003
X4              -0.13719    0.04856        -2.83       0.013
X6              1.075       1.404           0.77       0.457

S = 0.863504    R-Sq = 68.6%    R-Sq(adj) = 59.6%
PRESS = 19.0250    R-Sq(pred) = 42.73%

Analysis of Variance

Source          DF          SS             MS         F         P
Regression      4           22.7821        5.6955     7.64      0.002
Residual Error  14          10.4389        0.7456
Total           18          33.2211
```

Thus, we have

(a) $\hat{Y} = 0.41 + 0.715\,X_1 + 0.485\,X_3 - 0.137\,X_4 + 1.07\,X_6$

(b) From the ANOVA table, we have

$$\hat{\sigma}^2 = 0.7456.$$

7. We know that variance-covariance matrix of parameter vector β is given by

$$Var(\hat{\beta}) = (X'X)^{-1}\sigma^2$$

Now from the solution of 5(b), we have

$$\hat{\sigma}^2 = 0.7456.$$

Thus, an estimate of variance-covariance matrix of parameter vector $\beta[5 \times 1]$ is given by

$$Estimate \text{ of } \left(Var(\hat{\beta})\right) = (X'X)^{-1}\hat{\sigma}^2$$

Now from the estimate of variance-covariance matrix of parameter vector $\beta[5 \times 1]$, we can easily determine that the variance-covariance matrix of the estimators $\hat{\beta}_3$ and $\hat{\beta}_4$ in the model

$$E(Y) = \beta_0 + \beta_1 X_1 + \beta_3 X_3 + \beta_4 X_4 + \beta_6 X_6$$

is given by

$$\begin{pmatrix} 0.024713 & 0.001563 \\ 0.001563 & 0.003163 \end{pmatrix}(0.7456).$$

9. From the MINITAB output in Problem 5, the ANOVA table is as shown below.

		ANOVA			
Source	DF	SS	MS	F	P
Regression	4	22.7821	5.6955	7.64	0.002
Residual Error	14	10.4389	0.7456		
Total	18	33.2211			

Thus, we obtain

$$R^2 = \frac{SSR}{SS_{Total}} = \frac{22.7821}{33.2211} = 68.6\%.$$

Now from equation (16.3.20), we have

$$R^2_{adj} = 1 - \frac{n-1}{n-(k+1)}\frac{SSE}{SS_{Total}}$$

Substituting the values of n, k, SSE, and SS_{Total}, we obtain

$$R^2_{adj} = 1 - \frac{18}{14} \times \frac{10.4389}{33.2211} = 1 - \frac{187.9002}{465.0954} = 1 - 0.404 = 59.6\%.$$

11. Using the results of Problem 7 of Section 16.3 and Problem 10 above, it can be verified that

(a) (From the solution of Problem 7) As can easily be seen,

$$\hat{\sigma}_{\hat{\beta}_2} = 0.2633 \quad \hat{\sigma}_{\hat{\beta}_3} = 0.1359, \quad \hat{\sigma}_{\hat{\beta}_4} = 0.05339, \quad \hat{\sigma}_{\hat{\beta}_5} = 2.395$$

We know that the test statistic to test the hypothesis $H_0 : \beta_i = 0$ Vs $H_1 : \beta_i \neq 0$, is given by

$$\frac{\hat{\beta}_i - 0}{\hat{\sigma}_{\hat{\beta}_i}}$$

Substituting the values of $\hat{\beta}_i$ and $\hat{\sigma}_{\hat{\beta}_i}$, $i = 2, 3, 4, 5$, we find that the observed values of test statistics are -0.11, 3.83, -1.91 and -0.90, for $i = 2, 3, 4$, and 5, respectively. Now, we have $n = 19, k + 1 = 5$, so that $t_{n-5:.025} = t_{14:.025} = 2.145$. Hence, for $i = 2, 4$ and 5, we fail to reject the null hypothesis.

(b) A 95% confidence interval for β_i is given by

$$\hat{\beta}_i \pm t_{14:.025}\hat{\sigma}_{\hat{\beta}_i}$$

Thus, 95% confidence intervals for $\hat{\beta}_i$, $i = 2, 3, 4, 5$, are given by (see the solution of Problem 7) $(-0.0297 \pm (2.145 \times 0.2633))$, $(0.5205 \pm (2.145 \times 0.1359))$, $(-0.10180 \pm (2.145 \times 0.05339))$, $(-2.161 \pm (2.145 \times 2.395))$, or (-0.0297 ± 0.56478), (0.5205 ± 0.29151), (-0.10180 ± 0.11452), (-2.161 ± 5.13727) respectively.

13. Using MINITAB it can be verified that the ANOVA Table for Problem 8 of Section 16.3 is as shown below.

Analysis of Variance

Source	DF	SS	MS	F	P
Regression	6	23.6286	3.9381	4.93	0.009
Residual Error	12	9.5924	0.7994		
Total	18	33.2211			

Thus, we obtain

$$R^2 = \frac{SSR}{SS_{Total}} = \frac{23.6286}{33.2211} = 71.1\%.$$

Now from equation (16.3.20), we have

$$R^2_{adj} = 1 - \frac{n-1}{n-(k+1)}\frac{SSE}{SS_{Total}}$$

Substituting $n = 19$, $k = 6$, $SSE = 9.5924$, and $SS_{Total} = 33.2211$, we obtain

$$R^2_{adj} = 1 - \frac{18}{12} \times \frac{9.5924}{33.2211} = 1 - \frac{172.6632}{398.6532} = 1 - 0.433 = 56.7\%.$$

15. An answer to the question whether or not we should use the model developed in Problem 5 above to find a 95% confidence interval for $E(Y)$ and/or a 95% prediction interval for the pull strength Y at $X_1 = 8.0$, $X_3 = 38$, $X_4 = 80$, and $X_6 = 3.2$ is **No**. Because the given values of the predictor variables fall outside the experimental region, where we do not know what is the shape of the model. That is, we do not know whether or not the developed model outside the experimental region holds or not.

17. In this problem we fit the model $E(Y) = \beta_0 + \beta_1 X_1 + \beta_2 X_2$, using MINITAB (**Stat > Regression > Regression**). We then obtain the following MINITAB output:

Regression Analysis: Y versus X1, X2

```
The regression equation is
Y = 18.0 + 1.50 X1 + 0.877 X2

Predictor       Coef      SE Coef        T          P
Constant      18.0000      0.9790     18.39      0.000
X1             1.4997      0.3497      4.29      0.002
X2             0.8770      0.4294      2.04      0.071

S = 3.39142    R-Sq = 75.1%    R-Sq(adj) = 69.6%
PRESS = 199.686    R-Sq(pred) = 52.00%

Analysis of Variance
Source           DF        SS          MS         F              P
Regression        2      312.48      156.24     13.58          0.002
Residual Error    9      103.52       11.50
Total            11      416.00
```

Thus, from the above MINITAB output the fitted model is given by

$$\hat{Y} = 18.0 + 1.50\,X1 + 0.877\,X2$$

19. Using MINITAB (**Stat** > **Regression** > **General Regression**), we obtain the following output.
General Regression Analysis: Y versus X1, X2, X3

```
The regression equation is

Y = - 1.77 + 0.421 X1 + 0.222 X2 - 0.128 X3 - 0.0193 X1*X1 - 0.0074 X2*X2
+ 0.00082 X3*X3 - 0.0199 X1*X2 + 0.00915 X1*X3 + 0.00258 X2*X3

Predictor       Coef      SE Coef        T          P
Constant      -1.769       1.287      -1.37      0.188
X1             0.4208      0.2942      1.43      0.172
X2             0.2225      0.1307      1.70      0.108
X3            -0.12800     0.07025    -1.82      0.087
X1*X1         -0.01932     0.01680    -1.15      0.267
X2*X2         -0.00745     0.01205    -0.62      0.545
X3*X3          0.000824    0.001441    0.57      0.575
X1*X2         -0.01988     0.01204    -1.65      0.118
X1*X3          0.009151    0.007621    1.20      0.247
X2*X3          0.002576    0.007039    0.37      0.719

S = 0.0609233    R-Sq = 91.7%    R-Sq(adj) = 87.0%

Analysis of Variance

Source           DF        SS           MS         F         P
Regression        9      0.655671     0.072852   19.63     0.000
Residual Error   16      0.059386     0.003712
Total            25      0.715057

Unusual Observations
Obs.     X1       Y         Fit      SE Fit     Residual    St Resid
 1       7.3    0.2220    0.2727    0.0555     -0.0507      -2.02R
10       8.4    0.4560    0.3485    0.0310      0.1075       2.05R
25       7.3    0.2510    0.1285    0.0261      0.1225       2.23R
26       7.6    0.0000   -0.0246    0.0603      0.0246       2.80R

R denotes an observation with a large standardized residual.
```

From the above MINTAB output we obtain the following:

(a) $\hat{Y} = -1.77 + 0.421\ X_1 + 0.222\ X_2 - 0.128\ X_3 - 0.0193\ X_1^2 - 0.0074X_2^2 + 0.00082\ X_3^2 - 0.0199\ X_1X_2 + 0.00915$
$X_1X_3 + 0.00258\ X_2X_3$.

(b) From the ANOVA table the p-value for regression is zero, and R^2 value is quite high. We conclude that the model is significant. Note that the above output shows that none of the individual predictor variables are significant. This often happens because some of the predictor variables may be significant, but in the presence of other terms in the model, may not be significant. For example by running stepwise regression we found X_1, X_3, and X_2^2 are highly significant.

(c)

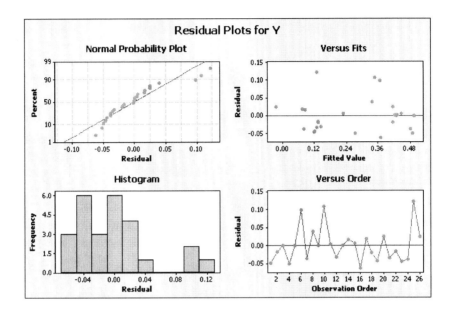

The above residual plots do not indicate any abnormalities about the assumptions, that is, the model is adequate.

(d) Here we first fit reduced model, that is, $E(Y) = \beta_0 + \beta_1X_1 + \beta_2X_2 + \beta_3X_3$. Thus, the fitted model is

Regression Analysis: Y versus X1, X2, X3

```
The regression equation is
Y = - 0.369 + 0.0865 X1 + 0.0244 X2 - 0.0286 X3


Predictor      Coef          SE Coef          T           P
Constant    -0.3693          0.1436        -2.57       0.017
X1           0.08651         0.01600         5.41       0.000
X2           0.02441         0.01023         2.39       0.026
X3          -0.028577        0.004477       -6.38       0.000


S = 0.0658146    R-Sq = 86.7%    R-Sq(adj) = 84.9%

PRESS = 0.174489    R-Sq(pred) = 75.60%


Analysis of Variance

Source            DF        SS          MS          F         P
Regression         3      0.61976     0.20659     47.69     0.000
Residual Error    22      0.09529     0.00433
Total             25      0.71506
```

```
Source  DF     Seq SS
X1       1    0.41529
X2       1    0.02802
X3       1    0.17645

Unusual Observations

Obs    X1       Y      Fit    SE Fit    Residual   St Resid
26    7.6   0.0000  -0.1130   0.0492      0.1130    2.58RX
```

Now to test the hypothesis that contribution of the second-order terms is zero at $\alpha = 0.05$ level of significance we use the test statistic

$$F = \frac{(SSR_1 - SSR_2)/r}{SSE/[n - (k+1)]}$$

Here, we have $n = 26$, $k = 9$, $r = 6$, $SSR_1 = 0.655671$, $SSR_2 = 0.61976$, and $SSE = 0.059386$. Substituting these values the observed value of the test statistic is 1.6125, which is less than the Table value $F_{6,16;05} = 2.74$. Hence, we fail to reject the null hypothesis that the contribution of the second-order terms is insignificant. p-value $= 0.2076$. However, note that in part (b) we saw that the term X_2^2 is highly significant.

21. **(a)** Using MINITAB we obtain the following fitted model.

$\hat{Y} = -0.259 + 0.0782\,X_1 + 0.121\,X_2 - 0.110\,X_3 - 0.0124\,X_1X_2 + 0.00842\,X_1X_3 + 0.00233\,X_2X_3$

(b) The Mallows' C_p, PRESS and R^2_{Pred} statistics are 7.0, 0.429932, and 39.87 respectively. Comparing these statistics with those obtained in Problem 20 (b), we find that each statistic indicates that the model fitted in Problem 20, is a better model.

23. From equations (16.4.17) and (16.4.18) we know that the residuals and their covariance matrix are given by $(I - H)Y$ and $(I - H)\,\sigma^2$ respectively, where matrix H is given in the solution of Problem 22 above. By substituting the value of $\hat{\sigma} = 0.0658$, (from the solution of Problem 20) and the matrix H given above one can easily find the residuals and their covariance matrix. Because of the large size of the hat matrix we do not present theses results here.

25. Using the information in this problem it can be seen that the least square normal equations for the regression model

$$Y_i = \beta_0 + \beta_1 X_{i1} + \beta_2 X_{i2} + \varepsilon_i \qquad i = 1, 2, \cdots, 13$$

are given by (see (16.3.8))

$$\begin{pmatrix} 13 & 97 & 626 \\ 97 & 1139 & 4922 \\ 626 & 4922 & 33050 \end{pmatrix} \times \begin{pmatrix} \hat{\beta}_0 \\ \hat{\beta}_1 \\ \hat{\beta}_2 \end{pmatrix} = \begin{pmatrix} 1240.5 \\ 10032 \\ 62027.8 \end{pmatrix}$$

$$\begin{pmatrix} \hat{\beta}_0 \\ \hat{\beta}_1 \\ \hat{\beta}_2 \end{pmatrix} = \begin{pmatrix} 13 & 97 & 626 \\ 97 & 1139 & 4922 \\ 626 & 4922 & 33050 \end{pmatrix}^{-1} \times \begin{pmatrix} 1240.5 \\ 10032 \\ 62027.8 \end{pmatrix} = \begin{pmatrix} 0.902623 & -0.0083871 & -0.0158475 \\ -0.008387 & 0.0025411 & -0.0002196 \\ -0.015848 & -0.0002196 & 0.0003631 \end{pmatrix} \times \begin{pmatrix} 1240.5 \\ 10032 \\ 62027.8 \end{pmatrix}$$

$$= \begin{pmatrix} 52.5773 \\ 1.4683 \\ 0.6623 \end{pmatrix}$$

Thus, we have

$$\hat{\beta}_0 = 52.5773, \hat{\beta}_1 = 1.4683, \hat{\beta}_2 = 0.6623$$

Hence, the fitted model is

$$\hat{Y} = 52.5773 + 1.4683X_1 + 0.6623X_2$$

Now from equation (16.3.16), we know that

$$SSE = Y'Y - \hat{\beta}'X'Y$$

$$= 121088 - \begin{pmatrix} 52.5773 \\ 1.4683 \\ 0.6623 \end{pmatrix}' \times \begin{pmatrix} 1240.5 \\ 10032 \\ 62027.8 \end{pmatrix}$$

$$= 121088 - 121030.1855$$

$$= 57.8145$$

Now $n = 13$, $k = 2$, so that $n - (k + 1) = 10$. Thus, we have

$$\hat{\sigma}^2 = SSE/(n - 3) = 57.8145/10 = 5.782.$$

(b) Now using equations (16.3.35) we know that a confidence interval for the expected response $(E(Y|X_0))$ at X_0 with confidence coefficient $(1 - \alpha)$ is given by

$$\left(\hat{Y}_0 \pm t_{n-(k+1);\alpha/2}\hat{\sigma}\sqrt{X_0(X'X)^{-1}X_0'}\right),$$

Now, it can easily be seen that at $X_0 = (1, X_1 = 7 \ X_2 = 50)$, $\hat{Y}_0 = 95.9664$
Substituting the values

$$\hat{Y} = 95.9664, \quad t_{10} = 2.228, \quad X_0 = (1, \quad X_1 = 7 \quad X_2 = 50), \quad \hat{\sigma} = 2.4046$$

$$(X'X)^{-1} = \begin{pmatrix} 0.902623 & -0.0083871 & -0.0158475 \\ -0.0083871 & 0.0025411 & -0.0002196 \\ -0.0158475 & -0.0002196 & 0.0003631 \end{pmatrix},$$

we obtain a 95% confidence interval for $E(Y|X_1 = 7, X_2 = 50)$ as

$$\left(95.9664 \pm 2.228(2.4046)\sqrt{0.0791}\right) = (95.9664 \pm 1.5068) = (94.4596, 97.4732)$$

27. Using MINITAB (**Stat** > **Regression** > **Regression**), we obtain the following output.

Regression Analysis: Y versus X1, X2, X3, X4, X5

```
The regression equation is

Y = - 0.0784 + 0.000044 X1 + 0.00245 X2 + 0.0183 X3 + 0.00779 X4 - 0.00313 X5

Predictor      Coef        SE Coef        T          P
Constant     -0.07837     0.07360      -1.06      0.297
X1            0.00004392  0.00003627    1.21      0.237
X2            0.0024544   0.0002082    11.79      0.000
X3            0.01835     0.02014       0.91      0.371
X4            0.007786    0.001349      5.77      0.000
X5           -0.003134    0.008053     -0.39      0.700

S = 0.00227199       R-Sq = 96.9%     R-Sq(adj) = 96.3%
PRESS = 0.000248040     R-Sq(pred) = 94.27%
```

```
Analysis of Variance
Source              DF          SS           MS           F          P
Regression           5    0.00419226   0.00083845    162.43      0.000
Residual Error      26    0.00013421   0.00000516
Total               31    0.00432647
```

```
Unusual Observations

Obs      X1             Y          Fit      SE Fit     Residual    St Resid
29     1650     0.040000    0.033578    0.001040     0.006422       3.18R
32     1700     0.068000    0.069528    0.001962    -0.001528      -1.33X
```

```
R denotes an observation with a large standardized residual.
X denotes an observation whose X value gives it large leverage.
```

```
Predicted Values for New Observations (also, see Problem 28 below)

New Obs.      Fit        SE Fit         95% CI                   95% PI
1         0.027942     0.001137   (0.025605, 0.030279)   (0.022720, 0.033164)
```

```
Values of Predictors for New Observations (also, see Problem 28 below)

New Obs.      X1        X2       X3       X4       X5
1           1675      2.25     1.12     1.20     0.850
```

From the above MINITAB output we have the following:

(a) $\hat{Y} = -0.0784 + 0.000044\ X_1 + 0.00245\ X_2 + 0.0183\ X_3 + 0.00779\ X_4 - 0.00313\ X_5$

(b) From the ANOVA table p-value for the regression is 0.000. Moreover the R^2 is very high. Hence we can conclude that the fitted model is a good fit. It is important that you check the adequacy of the model. The residual plot below shows no abnormalities with the model assumptions.

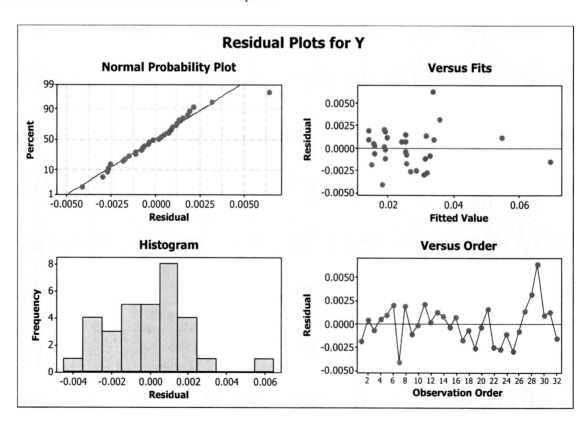

29. Using MINITAB (**Stat** > **Regression** > **General Regression**), we obtain the following output.

General Regression Analysis: Y versus X1* X2, X2*X3

```
Regression Equation
Y = 0.0124683 - 9.81655e-006 X1*X2 + 0.0198779 X2*X3
```

```
Coefficients
Term          Coef        SE Coef          T            P
Constant    0.0124683    0.0009618      12.9638       0.000
X1*X2      -0.0000098    0.0000016      -6.0468       0.000
X2*X3       0.0198779    0.0027732       7.1678       0.000
```

```
Summary of Model
S = 0.00263134        R-Sq = 95.36%        R-Sq(adj) = 95.04%
PRESS = 0.000365400   R-Sq(pred) = 91.55%
```

```
Analysis of Variance

Source          DF      Seq SS      Adj SS      Adj MS         F          P
Regression       2    0.0041257   0.0041257   0.0020628    297.927    0.0000000
  X1*X2          1    0.0037699   0.0002532   0.0002532     36.564    0.0000014
  X2*X3          1    0.0003557   0.0003557   0.0003557     51.378    0.0000001
Error           29    0.0002008   0.0002008   0.0000069
  Lack-of-Fit   16    0.0001807   0.0001807   0.0000113      7.318    0.0004104
  Pure Error    13    0.0000201   0.0000201   0.0000015
Total           31    0.0043265
```
Note that the lack-of-fit degree of freedom we get from the repeated observations.

```
Fits and Diagnostics for Unusual Observations

Obs     Y         Fit        SE Fit      Residual     St Resid
28    0.039   0.0331784   0.0006448    0.0058216     2.28197 R
31    0.056   0.0523399   0.0014773    0.0036601     1.68083 X
32    0.068   0.0714783   0.0021770   -0.0034783    -2.35325 R X
```

R denotes an observation with a large standardized residual.
X denotes an observation whose X value gives it large leverage.

```
Predicted Values for New Observations

New Obs.      Fit         SE Fit              95% CI                      95% PI
      1    0.0266933   0.0005565   (0.0255551, 0.0278316)   (0.0211926, 0.0321941)
```

Values of Predictors for New Observations (the "new observations" is at the values
of X_1, X_2, X_3 specified below, and the fitted is given above for these values of X_1, X_2, X_3)

```
New Obs.      X1       X2       X3
1            1670     2.2      1.15
```

From the above MINITAB output, we have the following:

(a) The fitted model is

$$\hat{Y} = 0.0124683 - (9.81655\mathrm{E}^{-6})X_1 X_2 + 0.0198779\, X_2 X_3$$

(b) $R^2_{pred} = 91.55\%$, which is quite high. Checking the p-value for the regression we find it is 0.000 which indicates the fit is quite good. However, if we look at the p-value for the lack of fit from the ANOVA table then we note it is only .0004, which indicates that the lack of fit is quite significant. This means that although the fitted model is quite good but it may lack some other terms in the model. Of course, it is also important to check the adequacy of the model which we leave for the reader to check.

31. In Problem 5 of the Practice Problems for Sections 16.6 and 16.7 the fitted model is

$$\hat{Y} = 0.148 - 0.00725\, X_3 + 1.59\, X_1^2$$

Thus, from the fitted model and the data in Problem 19 we can determine that if fitting Y on X_3, X_1^2, the $X[26 \times 3]$ matrix is given by

$$X = \begin{pmatrix}
1 & 0.0 & 53.29 \\
1 & 0.3 & 75.69 \\
1 & 1.0 & 77.44 \\
1 & 0.2 & 65.61 \\
1 & 1.0 & 81.00 \\
1 & 2.8 & 75.69 \\
1 & 1.0 & 86.49 \\
1 & 3.4 & 57.76 \\
1 & 0.3 & 100.00 \\
1 & 4.1 & 70.56 \\
1 & 2.0 & 86.49 \\
1 & 7.1 & 59.29 \\
1 & 2.0 & 96.04 \\
1 & 6.8 & 53.29 \\
1 & 6.6 & 72.25 \\
1 & 5.0 & 90.25 \\
1 & 7.8 & 54.76 \\
1 & 7.7 & 60.84 \\
1 & 8.0 & 59.29 \\
1 & 4.2 & 106.09 \\
1 & 8.5 & 60.84 \\
1 & 6.6 & 50.41 \\
1 & 9.5 & 59.29 \\
1 & 10.9 & 54.76 \\
1 & 5.2 & 53.29 \\
1 & 20.7 & 57.76
\end{pmatrix}$$

Now from equation (16.3.33), the estimated variance of $\left(\hat{Y}|X_0\right)$ is given by

$$\hat{\sigma}^2 (X_0(X'X)^{-1}X_0')$$

From the solution of Problem 5 of Practice Problems for Sections 16.6 and 16.7, we have

$$\hat{\sigma}^2 = 0.00048.$$

Here, $X_0 = (1, X_3, X_1^2) = (1, 6, 72.25)$. Thus, substituting the values of $\hat{\sigma}^2$, X_0, and $(X'X)^{-1}$, we obtain

$$\text{Var}(\hat{Y}|X_0) = 0.00002064.$$

(b) From the MINITAB output in Problem 5 of Practice Problems for Sections 16.6 and 16.7, we have (β_1, β_2 are regression coefficients of X_3, X_1^2)

$$\hat{\beta}_1 = -0.007246, \hat{\beta}_2 = 1.58934$$

and the standard errors for each of these regression coefficients are

$$\hat{\sigma}_{\hat{\beta}_1} = 0.001415, \hat{\sigma}_{\hat{\beta}_2} = 0.07201.$$

A 95% confidence interval for regression coefficient $\beta_i, i = 1, 2$ is given by ($n = 26$, $k = 2$)

$$\hat{\beta}_i \pm t_{n-(k+1):\alpha/2} \hat{\sigma}_{\hat{\beta}_i}$$

Now substituting the $\hat{\beta}_i$'s and $\hat{\sigma}_{\hat{\beta}_i}$'s we obtain the confidence intervals, which are as follows:

$$(-0.007246 \pm 0.002928), (1.58934 \pm 0.14899).$$

33. Using MINITAB (**Stat** > **Regression** > **Binary Logistic Regression**), we obtain the following output.

Binary Logistic Regression: Y versus X1, X2, X3

```
Link Function:    Logit
Response Information
```

Variable	Value	Count	
Y	1	6	(Event)
	0	9	
	Total	15	

```
Logistic Regression Table
```

Predictor	Coef	SE Coef	Z	P	Odds Ratio	95% CI Lower	95% CI Upper
Constant	-53.0928	28.3423	-1.87	0.061			
X1	0.114493	0.114737	1.00	0.318	1.12	0.90	1.40
X2	1.13665	0.592392	1.92	0.055	3.12	0.98	9.95
X3							
1	-2.44767	2.27096	-1.08	0.281	0.09	0.00	7.41

```
Log-Likelihood = -5.130
Test that all slopes are zero: G = 9.931, DF = 3, P-Value = 0.019
Goodness-of-Fit Tests
```

Method	Chi-Square	DF	P
Pearson	9.8549	11	0.543

```
Table of Observed and Expected Frequencies:
(See Hosmer-Lemeshow Test for the Pearson Chi-Square Statistic)
```

						Group					
Value	1	2	3	4	5	6	7	8	9	10	Total
1											
Obs	0	0	0	0	1	0	1	1	1	2	6
Exp	0.0	0.1	0.1	0.2	0.2	0.6	0.5	1.6	0.9	1.9	
0											
Obs	1	2	1	2	0	2	0	1	0	0	9
Exp	1.0	1.9	0.9	1.8	0.8	1.4	0.5	0.4	0.1	0.1	
Total	1	2	1	2	1	2	1	2	1	2	15

Interpretation of the MINITAB output

From the logistic regression table we see that none of the predictor variables are significant when all of them are considered together but we may consider X_2 as being almost significant.

$$\hat{\eta} = -53.0928 + 0.114493X_1 + 1.13665X_2 - 2.44767X_3$$

That is,

$$\hat{\eta}(X_1, X_2, X_3) = -53.0928 + 0.114493X_1 + 1.13665X_2 - 2.44767X_3$$

$$\hat{\eta}(X_1, X_2 + 1, X_3) = -53.0928 + 0.114493X_1 + 1.13665(X_2 + 1) - 2.44767X_3$$

By taking the difference of the last two expressions, we obtain

$$\hat{\eta}(X_1, X_2 + 1, X_3) - \hat{\eta}(X_1, X_2, X_3) = 1.13665$$

Now using equation (16.8.4) we know that $\hat{\eta}$ is the estimate of log-odds, at (X_1, X_2, X_3), which implies that

$$\ln(odds(X_1, X_2 + 1, X_3)) - \log(odds(X_1, X_2 + 1, X_3)) = \ln((odds(X_1, X_2 + 1, X_3))/(odds(X_1, X_2, X_3))) = 1.13665$$

that is,

$$odds(X_1, X_2 + 1, X_3)/odds(X_1, X_2, X_3) = e^{1.13665} = 3.1183.$$

This can be interpreted (based on these data) as that the odds that a woman getting breast cancer with one year increase in age for first pregnancy have increased by three fold. (Note: same interpretation is not applicable for the predictor variable X_3 since it is a qualitative variable).

35. Using MINITAB (**Stat** > **Regression** > **Binary Logistic Regression),** we obtain the following output.

Binary Logistic Regression: Y versus X1, X2

```
Link Function: Logit

Response Information

Variable    Value    Count
Y               1        9    (Event)
                0        6
            Total       15

Logistic Regression Table
```

Predictor	Coef	SE Coef	Z	P	Odds Ratio	95% CI Lower	Upper
Constant	-6.61146	3.67291	-1.80	0.072			
X1	0.497121	0.262557	1.89	0.058	1.64	0.98	2.75
X2	0.0507921	0.0860793	0.59	0.555	1.05	0.89	1.25

Log-Likelihood = -6.281
Test that all slopes are zero: G = 7.629, DF = 2, P-Value = 0.022

Goodness-of-Fit Tests

Method	Chi-Square	DF	P
Pearson	11.0780	12	0.522

Table of Observed and Expected Frequencies:
(See Hosmer-Lemeshow Test for the Pearson Chi-Square Statistic)

					Group						
Value	1	2	3	4	5	6	7	8	9	10	Total
1											
Obs	0	1	0	1	0	1	1	2	1	2	9
Exp	0.1	0.4	0.3	0.7	0.5	1.5	0.9	1.8	0.9	2.0	
0											
Obs	1	1	1	1	1	1	0	0	0	0	6
Exp	0.9	1.6	0.7	1.3	0.5	0.5	0.1	0.2	0.1	0.0	
Total	1	2	1	2	1	2	1	2	1	2	15

From the above MINITAB output we have the following:

(a) The logit regression model is given by

$$\hat{\eta} = -6.61146 + 0.497121X_1 + 0.0507921X_2$$

(b) Since the p-values are greater than 0.05 the level of significance, we do not reject the null hypotheses. However, the predictor variable X_1 is marginally significant.

37. Using MINITAB (**Stat** > **Regression** > **General Regression Analysis**), we obtain the following output.

General Regression Analysis: Y versus X1, X2, X3, X4

Regression Equation

X4

0 Y = -97.205 + 0.0415982 X1 + 12.2375 X2 + 3.92213 X3
1 Y = -86.8758 + 0.0415982 X1 + 12.2375 X2 + 3.92213 X3

Coefficients

Term	Coef	SE Coef	T	P
Constant	-92.0404	95.8258	-0.96050	0.359
X1	0.0416	2.0048	0.02075	0.984
X2	12.2375	6.1543	1.98843	0.075
X3	3.9221	9.5509	0.41066	0.690
X4				
0	-5.1646	7.7876	-0.66318	0.522

Summary of Model

```
S = 26.3837       R-Sq = 58.75%   R-Sq(adj) = 42.25%
PRESS = 14650.4   R-Sq(pred) = 13.18%
```

Analysis of Variance

Source	DF	Seq SS	Adj SS	Adj MS	F	P
Regression	4	9913.4	9913.43	2478.36	3.56036	0.047035
X1	1	22.1	0.30	0.30	0.00043	0.983854
X2	1	9302.5	2752.27	2752.27	3.95387	0.074816
X3	1	282.7	117.39	117.39	0.16864	0.689986
X4	1	306.1	306.15	306.15	0.43981	0.522208
Error	10	6961.0	6960.97	696.10		
Total	14	16874.4				

Fits and Diagnostics for Unusual Observations

Obs	Y	Fit	SE Fit	Residual	St Resid
4	38	94.9917	12.8406	-56.9917	-2.47272 R

R denotes an observation with a large standardized residual.
Predicted Values for New Observations

New Obs.	Fit	SE Fit	95% CI	95% PI
1	97.2135	13.9385	(66.1566, 128.270)	(30.7276, 163.699)

Values of Predictors for New Observations

New Obs.	X1	X2	X3	X4
1	24	14	3	1

From the above MINITAB output we have the following:

(a) The fitted regression model $E(Y|X) = \beta_0 + \beta_1 X_1 + \beta_2 X_2 + \beta_3 X_3 + \beta_4 X_4$ is

$$\text{For Males} : \hat{Y} = -97.205 + 0.0415982\,X_1 + 12.2375\,X_2 + 3.92213\,X_3$$

$$\text{For Females} : \hat{Y} = -86.8758 + 0.0415982\,X_1 + 12.2375\,X_2 + 3.92213\,X_3$$

(b) Using the ANOVA table in the above MINITAB output the p-value for the regression is 0.047035, which is less than 5% the level of significance. Hence the fitted model is significant.

(c) The 95% confidence interval for mean stress level (**Y**) and 95% prediction interval for stress level (**Y**) are

$$(66.1566, 128.270) \quad \text{and} \quad (30.7276, 163.699)$$

respectively.

Note: *Since in this problem the categorical variable has two categories, in the design matrix the MINITAB uses 1 and −1 respectively for male and female, irrespective of what we used, i.e., 0, and 1 to denote these categories.*

39. From the above MINITAB output in Problem 38, we obtain the following 95% confidence and prediction intervals.

(a) A 95% confidence interval for $E(Y|X_0 = (24, 14, 1))$, is given by

$$(71.3893, 115.419)$$

(b) A 95% prediction interval for Y at $X_0 = (24, 14, 1)$, is given by

$$(33.3880, 153.420)$$

Comparing the two intervals we find as expected that the prediction interval is much larger than the confidence interval.

17

ANALYSIS OF VARIANCE

PRACTICE PROBLEMS FOR SECTION 17.2

1. The result follows from Theorem 17.2.1, since

$$rank(X') = rank(X' : X')$$

3. For the given experiment the design model is

$$Y_{ij} = \mu + \beta_j + \epsilon_{ij} \qquad i = 1, 2; \quad j = 1, 2$$

so that parameter vector is

$$\eta = \begin{pmatrix} \mu \\ \beta_1 \\ \beta_2 \end{pmatrix}$$

and the transpose of the matrix X is

$$X' = \begin{pmatrix} 1 & 1 & 1 & 1 \\ 1 & 1 & 0 & 0 \\ 0 & 0 & 1 & 1 \end{pmatrix}$$

and is of rank 2. We now show that a linear function $c_1\beta_1 + c_2\beta_2$ such that $c_1 + c_2 = 0$ is linearly estimable. The function $c_1\beta_1 + c_2\beta_2$ can be written as $t'\eta$ where $t' = (0 \quad c_1 \quad c_2)$. Now

the rank of $(X' : t)$, i.e., $\qquad (X' : t) = \begin{pmatrix} 1 & 1 & 1 & 1 & 0 \\ 1 & 1 & 0 & 0 & c_1 \\ 0 & 0 & 1 & 1 & c_2 \end{pmatrix}$

Solutions Manual to Accompany Statistics and Probability with Applications for Engineers and Scientists, Bhisham C. Gupta and Irwin Guttman.
© 2014 John Wiley & Sons, Inc. Published 2014 by John Wiley & Sons, Inc.

is again 2, since $c_1 + c_2 = 0$ the sum of the last two rows is equal to the first row, so that $rank(X') = rank(X' : t)$. Hence, the linear function $c_1\beta_1 + c_2\beta_2$ is linearly estimable.

5. No. Since by Theorem 17.2.3, the maximum number of linearly estimable functions is exactly equal to the rank of the design matrix, which in this case is 2.

PRACTICE PROBLEMS FOR SECTION 17.3

1. MINITAB (One-way ANOVA): Yield versus Method

```
Source    DF      SS       MS       F        P
Method     3    111.83    37.28    24.23    0.000
Error     12     18.47     1.54
Total     15    130.30
S = 1.240         R-Sq = 85.83%      R-Sq(adj) = 82.29%
```

In this problem, the model is:

$$y_{ij} = \mu + \delta_j + \varepsilon_{ij}; \qquad i = 1, \ldots, n_j; j = 1, \ldots, 4, \sum_j \delta_j = 0$$

The hypothesis we wish to test is

$$H_0 : \delta_1 = \delta_2 = \delta_3 = \delta_4 \quad Versus \quad H_1 : not \; all \; equal$$

Since the $p - value = 0.000 < \alpha(= .05)$ we reject the hypothesis of equal means and conclude that differences between catalytic methods are significant.

3. (a) MINITAB (One-way ANOVA): Yield versus Variety

```
Source    DF      SS       MS       F        P
Variety    2     28.45    14.23    5.69     0.025
Error      9     22.51     2.50
Total     11     50.96
S = 1.581         R-Sq = 55.84%      R-Sq(adj) = 46.02%
```

```
                                   Individual 95% CIs For Mean Based on Pooled StDev
Level   N     Mean    StDev    -+---------+---------+---------+--------
I       4    41.650   1.756     (--------*--------)
II      4    45.250   1.964                         (--------*--------)
III     4    42.475   0.750         (--------*--------)
                               -+---------+---------+---------+--------
                              40.0      42.0      44.0      46.0
Pooled StDev = 1.581
```

Since $p - value = 0.025 < .05 = \alpha$, we reject the hypothesis of equal means and conclude that differences between varieties are significant.

(b) Since in part (a) we reject the null hypothesis of equal means, we proceed to estimate the effects of different varieties. These are:

$$\hat{\delta}_1 = 41.650 - 43.125 = -1.475, \quad \hat{\delta}_2 = 45.250 - 43.125 = 2.125, \quad \hat{\delta}_3 = 42.475 - 43.125 = -0.650.$$

5. (a) MINITAB (One-way ANOVA): Observations versus Training Program

```
Source             DF      SS      MS       F       P
Training Program    4    16.67    4.17    1.94    0.147
Error              18    38.63    2.15
Total              22    55.30
```

$S = 1.465$ R-Sq $= 30.14\%$ R-Sq(adj) $= 14.62\%$

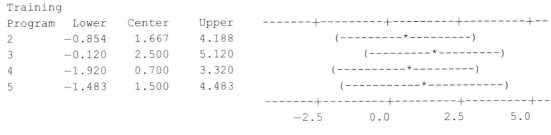

```
                                 Individual 90% CIs For Mean Based on
                                 Pooled StDev
Level    N      Mean    StDev    -----+---------+---------+---------+----
1        4     6.500    1.291      (-------*--------)
2        6     8.167    1.472             (-----*------)
3        5     9.000    1.225                (-------*-------)
4        5     7.200    0.837        (-------*-------)
5        3     8.000    2.646           (--------*---------)
                                 -----+---------+---------+---------+----
                                    6.0       7.5       9.0      10.5
```

Pooled StDev $= 1.465$

Since $p - value = 0.147 > .10 = \alpha$, we do not reject the null hypothesis of equal means and conclude that differences between training programs are not significant.

(b) Tukey 90% Simultaneous Confidence Intervals for All Pairwise Comparisons among Levels of Training Programs are:

Training Program $= 1$ subtracted from:

```
Training
Program   Lower    Center    Upper    -------+---------+---------+---------+--
2        -0.854    1.667     4.188            (---------*---------)
3        -0.120    2.500     5.120               (---------*---------)
4        -1.920    0.700     3.320         (----------*---------)
5        -1.483    1.500     4.483           (-----------*-----------)
                                         -------+---------+---------+---------+--
                                            -2.5       0.0       2.5       5.0
```

Training Program $= 2$ subtracted from:

```
Training
Program   Lower    Center    Upper    -------+---------+---------+---------+--
3        -1.532    0.833     3.198            (--------*---------)
4        -3.332   -0.967     1.398     (--------*---------)
5        -2.928   -0.167     2.595        (----------*----------)
                                         -------+---------+---------+---------+--
                                            -2.5       0.0       2.5       5.0
```

Training Program $= 3$ subtracted from:

```
Training
Program   Lower    Center    Upper    -------+---------+---------+---------+--
4        -4.270   -1.800     0.670     (---------*---------)
5        -3.852   -1.000     1.852      (----------*----------)
                                         -------+---------+---------+---------+--
                                            -2.5       0.0       2.5       5.0
```

```
Training Program = 4 subtracted from:

Training
Program Lower   Center   Upper
5      -2.052   0.800    3.652
```
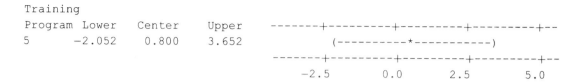

From the above, we see that all the confidence intervals for the multiple comparisons contain zero (except we may note that training program 1 and 3 are different but only very marginally.

Remark: The marginal difference may be due to the fact that since sample size are not equal the confidence level may be greater than 90% or significance level may be less than 10%. In other words, they are marginally different at significance level less than 10%).

7. In this problem we obtain the pairwise confidence intervals using Bonferroni method. Here, for each confidence interval we use $t_{12;\alpha/12} = 3.1527(N - a = 12)$.

$$\delta_2 - \delta_1 : \quad (3.750 \pm 2.987) = (0.763, 6.737)$$
$$\delta_3 - \delta_1 : \quad (1.275 \pm 2.766) = (-1.491, 4.041)$$
$$\delta_4 - \delta_1 : \quad (6.530 \pm 2.625) = (3.905, 9.155)$$
$$\delta_3 - \delta_2 : \quad (-2.475 \pm 2.987) = (-5.462, 0.512)$$
$$\delta_4 - \delta_2 : \quad (2.780 \pm 2.856) = (-0.076, 5.636)$$
$$\delta_4 - \delta_3 : \quad (5.255 \pm 2.625) = (2.63, 7.88)$$

Thus, the effects of Method I are significantly different from Method II and Method IV, and the effect of Method III is significantly different from Method IV. Note that Bonferroni confidence intervals are slightly narrower than those obtained in Problem 2 using the S-method. However, our conclusions remain the same.

9. In this problem, we wish to determine 99% pairwise confidence intervals using Tukey's method. These confidence intervals are as given below.

```
Tukey 99% Simultaneous Confidence Intervals
All Pairwise Comparisons among Levels of Variety

Variety = I subtracted from:

Variety Lower   Center   Upper
II     -0.694   3.600    7.894
III    -3.469   0.825    5.119
```
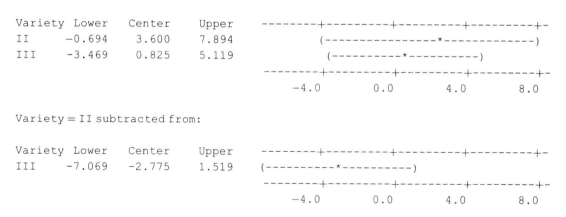

```
Variety = II subtracted from:

Variety Lower   Center   Upper
III    -7.069   -2.775   1.519
```

Since all these confidence intervals contain zero, so we conclude that at 1% level of significance we do not reject the null hypothesis of equal means.

11. Here we interested in finding the pairwise confidence intervals using the S-method.

$$\delta_2 - \delta_1 : \quad (3.600 \pm 4.613) = (-1.013, 8.213)$$
$$\delta_3 - \delta_1 : \quad (0.825 \pm 4.613) = (-3.788, 5.438)$$
$$\delta_3 - \delta_2 : \quad (-2.775 \pm 4.613) = (-7.388, 1.838)$$

Since all the confidence intervals contain zero, we do not reject the null hypothesis of equal means. Our conclusion in this problem is the same as in Problems 9 and 10.

13. In this problem we obtain pairwise confidence intervals using Bonferroni method. Here, for each confidence interval we use $t_{18;\alpha/20} = 3.1966, (N - a = 18)$. From Problem 5, we have $MS_E = 2.15$.

$$\delta_2 - \delta_1 : \quad (1.667 \pm 3.026) = (-1.359, 4.693)$$
$$\delta_3 - \delta_1 : \quad (2.500 \pm 3.144) = (-0.644, 5.644)$$
$$\delta_4 - \delta_1 : \quad (0.700 \pm 3.144) = (-2.444, 3.844)$$
$$\delta_5 - \delta_1 : \quad (1.500 \pm 3.580) = (-2.080, 5.080)$$
$$\delta_3 - \delta_2 : \quad (0.833 \pm 2.838) = (-2.005, 3.671)$$
$$\delta_4 - \delta_2 : \quad (-.0967 \pm 2.838) = (-3.895, 1.871)$$
$$\delta_5 - \delta_2 : \quad (-0.167 \pm 3.314) = (-3.481, 3.147)$$
$$\delta_4 - \delta_3 : \quad (-1.800 \pm 2.964) = (-4.764, 1.164)$$
$$\delta_5 - \delta_3 : \quad (-1.000 \pm 3.423) = (-4.423, 2.423)$$
$$\delta_5 - \delta_4 : \quad (0.800 \pm 3.423) = (-2.623, 4.223)$$

Here we note that all the confidence intervals contain zero. Hence, we may conclude that based on these data all training programs are equally good. Our conclusions in this problem and Problem 12 are the same.

PRACTICE PROBLEMS FOR SECTION 17.4

1. MINITAB (Two-way ANOVA): Observations versus Plants, Coatings

Source	DF	SS	MS	F	P
Plants	2	0.04667	0.023333	0.08	0.928
Coatings	3	0.22000	0.073333	0.24	0.866
Error	6	1.84000	0.306667		
Total	11	2.10667			

S = 0.5538   R-Sq = 12.66%   R-Sq(adj) = 0.00%

P-values for both coatings (treatments) and block (plants) are greater than $0.05 = \alpha$. Hence, we do not reject the null hypothesis of all coatings have same effect and all bocks have same effects.

MINITAB (Friedman Test): Observations versus Coatings blocked by Plants

S = 1.00   DF = 3   P = 0.801

Coatings	N	Est Median	Sum of Ranks
A	3	4.9000	7.0
B	3	5.1000	8.0
C	3	4.8000	6.0
D	3	5.4000	9.0

Grand median = 5.0500

Using the Friedman Test we make the same conclusion as in part (a), however, the Friedman Test gives lower p-value (.801), which is slightly lower than (.866) obtained in the ANOVA table.

3. MINITAB (Two-way ANOVA): Observations versus Chemicals, Machines (one missing obs.)

```
Source      DF      SS        MS      F       P
Chemicals    4    164.3    41.0750   3.13   0.056
Machines     3     76.2    25.4000   1.94   0.177
Error       12    157.3    13.1083
Total       19    397.8

S = 3.621    R-Sq = 60.46%    R-Sq(adj) = 37.39%
```

The ANOVA table shows that at significance level $\alpha = 0.10$ the chemicals are significantly different. However, machines are not significantly different. The residual plots given below do not present any abnormalities.

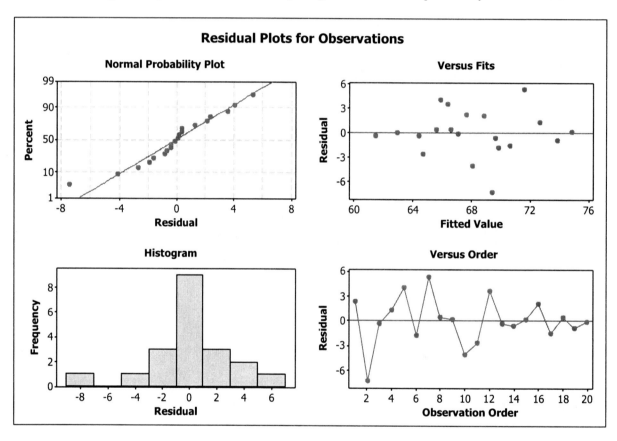

5. MINITAB (Friedman Test): Observations versus Chemicals blocked by Machines

	M_1	M_2	M_3	M_4	Total
C_1	70	68	62	71	271
C_2	62	77	70	69	278
C_3	64	67	61	66	258
C_4	74	75	69	73	291
C_5	70	64	75	67	276
Total	340	351	337	346	1374

```
S = 8.25   DF = 4   P = 0.083
S = 8.35   DF = 4   P = 0.079 (adjusted forties)
```

```
                            Sum of
Chemicals    N    Est Median    Ranks
1            4        68.800     12.5
2            4        71.900     14.0
3            4        65.800      6.0
4            4        73.700     18.0
5            4        67.300      9.5
```

```
Grand median = 69.500
```

Since the p-value is less than 0.10 the level of significance, the chemical are significantly different. The Friedman Test does not give any results regarding the blocks.

7. **MINITAB (Friedman Test): Observations versus Chemicals blocked by Machines**

```
S = 8.25   DF = 4   P = 0.083
S = 8.35   DF = 4   P = 0.079 (adjusted for ties)
```

```
                            Sum of
Chemicals    N    Est Median    Ranks
1            4        68.844     12.5
2            4        71.889     14.0
3            4        65.657      6.0
4            4        73.755     18.0
5            4        67.245      9.5
```

```
Grand median = 69.478
```

Again, since the p-value is less than 0.10 the level of significance, the chemicals are significantly different. Friedman Test does not give any results regarding the blocks. Otherwise, our conclusion in this problem is the same as in Problem 4.

PRACTICE PROBLEMS FOR SECTION 17.5

The mathematical model for the experiment in this example is

$$Y_{ijk} = \mu + \alpha_i + \beta_j + \gamma_{ij} + \varepsilon_{ijk}, \quad i = 1, 2, 3; j = 1, 2, 3, 4; k = 1, 2.$$

with

$$\sum_{i=1}^{3} \alpha_i = 0, \quad \sum_{j=1}^{4} \beta_j = 0, \quad \sum_{i=1}^{3} \gamma_{ij} = \sum_{j=1}^{4} \gamma_{ij} = 0; \varepsilon_{ijk} \sim N(0, \sigma^2), \text{ and } \varepsilon_{ijk}\text{'s are independent.}$$

MINITAB (Two-way ANOVA): Observations versus Temperatures, Raw Mareial

Source	DF	SS	MS	F	P
Temperatures	2	68.583	34.2917	3.04	0.086
Raw Material	3	36.458	12.1528	1.08	0.396
Interaction	6	167.417	27.9028	2.47	0.086
Error	12	135.500	11.2917		
Total	23	407.958			

```
S = 3.360    R-Sq = 66.79%    R-Sq(adj) = 36.34%
                        Individual 95% CIs For Mean Based on Pooled StDev
Temperatures       Mean   +---------+---------+---------+---------
1                75.000               (---------*---------)
2                72.625     (----------*---------)
3                76.750                   (---------*---------)
                        +---------+---------+---------+---------
                        70.0      72.5      75.0      77.5


                        Individual 95% CIs For Mean Based on
Raw                      Pooled StDev
Material          Mean   --------+---------+---------+---------+-
1               73.5000     (-----------*-----------)
2               76.3333         (-----------*-----------)
3               73.6667     (-----------*-----------)
4               75.6667         (-----------*-----------)
                        --------+---------+---------+---------+-
                        72.5      75.0      77.5      80.0
```

3. In this problem the hypothesis that we wish to test is

$$H_0 : \alpha_1 = \alpha_2 = \alpha_3 \quad versus \quad H_1 : all\ \alpha_i\ are\ not\ equal$$

Again, in this case the p-value is greater than 0.05 the level of significance, so that we do not reject the null hypothesis.

5. General Linear Model: Obs. versus Raw Material, Temp., Blocks

```
Factor          Type    Levels    Values
Raw Material    fixed      4      1, 2, 3, 4
Temp.           fixed      3      1, 2, 3
Blocks          fixed      2      1, 2
```

Analysis of Variance for Obs, using Adjusted SS for Tests

Source	DF	Seq SS	Adj SS	Adj MS	F	P
Raw Material	3	36.46	36.46	12.15	0.99	0.434
Temp.	2	68.58	68.58	34.29	2.78	0.105
Blocks	1	0.04	0.04	0.04	0.00	0.955
Raw Material*Temp.	6	167.42	167.42	27.90	2.27	0.114
Error	11	135.46	135.46	12.31		
Total	23	407.96				

```
S = 3.50919    R-Sq = 66.80%    R-Sq(adj) = 30.57%
```

Since the p-values for interaction and for raw materials, temperatures and blocks are all greater than 0.05 the level of significance, we do not reject any of the null hypotheses. In summary, the interactions and the main effects are not significant. Also, the plants are not significantly different.

7. The model for this experiment is

$$y_{ijk} = \mu + \alpha_i + \beta_j + \theta_k + \gamma_{ij} + \varepsilon_{ijk},$$

$$\sum_{i=1}^{2} \alpha_i = \sum_{j=1}^{4} \beta_j = \sum_{k=1}^{3} \theta_k = \sum_{i} \gamma_{ij} = \sum_{j} \gamma_{ij} = 0; \varepsilon_{ijk} \sim N(0, \sigma^2),$$

	Block 1			
Plate Temperature	L_1	L_2	L_3	L_4
$T_1(550°F)$	3774	4710	4176	4540
$T_2(600°F)$	4216	3828	4122	4484

	Block 2			
Plate Temperature	L_1	L_2	L_3	L_4
$T_1(550°F)$	4364	4180	4140	4530
$T_2(600°F)$	4524	4170	4280	4332

	Block 3			
Plate Temperature	L_1	L_2	L_3	L_4
$T_1(550°F)$	4374	4514	4398	3964
$T_2(600°F)$	4136	4180	4226	4390

Here we first find the block totals. That is

$$T_{..1} = 33850, \quad T_{..2} = 34520, \quad T_{..3} = 34182; G = 102552.$$

Hence,

$$SS_{bl} = \frac{1}{8}((33850)^2 + (34520)^2 + (34182)^2) - (102552)^2/24 = 28057.$$

From Problem 6 above, we then have that
$SS_{treat} = SS_A + SS_B + SS_{AB} = 25091 + 85943 + 253668 = 364702$
Also, $SS_{total} = 1189776$
Hence,

$$SS_E = SS_{total} - SS_{treat} - SS_{bl} = 1189776 - 364702 - 28057 = 797017.$$

We summarize the above results in the following ANOVA table.

	ANOVA				
Source	DF	SS	MS	F	P-value
Temperature	1	25091	25091	.4407	.5176
Current	3	85943	28647.6	.5032	.6832
Interaction	3	253668	84556	1.485	.2616
Block	2	28057	14028.5	.2464	.7849
Error	14	797017	56929.8		
Total	23	1189776			

From the above table, we first see that the p-value for the interaction is $P_{\text{interaction}} = 0.2616 > 0.05$, so that the interactions between temperature and current are not significant. Also,

$P_{\text{temperature}} = 0.5176 > 0.05$, so that the effects of temperature are not significant.

$P_{\text{current}} = 0.6832 > 0.05$, so that the effects of current are not significant.

$P_{block} = 0.7849 > 0.05$, so that the effects of blocks are not significant.

9. (a) MINITAB (Two-way ANOVA): Time versus Cloth, Machine

```
Source        DF      SS        MS        F        P
Cloth          2    56.953   28.4764    7.94    0.002
Machine        8    76.887    9.6109    2.68    0.026
Interaction   16    69.567    4.3479    1.21    0.320
Error         27    96.889    3.5885
Total         53   300.295


S = 1.894    R-Sq = 67.74%    R-Sq(adj) = 36.67%
```

(b) Since $P_{\text{interaction}} = 0.320 > 0.05$, interaction effects are not significant.

(c) $F_{\text{cloth}} = 7.94, F_{\text{machine}} = 2.68.$
   Since $P_{\text{cloth}} = 0.002 < 0.05$, cloth effects are significant.
   Since $P_{\text{machine}} = 0.026 < 0.05$, machine effects are significant.

PRACTICE PROBLEMS FOR SECTION 17.6

1. Using the method discussed in this section we obtain the following ANOVA table.

		ANOVA			
Source	DF	SS	MS	F-Ratio	P-value
Temperature	3	69.25	23.08	5.3303	0.0396
Catalyst	3	14.25	4.75	1.0970	0.4201
Reaction. Time	3	36.25	12.08	2.7898	0.1318
Error	6	26.0	4.33		
Total	15	145.75			

3. The ANOVA table for the data in this problem is given below. The model used for the experiment is:

$$y_{ijk} = \mu + \alpha_i + \beta_j + \gamma_k + \varepsilon_{ijk}; \quad (i,j,k) \in \Omega, i = 1,2,3; j = 1,2,3; k = 1,2,3$$

with

$$\sum_i \alpha_i = \sum_j \beta_j = \sum_k \gamma_k = 0; \varepsilon_{ijk} \sim N(0,\sigma^2), \text{and } \varepsilon_{ijk}\text{'s are independent}$$

		ANOVA			
Source	DF	SS	MS	F-Ratio	P-value
Operators	2	8.22	4.11	1.9433	0.3398
Machines	2	22.22	11.11	5.2530	0.1599
Bleach	2	2.88	1.44	0.6827	0.5943
Error	2	4.23	2.115		
Total	8	37.55			

$$S = 1.45297 \quad \text{R-Sq} = 88.76\% \quad \text{R-Sq(adj)} = 55.03\%$$

Since all p-values are greater than the level of significance 0.05, none of the hypotheses that operators are different, machines are different, or bleach levels are different are rejected. In other words, operators, machines or bleach levels do not have a significant effect on the shearing strength of the paper.

PRACTICE PROBLEMS FOR SECTION 17.7

1. The model used for this type of experiment is

$$y_{ijk} = \mu + \alpha_i + \beta_j + \gamma_{ij} + \varepsilon_{ijk}, \quad i = 1, 2, \cdots, 4; j = 1, 2, \cdots, 5; \text{and } k = 1, 2, 3$$

where

α_i's are unknown constants satisfying the side condition $\sum \alpha_i = 0$, and the β_j, γ_{ij} (for fixed i) and ε_{ijk} are independently and normally distributed with mean zero and variances $\sigma_\beta^2, \sigma_\gamma^2$ and σ_ε^2, respectively. Moreover, γ_{ij} satisfy the side condition $\sum_i \gamma_{ij} = 0$.

ANOVA				
Source	DF	SS	MS	E(MS)
A	3	$12 \sum (\bar{y}_{i\cdot\cdot} - \bar{y}_{\cdots})^2$	$MS_A = SS_A/4$	$\sigma_\varepsilon^2 + 3\sigma_\gamma^2 + 5\Sigma\alpha_i^2$
B	4	$15 \sum (\bar{y}_{\cdot j\cdot} - \bar{y}_{\cdots})^2$	$MS_B = SS_B/3$	$\sigma_\varepsilon^2 + 12\sigma_\beta^2$
AB	12	By subtraction	$MS_{AB} = SS_{AB}/12$	$\sigma_\varepsilon^2 + 3\sigma_\gamma^2$
Error	40	$\sum\sum\sum (y_{ijk} - \bar{y}_{ij\cdot})^2$	$MS_E = SS_E/40$	σ_ε^2
Total	59	$\sum y_{ijk}{}^2 - 60 \times \bar{y}_{\cdots}{}^2$		

The Hypotheses that we wish to test and the corresponding test statistics and their distributions under H_0 are:

Factor A:	$H_0 : \alpha_i = 0$	Vs $H_1 : not\ all\ \alpha_i = 0$	$MS_A/MS_{AB} \sim F_{3,12}$
Factor B:	$H_0 : \sigma_\beta^2 = 0$	Vs $H_1 : \sigma_\beta^2 \neq 0$	$MS_B/MS_E \sim F_{4,40}$
Int. AB:	$H_0 : \sigma_\gamma^2 = 0$	Vs $H_1 : \sigma_\gamma^2 \neq 0$	$MS_{AB}/MS_E \sim F_{12,40}$

3. Nested ANOVA: Obs. versus FWA, Rolls

```
Analysis of Variance for Obs.

Source     DF      SS         MS         F        P
FWA         2    192.8889    96.4444    2.380    0.148
Rolls       9    364.7500    40.5278    1.288    0.294
Error      24    755.3333    31.4722
Total      35   1312.9722

Variance Components

                      % of
Source   Var Comp.   Total     StDev
FWA        4.660     11.90     2.159
Rolls      3.019      7.71     1.737
Error     31.472     80.39     5.610
Total     39.150               6.257
```

```
Expected Mean Squares column

1       FWA         (3) + 3(2) + 12(1)
2       Rolls       (3) + 3(2)
3       Error       (3)
```

Here the p-values for FWA and Rolls are greater than 5%, so we do not reject the any of the null hypotheses. That is, there is no significant variation among different types of FWA and there is no significant variation among the rolls within each FWA.

5. Results for: PRACTICE PROBLEM 17.7.5 MINITAB.MTW

ANOVA: GAIC versus Manufacturer, Clinic (Using Balanced ANOVA & Restricted model)

```
Factor                  Type      Levels    Values
Manufacturer            fixed        3      1, 2, 3
Clinic(Manufacturer)    random       4      1, 2, 3, 4


Analysis of Variance for GAIC

Source                  DF       SS        MS        F        P
Manufacturer             2    0.3608    0.1804     0.79    0.484
Clinic(Manufacturer)     9    2.0637    0.2293     1.83    0.163
Error                   12    1.5050    0.1254
Total                   23    3.9296


S = 0.354142    R-Sq = 61.70%    R-Sq(adj) = 26.59%
```

	Source	Variance component	Error term	Expected Mean Square for Each Term (using restricted model)
1	Manufacturer		2	(3) + 2 (2) + 8 Q[1]
2	Clinic(Manufacturer)	0.05194	3	(3) + 2 (2)
3	Error	0.12542		(3)

Here the p-values for manufacturers and clinics are greater than 0.05, so we do not reject any of the null hypotheses. That is, there is no difference among different manufacturers and there is no variation among the clinics within each manufacturer.

7. Since the department has only four instructors, so in this problem Instructors is a fixed factor.

Results for: PRACTICE PROBLEM 17.7.7 MINITAB.MTW

```
MTB > ANOVA 'Score' = Instructor Student(Instructor);
SUBC>   Random 'Student';
SUBC>   Restrict;
SUBC>   EMS.
```

ANOVA: Score versus Instructor, Student

```
Factor                  Type      Levels    Values
Instructor              fixed        4      1, 2, 3, 4
Student(Instructor)     random       3      1, 2, 3
```

Analysis of Variance for Score

Source	DF	SS	MS	F	P
Instructor	3	37.90	12.63	0.19	0.902
Student(Instructor)	8	539.83	67.48	1.14	0.360
Error	36	2126.75	59.08		
Total	47	2704.48			

S = 7.68612   R-Sq = 21.36%   R-Sq(adj) = 0.00%

	Source	Variance component	Error term	Expected Mean Square for Each Term (using restricted model)
1	Instructor		2	(3) + 4 (2) + 12 Q[1]
2	Student(Instructor)	2.101	3	(3) + 4 (2)
3	Error	59.076		(3)

Since $p_{Instructor} = 0.902 > 0.01$, so we do not reject the null hypothesis of instructors are equally good. $p_{students} = 0.360 > 0.01$, so we do not reject the null hypothesis of students same variation within each instructor. The students variation of the total variation is only $2.101/(2.101 + 59.076) = 3.4\%$.

Note: The restricted model is when sum of interaction effects at each level of a fixed factor is assumed zero (see model conditions for mixed effects models).

REVIEW PRACTICE PROBLEMS

1. MINITAB (One-way ANOVA): Weight versus Feed

Source	DF	SS	MS	F	P
Feed	3	12000	4000	11.65	0.000
Error	16	5492	343		
Total	19	17493			

S = 18.53   R-Sq = 68.60%   R-Sq(adj) = 62.71%

```
                                        Individual 95% CIs For Mean Based on
                                        Pooled StDev
Level   N      Mean    StDev   ---------+---------+---------+---------+
1       5     50.80    10.87   (-----*-----)
2       5     79.60    15.52                   (-----*----)
3       5     72.40    17.99            (-----*-----)
4       5    118.60    26.27                               (-----*----)
                                ---------+---------+---------+---------+
                                        60        90       120       150
```

Pooled StDev = 18.53

Tukey 99% Simultaneous Confidence Intervals
All Pairwise Comparisons among Levels of Feed

Individual confidence level = 99.79%
Feed = 1 subtracted from:

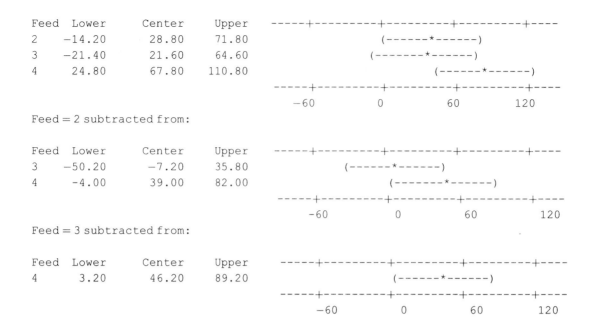

```
Feed   Lower    Center    Upper     -----+---------+---------+---------+----
2     -14.20    28.80    71.80                 (------*------)
3     -21.40    21.60    64.60                (-------*------)
4      24.80    67.80   110.80                       (------*------)
                                   -----+---------+---------+---------+----
                                      -60        0         60        120

Feed = 2 subtracted from:

Feed   Lower    Center    Upper     -----+---------+---------+---------+----
3     -50.20    -7.20    35.80           (------*------)
4      -4.00    39.00    82.00                  (-------*------)
                                   -----+---------+---------+---------+----
                                      -60        0         60        120

Feed = 3 subtracted from:

Feed   Lower    Center    Upper     -----+---------+---------+---------+----
4       3.20    46.20    89.20                      (------*------)
                                   -----+---------+---------+---------+----
                                      -60        0         60        120
```

Since $P = 0.000 \leq 0.01 = \alpha$, we reject the hypothesis of equal means and conclude that differences between feeds are significant.

The Tukey 99% simultaneous confidence intervals show that Feed 4 produced significantly different means than from Feed 1 and Feed 3.

3. We noted in Problem 2, that machine effects are not significant in both cases when days are assumed to have significant effects and when they do not have significant effects. So we do not find the confidence intervals for the given contrasts.

5. **(a)** **MINITAB (One-way ANOVA): Yield versus Method**

```
Source   DF        SS        MS        F          P
Method    2     22.38     11.19     6.59      0.015
Error    10     16.99      1.70
Total    12     39.37
```

$S = 1.303$ R-Sq $= 56.85\%$ R-Sq(adj) $= 48.22\%$

```
                                   Individual 95% CIs For Mean Based on
                                   Pooled StDev
Level   N      Mean     StDev     ------+---------+---------+---------+---
I       4    48.550     1.066     (---------*--------)
II      4    50.450     0.957               (--------*---------)
III     5    51.720     1.645                       (--------*-------)
                                  ------+---------+---------+---------+---
                                     48.0      49.5      51.0      52.5
```

Pooled StDev $= 1.303$

Assuming the use of a significance level of $\alpha = 0.05$, the p-value of 0.015 indicates that the effects due to the different catalytic methods are significantly different.

(b) With an overall mean of $\bar{y} = 50.35$, the estimated effects of the different catalytic methods are

$$\hat{\delta}_I = 48.55 - 50.35 = -1.80, \hat{\delta}_{II} = 50.45 - 50.35 = .10, \hat{\delta}_{III} = 51.72 - 50.35 = 1.37.$$

As noted in part (a) the p-value for the observed F-statistic is 0.015.

7. (a) MINITAB (One-way ANOVA): Illumination versus Technique

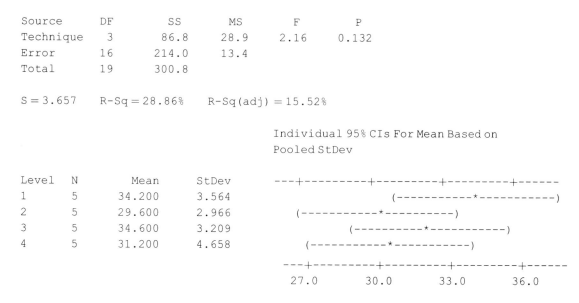

```
Source       DF        SS        MS       F       P
Technique     3      86.8      28.9    2.16   0.132
Error        16     214.0      13.4
Total        19     300.8

S = 3.657    R-Sq = 28.86%    R-Sq(adj) = 15.52%

                                 Individual 95% CIs For Mean Based on
                                 Pooled StDev

Level  N      Mean    StDev    ---+---------+---------+---------+------
1      5    34.200    3.564                    (-----------*-----------)
2      5    29.600    2.966       (-----------*----------)
3      5    34.600    3.209           (----------*-----------)
4      5    31.200    4.658        (-----------*-----------)
                                 ---+---------+---------+---------+------
                                 27.0      30.0      33.0      36.0

Pooled StDev = 3.657
```

Since the p-value $= 0.132$, we conclude that the effects due to lighting technique are not significantly different.

(b) The hypothesis of lighting technique-effect zero is not rejected.

9. In this problem we wish to test, using t-test, the hypothesis

$$H_0 : \alpha_1 - \alpha_4 = 0 \quad Vs \quad H_1 : \alpha_1 - \alpha_4 \neq 0$$

From Problem 8, we obtain

$$\hat{\alpha}_1 - \hat{\alpha}_4 = 2.314 - 2.676 = -0.362$$

Thus the observed value of the test statistic is

$$Obs.value = \frac{-0.362}{0.2558\sqrt{\left(\frac{1}{5} + \frac{1}{5}\right)}} = -2.2375$$

which gives p-value of 0.045. We will reject the null hypothesis if we use $\alpha = .05$. It seems a contradiction to our conclusion in Problem 8, which, in fact, is not. Since if we take for each pair of effects of four brands of tires (total six pairs) $\alpha = .05$, then in Problem 8, the real level of significance is $1 - (0.95)^6 = 0.2649$. Hence, in this case we must use level of significance $\alpha = .05/6 = .00833$ (Bonferroni method) for overall level of significance to be 0.05.

11. MINITAB (Two-way ANOVA): Score versus Group, Method

```
Source    DF      SS       MS        F       P
Group      3    139.6    46.533    0.81    0.514
Method     4    449.5   112.375    1.95    0.167
Error     12    692.9    57.742
Total     19   1282.0
```

```
S = 7.599    R-Sq = 45.95%    R-Sq(adj) = 14.42%
```

```
                        Individual 95% CIs For Mean Based on
                        Pooled StDev
Method    Mean    ------+---------+---------+---------+---
1         80.75                       (---------*---------)
2         67.75       (----------*---------)
3         76.50            (----------*---------)
4         78.50              (---------*---------)
5         71.50          (---------*----------)
                  ------+---------+---------+---------+---
                      64.0      72.0      80.0      88.0
```

Since the p-value (0.167) for teaching methods is greater than $\alpha = .05$ the effects due to teaching methods are insignificant. The blocks here are "age groups".

13. MINITAB (Two-way ANOVA): Flies versus Block, Treatment

```
Source      DF      SS        MS       F       P
Block        6   196529   32754.8    5.42    0.006
Treatment    2     1184     591.8    0.10    0.907
Error       12    72564    6047.0
Total       20   270277
```

```
S = 77.76   R-Sq = 73.15%    R-Sq(adj) = 55.25%
```

Since the p-value for treatments is $0.907 > .05$, we conclude that the effects due to treatment are not significantly different.

Since the p-value for blocks is $<.05$, we conclude that the effects due to block are significantly different.

15. MINITAB (Two-way ANOVA): Length versus Disease, Group

Two-way ANOVA: Time versus Age Groups, Disease

```
Source        DF     SS       MS        F       P
Age Groups     3    750.5   250.167   17.32    0.000
Disease        3     78.5    26.167    1.81    0.215
Error          9    130.0    14.444
Total         15    959.0
```

```
S = 3.801   R-Sq = 86.44%    R-Sq(adj) = 77.41%
```

Since $p_{Disease} = 0.215 > .01$, the disease effects are not significantly different. However, $p_{Group} = 0.000 < .01$, effects due to age groups are significantly different. Here age groups are used as the blocks.

17. (a) MINITAB (Two-way ANOVA): Solids versus Analyst, Batch

```
Source     DF      SS         MS         F        P
Analyst     2    0.0311     0.0156      2.06    0.178
Batch       5   93.5561    18.7112   2476.49    0.000
Error      10    0.0756     0.0076
Total      17   93.6628
```

```
S = 0.08692    R-Sq = 99.92%    R-Sq(adj) = 99.86%
```

(b) Since $p_{analyst} = 0.178 > .05$, the effects due to the analysts are not significant.

(c) Since $p_{Batch} = 0.000 < 0.05$, so the effects due to the batches are significant.

```
                   Individual 95% CIs For Mean Based on
                   Pooled StDev
Batch    Mean      -----+---------+---------+---------+----
a       20.1000                                      (*)
b       14.8000              (*)
c       13.0333    (*)
d       17.8000                            (*)
e       16.1000                    (*
f       15.0000              (*)
                   -----+---------+---------+---------+----
                      14.0       16.0      18.0      20.0
```

$$\hat{\beta}_1 = 3.9611, \hat{\beta}_2 = -1.3389, \hat{\beta}_3 = 3.1056, \hat{\beta}_4 = 1.6611, \hat{\beta}_5 = -0.0389, \hat{\beta}_6 = -1.1389$$

(d) A contrast that can be used to test whether Supplier 1 is different from Supplier 2 is $\beta_a + \beta_b + \beta_c - \beta_d - \beta_e - \beta_f$. Since the effects due to analysts are not significant we can pool SS and the degrees of freedom from the analysts with the error SS, yielding $MS_E = 0.1067/12 = 0.008892$. Using the S-method, the contrast has a point estimate of

$$20.10 + 14.80 + 13.03 - 17.80 - 16.10 - 15.00 = -0.97$$

and a margin of error of

$$\theta_{\hat{\sigma}\hat{\psi}} = \sqrt{(5)(3.106)(0.008892)\frac{1^2 + 1^2 + 1^2 + (-1)^2 + (-1)^2 + (-1)^2}{3}} = 0.5264.$$

This gives a confidence interval of $(-1.50, -0.44)$, which does not contain "0". Since both the limits are negative, we conclude that the effect of using Supplier 1 has lower effect than using the Supplier 2.

(e) The effects of analysts were insignificant, but the effects of batches were significant. In fact, batches from Supplier 1 showed lower means than batches from Supplier 2.

19. (a) MINITAB (Two-way ANOVA): Corrosion versus Analyst, Metal

```
Source   DF      SS        MS        F       P
Analyst   3    6.3975    2.1325     1.90    0.184
Metal     4   62.1650   15.5412    13.82    0.000
Error    12   13.4950    1.1246
Total    19   82.0575
```

```
S = 1.060      R-Sq = 83.55%      R-Sq(adj) = 73.96%
```

Since $p_{Analyst} = 0.184 > .05$, we conclude that the effects of analysts are not significant.

Since $p_{metal} = 0.000 < .05$, we conclude that the effects of metals are significant.

(b)

```
                   Individual 95% CIs For Mean Based on
                   Pooled StDev
Metal    Mean      ---------+---------+---------+---------+
1        13.275    (----*-----)
2        13.550     (-----*-----)
3        14.800        (-----*-----)
4        17.775                          (-----*-----)
5        16.725                    (-----*----)
                   ---------+---------+---------+---------+
                       14.0      16.0      18.0      20.0
```

Since $\hat{\beta}_j = \bar{y}_{.j} - \bar{y}_{..}$, the metal effects are estimated to be:

$$\hat{\beta}_1 = -1.95, \hat{\beta}_2 = -1.675, \hat{\beta}_3 = -0.425, \hat{\beta}_4 = 2.550, \hat{\beta}_5 = 1.500.$$

21. (a) MINITAB (Two-way ANOVA): Time versus Poison, Treatment

Source	DF	SS	MS	F	P
Poison	2	1.00868	0.504340	22.73	0.000
Treatment	3	0.98979	0.329930	14.87	0.000
Interaction	6	0.27880	0.046467	2.09	0.078
Error	36	0.79873	0.022187		
Total	47	3.07600			

$S = 0.1490$ R-Sq $= 74.03\%$ R-Sq(adj) $= 66.10\%$

```
                   Individual 95% CIs For Mean Based on
                   Pooled StDev
Poison   Mean      -------+---------+---------+---------+--
I        0.617500                       (----*----)
II       0.531875                  (----*----)
III      0.276250   (----*----)
                   -------+---------+---------+---------+--
                      0.30      0.45      0.60      0.75
```

```
                   Individual 95% CIs For Mean Based on
                   Pooled StDev
Treatment  Mean    ------+---------+---------+---------+---
A        0.297500  (-----*-----)
B        0.676667                      (-----*-----)
C        0.392500        (-----·*-----)
D        0.534167              (-----*----)
                   ------+---------+---------+---------+---
                      0.30      0.45      0.60      0.75
```

(b) Since $p_{Interaction} = 0.078 > .05$, we do not reject the hypothesis that interactions effects are zero.

(c) Since $p_{Poison} = 0.000 < .05$ and $P_{Treatment} = 0.000 < .05$, so we reject the hypotheses that Poison effects are 0 and Treatment effects are 0.

23. (a) MINITAB (Two-way ANOVA): Byproduct versus Temperature, Catalyst

Source	DF	SS	MS	F	P
Temperature	2	0.032258	0.0161292	4.11	0.044
Catalyst	3	0.157433	0.0524778	13.37	0.000
Interaction	6	0.163542	0.0272569	6.94	0.002
Error	12	0.047100	0.0039250		
Total	23	0.400333			

$S = 0.06265$ $R\text{-}Sq = 88.23\%$ $R\text{-}Sq(adj) = 77.45\%$

(b) Since $p_{Interaction} = 0.002 < .05$, we conclude that the interaction effects are significant.

(c)-(d) Since the interactions are significant, we cannot test hypotheses of Temperature and Catalysts effects are zero. One way to achieve this goal is to test the Temperature effects at each level of Catalyst and test the Catalyst effects at each level of Temperature.

25. MINITAB (General Linear Model): Observation versus Block, A, B

Factor	Type	Levels	Values
Block	fixed	2	1, 2
A	fixed	3	1, 2, 3
B	fixed	3	1, 2, 3

Analysis of Variance for Observation, using Adjusted SS for Tests

Source	DF	Seq SS	Adj SS	Adj MS	F	P
Block	1	2.98	2.98	2.98	3.28	0.107
A	2	301.44	301.44	150.72	165.86	0.000
B	2	774.23	774.23	387.12	425.99	0.000
A*B	4	1.58	1.58	0.39	0.43	0.781
Error	8	7.27	7.27	0.91		
Total	17	1087.50				

$S = 0.953278$ $R\text{-}Sq = 99.33\%$ $R\text{-}Sq(adj) = 98.58\%$

$S = 0.858632$ $R\text{-}Sq = 99.19\%$ $R\text{-}Sq(adj) = 98.85\%$

Since $p_{Block} = 0.107 > .05$, we conclude that the block effect are not significant.
Since $p_{Interaction} = 0.781 > .05$, we conclude that the interactions are not significant.
Since $p_A = 0.000 < .05$, and $p_B = 0.000 < .05$, we conclude that the effects of factor A and B are significant.

27. (a) MINITAB
   General Linear Model: MPG versus Vehicle, Run, Driver

Factor	Type	Levels	Values
Vehicle	fixed	4	A, B, C, D
Run	fixed	4	1, 2, 3, 4
Driver	fixed	4	a, b, c, d

```
Analysis of Variance for MPG, using Adjusted SS for Tests

Source      DF      Seq SS      Adj SS      Adj MS        F         P
Vehicle      3     0.06395     0.06395     0.02132      0.51     0.689
Run          3     0.25135     0.25135     0.08378      2.01     0.214
Driver       3     1.61850     1.61850     0.53950     12.93     0.005
Error        6     0.25040     0.25040     0.04173
Total       15     2.18420


S = 0.204287    R-Sq = 88.54%    R-Sq(adj) = 71.34%
```

(b) Since $p_{Drivers} = .005 < .05$, we conclude that the effects of drivers are significant. The estimated effects of the drivers are:

$$\hat{\delta}_a = 9.18 - 9.28 = -0.10, \hat{\delta}_b = 9.76 - 9.28 = 0.48, \hat{\delta}_c = 8.88 - 9.28 = -0.40, \hat{\delta}_d = 9.30 - 9.28 = 0.02.$$

(c) Since $p_{Vehicle} = 0.698$, we conclude that the effects of the vehicles are not significant.

29.

Source	Sums of Squares	Degrees of Freedom	Mean Sum of Squares	*F-ratio*
Between rows	58.1	5	11.62	16.366
Between columns	78.6	5	15.72	22.14
Between treatments	81.3	5	16.26	22.90
Error	14.2	20	0.71	
Total	232.2	35		

It can easily be seen that the p-values between rows, between columns, and between treatments are 0.000. Hence, we conclude that the effects of rows, columns, and treatments are all significant at any level of significance..

31. MINITAB (BALNCED ANOVA): Force versus Machine, Sample A (Machine)

ANOVA: Data versus Machine, Samples

```
Factor            Type      Levels    Values
Machine           fixed        3      1, 2, 3
Samples(Machine)  random       4      1, 2, 3, 4


Analysis of Variance for Data

Source            DF        SS        MS        F        P
Machine            2    645.500   322.750    50.74    0.000
Samples(Machine)   9     57.250     6.361     1.36    0.259
Error             24    112.000     4.667
Total             35    814.750


S = 2.16025    R-Sq = 86.25%    R-Sq(adj) = 79.95%


                                         Expected Mean Square
                    Variance    Error    for Each Term (using
   Source           component   term     restricted model)
1  Machine                        2      (3) + 3 (2) + 12 Q[1]
2  Samples(Machine)   0.5648      3      (3) + 3 (2)
3  Error              4.6667             (3)
```

Since $p_{Machines} = 0.000 < 0.01$, we reject the null hypothesis of equal machines effects.

Since $p_{Sample} = 0.259 > 0.01$, we do not reject the null hypothesis of no sample variation within each machine.

Note: The restricted model is when sum of interaction effects at each level of a random factor is assumed zero (see model conditions for mixed effect model).

33. MINITAB (Using Balanced ANOVA): Rate versus Process, Batch(Process)
ANOVA: Data versus Process, Batch(Process)

Factor	Type	Levels	Values
Process	fixed	3	1, 2, 3
Batch(Process)(Process)	random	4	1, 2, 3, 4

Analysis of Variance for Data

Source	DF	SS	MS	F	P
Process	2	676.06	338.03	1.46	0.281
Batch(Process)(Process)	9	2077.58	230.84	12.20	0.000
Error	24	454.00	18.92		
Total	35	3207.64			

$S = 4.34933$ R-Sq $= 85.85\%$ R-Sq(adj) $= 79.36\%$

	Source	Variance component	Error term	Expected Mean Square for Each Term (using restricted model)
1	Process		2	(3) + 3 (2) + 12 Q[1]
2	Batch(Process)	70.64	3	(3) + 3 (2)
3	Error	18.92		(3)

For using balanced ANOVA in MINITAB, batches are renamed as 1, 2, 3, 4 for each process.

Since $p_{Process} = 0.281 > 0.01$, we do not reject the null hypothesis of equal effects among processes.

Since $p_{Batch} = 0.000 > 0.01$, we reject the null hypothesis of no variations among the batches within the same process. Note that 78.9% of the total variation is due to batches within each process.

Note: The restricted model is when sum of interaction effects at each level of a random factor is assumed zero (see model conditions for mixed effect model).

35. (a) MINITAB (Nested ANOVA): Epinephrine versus Anesthesia, Mouse

Analysis of Variance for Epinephrine

Source	DF	SS	MS	F	P
Anesthesia	2	0.0021	0.0010	0.111	0.896
Mouse	9	0.0832	0.0092	0.191	0.991
Error	12	0.5806	0.0484		
Total	23	0.6659			

Expected Mean Squares

1	Anesthesia	1.00(3) + 2.00(2) + 8.00(1)
2	Mouse(Anesthesia)	1.00(3) + 2.00(2)
3	Error	1.00(3)

(b) Since $p_{anesthesia} = 0.896 > .05$, we do not reject the null hypothesis of no variation among anesthesia.

(c) Since $p_{\text{mouse}} = 0.991 > .05$, we do not reject the null hypothesis of no variation among mice within the each anesthesia.

37. MINITAB (One-way ANOVA): Heart Rate versus Age Group

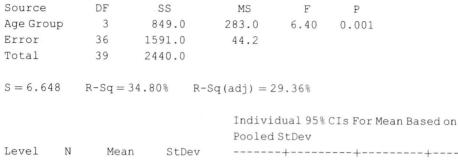

```
Source      DF      SS       MS      F      P
Age Group    3     849.0    283.0   6.40   0.001
Error       36    1591.0     44.2
Total       39    2440.0

S = 6.648    R-Sq = 34.80%    R-Sq(adj) = 29.36%
```

```
                                    Individual 95% CIs For Mean Based on
                                    Pooled StDev
Level   N     Mean     StDev       -------+---------+---------+---------+--
1      10    53.800    9.199                (------*------)
2      10    60.100    5.087                        (------*------)
3      10    50.200    6.730            (------*------)
4      10    47.900    4.581        (------*------)
                                    -------+---------+---------+---------+--
                                        48.0      54.0      60.0      66.0
```

Pooled StDev = 6.648

Tukey 95% Simultaneous Confidence Intervals
All Pairwise Comparisons among Levels of Age Group

Individual confidence level = 98.93%

Age Group = 1 subtracted from:

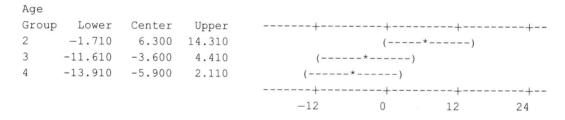

```
Age
Group   Lower   Center   Upper    -------+---------+---------+---------+--
2      -1.710    6.300   14.310                    (-----*------)
3     -11.610   -3.600    4.410          (------*------)
4     -13.910   -5.900    2.110            (------*------)
                                  -------+---------+---------+---------+--
                                       -12        0        12        24
```

Age Group = 2 subtracted from:

```
Age
Group   Lower   Center   Upper    -------+---------+---------+---------+--
3     -17.910   -9.900   -1.890         (------*-----)
4     -20.210  -12.200   -4.190     (------*------)
                                  -------+---------+---------+---------+--
                                       -12        0        12        24
```

Age Group = 3 subtracted from:

```
Age
Group   Lower   Center   Upper    -------+---------+---------+---------+--
4     -10.310   -2.300    5.710           (------*------)
                                  -------+---------+---------+---------+--
                                       -12        0        12        24
```

Since $p_{Age\ Group} = 0.001 < .05$, the effects among age groups are significantly different. Also, the Tukey method indicates that group 2 differs from groups 3 and 4. Estimates of effects due to age groups are $\hat{\alpha}_1 = 0.8, \hat{\alpha}_2 = 7.1, \hat{\alpha}_3 = -2.8, \hat{\alpha}_4 = -5.1$.

39. MINITAB (Kruskal-Wallis Test): Ratio versus Preparation

```
Kruskal-Wallis Test on Ratio

Preparation      N       Median       Total Ranks (T.j)
A                4       1.176            17.5
B                4       1.176            40.5
C                4       1.176            26.5
D                4       1.176            51.5
Overall         16                        136
```

Now the observed value of the Kruskal-Wallis test statistic (distributed as Chi-Square) is found to be

$$H = \frac{12}{N(N+1)} \sum_{j=1}^{a} \frac{T_j^2}{n_j} - 3(N+1) = 7.47$$

Hence, we have

$$H = 7.47, DF = 3, p\text{-value} = 0.058.$$

Since p- value is greater than $\alpha = .05$, we do not reject the null hypothesis of equal means among preparations.

41. MINITAB (Friedman Test): Wear versus Brand blocked by Car

```
S = 3.96    DF = 3    P = 0.266

Tire                            Sum of
Brand        N    Est Median    Ranks
1            5      2.4050        8.0
2            5      2.4850       12.0
3            5      2.4725       15.0
4            5      2.7575       15.0
```

From the MINITAB output above the p-value is 0.266, which greater than 0.05, the level of significance, and we conclude that at the 5% level of significance the effects among the brands are not significant. The Friedman Test does not give any results regarding the blocks.

43. MINITAB (Friedman Test): Length versus Disease blocked by Group

```
S = 3.90    DF = 3    P = 0.272
                                Sum of
Disease      N    Est Median    Ranks
Cancer       4      43.969        7.0
Hip          4      51.094       14.0
Stroke       4      45.844        9.0
TB           4      47.469       10.0

Grand median = 47.094
```

From the MINITAB output above the p-value is .272, which is greater than 0.05, the level of significance. Thus, we conclude that the effects among diseases are not significantly different. Friedman Test does not give any results regarding the blocks. Note that the MINITAB lists the treatments in an alphabetic order.

18

THE 2^k FACTORIAL DESIGNS

PRACTICE PROBLEMS FOR SECTION 18.2

1. A factorial experiment is undertaken when the effect on a response Y of *several* factors, say $A, B, C, \cdots$ is investigated and the factors are *varied together*. Commonly, each factor is studied at several levels, e.g., $A_1, A_2, \cdots, A_a$ are a levels of factor A, $B_1, B_2, \cdots, B_b$ are b levels of factor B, $C_1, C_2, \cdots, C_c$ are c levels of factor C, and so on. Then the use of $a \times b \times c \times \cdots$ treatments (experiments), say $A_i \times B_j \times C_k \times \cdots$, $i = 1, 2, \cdots, a$; $j = 1, 2, \cdots, b; k = 1, 2, \cdots, c; \cdots$ to observe the response $Y_{ijk\cdots}$ constitute the factorial experiment for this investigation.

3. A set of orthogonal contrasts of (Y_1, Y_2, Y_3, Y_4) is:

$$L_1 = Y_1 - Y_2 - Y_3 + Y_4 = \sum_{i=1}^{4} c_i Y_i$$

$$L_2 = -Y_1 + Y_2 - Y_3 + Y_4 = \sum_{i=1}^{4} d_i Y_i$$

$$L_3 = -Y_1 - Y_2 + Y_3 + Y_4 = \sum_{i=1}^{4} e_i Y_i.$$

We have that $\sum_{i=1}^{4} c_i = \sum_{i=1}^{4} d_i = \sum_{i=1}^{4} e_i = 0$ so that L_i's are contrasts in (Y_1, Y_2, Y_3, Y_4), and also we have that $\sum_{i=1}^{4} c_i d_i = \sum_{i=1}^{4} d_i e_i = \sum_{i=1}^{4} c_i e_i = 0$ so that $L_i, i = 1, 2, 3$ are a set of mutually orthogonal contrasts. Maximum three orthogonal contrasts in (Y_1, Y_2, Y_3, Y_4).

5. Suppose $n = 4$ and $(Y_1, \cdots, Y_4)$ observed to be (11, 9, 17, 13) respectively. Further, we suppose that $L = Y_1 + Y_2 - Y_3 - Y_4$, so that $L = 11 + 9 - 17 - 13 = -10$.

Note that $\Sigma c_i^2 = 1^2 + 1^2 + (-1)^2 + (-1)^2 = 4$, and in this case it can be seen that an unbiased estimator of σ^2 is observed to be $S^2 = 11.67$.

To test $H_0 : E(L) = 0$ Vs $H_1 : E(L) \neq 0$ we know from Problem 4 that under the null hypothesis the test statistic

$$\frac{L}{S\sqrt{\Sigma c_i^2}} = \frac{L}{S\sqrt{4}} = \frac{L}{2S}$$

Solutions Manual to Accompany Statistics and Probability with Applications for Engineers and Scientists, Bhisham C. Gupta and Irwin Guttman.
© 2014 John Wiley & Sons, Inc. Published 2014 by John Wiley & Sons, Inc.

is distributed as t_3. Thus we would reject $H_0 : E(L) = 0$ in favor of $H_1 : E(L) \neq 0$ at significance level $\alpha = .05$ if

$$\left|\frac{L}{2S}\right| > t_{n-1,\alpha/2} = t_{3;.025} = 3.182$$

But the observed value of the test statistic in this problem is

$$\left|\frac{L}{2S}\right| = \left|\frac{-10}{2\sqrt{11.67}}\right| = 1.464 < 3.182,$$

so that we do not reject $H_0 : E(L) = 0$.

7. The answer varies from field to field. There are several examples and problems from different fields that are given in this chapter.

PRACTICE PROBLEMS FOR SECTION 18.3

1.

	Notation 1	Notation 2				Notation 3			
	1	0	0	0	0	-1	-1	-1	-1
	a	1	0	0	0	1	-1	-1	-1
	b	0	1	0	0	-1	1	-1	-1
	ab	1	1	0	0	1	1	-1	-1
	c	0	0	1	0	-1	-1	1	-1
	ac	1	0	1	0	1	-1	1	-1
	bc	0	1	1	0	-1	1	1	-1
	abc	1	1	1	0	1	1	1	-1
	d	0	0	0	1	-1	-1	-1	1
	ad	1	0	0	1	1	-1	-1	1
	bd	0	1	0	1	-1	1	-1	1
	abd	1	1	0	1	1	1	-1	1
	cd	0	0	1	1	-1	-1	1	1
	acd	1	0	1	1	1	-1	1	1
	bcd	0	1	1	1	-1	1	1	1
	$abcd$	1	1	1	1	1	1	1	1

Note that notation 3 used above is equivalent to using the notation "-" and "+".

3. For a 2^4 factorial experiment, there are

$$\binom{4}{2} = 6 \text{ two factor interactions}$$

$$\binom{4}{3} = 4 \text{ three factor interactions}$$

$$\binom{4}{4} = 1 \text{ four factor interaction.}$$

5. The design matrix for a 2^3 factorial design is given below.

$$\begin{array}{cccccccc} \mu & A & B & C & AB & AC & BC & ABC \end{array}$$

$$\begin{pmatrix} 1 & -1 & -1 & -1 & 1 & 1 & 1 & -1 \\ 1 & 1 & -1 & -1 & -1 & -1 & 1 & 1 \\ 1 & -1 & 1 & -1 & -1 & 1 & -1 & 1 \\ 1 & 1 & 1 & -1 & 1 & -1 & -1 & -1 \\ 1 & -1 & -1 & 1 & 1 & -1 & -1 & 1 \\ 1 & 1 & -1 & 1 & -1 & 1 & -1 & -1 \\ 1 & -1 & 1 & 1 & -1 & -1 & 1 & -1 \\ 1 & 1 & 1 & 1 & 1 & 1 & 1 & 1 \end{pmatrix}$$

7. In this problem we are dealing with a 2^3 factorial experiment with 4 replications. We have seen in Problems 4 and 6 above that estimates of main effects and/or interactions are of the form

$$\text{Est.} = \overline{Y}_+ - \overline{Y}_-,$$

and here $\overline{Y}_+$ *and* $\overline{Y}_-$ are means of 16 observations generated using the plus and minus signs, respectively [note that the number $16 = r \times 2^{k-1} = 4 \times 2^{3-1} = 4 \times 2^2$]. Since all observations are assumed to be independent, we have

$$\text{Var (Est.)} = Var(\overline{Y}_+ - \overline{Y}_-) = Var(\overline{Y}_+) + Var(\overline{Y}_-) = \sigma^2/16 + \sigma^2/16 = \sigma^2/8.$$

Or the estimate of the standard error is $\hat{\sigma}/\sqrt{8} = \sqrt{MSE/8}$.
In general, we have

$$\text{Var (Est.)} = Var(\overline{Y}_+ - \overline{Y}_-) = Var(\overline{Y}_+) + Var(\overline{Y}_-) = \sigma^2/r2^{k-1} + \sigma^2/r2^{k-1} = \sigma^2/r2^{k-2}.$$

Note that the estimate of $\text{Var (EST.)} = Var(\overline{Y}_+) + Var(\overline{Y}_-)$ is

$$\begin{aligned} V\hat{a}r(EST.) &= V\hat{a}r(\overline{Y}_+) + V\hat{a}r(\overline{Y}_-) \\ &= \hat{\sigma}^2/r2^{k-1} + \hat{\sigma}^2/r2^{k-1} = \hat{\sigma}^2/r2^{k-2}, \end{aligned}$$

where $\hat{\sigma}^2 = MSE = \sum\limits_{i=1}^{2^k} \sum\limits_{j=1}^{r} (y_{ij} - \bar{y}_{i.})^2/(r-1)2^k, \bar{y}_{i.} = T_i/r$.

We may write $V\hat{a}r(EST.) = MSE/r2^{k-2}$
Or The standard error is $\sqrt{(MSE/r2^{k-2})}$.

Note that the degrees of freedom for MSE is $(r-1)2^k$, since each treatment is replicated r times, which provides $(r-1)$ degrees of freedom for MSE, and there are 2^k treatments.

9. In this we are interested in determining 99% confidence intervals for the main effects. Suppose we first determine a confidence interval for the main effect A. Using the result of Problem 7, we can conclude that

$$(\hat{A} - A)/\sqrt{(MSE/8}$$

has t distribution with $24[=(r-1)2^k = (4-1)2^3]$ degrees of freedom, where $\hat{\sigma}^2 = S^2 = MSE$. Hence, 99% confidence interval for the main effect A is given by

$$(\hat{A} \pm t_{24;.005}(\sqrt{MSE/8})) = (\hat{A} \pm 2.797(\sqrt{MSE/8})) = (\hat{A} \pm 0.989\sqrt{MSE}).$$

or

$$((\overline{Y}_{+}^{(A)} - \overline{Y}_{-}^{(A)}) \pm t_{24;.005}(\sqrt{MSE/8})) = ((\overline{Y}_{+}^{(A)} - \overline{Y}_{-}^{(A)}) \pm 2.797(\sqrt{MSE/8}))$$
$$= ((\overline{Y}_{+}^{(A)} - \overline{Y}_{-}^{(A)}) \pm 0.989\sqrt{MSE}).$$

Similarly, we can determine confidence intervals for main effects B and C. Thus, confidence intervals for main effects B and C are

$$((\overline{Y}_{+}^{(B)} - \overline{Y}_{-}^{(B)}) \pm t_{24;.005}(\sqrt{MSE/8})) = ((\overline{Y}_{+}^{(B)} - \overline{Y}_{-}^{(B)}) \pm 2.797(\sqrt{MSE/8}))$$
$$= ((\overline{Y}_{+}^{(B)} - \overline{Y}_{-}^{(B)}) \pm 0.989\sqrt{MSE}),$$

$$((\overline{Y}_{+}^{(C)} - \overline{Y}_{-}^{(C)}) \pm t_{24;.005}(\sqrt{MSE/8})) = ((\overline{Y}_{+}^{(C)} - \overline{Y}_{-}^{(C)}) \pm 2.797(\sqrt{MSE/8}))$$
$$= ((\overline{Y}_{+}^{(C)} - \overline{Y}_{-}^{(C)}) \pm 0.989\sqrt{MSE}),$$

respectively.

PRACTICE PROBLEMS FOR SECTION 18.4

1. MINITAB (Factorial Fit: Responses versus A, B, C, D

Estimated Effects and Coefficients for Responses (coded units)

Term	Effect	Coef
Constant		17.556
A	8.812	4.406
B	-2.537	-1.269
C	-1.112	-0.556
D	0.087	0.044
A*B	-0.737	-0.369
A*C	1.038	0.519
A*D	0.238	0.119
B*C	1.037	0.519
B*D	0.287	0.144
C*D	-0.137	-0.069
A*B*C	0.188	0.094
A*B*D	-0.063	-0.031
A*C*D	1.362	0.681
B*C*D	0.263	0.131
A*B*C*D	-0.237	-0.119

Analysis of Variance for Responses (coded units)

Source	DF	Seq SS	Adj SS	Adj MS	F	P
Main Effects	4	341.378	341.378	85.344	*	*
A	1	310.641	310.641	310.641	*	*
B	1	25.756	25.756	25.756	*	*
C	1	4.951	4.951	4.951	*	*
D	1	0.031	0.031	0.031	*	*
2-Way Interactions	6	11.419	11.419	1.903	*	*
A*B	1	2.176	2.176	2.176	*	*
A*C	1	4.306	4.306	4.306	*	*

A*D	1	0.226	0.226	0.226	*	*
B*C	1	4.306	4.306	4.306	*	*
B*D	1	0.331	0.331	0.331	*	*
C*D	1	0.076	0.076	0.076	*	*
3-Way Interactions	4	7.857	7.857	1.964	*	*
A*B*C	1	0.141	0.141	0.141	*	*
A*B*D	1	0.016	0.016	0.016	*	*
A*C*D	1	7.426	7.426	7.426	*	*
B*C*D	1	0.276	0.276	0.276	*	*
4-Way Interactions	1	0.226	0.226	0.226	*	*
A*B*C*D	1	0.226	0.226	0.226	*	*
Residual Error	0	*	*	*		
Total	15	360.879				

Estimates of all main effects and interactions are given in the MINITAB output. Note the ANOVA table is incomplete since we do not have any degrees of freedom for estimating the error variance.

(a) From the above MINITAB printout we note that the only significant effects seems to be A, B, C, AC, BC, and ACD. However, dimension A seems to be the most critical in influencing the response.

(b) The variance of effects estimate is given by $\sigma^2/r2^{k-2}$. Hence, in this problem the variance of effects estimate is equal to 1 or the standard error is equal to 1. The 95% confidence intervals for the significant effects are given by

$$
\begin{aligned}
&For\ A: &&(8.812 \pm 1.96)\\
&For\ B: &&(-2.537 \pm 1.96)\\
&For\ C: &&(-1.112 \pm 1.96)\\
&For\ AC: &&(1.038 \pm 1.96)\\
&For\ BC: &&(1.037 \pm 1.96)\\
&For\ ACD: &&(1.362 \pm 1.96)
\end{aligned}
$$

(c) The normal plot of the effects is shown below. It indicates that the only significant effects are A, B, C, AB, BC, and ACD.

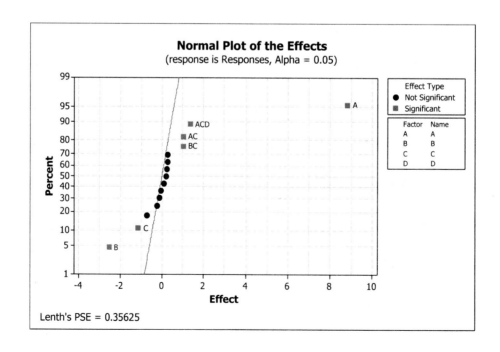

(d) The ANOVA table assuming the non-significant effects equal to zero and estimating the error variance by pooling the sum of squares of the non-significant interactions is the following:

Analysis of Variance for Responses (coded units)

Source	DF	Seq SS	Adj SS	Adj MS	F	P
Main Effects	4	341.378	341.378	85.344	197.04	0.000
A	1	310.641	310.641	310.641	717.21	0.000
B	1	25.756	25.756	25.756	59.46	0.000
C	1	4.951	4.951	4.951	11.43	0.010
D	1	0.031	0.031	0.031	0.07	0.797
2-Way Interactions	2	8.611	8.611	4.306	9.94	0.007
A*C	1	4.306	4.306	4.306	9.94	0.014
B*C	1	4.306	4.306	4.306	9.94	0.014
3-Way Interactions	1	7.426	7.426	7.426	17.14	0.003
A*C*D	1	7.426	7.426	7.426	17.14	0.003
Residual Error	8	3.465	3.465	0.433		
Total	15	360.879				

From the above ANOVA, we obtain

$$\hat{\sigma}^2 = MSE = 0.433$$

Thus, the standard error of the estimates of 6 significant effects is estimated to be $\sqrt{0.433/4} = 0.3290$. Therefore, The 95% confidence intervals for the significant effects, including the most crucial critical dimension A, are given by

$$A: \quad (8.812 \pm 2.306 \times 0.3290) = (8.812 \pm 0.7587)$$
$$B: \quad (-2.537 \pm 2.306 \times 0.3290) = (-2.537 \pm 0.7587)$$
$$C: \quad (-1.112 \pm 2.306 \times 0.3290) = (-1.112 \pm 0.7587)$$
$$AC: \quad (1.038 \pm 2.306 \times 0.3290) = (1.038 \pm 0.7587)$$
$$BC: \quad (1.037 \pm 2.306 \times 0.3290) = (1.037 \pm 0.7587)$$
$$ACD: \quad (1.362 \pm 2.306 \times 0.3290) = (1.362 \pm 0.7587)$$

3. (a) We know that the variance of the estimates of effects is given by $\sigma^2/r2^{k-2}$ or the standard error is given by $\sigma/\sqrt{r2^{k-2}}$. Now using MSE as an estimate of the error variance, we obtain the standard error of the effects is equal to

$$\hat{\sigma}/\sqrt{r2^{k-2}} = 4.272/\sqrt{2 \times 2^{4-2}} = 1.5104.$$

(b) The 95% confidence intervals for the factor effects are given by $(t_{16,.025} = 2.120)$

$$[(\text{Estimate of factor effect}) \pm 2.120 \times 1.5104]$$

or

$$[(\text{Estimate of factor effect}) \pm 3.202]$$

(c) Substituting the values of estimates of factor effects in

$$[(\text{Estimate of factor effect}) \pm 3.202]$$

we can easily verify that the only confidence interval that does not contain zero is that of the main effect D. This implies that the only significant effect is the main effect D. This matches with our conclusion in Problem 2.

5. MINITAB (Factorial Fit): Yield versus A, B, C, D

(a) See solution of Problem 4 above. Estimate of the error variance is given by

$$\hat{\sigma}^2 = MSE = 3.062.$$

(b) From the MINITAB output we determine the main effects B, C, and D are significant at the 5% level of significance, since the p-values are less than .05. The standard error of effects is given by $\sigma^2/r2^{k-2}$. Hence, an estimated standard error of the factor effects is given by ($r = 1$, $k = 4$)

$$\text{Standard error of factor effects} = \sqrt{\hat{\sigma}^2/r2^{k-2}} = \sqrt{3.062/4} = 0.8750$$

(c) The 99% confidence intervals for the main effects B, C, and D are given by ($t_{4;.005} = 4.604$)

$$[\textit{Estimated effect} \pm t_{4,.005} \times \text{Standard Error}]$$
$$B: (11.125 \pm 3.525)$$
$$C: (7.875 \pm 3.525)$$
$$D: (5.875 \pm 3.525)$$

Note the that none of the 99% confidence for the main effect B, C, and D contains zero, which means all of them are of significant 1% level of significance.

PRACTICE PROBLEMS FOR SECTION 18.5

1. In this problem we have a 2^4 design and we wish to create two blocks using the interaction ABCD as a design generator. Since in this case the interaction ABCD is confounded with the block effects, the contrast between the two blocks is the same as the contrast for the interaction ABCD. In other words, one block consist of 8 runs of the 2^4 factorial which contains a plus sign in the four factor interaction contrast ABCD, while the second block consists of the eight runs possessing a minus sign in the four factor interaction. Thus, the treatments in the two blocks in this case are:

$$\text{Block 1(Principle block): } 1, ab, ac, bc, ad, bd, cd, abcd$$
$$\text{Block 2: } a, b, \quad c, \quad d, \quad abc, abd, acd, bcd$$

(b) Since using the principal block generated by the interaction ABCD is equivalent to using $^1/_2$ replication using ABCD as the defining contrast. This means all main effects are confounded with three factor interactions and two factor interactions are confounded with each other. Thus, to analyze the data we assume ABC, ABD, ACD, BCD, BC, BD, and CD are negligible (also, see the alias structure given below). Thus, the analysis of the 1/2 replication with each treatment replicated 2 times is as follows:

Factorial Fit: Yield versus A, B, C, D

```
Estimated Effects and Coefficients for Yield (coded units)
```

Term	Effect	Reg. Coef	SE Coef	T	P
Constant		56.500	1.008	56.06	0.000
A	-2.250	-1.125	1.008	-1.12	0.297
B	-1.000	-0.500	1.008	-0.50	0.633
C	-1.500	-0.750	1.008	-0.74	0.478
D	7.250	3.625	1.008	3.60	0.007
A*B	2.750	1.375	1.008	1.36	0.210
A*C	3.750	1.875	1.008	1.86	0.100
A*D	1.000	0.500	1.008	0.50	0.633

```
S = 4.03113     PRESS=520
R-Sq = 71.98%   R-Sq(pred) = 0.00%   R-Sq(adj) = 47.47%
```

```
Analysis of Variance for Yield (coded units)
```

Source	DF	Seq SS	F	P
Main Effects	4	243.500	3.75	0.053
A	1	20.250	1.25	0.297
B	1	4.000	0.25	0.633
C	1	9.000	0.55	0.478
D	1	210.250	12.94	0.007
2-Way Interactions	3	90.500	1.86	0.215
A*B	1	30.250	1.86	0.210
A*C	1	56.250	3.46	0.100
A*D	1	4.000	0.25	0.633
Residual Error	8	130.000	16.250	
Pure Error	8	130.000	16.250	
Total	15	464.000		

From the ANOVA table we see that only significant effect is the main effect of factor D. This is also confirmed by the normal plots. The two factor interactions AB, AC, and AD are confounded with CD, BD, and BC respectively. All three factor interactions are confounded with main effects.

Alias Structure

```
I + A*B*C*D
A + B*C*D
B + A*C*D
C + A*B*D
D + A*B*C
A*B + C*D
A*C + B*D
A*D + B*C
```

Effects Normal Plot for Yield

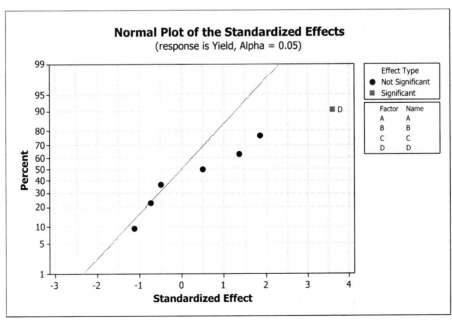

The residual plot given below does not indicate any abnormality about our assumptions for the model considered in this problem.

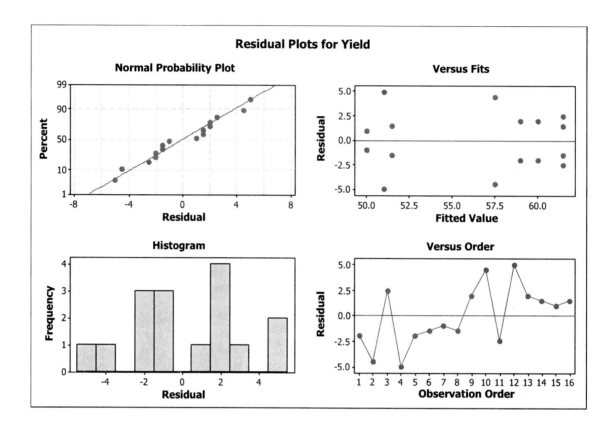

3. In this problem we have a 2^5 design and we wish to create four blocks using the interaction ABCD and BCE as design generators. Here, there are three degrees of freedom associated with 4 blocks, but the interaction ABCD and BCE count for only two degrees of freedom. Hence, one additional degree of freedom must be confounded with blocks. This degree of freedom corresponds to the generalized interaction (ABCD)×(BCE) = ADE. Thus, three interactions that are confounded with blocks are ABCD, BCE and ADE. The four blocks obtained by using ABCD and BCE as design generators consists of 8 treatments each, which are such that they have $(-,-),(+,-),(-,+)$, and $(+,+)$ in the contrasts (ABCD, BCE).

Thus writing the two columns of the design matrix corresponding to the interactions ABCD and BCE, we can see that the desired blocks are

1. $(-,-)$ 1, ad, bc, abcd, abe, ace, cde, bde
2. $(+,-)$ a, d, abc, bcd, be, abde, ce, acde
3. $(-,+)$ b, abd, c, acd, ae, de, abce, bcde
4. $(+,+)$ ab, ac, bd, cd, e, ade, bce, abcde.

We can obtain these blocks by using MINITAB as follows:

- Select **Stat > DOE > Factorial > Create Factorial Design**
- Select **2-level factorial (Specific Generators)**
- Enter number of factors
- Select appropriate fraction (**e.g., 1/2 fraction or full fraction**)
- Select **generators** and then in the new dialog box enter the desired generators in the second window.
- click **OK.** The desired design appears in the **worksheet.**

5. Factorial Fit: Observations versus Block, A, B, C, D

Estimated Effects and Coefficients for Observations (coded units)

Term	Effect	Coef
Constant		17.556
Block		-0.094
A	8.812	4.406
B	-2.537	-1.269
C	-1.112	-0.556
D	0.087	0.044
A*B	-0.737	-0.369
A*C	1.038	0.519
A*D	0.238	0.119
B*C	1.037	0.519
B*D	0.287	0.144
C*D	-0.137	-0.069
A*B*D	-0.062	-0.031
A*C*D	1.363	0.681
B*C*D	0.263	0.131
A*B*C*D	-0.238	-0.119

Analysis of Variance for Observations (coded units)

Source	DF	Seq SS	F	P
Blocks	1	0.141	*	*
Main Effects	4	341.377	*	*
A	1	310.641	*	*
B	1	25.756	*	*
C	1	4.951	*	*
D	1	0.031	*	*
2-Way Interactions	6	11.419	*	*
A*B	1	2.176	*	*
A*C	1	4.306	*	*
A*D	1	0.226	*	*
B*C	1	4.306	*	*
B*D	1	0.331	*	*
C*D	1	0.076	*	*
3-Way Interactions	3	7.717	*	*
A*B*D	1	0.016	*	*
A*C*D	1	7.426	*	*
B*C*D	1	0.276	*	*
4-Way Interactions	1	0.226	*	*
A*B*C*D	1	0.226	*	*
Residual Error	0	*		
Total	15	360.879		

Note in this problem, as in Problem 4, we cannot test any hypothesis about the various effects since there is no degrees of freedom for error. This, of course, can be done by assuming some of the higher order interactions to be negligible and using corresponding sum of squares and degrees of freedom for error. Partial Anova is given above. Further, note that the normal plot, however, shows that the effects A, B, C, and ACD are significant at 5% level.

Effects Plot for Observations

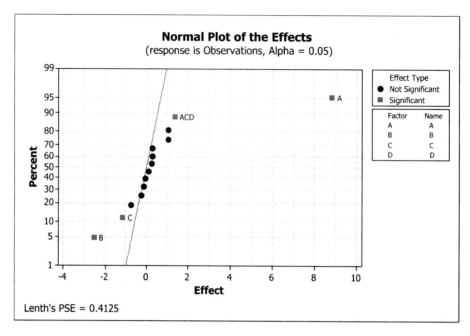

Thus, assuming interactions ABD, BCD, and ABCD negligible and reanalyzing the data we get the following output, including the new normal plot for interactions:

Factorial Fit: Yield versus Block, A, B, C, D

Estimated Effects and Coefficients for Yield (coded units)

Term	Effect	Coef	SE Coef	T	P
Constant		17.556	0.1038	169.18	0.000
Block		-0.094	0.1038	-0.90	0.433
A	8.812	4.406	0.1038	42.46	0.000
B	-2.537	-1.269	0.1038	-12.23	0.001
C	-1.112	-0.556	0.1038	-5.36	0.013
D	0.087	0.044	0.1038	0.42	0.702
A*B	-0.737	-0.369	0.1038	-3.55	0.038
A*C	1.038	0.519	0.1038	5.00	0.015
A*D	0.238	0.119	0.1038	1.14	0.336
B*C	1.037	0.519	0.1038	5.00	0.015
B*D	0.287	0.144	0.1038	1.39	0.260
C*D	-0.137	-0.069	0.1038	-0.66	0.555
A*C*D	1.363	0.681	0.1038	6.56	0.007

S = 0.415080   PRESS = 14.7022
R-Sq = 99.86%   R-Sq(pred) = 95.93% R-Sq(adj) = 99.28%

Analysis of Variance for Yield (coded units)

Source	DF	Seq SS	Adj SS	Adj MS	F	P
Blocks	1	0.141	0.141	0.141	0.82	0.433
Main Effects	4	341.377	341.377	85.344	495.35	0.000
A	1	310.641	310.641	310.641	1802.99	0.000
B	1	25.756	25.756	25.756	149.49	0.001
C	1	4.951	4.951	4.951	28.73	0.013
D	1	0.031	0.031	0.031	0.18	0.702
2-Way Interactions	6	11.419	11.419	1.903	11.05	0.037
A*B	1	2.176	2.176	2.176	12.63	0.038
A*C	1	4.306	4.306	4.306	24.99	0.015
A*D	1	0.226	0.226	0.226	1.31	0.336
B*C	1	4.306	4.306	4.306	24.99	0.015
B*D	1	0.331	0.331	0.331	1.92	0.260
C*D	1	0.076	0.076	0.076	0.44	0.555
3-Way Interactions	1	7.426	7.426	7.426	43.10	0.007
A*C*D	1	7.426	7.426	7.426	43.10	0.007
Residual Error	3	0.517	0.517	0.172		
Total	15	360.879				

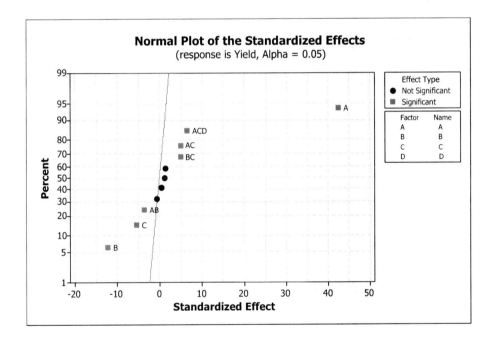

The above output and the normal plot show that main effects A, B, C, and two factor interactions AB, AC, BC, and three factor interaction ACD are significant at the 5% level of significance. Moreover, R-square, R-square (pred) are very high and PRESS Statistic is quite low, so the fitted model is quite valuable.

7. Here we present detailed calculations of the Yates Method. For the actual method refer to Section 18.5.2. Clearly, the results obtained by using Yates' Method match with those obtained using MINITAB in Problem 1 of Section 18.4. It can easily be verified that the effects which are significant at 5% level are A, B, C, AC. BC, and ACD.

Treatments	Observations	(1)	(2)	(3)	(4)	Effects (4)/8	Sums of Squares $(4)^2/16$
1	14.8	39.6	72.0	140.1	280.9	17.55*	
a	24.8	32.4	68.1	140.8	70.5	8.812	310.6406
b	12.3	36.1	72.9	34.3	−20.3	−2.537	25.7556
ab	20.1	32.0	67.9	36.2	−5.9	−0.737	2.1756
c	13.8	40.0	17.8	−11.3	−8.9	−1.112	4.9506
ac	22.3	32.9	16.5	−9.0	8.3	1.038	4.3056
bc	12.0	34.9	13.3	−2.7	8.3	1.038	4.3056
abc	20.0	33.0	22.9	−3.2	1.5	0.187	0.1406
d	16.3	10.0	−7.2	−3.9	0.7	0.087	0.0306
ad	23.7	7.8	−4.1	−5.0	1.9	0.238	0.2256
bd	13.5	8.5	−7.1	−1.3	2.3	0.287	0.3306
abd	19.4	8.0	−1.9	9.6	−0.5	−0.062	0.0156
cd	11.3	7.4	−2.2	3.1	−1.1	−0.137	0.0756
acd	23.6	5.9	−0.5	5.2	10.9	1.363	7.4256
bcd	11.2	12.3	−1.5	1.7	2.1	0.263	0.2756
abcd	21.8	10.6	−1.7	−0.2	−1.9	−0.238	0.2256

* The first entry in column 4 is found using Yates method is the grand mean if divided by $2^4 = 16$, and other entries of column 4 give the estimates of the effects when divided by $2^{4-1} = 8$.

These result match with the results obtained using MINITAB in Problem 1 of Section 18.4.

PRACTICE PROBLEMS FOR SECTION 18.6

1. MINITAB (Factorial Fit): Dimension versus A, B, C, D

See the alias structure below that assumes when we are using the $\frac{1}{2}$ replication of the 2^4 designs in factors A, B, C, D. To create $\frac{1}{2}$ replication we use the four factor interaction ABCD as a defining contrast

```
Estimated Effects and Coefficients for Dimension (coded units)

Term        Effect    Coef
Constant              17.4375
A           9.0750    4.5375
B          -1.1750   -0.5875
C          -1.1750   -0.5875
D           0.2750    0.1375
A*B        -0.8750   -0.4375
A*C         1.3250    0.6625
A*D         1.2750    0.6375

S = *       PRESS = *
```

```
Analysis of Variance for Dimension (coded units)

Source              DF    Seq SS    Adj SS    Adj MS    F    P
Main Effects         4   170.385   170.385    42.596    *    *
A                    1   164.711   164.711   164.711    *    *
B                    1     2.761     2.761     2.761    *    *
C                    1     2.761     2.761     2.761    *    *
D                    1     0.151     0.151     0.151    *    *
2-Way Interactions   3     8.294     8.294     2.765    *    *
A*B                  1     1.531     1.531     1.531    *    *
A*C                  1     3.511     3.511     3.511    *    *
A*D                  1     3.251     3.251     3.251    *    *
Residual Error       0        *         *         *
Total                7   178.679

Alias Structure
I + A*B*C*D
A + B*C*D
B + A*C*D
C + A*B*D
D + A*B*C
A*B + C*D
A*C + B*D
A*D + B*C
```

In this problem we cannot test any hypothesis since there is no error degrees of freedom available for residual error. This, of course, can be done by assuming that some of the higher order interactions are negligible and then using corresponding sum of squares and degrees of freedom for error. Without doing this partial ANOVA is given above. Further, note that the normal plot shows that the only effect that is significant is due to factor A. We reanalyzed the data by assuming two factor interactions as negligible. The new ANOVA showed that the main effect A is the only effect that is significant at 5% level of significance.

Effects Plot for Dimension

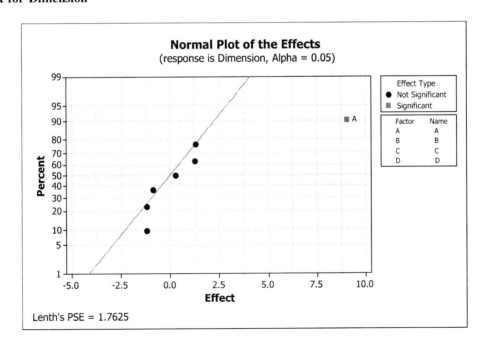

3. Suppose factor D is unimportant. So we construct a design using three factors, namely, A B, and C and then analyze and then reintroduce the factor D as D=ABC. The new design which is simply a $1/2$ replication of a 2^4 design, its ANOVA table, factor effects and normal probability plot for the effects are given below.

A	B	C	D	Observations
−1	−1	−1	−1	68
1	1	−1	−1	81
1	−1	1	−1	66
−1	1	1	−1	79
1	−1	−1	1	73
−1	1	−1	1	75
−1	−1	1	1	81
1	1	1	1	92

Notice that the treatments above are such that multiplying levels used for A, B, C, D, gives $+1$ (or $D = ABC$), so that the generator is $I = ABCD$. The confounding pattern or the alias structure is easily found to be

$$I = ABCD, A = BCD, B = ACD, C = ABC, D = ABC, AB = CD, AC = BD, AD = BC$$

Note that the ANOVA table given below is not complete since there is no degree of freedom available for residual error. Try to reanalyze these data by assuming that all two factor interactions are negligible. The normal probability plot indicates that no effect is significant at the 5% level of significance.

MINITAB (Factorial Fit): Observations versus A, B, C, D

```
Estimated Effects and Coefficients for Observations (coded units)
Term       Effect     Coef
Constant              76.875
A          2.250      1.125
B          9.750      4.875
C          5.250      2.625
D          6.750      3.375
A*B        7.250      3.625
A*C       -3.250     -1.625
A*D        2.250      1.125
```

```
Analysis of Variance for Observations (coded units)

Source              DF    Seq SS    Adj SS    Adj MS    F    P
Main Effects        4     346.50    346.50    86.63     *    *
A                   1      10.13     10.13    10.13     *    *
B                   1     190.13    190.13   190.13     *    *
C                   1      55.12     55.13     5.13     *    *
D                   1      91.13     91.13    91.13     *    *
2-Way Interactions  3     136.37    136.37    45.46     *    *
A*B                 1     105.12    105.13   105.13     *    *
A*C                 1      21.13     21.13    21.13     *    *
A*D                 1      10.12     10.12    10.12     *    *
Residual Error      0         *         *         *
Total               7     482.88
```

Effects Plot for Observations

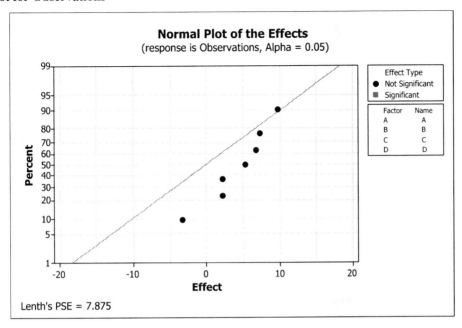

Note that in this problem we cannot estimate main effects without assuming three factor interactions to be zero. In order to perform testing of hypothesis about the main effects we have to assume that some of the two factor interactions are zero. However, the normal plot indicates that none of the effects is significant at 5% level of significance.

5.

Effects Plot for Observations

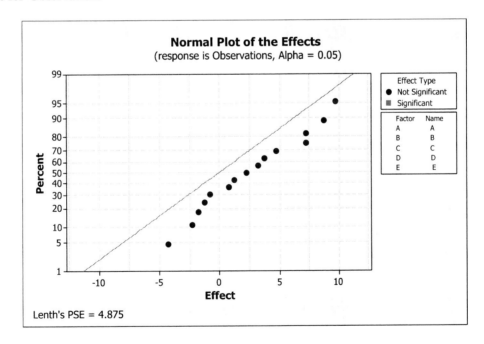

From the above normal probability plot we notice that none of the effects is significant at the 5% level of significance.

7. Assuming all two factor and higher order interactions equal to zero we have two degrees of freedom to estimate the error variance and the corresponding sun of squares is 98. Thus, using the result of Problem 6, we obtain the following ANOVA:

ANOVA					
Source	DF	SS	MS	F-Ratio	p-value
A	1	8.00	8.00	0.1633	0.7252
B	1	84.5	84.5	1.7244	0.3196
C	1	84.5	84.5	1.7244	0.3196
D	1	312.5	312.5	6.3776	0.1274
E	1	220.5	220.5	4.5	0.1680
Error	2	98.00	49.00		
Total	7	808.0			

Clearly, the above ANOVA table shows that none of the main effects at the 5% level is significant. These conclusions match with those we obtained using a normal probability plot of the effects in Problem 6.

9. The design using factors A_1, A_2, and A_5, and the corresponding data is as follows:

A_1	A_2	A_5	Observations
−1	−1	−1	13.0 13.4
1	−1	−1	17.5 18.1
−1	1	−1	10.6 10.2
1	−1	1	15.1 14.8
−1	−1	1	20.2 19.5
1	−1	1	19.1 19.2
−1	1	1	17.3 16.9
1	1	1	15.7 15.7

MINITAB(Factorial Fit): Observations versus A1, A2, A5

```
Estimated Effects and Coefficients for Observations (coded units)

Term          Effect    Coef    SE Coef      T        P
Constant              16.019   0.07474   214.33   0.000
A1             1.763    0.881   0.07474    11.79   0.000
A2            -2.962   -1.481   0.07474   -19.82   0.000
A5             3.862    1.931   0.07474    25.84   0.000
A1*A2         -0.188   -0.094   0.07474    -1.25   0.245
A1*A5         -2.813   -1.406   0.07474   -18.82   0.000
A2*A5         -0.138   -0.069   0.07474    -0.92   0.385
A1*A2*A5      -0.162   -0.081   0.07474    -1.09   0.309

S = 0.298957    PRESS = 2.86
R-Sq = 99.49%   R-Sq(pred) = 97.96%  R-Sq(adj) = 99.04%
```

Analysis of Variance for Observations (coded units)

Source	DF	Seq SS	Adj SS	Adj MS	F	P
Main Effects	3	107.207	107.207	35.7356	399.84	0.000
A1	1	12.426	12.426	12.4256	139.03	0.000
A2	1	35.106	35.106	35.1056	392.79	0.000
A5	1	59.676	59.676	59.6756	667.70	0.000
2-Way Interactions	3	31.857	31.857	10.6190	118.81	0.000
A1*A2	1	0.141	0.141	0.1406	1.57	0.245
A1*A5	1	31.641	31.641	31.6406	354.02	0.000
A2*A5	1	0.076	0.076	0.0756	0.85	0.385
3-Way Interactions	1	0.106	0.106	0.1056	1.18	0.309
A1*A2*A5	1	0.106	0.106	0.1056	1.18	0.309
Residual Error	8	0.715	0.715	0.0894		
Pure Error	8	0.715	0.715	0.0894		
Total	15	139.884				

From the above ANOVA table and normal probability plot (shown below) of the effects we determine that the all the main effects and two factor interaction A_1A_5 are significant at any level of significance. These conclusions are the same as arrived in Problem 8. Moreover, note that the R-square, and R-square(pred) values are exceptionally high and the PRESS statistic is very low, which indicate that the fitted model is very good.

Effects Plot for Observations

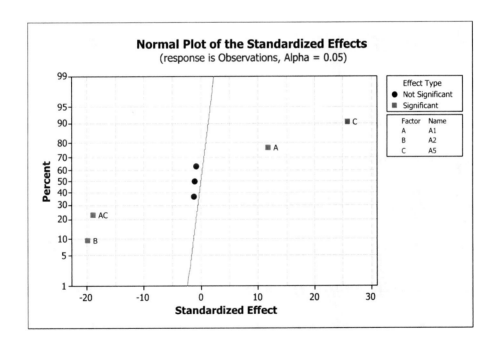

REVIEW PRACTICE PROBLEMS

1. (a) Let $(t_1t_2t_3t_4)$ denote the treatment used to generate the observation Y.

$$Y_{ijk} = \mu + \sum_i \alpha_i x_i + \sum_{i<j}^{i} \sum_j \alpha_{ij} x_i x_j + \beta_k + \varepsilon_{ijk}, i,j = 1,2, k = 1,2,3,4,5$$

where α_i, α_{ij} are main effects, two factor interaction and β_k is the batch effect, and $x_i = 1 \ or -1$ according as the ith factor is at the higher or at the lower level and ε_{ijk} are random errors that are independent and normally distributed with mean 0 and common variance σ^2.

MINITAB(Factorial Fit): Yield versus Batch, pH, Temperature

Estimated Effects and Coefficients for Yield (coded units)

Term	Effect	Coef	SE Coef	T	P
Constant		69.200	1.200	57.68	0.000
Batch 1		12.050	2.399	5.02	0.000
Batch 2		-3.450	2.399	-1.44	0.176
Batch 3		-0.450	2.399	-0.19	0.854
Batch 4		-8.200	2.399	-3.42	0.005
pH	51.800	25.900	1.200	21.59	0.000
Temperature	20.200	10.100	1.200	8.42	0.000
pH*Temperature	-1.200	-0.600	1.200	-0.50	0.626

S = 5.36501   PRESS = 959.444
R-Sq = 97.93%   R-Sq(pred) = 94.26% R-Sq(adj) = 96.73%

Analysis of Variance (ANOVA) for Yield (coded units)

Source	DF	Seq SS	Adj SS	Adj MS	F	P
Batch	4	898.2	898.2	224.6	7.80	0.002
Main Effects	2	15456.4	15456.4	7728.2	268.50	0.000
pH	1	13416.2	13416.2	13416.2	466.11	0.000
Temperature	1	2040.2	2040.2	2040.2	70.88	0.000
2-Way Interactions	1	7.2	7.2	7.2	0.25	0.626
pH*Temperature	1	7.2	7.2	7.2	0.25	0.626
Residual Error	12	345.4	345.4	28.8		
Total	19	16707.2				

Unusual Observations for Yield

Obs	StdOrder	Yield	Fit	SE Fit	Residual	St Resid
10	10	94.000	85.150	3.393	8.850	2.13R

R denotes an observation with a large standardized residual.

(b) We first find the sum of squares due to treatments, which consists of the sum of squares for the pH effect, temperature effect and interaction (ph×temperature). That is,

$$SS_{treat} = 13416.2 + 2040.2 + 7.2 = 15463.6$$

$$SSE = 345.4$$

Thus, the hypothesis that there are no treatment effects is rejected if the observed value of the test statistic $F > F_{3,12;.05} = 3.4903$. Since the observed value of $F_{treat} = MS_{treat}/MS_E$ is

$$F_{treat} = \frac{15463.6/3}{345.4/12} = 149.10 > 3.4903,$$

so that the null hypothesis of no treatment effects is rejected.

(c) The hypothesis that there are no batch differences ($\beta_k = 0$) is rejected if observed value of test statistic $F_{batch} > F_{4,12;.05} = 3.2592$. Since the observed value is

$$F_{batch} = \frac{898.2/4}{345.4/12} = 7.80 > 3.2592,$$

so that the null hypothesis $\beta_k = 0$ is rejected.

(d) From the above printout, we obtain

$$\text{ph effect} = 51.8, \text{Temperature effect } 20.2$$

(e) See the ANOVA given above in part (a)

(f) $\hat{\sigma}^2 = \text{MSE} = S^2 = 28.8$. The 95% confidence interval for the interaction effect (pH*Temp) is given by

$$\left[(\text{interaction effect}) \pm t_{12;.025} \frac{S}{\sqrt{r2^{k-2}}} \right] = \left[-1.2 \pm 2.179 \frac{5.36}{\sqrt{5 \times 2^0}} \right] = (-1.2 \pm 5.23) = (-6.43, 4.03).$$

which contains zero, therefore, the null hypothesis of no interaction is not rejected. This fact is also supported by the ANOVA table.

3. (a)

YATES' ALGORITHM					
Treatment	Totals	(1)	(2)	(3)	Effects
$-1 -1 -1$	7.98	17.24	35.32	67.11	$\hat{\mu} = 2.80$
$1 -1 -1$	9.26	18.08	31.79	3.63	0.3025
$-1\ \ 1 -1$	8.51	15.67	2.34	1.29	0.1075
$1\ \ 1 -1$	9.57	16.12	1.29	0.09	0.0075
$-1 -1\ \ 1$	7.59	1.28	0.84	-3.53	-0.2942
$1 -1\ \ 1$	8.08	1.06	0.45	-1.05	-0.0875
$-1\ \ 1\ \ 1$	7.66	0.49	-0.22	-0.39	-0.0325
$1\ \ 1\ \ 1$	8.46	0.80	0.31	0.53	0.0442

These effects are confirmed in the Minitab output. The estimates of effects except the general mean are equal to Column(3)/12.

MINITAB(Factorial Fit): Purity versus Block, Feed, Ratio, Temperature

Estimated Effects and Coefficients for Purity (coded units)

Term	Effect	Coef	SE Coef	T	P
Constant		2.7963	0.01310	213.40	0.000
Block 1		-0.0662	0.01853	-3.58	0.003
Block 2		-0.0725	0.01853	-3.91	0.002
Feed	0.3025	0.1513	0.01310	11.54	0.000
Ratio	0.1075	0.0537	0.01310	4.10	0.001
Temperature	-0.2942	-0.1471	0.01310	-11.22	0.000
Feed*Ratio	0.0075	0.0038	0.01310	0.29	0.779
Feed*Temperature	-0.0875	-0.0437	0.01310	-3.34	0.005

```
Ratio*Temperature        -0.0325   -0.0162   0.01310  -1.24   0.235
Feed*Ratio*Temperature   0.0442     0.0221   0.01310   1.69   0.114
```

S = 0.0641937   PRESS = 0.169543
R-Sq = 96.13%   R-Sq(pred) = 88.63% R-Sq(adj) = 93.64%

Analysis of Variance for Purity (coded units)

Source	DF	Seq SS	Adj SS	Adj MS	F	P
Blocks	2	0.23117	0.23117	0.115587	28.05	0.000
Main Effects	3	1.13758	1.13758	0.379193	92.02	0.000
Feed	1	0.54904	0.54904	0.549038	133.23	0.000
Ratio	1	0.06934	0.06934	0.069337	16.83	0.001
Temperature	1	0.51920	0.51920	0.519204	125.99	0.000
2-Way Interactions	3	0.05261	0.05261	0.017537	4.26	0.025
Feed*Ratio	1	0.00034	0.00034	0.000338	0.08	0.779
Feed*Temperature	1	0.04594	0.04594	0.045937	11.15	0.005
Ratio*Temperature	1	0.00634	0.00634	0.006337	1.54	0.235
3-Way Interactions	1	0.01170	0.01170	0.011704	2.84	0.114
Feed*Ratio*Temperature	1	0.01170	0.01170	0.011704	2.84	0.114
Residual Error	14	0.05769	0.05769	0.004121		
Total	23	1.49076				

(b) See the above MINITAB printout.

(c) Since the p-value for blocks is 0.000, we conclude that there is significant differences between replicates (blocks).

(d) The contrast of interest is $\tau = \beta_1 + \beta_2 - 2\beta_3$, where β_j is the effects of block $j, j = 1, 2\ 3$. The estimate $\hat{\tau}$ is given by

$$\hat{\tau} = \Sigma c_i \bar{y}_i = (21.84/8) + (21.79/8) - 2(23.48/8) = -0.41625$$

The contrast sum of squares is given by

$$\frac{8 \times (\Sigma c_i \bar{y}_i)^2}{\Sigma c_i^2} = \frac{8 \times (0.41625)^2}{6} = 0.2310$$

so that the observed value of test statistic is

$$F_\tau = \frac{0.2310}{0.00412} = 56.07 > F_{1,14;.05} = 4.6001.$$

Hence, we reject the null hypothesis that the contrast effect is zero.

(e) The 95% confidence interval for "ADN feed rate" is

$$[0.3025 \pm t_{14;.025} \sqrt{S^2/r2^{k-2}}] = [0.3025 \pm 2.145 \times \sqrt{0.00412/6}]$$
$$= (0.3025 \pm 0.0562) = (0.2463, 0.3587).$$

5. MINITAB(Factorial Fit): Thickness versus Block, A, B, C, D

Here, since each block represents the principal block generated by the interaction ABCD, MINITAB automatically removes all three factor interactions that are confounded with the main effects and 3-two factor interactions which were confounded with the remaining two factor interactions. For example, A was confounded with BCD and AB was confounded with CD.

Estimated Effects and Coefficients for Thickness (coded units)

Term	Effect	Coef	SE Coef	T	P
Constant		3.76812	0.1196	31.50	0.000
Block		-0.06687	0.1196	-0.56	0.594
A	0.36125	0.18062	0.1196	1.51	0.175
B	-0.02375	-0.01187	0.1196	-0.10	0.924
C	-0.05125	-0.02563	0.1196	-0.21	0.836
D	0.04875	0.02438	0.1196	0.20	0.844
A*B	0.00625	0.00313	0.1196	0.03	0.980
A*C	-0.17625	-0.08812	0.1196	-0.74	0.485
A*D	0.01875	0.00938	0.1196	0.08	0.940

S = 0.478479   PRESS = 8.37273
R-Sq = 31.64%   R-Sq(pred) = 0.00%   R-Sq(adj) = 0.00%

Analysis of Variance for Thickness (coded units)

Source	DF	Seq SS	Adj SS	Adj MS	F	P
Blocks	1	0.07156	0.07156	0.071556	0.31	0.594
Main Effects	4	0.54427	0.54427	0.136069	0.59	0.678
A	1	0.52201	0.52201	0.522006	2.28	0.175
B	1	0.00226	0.00226	0.002256	0.01	0.924
C	1	0.01051	0.01051	0.010506	0.05	0.836
D	1	0.00951	0.00951	0.009506	0.04	0.844
2-Way Interactions	3	0.12582	0.12582	0.041940	0.18	0.905
A*B	1	0.00016	0.00016	0.000156	0.00	0.980
A*C	1	0.12426	0.12426	0.124256	0.54	0.485
A*D	1	0.00141	0.00141	0.001406	0.01	0.940
Residual Error	7	1.60259	1.60259	0.228942		
Total	15	2.34424				

(a) For estimates see the above MINITAB printout. The normal probability plot of the estimated effects given below. The ANOVA table as well as the normal probability plot indicate that none of the effects is significant at 5% level of significance.

(b) See the ANOVA table given above, $\hat{\sigma}^2 = S^2 = 0.228942$.

(c) We do not reject the null hypothesis of no block effects since from the above ANOVA table we determine that the p-value is $0.594 > 0.05$.

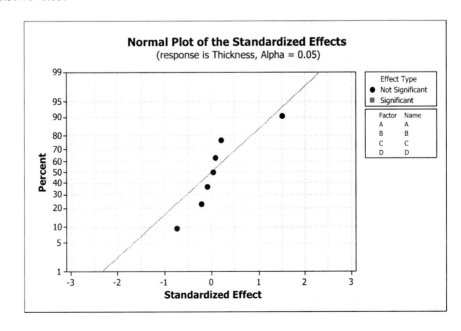

7. In this example we use the six factor interaction ABCDEF as a design generator. The desired half-fraction can be obtained by using MINITAB as follows:

- Select **Stat > DOE > Factorial > Create Factorial Design**
- Select **2-level factorial (Specific Generators)**
- Enter number of factors (number of factors $- 1 = 6 - 1 = 5$)
- Select appropriate fraction (**e.g., 1/2 fraction or full fraction**)
- Select **generators** and then in the new dialog box enter the desired generators ABCDEF in the first window as F = ABCDE.
- Click **OK** the desired design given below appears in the **worksheet**

  Alias Structure is as follows:

  I + ABCDEF, A+ BCDEF, B+ ACDEF, C+ ABDEF, D+ ABCEF, E+ ABCDF,

  F+ ABCDE, AB+ CDEF, AC+ BDEF, AD+ BCEF, AE+ BCDF, AF+ BCDE,

  BC+ ADEF, BD+ ACEF, BE+ ACDF, BF+ ACDE, CD+ ABEF, CE+ ABDF,

  CF+ ABDF, DE+ ABCF, DF+ ABCE, EF+ ABCD, ABC+DEF, ABD+CEF, ABE+CDF,

  ABF+CDE, ACD+BEF, ACE+BDF, ACF+BDE, ADE+BCF, ADF+BCE, AEF+ BCD.

$$
\begin{pmatrix}
A & B & C & D & E & F \\
1 & 1 & 1 & 1 & -1 & -1 \\
1 & 1 & 1 & -1 & -1 & 1 \\
-1 & -1 & 1 & -1 & -1 & 1 \\
1 & -1 & -1 & -1 & 1 & -1 \\
-1 & 1 & 1 & -1 & 1 & 1 \\
1 & -1 & 1 & -1 & 1 & 1 \\
1 & 1 & 1 & -1 & 1 & -1 \\
1 & 1 & -1 & -1 & 1 & 1 \\
-1 & -1 & -1 & -1 & 1 & 1 \\
-1 & 1 & -1 & 1 & 1 & 1 \\
1 & -1 & -1 & 1 & 1 & 1 \\
-1 & -1 & -1 & 1 & -1 & 1 \\
-1 & 1 & 1 & -1 & -1 & -1 \\
1 & 1 & -1 & -1 & -1 & -1 \\
-1 & 1 & -1 & -1 & -1 & 1 \\
1 & -1 & 1 & 1 & -1 & 1 \\
-1 & 1 & 1 & 1 & -1 & 1 \\
-1 & -1 & 1 & 1 & -1 & -1 \\
-1 & 1 & -1 & -1 & 1 & -1 \\
1 & 1 & -1 & 1 & -1 & 1 \\
-1 & -1 & -1 & -1 & -1 & -1 \\
-1 & 1 & 1 & 1 & 1 & -1 \\
1 & -1 & 1 & -1 & -1 & -1 \\
1 & 1 & -1 & 1 & 1 & -1 \\
1 & -1 & -1 & -1 & -1 & 1 \\
1 & -1 & -1 & 1 & -1 & -1 \\
-1 & -1 & 1 & -1 & 1 & -1 \\
-1 & -1 & 1 & 1 & 1 & 1 \\
-1 & 1 & -1 & 1 & -1 & -1 \\
1 & 1 & 1 & 1 & 1 & 1 \\
1 & -1 & 1 & 1 & 1 & -1 \\
-1 & -1 & -1 & 1 & 1 & -1 \\
\end{pmatrix}
$$

9. MINITAB(Factorial Fit): Thickness versus A, B, C, D

Estimated Effects and Coefficients for Thickness (coded units)

Term	Effect	Coef	SE Coef	T	P
Constant		3.76812	0.1144	32.95	0.000
A	0.36125	0.18062	0.1144	1.58	0.153
B	-0.02375	-0.01187	0.1144	-0.10	0.920
C	-0.05125	-0.02563	0.1144	-0.22	0.828
D	0.04875	0.02438	0.1144	0.21	0.837
A*B	0.00625	0.00313	0.1144	0.03	0.979
A*C	-0.17625	-0.08812	0.1144	-0.77	0.463
A*D	0.01875	0.00938	0.1144	0.08	0.937

S = 0.457459   PRESS = 6.6966
R-Sq = 28.58%   R-Sq(pred) = 0.00%   R-Sq(adj) = 0.00%

Analysis of Variance for Thickness (coded units)

Source	DF	Seq SS	Adj SS	Adj MS	F	P
Main Effects	4	0.54427	0.54427	0.136069	0.65	0.643
A	1	0.52201	0.52201	0.522006	2.49	0.153
B	1	0.00226	0.00226	0.002256	0.01	0.920
C	1	0.01051	0.01051	0.010506	0.05	0.828
D	1	0.00951	0.00951	0.009506	0.05	0.837
2-Way Interactions	3	0.12582	0.12582	0.041940	0.20	0.893
A*B	1	0.00016	0.00016	0.000156	0.00	0.979
A*C	1	0.12426	0.12426	0.124256	0.59	0.463
A*D	1	0.00141	0.00141	0.001406	0.01	0.937
Residual Error	8	1.67415	1.67415	0.209269		
Pure Error	8	1.67415	1.67415	0.209269		
Total	15	2.34424				

Alias Structure
I + A*B*C*D
A + B*C*D
B + A*C*D
C + A*B*D
D + A*B*C
A*B + C*D
A*C + B*D
A*D + B*C

In this problem when estimating the main effects, it is necessary to assume that all three factor interactions are zero, since main effects and three factor interactions are confounded with each other. From the ANOVA table given above it is clear that none of the main effects or two factor interactions (assuming BC, BD, and CD negligible) is significant at the 5% level of significance, since all the corresponding p-values are greater than 0.05.

11. MINITAB (Factorial Fit): Yield versus A, B, C, D

Estimated Effects and Coefficients for Yield (coded units)

Term	Effect	Coef
Constant		21.063
A	3.625	1.812
B	-0.875	-0.437
C	2.625	1.312
D	2.875	1.437
A*B	-2.875	-1.438

```
A*C          3.625     1.813
A*D          0.875     0.438
B*C         -0.375    -0.188
B*D         -3.125    -1.562
C*D         -0.125    -0.063
A*B*C        2.125     1.063
A*B*D        0.375     0.187
A*C*D       -0.625    -0.312
B*C*D        0.875     0.437
A*B*C*D      0.875     0.437
```

Analysis of Variance for Yield (coded units)

Source	DF	Seq SS	Adj SS	Adj MS	F	P
Main Effects	4	116.250	116.250	29.0625	*	*
A	1	52.563	52.563	52.5625	*	*
B	1	3.062	3.063	3.0625	*	*
C	1	27.562	27.562	27.5625	*	*
D	1	33.063	33.062	33.0625	*	*
2-Way Interactions	6	128.375	128.375	21.3958	*	*
A*B	1	33.063	33.063	33.0625	*	*
A*C	1	52.563	52.563	52.5625	*	*
A*D	1	3.063	3.063	3.0625	*	*
B*C	1	0.562	0.563	0.5625	*	*
B*D	1	39.062	39.062	39.0625	*	*
C*D	1	0.063	0.063	0.0625	*	*
3-Way Interactions	4	23.250	23.250	5.8125	*	*
A*B*C	1	18.063	18.063	18.0625	*	*
A*B*D	1	0.562	0.562	0.5625	*	*
A*C*D	1	1.562	1.562	1.5625	*	*
B*C*D	1	3.062	3.062	3.0625	*	*
4-Way Interactions	1	3.062	3.062	3.0625	*	*
A*B*C*D	1	3.062	3.062	3.0625	*	*
Residual Error	0	*	*	*		
Total	15	270.937				

Note that the above ANOVA is incomplete since there is no degree of freedom for Residual error.

Estimated Coefficients for Yield using data in uncoded units

```
Term         Coef
Constant   21.0625
A           1.81250
B          -0.437500
C           1.31250
D           1.43750
A*B        -1.43750
A*C         1.81250
A*D         0.437500
B*C        -0.187500
B*D        -1.56250
C*D        -0.0625000
A*B*C       1.06250
A*B*D       0.187500
A*C*D      -0.312500
B*C*D       0.437500
A*B*C*D     0.437500
```

(a) Estimates of various effects are given above in the MINITAB printout. The normal probability plot of the estimated effects is shown below. The normal probability plot indicates that the main effect A and interaction AC are significant at the 5% level of significance.

(b) In this Problem without assuming some of the higher order interaction to be zero we cannot have a complete ANOVA table since we cannot estimate the error variance. However, a partial ANOVA table is given above.

Effects Plot for Yield

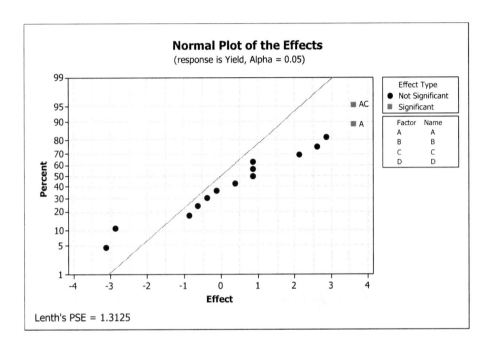

13. MINITAB Factorial Fit: Yield versus A, B, C, D

As in Problem 12, we let A, B, C, D stand for the factors, breed, diet, protein and methods, respectively.

Estimated Effects and Coefficients for Yield (coded units)

Term	Effect	Coef
Constant		392.88
A	-12.75	-6.38
B	11.25	5.63
C	13.75	6.87
D	-49.75	-24.87
A*B	-3.25	-1.62
A*C	-4.75	-2.38
A*D	-14.25	-7.12
B*C	-25.75	-12.87
B*D	36.75	18.37
C*D	-13.75	-6.88
A*B*C	23.75	11.87
A*B*D	6.25	3.13
A*C*D	27.75	13.87
B*C*D	-8.25	-4.13
A*B*C*D	-14.75	-7.37

Analysis of Variance for Yield (coded units)

Source	DF	Seq SS	Adj SS	Adj MS	F	P
Main Effects	4	11813.0	11813.0	2953.25	*	*
A	1	650.2	650.3	650.25	*	*
B	1	506.2	506.3	506.25	*	*
C	1	756.2	756.3	756.25	*	*
D	1	9900.3	9900.2	9900.25	*	*
2-Way Interactions	6	9755.5	9755.5	1625.92	*	*
A*B	1	42.2	42.2	42.25	*	*
A*C	1	90.3	90.3	90.25	*	*
A*D	1	812.3	812.2	812.25	*	*
B*C	1	2652.2	2652.2	2652.25	*	*
B*D	1	5402.2	5402.2	5402.25	*	*
C*D	1	756.2	756.2	756.25	*	*
3-Way Interactions	4	5765.0	5765.0	1441.25	*	*
A*B*C	1	2256.2	2256.2	2256.25	*	*
A*B*D	1	156.3	156.3	156.25	*	*
A*C*D	1	3080.2	3080.2	3080.25	*	*
B*C*D	1	272.2	272.3	272.25	*	*
4-Way Interactions	1	870.2	870.2	870.25	*	*
A*B*C*D	1	870.2	870.2	870.25	*	*
Residual Error	0	*	*	*		
Total	15	28203.7				

(a) Estimates of various effects and a partial ANOVA table are given above in the MINITAB printout. The normal probability plot of estimated effects is shown below. The normal probability plot indicates that none of the effects are significant at the 5% level of significance.

Effects Plot for Yield

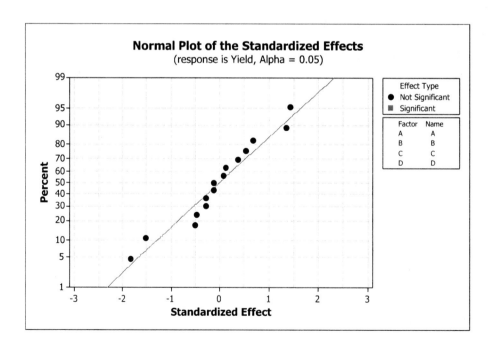

15. MINITAB (Factorial Fit: Yield versus M, N, P, K

Estimated Effects and Coefficients for Yield (coded units)

Term	Effect	Coef	SE Coef	T	P
Constant		67.375	0.9458	71.24	0.000
M	4.625	2.313	0.9458	2.45	0.026
N	-5.250	-2.625	0.9458	-2.78	0.014
P	-1.625	-0.813	0.9458	-0.86	0.403
K	1.500	0.750	0.9458	0.79	0.439
M*N	2.375	1.187	0.9458	1.26	0.227
M*P	-0.250	-0.125	0.9458	-0.13	0.897
M*K	-1.875	-0.938	0.9458	-0.99	0.336
N*P	-7.375	-3.688	0.9458	-3.90	0.001
N*K	5.750	2.875	0.9458	3.04	0.008
P*K	-3.375	-1.688	0.9458	-1.78	0.093
M*N*P	-1.250	-0.625	0.9458	-0.66	0.518
M*N*K	-0.875	-0.437	0.9458	-0.46	0.650
M*P*K	-6.250	-3.125	0.9458	-3.30	0.004
N*P*K	-2.375	-1.188	0.9458	-1.26	0.227
M*N*P*K	1.500	0.750	0.9458	0.79	0.439

S = 5.35023   PRESS = 1832
R-Sq = 78.67%   R-Sq(pred) = 14.69% R-Sq(adj) = 58.68%

Analysis of Variance for Yield (coded units)

Source	DF	Seq SS	Adj SS	Adj MS	F	P
Main Effects	4	430.75	430.750	107.688	3.76	0.024
M	1	171.13	171.125	171.125	5.98	0.026
N	1	220.50	220.500	220.500	7.70	0.014
P	1	21.13	21.125	21.125	0.74	0.403
K	1	18.00	18.000	18.000	0.63	0.439
2-Way Interactions	6	864.50	864.500	144.083	5.03	0.004
M*N	1	45.12	45.125	45.125	1.58	0.227
M*P	1	0.50	0.500	0.500	0.02	0.897
M*K	1	28.13	28.125	28.125	0.98	0.336
N*P	1	435.13	435.125	435.125	15.20	0.001
N*K	1	264.50	264.500	264.500	9.24	0.008
P*K	1	91.13	91.125	91.125	3.18	0.093
3-Way Interactions	4	376.25	376.250	94.063	3.29	0.038
M*N*P	1	12.50	12.500	12.500	0.44	0.518
M*N*K	1	6.13	6.125	6.125	0.21	0.650
M*P*K	1	312.50	312.500	312.500	10.92	0.004
N*P*K	1	45.13	45.125	45.125	1.58	0.227
4-Way Interactions	1	18.00	18.000	18.000	0.63	0.439
M*N*P*K	1	18.00	18.000	18.000	0.63	0.439
Residual Error	16	458.00	458.000	28.625		
Pure Error	16	458.00	458.000	28.625		
Total	31	2147.50				

Estimates of the various effects and ANOVA table are given above in the MINITAB printout. The normal probability plot of estimated effects is shown below. The normal probability plot as well as the ANOVA table indicates that the main effects M and N, and interactions NP, NK, and MPK are significant at the 5% level of significance. The estimate of the error variance is $\hat{\sigma}^2 = MSE = 28.625$.

Effects Plot for Yield

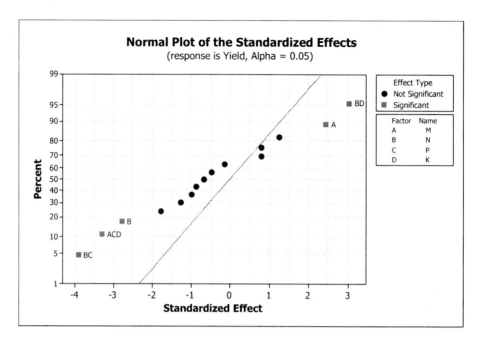

17. In this problem we denote depth and width by D and W respectively, so that we are having a 2^{6-2} experiment in factors M, N, P, K, D, W. Further, we consider the four factor interactions MNPD and MKDW as design generators. This results in D and W to be confounded with those interactions which are not significant. The generalized interaction of defining generators MNPD and MKDW is NKPW.

MINITAB (Factorial Fit): Yield versus M, N, P, K, D, W

Estimated Effects and Coefficients for Yield (coded units)

Term	Effect	Coef	SE Coef	T	P
Constant		67.375	0.9458	71.24	0.000
M	4.625	2.313	0.9458	2.45	0.026
N	-5.250	-2.625	0.9458	-2.78	0.014
P	-1.625	-0.813	0.9458	-0.86	0.403
K	1.500	0.750	0.9458	0.79	0.439
D	-1.250	-0.625	0.9458	-0.66	0.518
W	-2.375	-1.187	0.9458	-1.26	0.227
M*N	2.375	1.188	0.9458	1.26	0.227
M*P	-0.250	-0.125	0.9458	-0.13	0.897
M*K	-1.875	-0.938	0.9458	-0.99	0.336
M*D	-7.375	-3.688	0.9458	-3.90	0.001
M*W	1.500	0.750	0.9458	0.79	0.439
N*K	5.750	2.875	0.9458	3.04	0.008
N*W	-3.375	-1.687	0.9458	-1.78	0.093
M*N*K	-0.875	-0.438	0.9458	-0.46	0.650
M*N*W	-6.250	-3.125	0.9458	-3.30	0.004

S=5.35023   PRESS = 1832
R-Sq = 78.67%   R-Sq(pred) = 14.69% R-Sq(adj) = 58.68%

Analysis of Variance for Yield (coded units)

Source	DF	Seq SS	Adj SS	Adj MS	F	P
Main Effects	6	488.38	488.375	81.396	2.84	0.044

M	1	171.13	171.125	171.125	5.98	0.026
N	1	220.50	220.500	220.500	7.70	0.014
P	1	21.13	21.125	21.125	0.74	0.403
K	1	18.00	18.000	18.000	0.63	0.439
D	1	12.50	12.500	12.500	0.44	0.518
W	1	45.12	45.125	45.125	1.58	0.227
2-Way Interactions	7	882.50	882.500	126.071	4.40	0.007
M*N	1	45.12	45.125	45.125	1.58	0.227
M*P	1	0.50	0.500	0.500	0.02	0.897
M*K	1	28.13	28.125	28.125	0.98	0.336
M*D	1	435.13	435.125	435.125	15.20	0.001
M*W	1	18.00	18.000	18.000	0.63	0.439
N*K	1	264.50	264.500	264.500	9.24	0.008
N*W	1	91.12	91.125	91.125	3.18	0.093
3-Way Interactions	2	318.62	318.625	159.312	5.57	0.015
M*N*K	1	6.12	6.125	6.125	0.21	0.650
M*N*W	1	312.50	312.500	312.500	10.92	0.004
Residual Error	16	458.00	458.000	28.625		
Pure Error	16	458.00	458.000	28.625		
Total	31	2147.50				

```
Alias Structure
I + M*N*P*D + M*K*D*W + N*P*K*W
M + N*P*D + K*D*W + M*N*P*K*W
N + M*P*D + P*K*W + M*N*K*D*W
P + M*N*D + N*K*W + M*P*K*D*W
K + M*D*W + N*P*W + M*N*P*K*D
D + M*N*P + M*K*W + N*P*K*D*W
W + M*K*D + N*P*K + M*N*P*D*W
M*N + P*D + M*P*K*W + N*K*D*W
M*P + N*D + M*N*K*W + P*K*D*W
```

Effects Plot for Yield

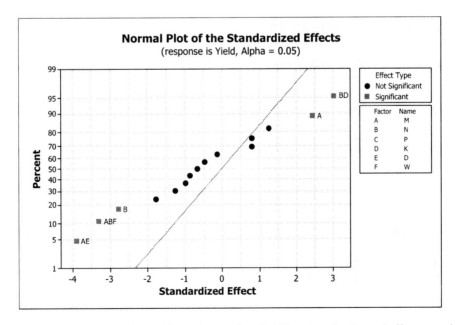

Estimates of the various effects, ANOVA table, and normal probability plot of estimated effects are given above in the MINITAB printout. The normal probability plot of estimated effects as well as the ANOVA table indicate that the main effects M and N, and interactions NK, MD, and MNW are significant at the 5% level of significance.

19. Factorial Fit: Observation versus A, B, C, D

Estimated Effects and Coefficients for Observation (coded units)

Term	Effect	Coef
Constant		27.750
A	-0.000	-0.000
B	5.000	2.500
C	8.000	4.000
D	5.000	2.500
A*B	1.500	0.750
A*C	2.500	1.250
A*D	-3.500	-1.750

Analysis of Variance for Observation (coded units)

Source	DF	Seq SS	Adj SS	Adj MS	F	P
Main Effects	4	228.000	228.000	57.000	*	*
A	1	0.000	0.000	0.000	*	*
B	1	50.000	50.000	50.000	*	*
C	1	128.000	128.000	128.000	*	*
D	1	50.000	50.000	50.000	*	*
2-Way Interactions	3	41.500	41.500	13.833	*	*
A*B	1	4.500	4.500	4.500	*	*
A*C	1	12.500	12.500	12.500	*	*
A*D	1	24.500	24.500	24.500	*	*
Residual Error	0	*	*	*		
Total	7	269.500				

Alias Structure

I + A*B*C*D
A + B*C*D
B + A*C*D
C + A*B*D
D + A*B*C
A*B + C*D
A*C + B*D
A*D + B*C

Effects Plot for Observation

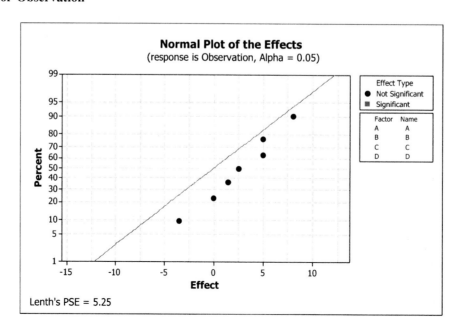

Normal Plot of the Effects
(response is Observation, Alpha = 0.05)

Lenth's PSE = 5.25

The normal probability plot of the estimated effects indicates that none of the effects is significant at 5% of significance. However, assuming all two factor and higher order interactions to be zero, we obtain the following ANOVA.
ANOVA

```
Analysis of Variance for Observation (coded units)

Source          DF    Seq SS    Adj SS    Adj MS      F        P
Main Effects     4   228.000   228.000    57.000      *
A                1     0.000     0.000     0.000    0.000    1.000
B                1    50.000    50.000    50.000    3.703    0.1500
C                1   128.000   128.000   128.000    9.481    0.0542
D                1    50.000    50.000    50.000    3.703    0.1500
Residual Error   3    41.500    13.500    13.500
Total            7   269.500
```

Note that none of the main effects are significant at the 5% level of significance but note that the main effect C comes very close to be significant at the 5% level of significance.

21. The design plan of a 2^6 design in 8 blocks using design generators ABCD, CDEF, and ACDE is as given below:

$$
\begin{pmatrix}
\text{Bl.} & \text{A} & \text{B} & \text{C} & \text{D} & \text{E} & \text{F} \\
1 & -1 & -1 & 1 & -1 & -1 & -1 \\
1 & -1 & -1 & -1 & 1 & -1 & -1 \\
1 & -1 & 1 & -1 & -1 & 1 & -1 \\
1 & -1 & 1 & 1 & 1 & 1 & -1 \\
1 & 1 & -1 & -1 & -1 & -1 & 1 \\
1 & 1 & -1 & 1 & 1 & -1 & 1 \\
1 & 1 & 1 & 1 & -1 & 1 & 1 \\
1 & 1 & 1 & -1 & 1 & 1 & 1 \\
2 & -1 & 1 & 1 & -1 & -1 & -1 \\
2 & -1 & 1 & -1 & 1 & -1 & -1 \\
2 & -1 & -1 & -1 & -1 & 1 & -1 \\
2 & -1 & -1 & 1 & 1 & 1 & -1 \\
2 & 1 & 1 & -1 & -1 & -1 & 1 \\
2 & 1 & 1 & 1 & 1 & -1 & 1 \\
2 & 1 & -1 & 1 & -1 & 1 & 1 \\
2 & 1 & -1 & -1 & 1 & 1 & 1 \\
3 & 1 & -1 & -1 & -1 & -1 & -1 \\
3 & 1 & -1 & 1 & 1 & -1 & -1 \\
3 & 1 & 1 & 1 & -1 & 1 & -1 \\
3 & 1 & 1 & -1 & 1 & 1 & -1 \\
3 & -1 & -1 & 1 & -1 & -1 & 1 \\
3 & -1 & -1 & -1 & 1 & -1 & 1 \\
3 & -1 & 1 & -1 & -1 & 1 & 1 \\
3 & -1 & 1 & 1 & 1 & 1 & 1 \\
4 & 1 & 1 & -1 & -1 & -1 & -1 \\
4 & 1 & 1 & 1 & 1 & -1 & -1 \\
4 & 1 & -1 & 1 & -1 & 1 & -1 \\
4 & 1 & -1 & -1 & 1 & 1 & -1 \\
4 & -1 & 1 & 1 & -1 & -1 & 1 \\
4 & -1 & 1 & -1 & 1 & -1 & 1 \\
4 & -1 & -1 & -1 & -1 & 1 & 1 \\
4 & -1 & -1 & 1 & 1 & 1 & 1
\end{pmatrix}
$$

$$\begin{pmatrix}
\text{Bl.} & \text{A} & \text{B} & \text{C} & \text{D} & \text{E} & \text{F} \\
5 & 1 & 1 & 1 & -1 & -1 & -1 \\
5 & 1 & 1 & -1 & 1 & -1 & -1 \\
5 & 1 & -1 & -1 & -1 & 1 & -1 \\
5 & 1 & -1 & 1 & 1 & 1 & -1 \\
5 & -1 & 1 & -1 & -1 & -1 & 1 \\
5 & -1 & 1 & 1 & 1 & -1 & 1 \\
5 & -1 & -1 & 1 & -1 & 1 & 1 \\
5 & -1 & -1 & -1 & 1 & 1 & 1 \\
6 & 1 & -1 & 1 & -1 & -1 & -1 \\
6 & 1 & -1 & -1 & 1 & -1 & -1 \\
6 & 1 & 1 & -1 & -1 & 1 & -1 \\
6 & 1 & 1 & 1 & 1 & 1 & -1 \\
6 & -1 & -1 & -1 & -1 & -1 & 1 \\
6 & -1 & -1 & 1 & 1 & -1 & 1 \\
6 & -1 & 1 & 1 & -1 & 1 & 1 \\
6 & -1 & 1 & -1 & 1 & 1 & 1 \\
7 & -1 & 1 & -1 & -1 & -1 & -1 \\
7 & -1 & 1 & 1 & 1 & -1 & -1 \\
7 & -1 & -1 & 1 & -1 & 1 & -1 \\
7 & -1 & -1 & -1 & 1 & 1 & -1 \\
7 & 1 & 1 & 1 & -1 & -1 & 1 \\
7 & 1 & 1 & -1 & 1 & -1 & 1 \\
7 & 1 & -1 & -1 & -1 & 1 & 1 \\
7 & 1 & -1 & 1 & 1 & 1 & 1 \\
8 & -1 & -1 & -1 & -1 & -1 & -1 \\
8 & -1 & -1 & 1 & 1 & -1 & -1 \\
8 & -1 & 1 & 1 & -1 & 1 & -1 \\
8 & -1 & 1 & -1 & 1 & 1 & -1 \\
8 & 1 & -1 & 1 & -1 & -1 & 1 \\
8 & 1 & -1 & -1 & 1 & -1 & 1 \\
8 & 1 & 1 & -1 & -1 & 1 & 1 \\
8 & 1 & 1 & 1 & 1 & 1 & 1 \\
\end{pmatrix}$$

23. MINITAB (Factorial Fit: Chem. Prod. versus A, B, C, D.

Here, we have used the principal block of the half fraction generated by $I = ABCD$. The alias structure given below shows that by letting three factor interactions to be zero we can estimate all the main effects.

```
Estimated Effects and Coefficients for Chem. Prod. (coded units)

Term         Effect        Coef
Constant                 22.6125
A            0.0750        0.0375
B            1.6250        0.8125
C           -0.4250       -0.2125
D            2.9250        1.4625
A*B         -0.5250       -0.2625
A*C          0.3250        0.1625
A*D          0.1750        0.0875

Analysis of Variance for Chem. Prod. (coded units)

Source              DF    Seq SS     Adj SS     Adj MS     F    P
Main Effects         4   22.7650    22.7650     5.6912     *    *
A                    1    0.0112     0.0112     0.0112     *    *
B                    1    5.2812     5.2812     5.2812     *    *
C                    1    0.3613     0.3613     0.3613     *    *
D                    1   17.1112    17.1112    17.1112     *    *
2-Way Interactions   3    0.8237     0.8237     0.2746     *    *
A*B                  1    0.5512     0.5512     0.5512     *    *
A*C                  1    0.2112     0.2112     0.2112     *    *
A*D                  1    0.0612     0.0612     0.0612     *    *
Residual Error       0       *          *          *
Total                7   23.5887

Alias Structure
I + A*B*C*D, A + B*C*D, B + A*C*D, C + A*B*D, D + A*B*C, A*B + C*D, A*C + B*D,
A*D + B*C
```

Estimates of all main effects assuming three factor interactions zero are given in the above MINITAB output. From the normal probability plot of the effects given below we note that the main effect D is significant at the 5% level of significance.

No, it is not possible to estimate all the main effects by selecting only four observations. Since there are five unknown parameters, including the general mean and the four main effects.

Effects Plot for Chem. Prod.

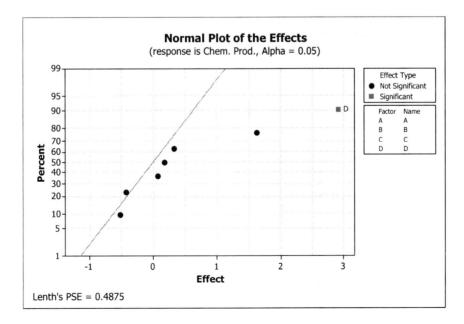

25. MINITAB (Factorial Fit: Y versus T, C, S, M

Estimated Effects and Coefficients for Y (coded units)

Term	Effect	Coef
Constant		66.94
T	-1.13	-0.56
C	7.37	3.69
S	-1.87	-0.94
M	-27.62	-13.81
T*C	1.88	0.94
T*S	4.13	2.06
T*M	-3.13	-1.56
C*S	-3.88	-1.94
C*M	4.87	2.44
S*M	-1.37	-0.69
T*C*S	-4.37	-2.19
T*C*M	-1.12	-0.56
T*S*M	2.13	1.06
C*S*M	1.62	0.81
T*C*S*M	-3.37	-1.69

Analysis of Variance for Y (coded units)

Source	DF	Seq SS	Adj SS	Adj MS	F	P
Main Effects	4	3289.25	3289.25	822.31	*	*
T	1	5.06	5.06	5.06	*	*
C	1	217.56	217.56	217.56	*	*
S	1	14.06	14.06	14.06	*	*
M	1	3052.56	3052.56	3052.56	*	*
2-Way Interactions	6	283.88	283.88	47.31	*	*
T*C	1	14.06	14.06	14.06	*	*
T*S	1	68.06	68.06	68.06	*	*
T*M	1	39.06	39.06	39.06	*	*
C*S	1	60.06	60.06	60.06	*	*
C*M	1	95.06	95.06	95.06	*	*
S*M	1	7.56	7.56	7.56	*	*
3-Way Interactions	4	110.25	110.25	27.56	*	*
T*C*S	1	76.56	76.56	76.56	*	*
T*C*M	1	5.06	5.06	5.06	*	*
T*S*M	1	18.06	18.06	18.06	*	*
C*S*M	1	10.56	10.56	10.56	*	*
4-Way Interactions	1	45.56	45.56	45.56	*	*
T*C*S*M	1	45.56	45.56	45.56	*	*
Residual Error	0	*	*	*		
Total	15	3728.94				

Effects Plot for Y

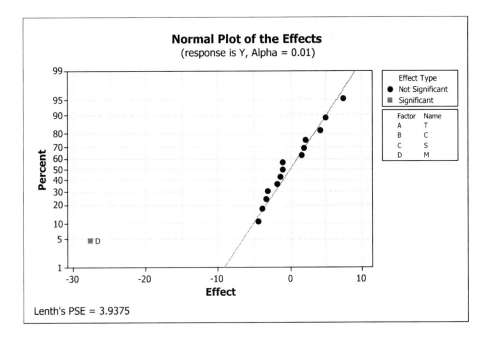

Lenth's PSE = 3.9375

Estimates of all main effects and interaction are given in the above MINITAB output. From the normal probability plot of the estimated effects given above we note that the main effects M is the only significant effect at the 1% level of significance.

Now assuming three and four factor interactions to be zero we get the following MINITAB output.

Factorial Fit: Scores versus T, C, S, M

Estimated Effects and Coefficients for Scores (coded units)

Term	Effect	Coef	SE Coef	T	P
Constant		66.94	1.396	47.96	0.000
T	-1.12	-0.56	1.396	-0.40	0.704
C	7.37	3.69	1.396	2.64	0.046
S	-1.88	-0.94	1.396	-0.67	0.532
M	-27.62	-13.81	1.396	-9.90	0.000
T*C	1.88	0.94	1.396	0.67	0.532
T*S	4.13	2.06	1.396	1.48	0.199
T*M	-3.13	-1.56	1.396	-1.12	0.314
C*S	-3.87	-1.94	1.396	-1.39	0.224
C*M	4.87	2.44	1.396	1.75	0.141
S*M	-1.37	-0.69	1.396	-0.49	0.643

S = 5.58234   PRESS = 1595.52
R-Sq = 95.82%   R-Sq(pred) = 57.21%   R-Sq(adj) = 87.46%

Analysis of Variance for Scores (coded units)

Source	DF	Seq SS	Adj SS	Adj MS	F	P
Main Effects	4	3289.25	3289.25	822.31	26.39	0.001
T	1	5.06	5.06	5.06	0.16	0.704

C	1	217.56	217.56	217.56	6.98	0.046
S	1	14.06	14.06	14.06	0.45	0.532
M	1	3052.56	3052.56	3052.56	97.96	0.000
2-Way Interactions	6	283.88	283.88	47.31	1.52	0.332
T*C	1	14.06	14.06	14.06	0.45	0.532
T*S	1	68.06	68.06	68.06	2.18	0.199
T*M	1	39.06	39.06	39.06	1.25	0.314
C*S	1	60.06	60.06	60.06	1.93	0.224
C*M	1	95.06	95.06	95.06	3.05	0.141
S*M	1	7.56	7.56	7.56	0.24	0.643
Residual Error	5	155.81	155.81	31.16		
Total	15	3728.94				

The ANOVA table shows that the only main effects that are significant at 5% level of significance are C and M, the other two main effects and all two factor interactions are not insignificant at the 5% level of significance.

19

RESPONSE SURFACES

PRACTICE PROBLEMS FOR SECTION 19.2

1. MINITAB (Response Surface Regression): Responses versus X1, X2, X3

The analysis was done using coded units.

Estimated Regression Coefficients for Responses

Term	Coef	SE Coef	T	P
Constant	50.8583	1.279	39.756	0.000
X1	-0.1250	1.567	-0.080	0.938
X2	1.2500	1.567	0.798	0.448
X3	1.5500	1.567	0.989	0.351

S = 4.43148   PRESS = 484.351
R-Sq = 16.85%   R-Sq(pred) = 0.00%   R-Sq(adj) = 0.00%

Analysis of Variance for Responses

Source	DF	Seq SS	Adj SS	Adj MS	F	P
Regression	3	31.845	31.845	10.6150	0.54	0.668
Linear	3	31.845	31.845	10.6150	0.54	0.668
X1	1	0.125	0.125	0.1250	0.01	0.938
X2	1	12.500	12.500	12.5000	0.64	0.448
X3	1	19.220	19.220	19.2200	0.98	0.351
Residual Error	8	157.104	157.104	19.6380		
Lack-of-Fit	5	134.937	134.937	26.9873	3.65	0.158
Pure Error	3	22.167	22.167	7.3892		
Total	11	188.949				

Estimated Regression Coefficients for Responses using data in uncoded units

```
Term        Coef
Constant    50.8583
X1          -0.125000
X2          1.25000
X3          1.55000
```

(a) From the MINITAB output, we obtain the first-order fitted model as

$$\hat{Y} = 50.8583 - 0.125x_1 + 1.25x_2 + 1.55x_3.$$

(b) "Lack of Fit" MS $= 26.9873$, "Pure Error" MS $= 7.3892$

(c) See the MINITAB output above. The "lack of fit" statistic $F_L = MS_L/MS_{PureError}$ is observed to be 3.65 with p-value of $0.158 > 0.05$, so that we conclude that there is no evidence of lack of fit.

3. MINITAB (Response Surface Regression: Responses versus X1, X2, X3, X4

The analysis was done using coded units.
Estimated Regression Coefficients for Responses

```
Term        Coef       SE Coef     T          P
Constant    25.2500    0.7134      35.394     0.000
X1          1.3750     0.8737      1.574      0.160
X2          2.1250     0.8737      2.432      0.045
X3          1.1250     0.8737      1.288      0.239
X4          -0.1250    0.8737      -0.143     0.890
```

```
S = 2.47126      PRESS = 62.3008
R-Sq = 58.99%    R-Sq(pred) = 40.24%    R-Sq(adj) = 35.56%
```

Analysis of Variance for Responses

Source	DF	Seq SS	Adj SS	Adj MS	F	P
Regression	4	61.500	61.5000	15.3750	2.52	0.135
Linear	4	61.500	61.5000	15.3750	2.52	0.135
X1	1	15.125	15.1250	15.1250	2.48	0.160
X2	1	36.125	36.1250	36.1250	5.92	0.045
X3	1	10.125	10.1250	10.1250	1.66	0.239
X4	1	0.125	0.1250	0.1250	0.02	0.890
Residual Error	7	42.750	42.7500	6.1071		
Lack-of-Fit	4	4.750	4.7500	1.1875	0.09	0.978
Pure Error	3	38.000	38.0000	12.6667		
Total	11	104.250				

Estimated Regression Coefficients for Responses using data in uncoded units

```
Term        Coef
Constant    25.2500
X1          1.37500
X2          2.12500
X3          1.12500
X4          -0.125000
```

(a) From the MINITAB output, we obtain the first-order fitted model as

$$\hat{Y} = 25.25 + 1.375x_1 + 2.125x_2 + 1.125x_3 - 0.125X_4.$$

(b) "Lack of Fit" MS $= 1.1875$, "pure Error" $= 12.6667$

(c) For ANOVA see the MINITAB output above. Since p-value is 0.978, so there is no evidence of lack of fit.

5. MINITAB (Response Surface Regression: Responses versus X1, X2, X3, X4

The analysis was done using coded units.

Estimated Regression Coefficients for Responses

Term	Coef	SE Coef	T	P
Constant	24.8750	0.2394	103.924	0.000
X1	1.3750	0.2394	5.745	0.010
X2	2.1250	0.2394	8.878	0.003
X3	1.1250	0.2394	4.700	0.018
X4	-0.1250	0.2394	-0.522	0.638

S = 0.677003   PRESS = 9.77778
R-Sq = 97.81%   R-Sq(pred) = 84.45%   R-Sq(adj) = 94.90%

Analysis of Variance for Responses

Source	DF	Seq SS	Adj SS	Adj MS	F	P
Regression	4	61.5000	61.5000	15.3750	33.55	0.008
Linear	4	61.5000	61.5000	15.3750	33.55	0.008
X1	1	15.1250	15.1250	15.1250	33.00	0.010
X2	1	36.1250	36.1250	36.1250	78.82	0.003
X3	1	10.1250	10.1250	10.1250	22.09	0.018
X4	1	0.1250	0.1250	0.1250	0.27	0.638
Residual Error	3	1.3750	1.3750	0.4583		
Total	7	62.8750				

Estimated Regression Coefficients for Responses using data in uncoded units

Term	Coef
Constant	24.8750
X1	1.37500
X2	2.12500
X3	1.12500
X4	-0.125000

Again, we note that the fitted model has changed significantly. There is no "pure error" degree of freedom available. This means we cannot test the adequacy of the fitted model. However, the ANOVA table indicates the overall first-order model is a good fit and each regression coefficient (except β_4) is significantly different from zero.

(a) The fitted model is

$$\hat{Y} = 24.875 + 1.375x_1 + 2.125x_2 + 1.125x_3 - 0.125X_4$$

(b) The alias structure is easily seen to be

$$\beta_{12} + \beta_{34}; \beta_{13} + \beta_{24}; \beta_{14} + \beta_{23}$$

The second-order model considered here is

$$E(Y) = \beta_0 + \sum_{i=1}^{4} \beta_i X_i + \beta_{12} X_1 X_2 + \beta_{13} X_1 X_3 + \beta_{14} X_1 X_4$$

When fitted to the factorial points of the data in Problem 3, we obtain

$$\hat{Y} = 25.00 + 1.375 x_1 + 2.125 x_2 + 1.125 x_3 - 0.125 X_4 + 0.125 x_1 x_2 + 0.125 x_1 x_3 + 0.375 x_1 x_4$$

(c) See part (a) and the MINITAB output given above.

(d) The ANOVA table for this part is given below.

Analysis of Variance (ANOVA) for Responses (coded units)

Source	DF	Seq SS	Adj SS	Adj MS	F	P
Main Effects	4	61.5000	61.5000	15.3750	33.55	0.008
X1	1	15.1250	15.1250	15.1250	33.00	0.010
X2	1	36.1250	36.1250	36.1250	78.82	0.003
X3	1	10.1250	10.1250	10.1250	22.09	0.018
X4	1	0.1250	0.1250	0.1250	0.27	0.638
Residual Error	3	1.3750	1.3750	0.4583		
Total	7	62.8750				

Clearly, the above ANOVA table indicates that $\hat{\beta}_1, \hat{\beta}_2, \hat{\beta}_3$ are significant at the 5% level of significance, so that main effects A, B, C are significant. However, $\hat{\beta}_4$ is not significant at the 5% level of significance, that is, main effect D is not significant.

PRACTICE PROBLEMS FOR SECTION 19.3

1. **MINITAB (Response Surface Regression: Responses versus Block, x_1, x_2)**

The analysis was done using coded units.

Estimated Regression Coefficients for Responses

Term	Coef	SE Coef	T	P
Constant	32.2502	1.2447	25.910	0.000
Block	0.5000	0.7186	0.696	0.518
x_1	2.2678	0.8802	2.576	0.050
x_2	0.5303	0.8802	0.603	0.573
x_1^2	-0.1877	0.9842	-0.191	0.856
x_2^2	0.0624	0.9842	0.063	0.952
$x_1 x_2$	5.0000	1.2447	4.017	0.010

S = 2.48944   PRESS = 243.060
R-Sq = 82.56%   R-Sq(pred) = 0.00%   R-Sq(adj) = 61.63%

Analysis of Variance for Responses

Source	DF	Seq SS	Adj SS	Adj MS	F	P
Blocks	1	3.000	3.000	3.000	0.48	0.518
Regression	5	143.680	143.680	28.736	4.64	0.059
Linear	2	43.388	43.388	21.694	3.50	0.112
x_1	1	41.139	41.139	41.139	6.64	0.050
x_2	1	2.250	2.250	2.250	0.36	0.573

Square	2	0.292	0.292	0.146	0.02	0.977
x_1^2	1	0.267	0.225	0.225	0.04	0.856
x_2^2	1	0.025	0.025	0.025	0.00	0.952
Interaction	1	100.000	100.000	100.000	16.14	0.010
$x_1 x_2$	1	100.000	100.000	100.000	16.14	0.010
Residual Error	5	30.986	30.986	6.197		
Lack-of-Fit	3	24.486	24.486	8.162	2.51	0.298
Pure Error	2	6.500	6.500	3.250		
Total	11	177.667				

Note that the each replicated treatment provides degrees of freedom for pure error, i.e., (number of replications -1). In this problem, the two center points in block I and Bock II provide one degree of freedom each, so that we have total of two degrees of freedom for the pure error.

(a) The fitted model is

$$\hat{Y} = 32.2502 + 2.2678x_1 + 0.5303x_2 - 0.1877x_1^2 + 0.0624x_2^2 + 5.000x_1 x_2$$

(b) For ANOVA table see the MINITAB output $\hat{\sigma}^2 = 3.250$.

(c) From the ANOVA table we can note that the pure quadratic terms are not significant at the 5% level of significance. However, the interaction term is significant at any level greater than the .01 level of significance.

3. MINITAB (Response Surface Regression): Responses versus X1, X2, X3

```
The analysis was done using coded units.
Estimated Regression Coefficients for Responses
```

Term	Coef	SE Coef	T	P
Constant	35.0000	2.974	11.769	0.000
X1	0.5000	3.642	0.137	0.894
X2	0.7500	3.642	0.206	0.842
X3	-0.2500	3.642	-0.069	0.947

```
S = 10.3017      PRESS = 2414.50
R-Sq = 0.82%     R-Sq(pred) = 0.00%     R-Sq(adj) = 0.00%
```

```
Analysis of Variance for Responses
```

Source	DF	Seq SS	Adj SS	Adj MS	F	P
Regression	3	7.000	7.000	2.333	0.02	0.995
Linear	3	7.000	7.000	2.333	0.02	0.995
X1	1	2.000	2.000	2.000	0.02	0.894
X2	1	4.500	4.500	4.500	0.04	0.842
X3	1	0.500	0.500	0.500	0.00	0.947
Residual Error	8	849.000	849.000	106.125		
Lack-of-Fit	5	637.000	637.000	127.400	1.80	0.333
Pure Error	3	212.000	212.000	70.667		
Total	11	856.000				

(a) The first-order fitted model is

$$\hat{Y} = 35.00 + 0.50x_1 + 0.75x_2 - 0.25x_3$$

(b) "Pure error" MS $= 70.667$, "Lack of fit" MS $= 127.400$

(c) For the ANOVA table see the MINITAB output above. For the adequacy of the model we note that the p-value of the lack of fit is 0.333, which is quite large (greater than 0.05). Hence the fit of the first-order model is declared adequate.

Further, note that even though we declared the model is adequate but referring to the values of R-sq, PRESS, and R-sq(pred) the model does not seem to be very useful.

5. MINITAB (Response Surface Regression): Y versus Blocks, X1, X2, X3

```
The analysis was done using uncoded units.
```

```
Estimated Regression Coefficients for Y
```

Term	Coef	SE Coef	T	P
Constant	31.9814	2.079	15.382	0.000
Block 1	2.3519	1.771	1.328	0.217
Block 2	-1.4814	1.771	-0.836	0.425
X1	1.4669	1.461	1.004	0.342
X2	-0.9720	1.461	-0.665	0.523
X3	-1.4478	1.461	-0.991	0.348
X1*X1	-2.0000	1.349	-1.483	0.172
X2*X2	-2.6667	1.349	-1.977	0.079
X3*X3	-0.8333	1.349	-0.618	0.552
X1*X2	2.2500	1.933	1.164	0.274
X1*X3	-4.0000	1.933	-2.069	0.068
X2*X3	-1.0000	1.933	-0.517	0.617

```
S = 5.46740      PRESS = 2644.81
R-Sq = 63.80%    R-Sq(pred) = 0.00%   R-Sq(adj) = 19.56%
```

```
Analysis of Variance for Y
```

Source	DF	Seq SS	Adj SS	Adj MS	F	P
Blocks	2	52.849	52.852	26.426	0.88	0.446
Regression	9	421.357	421.357	46.817	1.57	0.257
Linear	3	72.694	72.694	24.231	0.81	0.519
X1	1	30.124	30.124	30.124	1.01	0.342
X2	1	13.227	13.227	13.227	0.44	0.523
X3	1	29.343	29.343	29.343	0.98	0.348
Square	3	172.162	172.162	57.387	1.92	0.197
X1*X1	1	49.307	65.735	65.735	2.20	0.172
X2*X2	1	111.444	116.865	116.865	3.91	0.079
X3*X3	1	11.411	11.411	11.411	0.38	0.552
Interaction	3	176.500	176.500	58.833	1.97	0.189
X1*X2	1	40.500	40.500	40.500	1.35	0.274
X1*X3	1	128.000	128.000	128.000	4.28	0.068
X2*X3	1	8.000	8.000	8.000	0.27	0.617
Residual Error	9	269.032	269.032	29.892		
Lack-of-Fit	5	201.865	201.865	40.373	2.40	0.208
Pure Error	4	67.167	67.167	16.792		
Total	20	743.238				

(a) The second-order fitted model is

$$\hat{Y} = 31.98 + 1.47x_1 - 0.97x_2 - 1.45x_3 - 2.00x_1^2 - 2.67x_2^2 - 0.83x_3^2 + 2.25x_1x_2 \\ - 4.00x_1x_3 - 1.00x_2x_3$$

The p-value for lack of fit is $0.208 > 0.05$, so we do not reject the null hypothesis of no lack of fit. In other words, there is no evidence of lack of fit of the complete second order model.

(b) From the ANOVA table, see the MINITAB output above, we note that $\hat{\sigma}^2 = 16.792$.

(c) From the ANOVA table we note that the p-value for the second order terms is greater than 0.05, the level of significance. Hence, the second-order terms are not significant at 5% level. Also, the p-value for the first-order terms is $0.519 > 0.05$. Hence, the first order terms are not significant. Also the p-values for the individual terms are $> .05$. Further, note that even though we declared the model is adequate but referring to the values of R-sq, PRESS, and R-sq(pred) the model does not seem to be very useful.

PRACTICE PROBLEMS FOR SECTION 19.4

1. From the analysis of Problem 1 of Section 19.3, we see that the fit of the second-order model is quite adequate. We now determine the nature of the stationary point and the fitted response surface:

$$\hat{Y} = 32.2502 + 2.2678x_1 + 0.5303x_2 - 0.1877x_1^2 + 0.0624x_2^2 + 5.000x_1x_2$$

We first find the stationary point (x_1^0, x_2^0). It is well known that the stationary point is the solution of the following set of equations,

$$\frac{\partial \hat{Y}}{\partial x_1} = 0, \quad \frac{\partial \hat{Y}}{\partial x_2} = 0,$$

The two equations are:

$$-0.3754x_1^0 + 5.00x_2^0 = -2.2678$$
$$5.00x_1^0 + 0.1248x_2^0 = -0.5303$$

In matrix notation, this set of equations can be written as

$$\begin{bmatrix} -0.3754 & 5.00 \\ 5.00 & 0.1248 \end{bmatrix} \begin{bmatrix} x_1^0 \\ x_2^0 \end{bmatrix} = \begin{bmatrix} -2.2678 \\ -0.5303 \end{bmatrix}$$

or

$$\begin{bmatrix} x_1^0 \\ x_2^0 \end{bmatrix} = \frac{1}{2} \times \begin{bmatrix} -0.1877 & 2.5 \\ 2.5 & 0.0624 \end{bmatrix}^{-1} \times \begin{bmatrix} -2.2678 \\ -0.5303 \end{bmatrix} = \begin{bmatrix} -0.0946 \\ -0.4607 \end{bmatrix}$$

Thus, the stationary point is $(-0.0946, -0.4607)$.

The value of the estimated response $\hat{Y}$ at the stationary point is

$$\hat{y}_0 = 32.02$$

Thus, if we now shift our origins to the stationary point, that is

$$x_1 = u_1 - 0.0946, x_2 = u_2 - 0.4607.$$

where (u_1, u_2) are the new coordinates, then the fitted polynomial can be written as

$$\hat{Y} = 32.02 - 0.1877u_1^2 + 0.0624u_2^2 + 5u_1u_2.$$

This in turn can be reduced to its canonical form by determining the eigen-values of the matrix

$$\mathbf{B} = \begin{bmatrix} -0.1877 & 2.5 \\ 2.5 & 0.0624 \end{bmatrix}$$

The eigen-values of this matrix are found to be

$$-2.57, 2.44$$

Thus, the canonical form of the second-order fitted model or the response function is

$$\hat{Y} = 32.02 - 2.57w_1^2 + 2.44w_2^2$$

Since one eigen-value is negative and one eigen-value is positive, the stationary point is a saddle point and the fitted surface is a saddle. The contour plot for the fitted response surface is shown below. It confirms our conclusion that the stationary point is a saddle point.

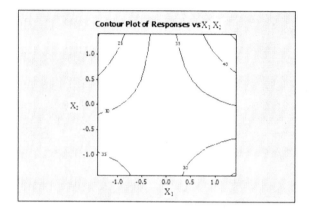

3. From the solution of Problem 2 of Section 19.3, the fitted model is

$$\hat{Y} = 35.5998 + 1.2782x_1 - 1.1616x_2 + 0.5924x_3 - 0.7987x_1^2 - 0.6219x_2^2$$
$$+ 0.6155x_3^2 - 2.2500x_1x_2 + 7.2500x_1x_3 - 4.5000x_2x_3$$

It can easily be seen that the matrix **B** is

$$B = \begin{bmatrix} -0.7987 & -1.1250 & 3.6250 \\ -1.1250 & -0.6219 & -2.2500 \\ 3.6250 & -2.2500 & 0.6155 \end{bmatrix}$$

and the stationary point is $(-0.0755, -0.0468, -0.2075)$. Further, the estimated value of Y at the stationary point is 35.7719. The eigen-values of the B matrix are found to be 4.73, -3.91, and -1.63. Thus the canonical form of the fitted model is

$$\hat{Y} = 35.77 + 4.73w_1^2 - 3.91w_2^2 - 1.63w_3^2$$

Since one eigen-value is positive and other two are negative, so the stationary point is saddle point or a minimax point and the response surface is a saddle. The contour plots for the fitted response surface are shown below. They confirm our conclusion that the stationary point is a minimax point.

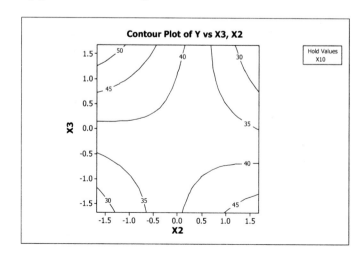

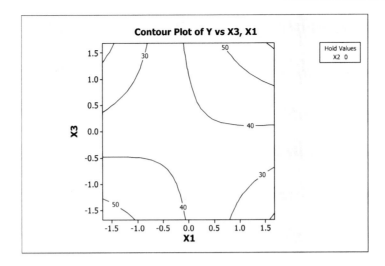

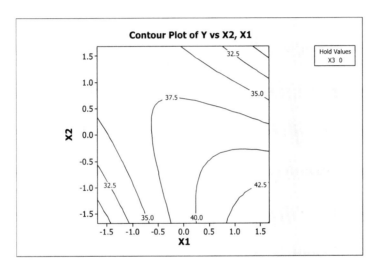

5. MINITAB (**Response Surface Regression**): **Y versus Blocks, X1, X2, X3**

The design that we use to fit the first order model is the following:

$$
\begin{pmatrix}
Bl. & X1 & X2 & X3 \\
1 & -1 & -1 & -1 \\
1 & 1 & 1 & -1 \\
1 & 1 & -1 & 1 \\
1 & -1 & 1 & 1 \\
1 & 0 & 0 & 0 \\
1 & 0 & 0 & 0 \\
2 & 1 & -1 & -1 \\
2 & -1 & 1 & -1 \\
2 & -1 & -1 & 1 \\
2 & 1 & 1 & 1 \\
2 & 0 & 0 & 0 \\
2 & 0 & 0 & 0
\end{pmatrix}
\begin{pmatrix}
Y \\
29 \\
39 \\
27 \\
21 \\
35 \\
33 \\
36 \\
18 \\
31 \\
21 \\
23 \\
32
\end{pmatrix}
$$

The analysis was done using coded units.

Estimated Regression Coefficients for Y

Term	Coef	SE Coef	T	P
Constant	28.750	1.738	16.544	0.000
Block	1.917	1.738	1.103	0.307
X1	3.000	2.128	1.410	0.202
X2	-3.000	2.128	-1.410	0.202
X3	-2.750	2.128	-1.292	0.237

S = 6.01981   PRESS = 989.687
R-Sq = 49.49%   R-Sq(pred) = 0.00%   R-Sq(adj) = 20.63%

Analysis of Variance for Y

Source	DF	Seq SS	Adj SS	Adj MS	F	P
Blocks	1	44.08	44.08	44.08	1.22	0.307
Regression	3	204.50	204.50	68.17	1.88	0.221
Linear	3	204.50	204.50	68.17	1.88	0.221
X1	1	72.00	72.00	72.00	1.99	0.202
X2	1	72.00	72.00	72.00	1.99	0.202
X3	1	60.50	60.50	60.50	1.67	0.237
Residual Error	7	253.67	253.67	36.24		
Lack-of-Fit	5	211.17	211.17	42.23	1.99	0.368
Pure Error	2	42.50	42.50	21.25		
Total	11	502.25				

(a) The design we used is consist of 8 factor points and four center points in block 1 and block 2 of Problem 5 of Section 19.3. The fitted model is

$$\hat{Y} = 28.75 + 3.00x_1 - 3.00x_2 - 2.75x_3$$

From the ANOVA table given in MINITAB output we note that the p-value for "lack of fit" is $0.368 > 0.05$, so that there is no evidence of lack of fit. Also, the regression coefficients, $\hat{\beta}_1, \hat{\beta}_2$, and $\hat{\beta}_3$ are not very small. Thus, our next step is to determine the path of steepest ascent. From the fitted equation, we have the changes in x_i's, $i = 1, 2$, and 3 (in the units of the design) along the *path of steepest ascent* are proportional to

$$\hat{\beta}_1 = 3.0, \hat{\beta}_2 = -3.00, \text{and } \hat{\beta}_3 = -2.75,$$

respectively. Since in this problem we do not have x_i's in their *original units*, so the points on the steepest ascent are given in terms of design units.

Now we select one of the variables, say x_1 as the standard variable and then calculate the changes in other variables x_2 and x_3, which would correspond to 1 design unit change in x_1. Thus, the changes in x_1, x_2 and x_3, are 1, $-1 = -3.00/3.0$, and $-.917 = -2.75/3.0$ in design units, respectively.

Now, to obtain the path of steepest ascent, we start at the center point $(0, 0, 0)$. For each 1 design unit increase in x_1, the variables x_2 and x_3 are decreased by 1 and .917 design units, respectively. Table below gives the various points on the path of steepest ascent. The estimated values of Y at these points are calculated from the fitted first-order model and are also given.

Initial point	x_1	x_2	x_3	$\hat{Y}$
	0	0	0	28.750
1	1	−1	−0.917	37.272
2	2	−2	−1.834	45.794
3	3	−3	−2.751	54.316
4	4	−4	−3.668	62.838
5	5	−5	−4.585	71.360

REVIEW PRACTICE PROBLEMS

1. From the solution for Problem 1 of Section 19.2, the fitted model is

$$\hat{Y} = 50.8583 - 0.125x_1 + 1.25x_2 + 1.55x_3$$

From the ANOVA table given in MINITAB output we note that the p-value for "lack of fit" is $0.158 > 0.05$, so that the fit of the first-order model is deemed to be quite adequate. Also, the regression coefficients, $\hat{\beta}_1, \hat{\beta}_2$, and $\hat{\beta}_3$ are not very small. Thus, our next step is to determine the path of steepest ascent. From the fitted equation, we have the changes in x_i's, $i = 1, 2$, and 3 (in design units) along the *path of steepest ascent* are proportional to

$$\hat{\beta}_1 = -0.125, \hat{\beta}_2 = 1.25, \text{ and } \hat{\beta}_3 = 1.55,$$

respectively. Thus, in the original units the changes in X_i's are proportional to

$$X_1 = -.125 \times 10 = -1.25, X_2 = 1.25 \times 4 = 5, X_3 = 1.55 \times 2 = 3.10$$

Now we select one of the variables, say X_1 as the standard variable and then calculate the changes in other variables X_2 and X_3, which would correspond to 1 original unit change in X_1. Thus, the changes in X_2 and X_3 are 1, $-4.0 = 5/-1.25$, and $-2.48 = 3.10/-1.25$ units, respectively.

Now, to obtain the path of steepest ascent, we start at the center point $(x_1, x_2, x_3) = (0, 0, 0)$, which is the point in original units $(X_1, X_2, X_3) = (132, 28, 8)$. For each 1 design unit increase in X_1, the variables X_2 and X_3 are decreased by -4.0 and -2.48 units, respectively. The table below gives the various points on the path of steepest ascent. The estimated values of Y at these points are calculated from the fitted first-order model and are given in the table below. Note that to calculate the predicted values, all the levels must be converted into design units.

	Points on the path of Steepest Ascent		
	X_1	X_2	X_3
Initial Point (center point)	132	28	8.0
1	133	24	5.52
2	134	20	3.04
3	135	16	0.56
4	136	12	−1.92
5	137	8	−4.40

Predicted Responses at the Points (in design units) Along the steepest Ascent

Initial point	x_1	x_2	x_3	$\hat{Y}$
	0	0	0	50.8583
1	0.1	-1	-1.24	47.6738
2	0.2	-2	-2.48	44.4893
3	0.3	-3	-3.72	41.3048
4	0.4	-4	-4.96	38.1203
5	0.5	-5	-6.20	34.9358

(3) (a)

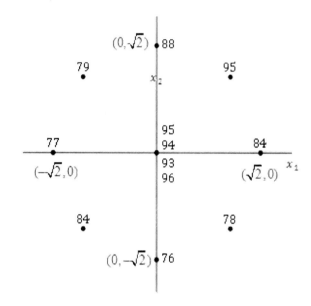

(b) MINITAB (Response Surface Regression): Observations versus X1, X2

```
Estimated Regression Coefficients for Observations

Term        Coef      SE Coef      T          P
Constant    86.583    2.200        39.364     0.000
X1          2.488     2.694        0.923      0.380
X2          3.622     2.694        1.344      0.212

S = 7.61948      PRESS = 904.638
R-Sq = 22.81%    R-Sq(pred) = 0.00%    R-Sq(adj) = 5.66%

Analysis of Variance for Observations

Source           DF    Seq SS     Adj SS     Adj MS     F        P
Regression       2     154.408    154.408    77.204     1.33     0.312
  Linear         2     154.408    154.408    77.204     1.33     0.312
    X1           1     49.499     49.499     49.499     0.85     0.380
    X2           1     104.909    104.909    104.909    1.81     0.212
Residual Error   9     522.509    522.509    58.057
  Lack-of-Fit    6     517.509    517.509    86.251     51.75    0.004
  Pure Error     3     5.000      5.000      1.667
Total            11    676.917
```

The fitted first order model is: $\hat{Y} = 86.583 + 2.488x_1 + 3.622x_2$.

From the ANOVA table we note that the p-value for lack of fit is 0.004 so that we reject the null hypothesis of no lack of fit. In other words, the first-order model is not adequate. The residual plots also indicate that a second-order model is likely to provide a fit better.

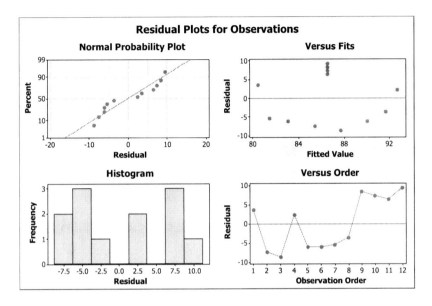

(c) MINITAB (Response Surface Regression): Observations versus X1, X2

Here, we assume the model $E(Y) = \beta_0 + \sum_{j=1}^{2} \beta_j x_j + \sum_{j=1}^{2} \beta_{jj} x_j^2 + \beta_{12} x_1 x_2$.

Estimated Regression Coefficients for Observations

Term	Coef	SE Coef	T	P
Constant	94.500	0.9844	95.998	0.000
X1	2.488	0.6961	3.574	0.012
X2	3.622	0.6961	5.202	0.002
X1*X1	-6.313	0.7784	-8.111	0.000
X2*X2	-5.563	0.7784	-7.147	0.000
X1*X2	5.500	0.9844	5.587	0.001

```
S = 1.96879      PRESS = 138.715
R-Sq = 96.56%    R-Sq(pred) = 79.51%    R-Sq(adj) = 93.70%
```

Analysis of Variance for Observations

Source	DF	Seq SS	Adj SS	Adj MS	F	P
Regression	5	653.660	653.660	130.732	33.73	0.000
Linear	2	154.408	154.408	77.204	19.92	0.002
X1	1	49.499	49.499	49.499	12.77	0.012
X2	1	104.909	104.909	104.909	27.07	0.002
Square	2	378.252	378.252	189.126	48.79	0.000
X1*X1	1	180.277	254.974	254.974	65.78	0.000
X2*X2	1	197.975	197.975	197.975	51.08	0.000
Interaction	1	121.000	121.000	121.000	31.22	0.001
X1*X2	1	121.000	121.000	121.000	31.22	0.001
Residual Error	6	23.257	23.257	3.876		

Lack-of-Fit	3	18.257	18.257	6.086	3.65	0.158
Pure Error	3	5.000	5.000	1.667		
Total	11	676.917				

the fitted model is

$$\hat{Y} = 94.5 + 2.488x_1 + 3.622x_2 - 6.313x_1^2 - 5.563x_2^2 + 5.500x_1x_2$$

From the ANOVA table we note that the p-value for lack of fit is 0.158, so that we do not reject the null hypothesis of no lack of fit. In other words, the second-order model provides an adequate fit.

(d) The B matrix is

$$\mathbf{B} = \begin{bmatrix} -6.313 & 2.750 \\ 2.750 & -5.563 \end{bmatrix}$$

The stationary point is $(x_1^0, x_2^0) = (0.4318, 0.5390)$, and $\hat{Y} = (x_1^0, x_2^0)$ is $\hat{Y} = 96.0133$. Thus, the canonical form of the fitted second-order model is

$$\hat{Y} = 96.0133 - 8.713w_1^2 - 3.163w_2^2$$

Since both the eigen-values are negative the stationary point is a maximum point and the fitted response surface is a mound.

5.(a)

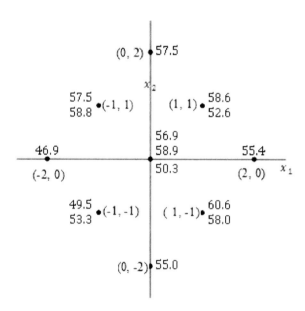

(b) MINITAB (Factorial Fit: Y versus X1, X2

The analysis was done using coded units.

Estimated Effects and Coefficients for Y (coded units)

Term	Effect	Coef	SE Coef	T	P
Constant		55.663	1.765	31.53	0.000
X1	3.575	1.788	1.765	1.01	0.358
X2	2.425	1.213	1.765	0.69	0.523

S = 4.99272   PRESS = 319.069
R-Sq = 23.04%   R-Sq(pred) = 0.00%   R-Sq(adj) = 0.00%

Analysis of Variance for Y (coded units)

Source	DF	Seq SS	Adj SS	Adj MS	F	P
Main Effects	2	37.32	37.32	18.66	0.75	0.520
X1	1	25.56	25.56	25.56	1.03	0.358
X2	1	11.76	11.76	11.76	0.47	0.523
Residual Error	5	124.64	124.64	24.93		
Lack of Fit	1	75.03	75.03	75.03	6.05	0.070
Pure Error	4	49.60	49.60	12.40		
Total	7	161.96				

Note that 4 degrees of freedom for the pure error are provided by the replication of the 2^2 factorial design, i.e., $4 = 4(2\text{-}1)$.

Since p-value of lack of fit is $0.07 > 0.05$, we do not reject the null hypothesis of no lack of fit, that is, there is no evidence of lack of fit. The first-order model is

$$\hat{Y} = 55.663 + 1.788x_1 + 1.213x_2$$

(c) MINITAB (Response Surface Regression): Y versus X1, X2

Here, we assume the model $E(Y) = \beta_0 + \sum_{j=1}^{2} \beta_j x_j + \sum_{j=1}^{2} \beta_{jj} x_j^2 + \beta_{12} x_1 x_2$.

The analysis was done using coded units.

Estimated Regression Coefficients for Y

Term	Coef	SE Coef	T	P
Constant	56.0615	1.574	35.611	0.000
X1	3.9125	1.639	2.388	0.041
X2	1.8375	1.639	1.121	0.291
X1*X1	-4.3904	2.980	-1.473	0.175
X2*X2	0.7096	2.980	0.238	0.817
X1*X2	-12.2500	4.635	-2.643	0.027

S = 3.27712   PRESS = 205.142
R-Sq = 65.27%   R-Sq(pred) = 26.29% R-Sq(adj) = 45.98%

Analysis of Variance for Y

Source	DF	Seq SS	Adj SS	Adj MS	F	P
Regression	5	181.648	181.648	36.3297	3.38	0.054
Linear	2	74.736	74.736	37.3681	3.48	0.076
X1	1	61.231	61.231	61.2306	5.70	0.041
X2	1	13.506	13.506	13.5056	1.26	0.291
Square	2	31.881	31.881	15.9404	1.48	0.277
X1*X1	1	31.272	23.310	23.3099	2.17	0.175
X2*X2	1	0.609	0.609	0.6089	0.06	0.817
Interaction	1	75.031	75.031	75.0313	6.99	0.027
X1*X2	1	75.031	75.031	75.0313	6.99	0.027
Residual Error	9	96.656	96.656	10.7395		
Lack-of-Fit	3	6.544	6.544	2.1813	0.15	0.929
Pure Error	6	90.112	90.112	15.0186		
Total	14	278.304				

The p-value for "lack of fit" is 0.927, so we do not reject the null hypothesis of no lack of fit. In other words, the fit of the second-order model is adequate. Also, the residual plots indicate adequacy of the second-order model. The fitted second-order model is

$$\hat{Y} = 56.0615 + 3.9125x_1 + 1.8375x_2 - 4.3904x_1^2 + 0.7096x_2^2 - 12.2500x_1x_2.$$

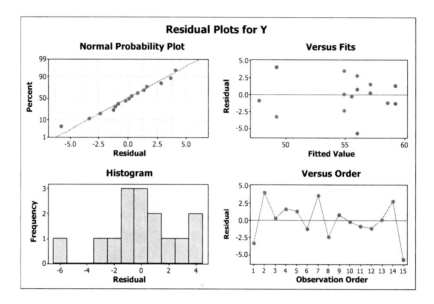

7. In this problem we have that $k = 3$ and the design T is a 1/2 replication of a 2^3 factorial experiment defined by the contrast $I = -ABC$. That is,

$$T = \begin{bmatrix} -1 & -1 & -1 \\ -1 & 1 & 1 \\ 1 & -1 & 1 \\ 1 & 1 & -1 \end{bmatrix}$$

where the upper and lower levels of each factor are coded to 1 and -1, respectively. We now show that the estimates of $\hat{\beta}_0, \hat{\beta}_1, \hat{\beta}_2$ and $\hat{\beta}_3$ are biased if the true model consists of first-order and two factor interactions terms as well. In other words, we assume the linear model and we fit

$$Y = \beta_0 x_0 + \beta_1 x_1 + \beta_2 x_2 + \beta_3 x_3 + \varepsilon,$$

but the true model is

$$Y = \beta_0 x_0 + \beta_1 x_1 + \beta_2 x_2 + \beta_3 x_3 + \beta_{12} x_1 x_2 + \beta_{13} x_1 x_3 + \beta_{23} x_2 x_3 + \varepsilon$$

Thus, we have

$$\gamma = \begin{bmatrix} \beta_0 \\ \beta_1 \\ \beta_2 \\ \beta_3 \end{bmatrix}, \quad X = \begin{matrix} x_0 & x_1 & x_2 & x_3 \\ \begin{bmatrix} 1 & -1 & -1 & -1 \\ 1 & -1 & 1 & 1 \\ 1 & 1 & -1 & 1 \\ 1 & 1 & 1 & -1 \end{bmatrix} \end{matrix}, \quad X_1 = \begin{matrix} x_1x_2 & x_1x_3 & x_2x_3 \\ \begin{bmatrix} 1 & 1 & 1 \\ -1 & -1 & 1 \\ -1 & 1 & -1 \\ 1 & -1 & -1 \end{bmatrix} \end{matrix}, \quad \text{and} \quad \gamma_1 = \begin{bmatrix} \beta_{12} \\ \beta_{13} \\ \beta_{23} \end{bmatrix}$$

The above means that

$$E(\hat{\gamma}) = \gamma + (X'X)^{-1}X'X_1\gamma_1 =$$

$$\gamma + \left[\begin{bmatrix} 1 & 1 & 1 & 1 \\ -1 & -1 & 1 & 1 \\ -1 & 1 & -1 & 1 \\ -1 & 1 & 1 & -1 \end{bmatrix}\begin{bmatrix} 1 & -1 & -1 & -1 \\ 1 & -1 & 1 & 1 \\ 1 & 1 & -1 & 1 \\ 1 & 1 & 1 & -1 \end{bmatrix}\right]^{-1} \times \begin{bmatrix} 1 & 1 & 1 & 1 \\ -1 & -1 & 1 & 1 \\ -1 & 1 & -1 & 1 \\ -1 & 1 & 1 & -1 \end{bmatrix} \times \begin{bmatrix} 1 & 1 & 1 \\ -1 & -1 & 1 \\ -1 & 1 & -1 \\ 1 & -1 & -1 \end{bmatrix}\begin{bmatrix} \beta_{12} \\ \beta_{13} \\ \beta_{23} \end{bmatrix}$$

$$= \gamma + \frac{1}{4}\begin{bmatrix} 1 & 0 & 0 & 0 \\ 0 & 1 & 0 & 0 \\ 0 & 0 & 1 & 0 \\ 0 & 0 & 0 & 1 \end{bmatrix}\begin{bmatrix} 0 & 0 & 0 \\ 0 & 0 & -4 \\ 0 & -4 & 0 \\ -4 & 0 & 0 \end{bmatrix}\begin{bmatrix} \beta_{12} \\ \beta_{13} \\ \beta_{23} \end{bmatrix} = \gamma + \begin{bmatrix} 0 & 0 & 0 \\ 0 & 0 & -1 \\ 0 & -1 & 0 \\ -1 & 0 & 0 \end{bmatrix}\begin{bmatrix} \beta_{12} \\ \beta_{13} \\ \beta_{23} \end{bmatrix} = \gamma - \begin{pmatrix} 0 \\ \beta_{23} \\ \beta_{13} \\ \beta_{12} \end{pmatrix}.$$

Thus, have that

$$E(\hat{\beta}_0) = \beta_0, \quad E(\hat{\beta}_1) = \beta_1 - \beta_{23}, \quad E(\hat{\beta}_2) = \beta_2 - \beta_{13}, \quad E(\hat{\beta}_3) = \beta_3 - \beta_{12}.$$

In other words, $\hat{\beta}_1, \hat{\beta}_2$ and $\hat{\beta}_3$ are biased estimators of β_1, β_2 and β_3, with bias of $-\beta_{23}, -\beta_{13}$, and $-\beta_{12}$, respectively.

9. MINITAB (Response Surface Regression): Y versus X1, X2, X3

The analysis was done using coded units.

Estimated Regression Coefficients for Y

```
Term         Coef      SE Coef        T          P
Constant    4.23083    0.04733      89.390      0.000
X1         -0.84625    0.05797     -14.599      0.000
X2         -0.16375    0.05797      -2.825      0.022
X3         -0.08875    0.05797      -1.531      0.164
```

```
S = 0.163957      PRESS = 0.576798
R-Sq = 96.54%     R-Sq(pred) = 90.73%     R-Sq(adj) = 95.25%
```

Analysis of Variance for Y

```
Source           DF    Seq SS     Adj SS     Adj MS       F         P
Regression        3    6.00664    6.00664    2.00221     74.48     0.000
  Linear          3    6.00664    6.00664    2.00221     74.48     0.000
    X1            1    5.72911    5.72911    5.72911    213.12     0.000
    X2            1    0.21451    0.21451    0.21451      7.98     0.022
    X3            1    0.06301    0.06301    0.06301      2.34     0.164
Residual Error    8    0.21505    0.21505    0.02688
  Lack-of-Fit     5    0.17645    0.17645    0.03529      2.74     0.218
  Pure Error      3    0.03860    0.03860    0.01287
Total            11    6.22169
```

(a) The first-order fitted model is

$$\hat{Y} = 4.231 - 0.846x_1 - 0.164x_2 - 0.089x_3$$

Since the p-value for the "lack of fit" is 0.218 the fit of first-order model is adequate.

(b) $S^2 = \hat{\sigma}^2 = 0.01287$; the three degrees of freedom that are used to estimate the error variance come from the observations at the four center points.

10. MINITAB (Response Surface Regression): Y versus X1, X2, X3

The analysis was done using coded units.

Estimated Regression Coefficients for Y

Term	Coef	SE Coef	T	P
Constant	4.29542	0.06446	66.640	0.000
X1	-1.48750	0.05497	-27.061	0.000
X2	-0.35917	0.05497	-6.534	0.000
X3	-0.19917	0.05497	-3.623	0.003
X1*X1	-0.26000	0.09521	-2.731	0.016
X2*X2	-0.08000	0.09521	-0.840	0.415
X3*X3	0.10000	0.09521	1.050	0.311
X1*X2	-0.32500	0.19042	-1.707	0.110
X1*X3	-0.10500	0.19042	-0.551	0.590
X2*X3	-0.37500	0.19042	-1.969	0.069

S = 0.134646   PRESS = 0.867103
R-Sq = 98.30%   R-Sq(pred) = 94.20%   R-Sq(adj) = 97.21%

Analysis of Variance for Y

Source	DF	Seq SS	Adj SS	Adj MS	F	P
Regression	9	14.6950	14.6950	1.6328	90.06	0.000
Linear	3	14.2879	14.2879	4.7626	262.70	0.000
X1	1	13.2759	13.2759	13.2759	732.29	0.000
X2	1	0.7740	0.7740	0.7740	42.69	0.000
X3	1	0.2380	0.2380	0.2380	13.13	0.003
Square	3	0.2784	0.2784	0.0928	5.12	0.013
X1*X1	1	0.2133	0.1352	0.1352	7.46	0.016
X2*X2	1	0.0451	0.0128	0.0128	0.71	0.415
X3*X3	1	0.0200	0.0200	0.0200	1.10	0.311
Interaction	3	0.1286	0.1286	0.0429	2.37	0.115
X1*X2	1	0.0528	0.0528	0.0528	2.91	0.110
X1*X3	1	0.0055	0.0055	0.0055	0.30	0.590
X2*X3	1	0.0703	0.0703	0.0703	3.88	0.069
Residual Error	14	0.2538	0.2538	0.0181		
Lack-of-Fit	5	0.1596	0.1596	0.0319	3.05	0.070
Pure Error	9	0.0942	0.0942	0.0105		
Total	23	14.9488				

(a) The fitted second-order model is

$$\hat{Y} = 4.30 - 1.49x_1 - 0.36x_2 - 0.20x_3 - 0.26x_1{}^2 - 0.08x_2{}^2 + 0.10x_3{}^2 - 0.33x_1x_2 - 0.11x_1x_3 - 0.38x_2x_3.$$

(b) $S^2 = \hat{\sigma}^2 = 0.0105$, the nine degrees of freedom that are used to estimate the error variance are counted for as follows:

three degrees of freedom from four center points + one degree of freedom that comes from each of the star points that are replicated twice.

(c) For testing the adequacy of the fitted model the test statistic used is given by $F_L = MS_{lack\ of\ fit}/MS_{Pure\ Error}$. The observed value of this test statistics is found to be

$$F = \frac{MS_{lack\ of\ fit}}{MS_{Pure\ error}} = \frac{0.0319}{0.0105} = 3.05 < F_{5,9;0.05} = 3.48166,$$

so that we do not reject the null hypothesis of no lack of fit or there is no evidence of lack of fit.

11. See solution of part (c) and MINITAB output of Problem 10.

13. Suppose the true quadratic model is

$$Y = \beta_0 + \sum_{i=1}^{4} \beta_i x_i + \sum_{i=1}^{4} \beta_{ii} x_i^2 + \sum_{\substack{i=1\ j=1 \ i<j}}^{4}\sum^{4} \beta_{ij} x_i x_j + \varepsilon;$$

Now suppose that $\overline{Y}$ is the average of the observations of the factorial points and $\overline{Y}_c$ is the average of the observations at the center. Then it follows from the fact $\hat{\beta}_0 = \overline{Y}$, and that

$$E(\overline{Y}) = \beta_0 + \beta_{11} + \beta_{22} + \beta_{33} + \beta_{44}.$$

and

$$E(\overline{Y}_c) = \beta_0 + \sum_{i=1}^{4} \beta_i(0) + \sum_{i=1}^{4} \beta_{ii}(0)^2 + \sum_{\substack{i=1\ j=1 \ i<j}}^{4}\sum^{4} \beta_{ij}(0)(0) = \beta_0,$$

Hence, we have that

$$E(\overline{Y} - \overline{Y}_c) = \beta_{11} + \beta_{22} + \beta_{33} + \beta_{44}$$

In other words, $\overline{Y} - \overline{Y}_c$ is an estimator of $\beta_{11} + \beta_{22} + \beta_{33} + \beta_{44}$. Note that $\overline{Y} - \overline{Y}_c$ is a contrast with each coefficient corresponding to the factorial points equal to $1/n$ and corresponding to the center points is $1/n_c$. The sum of squares due to $\beta_{11} + \beta_{22} + \beta_{33} + \beta_{44}$ with one degree of freedom is given by (See Chapter 18) $[C^2 = curvatue]$

$$C^2 = \frac{(contrast)^2}{r\sum_{i=1}^{n+n_c} c_i^2} = \frac{(\overline{Y} - \overline{Y}_c)^2}{\left(\sum\frac{1}{n^2} + \sum\frac{1}{n_c^2}\right)} = \frac{(\overline{Y} - \overline{Y}_c)^2}{\left(\frac{1}{n} + \frac{1}{n_c}\right)} = \frac{n \times n_c}{(n + n_c)} \times (\overline{Y} - \overline{Y}_c)^2$$

where r is the number of replications, which in this case is equal to 1. To test the hypothesis $H_0 : \beta_{11} + \beta_{22} + \beta_{33} + \beta_{44} = 0$, we compare C^2 with the mean square for pure error.

15. In this problem we are given the second-order fitted model is

$$\hat{Y} = 15.4 + 0.5x_1 - 1.2x_2 + 0.85x_3 + 2.6x_1x_2 - 1.8x_1x_3 + 2.1x_2x_3 + 3.2x_1^2 + 1.4x_2^2 + 2.7x_3^2$$

and we are interested in reducing the above fitted model to its canonical form and to describe the nature of the fitted response surface.

Here, we first determine the stationary point and then find the estimated value of Y at the stationary point. The matrix B for this model is

$$B = \begin{bmatrix} 3.20 & 1.30 & -0.90 \ 1.30 & 1.40 & 1.05 \ -0.90 & 1.05 & 2.70 \end{bmatrix}$$

The stationary point is $(x_1^0, x_2^0, x_3^0) = (-10.768, 18.688, -11.015)$. The estimated value of Y at the stationary point is -3.186 and the eigen-values of the matrix B are $(4.01, 3.26, 0.03)$. Thus, the canonical form of the fitted model is

$$\hat{Y} = -3.186 + 4.01w_1^2 + 3.26w_2^2 + 0.03w_3^2$$

Since all the eigen-values are positive the stationary point is a minimum point and the fitted response surface is a basin.

17. The estimated value of Y at "point 6" of the steepest path is found by substituting the point, i.e., $(6, -3.3, 5.70)$ in the fitted model. This gives the estimated response on "point 6" of the path to be

$$\hat{Y} = 50 + 2.2(6) - 1.2(-3.3) + 2.1(5.70) = 79.13.$$

19. **MINITAB (Response Surface Regression): Y versus X1, X2**

The analysis was done using coded units.

Estimated Regression Coefficients for Y

Term	Coef	SE Coef	T	P
Constant	21.775	0.8658	25.151	0.000
X1	1.578	0.8657	1.823	0.118
X2	-2.044	0.8657	-2.361	0.056
X1*X1	-1.450	1.3688	-1.059	0.330
X2*X2	1.200	1.3688	0.877	0.414
X1*X2	-1.350	1.7310	-0.780	0.465

S = 1.73154   PRESS = 113.378
R-Sq = 66.41%   R-Sq(pred) = 0.00%   R-Sq(adj) = 38.41%

Analysis of Variance for Y

Source	DF	Seq SS	Adj SS	Adj MS	F	P
Regression	5	35.560	35.560	7.1119	2.37	0.161
Linear	2	26.673	26.673	13.3365	4.45	0.065
X1	1	9.959	9.959	9.9593	3.32	0.118
X2	1	16.714	16.714	16.7137	5.57	0.056
Square	2	7.064	7.064	3.5321	1.18	0.370
X1*X1	1	4.760	3.365	3.3650	1.12	0.330
X2*X2	1	2.304	2.304	2.3041	0.77	0.414
Interaction	1	1.822	1.822	1.8225	0.61	0.465
X1*X2	1	1.822	1.822	1.8225	0.61	0.465
Residual Error	6	17.989	17.989	2.9982		
Lack-of-Fit	3	15.262	15.262	5.0873	5.60	0.096
Pure Error	3	2.728	2.728	0.9092		
Total	11	53.549				

(a) From the above MINITAB output we can state that the second-order fitted model is

$$\hat{Y} = 21.775 + 1.578x_1 - 2.044x_2 - 1.45x_1^2 + 1.20x_2^2 - 1.35x_1x_2$$

(b) From the ANOVA table given in MINITAB output we note that the p-value for "lack of fit" is $0.096 > 0.05$, so that we do not reject the null hypothesis of no lack of fit at the 5% level of significance. In other words, there is no evidence of lack of fit at the 5% level of significance.

(c) Here, we know that the second-order fitted model is

$$\hat{Y} = 21.775 + 1.578x_1 - 2.044x_2 - 1.45x_1^2 + 1.20x_2^2 - 1.35x_1x_2$$

and we are now interested in reducing the fitted model to its canonical form and describe the nature of the fitted response surface.

Here, we first determine the stationary point and then find the estimated value of Y at the stationary point. The matrix B for this model is

$$\mathbf{B} = \begin{bmatrix} -1.450 & -0.675 \\ -0.675 & 1.200 \end{bmatrix}$$

The stationary point is $(x_1^0, x_2^0) = (0.117, 0.917)$. The estimated value of Y at the stationary point is 20.930 and the eigen values of the matrix B are $(-1.612, 1.362)$. Thus, the canonical form of the fitted model is

$$\hat{Y} = 20.0843 - 1.612w_1^2 + 1.362w_2^2.$$

Since one of the eigen-value is positive and the other is negative the stationary point is a minimax point and the fitted response surface is a saddle. The plots of response surface and contours are given below. These plots confirm our conclusion.

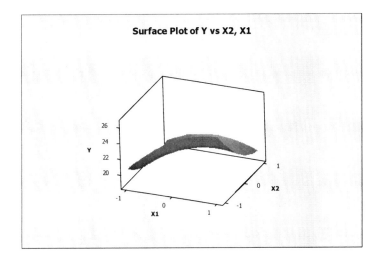

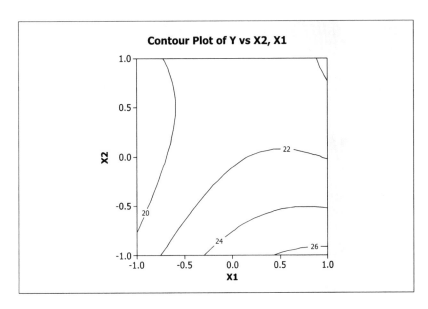

21. MINITAB (Response Surface Regression): Y versus X1, X2, X3

The analysis was done using coded units.

Estimated Regression Coefficients for Y

Term	Coef	SE Coef	T	P
Constant	64.5879	3.902	16.551	0.000
X1	6.7434	3.557	1.896	0.095
X2	1.3847	3.557	0.389	0.707
X3	0.6504	3.557	0.183	0.859
X1*X1	-2.7228	6.216	-0.438	0.673
X2*X2	-3.9728	6.216	-0.639	0.541
X3*X3	-1.7728	6.216	-0.285	0.783
X1*X2	5.3046	7.818	0.679	0.517
X1*X3	-19.4502	7.818	-2.488	0.038
X2*X3	-7.4265	7.818	-0.950	0.370

S = 7.81614   PRESS = 3473.43
R-Sq = 59.69%   R-Sq(pred) = 0.00%   R-Sq(adj) = 14.34%

Analysis of Variance for Y

Source	DF	Seq SS	Adj SS	Adj MS	F	P
Regression	9	723.71	723.709	80.412	1.32	0.355
Linear	3	230.83	230.834	76.945	1.26	0.351
X1	1	219.53	219.535	219.535	3.59	0.095
X2	1	9.26	9.257	9.257	0.15	0.707
X3	1	2.04	2.042	2.042	0.03	0.859
Square	3	31.50	31.500	10.500	0.17	0.912
X1*X1	1	5.09	11.720	11.720	0.19	0.673
X2*X2	1	21.44	24.951	24.951	0.41	0.541
X3*X3	1	4.97	4.968	4.968	0.08	0.783
Interaction	3	461.38	461.375	153.792	2.52	0.132
X1*X2	1	28.13	28.125	28.125	0.46	0.517
X1*X3	1	378.13	378.125	378.125	6.19	0.038
X2*X3	1	55.12	55.125	55.125	0.90	0.370
Residual Error	8	488.74	488.736	61.092		
Lack-of-Fit	5	448.07	448.066	89.613	6.61	0.075
Pure Error	3	40.67	40.670	13.557		
Total	17	1212.44				

(a) From the above MINITAB output we can state that the second-order fitted model is

$$\hat{Y} = 64.59 + 6.74x_1 + 1.38x_2 + 0.65x_3 - 2.72x_1^2 - 3.97x_2^2 - 1.77x_3^2 \\ + 5.30x_1x_2 - 19.45x_1x_3 - 7.43x_2x_3$$

(b) From the ANOVA table given in MINITAB output we note that the p-value for "lack of fit" is $0.075 > 0.05$, so that we do not reject the null hypothesis of no evidence of lack of fit at the 5% level of significance.

(c) The B matrix is

$$B = \begin{bmatrix} -2.72 & 2.65 & -9.72 \\ 2.65 & -3.97 & -3.72 \\ -9.72 & -3.72 & -1.77 \end{bmatrix}$$

and the stationary point is $(x_1^0, x_2^0, x_3^0) = (0.0176, -0.1073, 0.3125)$. The estimated value of Y at the stationary point is 64.6768. The eigen-values of matrix B are $(-12.0361, 9.0581, -5.4820)$

$$\hat{Y} = 64.6768 - 12.0361w_1^2 + 9.0581w_2^2 - 5.4820w_3^2$$

Since one eigen-value is positive and other two are negative, the stationary point is minimax and the response surface is a saddle. The response surface plot confirms our conclusion.

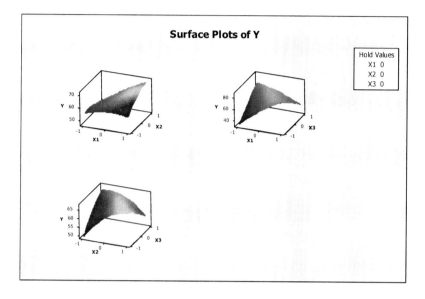

23. (a)

$$X = \begin{pmatrix} & x_0 & x_1 & x_2 & x_3 & x_4 \\ & 1 & -1 & -1 & -1 & -1 \\ & 1 & 1 & -1 & -1 & 1 \\ & 1 & -1 & 1 & -1 & 1 \\ & 1 & 1 & 1 & -1 & -1 \\ & 1 & -1 & -1 & 1 & 1 \\ & 1 & 1 & -1 & 1 & -1 \\ & 1 & -1 & 1 & 1 & -1 \\ & 1 & 1 & 1 & 1 & 1 \\ & 1 & 0 & 0 & 0 & 0 \\ & 1 & 0 & 0 & 0 & 0 \\ & 1 & 0 & 0 & 0 & 0 \\ & 1 & 0 & 0 & 0 & 0 \end{pmatrix}$$

(b) The design considered in part (a) is orthogonal since $X'X$ is a diagonal matrix, that is

$$X'X = \begin{pmatrix} 12 & 0 & 0 & 0 & 0 \\ 0 & 8 & 0 & 0 & 0 \\ 0 & 0 & 8 & 0 & 0 \\ 0 & 0 & 0 & 8 & 0 \\ 0 & 0 & 0 & 0 & 8 \end{pmatrix}$$

(c) The design is not variance optimal since $X'X \neq NI_p$ or $X'X \neq 12 \times I_5$.

CPSIA information can be obtained at www.ICGtesting.com
Printed in the USA
BVOW01n1608051113

335440BV00010B/24/P

Adobe® DREAMWEAVER® CS6
COMPREHENSIVE

Corinne L. Hoisington

Jessica L. Minnick

COURSE TECHNOLOGY
CENGAGE Learning™

SHELLY CASHMAN SERIES®

Australia • Brazil • Japan • Korea • Mexico • Singapore • Spain • United Kingdom • United States

COURSE TECHNOLOGY
CENGAGE Learning™

Adobe® Dreamweaver® CS6 Comprehensive
Corinne L. Hoisington, Jessica L. Minnick

Editor-in-Chief: Marie Lee

Executive Editor: Kathleen McMahon

Senior Product Manager: Emma F. Newsom

Associate Product Manager:
  Crystal Parenteau

Editorial Assistant: Sarah Ryan

Director of Marketing: Cheryl Costantini

Marketing Manager: Mark Linton

Marketing Coordinator: Benjamin Genise

Print Buyer: Julio Esperas

Director of Production: Patty Stephan

Content Project Manager: Matthew Hutchinson

Development Editor: Lisa Ruffolo

Copyeditor: Suzanne Huizenga

Proofreader: Kathy Orrino

Indexer: Rich Carlson

QA Manuscript Reviewers: Jeffrey Schwartz,
  Serge Palladino, Danielle Shaw,
  Susan Whalen

Art Director: GEX Publishing Services

Cover Designer: Lisa Kuhn, Curio Press, LLC

Cover Photo: Tom Kates Photography

Text Design: Joel Sadagursky

Compositor: PreMediaGlobal

For product information and technology assistance, contact us at
Cengage Learning Customer & Sales Support, 1-800-354-9706

For permission to use material from this text or product,
submit all requests online at **cengage.com/permissions**
Further permissions questions can be emailed to
permissionrequest@cengage.com

Adobe, the Adobe logos, and Dreamweaver are either registered trademarks or trademarks of Adobe Systems Incorporated in the United States and/or other countries. THIS PRODUCT IS NOT ENDORSED OR SPONSORED BY ADOBE SYSTEMS INCORPORATED, PUBLISHER OF DREAMWEAVER.

Library of Congress Control Number: 2012947746
ISBN-13: 978-1-133-52593-6
ISBN-10: 1-133-52593-8

Course Technology
20 Channel Center Street
Boston, Massachusetts 02210
USA

Cengage Learning is a leading provider of customized learning solutions with office locations around the globe, including Singapore, the United Kingdom, Australia, Mexico, Brazil, and Japan. Locate your local office at:
international.cengage.com/region

Cengage Learning products are represented in Canada by Nelson Education, Ltd.

For your course and learning solutions, visit **www.cengage.com**

To learn more about Course Technology,
visit **www.cengage.com/coursetechnology**

Purchase any of our products at your local college bookstore or at our preferred online store **www.cengagebrain.com**

Printed in the United States of America
1 2 3 4 5 6 7 17 16 15 14 13 12

Adobe DREAMWEAVER CS6 COMPREHENSIVE

Contents

Appendices

Preface

The Shelly Cashman Series® offers the finest textbooks in computer education. We are proud of the fact that our previous Dreamweaver® books have been so well received. With each new edition of our Dreamweaver® books, we make significant improvements based on the software and comments made by instructors and students. For this Adobe® Dreamweaver® CS6 text, the Shelly Cashman Series® development team carefully reviewed our pedagogy and analyzed its effectiveness in teaching today's Dreamweaver® student. Students today read less, but need to retain more. They need not only to be able to perform skills, but to retain those skills and know how to apply them to different settings. Today's students need to be continually engaged and challenged to retain what they're learning.

With this Adobe® Dreamweaver® CS6 text, we continue our commitment to focusing on the user and how they learn best.

Objectives of This Textbook

Adobe® Dreamweaver® CS6: Comprehensive is intended for a first course that offers an introduction to Dreamweaver® CS6 and creation of Web sites. No experience with a computer is assumed, and no mathematics beyond the high school freshman level is required. The objectives of this book are:

- To teach the fundamentals of Dreamweaver® CS6
- To expose students to proper Web site design and management techniques
- To acquaint students with the proper procedures to create Web sites suitable for coursework, professional purposes, and personal use
- To develop an exercise-oriented approach that allows learning by doing
- To introduce students to new input technologies
- To encourage independent study and provide help for those who are working independently

New to This Edition

HTM5 and CSS3

Engaging coverage of the latest HTML5 and CSS3 standards including style sheets which provide students with a solid understanding of professional Web design.

Professional Web Design

Explore creative designed centered solutions for creating a business and personal site that captures the attention of your targeted audience.

Mobile Web Site

Design a mobile Web site using a Web standards approach for delivering content beyond the desktop.

Web Accessibility

Integration of guidelines and standards for Web accessibility and disability access to the Web.

Social Networking

Coverage of social networking within a Web site to market business products and connect social trends.

The Shelly Cashman Approach

A Proven Pedagogy with an Emphasis on Project Planning

Each chapter presents a practical problem to be solved, within a project planning framework. The project orientation is strengthened by the use of Plan Ahead boxes, which encourage critical thinking about how to proceed at various points in the project. Step-by-step instructions with supporting screens guide students through the steps. Instructional steps are supported by the Q&A, Experimental Step, and BTW features.

A Visually Engaging Book that Maintains Student Interest

The step-by-step tasks, with supporting figures, provide a rich visual experience for the student. Call-outs on the screens that present both explanatory and navigational information provide students with information they need when they need to know it.

Supporting Reference Materials (Appendices and Quick Reference)

The appendices provide additional information about the Application at hand and include such topics as the Help Feature and customizing the application. With the Quick Reference, students can quickly look up information about a single task, such as creating a site, and find page references of where in the book the task is illustrated.

Integration of the World Wide Web

The World Wide Web is integrated into the Dreamweaver CS6 learning experience by (1) BTW annotations; and (2) BTW, Q&A, and Quick Reference Summary Web pages.

End-of-Chapter Student Activities

Extensive end-of-chapter activities provide a variety of reinforcement opportunities for students where they can apply and expand their skills. To complete some of these assignments, you will be required to use the Data Files for Students. Visit http://www.cengage.com/ct/studentdownload for detailed access instructions or contact your instructor for information about accessing the required files.

Instructor Resources

The Instructor Resources include both teaching and testing aids and can be accessed via CD-ROM or at **www.cengage.com/login**.

INSTRUCTOR'S MANUAL Includes lecture notes summarizing the chapter sections, figures and boxed elements found in every chapter, teacher tips, classroom activities, lab activities, and quick quizzes in Microsoft Word files.

SYLLABUS Easily customizable sample syllabi that cover policies, assignments, exams, and other course information.

FIGURE FILES Illustrations for every figure in the textbook in electronic form.

POWERPOINT PRESENTATIONS A multimedia lecture presentation system that provides slides for each chapter. Presentations are based on chapter objectives.

SOLUTIONS TO EXERCISES Includes solutions for all end-of-chapter and chapter reinforcement exercises.

TEST BANK & TEST ENGINE Test Banks include 112 questions for every chapter, featuring objective-based and critical thinking question types, and including page number references and figure references, when appropriate. Also included is the test engine, ExamView, the ultimate tool for your objective-based testing needs.

DATA FILES FOR STUDENTS Includes all the files that are required by students to complete the exercises. Visit www.cengage.com/ct/studentdownload for detailed instructions.

Learn Online

CengageBrain.com is the premier destination for purchasing or renting Cengage Learning textbooks, ebooks, eChapters and study tools, at a significant discount (eBooks up to 50% off Print). In addition, CengageBrain.com provides direct access to all digital products including eBooks, eChapters and digital solutions (i.e. CourseMate, SAM) regardless of where purchased. The following are some examples of what is available for this product on www.cengagebrain.com.

STUDENT COMPANION SITE Online practice opportunities and learning tools are available for no additional cost at www.cengagebrain.com can help reinforce chapter terms and concepts.

ADOBE DREAMWEAVER CS6 COURSEMATE CourseMate with ebook for Adobe Dreamweaver CS6 keeps today's students engaged and involved in the learning experience. Adobe Dreamweaver CS6 CourseMate includes an integrated, multi-media rich eBook, and a variety of interactive learning tools, including quizzes, activities, videos, and other resources that specifically reinforce and build on the concepts presented in the chapter. These interactive activities are tracked within CourseMate's Engagement Tracker, making it easy to assess students' retention of concepts. All of these resources enable students to get more comfortable using technology and help prepare students to use the Internet as a tool to enrich their lives. Available at the Comprehensive level in Spring 2013.

CourseNotes

Course Technology's CourseNotes are six-panel quick reference cards that reinforce the most important and widely used features of a software application in a visual and user-friendly format. CourseNotes serve as a great reference tool during and after the student completes the course. CourseNotes are available for software applications such as Adobe Photoshop CS6, Microsoft Office 2010, and Windows 7. Topic-based CourseNotes are available for Best Practices in Social Networking, Hot Topics in Technology, and Web 2.0. Visit www.cengage.com/ct/coursenotes to learn more!

About Our Covers

The Shelly Cashman Series is continually updating our approach and content to reflect the way today's students learn and experience new technology. This focus on student success is reflected on our covers, which feature real students from Naugatuck Valley Community College using the Shelly Cashman Series in their courses, and reflect the varied ages and backgrounds of the students learning with our books. When you use the Shelly Cashman Series, you can be assured that you are learning computer skills using the most effective courseware available.

Textbook Walk-Through

The Shelly Cashman Series Pedagogy: Project-Based — Step-by-Step — Variety of Assessments

Plan Ahead boxes prepare students to create successful projects by encouraging them to think strategically about what they are trying to accomplish before they begin working.

Step-by-step instructions now provide a context beyond the point-and-click. Each step provides information on why students are performing each task, or what will occur as a result.

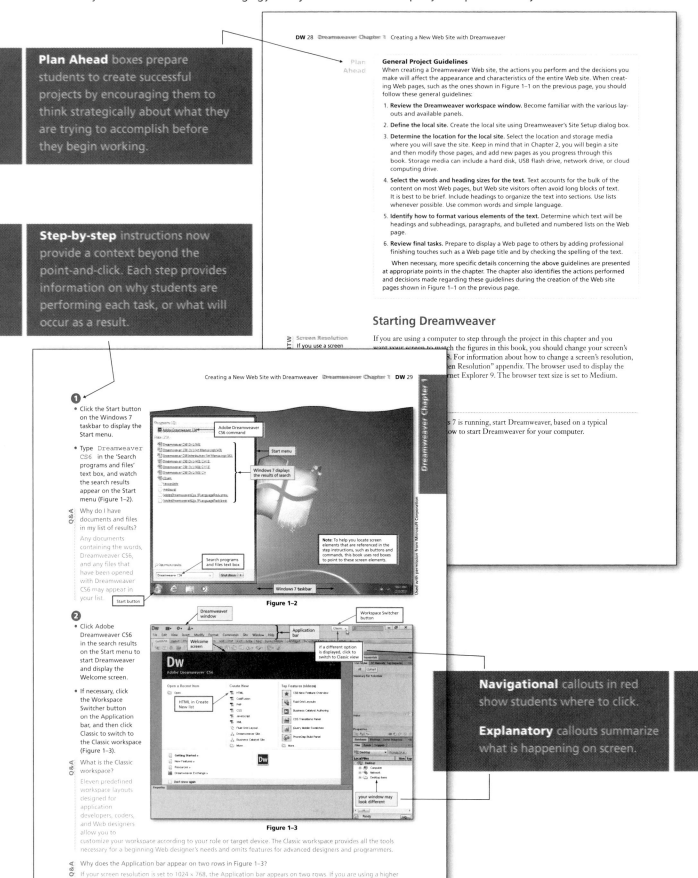

Plan Ahead

General Project Guidelines

When creating a Dreamweaver Web site, the actions you perform and the decisions you make will affect the appearance and characteristics of the entire Web site. When creating Web pages, such as the ones shown in Figure 1–1 on the previous page, you should follow these general guidelines:

1. **Review the Dreamweaver workspace window.** Become familiar with the various layouts and available panels.

2. **Define the local site.** Create the local site using Dreamweaver's Site Setup dialog box.

3. **Determine the location for the local site.** Select the location and storage media where you will save the site. Keep in mind that in Chapter 2, you will begin a site and then modify those pages, and add new pages as you progress through this book. Storage media can include a hard disk, USB flash drive, network drive, or cloud computing drive.

4. **Select the words and heading sizes for the text.** Text accounts for the bulk of the content on most Web pages, but Web site visitors often avoid long blocks of text. It is best to be brief. Include headings to organize the text into sections. Use lists whenever possible. Use common words and simple language.

5. **Identify how to format various elements of the text.** Determine which text will be headings and subheadings, paragraphs, and bulleted and numbered lists on the Web page.

6. **Review final tasks.** Prepare to display a Web page to others by adding professional finishing touches such as a Web page title and by checking the spelling of the text.

When necessary, more specific details concerning the above guidelines are presented at appropriate points in the chapter. The chapter also identifies the actions performed and decisions made regarding these guidelines during the creation of the Web site pages shown in Figure 1–1 on the previous page.

Starting Dreamweaver

If you are using a computer to step through the project in this chapter and you want your screen to match the figures in this book, you should change your screen's ... 8. For information about how to change a screen's resolution, ...en Resolution" appendix. The browser used to display the ...rnet Explorer 9. The browser text size is set to Medium.

Screen Resolution
If you use a screen

...s 7 is running, start Dreamweaver, based on a typical ...ow to start Dreamweaver for your computer.

Creating a New Web Site with Dreamweaver Dreamweaver Chapter 1 **DW 29**

1

- Click the Start button on the Windows 7 taskbar to display the Start menu.

- Type Dreamweaver CS6 in the 'Search programs and files' text box, and watch the search results appear on the Start menu (Figure 1–2).

Q&A Why do I have documents and files in my list of results?

Any documents containing the words, Dreamweaver CS6, and any files that have been opened with Dreamweaver CS6 may appear in your list.

Adobe Dreamweaver CS6 command

Start menu

Windows 7 displays the results of search

Note: To help you locate screen elements that are referenced in the step instructions, such as buttons and commands, this book uses red boxes to point to these screen elements.

Search programs and files text box

Start button

Windows 7 taskbar

Figure 1–2

Dreamweaver window

2

- Click Adobe Dreamweaver CS6 in the search results on the Start menu to start Dreamweaver and display the Welcome screen.

- If necessary, click the Workspace Switcher button on the Application bar, and then click Classic to switch to the Classic workspace (Figure 1–3).

Q&A What is the Classic workspace?

Eleven predefined workspace layouts designed for application developers, coders, and Web designers allow you to customize your workspace according to your role or target device. The Classic workspace provides all the tools necessary for a beginning Web designer's needs and omits features for advanced designers and programmers.

Application bar

Welcome screen

Workspace Switcher button

If a different option is displayed, click to switch to Classic view

HTML in Create New list

your window may look different

Figure 1–3

Q&A Why does the Application bar appear on two rows in Figure 1–3?

If your screen resolution is set to 1024 × 768, the Application bar appears on two rows. If you are using a higher screen resolution, the Application bar appears on one row at the top of the Dreamweaver window.

Used with permission from Microsoft Corporation

Navigational callouts in red show students where to click.

Explanatory callouts summarize what is happening on screen.

3

- Click HTML in the Create New list to create a blank HTML document.
- If the Dreamweaver window is not maximized, click the Maximize button next to the Close button on the Application bar to maximize the window.

Figure 1–4

- If the Design button is not selected, click the Design button on the Document toolbar to view the page design.
- If the Insert bar is not displayed, click Window on the Application bar and then click Insert to display the Insert panel (Figure 1–4).

Q&A

What if a message is displayed regarding default file types?
If a message is displayed, click the Close button.

Q&A

What is a maximized window?
A maximized window fills the entire screen. When you maximize a window, the Maximize button changes to a Restore Down button.

Other Ways

1. Double-click Dreamweaver icon on desktop, if one is present
2. Click Adobe Dreamweaver CS6 on Start menu
3. Click Adobe Dreamweaver icon on taskbar, if present

MAC For a detailed example of starting Dreamweaver using the Mac operating system, refer to the "To Start Adobe Dreamweaver CS6" steps in the For Mac Users appendix at the end of this book.

Dreamweaver Window

window open, take time to tour your new Web design weaver workspace lets you view documents and object lbar buttons for the most common operations so that you to your documents. In Dreamweaver, a document is an ed in a browser as a Web page. Figure 1–4 shows how the looks the first time you start Dreamweaver after installation ork efficiently, you should learn the basic terms and concepts rkspace, and understand how to choose options, use inspectors nces that best fit your work style.

Workspace

workspace is an integrated environment in which the anels are incorporated into one large application window. tomized workspace, Dreamweaver provides the Web site workspace layouts as shown in the Workspace Switcher

---

To View the Site in Live View

Live view provides a realistic rendering of what your page will look like in a browser, but lets you make any necessary changes without leaving Dreamweaver. The following steps display the page in Live view.

1

- Click the Live button on the Document toolbar to display a Live view of the site (Figure 2–67).

Experiment

- Click the Code, Split, and Design buttons to view the page in different views. When you are finished, click the Live button to return to Live view.

2

- Click the Live button again to return to Design view.

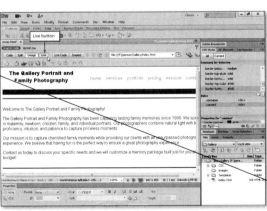

Figure 2–67

Other Ways

1. Press ALT+F11

To View the Site in the Browser

The following steps preview the Gallery Web site home page using Internet Explorer.

1. Click the 'Preview/Debug in browser' button on the Document toolbar.
2. Click Preview in IExplore in the Preview/Debug in browser list to display the Gallery Web site in the Internet Explorer browser.
3. Click the Internet Explorer Close button to close the browser.

To Quit Dreamweaver

The following steps quit Dreamweaver and return control to the operating system.

1. Click the Close button on the right side of the Application bar to close the window.
2. If Dreamweaver displays a dialog box asking you to save changes, click the No button.

Textbook Walk-Through

Chapter Summary

In this chapter, you have learned how to create a Web site template using the design building blocks of CSS style sheets. You defined a new Web site, created a Dreamweaver Web Template, and defined regions of the page using a style sheet. You defined each region with CSS rules that provided consistent site formatting. You also learned how to create an editable region within a template. You added a page to the site based on the template and placed text in the editable region. You displayed the completed Web page in Live view and in a browser. The following tasks are all the new Dreamweaver skills you learned in this chapter:

1. Create a New Site (DW 86)
2. Create Folders for the Image and CSS Files (DW 87)
3. Create a Blank HTML Template (DW 88)
4. Save the HTML Page as a Template (DW 90)
5. Add a Div Tag (DW 93)
6. Select CSS Rule Definitions (DW 97)
7. Add the Logo Div Tag and Define Its CSS Rules (DW 100)
8. Add the Navigation Div Tag and Define Its CSS Rules (DW 104)
9. Add the Image Div Tag and Define Its CSS Rules (DW 107)
10. Add the Content Div Tag and Define Its CSS Rules (DW 109)
11. Add the Footer Div Tag and Define Its CSS Rules (DW 111)
12. Create an Editable Region (DW 113)
13. Close the Template (DW 115)
14. Create a Page from a Template (DW 115)
15. View the Site in Live View (DW 118)

Apply Your Knowledge

Reinforce the skills and apply the concepts you learned in this chapter.

Creating a New Web Page Template

Instructions: First, create a new HTML5 Web page template and save it. Next, insert div tags in the template and create a new external style sheet file. Finally, add new CSS rules so that the completed template in Live view looks like Figure 2–68. The CSS rule definitions for the template are provided in Table 2–2.

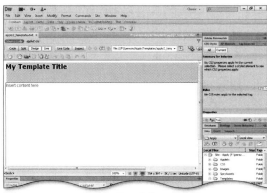

Figure 2–68

Continued >

Chapter Summary includes a concluding paragraph, followed by a listing of the tasks completed within a chapter together with the pages on which the step-by-step, screen-by-screen explanations appear.

Apply Your Knowledge usually requires students to open and manipulate a file from the Data Files that parallels the activities learned in the chapter.

Extend Your Knowledge

Extend the skills you learned in this chapter and experiment with new skills. You may need to use Help to complete the assignment.

Attaching an External Style Sheet to a Web Page

Note: To complete this assignment, you will be required to use the Data Files for Students. Visit www.cengage.com/ct/studentdownload for detailed instructions or contact your instructor for information about accessing the required files.

Instructions: A volunteer service organization wants to create a Web site using style sheets. You are creating a Web page for the organization that explains the difference between internal and external style sheets. Apply styles to a page by attaching an external style sheet to an existing Web page. The completed Web page is shown in Figure 2–69.

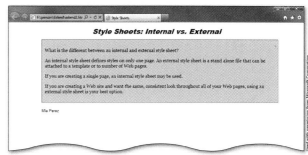

Figure 2–69

Perform the following tasks:

1. Use Windows Explorer to copy the CSS folder and the extend2.html file from the Chapter 02\ Extend folder into the *your last name and first initial*\Extend folder (the folder named perezm\ Extend, for example, which you created in Chapter 1).

2. Start Dreamweaver and open extend2.html.

3. On the CSS Styles panel, click the Attach Style Sheet button to display the Attach External Style Sheet dialog box. (*Hint*: Point to the buttons in the lower-right part of the CSS Styles panel to find the Attach Style Sheet button.) Click the Browse button in the Attach External Style Sheet dialog box, and then navigate to find and then select the extend2.css file located in the CSS folder in the Extend root folder. Add the CSS file as a link, and then accept your changes.

4. ~~Replace the text "Your name here"~~ .... with your name.
   .... ets.
   .... your document in your browser. Compare your document to
   .... changes and then save your changes.
   .... at specified by your instructor.

Make It Right

Analyze a Web page and correct all errors and/or improve the design.

Formatting and Checking the Spelling of a Web Page

Note: To complete this assignment, you will be required to use the Data Files for Students. Visit www.cengage.com/ct/studentdownload for detailed instructions, or contact your instructor for information about accessing the required files.

Instructions: Start Dreamweaver. You are working with a neighborhood association that wants to learn about Dreamweaver. An association member created a Web page and asks you to improve it. First, you will enhance the look of the Web page by applying styles, aligning text, and adding bullets. Next, you will make the Web page more useful by adding links to text and inserting a document title. Finally, you will check the spelling. The modified Web page is shown in Figure 1–65.

Figure 1–65

Perform the following tasks:

1. Use the Site Setup dialog box to define a new local site in the *your last name and first initial* folder (the F:\perezm folder, for example). Enter `Right` as the new site name and save it.

2. Click the 'Browse for folder' icon to navigate to your USB flash drive. Create a new subfolder within the *your last name and first initial* folder and name it `Right`. Open the folder and then select it as the local site folder. Save the Right site.

3. Using Windows Explorer, copy the right1.html file from the Chapter 01\Right folder into the

---

Extend Your Knowledge projects at the end of each chapter allow students to extend and expand on the skills learned within the chapter. Students use critical thinking to experiment with new skills to complete each project.

Make It Right projects call on students to analyze a file, discover errors in it, and fix them using the skills they learned in the chapter.

Textbook Walk-Through

16. Replace the text, Your name here., with your name.

17. Title the document `CSS Rules`.

18. Save your changes and then view the Web page in your browser. Compare your page to Figure 2–70. Make any necessary changes and then save your changes.

19. Submit the document in the format specified by your instructor.

In the Lab

Design and/or create a Web document using the guidelines, concepts, and skills presented in this chapter. Labs are listed in order of increasing difficulty.

Lab 1: Designing a New Template for the Healthy Lifestyle Web Site

Note: To complete this assignment, you will be required to use the Data Files for Students. Visit www.cengage.com/ct/studentdownload for detailed instructions or contact your instructor for information about accessing the required files.

Problem: In an effort to reduce health insurance costs, your company wants to provide resources for living a healthy lifestyle. You have been asked to create an internal Web site for your company with information about how to live a healthy lifestyle. This Web site will be used as a resource by employees at your company. You thoughtfully have planned the design of the Web site and now are ready to create a template for the site.

Define a new Web site and create a new HTML5 template. Use div tags and CSS rules in your template design. The template in Live view is shown in Figure 2–71, and the final Web page is shown in Figure 2–72. The CSS rule definitions for the template are provided in Table 2–4.

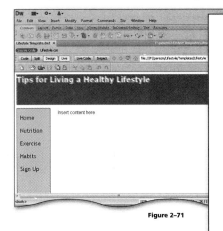

Figure 2–71

In the Lab assignments require students to utilize the chapter concepts and techniques to solve problems on a computer.

Table 1–3 Business Plan Tips	
Section	**Text**
Heading	Business Plan Tips
Paragraph 1	Understand your target audience. Identify demographics of your target audience.
Paragraph 2	Identify your industry and describe your products and services. Identify the benefits of your products and services.
Paragraph 3	Define your business model. How many employees do you need?
Paragraph 4	Describe how you will market your products and services. What is the best way to reach your target audience?
Paragraph 5	How much money do you need to get started? How will you obtain capital?

1. Start Dreamweaver. Use the Site Setup dialog box to define a new local site using `BusinessPlan` as the site name.

2. Click the 'Browse for folder' icon and then create a new subfolder in the *your last name and first initial* folder named `BusinessPlan`. Open and select the BusinessPlan folder, and then save the site.

3. On the Dreamweaver Welcome screen, click More in the Create New list. Create a blank HTML page using the 1 column liquid, centered layout, and HTML5 as the DocType. Save the new Web page using `index.html` as the file name.

4. Enter `Business Plan` as the Web page title.

5. Delete all the text on the page, replacing it with the text shown in Table 1–3. Press the ENTER key after typing each line or paragraph.

6. Press the ENTER key two times after typing the last sentence. Type your first and last names.

7. Select all of the text below the heading and use the Ordered List button in the Property inspector to create a numbered list for the text.

8. Save your changes and then view your document in your browser. Compare your document to Figure 1–68. Make any necessary changes and then save your work.

9. Submit the document in the format specified by your instructor.

Cases & Places exercises call on students to create open-ended projects that reflect academic, personal, and business settings.

Cases and Places

Apply your creative thinking and problem solving skills to design and implement a solution.

1: Protecting Yourself from Identity Theft

Personal

You recently read an article about the growing trend of identity theft. The article has prompted you to create an educational Web site on how people can protect themselves from identity theft. Define a new local site in the *your last name and first initial* folder and name it Protect Yourself from Identity Theft. Name the new subfolder Theft. Create and save a new HTML Web page. Name the file protect. Use your browser to conduct some research on ways to protect yourself from identity theft. Create a heading for your Web page. Apply the Heading 1 format to the title and center-align the title on the page. Below the heading, create an unordered list of 10 different ways to protect yourself from identity theft. Title the document ID Theft. Check the spelling in the document. Include your name at the bottom of the page. Submit the document in the format specified by your instructor.

Continued >

Web Site Development and Adobe Dreamweaver CS6

mozcann/iStockphoto.com

Adobe product screenshot(s) reprinted with permission from Adobe Systems Incorporated

Objectives

You will have mastered the material in this chapter when you can:

- Identify the 10 types of Web sites

- Define Web browsers and identify their main features

- Discuss how to connect to the Internet using an ISP or a data plan

- Discuss how to plan a Web site for your targeted audience

- Design a Web page with a focal point and appropriate colors, text, and images

- Design a site using accessibility guidelines

- Discuss the role of social networking within Web sites

- Identify the steps in hosting a Web site

- Understand how a Web project is managed

- Discuss how to test, publish, and maintain a Web site

- Recognize the HTML versions and the programming languages of the Web

- Discuss the advantages of using Web page authoring programs such as Dreamweaver

Web Site Development and Adobe Dreamweaver CS6

Web Site Planning and Design Basics

You make judgments about a Web site's visual appeal within seconds of opening a page on the site. This first impression influences subsequent judgments about the site's credibility and your willingness to make purchases. A Web site provides prospective customers with distinct impressions of a business through its online presence. A clean, professional design with intuitive navigation keeps you on a Web page instead of pressing the Back button in the browser. When you walk into a local physical store, you are more apt to make a purchase if the store is well organized, sells quality products, and establishes a positive shopping experience. Similarly, a good Web site is an online storefront providing a friendly environment with the same high level of attention to design and detail as a physical store.

To make that quality first impression, carefully plan your Web site before actually building it by defining its purpose and intended target audience. For example, the goal of a Web site for a local bed and breakfast inn is to encourage travelers to book a reservation at the inn. The site should provide the necessary information to book a stay: contact information, location, photos of the inn, rates, a neighborhood description, proximity to attractions, and a list of amenities, while also conveying a visual image representative of the inn's style. A well-planned site clearly communicates pertinent information to your target audience. Depending on the purpose, the site's goals may also include posting opinions, sharing personal interests, creating a social community, or generating revenue.

Types of Web Sites

Web sites can be classified into 10 categories according to their purpose: search engine/portal, social network, business, informational/educational, news, personal, blog, Web 2.0, e-commerce, and entertainment, as shown in Figure I–1. A search engine or a **portal** Web site (Figure I–1a) provides a variety of Internet services from a single, convenient location. Most portals offer free services such as search engines; e-mail; images; local, national, and worldwide news; sports; weather; reference tools; maps; stock quotes; newsgroups; and calendars. Popular search engines include Google, Bing, Yahoo!, and WolframAlpha. A **social network** (Figure I–1b) is an online community such as Facebook, LinkedIn, or Twitter that encourages members to share their interests, stories, photos, music, and videos with other members. A **business** Web site (Figure I–1c) displays content that markets or sells products or services. An **informational/educational** Web site (Figure I–1d) contains factual information, such as research and statistics. Web pages for governmental agencies, schools, and nonprofit organizations are examples of informational Web pages. A **news** Web site (Figure I–1e) features news articles and information about current events. By contrast, a **personal** Web site (Figure I–1f) is published by an individual or a family and is not associated with any organization. A **blog** (Figure I–1g), short for Weblog, uses a regularly updated journal format to reflect the interests, opinions, and personality of the author and sometimes of site visitors. Blogs can be created using free services such as Blogger.com. A **Web 2.0** site (Figure I–1h) shares user-created content with site visitors. Popular Web 2.0 sites include Erly, Flickr, Pinterest, and Animoto.

(a) search engine/portal

(b) social network

(c) business

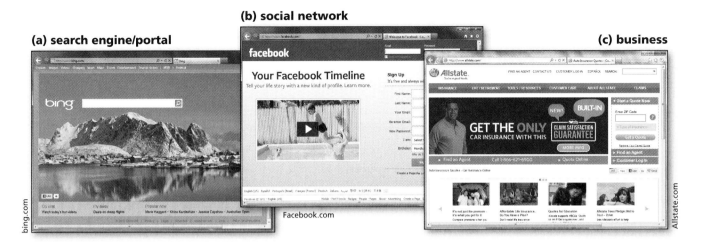

Facebook.com

bing.com

Allstate.com

(d) informational/educational

(e) news

(f) personal

www.nlm.nih.gov

bbc.com

www.stallardfamily.org

(g) blog

(h) Web 2.0

(i) e-commerce

blogger.com

erly.com

amazon.com

pogo.com

(j) entertainment

Figure I–1

E-commerce Web sites (Figure I–1i) such as Amazon, eBay, and Expedia specialize in generating revenue through sales and auctions of products and services. An **entertainment** Web site (Figure I–1j) offers an interactive and engaging environment and may contain music, video, sports, games, and other similar features. As you progress through this book, you will learn more about different types of Web pages.

Browser Considerations

The medium through which you access Web pages is quickly changing. Web pages are opened in a **browser**, a software application that displays content from the **World Wide Web (WWW)**. Browsers are the interface through which you view sites on your own Internet-connected device. You can access your favorite browser on a tablet (a slate device such as an iPad or a Windows or Android tablet), laptop or desktop computer (PC or Mac), or smartphone. The more popular Web browser programs, shown in Figure I–2, are Microsoft Internet Explorer, Mozilla Firefox, Apple Safari, Google Chrome, and Opera.

(a) Internet Explorer

(b) Mozilla Firefox

(c) Apple Safari

(d) Google Chrome

(e) Opera

Figure I–2

This book uses Internet Explorer (Figure I–3) as the primary browser. Using the browser's Tools menu, you can designate any page on the Web as the home page. Important features of Internet Explorer are summarized in Table I–1.

Entering a unique address or **Uniform Resource Locator (URL)** into a browser retrieves a Web page from a Web server, interprets the code within the page, and then displays the page within the browser. **Hypertext Transfer Protocol (HTTP)** is the protocol, or standard, that enables the host computer to transfer data to the client computer. The Web consists of a system of global network servers, also known as **Web servers**, that support specially formatted documents and provide a means for sharing these resources with many people at the same time. As you design a Web page, be aware that different browsers do not display pages identically. During the design and testing phases, ensure that your Web site looks and functions in the ways you intended. What you see when you surf the Web is your browser's interpretation of the underlying code, so testing your page in various browsers helps you create a consistent, positive experience.

Figure I–3

Table I–1 Internet Explorer Features	
Feature	**Description**
Address bar and search box	Displays the Web site address, or URL, of the Web page you are viewing or searches for the term entered on a search provider site such as Bing.
Web page tab	Provides the option to use tabs to switch from one site to another in a single browser window. The Web page tab also displays the title of the Web page.
Home button	Opens the default home page.
Favorites button	Lets you view favorites, feeds, and history.
Tools button	Provides access to print, zoom, and safety features and lets you view downloads and manage add-ons.
Document window	Displays the Web page content.

Although nearly all Web pages have unique characteristics, most share the same basic design elements. On most Web pages, you find headings or titles, text, pictures or images, background enhancements, and hyperlinks. A **hyperlink**, or link, can connect to another place in the same Web page or site — or a page on an entirely different Web site or document on a server in another city or country. A Web page can contain a variety of link types such as hyperlinks, image links, and social network links to Facebook and Twitter as shown in Figure I–4. Clicking a link displays the Web page associated with the link in a browser window. Linked pages can appear in the same browser window, in a new tab, or in a separate browser window, depending on the code associated with the link.

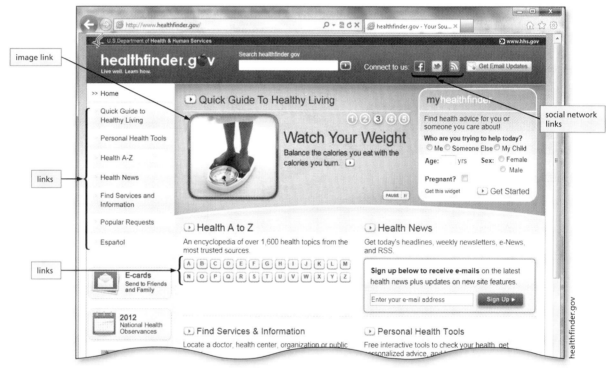

Figure I–4

Most Web pages are part of a **Web site**, which is a group of related Web pages that are linked together. Most Web sites contain a home page, which is generally the first Web page visitors see when they enter the site. A **home page** (also called an **index page**) typically provides information about the Web site's purpose and content. Most Web sites also contain additional content and pages. An individual, a company, or an organization owns and manages each Web site.

To access the Web, you must connect through a regional or national **Internet service provider (ISP)** using a wireless or wired connection, or a mobile data plan for your phone, tablet, or laptop. Figure I–5 illustrates ways to access the Internet using a service provider or data plan. An ISP provides temporary connections to individuals or businesses through its permanent Internet connection.

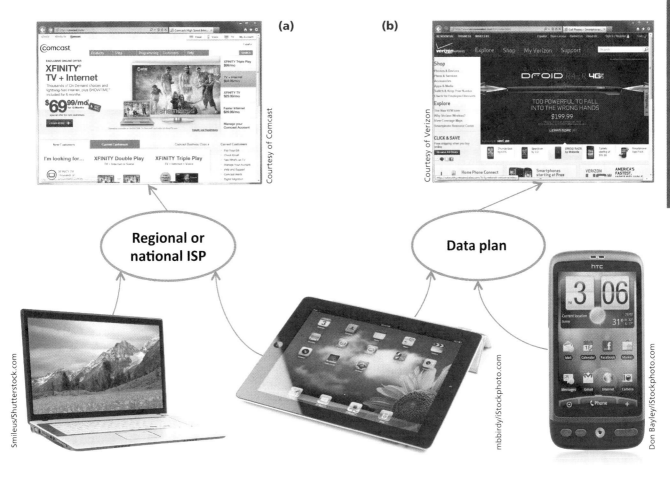

(a)

(b)

High-speed connectivity through DSL, cable modem, or satellite (wireless or wired)

Wireless connectivity through ISP or data plan

Smartphone data plan

Figure I–5

Web Site Navigation

If visitors to your Web site do not find what they are looking for quickly and easily, they will simply go elsewhere. Predicting how a visitor will access a Web site or at what point the visitor will enter the Web site structure is not possible. Visitors can arrive at any page within your site by a variety of ways, including clicking a hyperlink from another page or Facebook posting, using a search engine, typing a Web address directly, and so on. Therefore, every page of your Web site must provide clear answers to the three questions your visitors will ask: Where am I? Where do I go from here? How do I get to the home page? Once a visitor arrives at a Web site, navigation, the pathway through your site, must be clear and intuitive. Good site navigation uses a consistent and uniform layout for each page. For example, if one page provides a link at the bottom of the window to return to the home page, then all pages should provide navigation to the home page in the same location. Using intuitive navigation greatly enhances the usability of your Web site, which leads to higher user satisfaction and return rates. Every page within your site must provide the visitor with a sense of location, or context within the site.

BTW

Web Page Navigation Charts
To create a visual representation of your site navigation for planning purposes, consider using an organizational chart included in Microsoft Word or PowerPoint. For larger, more complex Web sites, you can chart and organize your content using Microsoft Visio or SmartDraw.

BTW

BTWs
For a complete list of the BTWs found in the margins of this book, visit the Dreamweaver BTW chapter resource on the student companion site located at www. cengagebrain.com.

Mobile Site Access

Using a mobile device with a monthly **data plan** subscription, you can connect to a wireless data network to access the Web, check e-mail, stream music, and download applications (apps). Terms of the data plan are set by the service provider, such as AT&T and Verizon, and typically limit the user to a certain amount of bandwidth per month based on a tiered rate plan (Figure I–5).

An **AirCard**, a type of wireless broadband modem, enables users to connect their laptop to the Internet with their data plan when not connected to a local Wi-Fi (Wireless Fidelity) network. Mobile devices come equipped with default browsers such as Safari for the iOS platform, a built-in Android browser for Android tablets and phones, and Internet Explorer for the Windows tablet or phone. In addition, a wide variety of free browsers are available for download.

Savvy businesses are investing in a mobile presence by creating mobile optimized Web pages based on each mobile device's capabilities and screen size. Using a mobile device with a data plan, you can browse the Internet while commuting on trains, racing through airports, or standing in line. Their small size gives users instant access, from anywhere at any time, as an alternative to sitting at a desk, tethered to a traditional desktop computer. Mobile Web browsers can display the content of most Web sites, but pages that are coded specifically to display on a mobile device provide a more personal, mobile-friendly experience, as shown in Figure I–6. By customizing a Web site for a phone or tablet, navigation is optimized for the mobile world, including an automatic touch interaction to directly call a company or instantly display a map and directions to your meeting.

Figure I–6

Planning a Web Site

The first step in building your first Web site is to design a detailed plan to ensure success. Defining your site's purpose, its target audience, the intended Web platform, and the proposed design is a crucial aspect of Web development. You need to make sure that visitors are immediately drawn into your site through captivating content, a compelling call to action, ease of use, and a sense of community.

Planning Basics

Those who rush into publishing their Web site without proper planning usually design sites that are unorganized and difficult to navigate. Visitors to this type of Web site often lose interest quickly and do not return. As you begin planning your Web site, consider the following guidelines to ensure that you set and attain realistic goals.

Purpose and Goal   Determine the purpose and goal of your Web site. Create a focus by developing a mission statement, which conveys the intention of the Web site. Consider the 10 basic types of Web sites mentioned previously. Will your Web site consist of one basic type or a combination of two or more types? For example, a business Web site's purpose may be to market new products and services or provide customer support. By focusing on the goals that you hope to achieve, you can effectively plan a site that fits your organization's business model.

Target Audience   Knowing your target audience is essential to good design because the needs of your intended visitors help shape the content of a site as well as its look and feel. Figure I–7 shows the Web site for a popular outdoor company named REI that is customized to its target audience. The intended audience at the REI site is anyone who enjoys nature and outdoor adventures. Notice the image on the REI home page during the month of January is customized to the time of the year. Visitors in the winter are most likely focused on snow activities including sledding and skiing, and interested in clothing that provides warmth during the winter season. Easy navigation using a wide variety of links is provided for the major categories of products sold at REI. A search tool at the top of the site enables quick and easy navigation to the desired products. Knowing the information that your target audience is searching for simplifies their purchasing experience at this site. Creating a welcoming, easy-to-navigate experience is vital to any site, so consider the characteristics of your target audience such as interest, gender, education, age range, income, profession/job field, and computer proficiency.

Multiplatform Display   Where will your target audience view your Web site? Will it be displayed on a Mac laptop using Safari, a Windows tablet using Internet Explorer, an Android phone on a built-in browser, or a desktop PC using Mozilla Firefox? Planning for Web presentation involves verifying that your site will function in a variety

Figure I–7

of browsers based on the intended layout and in different screen resolutions. **Screen resolution** refers to the number of pixels in a display, such as 1280 × 800. The layout of a Web site can change depending on the user's screen resolution. Creating a multi-platform Web page, also called a **cross-platform site,** that provides a similar display experience across various screen sizes, resolutions, browsers, and devices is supported by a new code environment called HTML5 and CSS3, which are explained later in the chapter. Web developers must plan their sites to deploy on any device without worrying that the device itself will not support a particular graphic or effect used in the page.

Design

A Web site consists of more than information and links. A well-designed Web site creates a positive interaction with the user by focusing on a visual, aesthetically pleasing way to present information. Web users prefer a simple, clean, and functional design. Avoid a cluttered design that uses multiple fonts, inconsistent icons, flashing ads, and ubiquitous links. Consider the following Web design principles when creating a memorable site.

Focal Point A mixture of elements including text, colors, and images all compete for your attention when you open a Web page. A design element called a **focal point** provides a dominating element that captures your attention. In Figure I–8, the United States Department of Agriculture nutrition site uses a dominant image as a focal point to immediately draw your attention to and illustrate the purpose of this site — choosing a plate full of healthy foods. A focal point in your site may be an element such as a prominent header, company logo, or central product image. Notice the amount of white space in the site in Figure I–8. **White space** is the empty space around the focal point and other design elements that enables important aspects of the page to stand out. White space does not have to be white; it can be the background color of the page. Appropriate use of white space can lead the user's eye to important content. When a Web page lacks a clear focal point and white space, competing images, flashing text, and abundant text can confuse the user.

Color as a Design Tool Color can effectively convey information that adds interest and vitality to your site. Color should be aesthetically pleasing to your target audience and suit the content of the page. For example, in Figure I–7 on the previous page the gray and white colors fit the page's winter theme. When designing Web pages for a

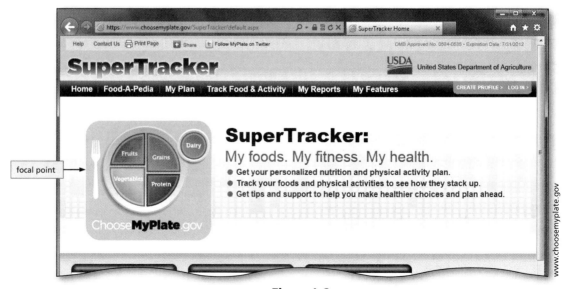

Figure I–8

business environment, consider the company's branding colors. In Figure I–9, the site for the Cengage Learning publishing company incorporates the colors in the logo shown in the upper-left corner of the page. The blue color scheme of this page reinforces the site's overall message and identity.

company logo

© Cengage Learning 2013

Figure I–9

The most common misuse of color is a background color that detracts from the readability of the text. How much time would you spend on a site featuring a bright yellow background and white text? A bright yellow background strains the eyes, making it impossible to read the text. Keep in mind that people use color to categorize objects in their everyday lives. Colors in the cool color group such as green and blue tend to have a calming effect. Similarly, many companies use color in their Web design to help users identify and categorize their brand with just one glance. For example, red is associated with power and energy, making it a good color for a sports car. Table I–2 describes the common colors and their related meanings.

Table I–2 Common Color Meanings	
Color	**Description**
Blue	Trust, security, conservative, technology (The most common color used on the Web)
Green	Nature, money, earth, health, good luck
White	Purity, cleanliness, innocence, precision
Red	Power, danger, passion, love, energy
Black	Sophistication, power, death, fear
Gray	Intellect, elegance, modesty
Orange	Energy, balance, warmth, brightness
Yellow	Cheer, optimism, joy, honesty
Brown	Reliability, earth, comfort
Purple	Mystery, spirituality, arrogance, royalty

Text Design   Like other Web elements, good **typography**, or the appearance and arrangement of the characters that make up your text, is vital to the success of your Web page. Adding text is not just about placing letters on a site, but rather using text as a page design element to improve readability. The introductory text of a Web page should instantly convey the purpose of the site. A Web site visitor will more likely read a short and well-structured introduction while skipping longer paragraphs of text. In the Web site for the Fish restaurant shown in Figure I–10, the shading on the left provides a focal point for the text. The introductory text is short, useful, and interesting. A call to action initiated by the text requests that you view the menus. The site clearly defines what its targeted audience wants to view first — the menus.

On an opening page, shorter text can captivate your audience. But when longer paragraphs are necessary to provide more detailed information, the text should be easy to read. A font face or font family is the typeface that will be applied to the text by a Web browser. Select fonts with good readability — especially when developing pages for smaller, mobile devices.

Image Design   Images provide an instant focal point to any site and give viewers an immediate cue about page content. In the Fish restaurant example in Figure I–10, the image of Asian food on the home page conveys the flavor of the cuisine and the ambiance of this restaurant. Even a cursory glance at this opening communicates a positive impression for that type of cuisine and atmosphere.

An image can stimulate visual interest, market a product, display a company logo, illustrate a process, or provide graphical information. Visitors' eyes are naturally drawn to photos of people. Creating a balance between the number of images and text elements is vital to achieving an uncluttered appearance. An image leaves a stronger impression because our brains are drawn to familiar, real-life objects rather than words alone.

Figure I–10

Accessibility Guidelines

In your planning phase, you must ensure your site is accessible to all users. Imagine if your school's Web site was not accessible to all students regardless of physical limitations. For schools and government entities it is not only unwise, but also likely in violation of federal law. Section 508, which was added as an amendment in 1998 to the Rehabilitation Act, requires that electronic and information technology that is developed by or purchased by federal agencies be accessible to people with disabilities (www.section508.gov). Businesses seek the largest possible audience and recognize the positive return on investment (ROI) for the extra costs of building a site accessible to all customers. Table I–3 categorizes the major types of disabilities. Each requires certain kinds of adaptations in the design of the Web content, but ultimately everyone benefits from helpful illustrations and clear navigation.

Table I–3 Disability Types and Design Strategies

Disability	Description	Design Strategies
Cognitive	Autism, learning disabilities, distractibility, inability to focus on large amounts of information	Keep text short. Break up text with headings. Use meaningful graphics.
Hearing	Deafness	Provide a transcription or summary of audio elements.
Motor	Inability to move mouse or use touch screen, slower response time	Minimize scrolling.
Visual	Blindness, poor vision, color blindness	Describe images in text (which may be read aloud by a screen reader). Use a large, easy-to-read font.

As a Web designer, removing barriers so people with disabilities have equal access to the Web is a moral and often a legal obligation under the Americans with Disabilities Act (ADA). For example, a visually impaired person often uses a screen reader, which is a software application that can vocalize screen content. A Web site image should contain information called **alternative text** describing the picture for the screen. The **World Wide Web Consortium (W3C)**, an international standards organization for the World Wide Web, provides Web standards, language specifications, and accessibility recommendations at www.w3.org to promote the growth of the Web. Forward-thinking Web developers plan their sites with accessibility in mind because it is the right thing to do. Accessibility needs to be an integral part of Web design planning rather than an afterthought. Throughout this text, accessibility is illustrated as each aspect of Web page development is covered.

Role of Social Networking

Planning a site extends beyond selecting content and following accessibility rules. Modern Web sites enable users to interact with one another and share information. These sites create a sense of community with two-way conversations between site visitors and business responses, including product reviews, targeted e-mail, a Facebook and Twitter marketing presence, YouTube links to new product video demonstrations, and blog feedback. These online interactions allow businesses to give their customers a voice and help them improve their products, services, and customer satisfaction. A Web site can provide customer reviews and ratings to help other customers. In addition, many sites incorporate Facebook pages and Twitter follower links by providing logos and inviting visitors to connect with them.

As shown in Figure I–11 on the next page, the publishing company Cengage Learning provides a Facebook and Twitter logo on its opening page to allow a user to "like" or "follow" its product line. When a visitor "likes" a business, he or she becomes a fan on Facebook, promoting that business to all of the visitor's personal contacts. Cengage Learning leverages Facebook and Twitter to provide instructors and students with the latest technology innovations, educational research, and e-book ventures. This social networking resource offers the customer a location to share learning success stories, ask questions, and post new ways to integrate classroom solutions. This presence creates an interwoven community of friends and colleagues who can quickly learn about the business from others who have liked or followed it.

Facebook Presence

Twitter Presence

©Cengage Learning 2013

Courtesy of Facebook

Courtesy of Twitter

Facebook link

Facebook and Twitter links

Figure I–11

Web Site Hosting

Creating a good Web site begins with planning the content and structure. But selecting a Web server host, which makes a Web site visible to the world through a unique URL, is another important consideration. Each Web site requires a Web server running continuously to deliver your Web pages to visitors quickly.

Obtain a Domain Name To allow visitors to access your Web site, you must obtain a domain name. Visitors access Web sites via an IP address or a domain name. An **IP address (Internet Protocol address)** is a number that uniquely identifies each computer or device connected to the Internet. A **domain name** is the text version of an IP address. The **Domain Name System (DNS)** is an Internet service that translates domain names into their corresponding IP addresses. The **Accredited Registrar Directory** provides a listing of domain name registrars accredited by the **Internet Corporation for Assigned Names and Numbers (ICANN)**. Your most difficult task likely will be to find a name that is not registered. Expect to pay approximately $8 to $50 per year for a domain name.

For example, a small hair salon named Shear Styles contacts a domain registrar, Network Solutions — which is shown in Figure I–12 — to create a site for its Web presence. The salon would like the URL www.shearstyles.com, but must first verify if that URL is available. As a Web domain registrar, Network Solutions determines if the site is available and the yearly cost of that domain name. The domain name ends with an extension such as .com (commercial entity), .net (network), .gov (government agency), .org (organization), or .edu (education) to represent the type of site. Domain names should be easy to recall or should reflect the organization's name so that people can easily find the site.

Obtain Server Space Locate an ISP that will host your Web site. Recall that an ISP is a business that has a permanent Internet connection. ISPs offer connections to individuals and companies for free or for a fee. Typically, an ISP for your home Internet connection provides a small amount of server space for free to host a personal site.

enter URL to determine if domain name is available

possible domain extensions

www.networksolutions.com

Figure I–12

If you select an ISP that provides free server space, your visitors will typically be subjected to advertisements and pop-up windows. Other options to explore for free or inexpensive server space include online communities, such as Bravenet (http://bravenet .com), Biz.ly (www.biz.ly), and webs.com (www.webs.com); and your educational institution's Web server. If the purpose of your Web site is to sell a product or service or to promote a professional organization, you should consider a fee-based ISP. Shop around to determine the best fit for your site. When selecting an ISP, consider the following questions and how they apply to your particular situation and Web site:

1. What is the monthly fee? Are setup fees charged?

2. How much server space is provided for the monthly fee? Is there unlimited storage? Can you purchase additional space? If so, how much does it cost?

3. How much bandwidth is available to download multimedia files?

4. Is your site hosted on a single dedicated server or cloud-hosted servers? (Cloud hosting forms a network of connected servers that are located in different locations across the world, providing multiple backup opportunities.)

5. What is the average server uptime on a monthly basis? What is the average server downtime?

6. What are the server specifications? Can the server handle heavy usage? Does it have battery backup power?

7. Are **server logs**, which keep track of the number of accesses, available?

8. What technical support does the ISP provide, and when is it available?

9. Does the server on which the Web site will reside have CGI and PHP scripting capabilities, and provide support for Active Server Pages (ASP), SQL Database, and File Transfer Protocol (FTP)?

10. Does the server on which the Web site will reside support e-commerce with **Secure Sockets Layer (SSL)** for encrypting confidential data such as credit card numbers? Are additional fees required for these capabilities?

Publish the Web Site You must publish, or upload, a finished Web site from your computer to a host server where your site will then be accessible to anyone on the Internet. Publishing, or uploading, is the process of transmitting all the files that constitute your Web site from your computer to the top directory, also called the root folder on the selected server or host computer. The files that make up your Web site can include Web pages, PDF documents, images, audio, video, animation, and others. You can use a variety of tools and methods to manage the upload task. Some of the more popular of these are **FTP (File Transfer Protocol)** and Web authoring programs such as Dreamweaver. These tools allow you to link to a remote server, enter a password, and then upload your files. Dreamweaver contains a built-in function similar to independent FTP programs.

Project Management

After completing the planning phase, the next step is to create, manage, and update the site. In most businesses, a **Webmaster** or Web project manager is in charge of delivering the site. Large commercial organizations employ a Web development team, shown in Figure I–13, which includes project managers, designers, programmers, a marketing group, a legal team to deal with copyright materials and permissions, editors, and strategic managers. The Webmaster works within the budget for site development, creates a schedule from start to end, and defines the quality of the work completed. The members of a Web team continue their roles throughout the development, testing, and maintenance of the site to keep the information current.

Goodluz/iStockphoto.com

Figure I–13

Test Web Pages

To test a Web page, you can use the Adobe BrowserLab at https:// browserlab.adobe.com to see how your pages are displayed in a variety of browsers and versions of browsers.

The best way to test a site is to use actual users. Each user independently reviews each page and records his or her feedback. In addition to the basic testing, safety testing with e-commerce sites is especially imperative where credit card numbers and personal information are part of the purchasing process. These security tests should report any possible vulnerabilities and recommendations to address them.

Testing the Site

A Web site's usability or ease of use is an integral part of the site's success. Due to the complexity of most Web sites with multimedia, interaction, and navigation, each page in a site must be tested on various browsers, operating systems, and platforms. Testing can take place at almost any stage of site development, but earlier is better. Among other things, the testing process verifies that the site is free of spelling and grammatical errors, all the links work correctly, and graphics appear as designed.

Maintaining the Site

An outdated Web site gives the impression that the site has been abandoned, making the visitor lose trust in the information on the site. In the long run, performing ongoing maintenance is usually less expensive than overhauling a site that is significantly out of date. Content on any site should be routinely reviewed for accuracy, currency, and alignment with the site's purpose. A Web site is a living entity, requiring the addition of updated images, topics, and videos at regular intervals. The site should be reviewed periodically for obsolete information and broken links. Use internal statistics from your ISP reports to track each visitor's behavior to determine popular pages within your site, how visitors found your site, their countries of origin, the browsers they are using, and the number of people visiting your site. Learning to use and apply the information derived from the server log will help you make your Web site successful.

Web Programming Languages

When you access a Web page with your browser, the Web server reads through a coded page line by line and executes the code to display the finished page. **Hypertext Markup Language (HTML)** is the language used to develop basic Web pages. To view the HTML source code for a Web page using Internet Explorer, right-click the Web page, and then click View source. Web site programming allows you to turn a simple, static HTML page into a dynamic Web page with the use of various programming languages added within the HTML code. Web programming languages, such as PHP and ASP, provide user interaction to fill in an online application, post your Facebook status, submit a dinner review at Yelp, purchase your textbooks online, or request an online Groupon coupon, for example.

HTML and XHTML

HTML, the first language of the Web developed in 1990 to organize information in a browser, continues to evolve. HTML code, or markup, consists of **tags** shown in Figure I–14, which are a set of symbols defined in HTML. Tags start with a left angle bracket (<) followed by an HTML keyword, and end with a right angle bracket (>). For example, the tags <p>, <h1>, and <table> represent paragraph, heading size 1, and table, respectively. Many tags have a start tag, or element, such as <h1> and an end tag, or element, such as </h1> to indicate where the formatting begins and ends.

HTML is useful for creating headings, paragraphs, lists, and so on, but is limited to these general types of formatting. **XHTML (Extensible Hypertext Markup Language)** is a rewritten version of HTML using XML (Extensible Markup Language) developed in 2000 and described later in the chapter. The difference between HTML and XHTML is minor, but the primary benefit is that XHTML is more widely accepted on mobile device platforms. Some XHTML elements such as a line break <br /> do not have an end element. Instead, the right angle bracket is

Load Testing

A Web site testing BTW process called load testing simulates the operation of hundreds or thousands of simultaneous visitors to determine how well a site performs under a heavy load.

photovibes/Shutterstock.com

Figure I–14

preceded by a space and forward slash. These are known as one-sided elements, or self-closing elements. XHMTL elements use lowercase tags in the Web page code source.

HTML5

The newest HTML standard is **HTML5**, representing the fifth major revision of the core Web language. Web developers use HTML5 to display their sites on a variety of smartphones, tablets, and computers without tailoring the code for specific hardware. HTML5 supports in-browser multimedia by adding simple tags to play audio and video elements. HTML5 enables developers to write less JavaScript code (a scripting Web language), which makes the site easier to code and update.

DHTML

HTML5 is combined with CSS3 (Cascading Style Sheets) and JavaScript to create more dynamic Web content. **CSS3** describes a template for the style and formatting of the page. The resulting combined code is called **DHTML**, which stands for Dynamic HTML, and creates movement or interactivity. DHTML is supported by modern browsers and creates a faster Web experience.

XML

The goal of HTML code is to display a Web page in a particular layout. But a different type of code, called **XML (Extensible Markup Language)**, is designed to transport data, not to display data. If a site visitor at a hotel chain Web site enters his or her home address to get directions from home to the hotel, the home address entered is supported and sent to the Web service using XML. In other words, XML code transports and stores data. Newer versions of Microsoft Office save their documents in XML, creating a direct way to carry these documents throughout the Web.

PHP

Embedded within HTML code, **PHP (Hypertext Preprocessor)** code is an open-source Web server-side scripting language that produces dynamic Web pages. A server-side scripting language means that when a visitor opens a Web page, the Web server processes the PHP code and then sends the results to the visitor's browser to be displayed within the HTML code. When you create an account to log on to a Web site, fill out a financial aid form online, search an online shopping site's database to determine if it has your size, or open a PDF form for your federal income taxes, the Web server processes the PHP code to administer each of these actions.

ASP

ASP (Active Server Pages) is very similar to PHP because both languages create dynamic Web pages. ASP leverages languages such as Visual Basic Script, JScript, and C#, among others, to create ASP code. A programmer can write code in Visual Basic using Visual Studio, and ASP code is generated automatically. ASP is a proprietary system owned by Microsoft.

jQuery

jQuery provides a full JavaScript Library that simplifies programming within an HTML document. If you have visited a site with a drag-and-drop interface, a drop-down calendar (Figure I–15), a horizontal slider, or text that could be expanded like an accordion, jQuery was most likely the code powering the effect. If your Web page requires a date picker calendar tool, jQuery provides code to paste directly in your HTML to create this widget. jQuery is a cross-browser, free open source library that contains thousands of animations, widgets, and plug-ins that cover a wide range of functionality.

Figure I–15

Web Page Authoring Programs

Web developers have several options for creating Web pages: a text editor, an HTML5 editor, software applications, or a **What You See Is What You Get (WYSIWYG)** text editor. A WYSIWYG text editor allows a user to view a document as it will appear in the final product and to edit the text, images, or other elements directly within that view. Microsoft Notepad (Figure I–16 on the next page), WordPad, and Notepad++ are each examples of a **text editor**. These simple, easy-to-use programs allow you to enter, edit, save, and print text. Software applications such as Microsoft Word and Microsoft Excel also provide a Save as Web Page command. This feature converts the application document into a file that Web browsers can display.

Web page authoring software allows you to see your design as you create it without writing code. Examples of a WYSIWYG text editor are programs such as Adobe Dreamweaver, Microsoft Expression Web, WordPress, Joomla, and Drupal. Technically, you do not need to know HTML to create Web pages in Dreamweaver; however, an understanding of HTML will help you if you need to alter Dreamweaver-generated code. If you know HTML, then you can make changes to the code in the code window (Figure I–17 on the next page) and Dreamweaver will accept the changes.

About Adobe Dreamweaver CS6

The standard in professional Web authoring, **Adobe Dreamweaver CS6** (Figure I–18 on the next page) is part of the Adobe Creative Suite, which includes Adobe Flash, Fireworks, Photoshop, Illustrator, InDesign, Acrobat, and other programs depending on the particular suite. Dreamweaver provides features that access these separate products. Dreamweaver makes it easy to get started and provides you with helpful tools to enhance your Web design and development experience. Working in a single environment, you create, build, and manage Web sites and Internet applications. Many of the new features in Dreamweaver CS6 focus on building mobile HTML5 applications for various platforms.

Dreamweaver contains coding tools and features that include references for HTML5, XML, CSS, and JavaScript, as well as code editors that allow you to edit the code directly using programming languages such as PHP, ASP, and jQuery. Using **Adobe Roundtrip technology**, Dreamweaver can import Microsoft Office or other software Web pages and delete the unused code. Downloadable extensions from the Adobe Web site make it easy to add functionality to any Web site. Examples of these extensions include shopping carts and online payment features.

HTML5 code in Notepad

```
<!DOCTYPE HTML>
<html>
<body>
<h1> Welcome to an HTML5 Page </h1>
<h2> This page displays a video </h2>|

<video width="320" height="240" controls="controls">
  <source src="movie.mp4" type="video/mp4" />
  <source src="movie.ogg" type="video/ogg" />
  <source src="movie.webm" type="video/webm" />
Your browser does not support the video tag.
</video>

</body>
</html>
```

Used with permission from Microsoft Corporation

Figure I–16

Dreamweaver HTML5 Code window

Figure I–17

Adobe Dreamweaver CS6

Figure I–18

Instead of writing individual files for every page, you can use a database to store content and then retrieve the content dynamically in response to a user's request. With this feature, you can update the information one time in one place, instead of manually editing many pages. Another key feature is **Cascading Style Sheets (CSS) styles**.

CSS3 (version 3) styles are collections of formatting definitions that affect the appearance of Web page elements. You can use CSS3 styles to format text, images, headings, tables, and so forth. Using CSS, you can make a formatting change in one place and update all the Web pages that contain that same formatting. CSS layouts are used to create a similar look and feel across Web sites. They also are used to reduce the amount of work and HTML code generated by consolidating display properties into a single file.

Dreamweaver provides the tools that help you author accessible content that complies with government guidelines. Accessibility is discussed in more detail as you progress through the book.

Dreamweaver allows you to easily publish Web sites to a local area network, which connects computers in a limited geographical area, or to the Web, so that anyone with Internet access can see them. The concepts and techniques presented in this book provide the tools you need to plan, develop, and publish professional Web sites.

Chapter Summary

In this chapter, you have learned how to plan, design, develop, test, publish, and maintain a Web site using accessibility guidelines. An overview of the basic types of Web pages also was presented. The Introduction furnished information on developing a Web site, including planning basics. The process of designing a Web site with a focal point and appropriate color, text, and images was discussed. Information about testing, publishing, and maintaining a Web site also was presented, including an overview of obtaining a domain name, acquiring server space, and uploading a Web site. Methods and tools used to create Web pages were introduced. A short overview of the languages used to create Web sites was explained. Finally, the advantages of using Dreamweaver in Web development were discussed. These advantages include a WYSIWYG text editor and a visual, customizable development environment using Cascading Style Sheets.

1. Web Site Planning and Design Basics (DW 2)
2. Planning a Web Site (DW 9)
3. Web Site Hosting (DW 15)
4. Project Management (DW 16)
5. Web Programming Languages (DW 18)
6. Web Page Authoring Programs (DW 20)

Apply Your Knowledge

Reinforce the skills and apply the concepts you learned in this chapter.

Planning a Web Site
Instructions: As discussed in this Introduction, you must plan and design your Web site thoughtfully before you create it. Use the information in Table I–4 to plan and design a Web site.

Table I–4 Planning a Web Site	
Web Site Element	**Questions to Ask**
Purpose and goal	What are the purpose and goal of your Web site? What type of Web site will you develop?
Target audience	Who is your target audience? For what purpose is the audience using your Web site?
Multiplatform display	Will your target audience view your Web site on a computer, tablet, or mobile device? Will you design a cross-platform site?
Focal point	What will you use to capture the attention of your target audience?
Color	What color(s) will you use to enhance your site? Refer to Table I–2 on page DW 12.
Text design	How will text be arranged on your pages?
Image design	What types of images will you use within your site?
Accessibility	How will your site accommodate people with disabilities?
Social networking	Will you incorporate social networking within your planning model?

Perform the following tasks:

1. Select a type of Web site to create, using Figure I–1 as a guide. Open your word processing program and answer the questions contained in Table I–4. Restate each question and use complete sentences for your answers.

2. Save the document with the file name Apply I-1_*lastname_firstname*. Submit the document in the format specified by your instructor.

Extend Your Knowledge

Extend the skills you learned in this chapter and experiment with new skills. You may need to use Help to complete the assignment.

Exploring Programming Languages

Instructions: As you learned in this Introduction, you can use many types of Web programming languages in the development of a Web site. This introduction discussed the following types of programming languages: HTML, XHTML, HTML5, DHTML, XML, PHP, ASP, and jQuery. Use the Web to identify other types of Web programming languages.

Perform the following tasks:

1. Open your Web browser and use a search engine to research three types of Web programming languages not discussed in this chapter.

2. Use your word processor to list the Web programming languages that you found and describe how each language can be used in the development of a Web site.

3. Save the document with the file name Extend I-1_*lastname_firstname*. Submit the document in the format specified by your instructor.

Make It Right

Analyze a Web site and suggest how to improve its design.

Improving Design

Instructions: Start your Web browser. Select and analyze two Web sites. Determine the focal point and typography for each Web site.

Use your word processing program to describe the focal point for each Web site and include a screen shot within your document of each site's home page. Describe the text and color design used in each Web site. Review Table I–3 on page DW 13. Discuss how the text and color design could be improved. Save your document as MIR_*lastname_firstname*. Submit the document in the format specified by your instructor.

In the Lab

Design and/or create a document using the guidelines, concepts, and skills presented in this chapter. Labs are listed in order of increasing difficulty.

Lab 1: Web Site Maintenance

Problem: Once a Web site is up and running, the Webmaster is responsible for maintaining the site. Assume you are the Webmaster for a state government agency. Develop a maintenance plan for the Web site.

Perform the following tasks:

1. Open your Web browser. Research tips on Web site maintenance.

2. Use your word processing program to describe your Web site maintenance plan.

3. Save the document with the file name Lab I-1_*lastname_firstname*. Submit the document in the format specified by your instructor.

In the Lab

Lab 2: Web Site Hosting

Problem: Determining the best ISP to host your site is vital to maintain a continuous site that attracts visitors. Assume you have developed a new Web site for a small online retail store. Conduct research on the Web to find a suitable ISP.

Perform the following tasks:

1. Start your browser. Use a search engine to find ISPs and research their services and costs.

2. Select two desirable ISPs and use your word processing program to compare their services and costs. Write a short summary explaining which ISP you would select and why.

3. Save the document with the file name Lab I-2_*lastname_firstname*. Submit the document in the format specified by your instructor.

In the Lab

Lab 3: Comparing Web Browsers

Problem: People use many different Web browsers to access the Internet. Typically, an individual will select a primary Web browser as his or her preferred browser based on its features and tools.

Perform the following tasks:

1. Open or download a Web browser that you do not normally use or have never used and examine its features and tools. Web browsers include:

   a. Microsoft Internet Explorer (http://windows.microsoft.com/en-US/internet-explorer/downloads/ie)

   b. Mozilla Firefox (www.mozilla.org/en-US/firefox/new/)

   c. Google Chrome (www.google.com/chrome/)

   d. Apple Safari (www.apple.com/safari/download/)

   e. Opera (www.opera.com/download/)

2. Open your preferred Web browser and visit a page you have visited previously. Examine the browser's features and tools and explore its ease of use.

3. Use your word processing program to identify your primary Web browser and the one you downloaded, and then compare and contrast the browsers' features and tools.

4. Describe the ease of use of the browser that you do not normally use, and explain why you would or would not use the browser as a secondary Web browser.

5. Save your document as Lab I-3_*lastname_firstname*. Submit the document in the format specified by your instructor.

Cases and Places

Apply your creative thinking and problem solving skills to design and implement a solution.

1: Determining Web Site Type

Personal

You are an expert on several video games and video game consoles. You have decided to design a Web site to share tips about certain video games and game consoles. Use a search engine to research the type of Web site you should design. Use your word processing program to write a one-page summary about the type of Web site you will design and why. Save the document as Case I-1_*lastname_firstname*. Use proper spelling and grammar. Submit the document in the format specified by your instructor.

2: Designing for Accessibility

Academic

You are the Webmaster for a small community college. As such, you must make the college's Web site accessible for students with disabilities. Start your browser and research various ways you can accomplish this task. Write a two-page summary of at least three ways you can make the Web site accessible for students with disabilities. Use proper spelling and grammar. Save the document as Case I-2_*lastname_firstname*. Submit the document in the format specified by your instructor.

3: Social Networking for a Business

Professional

Assume you are the owner of a local pest control business. You currently have a Web site and are now considering joining the world of social networking, but are unsure about the best social networking option. Start your browser and use a search engine to do some research. Use your word processing program to write a two-page summary about the information you found. Include a list of pros and cons of a business using social media. Use proper spelling and grammar. Save the document as Case I-3_*lastname_firstname*. Submit the document in the format specified by your instructor.

1 Creating a New Web Site with Dreamweaver

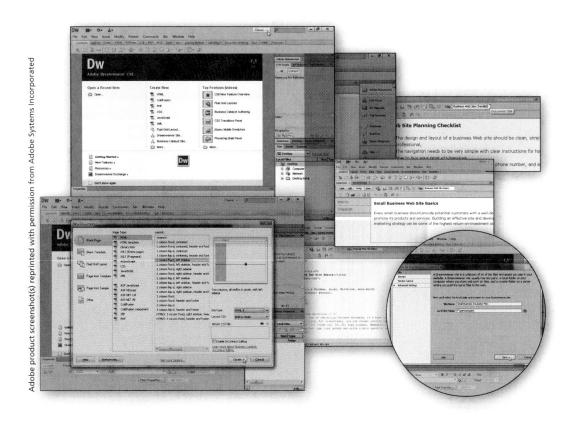

Adobe product screenshot(s) reprinted with permission from Adobe Systems Incorporated

Objectives

You will have mastered the material in this chapter when you can:

- Start Dreamweaver and customize the Dreamweaver workspace
- Describe the Dreamweaver workspace
- Show and hide panels
- Create a Dreamweaver Web site using a template
- Define a local site
- Add text to a Web page

- Change the format of the text headings
- Add links to a Web site
- Create an unordered list
- Save a Web site
- Check spelling
- Preview a Web site in a browser
- Use Dreamweaver Help

1 | Creating a New Web Site with Dreamweaver

What Is Dreamweaver CS6?

Adobe Dreamweaver CS6, the preferred professional Web site creation and management software, provides a rich user interface with powerful tools. Dreamweaver can be used to design a Web site that is displayed in any browser on multiple platforms, including PC and Mac computers, kiosks, tablets, and smartphones. Dreamweaver's icon-driven menus and detailed panels make it easy for users to add text, images, multimedia files, and links without typing one line of code. Dreamweaver creates code that reflects selections made in the user interface and provides content structure when rendering the page in the browser.

The Adobe Dreamweaver user interface is consistent across all Adobe authoring tools. This consistency allows for easy integration with other Adobe Web-related programs such as Adobe Flash, Photoshop, Illustrator, and Fireworks. Dreamweaver CS6 is part of the **Adobe Creative Suite 6**, a collection of graphic design, video editing, and Web development applications published by Adobe Systems. Dreamweaver CS6 runs on multiple operating systems, including Windows 8, Windows 7, Windows Vista, Windows XP, and Mac OS X. This text uses Dreamweaver CS6 on the PC platform, running the Windows 7 operating system.

Project Planning Guidelines

> The process of developing a Web site that communicates specific information requires careful analysis and planning. Start by identifying the purpose and audience of the Web site and developing a Web page design. If you are working with a client, ask your client to clearly express his or her expectations, such as who will visit the site and how they will use it. The Web page design contributes to the look and feel of the Web site, which includes the amount of text displayed on each page and the format of the text. Details of these guidelines are provided in the "Project Planning Guidelines" appendix. Each chapter in this book provides practical business applications of these planning guidelines.

Project — Small Business Incubator Web Site Plan

You can use Dreamweaver CS6 to produce Web sites such as the Small Business Incubator Web site shown in Figure 1–1. A business incubator is a program that supports start-up companies by providing resources such as office space, and services such as business advice and networking opportunities. A business incubator in Condor, California plans to create the site shown in Figure 1–1 to highlight best practices for small businesses that design their own Web sites. The two-page Small Business Incubator Web site includes the index, or home, page for the Web site and introduces the design elements. The page includes a simple navigation bar in the left column and a main heading followed by a short informational paragraph in the right column. The second page displays a checklist of best practices for designing any small business Web site.

The project in this chapter uses a built-in Dreamweaver layout to create a simple HTML5 page named index as shown in Figure 1–1a. Recall that HTML5 is the most recent standard of Hypertext Markup Language (HTML), the core language for creating Web pages. After entering text into the index page, you will create a second page (Figure 1–1b) named checklist, which includes a bulleted list of best practices for small business Web site planning. Creating a two-page Web site requires a basic understanding of the Dreamweaver user interface, layouts, links, heading sizes, and bullets.

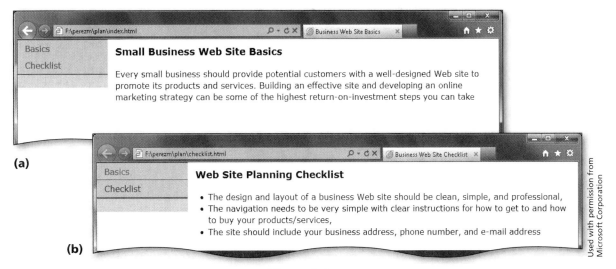

(a)

(b)

Used with permission from Microsoft Corporation

Figure 1–1

Overview

As you read this chapter, you will learn how to create the Web page project shown in Figure 1–1 by performing these general tasks:

- Customize the workspace.
- Create a new Dreamweaver HTML5 Web site with two columns.
- Enter text in the Web page.
- Change the format of the headings.
- Add links.
- Save the document.
- Add a second HTML5 page.
- Create a bulleted list.
- Check spelling.
- Preview the Web site in a browser.
- Save the Web site.

General Project Guidelines

When creating a Dreamweaver Web site, the actions you perform and the decisions you make will affect the appearance and characteristics of the entire Web site. When creating Web pages, such as the ones shown in Figure 1–1 on the previous page, you should follow these general guidelines:

1. **Review the Dreamweaver workspace window.** Become familiar with the various layouts and available panels.

2. **Define the local site.** Create the local site using Dreamweaver's Site Setup dialog box.

3. **Determine the location for the local site.** Select the location and storage media where you will save the site. Keep in mind that in Chapter 2, you will begin a site and then modify those pages, and add new pages as you progress through this book. Storage media can include a hard disk, USB flash drive, network drive, or cloud computing drive.

4. **Select the words and heading sizes for the text.** Text accounts for the bulk of the content on most Web pages, but Web site visitors often avoid long blocks of text. It is best to be brief. Include headings to organize the text into sections. Use lists whenever possible. Use common words and simple language.

5. **Identify how to format various elements of the text.** Determine which text will be headings and subheadings, paragraphs, and bulleted and numbered lists on the Web page.

6. **Review final tasks.** Prepare to display a Web page to others by adding professional finishing touches such as a Web page title and by checking the spelling of the text.

When necessary, more specific details concerning the above guidelines are presented at appropriate points in the chapter. The chapter also identifies the actions performed and decisions made regarding these guidelines during the creation of the Web site pages shown in Figure 1–1 on the previous page.

Starting Dreamweaver

If you are using a computer to step through the project in this chapter and you want your screen to match the figures in this book, you should change your screen's resolution to 1024×768. For information about how to change a screen's resolution, read the "Changing Screen Resolution" appendix. The browser used to display the Web page figures is Internet Explorer 9. The browser text size is set to Medium.

To Start Dreamweaver

The following steps, which assume Windows 7 is running, start Dreamweaver, based on a typical installation. You may need to ask your instructor how to start Dreamweaver for your computer.

1

- Click the Start button on the Windows 7 taskbar to display the Start menu.

- Type `Dreamweaver CS6` in the 'Search programs and files' text box, and watch the search results appear on the Start menu (Figure 1–2).

Used with permission from Microsoft Corporation

Q&A

Why do I have documents and files in my list of results?

Any documents containing the words, Dreamweaver CS6, and any files that have been opened with Dreamweaver CS6 may appear in your list.

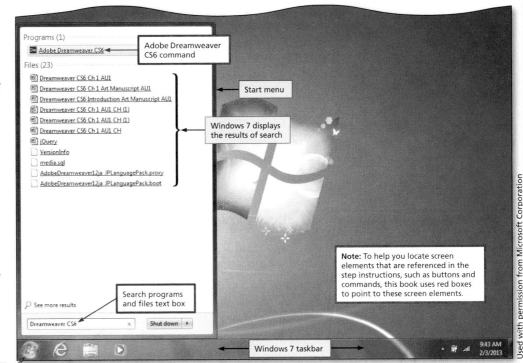

Figure 1–2

2

- Click Adobe Dreamweaver CS6 in the search results on the Start menu to start Dreamweaver and display the Welcome screen.

- If necessary, click the Workspace Switcher button on the Application bar, and then click Classic to switch to the Classic workspace (Figure 1–3).

Q&A

What is the Classic workspace?

Eleven predefined workspace layouts designed for application developers, coders, and Web designers allow you to

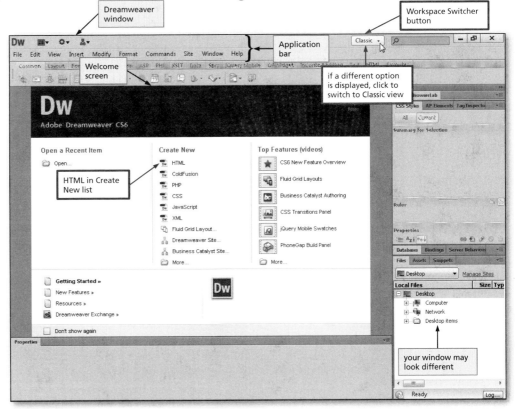

Figure 1–3

customize your workspace according to your role or target device. The Classic workspace provides all the tools necessary for a beginning Web designer's needs and omits features for advanced designers and programmers.

Q&A

Why does the Application bar appear on two rows in Figure 1–3?

If your screen resolution is set to 1024 × 768, the Application bar appears on two rows. If you are using a higher screen resolution, the Application bar appears on one row at the top of the Dreamweaver window.

3

- Click HTML in the Create New list to create a blank HTML document.

- If the Dreamweaver window is not maximized, click the Maximize button next to the Close button on the Application bar to maximize the window.

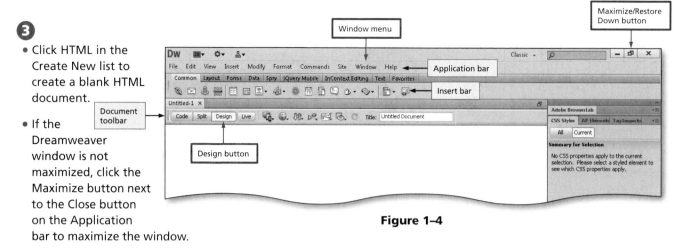

Figure 1–4

- If the Design button is not selected, click the Design button on the Document toolbar to view the page design.

- If the Insert bar is not displayed, click Window on the Application bar and then click Insert to display the Insert panel (Figure 1–4).

Q&A What if a message is displayed regarding default file types?

If a message is displayed, click the Close button.

Q&A What is a maximized window?

A maximized window fills the entire screen. When you maximize a window, the Maximize button changes to a Restore Down button.

Other Ways

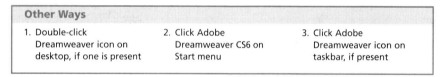

1. Double-click Dreamweaver icon on desktop, if one is present
2. Click Adobe Dreamweaver CS6 on Start menu
3. Click Adobe Dreamweaver icon on taskbar, if present

MAC For a detailed example of starting Dreamweaver using the Mac operating system, refer to the "To Start Adobe Dreamweaver CS6" steps in the For Mac Users appendix at the end of this book.

BTW

Q&As
For a complete list of the Q&As found in many of the step-by-step sequences in this book, visit the Dreamweaver CS6 Q&A chapter resource on the student companion site located at www.cengagebrain.com.

BTW

By the Way Boxes
For a complete list of the BTWs found in the margins of this book, visit the Dreamweaver BTW chapter resource on the student companion site located at www.cengagebrain.com.

Touring the Dreamweaver Window

With the Dreamweaver window open, take time to tour your new Web design environment. The Dreamweaver workspace lets you view documents and object properties. It provides toolbar buttons for the most common operations so that you quickly can make changes to your documents. In Dreamweaver, a document is an HTML file that is displayed in a browser as a Web page. Figure 1–4 shows how the Dreamweaver workspace looks the first time you start Dreamweaver after installation on most computers. To work efficiently, you should learn the basic terms and concepts of the Dreamweaver workspace, and understand how to choose options, use inspectors and panels, and set preferences that best fit your work style.

Dreamweaver Workspace

The **Dreamweaver workspace** is an integrated environment in which the Document window and panels are incorporated into one large application window. To create an efficient, customized workspace, Dreamweaver provides the Web site developer with 11 preset workspace layouts as shown in the Workspace Switcher list in Figure 1–5: App Developer, App Developer Plus, Business Catalyst, Classic, Coder, Coder Plus, Designer, Designer Compact, Dual Screen, Fluid Layout, and Mobile Applications. These workspaces provide different arrangements of panels: Depending on the workspace, some panels are hidden and some appear in different locations in the Dreamweaver window. Each workspace is designed for a different type of Dreamweaver user. For example, programmers who work primarily with HTML and other languages generally select the Coder or App Developer workspace. The Dual Screen option requires two monitors, with the Document window and Property inspector displayed on one monitor, and the panels displayed on a secondary monitor. The Classic workspace contains a visually integrated workspace and is ideal for beginners and nonprogrammers. Select the Mobile Applications view if you want to build an application intended for deployment on a tablet or smartphone device. The projects and exercises in this book use the Classic workspace.

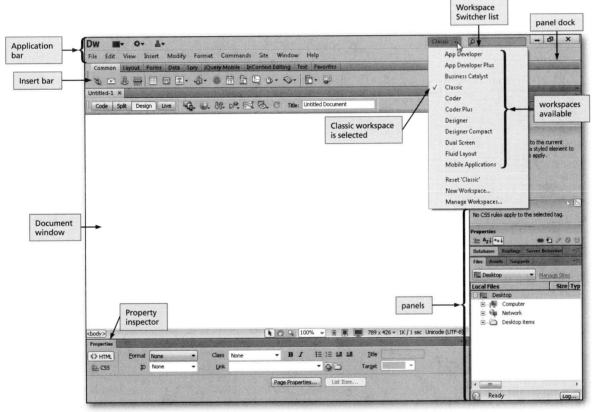

Figure 1–5

In Classic view, the Application bar is at the top of the window, the Insert panel is displayed below the Application bar, the Document window is in the center with the panel dock and panels on the right, and the Property inspector is located at the bottom of the window as shown in Figure 1–5. The following list describes the components of the Dreamweaver workspace:

Application Bar The **Application bar** displays the Dreamweaver menu names and buttons for working with the window layout, extending Dreamweaver, managing sites, switching the workplace layout, searching for help, and manipulating the window. When you point to a menu name on the Application bar, the menu name is selected. When you click a menu name, the corresponding menu is displayed. Figure 1–6 on the next page shows the Edit menu.

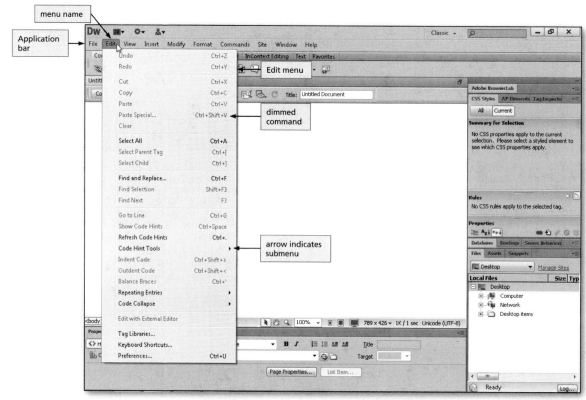

Figure 1–6

The menus contain lists of common actions for performing tasks such as opening, saving, modifying, previewing, and inserting data in your Web page. The menus may display some commands that appear gray, or dimmed — which indicates they are not available for the current selection — instead of black.

Insert Bar Below the Application bar, the **Insert bar** (Figure 1–7), also called the **Insert panel**, allows quick access to frequently used commands.

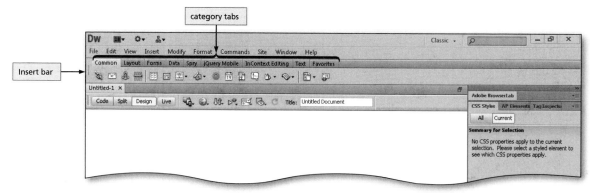

Figure 1–7

BTW

Switching from Insert Bar to Panel
If you drag the Insert bar to another part of the Dreamweaver window, it is displayed as a vertical panel instead of a horizontal bar.

You use the buttons on the Insert bar to insert various types of objects — such as images, tables, links, and dates — into a Web document. As you insert each object, a dialog box allows you to set and manipulate specific attributes of the object. The buttons on the Insert bar are organized into nine categories, such as Common and Layout, which you can access through tabs. Some categories also have buttons with pop-up menus. When you select an option from a pop-up menu, it becomes the default action for the button. When you start Dreamweaver, the category in which you last were working is displayed on the Insert bar.

Document Tab, Document Toolbar, and Document Window The **document tab** displays the Web page name, which is Untitled-1 for the first Web page you create in a Dreamweaver session, as shown in Figure 1–8. (The "X" is the Close button for the document tab.) The **Document toolbar** contains buttons that provide different views of the Document window (e.g., Code, Split, and Design), and some common operations, such as Preview/Debug in Browser, Refresh Design View, View Options, Visual Aids, and Check Browser Compatibility. The **Document window** displays the current document as you create and edit it.

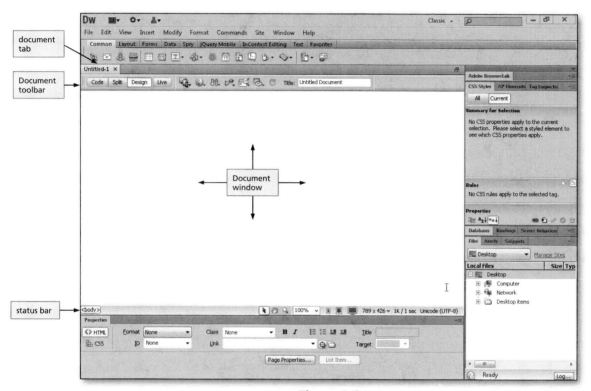

Figure 1–8

When you make changes to a document, Dreamweaver places an asterisk following the file name in the document tab, indicating that the changes have not been saved. The asterisk is removed after the document is saved. The file path leading to the document's location is displayed to the far right of the document tab.

Status Bar The **status bar**, located below the Document window (Figure 1–9), provides additional information about the document you are creating.

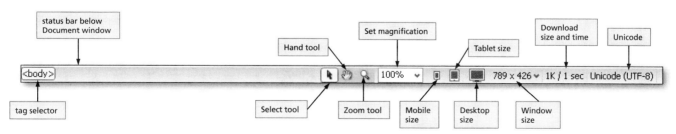

Figure 1–9

The status bar displays the following options:

- **Tag selector:** Click any tag in the hierarchy to select that tag and all its contents.
- **Select tool:** Use the Select tool to return to default editing after using the Zoom or Hand tool.
- **Hand tool:** To pan a page after zooming, use the Hand tool to drag the page.
- **Zoom tool:** Available in Design view or Split view, you can use the Zoom tool to check the pixel accuracy of graphics or to better view the page.
- **Set magnification:** Use the Set magnification context menu to change the view from 6% to 6400%; default is 100%.
- **Mobile size:** Set the Document window to mobile size values, such as for smartphones.
- **Tablet size:** Set the Document window to tablet size values.
- **Desktop size:** Set the Document window to desktop size values.
- **Window size:** Set the Window size value, which includes the window's current dimensions (in pixels). Click this value to display the Window size pop-up menu.
- **Download size and download time:** Refer to this area for the size and estimated download time of the current page. Dreamweaver CS6 calculates the size based on the entire contents of the page, including all linked objects such as images and plug-ins.
- **Unicode (UTF-8):** Refer to this area for the type of text encoding. Unicode is an industry standard that allows computers to consistently represent and manipulate text expressed in most of the world's writing systems.

Property Inspector The **Property inspector**, docked at the bottom of the Document window, provides properties such as the color or font style of a selected object or text in the document. Figure 1–10 shows the default Property inspector when an object is not selected.

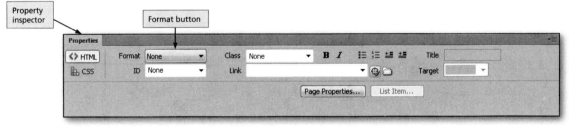

Figure 1–10

The Property inspector enables you to view and change a variety of properties for the selected object or text. The Property inspector is context sensitive, meaning it changes based on the selected object, which can include text, tables, images, and other objects. For example, to change the format of a selected heading, you click the Format button in the Property inspector and then select the new format in the list.

Panel Groups Within the panel dock shown in Figure 1–11 on the next page, related panels are displayed in a single panel group with individual tabs. A **panel dock** is a fixed area at the left or right edge of the workspace that hosts a panel group. A **panel** displays a collection of related tools, settings, and options. A **panel group** is a set of related panels docked together below one heading. A panel group typically contains three panels. Each panel provides a wide variety of tools to assist in developing and managing a Web site. For example, the Files panel is used to view and manage the files in your Dreamweaver site.

BTW

Dreamweaver Panels
Dreamweaver has many panels, inspectors, and bars. To open any of them, click Window on the Application bar.

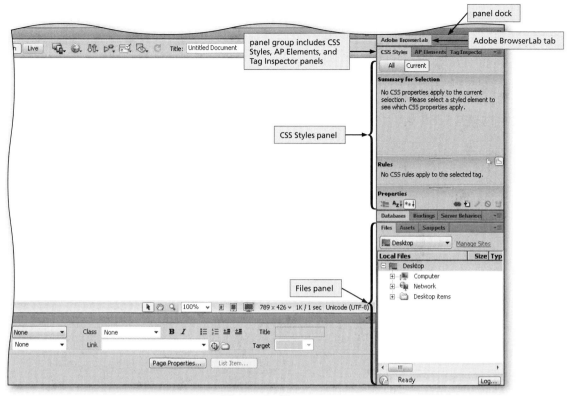

Figure 1–11

Displaying Document Views

The Document toolbar contains buttons that display different views of an active Web page.

Code View In **Code view**, the Document window displays the HTML, CSS, JavaScript, and other server-side (Web programming) language code within a Web page. Figure 1–12 shows the completed source code from the Small Business Incubator Plan Web site. The different parts of the code are associated with certain colors, making it easier to code by hand.

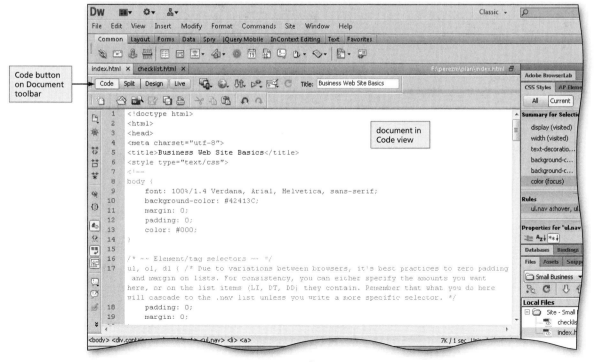

Figure 1–12

Split View Split view displays both the source code and the document design simultaneously. Figure 1–13 shows the completed chapter project Web page in Split view. If you add text in the document design pane, the source code is updated immediately.

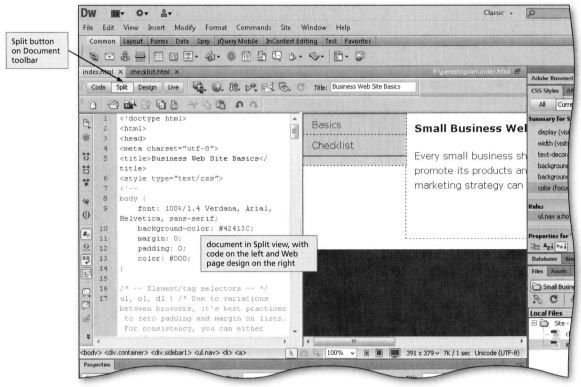

Figure 1–13

Design View The design environment, where you assemble your Web page elements and design your page, is called **Design view**, as shown in the completed project in Figure 1–14.

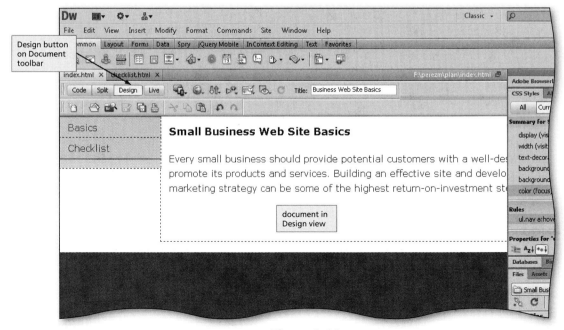

Figure 1–14

Live View **Live view** displays an interactive, browser-based view of the document. This view looks similar to the Design view except it does not support editing functions.

Opening and Closing Panels

The Dreamweaver panels help you organize and modify content using commands and functions. You can customize the workspace to display only the panels you want. Drag a panel by its title bar to move it from its default location and position it where you like, optimizing your Dreamweaver environment. Moving and hiding panels makes it easy to access the panels you need without cluttering your workspace. Each time you start Dreamweaver, the workspace is displayed in the same layout from the last time you used Dreamweaver.

Throughout the workspace, you can open and close the panel groups and display or hide other Dreamweaver features as needed, or move a panel to another location. You use the Window menu to open a panel and its group. Closing unused panels provides an uncluttered workspace in the Document window. You can use the panel options button or a panel's context menu (or shortcut menu) to close the panel or its group. You also can collapse a panel so it takes up less space in the panel dock by double-clicking the panel tab, or you can collapse the entire panel dock so it takes up less space in the Dreamweaver window. In either case, you can expand a panel, panel group, or panel dock to display one or more full panels again.

To Show, Hide, and Move Panels

The following steps show, hide, and move panels.

- Double-click the Properties tab below the Document window to collapse the Property inspector (Figure 1–15).

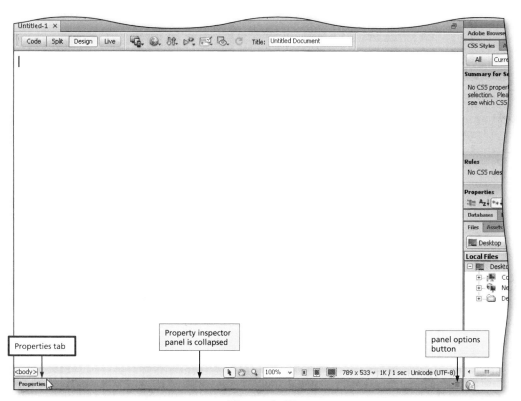

Figure 1–15

• Click the Collapse to Icons button in the panel dock to collapse all the panel groups (Figure 1–16).

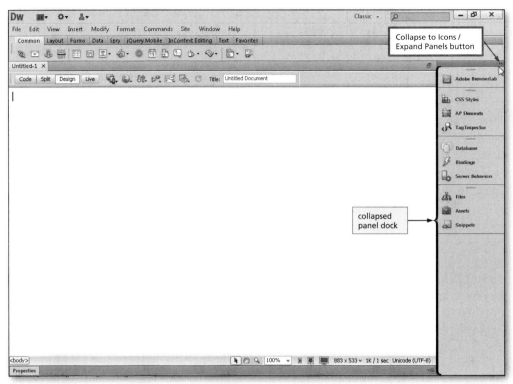

Figure 1–16

• Click the Expand Panels button to expand the panel groups.

Q&A What happened to the Collapse to Icons button?

The Collapse to Icons button and Expand Panels button are in the same location. After you collapse the panels, the button changes to the Expand Panels button so you can expand the panels again.

• Click the Properties tab to expand the Property inspector (Figure 1–17).

Q&A What is the fastest way to open and close panels?

The fastest way to open and close panels in Dreamweaver is to use the F4 key, which opens or closes all panels and inspectors at one time.

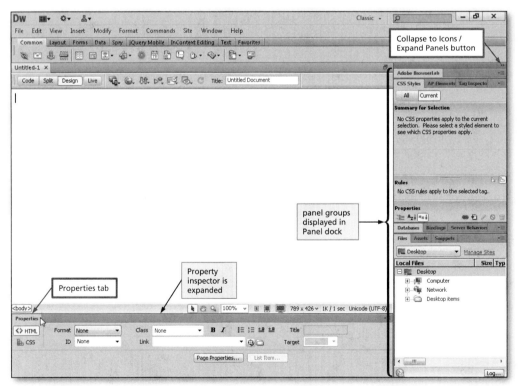

Figure 1–17

- Drag the Adobe
  BrowserLab panel
  by its tab to the
  center of the screen
  to move the panel
  to a new location
  (Figure 1–18).

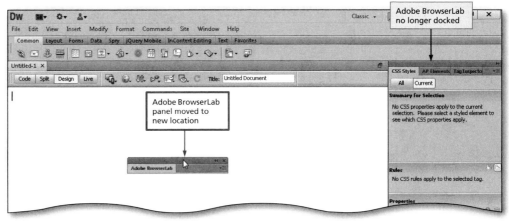

Figure 1–18

Other Ways

1. Click panel options
   button, click Close or click
   Close Tab Group

2. Right-click panel, click
   Close or click Close Tab
   Group

3. Right-click panel, click
   Minimize or click Expand
   Panel

4. Press F4

To Reset the Classic Workspace

After collapsing, expanding, and moving panels, you may want to return the workspace to its default settings. The default workspace, called Classic, displays commonly used panels. The following steps reset the Classic workspace.

- Click the Workspace
  Switcher button
  on the Application
  bar to display the
  Workspace Switcher
  menu (Figure 1–19).

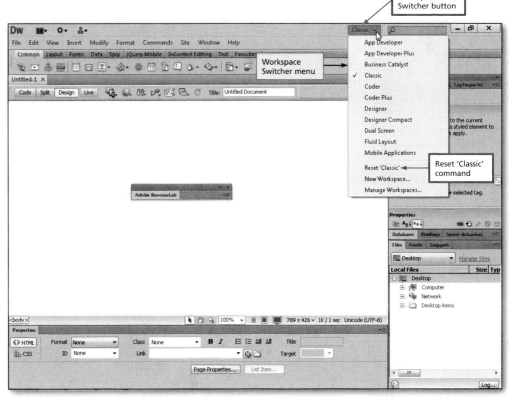

Figure 1–19

- Click Reset 'Classic' to restore the workspace to its default settings (Figure 1–20).

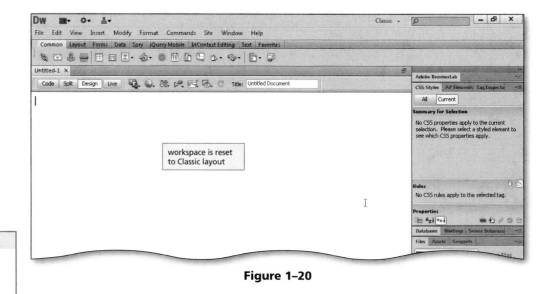

Other Ways

1. On Window menu, point to Workspace Layout, click Reset 'Classic'

Figure 1–20

To Display the Standard Toolbar

In the Classic workspace, Dreamweaver can display three toolbars: Style Rendering, Document, and Standard. You can choose to display or hide the toolbars by clicking View on the Application bar and then pointing to Toolbars. If a toolbar name has a check mark next to it, it is displayed in the window. To hide the toolbar, click the name of the toolbar so it no longer is displayed.

The Standard toolbar is not displayed by default in the Document window when Dreamweaver starts. As with other toolbars and panels, you can dock or undock and move the Standard toolbar so it can be displayed in a different location on your screen. Drag any toolbar by its selection handle to undock and move the toolbar.

The following steps display the Standard toolbar.

- Click View on the Application bar to display the View menu.

- If necessary, click the down-pointing arrow at the bottom of the View menu to scroll the menu.

- Point to Toolbars to display the Toolbars submenu (Figure 1–21).

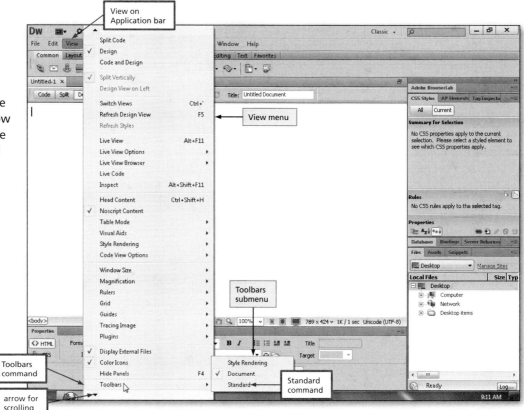

Figure 1–21

2

• Click Standard on the Toolbars submenu to display the Standard toolbar (Figure 1–22).

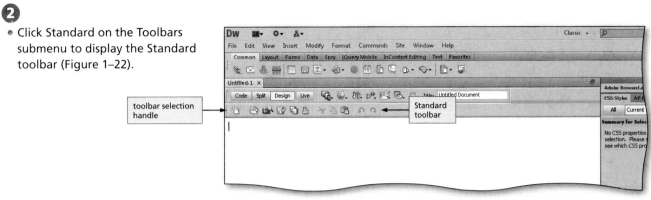

Figure 1–22

Other Ways

1. Right-click blank area on toolbar, click Standard

To Access Preferences

In addition to creating a customized environment by selecting a workspace; collapsing, expanding, and moving the panels; and adding toolbars, you can further customize your environment by adjusting your preferences. Dreamweaver's work **preferences** are options that modify the appearance of the workspace, user interactions, accessibility features, and default settings such as font, file types, browsers, code coloring, and code hints. Dreamweaver offers you the flexibility to shape your Web page tools and your code output.

The following steps access Dreamweaver's preferences.

1

• Click Edit on the Application bar to display the Edit menu (Figure 1–23).

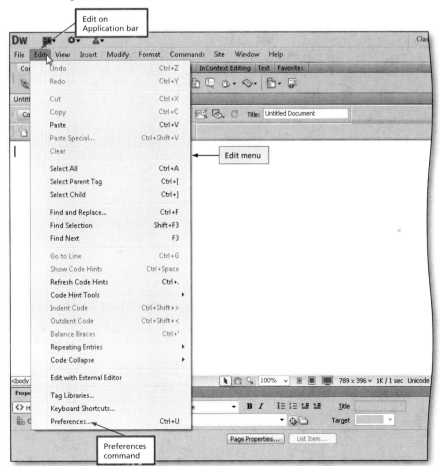

Figure 1–23

● Click Preferences to display the Preferences dialog box (Figure 1–24).

Q&A
How do I view the options within each category of preferences?

You can view each option by selecting the category in the left pane to display the options for that category in the main area of the dialog box.

● Click the OK button (Preferences dialog box) to close the dialog box.

Other Ways

1. Press CTRL+U

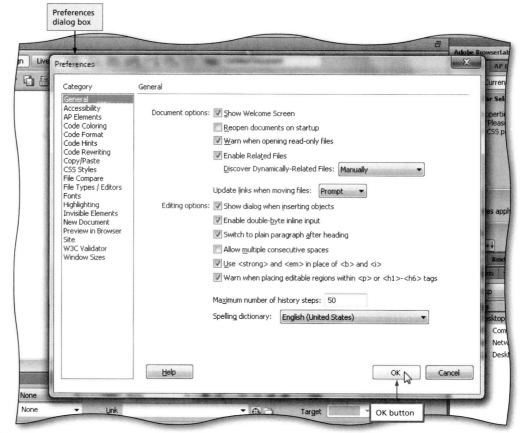

Figure 1–24

Understanding HTML5

Adobe Dreamweaver CS6 supports HTML5, the newest version of HTML. Enhanced features in HTML5 are streamlining Web site design and development so you can create more engaging Web sites, increase Web security, and deploy to multiple devices. You probably have viewed a Web site on a smartphone that was difficult to navigate because the site was not specifically designed for a mobile device. With Dreamweaver and HTML5, you can create an interactive site that will be rendered correctly on a variety of smartphones, tablets, and traditional computers. HTML5 includes native support of video without the use of a plug-in, such as Adobe Flash Player, so that videos play seamlessly on Apple devices, such as an iPad.

When a Dreamweaver page is created, an option called DocType is placed in the first line of HTML code. The **DocType** declaration is not a formatting tag, but rather an instruction to the Web browser indicating in what version of the markup language the page is written. The DocType declaration refers to a **Document Type Definition (DTD)**, which specifies the rules for the markup language so browsers can render the content correctly. For the projects in this book, the DocType will be set to HTML5 to access the latest features of HTML. The first line in an HTML5 document reads: <!DOCTYPE html>.

Understanding CSS3

Working hand in hand with HTML5, Dreamweaver CS6 also supports CSS3, Cascading Style Sheets. CSS3 is the new standard for Web site presentation. Web pages can be designed in two ways:

1. A Web page can contain layout and content information combined in a single HTML file. This option is not recommended because it is difficult to update and maintain such a Web page on larger sites.

2. A Web page can separate content from the layout in two separate files. The content information is coded in the HTML file, but the HTML file does not contain information about how that information is displayed. The appearance, or layout, is stored in a separate file called a CSS file.

A **CSS file** is a simple text file containing style rules that control the appearance of a Web page. Using CSS styles, you can control font size, font color, background, and many other attributes of a Web page, thus reducing a page's file size. In a large Web site that does not use CSS styles, making a formatting revision such as changing the font is very time consuming because each page must be changed individually to use the new font. By using CSS, you can change the font in one line of a CSS file and have the font automatically update throughout the entire site.

In this chapter, a simple text-only site about planning a Web site for a small business displays two basic pages created with a built-in Dreamweaver template that uses CSS. A **template** is a predesigned layout used to create pages with placeholder content. Dreamweaver templates provide a framework for designing a professional page that includes background colors, fonts, and a layout controlled by built-in CSS auto-generated code. In the chapter project, a predefined template with two fixed columns and a left sidebar displaying two vertical columns is used for the layout. Typically, the first vertical column displays a navigation menu while the second vertical column shows the main content of the page. Any predefined style within the template can be customized by changing the corresponding CSS settings.

Break Point: If you wish to take a break, this is a good place to do so. You can quit Dreamweaver now. To resume at a later time, start Dreamweaver, and continue following the steps from this location forward.

Creating a New Site

After touring the Dreamweaver environment, you are ready to define a local site for the Small Business Incubator Web site. When you define a site, you create the folder that will contain the files and any subfolders for the site. The site consists of two pages of text that provide information on basic Web design best practices for small businesses that want to create a Web presence.

Defining a Local Site

Web design and Web site management are two important skills that a Web developer must possess and apply. Dreamweaver CS6 is a site creation and management tool. To use Dreamweaver effectively, you first must define the local site. After a Web site is developed within the local site location, the site can be published to a remote server for access by others on the Internet.

The general definition of a **site**, or Web site, is a set of linked documents with shared attributes, such as related topics, a similar design, or a shared purpose. In Dreamweaver, the term, site, can refer to any of the following:

- **Web site:** A set of pages on a server that are viewed through a Web browser by a site visitor

- **Remote site:** Files on the server that make up a Web site, from the author's point of view rather than a visitor's
- **Local site:** Files on your computer that correspond to the files on the remote site (You edit the files on your computer, often called the local computer, and then upload them to the remote site.)
- **Dreamweaver site definition:** A set of defining characteristics for a local site, plus information on how the local site corresponds to a remote site

All Dreamweaver Web sites begin with a local root folder. As you become familiar with Dreamweaver and complete the chapters in this book, you will find references to a **local site folder**, **local root folder**, **root folder**, and **root**. These terms are interchangeable. This folder is no different from any other folder on your computer's hard drive or other storage media, except in the way Dreamweaver views it. By default, Dreamweaver searches for Web pages, links, images, and other files in the designated root folder. Within the root folder, you can create additional folders and subfolders to organize images and other objects. A **subfolder** is a folder inside another folder. Dreamweaver displays only the files in the root folder and its subfolders when you preview the Web site in a Web browser.

Dreamweaver provides two options to define a site and create the hierarchy: You can create the root folder and any subfolders, or create the pages and then create the folders when saving the files. In this book, you create the root folder and subfolders, and then create the Web pages.

Plan
Ahead

Determine the location for the local site
Before you create a Web site, you need to determine where you will save the site and its files.

- If you plan to work on your Web site in various locations or on more than one computer, you should create your site on removable media, such as a USB flash drive. The Web sites in this book use a USB flash drive because these drives are portable and can store a lot of data.
- If you always work on the same computer, you probably can create your site on the computer's hard drive. However, if you are working in a computer lab, your instructor or the lab supervisor might instruct you to save your site in a particular location on the hard drive or on removable media such as a USB flash drive. (This book assumes the Web site files are stored on a USB flash drive.)

Creating the Local Root Folder and Subfolders

You can use several options to create and manage your local root folder and subfolders, including Dreamweaver's Files panel, Dreamweaver's Site Setup feature, and Windows file management. In this book, you use the most common ways to manage files and folders: Dreamweaver's Site Setup feature to create the local root folder and subfolders, the Files panel to manage and edit your files and folders, and Windows file management to download and copy the data files.

To organize and create a Web site and understand how you access Web documents, you need to understand paths and folders. The term, path, sometimes is confusing for new users of the Web. It is, however, a simple concept: A **path** is the succession of folders that must be navigated to get from one folder to another. Because folders sometimes are referred to as **directories,** the two terms are often used interchangeably.

A typical path structure containing Web site files has a **master folder**, called the **root**, and is designated by the backslash symbol (\ ) in the path notation that appears in the Dreamweaver window. This root folder contains all of the other subfolders or nested folders. Further, each subfolder may contain additional subfolders or nested folders. On most sites, the root folder includes a subfolder for images.

For this book, you first will create a local root folder using your last name and first initial. Examples in this book use Mia Perez as the Web site author. Thus, Mia's local root folder is perezm and is located on drive F (a USB drive, which might have a different drive letter on your computer). Next, you will create a subfolder named plan for the Web site you create in this chapter. You will store related files and subfolders within the plan folder. When you navigate through this folder hierarchy, you are navigating along the path. The path to the Small Business Incubator Web site is F:\perezm\plan\. In all references to F:\perezm, substitute your last name and first initial and your drive location.

Using Site Setup to Create a Local Site

You create a local site using Dreamweaver's Site Setup dialog box, which provides four categories of settings. For the Web site you create in the chapter, you only need to work in the Site category, where you enter the name of your site and the path to the local site folder. For example, you will use Small Business Incubator Plan as the site name and F:\perezm\plan\ as the path to the local site. You can select the location of the local site folder instead of entering its path.

After you complete the site definition, the folder hierarchy structure is displayed in the Dreamweaver Local Files list on the Files panel. This hierarchy structure is similar to the Windows file organization. The **Local Files** list provides a view of the devices and folders on your computer, and shows how these devices and folders are organized.

BTW

Dreamweaver Help
At any time while using Dreamweaver, you can find answers to questions and display information about various topics through Dreamweaver Help. Used properly, this form of assistance can increase your productivity and reduce your frustrations by minimizing the time you spend learning how to use Dreamweaver. For instruction about Dreamweaver Help and exercises that will help you gain confidence in using it, read the "Adobe Dreamweaver CS6 Help" appendix at the end of this book.

To Quit and Restart Dreamweaver

The following step quits Dreamweaver and then restarts the program to display the Welcome screen.

1

- Click the Close button in the upper-right corner of the Dreamweaver window to quit Dreamweaver after touring the interface.

- Click the No button (Dreamweaver dialog box) if asked to save changes to Untitled-1.

- Click the Start button on the Windows 7 taskbar to display the Start menu.

- Type `Dreamweaver CS6` in the 'Search programs and files' text box.

- Click Adobe Dreamweaver CS6 in the search results on the Start menu to start Dreamweaver and display the Welcome screen (Figure 1–25).

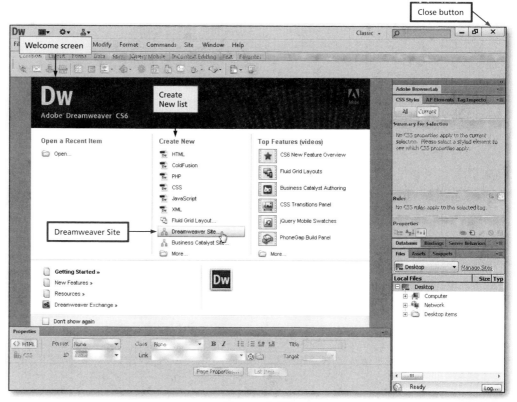

Figure 1–25

To Use Site Setup to Create a Local Site

The following steps define a local site by telling Dreamweaver where you plan to store local files. A USB drive is used for all projects and exercises in this book. If you are saving your sites in another location or on removable media, substitute that location for Removable Disk (F:).

- Click Dreamweaver Site in the Create New list to display the Site Setup dialog box (Figure 1–26).

Q&A

Should the name that appears in the Site Name text box be Unnamed Site 2?

Not necessarily. Your site number may be different.

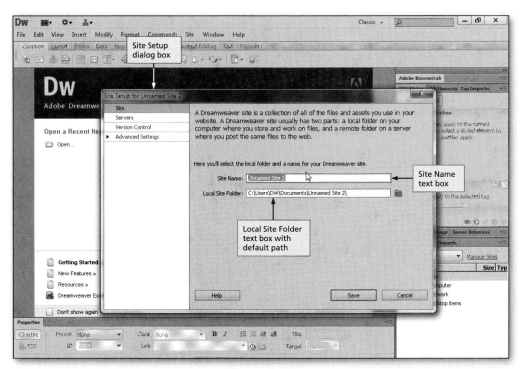

Figure 1–26

- Type Small Business Incubator Plan in the Site Name text box to name the site (Figure 1–27).

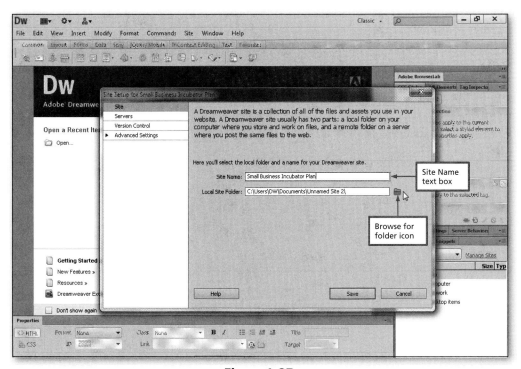

Figure 1–27

3

- Click the 'Browse for folder' icon to display the Choose Root Folder dialog box.

- Click the Select Box arrow to display locations on your system (Figure 1–28).

Q&A

Do I have to save to a USB flash drive?

No. You can save to any device or folder. A folder is a specific location on a storage medium. You can save to the default folder or a different folder.

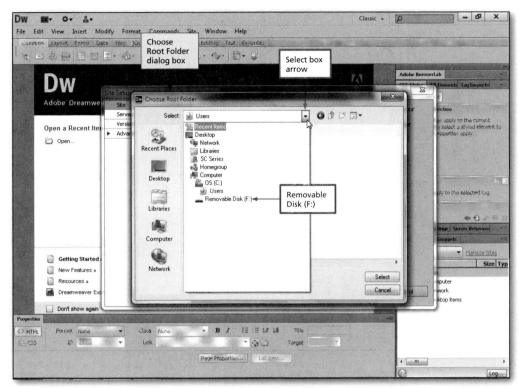

Figure 1–28

4

- Click Removable Disk (F:) in the list, or the name of your storage location.

Q&A

What if my USB flash drive has a different name or letter?

It is very likely that your USB flash drive has a different name and drive letter, and is connected to a different port. Verify that the device in the Select text box is correct.

- Click the Create New Folder button to create a folder for your local site (Figure 1–29).

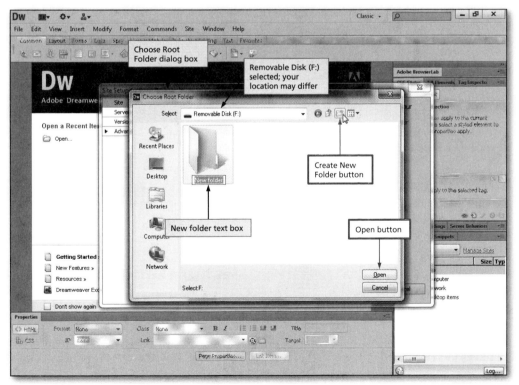

Figure 1–29

5

- For the root folder name, type your last name and first initial (with no spaces between your last name and initial) in the New folder text box. For example, type `perezm`.

- Press the ENTER key to rename the new folder (Figure 1–30).

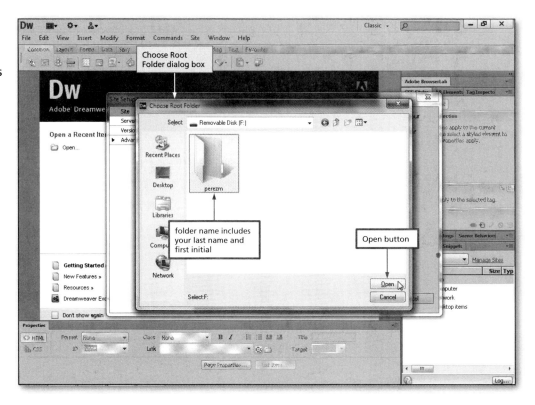

Figure 1–30

6

- Click the Open button to open the root folder.

- Click the Create New Folder button in the Choose Root Folder dialog box to create a folder for the Small Business Incubator Plan site within the folder with your last name and first initial.

- Type `plan` as the name of the new folder, press the ENTER key, and then click the Open button to create the plan subfolder and open it.

- Click the Select button to select the plan folder for the new site and display the Site Setup dialog box (Figure 1–31).

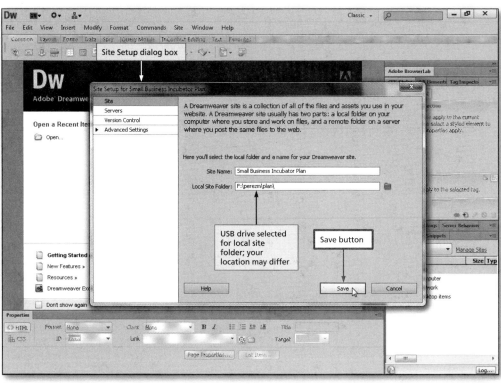

Figure 1–31

Q&A Why should I create a folder on the drive for my Web site?

Organizing your Web site folders now will save you time and prevent problems later.

Q&A Which files will I store in the plan folder?

The plan folder will contain all the files for the Small Business Incubator Plan site. In other words, the plan folder is the local root folder for the Web site.

7

- Click the Save button in the Site Setup dialog box to save the site settings and display the Small Business Incubator Plan site hierarchy on the Files panel (Figure 1–32).

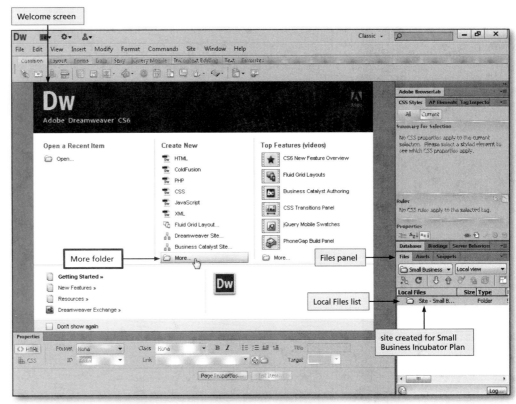

Figure 1–32

Selecting a Predefined Template

Dreamweaver CS6 provides templates with built-in layouts to help you quickly create a Web page. The layouts are predesigned pages with placeholder content. The placeholder content is replaced with your own formatted content.

Plan Ahead

Select the words and fonts for the text

Most informational Web pages start with a heading, include paragraphs of text and one or more lists, and then end with a closing line. Before you add text to a Web page, consider the following guidelines for organizing and formatting text:

- **Headings:** Start by identifying the headings you will use. Determine which headings are for main topics (Heading 1) and which are for subtopics (Heading 2 or 3).

- **Paragraphs:** For descriptions or other information, include short paragraphs of text. To emphasize important terms, format them as bold or italic.

- **Lists:** Use lists to organize key points, a sequence of steps, or other information you want to highlight. If amount or sequence matters, number each item in a list. Otherwise, use a bullet (a dot or other symbol that appears at the beginning of the paragraph).

- **Closing:** The closing is usually one sentence that provides information of interest to most Web page viewers, or that indicates where people can find more information about your topic.

The template used in the chapter project displays two fixed columns with an earth tone color design and a DocType defined as HTML5. Dreamweaver provides two types of predefined layouts: fixed and liquid. In a **fixed layout**, the values for the overall width, as well as any columns within the page, are written using pixel units in the CSS file. In a **liquid layout**, the values for the overall width, as well as any columns within the page, are written using percentages in the CSS file. A fixed layout offers a greater measure of control to align items within the fixed columns because the layout is not resized when the site visitor resizes his or her browser window.

To Select a Template Layout

The following steps create a two-column fixed layout with a left sidebar and a DocType defined as an HTML5 page.

1

● Click the More folder icon on the Welcome screen to display the New Document dialog box (Figure 1–33).

Q&A
What if I do not see a Welcome screen?

Click File on the Application bar and select New to display the New Document dialog box.

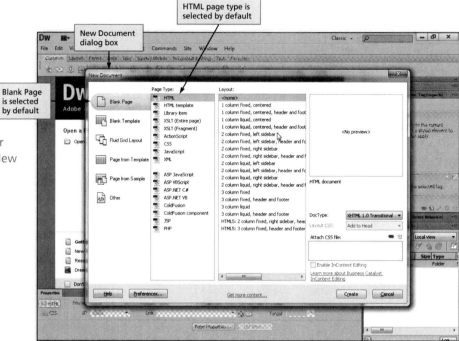

Figure 1–33

2

● Click '2 column fixed, left sidebar' in the Layout column in the New Document dialog box to display a preview of a Web page with two columns.

● Click the DocType button, and then click HTML 5 to change the DocType to HTML 5 (Figure 1–34).

Q&A
Do I need to select Blank Page in the left pane and HTML in the Page Type column?

No. Those options are selected by default in the New Document dialog box.

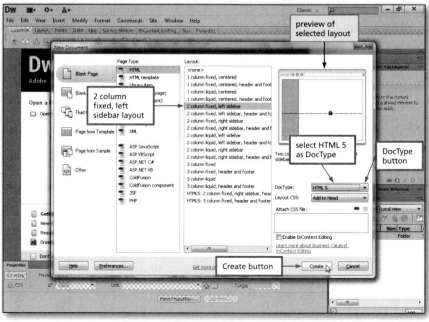

Figure 1–34

3

- Click the Create button to display the two-column fixed layout page in the Document window (Figure 1–35).

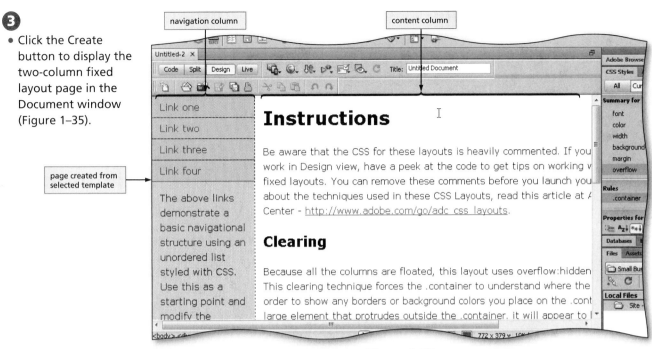

Figure 1–35

To Name and Save the Home Page

The **home page** is the starting point for a Web site. For most Web sites, the home page is named index. This name has special significance because most Web servers recognize index.html (or index.htm) as the default home page. If a folder contains multiple files, the browser determines that the first page to display on a site is the index file. Dreamweaver automatically adds the default extension .html to the file name. Documents with the .html or .htm extensions are displayed in Web browsers. If you have unsaved changes on a Web page, the document tab displays an asterisk after the html extension (index.html*). When you save the page, the asterisk disappears — confirming that your page is up to date in the saved file.

The following steps rename the untitled home page to index.html and then save it.

1

- Click File on the Application bar to display the File menu (Figure 1–36).

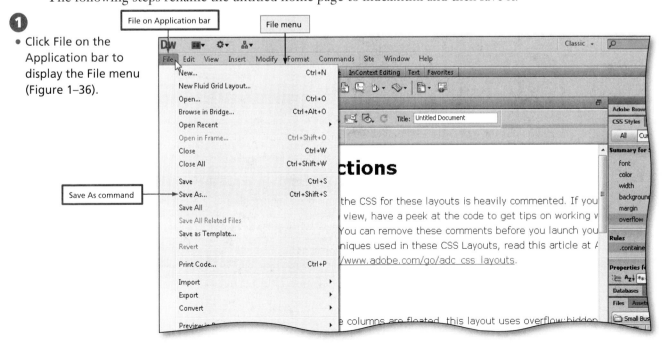

Figure 1–36

2

- Click Save As on the File menu to display the Save As dialog box (Figure 1–37).

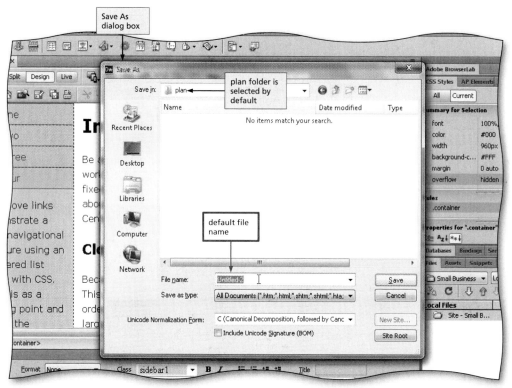

Figure 1–37

3

- If necessary, select the text in the File Name text box, and then type `index` to name the home page (Figure 1–38).

Q&A

Is it necessary to type the extension .html after the file name?

No. By default, Dreamweaver saves an HTML file with the extension .html.

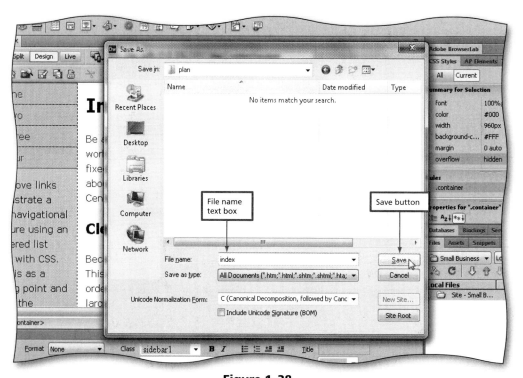

Figure 1–38

 For a detailed example of saving a Web page using the Mac operating system, refer to the "To Save a File in Dreamweaver" steps in the For Mac Users appendix at the end of the book.

4

- Click the Save button (Save As dialog box) to save the home page as index.html and to display the new file name on the document tab (Figure 1–39).

Q&A

What do the icons on the Files panel indicate?

A small device icon or folder icon is displayed next to each object listed on the Files panel. The device icon represents a device such as the Desktop or a disk drive, and the folder icon represents a folder. These icons may have an expand (plus sign) or collapse (minus sign) next to them indicating whether the device or folder contains additional folders. You click these icons to expand or collapse the view of the file hierarchy.

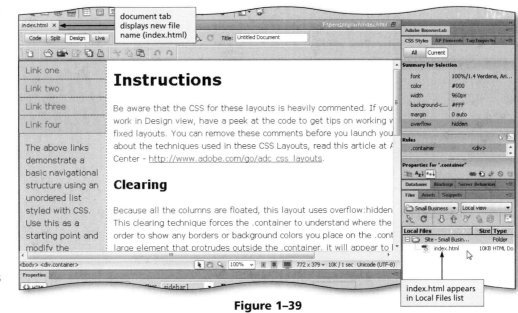

Figure 1–39

To Edit Navigation Link Text

The two columns in the template contain placeholder text instead of an empty page so that you can view how the page will look when displayed in a browser. The left column includes links for navigating from one page to another. Both pages in the Small Business Incubator Plan Web site should display only two links: a Basics link used to display the home page, and a Checklist link used to display the checklist page. The following steps change the link text for the first two links and delete the remaining links in the navigation column.

1

- Drag to select the text, Link one, in the left column of index.html and type `Basics` to change the text for the first link.

- Drag to select the text, Link two, and type `Checklist` to change the text for the second link (Figure 1–40).

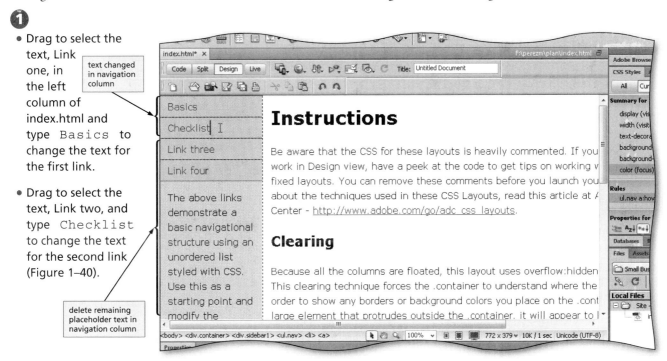

Figure 1–40

- Drag to select the text, Link three, and then press the DELETE key to remove the third link.
- Drag to select the text, Link four, and then press the DELETE key to remove the fourth link.
- Drag to select the paragraph below the links, and then press the DELETE key to remove the placeholder text (Figure 1–41).

placeholder text deleted from navigation pane

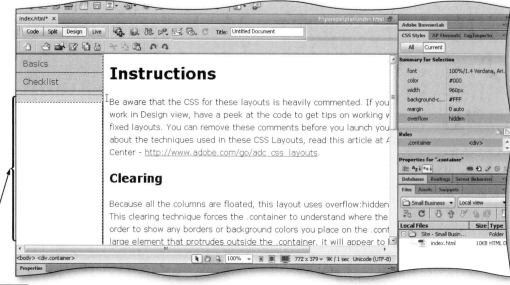

Figure 1–41

To Format Text Using Heading Styles

HTML defines a collection of font styles called **headings** to format the size of text. The advantage of using heading styles to format text that serves as a headline or title is that headings are displayed in relative sizes with Heading 1 <h1> being the largest heading and Heading 6 <h6> being the smallest heading. Any text formatted with Heading 1 is always larger than text formatted with Heading 2. **Formatting** involves setting heading styles, inserting special characters, and inserting or modifying other elements that enhance the appearance of the Web page. Dreamweaver provides three options for directly formatting text: the Format menu on the Application bar, the Text category on the Insert panel, and the Property inspector. In Chapter 2, you will format Web page text using CSS styles. To format the content title in the second column of the index page, you will use the text-related features of the Property inspector.

The Property inspector is one of the panels used most often when creating and formatting Web pages. Recall that it displays the properties, or characteristics, of the selected object. The object can be a table, text, an image, or some other item. The Property inspector is context sensitive, so its options change relative to the selected object.

The following steps edit the text and heading style in the content column.

- Drag to select the text, Instructions, in the right column of the index.html page, and then type Small Business Web Site Basics to change the content title (Figure 1–42).

updated title in content column

Figure 1–42

- Drag to select the paragraph below the new title, and then type `Every small business should provide potential customers with a well-designed Web site to promote its products and services. Building an effective site and developing an online marketing strategy can be some of the highest return-on-investment steps you can take.` to change the first paragraph.

- Drag to select the rest of the placeholder text in the content column, and then press the DELETE key to delete the text (Figure 1–43).

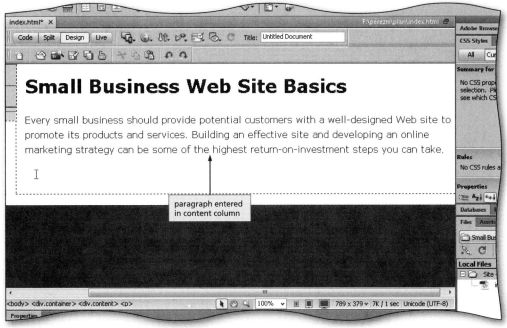

Figure 1–43

- Drag to select the content title, Small Business Web Site Basics.

- Click the Format button in the Property inspector to display styles to apply to the selected text (Figure 1–44).

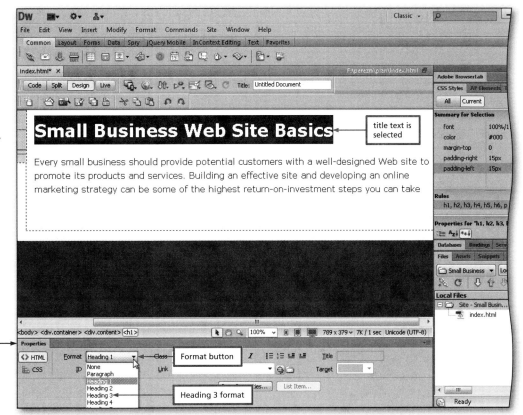

Figure 1–44

4

● Click Heading 3 to apply the Heading 3 style to the content title text (Figure 1–45).

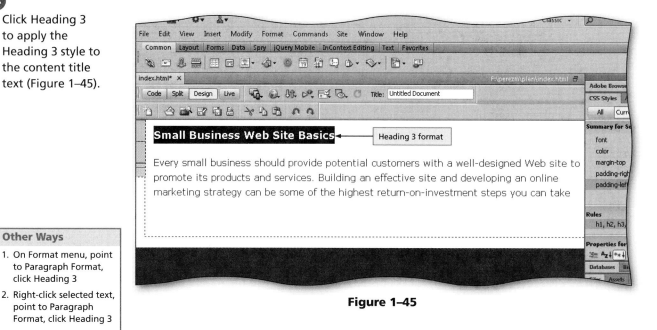

Other Ways

1. On Format menu, point to Paragraph Format, click Heading 3

2. Right-click selected text, point to Paragraph Format, click Heading 3

Figure 1–45

To Create a Link

Links allow users to move from page to page by clicking text or other objects. In the Property inspector, the Link box allows you to transform selected text or other objects into a hyperlink to a specified URL or Web page. You either can point to the hyperlink destination by dragging the Point to File button, or browse to a page in your Web site and select the file name. As you drag the Point to File button, Dreamweaver displays a line showing the connection between the Point to File button and the file link. In a browser, when the site visitor clicks the Basics link, the index page should open. The following steps create a hyperlink by dragging.

1

● Select the text, Basics, in the left column of index.html (Figure 1–46).

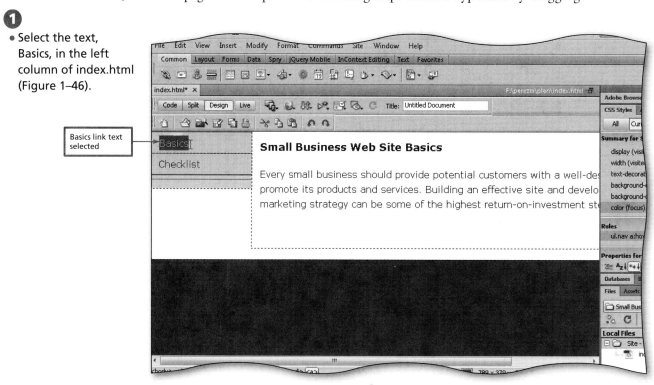

Figure 1–46

2

- Drag the Point to File button in the Property inspector to the index.html file on the Files panel to create a link to index.html (Figure 1–47).

3

- Release the mouse button to complete the link to index.html.

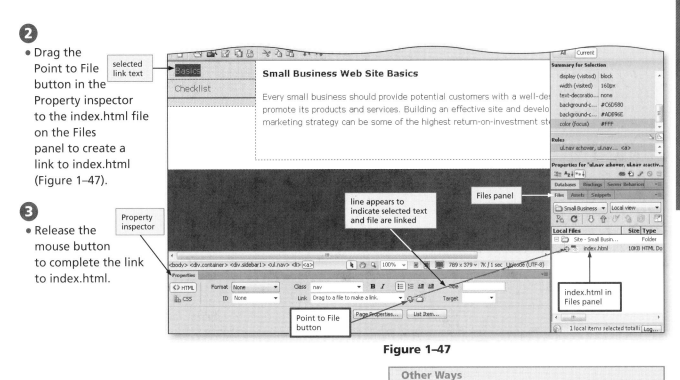

Figure 1–47

To Save the Home Page

The following steps save the changes to the home page, index.html.

1

- Click File on the Application bar to display the File menu (Figure 1–48).

2

- Click Save on the File menu to save the document.

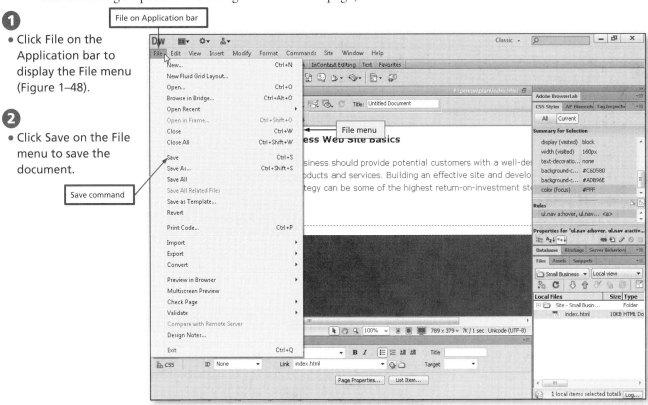

Figure 1–48

For a detailed example of saving a Web page using the Mac operating system, refer to the "To Save a File in Dreamweaver" steps in the For Mac Users appendix at the end of the book.

To Add a Second Page

The first page, index.html, is the model for the second page; in fact, a copy of the first page can serve as the second page. Creating a copy of the first page saves the design time of recreating the page.

The home page (index.html) and the second page (checklist.html) are identical except for the custom text in the content column of each page. As you add pages to your Web site, if your new page is similar to a previous page you created, first save the changes to the existing page, and then save the page again with the new page name, such as checklist.html. The following steps add a second page to the Web site.

- Click File on the Application bar to display the File menu.

- Click Save As on the File menu to display the Save As dialog box.

- Type checklist in File name text box to name the second page (Figure 1–49).

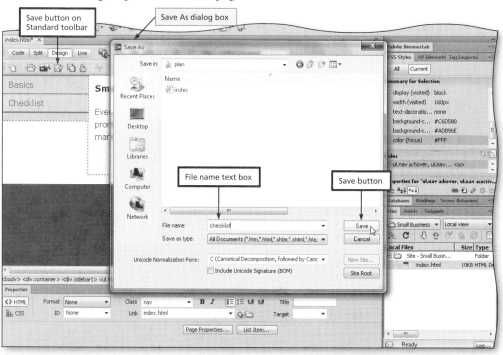

Figure 1–49

- Click the Save button in the Save As dialog box to save the second page as checklist.html (Figure 1–50).

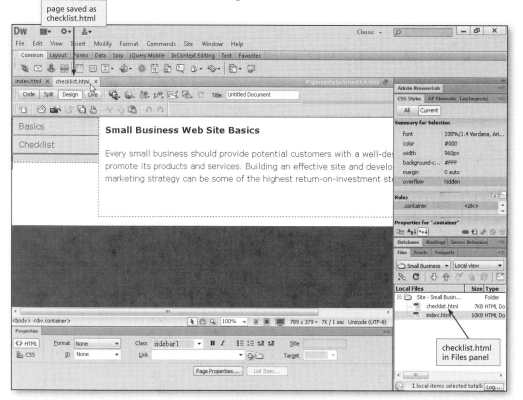

Figure 1–50

To Create an Unordered List

Using lists is a convenient way to group and organize information. Web pages can have three types of lists: ordered (numbered), unordered (bulleted), and definition. **Ordered lists** contain text preceded by numbered steps. **Unordered lists** contain text preceded by bullets (dots or other symbols) or image bullets. You use an unordered list if the items need not be listed in any particular order. **Definition lists** do not use leading characters such as bullet points or numbers. Glossaries and descriptions often use this type of list.

You can type a new list or you can create a list from existing text. When you select existing text and add bullets, the blank lines between the list items are deleted. The following steps edit the checklist.html page to include a new title and a bulleted list of design best practices for a business Web site.

1

- Select the title text, Small Business Web Site Basics, in checklist.html, and then type `Web Site Planning Checklist` to change the title text (Figure 1–51).

Figure 1–51

2

- Select the text in the paragraph below the title, and then press the DELETE key to delete the first paragraph.

- Type `The design and layout of a business Web site should be clean, simple, and professional` and then press the ENTER key to begin a new paragraph.

Figure 1–52

- Type `The navigation needs to be very simple with clear instructions for how to get to and how to buy your products/services` and then press the ENTER key to begin a new paragraph.

- Type `The site should include your business address, phone number, and e-mail address` (Figure 1–52).

- Drag to select the three checklist items (Figure 1–53).

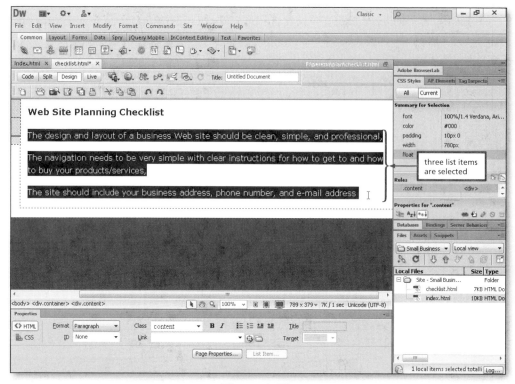

Figure 1–53

4

- In the Property inspector, click the Unordered List button to indent the text and add a bullet to each line.

- Click at the end of the third bulleted line to deselect the text (Figure 1–54).

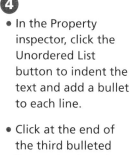

Q&A

How do I start a list with a different number or letter?

In the Document window, click the list item you want to change, click Format on the Application bar, point to List, and then click Properties. In the List Properties dialog box, select the options you want to define.

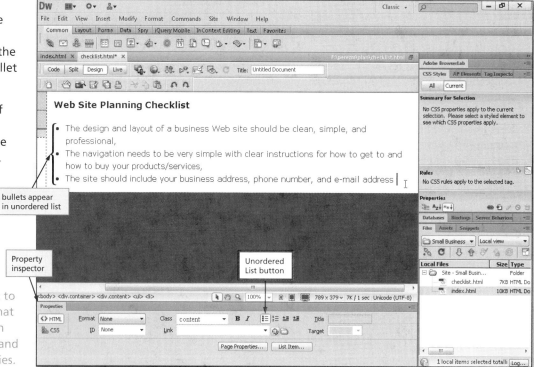

Figure 1–54

Other Ways

1. Format menu, point to List, click Unordered List
2. Right-click text, point to List, click Unordered List

To Add the Links to the Second Page

To complete the navigation, the text, Checklist, in the left column must be linked on both pages to checklist.html. The following steps link the navigation text to the second page.

- Drag the horizontal scroll bar at the bottom of the Document window to scroll to the left to display the left column.

- Select the text, Checklist, in the left column.

- Drag the Point to File button in the Property inspector to the checklist. html file on the Files panel to create a link to checklist.html (Figure 1–55).

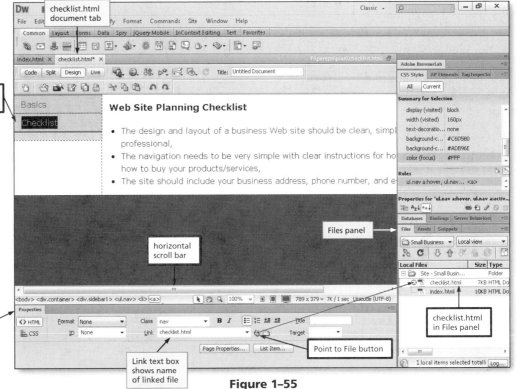

Figure 1–55

- Click the index.html document tab to display the index page.

- Drag to select the text, Checklist, in the left column of index.html.

- Drag the Point to File button in the Property inspector to the checklist.html file on the Files panel to create a link to checklist.html (Figure 1–56).

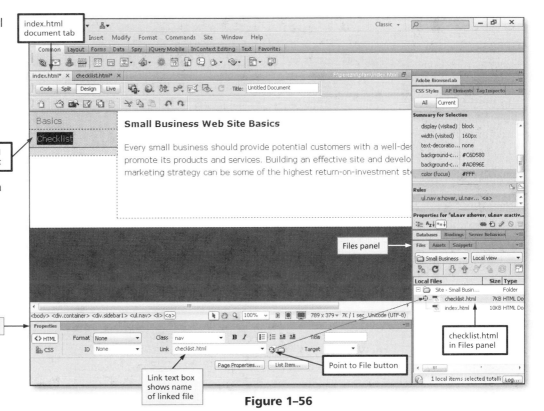

Figure 1–56

To Change the Web Page Title

A **Web page title** helps Web site visitors keep track of what they are viewing as they browse. It is important to give your Web page an appropriate title. When visitors to your Web page create bookmarks or add the Web page to their Favorites lists, they use the title for reference. If you do not title a page, the browser displays the page as Untitled Document in the browser tab, Favorites lists, and history lists. Because many search engines use the Web page title, you should create a descriptive and meaningful name. A document file name is not the same as the page title. The page title appears on the tab of the browser in Internet Explorer 7 and later versions.

The following steps change the Web page title of each page.

- With the index.html page still displayed, select the text, Untitled Document, in the Title text box on the Document toolbar to prepare to replace the text.

- Type Business Web Site Basics in the Title text box to enter a descriptive title for the Web page (Figure 1–57).

Figure 1–57

- Click the checklist. html document tab to display the checklist page.

- Select the text, Untitled Document, in the Title text box on the Document toolbar to prepare to replace the text.

- Type Business Web Site Checklist in the Title text box to enter a descriptive title for the Web page (Figure 1–58).

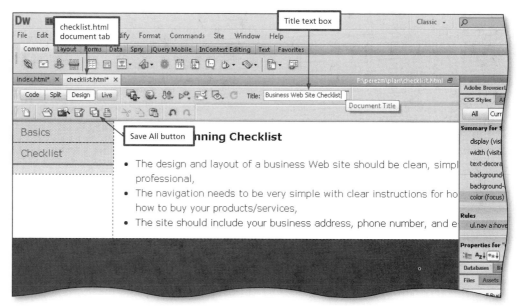

Figure 1–58

3

- Click the Save All button on the Standard toolbar to save both documents.

Q&A

What is the difference between the Save button and the Save All button?

When you click the Save button, you save changes only in the displayed document. When you click the Save All button, you save changes in all the open documents.

To Check Spelling

After you create a Web page, you should inspect the page visually for spelling errors. In addition, you can use Dreamweaver's Check Spelling command to identify possible misspellings. The Check Spelling command ignores HTML tags.

The following steps use the Check Spelling command to check the spelling of your entire document. Your Web page may contain different misspelled words depending on the accuracy of your typing.

- Click the index.html document tab.

- Click at the beginning of the document in the right column to position the insertion point.

- Click Commands on the Application bar to display the Commands menu (Figure 1–59).

Figure 1–59

- Click Check Spelling to display the Check Spelling dialog box.

- If a misspelled word is highlighted, click the correct spelling of the word in the Suggestions list or type the correct spelling, and then click the Change button.

Q&A What should I do if Dreamweaver highlights a proper noun or correctly spelled word as being misspelled?

Click Ignore or click Ignore All if proper nouns are displayed as errors.

- If necessary, continue to check the spelling and, as necessary, correct any misspelled word.

- Click the OK button when you are finished checking spelling.

- Click the Save button on the Standard toolbar to save the document.

Other Ways

1. Press SHIFT+F7

Previewing a Web Page in a Browser

After you have created a Web page, it is a good practice to test your Web page by previewing it in Web browsers to ensure that it is displayed correctly. Using this strategy helps you catch errors so you will not copy or repeat them.

Plan Ahead

Review final tasks
Before completing a Web page, perform the following tasks to make sure it is ready for others to view:

- Give your Web page a title.

- Check the spelling and proofread the text.

- Preview the page in one or more browsers so you can see how it looks when others open it.

BTW

Quick Reference
For a table that lists how to complete the tasks covered in this book using the mouse, menus, context menus, and keyboard, see the Quick Reference Summary at the back of the book or visit the Dreamweaver student companion site located at www.cengagebrain.com.

As you create your Web page, you should be aware of the variety of available Web browsers. HTML5 can help create more consistency across browsers, but the pages should be tested in each browser platform even if you are using HTML5. Each browser may display text, images, and other Web page elements differently. For this reason, you should preview your Web pages in more than one browser to make sure the browsers display your Web pages as you designed them. Be aware that visitors viewing your Web page might have earlier versions of these browsers. Dreamweaver also provides an option called Preview in Adobe BrowserLab that provides a free online comparison of your page in multiple browsers. **Adobe BrowserLab** is an online service that helps ensure your Web content is displayed as intended.

The Preview/Debug in browser button on the Document toolbar provides a list of all Web browsers currently installed on your computer. You can select a primary browser in the Preferences dialog box. Before previewing a document, save the document; otherwise, the browser will not display your most recent changes.

To Choose a Browser and Preview the Web Site

To select the browser you want to use for the preview, use the Preview/Debug in browser button on the Document toolbar. The following steps preview the Small Business Incubator Plan Web site using Internet Explorer and Firefox.

1
- Click the Preview/Debug in browser button on the Document toolbar to display a list of browsers installed on your computer (Figure 1–60).

Q&A Why do I have different browsers listed?

This list displays browsers that are installed on your local computer. If you want to test your page in other browsers, download and install multiple browsers.

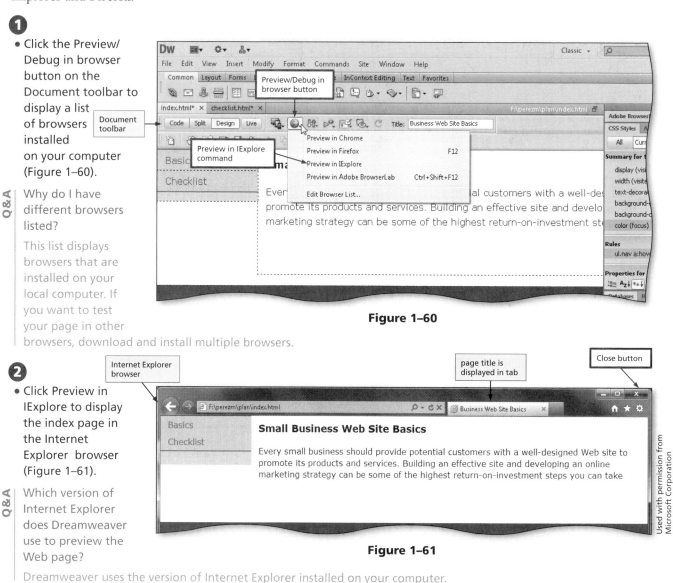

Figure 1–60

2
- Click Preview in IExplore to display the index page in the Internet Explorer browser (Figure 1–61).

Q&A Which version of Internet Explorer does Dreamweaver use to preview the Web page?

Dreamweaver uses the version of Internet Explorer installed on your computer.

Figure 1–61

- Click the Close button on the Internet Explorer title bar to close the browser.

- If the Firefox browser is installed on your system, click the Preview/Debug in browser button again, and then click Preview in Firefox to display the Web page in the Firefox browser.

- Click the Close button on the Firefox title bar to close the browser.

Other Ways

1. On File menu, point to Preview in Browser, click browser of choice

2. Right-click document, click Preview in Browser, click browser

3. F12, CTRL+F12

Dreamweaver Help

The built-in Help feature in Dreamweaver provides reference materials and other forms of assistance. When the main Help page opens, it connects to the Adobe Web site.

To Access Dreamweaver Help

The following step accesses Dreamweaver Help. You must be connected to the Web to complete this step.

1

- With the Dreamweaver program open, click Help on the Application bar to display the Help menu.

- Click Dreamweaver Help to access Adobe Community Help online.

- If necessary, double-click the title bar to maximize the window (Figure 1–62).

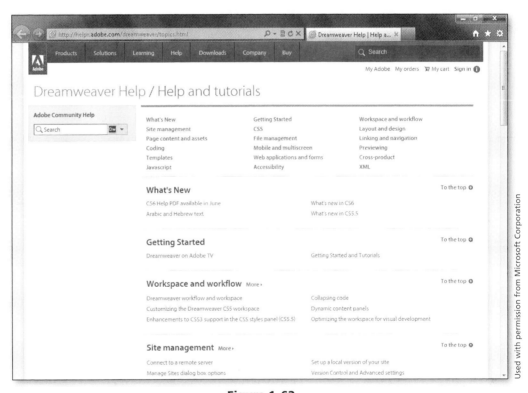

Figure 1–62

Used with permission from Microsoft Corporation

Other Ways

1. Press F1

To Quit Dreamweaver

The following step quits Dreamweaver and returns control to Windows.

- Click the Close button on the right side of the Application bar to close the window.

- If Dreamweaver displays a dialog box asking you to save changes, click the No button.

Other Ways

1. On File menu, click Exit

2. Press CTRL+Q

MAC For a detailed example of quitting Dreamweaver using the Mac operating system, refer to the "To Quit Dreamweaver" steps in the For Mac Users appendix at the end of this book.

Chapter Summary

In this chapter, you have learned how to start Dreamweaver, define a Web site, and create a Web page. You added a link and used Dreamweaver's Property inspector to connect to another page in the site. You also learned how to use an unordered list to organize information into a bulleted list. Once your Web page was completed, you learned how to save the Web page and preview it in a browser. To enhance your knowledge of Dreamweaver further, you learned the basics about Dreamweaver Help.

The following tasks are all the new Dreamweaver skills you learned, listed in the same order they were presented. For a list of keyboard commands for topics introduced in this chapter, see the Quick Reference for Windows at the back of this book. The list below includes all the new Dreamweaver skills you have learned in this chapter:

1. Start Dreamweaver (DW 28)
2. Show, Hide, and Move Panels (DW 37)
3. Reset the Classic Workspace (DW 39)
4. Display the Standard Toolbar (DW 40)
5. Access Preferences (DW 41)
6. Quit and Restart Dreamweaver (DW 45)
7. Use Site Setup to Create a Local Site (DW 46)
8. Select a Template Layout (DW 50)
9. Name and Save the Home Page (DW 51)
10. Edit Navigation Link Text (DW 53)
11. Format Text Using Heading Styles (DW 54)
12. Create a Link (DW 56)
13. Save the Home Page (DW 57)
14. Add a Second Page (DW 58)
15. Create an Unordered List (DW 59)
16. Add the Links to the Second Page (DW 61)
17. Change the Web Page Title (DW 62)
18. Check Spelling (DW 63)
19. Choose a Browser and Preview the Web Site (DW 64)
20. Access Dreamweaver Help (DW 65)
21. Quit Dreamweaver (DW 65)

Apply Your Knowledge

Reinforce the skills and apply the concepts you learned in this chapter.

Creating a New Web Page

Instructions: Start Dreamweaver. In this activity, you will define a local site, create a new Web page, and save the Web page. Next, you will give the page a title and then add text to the page. Finally, you will format the text and create an ordered list. The completed Web page is shown in Figure 1–63.

Figure 1–63

Perform the following tasks:

1. Use the Site Setup dialog box to define a new local site in the *your last name and first initial* folder (the F:\perezm folder, for example). Type `Apply` as the new site name.

2. Click the 'Browse for folder' icon and navigate to your USB flash drive. Create a new subfolder within the *your last name and first initial* folder and name it `Apply`. Open the folder, select it as the local site folder, and then save the Apply site.

3. Create a new HTML page. Save the new HTML page and type `apply1.html` as the file name.

4. Enter `Web Site Development` as the document title.

5. Type `Web Site Development Steps, Easy as 1-2-3` at the top of the page, and then press the ENTER key.

6. Type `The first step in developing a new Web site is to define a local site. A local site is defined in the Site Setup dialog box.` Press the ENTER key.

7. Type `Once the new site is defined, a new HTML page is created and saved. The new page is typically the home page and the home page is named index.html.` Press the ENTER key.

8. Type `Once the new HTML or home page is created, text is added to the page and then formatted.`

9. Press the ENTER key two times. Type your first and last names.

10. Apply the Heading 1 format to the first line of text.

11. Select paragraphs 2–4, and then click the Ordered List button in the Property inspector to make the text an ordered list.

12. Save your changes and then view your document in your browser. Compare your document to Figure 1–63. Make any necessary changes and then save your changes.

13. Submit the document in the format specified by your instructor.

Extend Your Knowledge

Extend the skills you learned in this chapter and experiment with new skills. You may need to use Help to complete the assignment.

Formatting a Web Page

Note: To complete this assignment, you will be required to use the Data Files for Students. Visit www.cengage.com/ct/studentdownload for detailed instructions, or contact your instructor for information about accessing the required files.

Instructions: Start Dreamweaver. A recreational soccer league wants to teach team managers how to create Web pages using Dreamweaver. The league president created a Web page and asks you to improve it. First, you will enhance the Web page by applying styles, aligning text, and indenting text. Finally, you will make the page more useful by adding links to text. The modified Web page is shown in Figure 1–64.

Continued >

Extend Your Knowledge *continued*

Figure 1–64

Perform the following tasks:

1. Use the Site Setup dialog box to define a new local site in the *your last name and first initial* folder (the F:\perezm folder, for example). Enter `Extend` as the new site name.

2. Click the 'Browse for folder' icon and navigate to your USB flash drive. Create a new subfolder within the *your last name and first initial* folder and name it `Extend`. Open the folder, select it as the local site folder, and then save the Extend site.

3. Using Windows Explorer, copy the extend1.html file from the Chapter 01\Extend folder into the *your last name and first initial*\Extend folder.

4. Return to Dreamweaver and then open the extend1.html document. (*Hint*: Click Open on the Welcome screen.)

5. Select the text in the first line, Customize Your Dreamweaver Interface, and apply the Heading 2 format.

6. Center-align the heading by using the Format menu on the Application bar.

7. Select the paragraph below the first line, starting with "You can customize" and ending with "customize your workspace." Use the Property inspector to apply italics to the text.

8. Select the three lines of text below the paragraph, starting with "Manage windows and panels" and ending with "Set general preferences for Dreamweaver". Indent these three lines of text by using the Blockquote button in the Property inspector or by using the Indent command on the Format menu.

9. In the last line of text, select "Dreamweaver Help". Using the Link text box in the Property inspector, link the following URL to the text: `http://helpx.adobe.com/ dreamweaver/topics.html`.

10. Place the insertion point at the end of the last sentence on the page. Press the ENTER key two times. Type your first and last names.

11. Save your changes and then view your document in your browser. Compare your document to Figure 1–64.

12. Click the Dreamweaver Help link to test your link. If the Adobe Dreamweaver Help site does not open, return to Dreamweaver and verify that you entered the correct URL in Step 9. Make any necessary changes and then save your changes.

13. Submit the document in the format specified by your instructor.

Make It Right

Analyze a Web page and correct all errors and/or improve the design.

Formatting and Checking the Spelling of a Web Page

Note: To complete this assignment, you will be required to use the Data Files for Students. Visit www.cengage.com/ct/studentdownload for detailed instructions, or contact your instructor for information about accessing the required files.

Instructions: Start Dreamweaver. You are working with a neighborhood association that wants to learn about Dreamweaver. An association member created a Web page and asks you to improve it. First, you will enhance the look of the Web page by applying styles, aligning text, and adding bullets. Next, you will make the Web page more useful by adding links to text and inserting a document title. Finally, you will check the spelling. The modified Web page is shown in Figure 1–65.

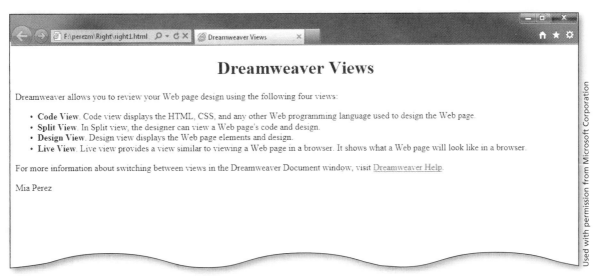

Figure 1–65

Perform the following tasks:
1. Use the Site Setup dialog box to define a new local site in the *your last name and first initial* folder (the F:\perezm folder, for example). Enter `Right` as the new site name and save it.

2. Click the 'Browse for folder' icon to navigate to your USB flash drive. Create a new subfolder within the *your last name and first initial* folder and name it `Right`. Open the folder and then select it as the local site folder. Save the Right site.

3. Using Windows Explorer, copy the right1.html file from the Chapter 01\Right folder into the *your last name and first initial*\Right folder.

4. Return to Dreamweaver and then open the right1.html document. (*Hint*: Click Open on the Welcome screen.)

5. Enter `Dreamweaver Views` as the Web page title.

6. Apply the Heading 1 format to the first line of text. Use the Format menu to center-align the heading on the page.

7. Apply bullets to the list of views, starting with "Code View" and ending with "Live View".

8. Bold each view name (Code View, Split View, Design View, and Live View).

9. Check the spelling using the Commands menu. Correct all misspelled words.

Continued >

Make It Right *continued*

10. In the last line of text, select "Dreamweaver Help". Using the Link text box in the Property inspector, link the following URL to the text: `http://helpx.adobe.com/ dreamweaver/topics.html`.

11. Place the insertion point at the end of the last sentence on the page. Press the ENTER key two times. Type your first and last names.

12. Save your changes and then view your document in your browser. Compare your document to Figure 1–65.

13. Click the Dreamweaver Help link to test your link. If it did not open the Adobe Dreamweaver Help site, return to Dreamweaver and verify that you entered the correct URL in Step 10. Make any necessary changes and then save your work.

14. Submit the document in the format specified by your instructor.

In the Lab

Design and/or create a Web site using the guidelines, concepts, and skills presented in this chapter. Labs are listed in order of increasing difficulty.

Lab 1: Creating a Family Reunion Web Site

Problem: It has been more than 10 years since the entire Hydes family celebrated together. Janna Hydes has decided to coordinate a family reunion and requests your assistance in spreading the word by creating a Web site. Janna has provided you with all of the details about the event, and she asks you to share the information on the site.

Define a new Web site, and create and format a Web page for the Hydes Family Reunion. The Web page as it is displayed in a browser is shown in Figure 1–66. The text for the Web site is provided in Table 1–1.

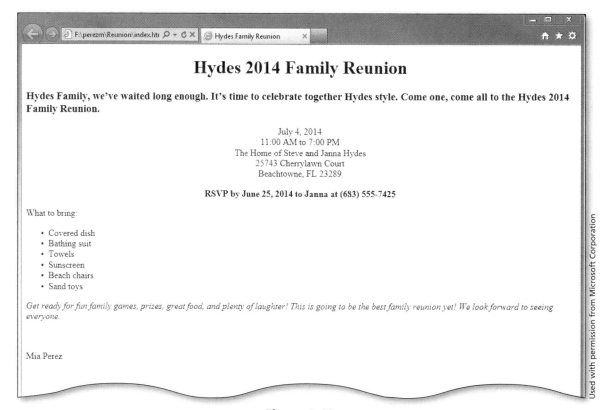

Figure 1–66

Table 1–1 Hydes Family Reunion

Section	Text
Heading	Hydes 2014 Family Reunion
Invitation paragraph	Hydes Family, we've waited long enough. It's time to celebrate together Hydes style. Come one, come all to the Hydes 2014 Family Reunion.
Item 1	July 4, 2014 11:00 AM to 7:00 PM The Home of Steve and Janna Hydes 25743 Cherrylawn Court Beachtowne, FL 23289
Item 2	RSVP by June 25, 2014 to Janna at (683) 555-7425
Item 3	What to bring: Covered dish Bathing suit Towels Sunscreen Beach chairs Sand toys
Closing paragraph	Get ready for fun family games, prizes, great food, and plenty of laughter! This is going to be the best family reunion yet! We look forward to seeing everyone.

Perform the following tasks:

1. Start Dreamweaver. Click Dreamweaver Site in the Create New list to display the Site Setup dialog box. Define a new local site by typing `Reunion` in the Site name text box.

2. Click the 'Browse for folder' icon to display the Choose Root Folder dialog box. The current path should be F:*your last name and first initial*\\ (substitute the drive letter as necessary). Create a new subfolder in the *your last name and first initial* folder. Type `Reunion` as the folder name, open and select the Reunion folder, and then save the site.

3. On the Welcome screen, click HTML in the Create New list to create a new HTML page. Save the page and type `index.html` as the file name.

4. Type `Hydes Family Reunion` in the Title text box.

5. Type the Web page text shown in Table 1–1. Press the ENTER key after typing the heading and invitation paragraph. For Item 1, press the SHIFT+ENTER keys to insert a line break after each line except the last line. Press the ENTER key after the last Item 1 line. Press the ENTER key after typing each line in Items 2 and 3. Type the closing paragraph.

6. Select the heading text and use the Property inspector to apply the Heading 1 format. Use the Format menu to center-align the heading.

7. Select the invitation paragraph and apply the Heading 3 format.

8. Select all the text for Item 1 and center-align it on the page.

9. Select the text for Item 2, use the Bold button in the Property inspector to apply the Bold style to the text, and then center-align it on the page.

10. Select all the items listed after "What to bring" and use the Unordered List button in the Property inspector to create an unordered bulleted list for the text.

11. Use the Italic button in the Property inspector to apply the Italic style to the closing paragraph.

12. Place the insertion point after the last sentence on the page. Press the ENTER key two times. Type your first and last names.

Continued >

In the Lab continued

13. Save your changes.

14. View your document in your browser and compare it to Figure 1–66. Make any necessary changes and then save your work.

15. Submit the document in the format specified by your instructor.

In the Lab

Lab 2: Creating a Recipe Web Site

Problem: Brooke Davis has acquired famous family recipes that have been passed down through generations. She wants to share recipes easily with other family members and friends. You talk to her about setting up a Web site to display the recipes. Because she likes the idea, she asks you to help her get started. You agree to develop the first two pages for her Web site.

Define a new Web site and use a fixed HTML layout to create two Web pages for Davis Family Recipes. The Web pages, as displayed in a browser, are shown in Figure 1–67. Text for the Web site recipe is provided in Table 1–2.

(a)

(b)

Figure 1–67

Table 1–2 Pork Tenderloin Recipe	
Section	**Text**
Heading	Timeless Pork Tenderloin
Ingredients section	Ingredients: 3-4 lb pork tenderloin 2 cloves garlic, chopped 1 teaspoon of kosher salt ¼ teaspoon of black pepper 1 tablespoon of olive oil
Directions section	Directions: Preheat the oven to 250 degrees. Cut numerous ½-inch slits on the top of the tenderloin. Rub the tenderloin with garlic and place garlic pieces into slits. In a small mixing bowl, combine salt, pepper, and olive oil. Rub the mixture on the tenderloin. Place the tenderloin in a shallow roasting pan and cook in the oven for 3 hours or until internal temperature reaches 160 degrees.

Perform the following tasks:

1. Start Dreamweaver. Use the Site Setup dialog box to define a new local site using `Davis Recipes` as the site name.

2. Click the 'Browse for folder' icon to display the Choose Root Folder dialog box. The current path should be F:*your last name and first initial*\\(substitute the drive letter as necessary). Create a new subfolder in the *your last name and first initial* folder. Enter `Recipes` as the folder name, open and select the Recipes folder, and then save the site.

3. On the Dreamweaver Welcome screen, click More in the Create New list. In the New Document dialog box, select Blank Page, Page Type: HTML, Layout: 2 column fixed, left sidebar. Set the DocType as HTML 5 and click the Create button. Save the new Web page using `index.html` as the file name.

4. Enter `Davis Family Recipes` as the Web page title.

5. Delete all of the text in the right column. Type `Davis Family Recipes` at the top of the right column. Press the ENTER key. (If the heading does not appear in the Heading 1 format, apply the Heading 1 format.)

6. Type `This site was created to share some famous family recipes that have been passed down over generations`. Press the ENTER key. Type `Click a recipe in the left navigation bar and start cooking!`

7. Press the ENTER key two times. Type your first and last names.

8. In the left navigation bar, replace the "Link one" text with `Home`. Replace the "Link two" text with `Timeless Pork Tenderloin`. Replace the "Link three" text with `Mom's Lasagna`. Replace the "Link four" text with `Mighty Meatballs`.

9. Delete all of the text below "Mighty Meatballs". Type `What are you hungry for tonight?` Use the Italic button in the Property inspector to apply italics to the text.

10. Save your changes.

11. To create a new document from the existing document, click File on the Application bar and then click Save As. Type `pork.html` as the file name and save the document.

12. Enter `Timeless Pork Tenderloin` as the Web page title.

13. Replace the text in the right column with the recipe text shown in Table 1–2. Press the ENTER key after typing each line or paragraph. (If the Timeless Pork heading does not appear in the Heading 1 format, apply the Heading 1 format.)

Continued >

In the Lab *continued*

14. Select the five items listed after "Ingredients" and use the Unordered List button in the Property inspector to create a bulleted list for the text.

15. Select the text, Home, in the upper-left column. Link it to the index.html page by dragging the Point to File button in the Property inspector to the index.html file in the Files panel.

16. Save your changes and close the pork.html file. The index.html document should be displayed. Select the "Timeless Pork Tenderloin" text in the left column. Link it to the pork.html page by dragging the Point to File button in the Property inspector to the pork.html file in the Files panel. Save your changes.

17. View the index.html page in your browser and compare it to Figure 1–67.

18. Test the link to Timeless Pork Tenderloin and the link to Home. If the links do not work, verify that you properly completed Steps 15 and 16. Make any necessary changes and then save your work.

19. Submit the documents in the format specified by your instructor.

In the Lab

Lab 3: Creating a Business Plan Tips Web Site

Problem: Tyler James is a business consultant who provides his clients with advice about starting their own businesses. He wants to create a Web site with tips for developing a solid business plan. He has hired you to develop his Web site.

Define a new Web site and use a liquid HTML layout to create a Business Plan Tips Web page. The Web page, as it is displayed in a browser, is shown in Figure 1–68. Text for the Web site is provided in Table 1–3.

Figure 1–68

Table 1–3 Business Plan Tips	
Section	**Text**
Heading	Business Plan Tips
Paragraph 1	Understand your target audience. Identify demographics of your target audience.
Paragraph 2	Identify your industry and describe your products and services. Identify the benefits of your products and services.
Paragraph 3	Define your business model. How many employees do you need?
Paragraph 4	Describe how you will market your products and services. What is the best way to reach your target audience?
Paragraph 5	How much money do you need to get started? How will you obtain capital?

1. Start Dreamweaver. Use the Site Setup dialog box to define a new local site using `BusinessPlan` as the site name.

2. Click the 'Browse for folder' icon and then create a new subfolder in the *your last name and first initial* folder named `BusinessPlan`. Open and select the BusinessPlan folder, and then save the site.

3. On the Dreamweaver Welcome screen, click More in the Create New list. Create a blank HTML page using the 1 column liquid, centered layout, and HTML5 as the DocType. Save the new Web page using `index.html` as the file name.

4. Enter `Business Plan` as the Web page title.

5. Delete all the text on the page, replacing it with the text shown in Table 1–3. Press the ENTER key after typing each line or paragraph.

6. Press the ENTER key two times after typing the last sentence. Type your first and last names.

7. Select all of the text below the heading and use the Ordered List button in the Property inspector to create a numbered list for the text.

8. Save your changes and then view your document in your browser. Compare your document to Figure 1–68. Make any necessary changes and then save your work.

9. Submit the document in the format specified by your instructor.

Cases and Places

Apply your creative thinking and problem solving skills to design and implement a solution.

1: Protecting Yourself from Identity Theft

Personal

You recently read an article about the growing trend of identity theft. The article has prompted you to create an educational Web site on how people can protect themselves from identity theft. Define a new local site in the *your last name and first initial* folder and name it Protect Yourself from Identity Theft. Name the new subfolder Theft. Create and save a new HTML Web page. Name the file protect. Use your browser to conduct some research on ways to protect yourself from identity theft. Create a heading for your Web page. Apply the Heading 1 format to the title and center-align the title on the page. Below the heading, create an unordered list of 10 different ways to protect yourself from identity theft. Title the document ID Theft. Check the spelling in the document. Include your name at the bottom of the page. Submit the document in the format specified by your instructor.

Continued >

Cases and Places *continued*

2: Creating a Web Site for a Literacy Promotion

Academic

You are in charge of the Literacy Committee at your university. Your mission is to promote reading for enjoyment to university students. You decide to create a Web site to inform students about the best-selling books. Define a new local site in the *your last name and first initial* folder and name it Reading for Fun. Name the new subfolder Read. Create and save a new HTML Web page. Name the file read. Use your browser to conduct some research on current best-selling books. Create a heading for your Web page. Apply the Heading 2 format to the title and center-align the title on the page. Below the heading, create an ordered list of the top 10 best-selling books. Bold three of the book titles. Title the document Reading for Fun. Check the spelling in the document. Include your name at the bottom of the page. Submit the document in the format specified by your instructor.

3: Creating a Web Site for an Accountant

Professional

Your friend is an accountant who wants to create a Web site to advertise his accountant services. You offer to help him develop a Web site. Define a new local site in the *your last name and first initial* folder and name it Accountant Services. Name the new subfolder Accountant. Create and save a new HTML Web page with a 2 column fixed, left sidebar layout. Be sure to select HTML 5 as the DocType. Name the file accountant. Delete the text in the right column. Create a heading for your Web page. Below the title, provide contact information for the accountant services. In the left column, provide links to Home, Services, Costs, and Schedule Appointment. Title the document Accountant Services. Check the spelling in the document. Delete the text below the links in the left column and type your name. Submit the document in the format specified by your instructor.

2 | Designing a Web Site Using a Template and CSS

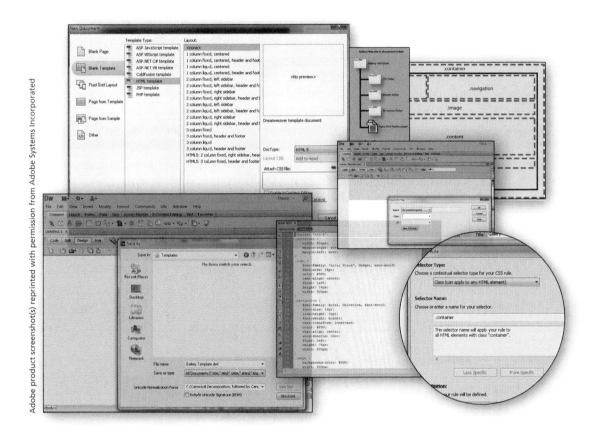

Objectives

You will have mastered the material in this chapter when you can:

- Describe the anatomy of a style sheet
- Describe the types of style sheets
- Create a Dreamweaver Web site using a template and CSS
- Save the HTML template as a .dwt file
- Define the regions of a Web page
- Describe the CSS categories

- Create a CSS style sheet
- Create a CSS rule
- Apply CSS rule definitions
- Create an editable region within a template
- Create a new page from a template
- Preview a Web page using Live view

2 | Designing a Web Site Using a Template and CSS

Designing Web Pages with CSS

By designing a Web page from scratch, you can create a page customized to your exact needs and preferences. A Web site's design provides the first impression of the credibility of an organization. Content is an important part of any site. But if the presentation of that content is not consistent, easy to navigate, and aesthetically pleasing, the site visitor will quickly leave. **CSS**, or **Cascading Style Sheets**, provides the style and layout for HTML content. CSS is the means by which a Web site's presentation is defined, styled, and modified. The newest version presented in the chapter is the CSS3 standard.

To understand the concept of CSS, imagine a chef at a cooking school teaching dozens of students. If the chef spends an hour with each student privately teaching a new cooking technique, the amount of time required to teach this same lesson over and over means it would take days to teach everyone. It makes more sense for the chef to teach the entire class the new cooking technique at once. Similarly, if your Web site has dozens of pages, changing the background color of every single page individually would be very time consuming. But with CSS, you can make one adjustment in the CSS file to display the new background color on every page of the site. CSS allows complete and total control over the style of an entire Web site.

Project — Custom Template and Style Sheet

Designing a professional Web site that appeals to your target audience begins with a plan to define the style of the site, which should suit its purpose and audience. A local family photographer has launched a new Web site using the latest HTML5 and CSS3 standards to market the company's photography services, as shown in Figure 2–1. The Gallery Portrait and Family Photography business (called Gallery for short) requires a site that meets the goals described in its business mission statement: "Our mission is to capture cherished family moments while providing our clients with an unsurpassed photography experience." The Gallery specializes in artistic family and individual portrait photos that capture the different personalities of each subject. You begin the Gallery site in this chapter, and then expand it in subsequent chapters to include information about the company's services, portfolio, pricing, session details, and contact information.

To create a unique and memorable design for the Gallery site, follow the same process that professional Web designers follow when building a site. The Gallery site uses a custom template to establish the layout of the Web pages. Attached to the template is an external style sheet, which uses CSS3 to define the design for each area of the site, including the logo, navigation, main content, and footer areas. An external CSS3 style sheet is the professional standard for styling content throughout an entire Web site. Finally, you create a page based on the template, and then customize the content to suit the first page of the site, shown in Figure 2–1.

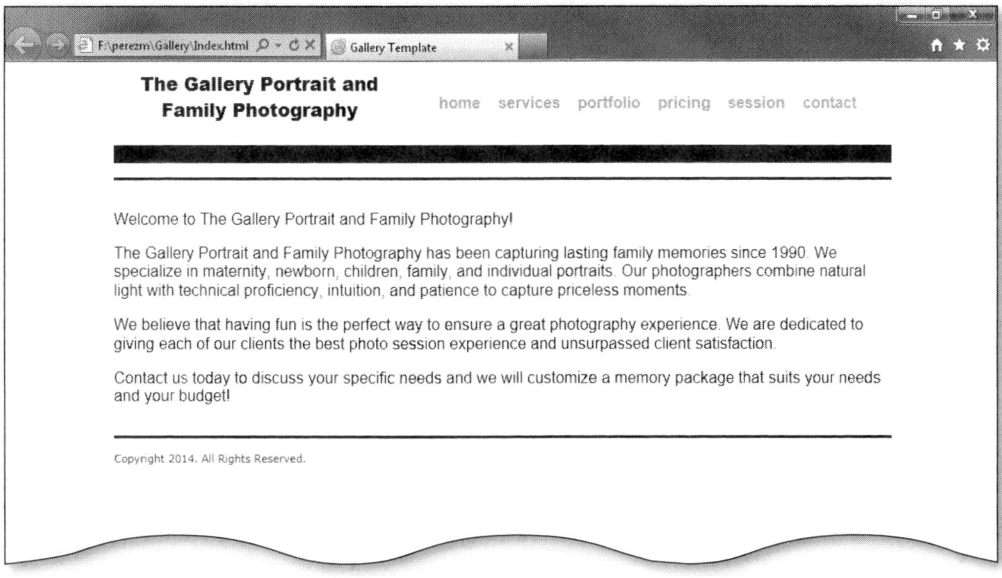

Figure 2–1

Overview

As you read this chapter, you will learn how to create the Web page project shown in Figure 2–1 by performing these general tasks:

- Create a new Dreamweaver HTML5 dynamic Web template.
- Add CSS and image folders.
- Create an external style sheet.
- Add CSS rules to the style sheet.
- Create an editable region in the template.
- Create a new page from the template.

General Project Guidelines

When creating a Dreamweaver Web site, the individual actions you perform and decisions you make will affect the appearance and characteristics of the entire Web site. When creating the opening page for a business Web site, as shown in Figure 2–1, you should follow these general guidelines:

1. **Create an HTML template.** Choose a blank HTML template so you can design a custom layout.

2. **Determine the layout and formatting of the Web site.** Before designing a CSS layout, carefully consider which site structure will convey your message effectively. Maintaining consistency in your page layout and design helps to ensure a productive user experience.

3. **Understand the anatomy of a CSS style sheet.** Recognize the elements of CSS styles.

4. **Create a CSS style sheet.** Define the element properties and values within the CSS style sheet.

   More specific details about these guidelines are presented at appropriate points throughout the chapter. The chapter also identifies the actions performed and decisions made regarding these guidelines during the creation of the Web site home page shown in Figure 2–1.

Plan
Ahead

Anatomy of a Style Sheet

The **World Wide Web Consortium (W3C)** — an international community where member organizations, a full-time staff, and the public work together to develop Web standards — mandates that CSS style sheets are the core of Web design. When you customize a Web site by creating, modifying, and applying CSS styles, all the Web pages in the site share a consistent look even as the content changes. A **style** is a rule that defines the appearance and position of text and graphics. A **style sheet** is a collection of styles that describes how to display the elements of an HTML document in a Web browser. You can develop CSS style sheets by entering code or by using Dreamweaver's CSS toolset. Designers typically define a style in the style sheet for a Web site and then apply the style to content in many locations throughout the site. Separating style from content means you can change a Web site's appearance easily. If you modify a CSS style, the site updates any content to which you applied that style to reflect the modifications.

A style sheet consists of **CSS rule definitions** that specify the layout and format properties that apply to an element, such as a heading, bullets, or a paragraph. For example, the heading at the top of each page could be defined in a CSS rule stating that this heading always appears as blue, 20-point, underlined text. The term, cascading, in Cascading Style Sheets refers to a sorting order that determines whether one style has precedence over another if two competing style rules affect the same content.

The CSS style sheet shown in Figure 2–2a lays the foundation for the design of the Web page in Figure 2–2b.

(a) Layout.css style sheet

(b) Home page for the Gallery Web site

Figure 2–2

Because a central style sheet provides the layout code for all pages within the site, style sheets originally were developed in the late 1990s to reduce the size of HTML files. In addition to smaller file sizes, CSS style sheets offer the following benefits, which have changed the architectural design of the Web:

- Faster download times because the styles are separated in a style sheet from the HTML code
- Reduced design expenses because one change in the style sheet updates the entire site
- Improved accessibility for site users who have disabilities
- Improved consistency in design and navigation throughout the site

Understanding the Structure of a Style

A style, also called a rule, uses CSS code to specify how to format an element or a section of a Web page. The anatomy of a style is shown in Figure 2–3. Although Dreamweaver automatically generates the CSS code you need after you make selections in the CSS rules dialog boxes, it is important to understand the elements of a style so you can make additional changes to the code.

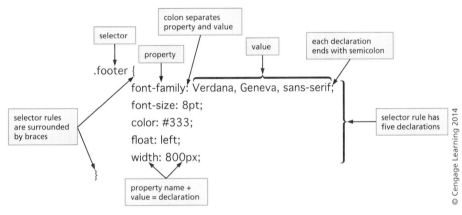

Figure 2–3

Selector   Every CSS style begins with a **selector**, which is the name of the style. The selector informs the browser that a special class or element is styled a certain way. Class selectors begin with a period, as in .content, and ID selectors begin with a hash symbol, as in #right. A **class** identifies a particular region or division on a page, such as the footer, that can be customized. For example, the style shown in Figure 2–3 formats a special class named .footer according to the properties listed within the curly braces. A selector can have multiple properties.

Declaration   Every CSS style also has a **declaration**, which defines the details of a style. In Figure 2–3, the first declaration indicates that the font-family for the .footer selector is Verdana, Geneva, sans-serif. A declaration includes a property and a value, and ends with a semicolon. The .footer selector has five declarations within the curly braces.

Property   A **property** identifies the type of formatting to apply, such as the font family, color, or width. In Figure 2–3, the first declaration is for the font-family property.

Value   The **value** of a property specifies the exact formatting to apply. Separate the property and value with a colon. In Figure 2–3 on the previous page, the width property is set to 800 px, indicating that the footer element should be 800 pixels wide. The font-family property has three values: Verdana, Geneva, and sans-serif. The Web site uses the first value, Verdana, if that font is available on the visitor's browser. If Verdana is not installed on the visitor's browser, the Web site uses the next value, Geneva, if that font is available. If the first two fonts are not available, the Web site uses the next value, sans-serif, as a generic font to format the text. Separate multiple values for a property by commas. If a value contains more than one word, place quotation marks around the value.

Declaration Block   The CSS style code found between the two curly braces is considered a declaration block within the style sheet.

Identifying Types of Style Sheets

The single selector displayed in Figure 2–3 applies only to the footers throughout the Web site. To create a more comprehensive rule set, use a style sheet to specify a collection of CSS rules as shown in Figure 2–4.

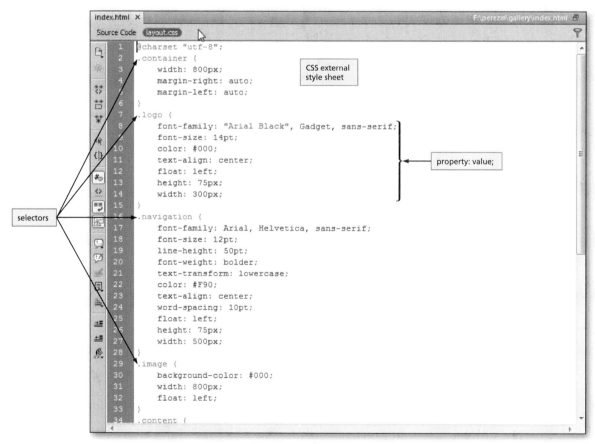

Figure 2–4

The style sheet in Figure 2–4 is an external style sheet. However, you can define CSS rules in an external style sheet, in an internal style sheet, or as inline styles. **External style sheets**, the most commonly used type of style sheet, allow you to store the code for the site styles in a separate document. Dreamweaver takes care of linking the external style sheet to the HTML file by automatically creating code, but you will strengthen your Web designer skills if you understand what is happening under the hood. If you are coding a Web page manually, you need to place a link tag in the HTML code (the content of the page) to attach the external style sheet to the page and apply the style definitions. You typically place code similar to the following example in the HTML <head> section of the page:

```
<link href="CSS/layout.css" rel="style sheet" type="text/css">
<style type="text/css">
```

The **link tag** forms a relationship between the current document and an external style sheet, which is named layout.css in this case. External style sheets are text files with a .css file name extension. An external style sheet is a perfect way to format a large Web site because the styles are applied automatically to all pages in the site. Site management is simple when using an external style sheet because you can update the style in one place in the style sheet and then apply the change throughout the site. Separating the page content from presentation makes it much easier to maintain the appearance of your site.

Another type of style sheet, called an **internal style sheet**, applies formatting styles within an HTML document. In a multiple-page Web site, a single page may have a unique layout that differs from the layouts specified in the external style sheet. You can use an internal style sheet to create a distinct look for this one page only. You embed internal styles in the <head> tag using the <style> tag, as shown in the following HTML code:

```
<head>
<style type="text/css">
body {background-color:navy;}
</style>
</head>
```

This style rule sets the background color for body text to navy. Internal styles take precedence over external styles. If an external style sheet sets the background color of the entire site to light blue, the internal style sheet overrides that rule and displays the background color as navy on this page only. The internal style overrides the external style because the internal style is specific to this single page. HTML pages with internal styles can take longer to load, but the advantage of having one page with its own style rules makes it worthwhile.

The third approach to defining styles is to include inline styles within your Web site. **Inline styles** allow you to insert a style rule within an HTML tag in an HTML page. For example, suppose you want to display only one heading in red. Add an inline style within the heading tag to format the heading, as shown in the following code:

```
<h1 style="color: red;">The Photographer's Gallery</h1>
```

When the inline style code is placed within a tag, it affects that tag only. The other heading styles are not affected by this inline style. Inline styles override external and internal style coding styles. Excessive inline styles can create a slow-loading page that is also difficult to edit, so consider using inline styles very rarely in your site development.

Creating a Dreamweaver Web Template

The design of the Gallery Web site should have a consistent look and feel throughout every page. Instead of using a ready-made design template to create a page as you did in Chapter 1, in this chapter, you create a custom template from scratch. A custom template provides a basic layout for the entire site using design elements that you specify. Templates are best used when you are creating a large site where every page shares the same design characteristics such as the logo, background, font, and arrangement. When you save an HTML page as a Dreamweaver Web Template, Dreamweaver creates a template folder at the root level of the local root folder and generates a **.dwt** file that becomes the design source for all the pages that you generate from it. The .dwt extension stands for Dreamweaver Web Template and is associated with a special type of Web document that adds structure and layout to a page.

Dreamweaver templates have a number of design layout regions, or divisions; some can be edited and others cannot. By creating a Dreamweaver Web Template, you can include editable, unlocked regions for adding content to the page. An **editable region** on a Web page is an area where other Web page authors can change the content. For example, you would locate a calendar of upcoming events in an editable region of a page so that anyone designing the page could modify or update the calendar as necessary to keep it current. A template can also have **noneditable regions**, which are sections with static, unchanging content. By using noneditable regions in a template, you prevent changes to certain areas, such as a navigation bar, and preserve the consistent layout of each page based on the template.

> **Create an HTML template**
> Before you create a Web site, you must determine the look and feel of each page in the site. By creating a common template for the site that can be applied to each page of the site, the entire Web site maintains a cohesive, consistent presentation. After creating a template, you use it to create Web pages that share the same layout, style, and content. Place this unvarying content in the noneditable regions of the template. The template also should include editable regions for elements that vary from page to page, such as the page heading and descriptive text. Attach a style sheet to the template so that all the pages created from the template use the same CSS styles.

Organizing the Site Structure

Carefully organizing a business or personal Web site from the start can save you frustration and problems with navigation. You save the Gallery site in a root folder on your USB drive. Within the root folder, you can create additional folders and subfolders to organize images, CSS files, templates, and other objects for the site. In this chapter, you create two folders within the Gallery site to hold the CSS and image resources necessary for the design of the project. After you define a site using Dreamweaver's Site Setup feature, you can create the folder hierarchy shown in Figure 2–5.

In Figure 2–5, the root folder for the Gallery site contains a CSS folder, an Images folder, a Templates folder, and the index.html file, representing the home page that appears when you open the site. An external style sheet named layout.css resides within the CSS folder. When you add images to the Gallery site in Chapter 3, you place them in the Images folder so that when you want to insert an image into a page, you know where to find it. Dreamweaver automatically creates the Templates folder when you save an HTML page as a template file with the .dwt extension. In this case, the Dreamweaver Web Template is named Gallery Template and is stored in the Templates folder.

Gallery Web site in the perezm folder

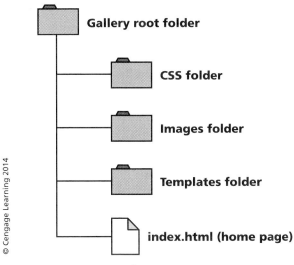

Gallery root folder

CSS folder

Images folder

Templates folder

index.html (home page)

© Cengage Learning 2014

Figure 2–5

To Start Dreamweaver

If you are stepping through this project on a computer and you want your screen to match the figures in this book, you should change your computer's resolution to 1024 × 768 and reset the Classic workspace. For more information about how to change the resolution on your computer, read the "Changing the Screen Resolution" appendix.

The following steps, which assume Windows 7 is running, start Dreamweaver based on a typical installation. You may need to ask your instructor how to start Dreamweaver for your system.

1 Click the Start button on the Windows 7 taskbar to display the Start menu and then type `Dreamweaver CS6` in the 'Search programs and files' box.

2 Click Adobe Dreamweaver CS6 in the list to start Dreamweaver and display the Welcome screen.

3 If the Dreamweaver window is not maximized, click the Maximize button next to the Close button on the Application bar to maximize the window (Figure 2–6).

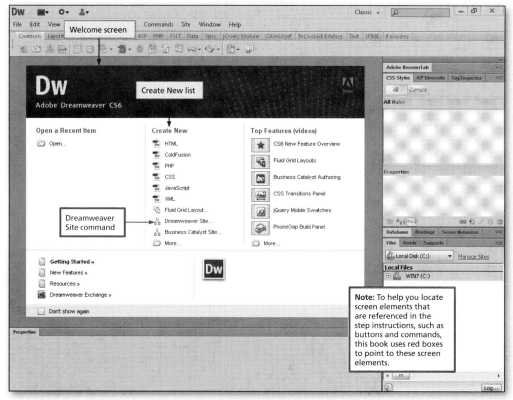

Figure 2–6

To Create a New Site

The following steps create a site named Gallery for the new photography studio Web site.

1

- Click Dreamweaver Site in the Create New list to display the Site Setup dialog box.

- Type `Gallery` in the Site name text box to name the site.

- Click the 'Browse for folder' icon to display the Choose Root Folder dialog box, and if necessary, navigate to your Removable Disk (F:) drive, click the root folder named with your last name and first initial (such as perezm), and then click the Open button to display the subfolders in the root folder.

- Click the Create New Folder button to create a folder, type `Gallery` as the name of the new folder, press the ENTER key, and then click the Open button to create the Gallery subfolder and open it.

- Click the Select button to display the Site Setup for Gallery dialog box (Figure 2–7).

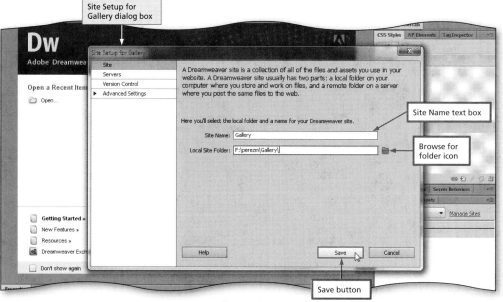

Figure 2–7

• Click the Save button to save the site settings and display the Gallery root folder in the Files panel (Figure 2–8).

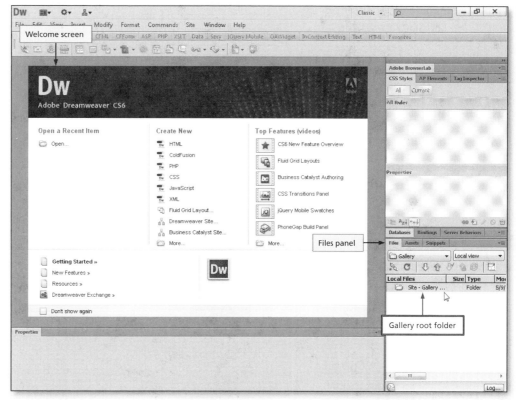

Figure 2–8

To Create Folders for the Image and CSS Files

The Gallery site uses an image file to display a background for the Web pages and a CSS file to set the Web page layout. To make it easy to manage files for a Web site, which can quickly grow to dozens of files, you should store the files in **resource folders,** which are subfolders in the root folder for the site. In each resource folder, store a single type of file. For example, store all the photos, line drawings, backgrounds, and other graphics in the Images folder. Store all the CSS style sheets in the CSS folder. The following steps create two resource folders for the image and CSS files.

• Right-click the Gallery root folder in the Files panel to display the folder's context menu (Figure 2–9).

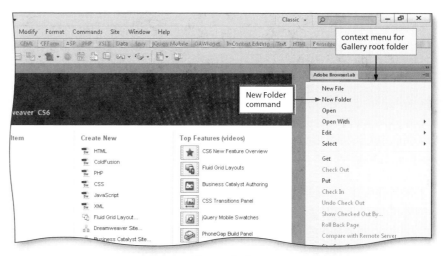

Figure 2–9

- Click New Folder on the context menu to create a new folder.

- Type `CSS` to name the first resource folder in the Gallery site, and then press the ENTER key to create the CSS folder (Figure 2–10).

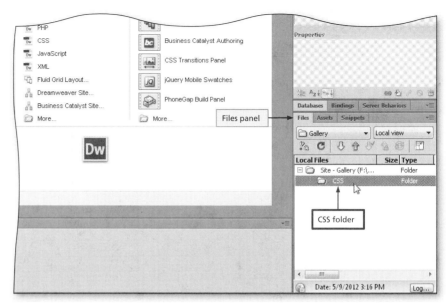

Figure 2–10

- Right-click the Gallery root folder in the Files panel to display the folder's context menu.

- Click New Folder on the context menu to create a new folder.

- Type `Images` to name the folder.

- Press the ENTER key to name the second resource folder (Figure 2–11).

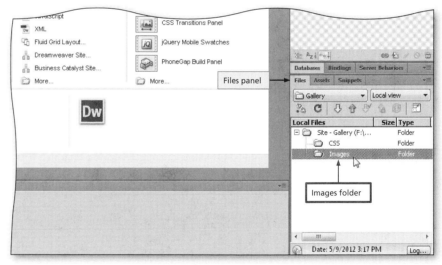

Figure 2–11

To Create a Blank HTML Template

Dreamweaver CS6 provides a blank HTML template with no predefined layout for use in developing a template from scratch. After defining a site, the next step is to design a consistent layout of elements that appear on each page of the site. Save the HTML template as a Dreamweaver Web Template so you can use it to create pages for the site. If you modify the template, you immediately update the design of all pages attached to the template.

To begin a new HTML layout, you use the New Document dialog box, which provides options for creating blank pages, blank templates with liquid and fixed layouts, and blank templates without any predefined layout. After creating a template, you save it with a specific name. By default, Dreamweaver stores the new template in a folder named Templates in the root folder of your site. The following steps create a new blank HTML template.

1

- Click the More folder on the Welcome screen to display the New Document dialog box (Figure 2–12).

What if I do not see a Welcome screen?

Instead of clicking the More folder on the Welcome screen, you can click File on the Application bar and then select New to display the New Document dialog box.

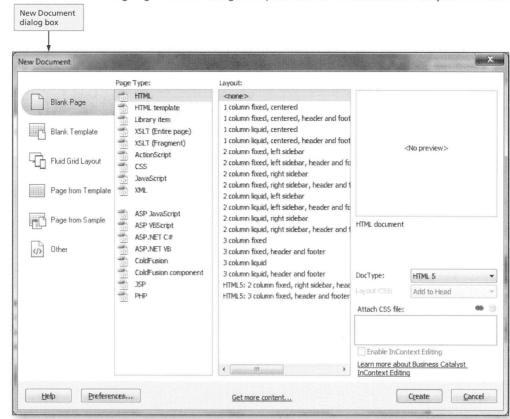

Figure 2–12

2

- Click Blank Template in the left pane to specify you are creating a blank template.

- Click HTML template in the Template Type list to select an HTML template.

- If necessary, click <none> in the Layout list to create a blank HTML template with no predefined layout.

- If necessary, click the DocType button and then click HTML 5 to set the DocType to HTML5 (Figure 2–13).

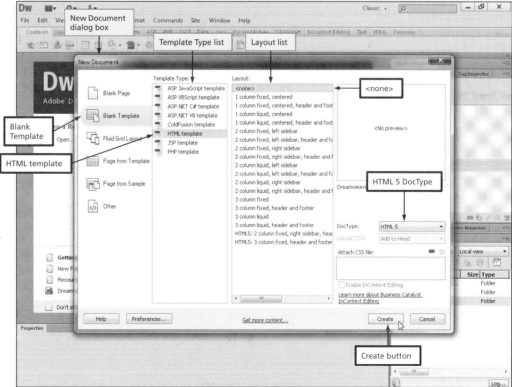

Figure 2–13

- Click the Create button to display the blank HTML template in the Document window (Figure 2–14).

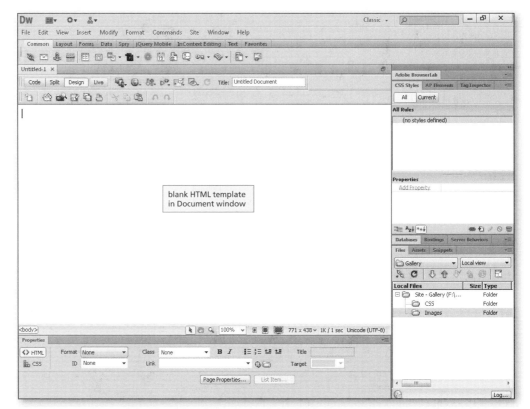

Figure 2–14

To Save the HTML Page as a Template

When you create a template, Dreamweaver uses the name of the site followed by *Template.dwt* as the name of the template, such as Gallery Template.dwt. When saving the template, a dialog box appears reminding you that the template does not have any editable regions. Recall that an editable region is an area in a template that contains text or other objects users can edit. For example, if every Web page based on the template will have a different main heading, include the heading in an editable region. You define editable regions later in this chapter. The following steps save the blank HTML template as a Dreamweaver Web Template.

1

- Click File on the Application bar to display the File menu (Figure 2–15).

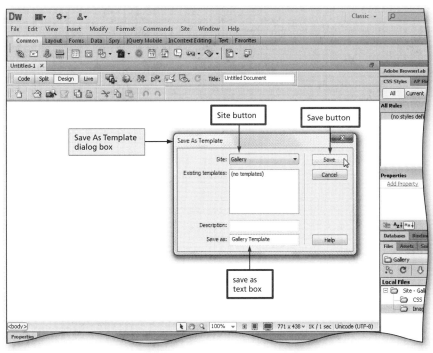

Figure 2–15

2

- Click Save on the File menu to display the Save As Template dialog box.

- If a warning dialog box is displayed, click the OK button to close the dialog box.

- If necessary, click the Site button to display the site list and then click Gallery to select the Gallery site.

- If necessary, select the placeholder text in the Save as text box, and then type `Gallery Template` to name the Dreamweaver Web template (Figure 2–16).

 Why is the Save As Template dialog box displayed when I select Save on the File menu?

Because this is the first time you are saving the template, Dreamweaver displays the Save As Template dialog box so you can specify a new name for the file.

Will the template include editable regions?

Yes. You create the editable regions later in this chapter and in other chapters.

Figure 2–16

3

- Click the Save button to save the Dreamweaver Web template as Gallery Template.dwt in a folder named Templates in the Gallery file hierarchy.

- Click the expand icon for the Templates folder in the Files panel to expand the Templates folder and display the Gallery Template file (Figure 2–17).

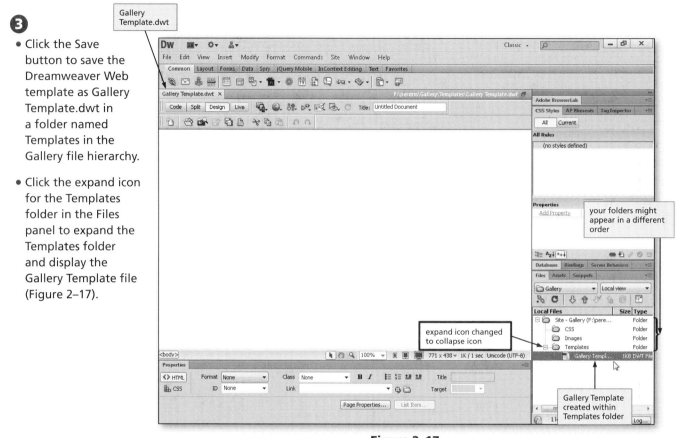

Figure 2–17

Adding CSS Styles

The Gallery Web site will use CSS styles to structure the layout of the pages. Instead of cluttering each HTML page in the site with individual style tags for each element, CSS rules style an element with specified properties such as font type, size, and color throughout the site.

Before you can take advantage of the power of CSS styles, you should organize the content of the Web pages and identify the sections to which you will apply certain types of CSS styles. The most common way to organize content involves dividing the page into different regions, or divisions. Figure 2–18 shows each region of the page sketched during the planning phase of the Web site project. These six regions of the page are the building blocks of the Gallery site design.

The easiest way to format elements within a site is by using specific CSS styles within each region. You can apply positioning and formatting styles to text, images, tables, and other elements in each region. The formatting styles determine the appearance of individual elements by making text in the navigation region, for example, bold, orange, 12-point lowercase Arial text. The positioning styles create simple to complex layouts, arranging blocks of text, graphics, or images on the page. During the design phase of a Web site project, designers often use a wireframe to sketch the layout of the page as shown in Figure 2–18. A **wireframe** is a block diagram that shows the main elements of a Web page layout as boxes with brief descriptions.

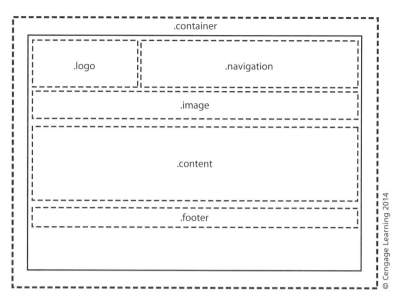

Figure 2–18

Each Dreamweaver CSS layout requires its own style sheet containing each region and the associated style rules necessary to make the layout work. An external style sheet declares style rules for each region, also called a class. Each class selector, identified with a beginning period, specifies a style for a group of elements.

Plan Ahead

Determine the layout and formatting of the Web site

When designing a Web site from scratch, you should consider the types of elements that you must format to create a consistent look across each page in the site. Consider each CSS rule that is needed to format every element:

• Determine the layout of the site, including the location of the logo, navigation, main content area, and footer.

• Plan the formatting necessary for each Web page element to determine the CSS rules for font styles, color, and heading size.

Recall that CSS is the current standard for formatting Web page elements. In addition, CSS is the standard for Web page layout. Instead of using HTML techniques, which involve tables or frames to structure content, CSS uses the **div tag**, an HTML tag that acts as a container for text, images, and other page elements. When laying out pages with CSS, you place the div tag around text, images, and other page elements to position the regions of content on the page. You can place other tags within a div tag.

BTW

Div Tags
To see examples of div tags used in CSS layouts, you can create a test page from a CSS layout listed in the New Document dialog box.

To Add a Div Tag

Because each region of the CSS layout is associated with a div tag, you need to insert a div tag for each region of content in the Web page or template. The Gallery site has six regions, so you insert six div tags. After naming a div tag, set the CSS rule definitions for the font, position, border, and other properties of that region. The outer region of the Web site, as shown in Figure 2–18, is called the .container class because it contains all the other regions. When adding a div tag, you specify where to store the CSS styles that apply to that region: in this document only, in an existing style sheet, or in a new style sheet. For the first div tag in a page or template, specify a new style sheet. Dreamweaver requests a name and location for the new .css file. In this case, name the new style sheet layout.css and store it in the CSS folder in the Gallery site. The following steps insert a div tag for the container region and save the CSS style sheet.

- On the Document toolbar, drag to select the text, Untitled Document, in the Title text box.

- Type `Gallery` as the title of the template page (Figure 2–19).

Figure 2–19

- Click the Document window, and then click the Insert Div Tag button on the Insert bar to display the Insert Div Tag dialog box (Figure 2–20).

Q&A What is the purpose of inserting a div tag?

You are inserting a div tag to create a division in the Gallery template.

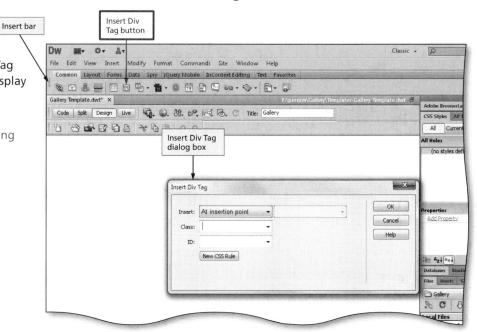

Figure 2–20

- If necessary, click the Class text box and then type `container` to name the div tag (Figure 2–21).

Q&A Do I also enter a name in the ID text box?

No. You create a div tag as a class or an ID. You can apply a class div tag to any other tag on the page. You can apply an ID div tag only once on a page.

Q&A Should I type a period before the class name?

No. A period is not necessary because Dreamweaver automatically places a period before the class name in the next dialog box.

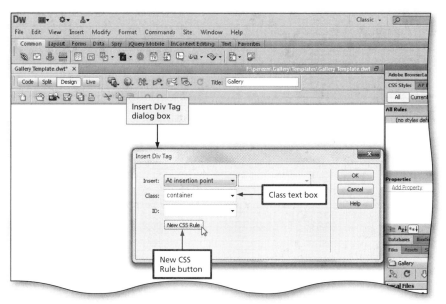

Figure 2–21

4

- Click the New CSS Rule button to display the New CSS Rule dialog box (Figure 2–22).

What is the purpose of the New CSS Rule dialog box?

Use this dialog box to add a CSS rule that defines the class you created; in this case, the container class.

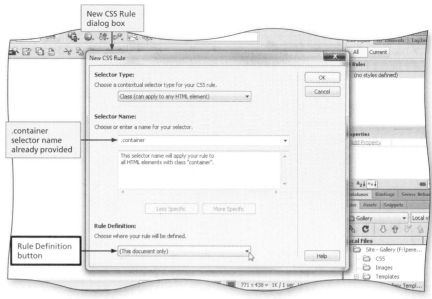

Figure 2–22

5

- Click the Rule Definition button and then click (New Style Sheet File) in the Rule Definition list to select the location in which you want to define the rule, which is in a new style sheet file (Figure 2–23).

When (New Style Sheet File) is selected, where does Dreamweaver save the CSS style sheet?

When you determine that the rule definition should be saved in a new style sheet, click the OK button to display a dialog box requesting where to save the .css file.

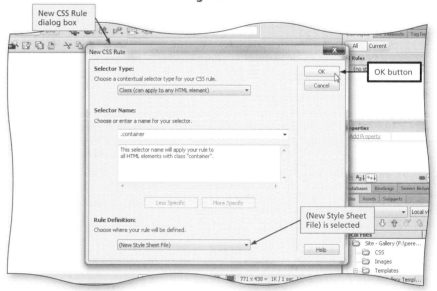

Figure 2–23

6

- Click the OK button to create a new CSS rule for the container region and to display the Save Style Sheet File As dialog box (Figure 2–24).

Figure 2–24

- Double-click the CSS folder in the Save Style Sheet File As dialog box to select the file location for the style sheet.

- Click the File name text box and then type layout to name the CSS style sheet within the CSS folder (Figure 2–25).

Q&A

Which file type should I select when I save the style sheet?

Dreamweaver automatically selects .css as the default file type.

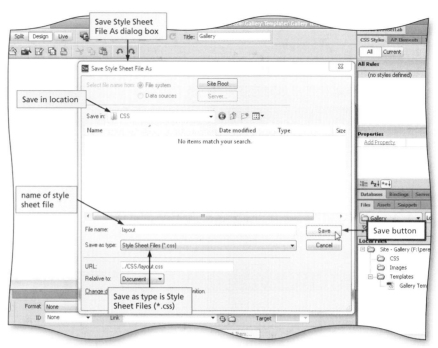

Figure 2–25

- Click the Save button to save the style sheet as layout.css and display the 'CSS Rule Definition for .container in layout.css' dialog box (Figure 2–26).

Q&A

Why is a CSS Rule Definition dialog box displayed after I save the style sheet?

You use this dialog box to specify the details of the styles to apply to the container region by selecting a formatting category and the appropriate value for each style rule.

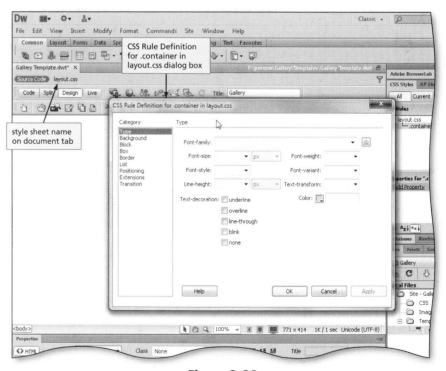

Figure 2–26

Other Ways

1. Click style sheet on CSS Styles panel, click New Rule button

Setting CSS Rule Definitions

The layout.css style sheet contains the CSS rules that define the styles in the Gallery Web site. Instead of entering code by hand, Dreamweaver CS6 provides a CSS Rule Definition dialog box that allows Web designers to define styles easily and effectively for CSS element rules. The CSS Rule Definition dialog box appears when you are creating or modifying styles.

BTW

Creating CSS Rules
Besides using the New CSS Rule button on the CSS Styles panel, you can click the Targeted Rule button on the Property inspector and then click New CSS Rule to create a CSS rule.

Plan Ahead

Create a CSS style sheet
To create an external style sheet, you need to define which rules to set in each division, or region, of the site. Six div tags with detailed CSS rules change the default settings of the font, color, margins, and other CSS properties. By defining styles in an external style sheet, you can apply the styles in any page in the site connected to that style sheet. Because editing a style in the external style sheet updates all instances of that style throughout the site, external style sheets are the most powerful and flexible way to use styles.

The CSS Rule Definition dialog box in Figure 2–26 consists of nine categories of style rules. Table 2–1 describes each category.

Table 2–1 Categories in the CSS Rule Definition Dialog Box

Category	Purpose
Type	Determines the appearance and format of text for the selected style
Background	Specifies the background color or background images to display as the page background
Block	Provides option styles to space and align text according to your custom settings
Box	Defines the spacing and placement of elements on a page, such as the location of an image within a defined region
Border	Specifies border styles, width, and color values for one or all edges of borders for text, images, and other Web elements
List	Defines list types, custom bullet images, and unique positioning selections
Positioning	Prescribes the placement of CSS elements within the page, which increases a designer's creative control over the appearance of a Web site
Extensions	Determines page breaks for printing and customizes the appearance of elements on the page
Transition	Enables animation changes in CSS values to occur smoothly over a specified duration

To Select CSS Rule Definitions

The .container class within the style sheet provides specifications for the width of the page and for the margin settings. The .container class is the default name for the large container that holds the other classes with the style sheet. In this case, you set the width of the page to 800px. **Pixels (px)** is the measurement unit for setting the dimensions of the container. The measurement value, such as 800, and the unit, such as px, typically are noted without a space between them. The **margin** determines the amount of space to maintain between the container region and the borders of the browser window. Here, you set the left and right margins to auto, which centers the container region horizontally within the browser window. The following steps define the CSS rule definitions for the .container class.

1

- Click Box in the Category list of the 'CSS Rule Definitions for .container in layout.css' dialog box to set the CSS rules for the layout of the container class (Figure 2–27).

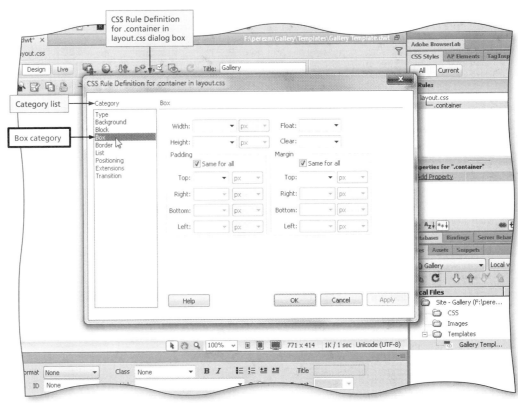

Figure 2–27

2

- Click the Width box and then type 800 to set a width of 800 pixels for the .container class.

- Click Same for all in the Margin section to remove the check mark from the 'Same for all' check box.

- Click the Right box arrow in the Margin section, and then click auto to set the right margin to center automatically within the browser window.

- Click the Left box arrow in the Margin section, and then click auto to set the left margin to center automatically within the browser window (Figure 2–28).

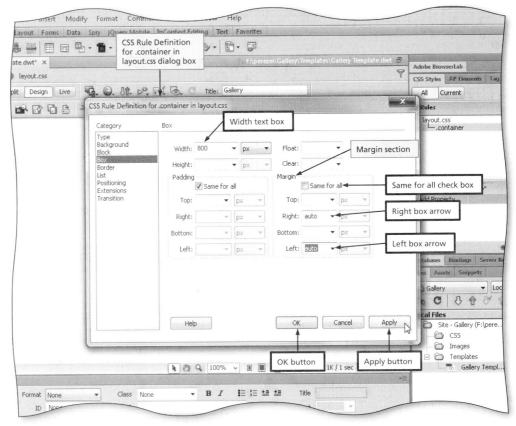

Figure 2–28

3

- Click the Apply button to apply the CSS rules for the .container class within the layout.css file.

- Click the OK button (CSS Rule Definition dialog box) to define the CSS rules for the .container class.

- Click the OK button (Insert Div Tag dialog box) to close the Insert Div Tag dialog box.

- If necessary, select the text 'Content for class "container" Goes Here' in the Document window, and then press the DELETE key to delete the text from the container region (Figure 2–29).

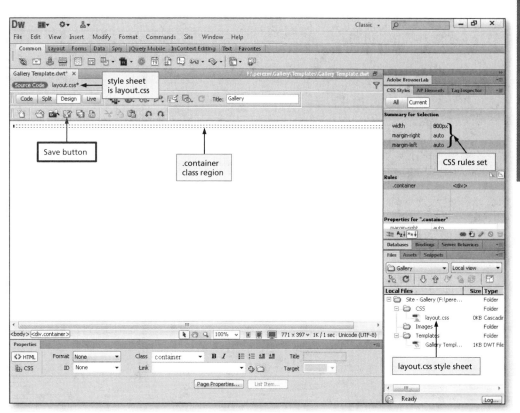

Figure 2–29

Q&A Why do I click the Apply button instead of the OK button in the CSS Rule Definition dialog box?

Click the Apply button to apply the styles in the current category. You can select another category and then set styles in that category. When you click the OK button, you apply all the selected styles in all the categories.

Q&A Why are dashed lines displayed in the Document window?

The dashed lines define the region set by the .container class. Later, you define other regions within the .container region.

Q&A Where are the classes listed for the style sheet?

The classes are listed on the CSS Styles panel on the right side of the Dreamweaver window.

4

- Click the Save button on the Standard toolbar to save your work.

Q&A What should I do if a dialog box notes that the template does not have any editable regions?

Click the OK button. You will add an editable region later in this chapter. Click the OK button each time this dialog box appears in this chapter.

Other Ways

1. Select class on CSS Styles panel, click Edit Rule button

Break Point: If you wish to take a break, this is a good place to do so. To resume at a later time, start Dreamweaver, if necessary, open the file called Gallery Template, and continue following the steps from this location forward.

To Add the Logo Div Tag and Define Its CSS Rules

A business typically uses its company logo, which can consist of an image and text, to create a recognizable reference to that business. In the Gallery site, the logo consists of the text, The Gallery Portrait and Family Photography, and an image. You add the image in the next chapter. Before you add the text, you define a layout region in the style sheet so you can control the placement and appearance of the region. In this case, you define a region named .logo in the layout.css style sheet to determine the custom arrangement of the Gallery logo. Next, you set the CSS rules for the region by defining the Type property such as font-family and font size, Block properties such as text alignment, and Box properties that set the size of the region. The **float property** determines where text and other objects should float around the region. In Chapter 3, you add an image to the template that should float to the left of the logo. The following steps define the CSS rule definitions for the .logo class.

- If necessary in the Gallery template, click within the container region.

- Click the Insert Div Tag button on the Insert bar to display the Insert Div Tag dialog box.

- If necessary, click the Class text box and then type logo to name the div tag (Figure 2–30).

Q&A

How can I tell if I am clicking within the container region?

If the insertion point is within the container region, <div.container> appears on the status bar.

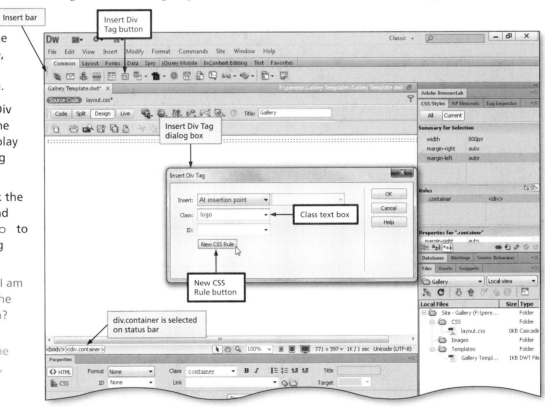

Figure 2–30

- Click the New CSS Rule button to display the New CSS Rule dialog box for adding a new CSS rule that defines the logo class (Figure 2–31).

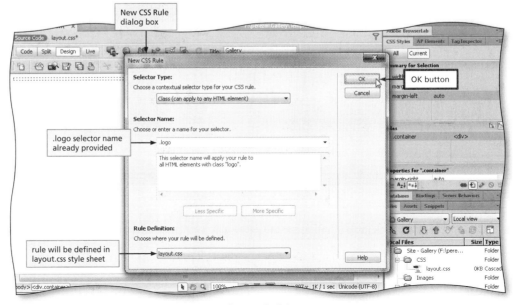

Figure 2–31

3

- Click the OK button to add the .logo selector to the layout.css style sheet and display the 'CSS Rule Definition for .logo in layout.css' dialog box (Figure 2–32).

Q&A Should I create another style sheet for the .logo class?

No. In this case, you save the six style classes within the same style sheet named layout.css.

Q&A How many classes can be added to the style sheet?

You can identify as many classes as you need to define the style of your site.

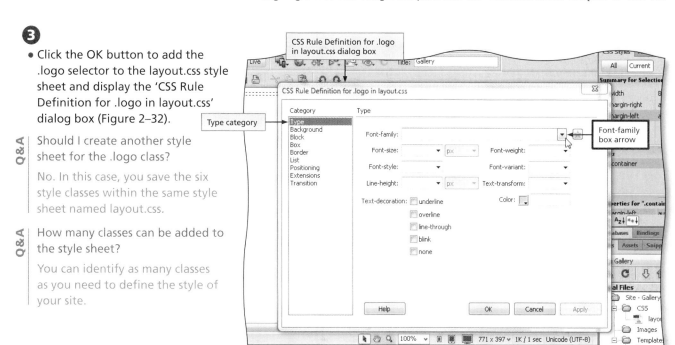

Figure 2–32

4

- If necessary, click Type in the Category list to display the Type options.

- Click the Font-family box arrow and then click 'Arial Black, Gadget, sans-serif' to set the font family for the .logo style.

- Click the Font-size box arrow and then click 14 to set the font size.

- Click the px button next to the Font-size box and then click pt to change the font size units from px to pt.

- Click the Color text box and then type #000 to change the font color to black (Figure 2–33).

Q&A Why is sans-serif or serif typically the last font listed in the Font-family font groupings?

The sans-serif and serif fonts are generic fonts that are displayed on any type of computer.

Q&A Are the px and pt font-size units basically the same?

The px unit measures fonts in pixels and the pt unit measures fonts in points. These units represent different sizes.

Figure 2–33

5

- Click the Apply button to apply the Type settings.

- Click Block in the Category list to display the Block options.

- Click the Text-align box arrow and then click center to set the text alignment for the .logo style (Figure 2–34).

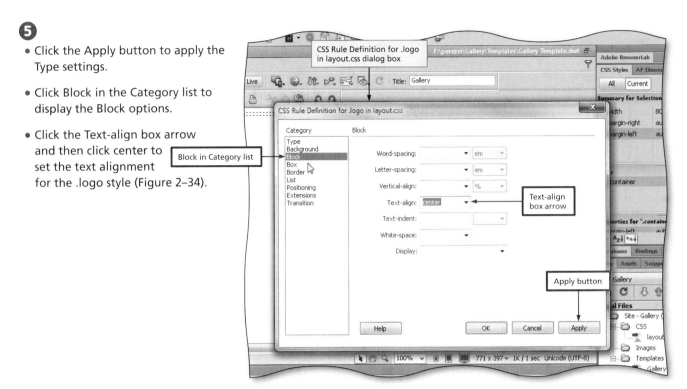

Figure 2–34

6

- Click the Apply button to apply the Block settings.

- Click Box in the Category list to display the Box options.

- Click the Width text box and then type 300 to set the width of the region containing the .logo style.

- Click the Height text box and then type 75 to set the height of the region.

- Click the Float box arrow and then click left to specify that other elements float to the left of elements in the .logo style (Figure 2–35).

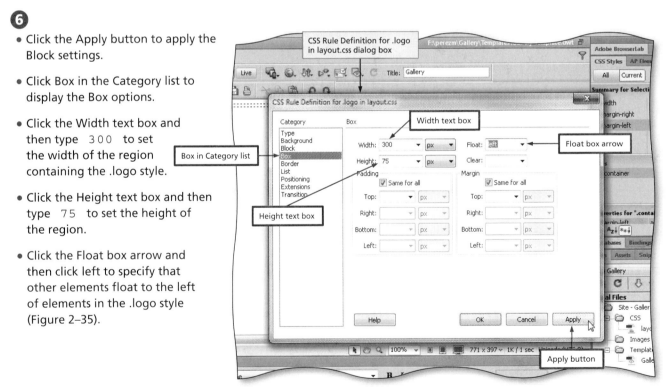

Figure 2–35

- Click the Apply button to apply the CSS rules for the .logo class within the layout.css file.

- Click the OK button (CSS Rule Definition dialog box) to define the CSS rules for the .logo class.

- Click the OK button (Insert Div Tag dialog box) to close the Insert Div Tag dialog box and display the .logo class region and its placeholder text (Figure 2–36).

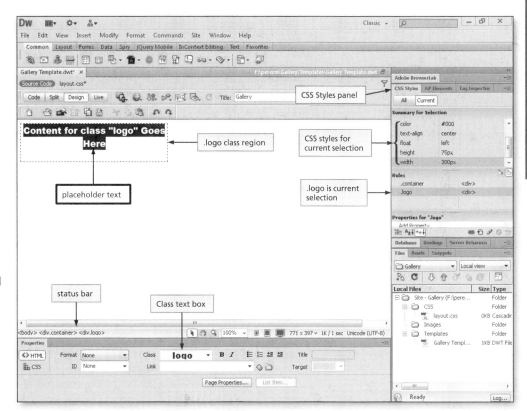

Figure 2–36

- If necessary, select the placeholder text 'Content for class "logo" Goes Here' in the Document window, and then type `The Gallery Portrait and Family Photography` to insert the logo text (Figure 2–37).

Figure 2–37

⑨

- Click the Save button on the Standard toolbar to save the template.

Q&A   What should I do if a Dreamweaver dialog box indicates the template does not have any editable regions?

Click the 'Don't warn me again' check box to insert a check mark, and then click the OK button. You specify editable regions later in the chapter.

To Add the Navigation Div Tag and Define Its CSS Rules

Most Web pages have a navigation region that contains links to other pages in the site. Including the navigation region in the Web site template means the navigation links appear in the same place and style on each page, which makes it easy for site visitors to find the links and navigate the site. Web sites typically include navigation links either horizontally across the top or vertically along the left side of each page. Many site designers prefer a horizontal navigation bar at the top of the page because it uses the full width of the page for content instead of crowding the left edge of the page with links, but either design can be effective as long as the navigation provides a strong visual focus. The navigation region for the Gallery site is displayed in the upper-right corner of the page. Similar to defining the logo region, you define the format and position of the navigation region and its contents. The following steps define the CSS rule definitions for the .navigation class.

1

- Click to the right of the .logo region within the document to move the insertion point to the div.container region.

- Click the Insert Div Tag button on the Insert bar to display the Insert Div Tag dialog box.

- If necessary, click the Class text box and then type `navigation` to name the div tag (Figure 2–38).

Q&A Why does the status bar display <body> <div.container>?

The container region is now selected instead of the logo region.

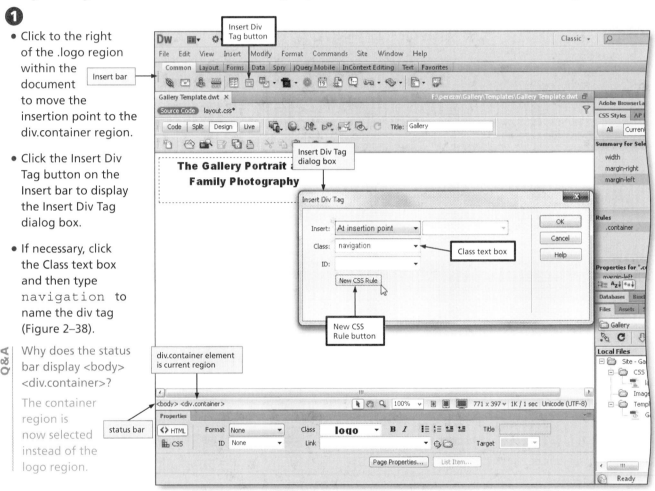

Figure 2–38

Q&A Why do I need to move the insertion point to the div.container region before creating the new navigation region?

You move the insertion point so you can create the new region in the div.container region, which you designed to hold all the other regions. Otherwise, you create the new navigation region in the logo region.

2

- Click the New CSS Rule button to display the New CSS Rule dialog box for adding a new CSS rule that defines the navigation class (Figure 2–39).

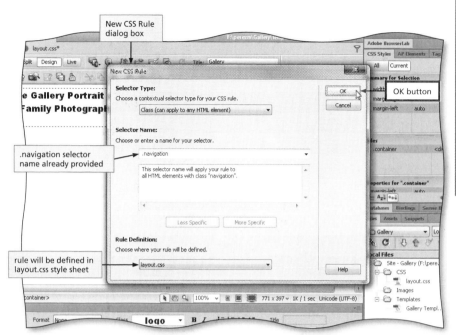

Figure 2–39

3

- Click the OK button to add the .navigation selector name to the layout.css file and display the 'CSS Rule Definition for .navigation in layout.css' dialog box.

- Click the Font-family box arrow and then click 'Arial, Helvetica, sans-serif' to set the font family for the .navigation style.

- Click the Font-size box arrow and then click 12 to set the font size for the .navigation style.

- Click the px button next to the Font-size box and then click pt to change the font size units from px to pt.

- Click the Line-height text box and then type 50 to set the line height to 50.

- Click the px button next to the Line-height box and then click pt to change the line height units from px to pt.

- Click the Font-weight box arrow and then click bolder to set the font to a bolder style.

- Click the Text-transform box arrow and then click lowercase to convert the text to lowercase letters.

- Click the Color text box and then type #F90 to change the style color to orange (Figure 2–40).

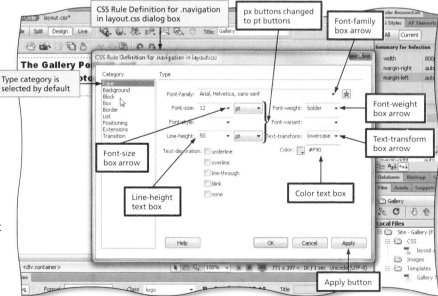

Figure 2–40

Q&A

What does the value, #F90, represent?

The value, #F90, is a hexadecimal number representing a color; in this case, orange. You can enter hexadecimal values in text boxes that request color values, or you can click the color palette button next to the text box to select a color.

- Click the Apply button to apply the Type settings.

- Click Block in the Category list to display the Block options.

- Click the Word-spacing text box, type 10, click the em button, and then click pt to change the word spacing in the .navigation style to 10 points.

- Click the Text-align box arrow and then click center to set the text alignment (Figure 2–41).

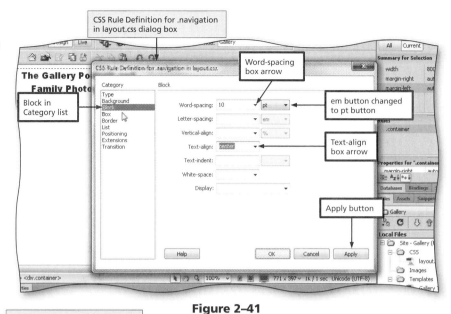

Figure 2–41

- Click the Apply button to apply the Block settings.

- Click Box in the Category list to display the Box options.

- Click the Width text box and then type 500 to set the width of the .navigation region.

- Click the Height text box and then type 75 to set the height of the .navigation region.

- Click the Float box arrow and then click left to change the float property (Figure 2–42).

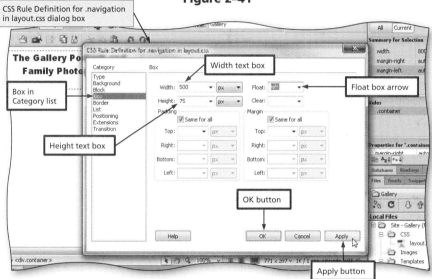

Figure 2–42

- Click the Apply button to apply the CSS rules for the .navigation class within the layout.css style sheet.

- Click the OK button (CSS Rule Definition dialog box) to define the CSS rules for the .navigation class.

- Click the OK button (Insert Div Tag dialog box) to close the Insert Div Tag dialog box and display the .navigation class region and its placeholder text.

- If necessary, select the text 'Content for class "navigation" Goes Here' and then type `home services portfolio pricing session contact` to enter the navigation text (Figure 2–43).

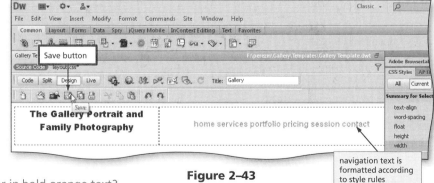

Figure 2–43

Q&A Why does the navigation text appear in bold orange text?

The style rules you just set are applied automatically to the navigation text.

- Click the Save button on the Standard toolbar to save the template.

To Add the Image Div Tag and Define Its CSS Rules

Most Web pages display one or more images to increase the visual appeal of the page. If one image or a certain type of image appears on each page, include an image region in the template so the image appears in the same place and style throughout the site. The pages in the Gallery site eventually will include family and portrait images as samples of photography the company has produced. To display these images consistently on each page, you add an image region to the template and then set its properties. The image region serves as a placeholder for the main images that will be displayed in the Gallery site. The following steps define the CSS rules of the .image class.

- If necessary, scroll right and then click to the right of the .navigation div within the document to move the insertion point to the div.container element.

- Click the Insert Div Tag button on the Insert bar to display the Insert Div Tag dialog box.

- In the Class text box, type `image` to name the div tag (Figure 2–44).

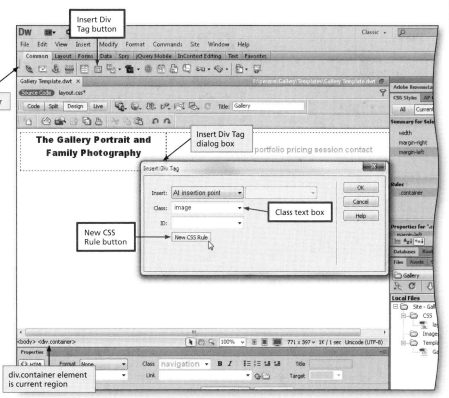

Figure 2–44

- Click the New CSS Rule button to display the New CSS Rule dialog box for adding a new CSS rule that defines the image class (Figure 2–45).

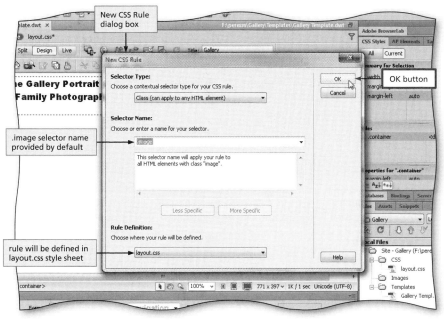

Figure 2–45

- Click the OK button to add the .image selector name to the layout.css style sheet and display the 'CSS Rule Definition for .image in layout.css' dialog box.

- Click Background in the Category list to display the Background options.

- Click the Background-color text box and then type #000 to change the background color of the .image region to black (Figure 2–46).

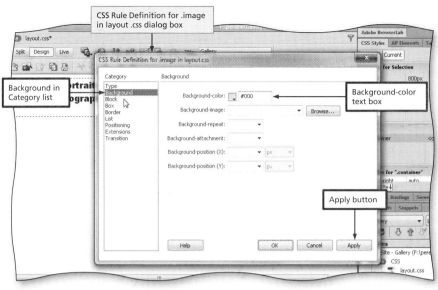

Figure 2–46

- Click the Apply button to apply the Background settings.

- Click Box in the Category list to display the Box options.

- In the Width text box, type 800 to set the width of the image region to 800 pixels.

- Click the Float box arrow and then click left to set the value of the float property (Figure 2–47).

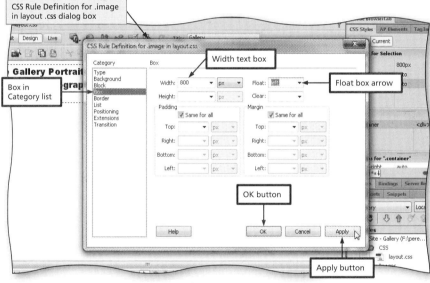

Figure 2–47

5

- Click the Apply button to apply the CSS rules for the .image class within the layout.css style sheet.

- Click the OK button (CSS Rule Definition dialog box) to define the CSS rules for the .image class.

- Click the OK button (Insert Div Tag dialog box) to close the Insert Div Tag dialog box and display the new image region (Figure 2–48).

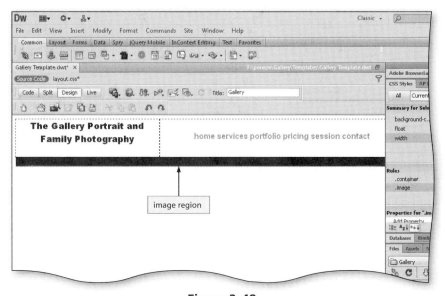

Figure 2–48

To Add the Content Div Tag and Define Its CSS Rules

One way to create a consistent look for a Web site is to include a heading or other text in the same style and place on each page. To achieve this consistency, add a region for the text content to the template, and then specify the style and placement properties in the style sheet. Because each page in the Gallery site will include content such as a heading or other text below the logo and navigation regions, you can add a content region to the template. On the opening page, the content region displays welcome text, but other pages will display different content. For now, you insert the content region with placeholder text and specify the style properties for the region. One new property to enter is **padding**, which specifies the amount of space between the text and its border. The following steps define the CSS rules for the .content class.

- If necessary, scroll right and then click to the right of the .image div within the document to move the insertion point to the div.container element.

- Click the Insert Div Tag button on the Insert bar to display the Insert Div Tag dialog box.

- In the Class text box, type content to name the div tag.

- Click the New CSS Rule button to display the New CSS Rule dialog box for adding a new CSS rule that defines the content class (Figure 2–49).

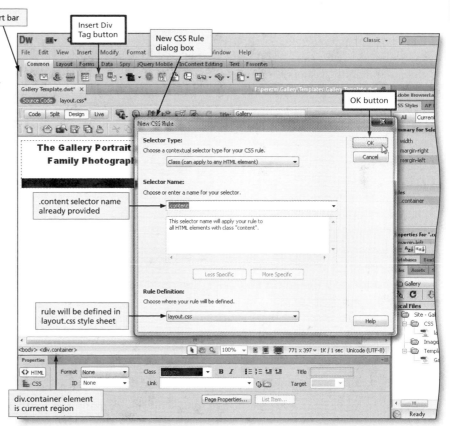

Figure 2–49

- Click the OK button to add the .content selector to the layout.css style sheet and display the 'CSS Rule Definition for .content in layout.css' dialog box.

- Click the Font-family box arrow and then click 'Arial, Helvetica, sans-serif' to set the font family.

- Click the Font-size box arrow, click 12, click the px button, and then click pt to change the units to pt and the font size to 12 points.

- Click the Color text box and then type #000 to change the background color of the .content region to black (Figure 2–50).

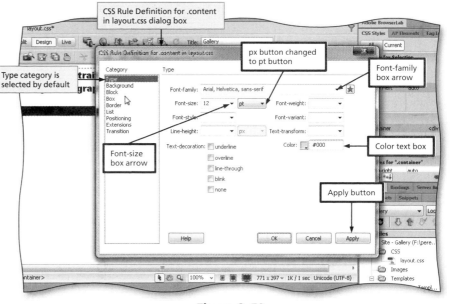

Figure 2–50

- Click the Apply button to apply the Type settings.
- Click Box in the Category list to display the Box options.
- Click the Width text box and then type 800 to set the width of the .content region.
- Click the Float box arrow and then click left to set the value of the float property.
- Click 'Same for all' in the Padding section to remove the check mark from the 'Same for all' check box.
- Click the Top text box in the Padding section, type 10, click the px button, and then click pt to change the units to pt and set the top padding to 10 points above the text.
- Click the Bottom text box in the Padding section, type 10, click the px button, and then click pt to change the units to pt and set the bottom padding.
- Click 'Same for all' in the Margin section to remove the check mark from the 'Same for all' check box.
- Click the Top text box in the Margin section, type 10, click the px button, and then click pt to change the units to pt and set the top margin to 10 points.
- Click the Bottom text box in the Margin section, type 10, click the px button, and then click pt to change the units to pt and set the bottom margin (Figure 2–51).

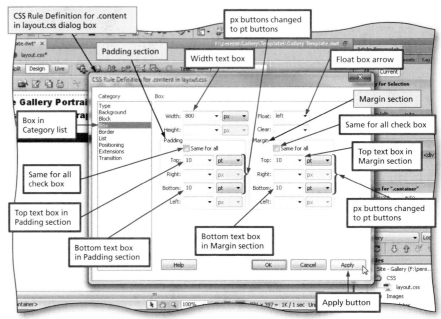

Figure 2–51

- Click the Apply button to apply the Box settings.
- Click Border in the Category list to display the Border options.
- Click 'Same for all' in the Style section to remove the check mark from the 'Same for all' check box.
- Click the Top box arrow, click solid, click the Bottom box arrow, and then click solid to select a solid style for the top and bottom borders.
- Click 'Same for all' in the Width section to remove the check mark from the 'Same for all' check box.
- Click the Top box arrow, click medium, click the Bottom box arrow, and then click medium to select a medium width for the top and bottom borders.
- Click 'Same for all' in the Color section to remove the check mark from the 'Same for all' check box.
- Click the Top text box, type #000, click the Bottom text box, and then type #000 to set the color of the top and bottom borders to black (Figure 2–52).

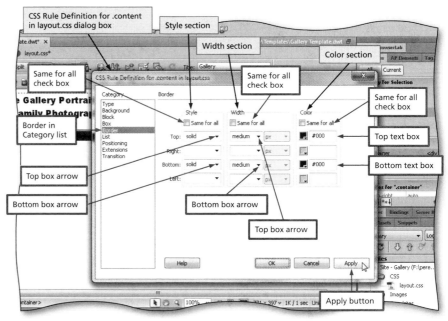

Figure 2–52

- Click the Apply button to apply the CSS rules for the .content class within the layout.css style sheet.

- Click the OK button (CSS Rule Definition dialog box) to define the CSS rules for the .content class.

- Click the OK button (Insert Div Tag dialog box) to close the Insert Div Tag dialog box and display the new content region (Figure 2–53).

- Click the Save button on the Standard toolbar to save your work.

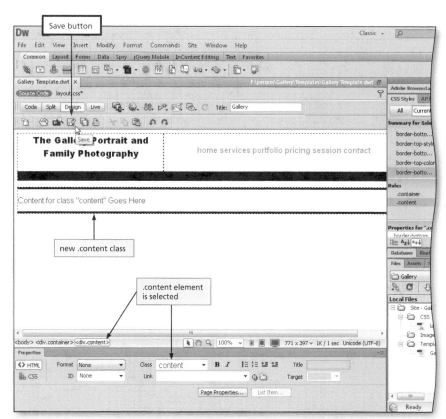

Figure 2–53

To Add the Footer Div Tag and Define Its CSS Rules

The footer of a site typically includes copyright or contact information. The Gallery site displays the year and the copyright information. The following steps define the CSS rules for the .footer class.

1

- If necessary, scroll right and then click to the right of the .content div within the document to move the insertion point to the div.container element.

- Click the Insert Div Tag button on the Insert bar to display the Insert Div Tag dialog box.

- In the Class text box, type `footer` to name the div tag.

- Click the New CSS Rule button to add a new CSS rule that defines the footer class (Figure 2–54).

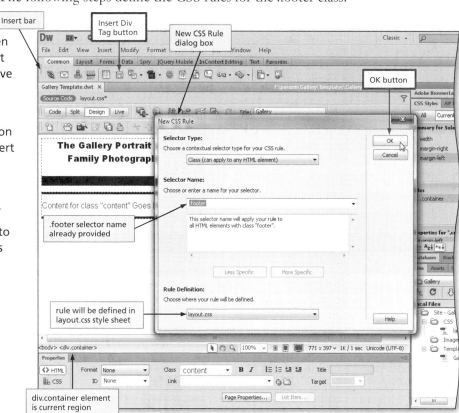

Figure 2–54

- Click the OK button to add the .footer selector name to layout.css and display the 'CSS Rule Definition for .footer in layout.css' dialog box.

- Click the Font-family box arrow and click 'Verdana, Geneva, sans-serif' to set the font family.

- Click the Font-size text box, type 8, click the px button, and then click pt to change the units to pt and the font size to 8 points.

- Click the Color text box and type #333 to change the font color of the .footer div region to dark gray (Figure 2–55).

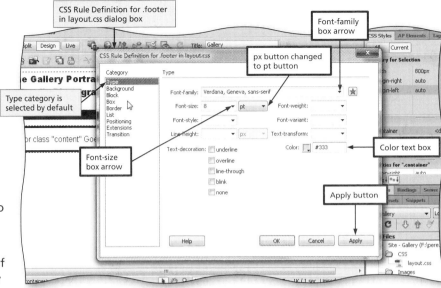

Figure 2–55

❸

- Click the Apply button to apply the Type settings.

- Click Box in the Category list to display the Box options.

- Click the Width text box and then type 800 to set a width of 800 pixels for the .footer class.

- Click the Float box arrow and then click left to set the value of the float property (Figure 2–56).

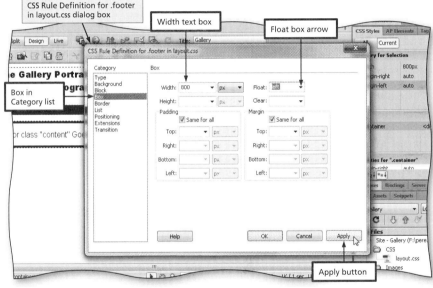

Figure 2–56

❹

- Click the Apply button to apply the CSS rules for the .footer class within the layout.css style sheet.

- Click the OK button (CSS Rule Definition dialog box) to define the CSS rules for the .footer class.

- Click the OK button (Insert Div Tag dialog box) to close the Insert Div Tag dialog box.

- If necessary, select the text 'Content for class "footer" Goes Here' and type Copyright 2014. All Rights Reserved. to enter the .footer class text (Figure 2–57).

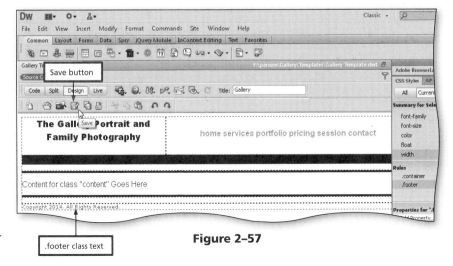

Figure 2–57

- Click the Save button on the Standard toolbar to save your work.

Creating an Editable Region of a Template

Recall that the two types of regions you can include in a template are editable and noneditable regions. Noneditable regions, also called locked regions, are the sections of a template that have static, unchanging content, such as a logo or a navigation bar. An editable region on a Web page is an area where other Web page authors can edit the content. As a Web developer, you create editable regions in a template to allow other authors to add or remove information without worrying that they will alter the page layout. This gives you control over the layout of the pages and the template itself. For example, if an element, such as a navigation bar, is exactly the same across the entire site, it should be in a noneditable region of the template. Dreamweaver inserts an element as a noneditable region by default, meaning that no one can edit its content in a Web page based on the template. If a content area contains different information on each page, that content should be an editable region of the template. Because Dreamweaver templates contain no editable regions by default, the next step after creating a template typically is to specify some regions as editable regions so each page in the site can display different content.

To Create an Editable Region

In the Gallery site, the logo, navigation, and footer regions should display the same content on each page. Therefore, they can remain noneditable regions in the template. However, the content region of each page in the Gallery site contains different text. You need to specify that the content region is an editable region in the template so you can display different text on each page. The following steps create an editable region in the template.

1

- Select the text 'Content for class "content" Goes Here' in the content region, and then press the DELETE key to delete the placeholder text.

- Click an edge of the content region to select the region (Figure 2–58).

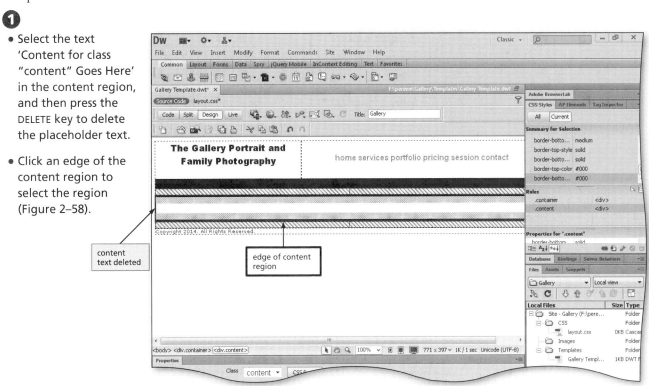

content text deleted

edge of content region

Figure 2–58

2

- Click Insert on the Application bar and then point to Template Objects to display the Template Objects submenu (Figure 2–59).

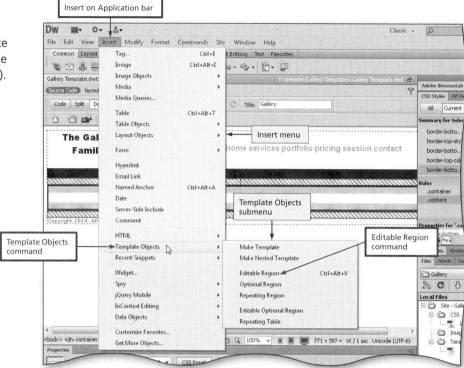

Figure 2–59

3

- Click Editable Region on the Template Objects submenu to display the New Editable Region dialog box.

- If necessary, select the text in the Name text box and then type `contentArea` to identify an editable region within the content region (Figure 2–60).

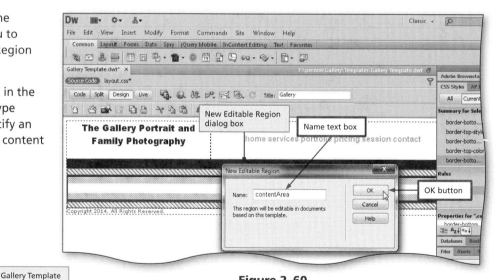

Figure 2–60

4

- Click the OK button to display the editable region.

- If necessary, click in the content region and then type `Insert page content here` to provide directions to add content to the editable region (Figure 2–61).

5

- Click the Save button on the Standard toolbar to save the completed template.

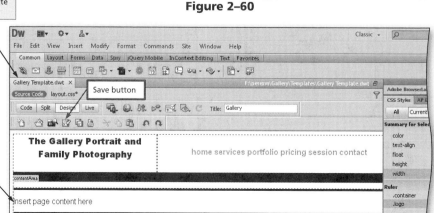

Figure 2–61

To Close the Template

The template is complete, so you can close it. The following steps close the template and display the Welcome screen again.

- Click File on the Application bar and then click Close to close the completed template and display the Welcome screen (Figure 2–62).

Q&A What should I do if an Adobe Dreamweaver CS6 dialog box is displayed?

Click the Yes button to save your changes to the style sheet.

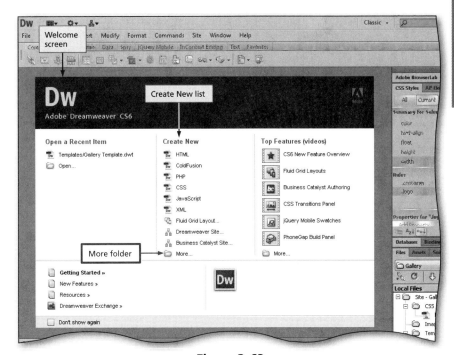

Figure 2–62

Other Ways

1. To add editable region, press CTRL+ALT+V

Creating a New Page from a Template

After creating and saving a template, you can use it to create Web pages on your site — just as you can create pages based on templates provided with Dreamweaver. To create a consistent site, be sure to associate each page with a template. As you create each new page of the site, you start with the template to establish the layout of the page. The noneditable regions of the template appear on the page as static, unchanging content. For example, if you add navigation, logo, and footer elements to a template, they appear on the Web page you create from the template and cannot be changed. Areas you specified as editable regions in the template appear on the Web page as content or areas you can change.

BTW

HTML and HTM File Extensions
Although Dreamweaver saves Web pages with a .html file extension by default, it can also save files with a .htm extension. Browsers recognize both file types as HTML files. You can set a preference to create pages using one extension or the other by clicking Edit on the Application bar, clicking Preferences, and then clicking the New Document category.

To Create a Page from a Template

To complete this assignment, you will be required to use the Data Files for Students. Visit www.cengage .com/ct/studentdownload for detailed instructions or contact your instructor for information about accessing the required files. The following steps open the Ch2_Home_Content file from the Data Files for Students.

The opening home page of the Gallery site is named index.html. In addition to the template elements, the index page uses the editable region of the template to display an opening message about the Gallery photography services. You add this message to the index.html page by copying text from a student data file named Ch2_Home_Page_Content.txt and pasting the text into index.html. The following steps create a new page from the template and add content to the editable region.

1

- Click More in the Create New list to display the New Document dialog box.

- Click Page from Template in the left pane of the New Document dialog box to create a page from the template.

- If necessary, click Gallery in the Site list to create a page for the Gallery site (Figure 2–63).

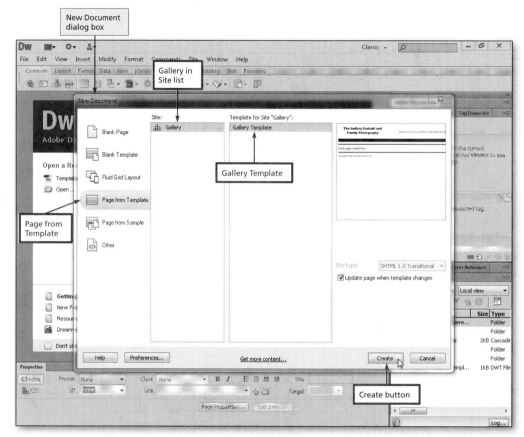

Figure 2–63

2

- Click the Create button to create a new page based on the template for the Gallery site.

- Click File on the Application bar and then click Save As to display the Save As dialog box.

- If necessary, select the text in the File name text box and then type `index` to name the new page created from the template (Figure 2–64).

Q&A

Which folder should I select to save the index.html file?

Save the home page, index.html, in the root Gallery folder, which is the folder displayed by default in the Save As dialog box. When a browser is directed to the Gallery folder, the browser displays index.html as the opening page.

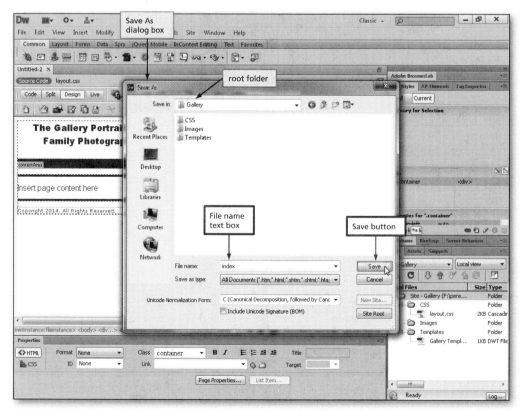

Figure 2–64

3

- Click the Save button to save and open index.html (Figure 2–65).

Q&A

When I move the mouse around the page, the pointer changes to a "not" symbol (circle with a line through it) over some parts of the page. What does this symbol represent in index.html?

The pointer displays a not symbol over the noneditable (locked) regions of the page.

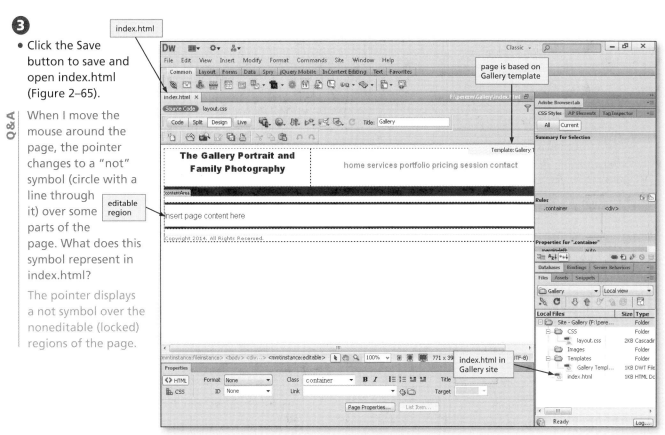

Figure 2–65

4

- If necessary, insert the drive containing your Data Files for Students into an available port. After a few seconds, if Windows displays a dialog box, click its Close button.

- With the Dreamweaver window open, click File on the Application bar, and then click Open to display the Open dialog box.

- In the Open dialog box, navigate to the storage location of the Data Files for Students.

- Click the Chapter 02 folder to display its contents, and then double-click the file, Ch2_Home_Page_Content, to open it.

- Select all the text in the file, click Edit on the Application bar, and then click Copy to copy the text.

- Click the index.html document tab, select the text, Insert page content here, on the index.html page, click Edit on the Application bar, and then click Paste to paste the text you copied (Figure 2–66).

5

- Close the Ch2_Home_Page_Content file.

- Click the Save All button on the Standard toolbar to save the site.

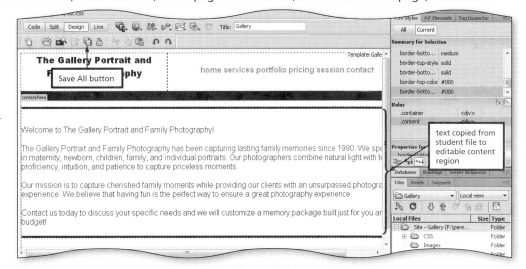

Figure 2–66

To View the Site in Live View

Live view provides a realistic rendering of what your page will look like in a browser, but lets you make any necessary changes without leaving Dreamweaver. The following steps display the page in Live view.

- Click the Live button on the Document toolbar to display a Live view of the site (Figure 2–67).

Experiment

- Click the Code, Split, and Design buttons to view the page in different views. When you are finished, click the Live button to return to Live view.

- Click the Live button again to return to Design view.

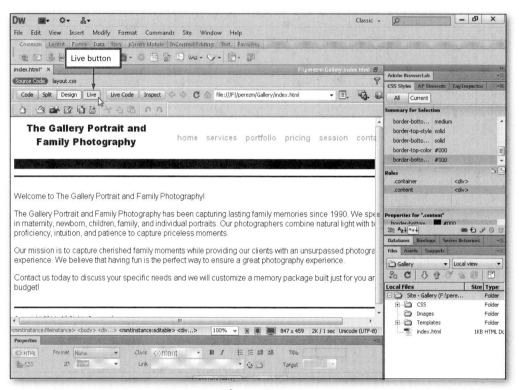

Figure 2–67

Other Ways

1. Press ALT+F11

To View the Site in the Browser

The following steps preview the Gallery Web site home page using Internet Explorer.

1. Click the 'Preview/Debug in browser' button on the Document toolbar.

2. Click Preview in IExplore in the Preview/Debug in browser list to display the Gallery Web site in the Internet Explorer browser.

3. Click the Internet Explorer Close button to close the browser.

To Quit Dreamweaver

The following steps quit Dreamweaver and return control to the operating system.

1. Click the Close button on the right side of the Application bar to close the window.

2. If Dreamweaver displays a dialog box asking you to save changes, click the No button.

Chapter Summary

In this chapter, you have learned how to create a Web site template using the design building blocks of CSS style sheets. You defined a new Web site, created a Dreamweaver Web Template, and defined regions of the page using a style sheet. You defined each region with CSS rules that provided consistent site formatting. You also learned how to create an editable region within a template. You added a page to the site based on the template and placed text in the editable region. You displayed the completed Web page in Live view and in a browser. The following tasks are all the new Dreamweaver skills you learned in this chapter:

1. Create a New Site (DW 86)
2. Create Folders for the Image and CSS Files (DW 87)
3. Create a Blank HTML Template (DW 88)
4. Save the HTML Page as a Template (DW 90)
5. Add a Div Tag (DW 93)
6. Select CSS Rule Definitions (DW 97)
7. Add the Logo Div Tag and Define Its CSS Rules (DW 100)
8. Add the Navigation Div Tag and Define Its CSS Rules (DW 104)
9. Add the Image Div Tag and Define Its CSS Rules (DW 107)
10. Add the Content Div Tag and Define Its CSS Rules (DW 109)
11. Add the Footer Div Tag and Define Its CSS Rules (DW 111)
12. Create an Editable Region (DW 113)
13. Close the Template (DW 115)
14. Create a Page from a Template (DW 115)
15. View the Site in Live View (DW 118)

Apply Your Knowledge

Reinforce the skills and apply the concepts you learned in this chapter.

Creating a New Web Page Template

Instructions: First, create a new HTML5 Web page template and save it. Next, insert div tags in the template and create a new external style sheet file. Finally, add new CSS rules so that the completed template in Live view looks like Figure 2–68. The CSS rule definitions for the template are provided in Table 2–2.

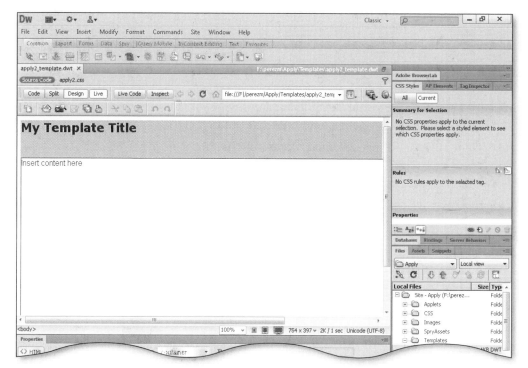

Figure 2–68

Continued >

Apply Your Knowledge *continued*

Perform the following tasks:

1. Use Windows Explorer to access your USB flash drive and create a new folder within the *your last name and first initial*\Apply folder (the folder named perezm\Apply, for example, which you created in Chapter 1). Name the folder `Templates`.

2. Start Dreamweaver. On the Dreamweaver Welcome screen, click the More folder. Create a new document as a blank template, HTML template, with no layout and DocType HTML5.

3. Save the new template in the Templates folder in the Apply root folder. Name the template `apply2_template.dwt`.

4. Insert a new div tag and name the class `container`.

5. Create a new CSS rule and specify that the rule definition will be defined in a new style sheet file. Name the style sheet file `apply2.css` and save it in a new folder named `CSS` in the root folder for the Apply site.

6. Define the CSS rule for the container class according to the settings provided in Table 2–2. Apply and accept your changes when you are finished.

7. Click after the placeholder text in the container and then press the ENTER key to insert a blank line.

8. Insert a div tag within the container. Name the class `header`. Refer to Table 2–2 to define the new CSS rule for the header in Apply2.css. Apply and accept your changes.

9. Replace text within the header div tag with `My Template Title`.

10. Insert a div tag below the header div tag but within the container div tag. Name the class `content`. Refer to Table 2–2 to define the new CSS rule for the content in apply2.css. Apply and accept your changes.

11. Delete the text within the content div tag and insert an editable region. Name the new editable region `Insert content here`.

12. Insert a div tag below the content div tag but within the container div tag. Name the class `footer`. Refer to Table 2–2 to define the new CSS rule for the footer in apply2.css. Apply and accept your changes.

13. Replace the text within the footer div tag with your name. Delete the placeholder text for the container region, and then press the BACKSPACE key to move the header to the top of the container.

14. Title the document `Apply2_Template`.

15. Save your changes and then view your document using Live view. Compare your document to Figure 2–68. Make any necessary changes and then save your changes.

16. Submit the document in the format specified by your instructor.

Table 2–2 CSS Rule Definitions for apply2_template

CSS Rule Definition for .container in apply2.css

Category	Property	Value
Box	Width	1000px
	Right Margin	auto
	Left Margin	auto

CSS Rule Definition for .header in apply2.css

Category	Property	Value
Type	Font-family	Verdana, Geneva, sans-serif
	Font-size	18pt
	Font-weight	bold
Background	Background-color	#9CC
Box	Width	1000px
	Height	75px

CSS Rule Definition for .content in apply2.css

Category	Property	Value
Type	Font-family	Arial, Helvetica, sans-serif
Box	Width	1000px
	Height	400px
Border	Style	solid, same for all
	Width	thin, same for all
	Color	#000, same for all

CSS Rule Definition for .footer in apply2.css

Category	Property	Value
Type	Font-family	Times New Roman, Times, serif
	Font-size	10pt
Box	Width	1000px

Extend Your Knowledge

Extend the skills you learned in this chapter and experiment with new skills. You may need to use Help to complete the assignment.

Attaching an External Style Sheet to a Web Page

Note: To complete this assignment, you will be required to use the Data Files for Students. Visit www.cengage.com/ct/studentdownload for detailed instructions or contact your instructor for information about accessing the required files.

Instructions: A volunteer service organization wants to create a Web site using style sheets. You are creating a Web page for the organization that explains the difference between internal and external style sheets. Apply styles to a page by attaching an external style sheet to an existing Web page. The completed Web page is shown in Figure 2–69.

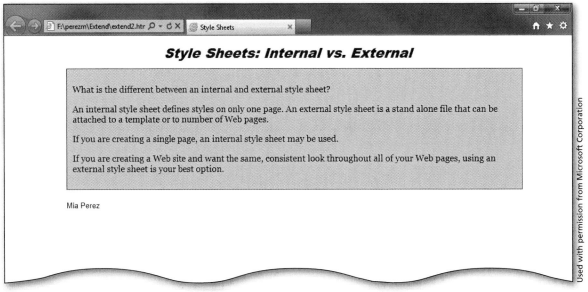

Figure 2–69

Perform the following tasks:

1. Use Windows Explorer to copy the CSS folder and the extend2.html file from the Chapter 02\ Extend folder into the *your last name and first initial*\Extend folder (the folder named perezm\ Extend, for example, which you created in Chapter 1).

2. Start Dreamweaver and open extend2.html.

3. On the CSS Styles panel, click the Attach Style Sheet button to display the Attach External Style Sheet dialog box. (*Hint*: Point to the buttons in the lower-right part of the CSS Styles panel to find the Attach Style Sheet button.) Click the Browse button in the Attach External Style Sheet dialog box, and then navigate to find and then select the extend2.css file located in the CSS folder in the Extend root folder. Add the CSS file as a link, and then accept your changes.

4. Replace the text, Your name here, with your name.

5. Title the document `Style Sheets`.

6. Save your changes and then view your document in your browser. Compare your document to Figure 2–69. Make any necessary changes and then save your changes.

7. Submit the document in the format specified by your instructor.

Make It Right

Analyze a Web page and correct all errors and/or improve the design.

Adding Div Tags and CSS Rule Definitions to a Web Page

Note: To complete this assignment, you will be required to use the Data Files for Students. Visit www.cengage.com/ct/studentdownload for detailed instructions or contact your instructor for information about accessing the required files.

Instructions: A bird-watching club is creating a Web site and wants to know the benefits of using CSS rules. You will create a Web page that lists these benefits. You also will create new div tags and add CSS rule definitions within an existing Web page. The CSS rule definitions for the Web page are provided in Table 2–3. The completed Web page is shown in Figure 2–70.

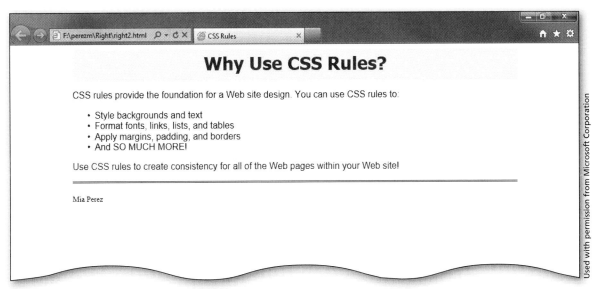

Why Use CSS Rules?

CSS rules provide the foundation for a Web site design. You can use CSS rules to:

- Style backgrounds and text
- Format fonts, links, lists, and tables
- Apply margins, padding, and borders
- And SO MUCH MORE!

Use CSS rules to create consistency for all of the Web pages within your Web site!

Mia Perez

Used with permission from Microsoft Corporation

Figure 2–70

Perform the following tasks:

1. Use Windows Explorer to copy the right2.html file from the Chapter 02\Right folder into the *your last name and first initial*\Right folder (the folder named perezm\Right, for example, which you created in Chapter 1).

2. Open right2.html in your browser to view it and note that it currently has no design and no styles applied. Close your browser.

3. Start Dreamweaver and open right2.html.

4. Select all of the text on the page.

5. Insert a new div tag and name the class `container`. Leave the insert type as the default 'Wrap around selection'.

6. Refer to Table 2–3 to define the new CSS rule for the class container. The Rule Definition will be for this document only. Apply and accept your changes.

7. Select the text, Why Use CSS Rules?

Continued >

Make It Right *continued*

Table 2–3 CSS Rule Definitions for right2.html		
CSS Rule Definition for .container		
Category	**Property**	**Value**
Box	Width	800px
	Right Margin	auto
	Left Margin	auto
CSS Rule Definition for .header		
Category	**Property**	**Value**
Type	Font-family	Tahoma, Geneva, sans-serif
	Font-size	24pt
	Font-weight	bold
	Color	#030
Background	Background-color	#FFC
Block	Text-align	center
Box	Width	800px
	Height	50px
	Top Padding	5px
CSS Rule Definition for .content		
Type	Font-family	Arial, Helvetica, sans-serif
	Font-size	12pt
Box	Width	800px
Border	Bottom Style	double
	Bottom Width	medium
	Bottom Color	#000
CSS Rule Definition for .footer		
Type	Font-size	10pt
Box	Width	800px
	Top Margin	20px

8. Insert a new div tag and name the class `header`. Leave the insert type as the default 'Wrap around selection'.

9. Refer to Table 2–3 to define the new CSS rule for the class header in this document only. Apply and accept your changes.

10. Select the text below Why Use CSS Rules? beginning with "CSS rules provide" and ending with "within your Web site!"

11. Insert a new div tag and name the class `content`. Leave the insert type as the default 'Wrap around selection'.

12. Refer to Table 2–3 to define the new CSS rule for the class content in this document only. Apply and accept your changes.

13. Select the text, Your name here.

14. Insert a new div tag and name the class `footer`. Leave the insert type as the default 'Wrap around selection'.

15. Refer to Table 2–3 to define the new CSS rule for the class footer in this document only. Apply and accept your changes.

16. Replace the text, Your name here., with your name.

17. Title the document `CSS Rules`.

18. Save your changes and then view the Web page in your browser. Compare your page to Figure 2–70. Make any necessary changes and then save your changes.

19. Submit the document in the format specified by your instructor.

In the Lab

Design and/or create a Web document using the guidelines, concepts, and skills presented in this chapter. Labs are listed in order of increasing difficulty.

Lab 1: Designing a New Template for the Healthy Lifestyle Web Site

Note: To complete this assignment, you will be required to use the Data Files for Students. Visit www.cengage.com/ct/studentdownload for detailed instructions or contact your instructor for information about accessing the required files.

Problem: In an effort to reduce health insurance costs, your company wants to provide resources for living a healthy lifestyle. You have been asked to create an internal Web site for your company with information about how to live a healthy lifestyle. This Web site will be used as a resource by employees at your company. You thoughtfully have planned the design of the Web site and now are ready to create a template for the site.

Define a new Web site and create a new HTML5 template. Use div tags and CSS rules in your template design. The template in Live view is shown in Figure 2–71, and the final Web page is shown in Figure 2–72. The CSS rule definitions for the template are provided in Table 2–4.

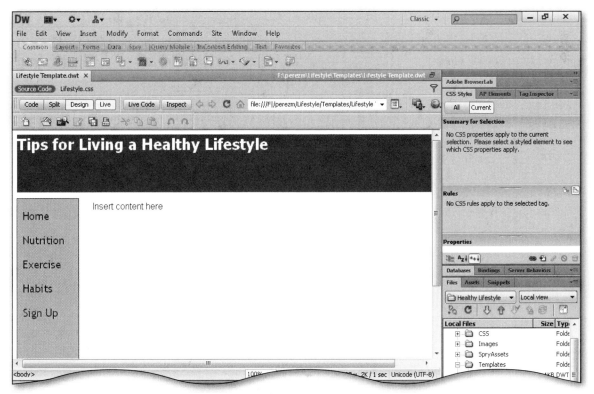

Figure 2–71

Continued >

In the Lab *continued*

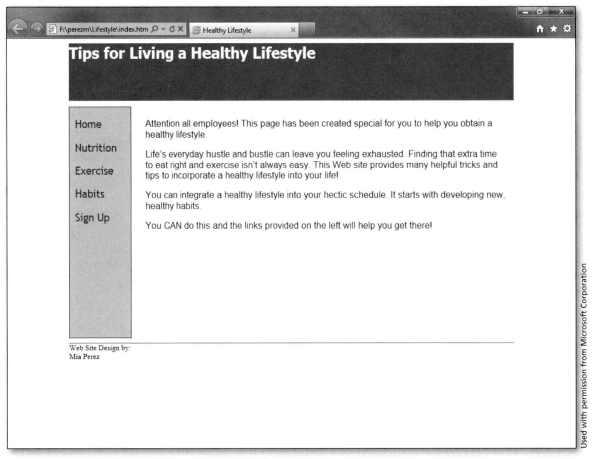

Figure 2–72

Perform the following tasks:

1. Start Dreamweaver. Click Dreamweaver Site in the Create New list of the Welcome screen to display the Site Setup dialog box. Type `Healthy Lifestyle` in the Site name text box.

2. Click the 'Browse for folder' icon to display the Choose Root Folder dialog box. Create a new subfolder in the *your last name and first initial* folder. Type `Lifestyle` as the folder name, select the folder, and then save the site.

3. On the Dreamweaver Welcome screen, click More in the Create New list. In the New Document dialog box, select Blank Template, Template Type: HTML template, Layout: <none>. Set the DocType to HTML5 and click the Create button. Save the new template using `Lifestyle Template` as the file name.

4. Use `Healthy Lifestyle` as the page title in the Title text box.

5. Insert a new div tag and name the class `container`.

6. Create a new CSS rule with a new style sheet file as the location for the rule definition. Name the style sheet file `Lifestyle.css`. Use the Create New Folder button in the Save Style Sheet File As dialog box to create a new folder and name it `CSS`. Save Lifestyle.css within the CSS folder.

7. Refer to Table 2–4 to define the CSS rule for the container. Apply and accept your changes.

8. Insert a blank line after the placeholder text in the container div tag.

9. Insert a div tag within the container. Name the class `header`.

Table 2–4 CSS Rule Definitions for Tips for Living a Healthy Lifestyle

CSS Rule Definition for .container in Lifestyle.css

Category	Property	Value
Box	Width	800px
	Right Margin	auto
	Left Margin	auto

CSS Rule Definition for .header in Lifestyle.css

Category	Property	Value
Type	Font-family	Tahoma, Geneva, sans-serif
	Font-size	20pt
	Font-weight	bold
	Color	#FFF
Background	Background-color	#039
Box	Width	800px
	Height	100px
	Bottom Margin	10px

CSS Rule Definition for .navigation in Lifestyle.css

Type	Font-family	Trebuchet MS, Arial, Helvetica, sans-serif
	Font-size	14pt
	Color	#333
Background	Background-color	#9CF
Box	Width	100px
	Height	400px
	Float	left
	Left Padding	10px
	Right Margin	10px
Border	Style	solid, same for all
	Width	thin, same for all
	Color	#003, same for all

CSS Rule Definition for .content in Lifestyle.css

Type	Font-family	Arial, Helvetica, sans-serif
	Font-size	12pt
Box	Width	650px
	Height	400px
	Float	left
	Padding	5px, same for all
	Left Margin	10px

CSS Rule Definition for .footer in Lifestyle.css

Type	Font-family	Times New Roman, Times, serif
	Font-size	10pt
Box	Width	800px
	Float	left
Border	Style	solid, same for all
	Width	thin, same for all
	Color	#333, same for all

Continued >

In the Lab *continued*

10. Create a new CSS rule. Refer to Table 2–4 to define the CSS rule for the header in Lifestyle. css. Apply and accept your changes.

11. Replace the text within the header div tag with `Tips for Living a Healthy Lifestyle`.

12. Insert a div tag after the header div tag but within the container div tag. Name the class `navigation`.

13. Create a new CSS rule. Refer to Table 2–4 to define the CSS rule for the navigation in Lifestyle.css. Apply and accept your changes.

14. Replace the text within the navigation div tag with the following list, pressing ENTER at the end of each line:

```
Home
Nutrition
Exercise
Habits
Sign Up
```

---

15. Insert a div tag to the right of the navigation div tag but within the container div tag. Name the class `content`.

16. Create a new CSS rule. Refer to Table 2–4 to define the CSS rule for the content in Lifestyle. css. Apply and accept your changes.

17. Delete the text within the content div tag and insert an editable region. Name the new editable region `Insert content here`.

18. Insert a div tag to the right of the content div tag but within the container div tag. Name the class `footer`.

19. Create a new CSS rule. Refer to Table 2–4 to define the CSS rule for the footer in Lifestyle.css. Apply and accept your changes.

20. Replace the text within the footer div tag with `Web Site Design by:`, press the SHIFT+ENTER keys, and then type your first and last names.

21. Delete the placeholder text for the container region, and then press the BACKSPACE key to move the header to the top of the container.

22. Save your changes and then view your template using Live view. Compare your template to Figure 2–71. Make any necessary changes and save your changes. Close the template.

23. Use the New Document dialog box to create a new Web page using the Lifestyle Template. Save the new page using `index` as the file name in the root folder of the Lifestyle site.

24. Replace the text, Insert content here, with text from the Lab1_Content.txt data file.

25. Save your changes and then view the Web page in your browser. Compare your page to Figure 2–72. Make any necessary changes and then save your changes.

26. Submit the documents in the format specified by your instructor.

In the Lab

Lab 2: Designing a New Template for Designs by Dolores

Note: To complete this assignment, you will be required to use the Data Files for Students. Visit www.cengage.com/ct/studentdownload for detailed instructions or contact your instructor for information about accessing the required files.

Problem: You are working as an intern for a Web site design company, Designs by Dolores. The owner, Dolores, is impressed with your Web site design knowledge and asks you to redesign her current site. You are excited about the opportunity to design your first Web site and begin by creating a template for the site.

Define a new Web site and create a new HTML5 template. Use div tags and CSS rules in your template design. The template in Live view is shown in Figure 2–73, and the final Web page is shown in Figure 2–74. The CSS rule definitions for the template are provided in Table 2–5.

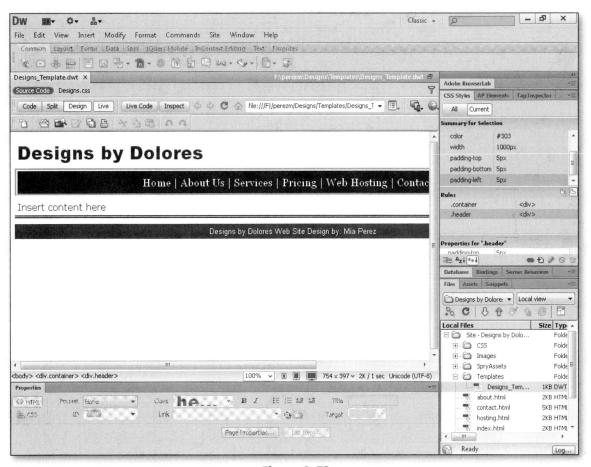

Figure 2–73

Continued >

In the Lab *continued*

Figure 2–74

Perform the following tasks:

1. Start Dreamweaver. Click Dreamweaver Site in the Create New list to display the Site Setup dialog box. Type `Designs by Dolores` in the Site name text box.

2. Click the 'Browse for folder' icon to display the Choose Root Folder dialog box. Create a new subfolder in the *your last name and first initial* folder. Use `Designs` as the folder name, and then save the folder and the site.

3. On the Dreamweaver Welcome screen, click More in the Create New list. Create a blank HTML5 template with no layout. Set the DocType to HTML5. Save the new template as `Designs_Template` in the Designs by Dolores site.

4. Use `Designs by Dolores` as the page title.

5. Insert a new div tag and name the class `container`.

6. Create a new CSS rule with a new style sheet file as the rule definition. Name the style sheet file `Designs.css`. Save Designs.css in a new folder named `CSS`.

7. Refer to Table 2–5 for the container CSS rule definition.

8. Insert a blank line after the placeholder text in the container div tag.

9. Insert a div tag within the container. Name the class `header`.

10. Create a new CSS rule. Refer to Table 2–5 for the header CSS rule definition.

11. Replace the text within the header div tag with `Designs by Dolores`.

12. Insert a div tag below the header div tag but within the container div tag. Name the class `navigation`.

13. Create a new CSS rule. Refer to Table 2–5 for the navigation CSS rule definition.

14. Replace the text within the navigation div tag with `Home | About Us | Services | Pricing | Web Hosting | Contact Us`.

15. Insert a div tag below the navigation div tag but within the container div tag. Name the class `content`.

Table 2–5 CSS Rule Definitions for Designs by Dolores

CSS Rule Definition for .container in Designs.css

Category	Property	Value
Box	Width	1000px
	Right Margin	Auto
	Left Margin	Auto

CSS Rule Definition for .header in Designs.css

Category	Property	Value
Type	Font-family	Arial Black, Gadget, sans-serif
	Font-size	24pt
	Color	#303
Box	Width	1000px
	Top Padding	5px
	Bottom Padding	5px
	Left Padding	5px

CSS Rule Definition for .navigation in Designs.css

Type	Font-family	Georgia, Times New Roman, Times, serif
	Font-size	14pt
	Color	#FFF
Background	Background-color	#303
Block	Text-align	center
Box	Width	1000px
	Top Padding	8px
	Bottom Padding	8px
Border	Style	groove, same for all
	Width	thick, same for all
	Color	#FFF, same for all

CSS Rule Definition for .content in Designs.css

Type	Font-family	Verdana, Geneva, sans-serif
	Font-size	12pt
Box	Width	1000px
	Bottom Padding	5px
	Left Padding	5px
	Top Margin	10px
	Bottom Margin	10px
Border	Bottom Style	double
	Bottom Width	thick
	Bottom Color	#303

CSS Rule Definition for .footer in Designs.css

Type	Font-family	Arial, Helvetica, sans-serif
	Font-size	10pt
	Color	#FFF
Background	Background-color	#333
Block	Text-align	center
Box	Width	1000px
	Top Padding	5px
	Bottom Padding	5px

Continued >

In the Lab *continued*

16. Create a new CSS rule. Refer to Table 2–5 for the content CSS rule definition.

17. Delete the text within the content div tag and insert an editable region. Name the new editable region `Insert content here`.

18. Insert a div tag below the content div tag but within the container div tag. Name the class `footer`.

19. Create a new CSS rule. Refer to Table 2–5 for the footer CSS rule definition.

20. Replace the text within the footer div tag with `Designs by Dolores Web Site Design by:` and then type your first and last names.

21. Delete the placeholder text for the container region, and then press the BACKSPACE key to move the header to the top of the container.

22. Save your changes and then view your template using Live view. Compare your template to Figure 2–73. Make and save any necessary changes, and then close the template.

23. Create a new Web page using the Designs Template. Save the new page using `index` as the file name in the root folder for the Designs site.

24. Replace the text, Insert content here, with text from the Lab2_Content.txt data file.

25. Save your changes and then view the Web page in your browser. Compare your page to Figure 2–74. Make any necessary changes and then save them.

26. Submit the documents in the format specified by your instructor.

In the Lab

Lab 3: Designing a New Template for Justin's Lawn Care Service

Note: To complete this assignment, you will be required to use the Data Files for Students. Visit www.cengage.com/ct/studentdownload for detailed instructions or contact your instructor for information about accessing the required files.

Problem: You have been hired to create a Web site for a new lawn care company, Justin's Lawn Care Service. You thoughtfully have planned the design of the Web site and now are ready to create a template for the site. You have met with Justin to discuss his needs for the Web site and are ready to start developing the site.

Define a new Web site and create a new HTML5 template. Use div tags and CSS rules in your template design. The template is shown in Figure 2–75, and the final Web page is shown in Figure 2–76. The CSS rule definitions for the template are provided in Table 2–6.

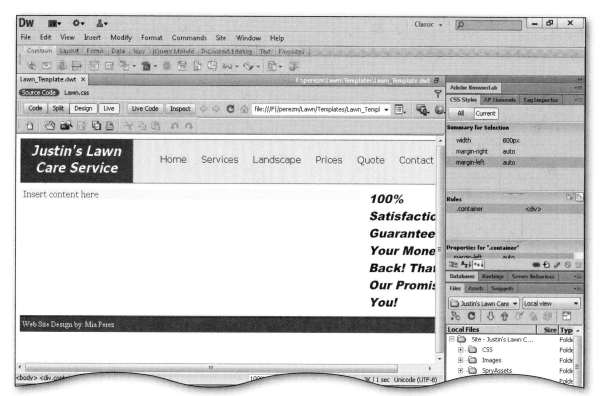

Figure 2–75

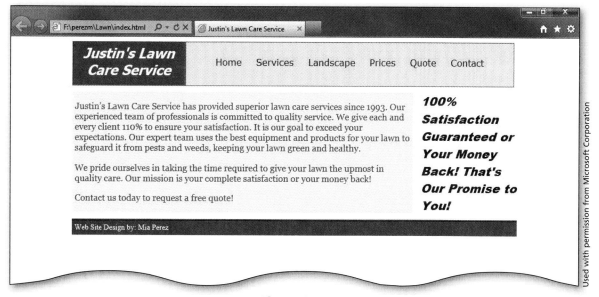

Figure 2–76

Perform the following tasks:

1. Start Dreamweaver. Use the Welcome screen to create a new site. Type `Justin's Lawn Care Service` as the site name.

2. Create a new subfolder named `Lawn` in the *your last name and first initial* folder, and then save the site.

Continued >

In the Lab *continued*

3. Create a blank HTML5 template with no layout. Save the new template as `Lawn_Template`.

4. Use `Justin's Lawn Care Service` as the page title.

5. Insert a new div tag and name the class `container`.

6. Create a new CSS rule with a new style sheet file named `Lawn.css`. Save Lawn.css in a new folder named `CSS`.

7. Refer to Table 2–6 for the container CSS rule definition.

8. Insert a blank line after the placeholder text in the container div tag.

9. Insert a div tag within the container. Name the class `header`.

10. Create a new CSS rule. Refer to Table 2–6 for the header CSS rule definition.

11. Replace the text within the header div tag with `Justin's Lawn Care Service.`

12. Insert a div tag to the right of the header div tag but within the container div tag. Name the class `navigation`.

13. Create a new CSS rule. Refer to Table 2–6 for the navigation CSS rule definition.

14. Replace the text within the navigation div tag with `Home Services Landscape Prices Quote Contact.`

15. Insert a div tag to the right of the navigation div tag but within the container div tag. Name the class `content`.

16. Create a new CSS rule. Refer to Table 2–6 for the content CSS rule definition.

17. Delete the text within the content div tag and insert an editable region. Name the new editable region `Insert content here.`

18. Insert a div tag to the right of the content div tag but within the container div tag. Name the class `sidebar`.

19. Create a new CSS rule. Refer to Table 2–6 for the sidebar CSS rule definition.

20. Replace the text within the sidebar div tag with `100% Satisfaction Guaranteed or Your Money Back! That's Our Promise to You!.`

21. Insert a div tag to the right of the sidebar div tag but within the container div tag. Name the class `footer`.

22. Create a new CSS rule. Refer to Table 2–6 for the footer CSS rule definition.

23. Replace the text within the footer div tag with `Web Site Design by:` and then type your first and last names.

24. Delete the placeholder text for the container region, and then press the BACKSPACE key to move the header to the top of the container.

25. Save your changes and then view the template using the Live view. Compare your template to Figure 2–75. Make and save any necessary changes, and then close the template.

26. Create a new Web page named `index` using the Lawn Template.

27. Replace the text, Insert content here, with text from the Lab3_content.txt data file.

28. Save your changes and then view the Web page in your browser. Compare your page to Figure 2–76. Make any necessary changes and then save your changes.

29. Submit the documents in the format specified by your instructor.

Table 2–6 CSS Rule Definitions for Justin's Lawn Care Service

CSS Rule Definition for .container in Lawn.css

Category	Property	Value
Box	Width	800px
	Right Margin	Auto
	Left Margin	Auto

CSS Rule Definition for .header in Lawn.css

Category	Property	Value
Type	Font-family	Tahoma, Geneva, sans-serif
	Font-size	18pt
	Font-weight	bold
	Font-style	italic
	Color	#FFF
Background	Background-color	#060
Block	Text-align	center
Box	Width	200px
	Float	left
	Height	70px
	Top Padding	5px
Border	Style	solid, same for all
	Width	thin, same for all
	Color	#960, same for all

CSS Rule Definition for .navigation in Lawn.css

Category	Property	Value
Type	Font-family	Verdana, Geneva, sans-serif
	Font-size	12pt
	Color	#030
Background	Background-color	#CFC
Block	Word-spacing	15pt
	Text-align	center
Box	Width	590px
	Float	left
	Height	50px
	Top Padding	25px
Border	Style	solid, same for all
	Width	thin, same for all
	Color	#960, same for all

CSS Rule Definition for .content in Lawn.css

Category	Property	Value
Type	Font-family	Georgia, Times New Roman, Times, serif
	Font-size	12pt
	Color	#360

Continued >

In the Lab continued

Table 2–6 CSS Rule Definitions for Justin's Lawn Care Service *(continued)*		
CSS Rule Definition for .content in Lawn.css *(continued)*		
Background	Background-color	#FFC
Box	Width	600px
	Float	left
	Right Padding	5px
	Left Padding	5px
	Top Margin	10px
	Right Margin	10px
	Bottom Margin	10px
CSS Rule Definition for .sidebar in Lawn.css		
Type	Font-family	Arial Black, Gadget, sans-serif
	Font-size	16pt
	Font-style	italic
	Color	#FFF
Box	Width	175px
	Float	right
	Top Margin	10px
	Bottom Margin	10px
CSS Rule Definition for .footer in Lawn.css		
Type	Font-family	Times New Roman, Times, serif
	Font-size	10pt
	Color	#FFF
Background	Background-color	#060
Box	Width	795px
	Float	left
	Top Padding	5px
	Bottom Padding	5px
	Left Padding	5px

Cases and Places

Apply your creative thinking and problem solving skills to design and implement a solution.

1: Creating a Web Site Template for Moving Tips

Personal

You recently moved out of your parents' home and, after realizing how much preparation is involved, you have decided to create a Web site with helpful tips and information about the moving process. Define a new local site in the *your last name and first initial* folder and name it

Moving Venture. Name the new subfolder Move. Create and save a new blank HTML template, layout none, and DocType HTML5. Name the file Move Template. Use a div tag to create a class container, and then insert the following class div tags within the container: header, navigation, content, and footer. Create a new CSS rule for each div tag and use a new style sheet file as the rule definition. Name the style sheet file move.css and save it to Removable Disk (F:)*your last name and first initial*\\Move\\CSS. The container box width should be between 800px and 1000px. Your CSS rule definitions for the other div tags should include a variety of category properties, such as font-family, font-weight, font-size, color, text-align, background, box width and height, and border style. Include a title in the header div tag. Make the navigation a left, vertical sidebar and include Home, Budget, Rentals, Tips, and Contact within the navigation bar. Include an editable region within the content div tag. Include your name in the footer. Title your document Moving Venture. Save the template. Create a new home page for the Web site using the template. Name the home page index. html. In the editable region, include a welcome paragraph that provides a mission statement and summary regarding the Web site's purpose. Submit the document in the format specified by your instructor.

2: Creating a Web Site Template for Student Campus Resources

Academic

You are a volunteer at your college campus library. The library provides print materials regarding various student activities, committees, and campus events. In an effort to reduce printing costs, the library has decided to develop a Web site with this information rather than printing numerous paper copies. You have been asked to design the Web site. Define a new local site in the *your last name and first initial* folder and name it Student Campus Resources. Name the new subfolder Campus. Create and save a new blank HTML template, layout none, and DocType HTML5. Name the file Campus Template. Use a div tag to create a class container, and then insert the following class div tags within the container: header, navigation, content, and footer. Create a new CSS rule for each div tag and use a new style sheet file as the rule definition. Name the style sheet file campus.css and save it to Removable Disk (F:)*your last name and first initial*\\Campus\\CSS. The container box width should be between 800px and 1000px. Your CSS rule definitions for the other div tags should include a variety of category properties, such as font-family, font-weight, font-size, color, text-align, background, box width and height, and border style. Include the text Student Campus Resources in the header div tag. Make the navigation horizontal, placed below the header, and include Home, Activities, Committees, Events, and Contact within the navigation bar. Include an editable region within the content div tag. Include your name in the footer. Title your document Student Campus Resources. Save the template. Create a new home page for the Web site using the template. Name the home page index.html. In the editable region, include a welcome paragraph that provides a mission statement and summary regarding the Web site purpose. Submit the document in the format specified by your instructor.

3: Creating a Web Site Template for French Villa Roast Café

Professional

You have been hired to design and develop a Web site for a local coffee shop, French Villa Roast Café. Define a new local site in the *your last name and first initial* folder and name it French Villa Roast Cafe. Name the new subfolder Cafe. Create and save a new blank HTML template, layout none, and DocType HTML5. Name the file Cafe Template. Use a div tag to create a class container, and then insert the following class div tags within the container: header, navigation, content, and footer. Create a new CSS rule for each div tag and use a new style sheet file as the rule definition. Name the style sheet file cafe.css and save it to Removable Disk (F:)*your last name and first initial*\\Cafe\\CSS. The container box width should be between 800px and 1000px. Your CSS

Continued >

Cases and Places *continued*

rule definitions for the other div tags should include a variety of category properties, such as font-family, font-weight, font-size, color, text-align, background, box width and height, and border style. Include the text, French Villa Roast Cafe, in the header div tag. Make the navigation horizontal, placed to the right of the header, and include Home, About, Menu, Rewards, and Contact within the navigation bar. Include an editable region within the content div tag. Include your name in the footer. Title your document French Villa Roast Cafe. Save the template. Create a new home page for the Web site using the template. Name the home page index.html. In the editable region, include a welcome paragraph that provides a mission statement and summary regarding the Web site purpose. Submit the document in the format specified by your instructor.

3 | Adding Graphics and Links

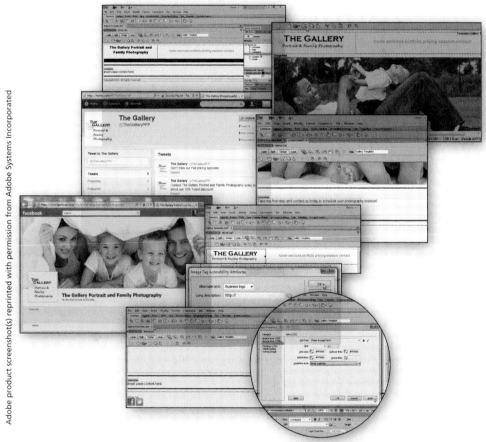

Objectives

You will have mastered the material in this chapter when you can:

- Modify a Dreamweaver template
- Edit a CSS rule
- Add graphics to a template
- Describe image file formats
- Insert images on a Web page
- Describe Dreamweaver's image accessibility features
- Create a Facebook and Twitter presence on the site
- Add images to an HTML page

- Describe the different types of links
- Add a relative link to a template
- Add an absolute link to a template
- Create an e-mail link
- Format a rollover link
- Add a CSS rule to an existing style sheet
- Add an image placeholder and replace it with an image

3 | Adding Graphics and Links

Introduction

A Web page that captures the attention of visitors includes appealing images and easy-to-follow navigation links to other pages within the site. After establishing the initial layout of a site with style sheets, you can add content, including images and links to other pages. The content is the information provided in the Web site, and it should be engaging, relevant, and appropriate to the audience. Some people in the audience may need assistance viewing the site if they have limited vision or other visual impairments, so accessibility issues also should be addressed when developing the site.

By captivating your audience with graphics, you motivate each user to follow the navigation links and investigate the message of your business or topic on your Web pages. A well-designed site includes images that convey the professionalism and focus of the site. As you select each image for a site, remember that a picture truly is worth a thousand words.

Project — Promotional Images

The images displayed on the site for The Gallery Portrait and Family Photography and shown in Figure 3–1 not only are crucial to the design of the site, but also convey the studio's artistic style of photography. Besides increasing the appeal of the Web pages, the images market the Gallery as a business, so they serve as promotion images. Chapter 3 uses Dreamweaver to add pages to the Gallery site and then enhance those pages by including promotion images. The Gallery owners already market the Gallery online through their pages on the Facebook and Twitter social networking sites. Adding Facebook and Twitter image links on the Gallery's Web pages increases customer awareness and brand loyalty.

(a) Home Page

(b) Services Page

(c) Portfolio Page

(d) Pricing Page

(e) Session Page

(f) Contact Page

Figure 3–1

Overview

As you read this chapter, you will learn how to create the Web page project shown in Figure 3–1 on the previous page by performing these general tasks:

- Modify a template.
- Add images to the site.
- Add pages to the site.
- Connect to social networks.
- Add relative, absolute, and e-mail links.
- Format links.
- Add a new CSS rule to an existing style sheet.
- Add an image placeholder.
- Replace an image placeholder.

Plan Ahead

> **General Project Guidelines**
>
> As you design any Web site, it is vital to consider several factors including the aesthetics of the graphics, the quality of the content, and the ease of the site's navigation. Web sites typically have a home page or an index page, but that does not necessarily mean that all visitors use it to enter the Web site. Generally, with most Web sites, the visitor can enter the site at any point that has a Web page address. This means each page requires links visitors can use to navigate to the other pages. As you modify the home page and add the pages shown in Figure 3–1 on the previous page, you should follow these general guidelines:
>
> 1. **Prepare images.** Select your images carefully to make sure they convey the look and feel of your site adequately. Each image placed on the Web must comply with copyright rules. Acquire and then organize your images within the Assets panel. Determine which image goes with which Web page.
>
> 2. **Consider accessibility.** Consider how people with accessibility concerns such as visual impairments can use the site and how the site can address these accessibility issues.
>
> 3. **Understand the use of social networking sites.** Recognize the value of marketing your site by linking to social networking sites such as Facebook and Twitter.
>
> 4. **Identify the navigation of the site.** Consider how each page is linked to other pages within the site. Links also can connect to outside sites and e-mail.
>
> More specific details about these guidelines are presented at appropriate points throughout the chapter. The chapter also identifies the actions performed and decisions made regarding these guidelines during the development of the pages within the site shown in Figure 3–1.

To Start Dreamweaver and Open the Gallery Site

Each time you start Dreamweaver, it opens to the last site displayed when you closed the program. The following steps start Dreamweaver and open the Gallery Web site.

1 Click the Start button on the Windows 7 taskbar to display the Start menu, and then type `Dreamweaver CS6` in the 'Search programs and files' box.

2 Click Adobe Dreamweaver CS6 in the list to start Dreamweaver.

3 If the Dreamweaver window is not maximized, click the Maximize button next to the Close button on the Application bar to maximize the window.

4 If the Gallery site is not displayed in the Files panel, click the Sites button on the Files panel toolbar and then click Gallery to display the files and folders in the Gallery site.

Modifying a Template

The Gallery Dreamweaver Template created in Chapter 1 uses <div> tags to form a number of locked regions and one editable region identified as the content. Recall that Web page designers can edit locked regions only inside the template itself. Editable regions are placeholders for content unique to each page created from the template. The template and other design documents can serve as the Web site **prototype**, a realistic representation of how the new Web site will look and function when it is fully developed. Prototypes can range from a wireframe layout drawing to a working model of the site before it undergoes final development. (A **wireframe** is a sketch that illustrates the arrangement of content on each Web page.) It is best to show a prototype to your customer and ask for his or her approval early in the design process because it is much easier to make changes during the design stages rather than in the final stages of site development.

For the Gallery site, the owners of the studio reviewed the prototype, which you created in Chapter 2. It displays a single photograph with a black background on each page. The Gallery owners want to place unique images on each page of the site to showcase more of the studio's fine photography. They also want to remove the black background and increase the amount of space provided for the images. To meet these objectives, you need to modify the template and insert a second editable region within the image <div> tag. After designing a Dreamweaver template, you can modify any portion of the template to provide more flexibility when updating the site.

BTW

Adobe Tools for Prototypes
You can use Adobe Fireworks or Photoshop to create prototypes and wireframes, especially if you plan to show these mock-ups to clients or others on your Web site development team.

To Open the Gallery Template

Before modifying the Gallery Template file, you must open it in Dreamweaver. The following steps start Dreamweaver and open the Gallery Template.

1 Start Dreamweaver as you usually do, and then click Open on the Dreamweaver Welcome screen to display the Open dialog box. If necessary, navigate to the Gallery site on Removable Disk (F:).

2 Double-click the Templates folder in the Open dialog box to display the contents of the Templates folder (Figure 3–2).

3 Double-click Gallery Template to display the template in the Document window.

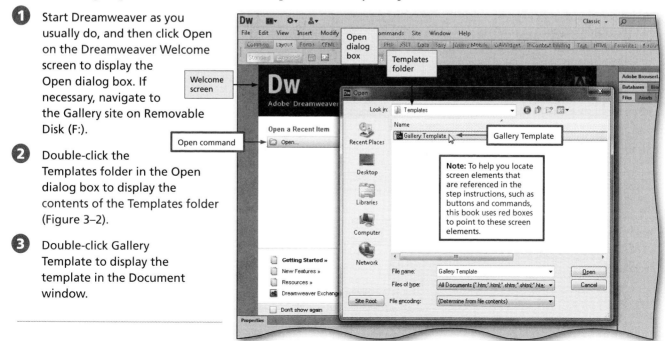

Figure 3–2

To Modify a Dreamweaver Template by Editing a CSS Rule

To edit the CSS rules established in a style sheet or document, you use the CSS Styles panel. You can display the CSS Styles panel in two modes. In **All mode**, the All Rules pane lists the CSS rules defined in the current document and in any style sheets attached to the document. Select a rule to display the CSS properties for that rule in the Properties pane. In **Current mode**, the CSS Styles panel shows style information for the current selection in the document, including CSS properties and rules. In either mode, the bottom of the CSS Styles panel contains buttons that allow you to alter the CSS rules. The button with a pencil icon is called the Edit Rule button, which you use to open a dialog box for editing the styles in the current document or the external style sheet. The layout.css style sheet defines all the styles in the Gallery Template. After opening the Gallery Template and selecting the image CSS rule in layout.css, you can use the CSS Styles panel to edit the background color and box height of the image container to meet the Gallery owner's objectives. The following steps modify the existing template to change CSS rules for the Gallery Web site.

1

- If necessary, click the All button on the CSS Styles panel to display the list of rules defined in layout.css (Figure 3–3).

Q&A What are the rules listed below layout.css in the CSS Styles panel?

These are the rules you defined in Chapter 2 to design and lay out the template for the Gallery Web site.

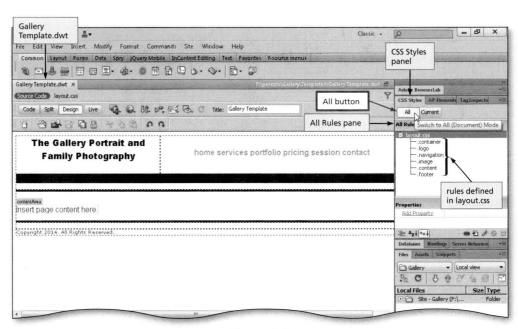

Figure 3–3

2

- Click .image in the All Rules pane to select the .image rule and display its properties in the Properties pane (Figure 3–4).

Q&A What part of the template does the .image rule format?

The .image rule determines the style of the image region, which is the area with the black background in Figure 3-4.

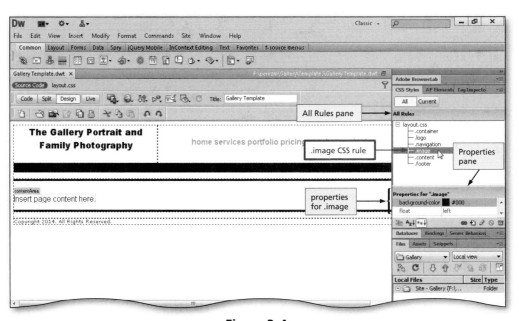

Figure 3–4

- Click the Edit Rule button on the CSS Styles panel to display the 'CSS Rule Definition for .image in layout.css' dialog box (Figure 3–5).

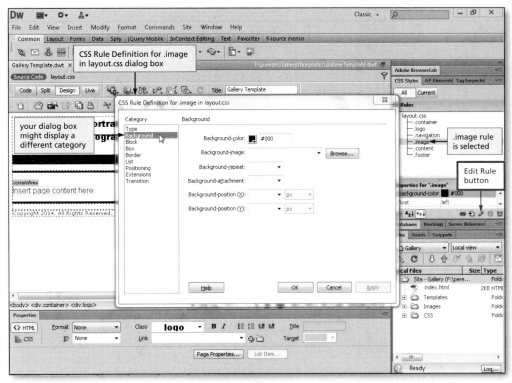

Figure 3–5

- If necessary, click Background in the Category list to display the Background options.

- Double-click the Background-color text box, delete #000, and then press the TAB key to remove the background color of the .image region (Figure 3–6).

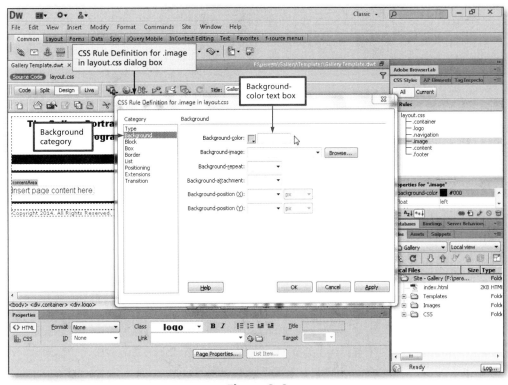

Figure 3–6

5
- Click Box in the Category list to display the Box options.

- Click the Height text box and then type 325 to set the height of the image placeholder to 325 pixels (Figure 3–7).

 6
- Click the Apply button in the 'CSS Rule Definition for .image in layout.css' dialog box to apply the CSS rules for .image in the layout.css file.

- Click the OK button in the 'CSS Rule Definition for .image in layout.css' dialog box to modify the CSS rules for .image.

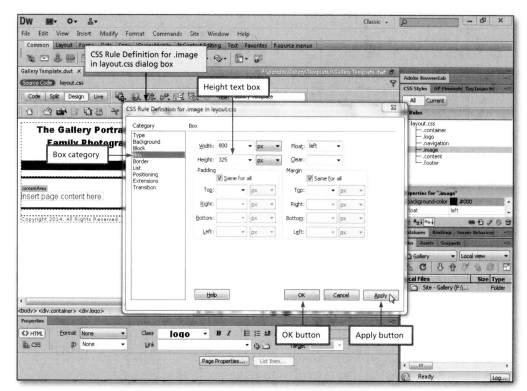

Figure 3–7

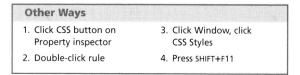

Other Ways

1. Click CSS button on Property inspector
2. Double-click rule
3. Click Window, click CSS Styles
4. Press SHIFT+F11

To Modify a Dreamweaver Template by Adding an Editable Region

When you defined the regions of the Gallery Template in Chapter 2, Dreamweaver inserted them as noneditable, or locked, regions by default. You changed the content <div> container to an editable region so that each page in the Gallery site could include different text. Now that the Gallery owners want to display different images on each page as well, you need to define the image <div> container in the template as an editable region. In a document, Dreamweaver outlines each editable region in blue and displays a small blue tab identifying the region's name. You can determine which regions are not editable in a document by moving the pointer around the Document window. The pointer changes to a "not" symbol (a circle with a line through it) when you point to a locked region. The pointer does not change when you are working in a template because you can modify locked regions in templates. The following steps modify the template to add an editable image region for the Gallery Web site.

①

- Select the text, Content for class "image" Goes Here, in the Gallery Template, and then press the DELETE key to delete the text.

- Click Insert on the Application bar to display the Insert menu, and then point to Template Objects to display the Template Objects submenu (Figure 3–8).

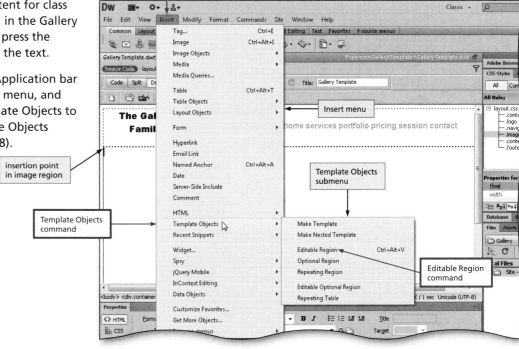

Figure 3–8

②

- Click Editable Region on the Template Objects submenu to display the New Editable Region dialog box.

- If necessary, select the text in the Name text box and then type imageArea to name the new editable region (Figure 3–9).

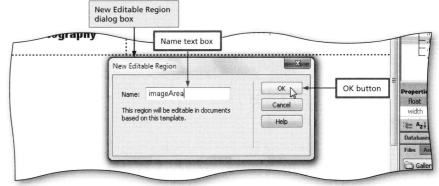

Figure 3–9

③

- Click the OK button in the New Editable Region dialog box to add the imageArea editable region to the template.

- Click the Save All button on the Standard toolbar to save the modified template and display the Update Template Files dialog box (Figure 3–10).

Q&A What is the purpose of the Update Template Files dialog box?

When you modify and then save a template, Dreamweaver displays the Update Template Files dialog box so you also can update all of the documents attached to the template. In this case, when you modify the Gallery Template, you can update the index.html document with the same changes.

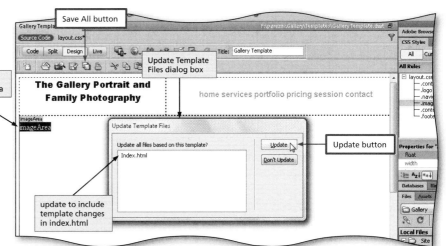

Figure 3–10

- Click the Update button to add the imageArea editable region to index.html and display the Update Pages dialog box (Figure 3–11).

5

- Click the Close button in the Update Pages dialog box to update the template and index.html.

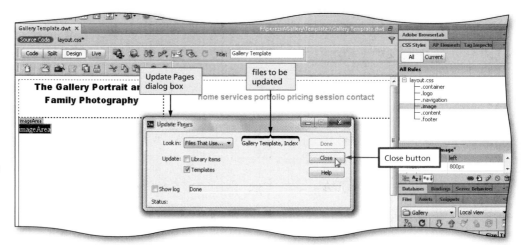

Figure 3–11

BTW

Editing Graphics
After you add a graphic to a Web page, you can use the graphic-editing tools on the Property inspector to fine-tune the image. For example, use the Crop tool to trim the image. Use the Sharpen button to increase the contrast of edges in the image.

Adding Graphics to the Web Site

The graphics that you select for a Web site have the power to create an emotional response in your audience. The best way to create interest in your Web site is to use images that complement the core message of the Web site. Images serve various purposes. For example, you can use photos to illustrate or support content, buttons to provide navigation, logos to identify a company or product, bullets to draw attention to text, mastheads to serve as title graphics, and drawings to add interest to a Web page background. The Gallery site should display photographs that represent the artistic family photography the studio provides. To include these photos in the Gallery site, you first must add the image files to the site's file structure.

Plan Ahead

Prepare images

Before you add images to a Web site, you must determine which images best support the site's mission to attract more traffic to your site. A personal Web site may include images of your friends, family members, or vacation settings taken with a digital camera. Business sites typically feature pictures of the products being sold. Keep the following guidelines in mind as you prepare images for a site:

- **Acquire the images.** To create your own images, you can take photos with a digital camera and store them in the JPEG format, use a scanner to scan your drawings and photos, or use a graphics editor such as Adobe Photoshop to create images. You also can download images from public domain Web sites, use clip art, or purchase images from stock photo collections. Be sure you have permission to reproduce the images you acquire from Web sites unless the images are clearly marked as being in the public domain.

- **Choose the right format.** Use JPEG files for photographic images and complicated graphics that contain color gradients and shadowing. Use GIF files for basic graphics, especially when you want to take advantage of transparency. You also can use PNG files for basic graphics, but not for photos.

(continued)

Prepare images *(continued)*

- **Keep the image file size small.** Use high-resolution images with an appropriate file size for faster loading. Because high-resolution image files are larger, and therefore take longer to download to a browser, use a graphics editor such as Adobe Photoshop to compress image files and reduce their file size without affecting quality. Background images in particular should have a small file size because they often appear on every page.

- **Check the dimensions.** Determine the dimensions of an image file in pixels. You can reduce the dimensions on the Web page by changing the width and height or by cropping the image. Enlarging images generally produces poor results.

As you select images, be aware of copyright laws. **Copyright** is the legal protection extended to the owners of original published and unpublished images and intellectual works. If you have not created the image yourself, you must obtain written authorization to use the image you intend to publish on your Web site unless the image is considered copyright-free. If you purchase images from stock photo collections, which are available at many Web sites, the rights to publish the images are included with your purchase. However, you should read the licensing agreement from each photo collection to determine under what conditions you can publish its images.

Understanding Image File Formats

Graphical images used on the Web fall into one of two broad categories: vector and bitmap. **Vector images** are composed of key points and paths that define shapes and coloring instructions, such as line and fill colors. A vector file contains a mathematical description of the image. The file describes the image to the computer, and the computer draws it. This type of image generally is associated with Adobe Flash, which is an animation program. One benefit of vector images is their small file size, particularly compared to the larger file sizes of bitmap images.

Bitmap images are the more common type of digital image file. A bitmap file maps, or plots, an image pixel by pixel. A **pixel**, or **picture element**, is the smallest point in a graphical image. Computer monitors display images by dividing the display screen into thousands (or millions) of pixels arranged in a **grid** of rows and columns. The pixels appear connected because they are so close together. This grid of pixels is a **bitmap**. The **bit-resolution** of an image is the number of bits used to represent each pixel. There are 8-bit images as well as 24- or 32-bit images, where each bit represents a pixel. An 8-bit image supports up to 256 colors, and a 24- or 32-bit image supports up to 16.7 million colors.

The three most common bitmap image file types that Web browsers support are JPEG, GIF, and PNG.

JPEG (.jpg) is an acronym for **Joint Photographic Experts Group**. JPEG files are the best format for photographic images because they can contain up to 16.7 million colors. **Progressive JPEG** is a new variation of the JPEG image format. This image format supports a gradually built display, which means the browser begins to build a low-resolution version of the full-sized JPEG image on the screen while the file is still downloading so visitors can view the image while the Web page downloads. Older browsers do not support progressive JPEG files.

GIF (.gif) is an acronym for **Graphics Interchange Format**. The GIF format uses 8-bit resolution, supports up to a maximum of 256 colors, and uses combinations of these 256 colors to simulate colors beyond that range. The GIF format is best for displaying images such as logos, icons, buttons, and other images with even colors and tones.

PNG (.png) stands for **Portable Network Graphics**. PNG, which is the native file format of Adobe Fireworks, is a GIF competitor and is used mostly for Web site images. All contemporary browsers support PNG files, though some older browsers do not support this format without a special plug-in.

When developing a Web site containing many pages, you should maintain a consistent, professional layout and design using images throughout all of the pages. The pages in a single site, for example, should use similar background colors or images, margins, and headings.

Adding Alt Text to Provide Accessibility

BTW

Alt Text and Screen Readers
People with visual impairments often use a screen reader to interact with Web pages. The screen reader recites the text provided as alt text to help users interpret the image.

People with visual impairments often use screen readers (speech synthesizers) that can read a text description aloud for each image and let users understand accompanying information about the images. Each image in a Web site should have **alternate text**, also called alt text, that assigns text to the image tag to describe the image. The **alt tag** is an HTML attribute that provides alternate text when nontextual elements, typically images, cannot be displayed. The alt text is considered an accessibility attribute because it provides access to everyone who visits your site. Alternate text always should describe the content of the image. For example, when the screen reader approaches a logo image, the alt tag text may be read as *Company image logo*.

Plan Ahead

> **Consider accessibility**
> After you select images for the site, consider what information the image is conveying as you create each alt tag (alternate text). The text should identify the same information that the image illustrates or communicates.

In addition to assisting people with visual impairments, alt tags can improve navigation when a graphics-intensive site is being viewed over a slow connection. Because the alt text appears before the page begins loading an image, site visitors can make navigation choices before graphics are fully rendered. Alt tags also determine how a search engine locates the content of your site. Search engines can only read text, so images with alt tags allow search engines to match the search description to the site's content, which may aid in search engine rankings. Alt tags are a required element for standards-based HTML coding.

To Copy Files into the Images Folder

Before adding an image to a site, you must add the image file to the file structure of the site. To complete this assignment, you will be required to use the Data Files for Students. Visit www.cengage.com/ct/student-download for detailed instructions or contact your instructor for information about accessing the required files. The following steps copy 12 files from the Data Files for Students to the Gallery site.

- If necessary, insert the drive containing your student data files into an available port. Use Windows Explorer to navigate to the storage location of the Data Files for Students.

- Double-click the Chapter 03 folder, and then double-click the Gallery folder to open the folders.

- Click the contact_image file, or the first file in the list, hold down the SHIFT key, and then click the twitter_image file, or the last file in the list, to select the images needed for the site (Figure 3–12).

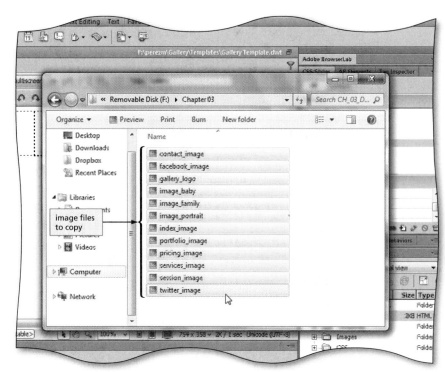

Figure 3–12

2

- Right-click the selected files, click Copy on the context menu, and then navigate to the *your last name and first initial* folder on Removable Disk F: to prepare to copy the files.

- Double-click the Gallery folder, and then double-click the Images folder to open the Images folder.

- Right-click anywhere in the open window, and then click Paste on the context menu to copy the files into the Images folder. Verify that the folder now contains 12 images (Figure 3–13).

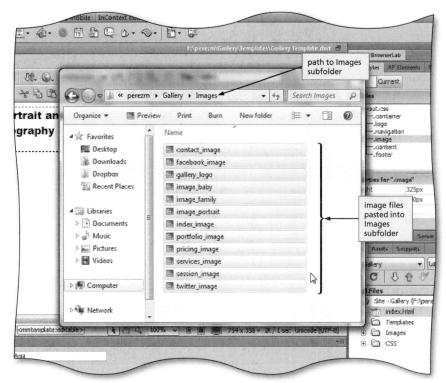

Figure 3–13

To Insert a Logo Image in the Template

Logos increase brand recognition and add visual appeal to any Web page. The Gallery's logo should appear in the upper-left corner of every page within the site to provide consistency to the layout. Instead of using a text logo, an image logo is available for the Gallery site. The following steps insert the Gallery logo into the Gallery Template.

1

- In Gallery Template.dwt, select the text, The Gallery Portrait and Family Photography, in the logo region and then press the DELETE key to delete the text.

- Click Insert on the Application bar to display the Insert menu (Figure 3–14).

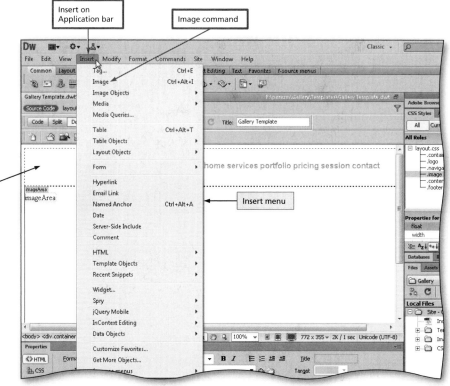

Figure 3–14

2

- Click Image on the Insert menu to display the Select Image Source dialog box.

- Double-click the Images folder to display the image files available.

- Click gallery_logo to select the logo image (Figure 3–15).

Q&A

How long will this image take to load in a browser?

The Select Image Source dialog box lists the file size and approximate download time below the Image preview.

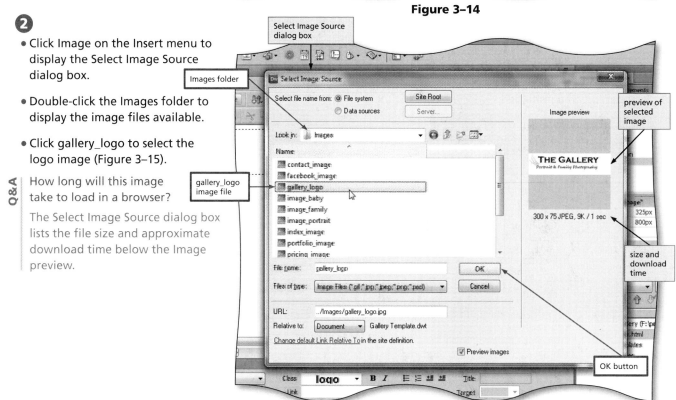

Figure 3–15

- Click the OK button in the Select Image Source dialog box to display the Image Tag Accessibility Attributes dialog box.

- In the Alternate text text box, type Business logo to add the alt tag necessary for accessibility (Figure 3–16).

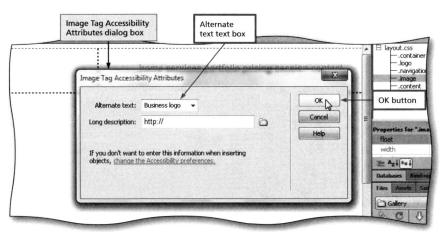

Figure 3–16

4

- Click the OK button to display the Gallery logo image in the logo region (Figure 3–17).

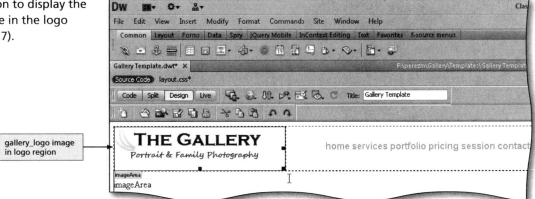

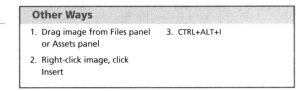

Figure 3–17

Other Ways
1. Drag image from Files panel or Assets panel 3. CTRL+ALT+I
2. Right-click image, click Insert

Marketing a Site with Facebook and Twitter

A **social networking site** is an online community in which members share their interests, ideas, and files such as photos, music, and videos with other registered users. Some social networking sites are purely social, while others have a business focus.

Understand the use of social networking sites	Plan Ahead
Social networking sites offer a way to promote products and services over the Internet to a larger target audience. Before placing a link to a social networking site on your Web site, a Facebook page and Twitter presence must be established with professional, business-generating content. To establish a presence, join Facebook and Twitter, and then follow the directions on each Web site to post text, images, and links to showcase your organization.	

Instead of advertising in a newspaper or magazine, many businesses target social networking sites, such as Facebook and Twitter, for their ads. **Facebook** is a social networking site that provides a platform to interact with customers and other businesses that are also members of Facebook. Visitors to a business-oriented Facebook page can engage with their favorite brands and receive product updates. Using Facebook, the Gallery site provides a more personalized, social experience. The Gallery owners have established the Facebook page shown in Figure 3–18.

Figure 3–18

Twitter is a social networking tool for posting very short updates, comments, or thoughts. If you want to receive posts, or tweets, from a Twitter member automatically, you can choose to become a follower of that member. Developing many followers is a goal of most business members. Using the Gallery's Twitter account, shown in Figure 3–19, the studio's owners can post information about special offers and photo packages, and links to the Gallery's Facebook page. Making a positive impression on your Twitter followers is invaluable when 'growing your business.

Figure 3–19

To Insert Social Networking Icons in the Template

Facebook and Twitter provide specific images for use as icons in other Web sites. Visitors to the Gallery Web site can click the Facebook icon to visit the Gallery's Facebook page, or they can click the Twitter icon to visit the Gallery's Twitter page. The following steps insert Facebook and Twitter icons in the footer of the Gallery Template.

- If necessary, scroll down in the Document window, and then click to the left of Copyright 2014 in the footer of the template to place the insertion point directly before the Copyright 2014 text.

- Type `Follow Us:` and then press the SHIFT+ENTER keys to add text to the footer (Figure 3–20).

Q&A What is the purpose of pressing the SHIFT+ENTER keys simultaneously?

Pressing the SHIFT+ENTER keys inserts a line break in the Web document. A line break starts a new line without adding blank space between the lines. Pressing the ENTER key creates a new paragraph. Browsers automatically add a blank line before and after a paragraph.

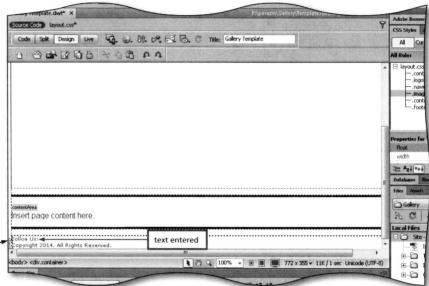

Figure 3–20

- Click Insert on the Application bar and then click Image on the Insert menu to display the Select Image Source dialog box.

- Click facebook_image in the Images folder to select the facebook_image file (Figure 3–21).

Q&A Does the Facebook icon automatically link to Facebook?

No. Later in this chapter, you add a link to the image to connect it to the Gallery's Facebook page.

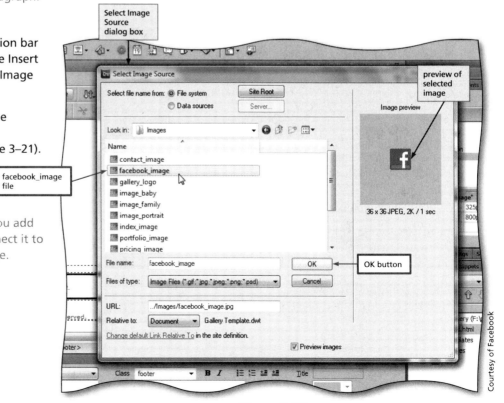

Figure 3–21

3

- Click the OK button in the Select Image Source dialog box to display the Image Tag Accessibility Attributes dialog box.

- In the Alternate text box, type `Facebook icon` to add the alt tag necessary for accessibility (Figure 3–22).

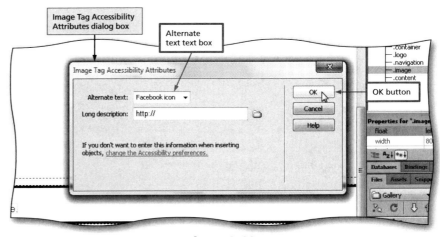

Figure 3–22

4

- Click the OK button to display the Facebook icon image in the footer region.

- Click to the right of the Facebook icon in the footer and then press the SPACEBAR to insert a space.

- Click Insert on the Application bar and then click Image to display the Select Image Source dialog box.

- Click twitter_image to select the Twitter icon in the Images folder (Figure 3–23).

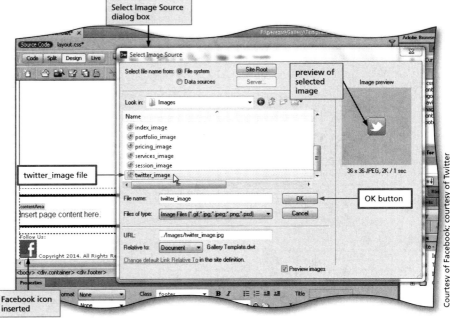

Figure 3–23

5

- Click the OK button in the Select Image Source dialog box to open the Image Tag Accessibility Attributes dialog box.

- In the Alternate text text box, type `Twitter icon` to add the alt tag necessary for accessibility.

- Click the OK button to display the Twitter icon image in the footer region (Figure 3–24).

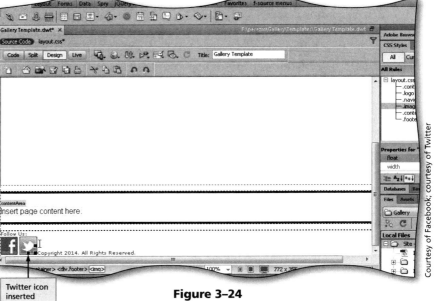

Figure 3–24

- Click to the left of the word, Copyright, in the footer to place the insertion point in front of that text.

- Press the ENTER key two times to create two blank lines between the Facebook and Twitter icons and the Copyright line.

- Click the Save All button on the Standard toolbar to save the template.

- Click Update in the Update Templates dialog box to add the icons to the Gallery Template and display the Update Pages dialog box.

- Click the Close button in the Update Pages dialog box to update the Gallery template and index.html (Figure 3–25).

- Click the Close button on the Gallery Template.dwt tab to close the template and display the Welcome screen.

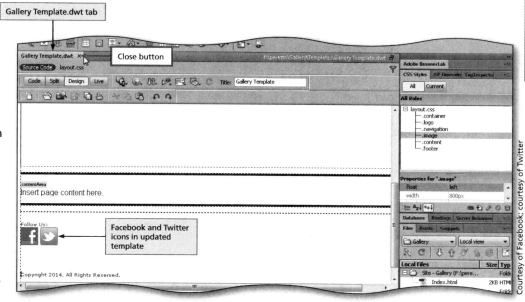

Figure 3–25

To Insert an Image on the Home Page

Because it is the first page most Web site visitors see, the home page must have enough visual interest to catch the attention of visitors and invite them to explore other pages. The following steps insert an image on index.html, the home page.

- Double-click index.html in the Files panel to open the index.html page.

- Select the text, imageArea, in the imageArea region and then press the DELETE key to delete the text (Figure 3–26).

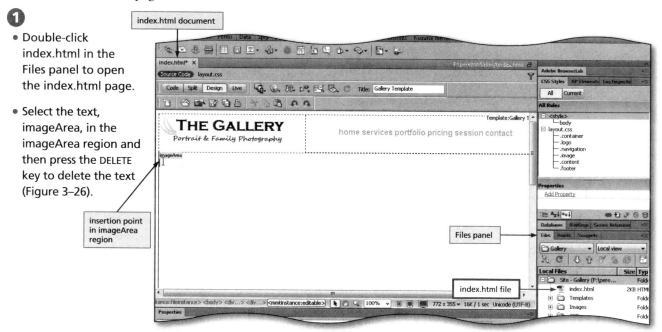

Figure 3–26

- Click Insert on the Application bar and then click Image on the Insert menu to display the Select Image Source dialog box.

- Click index_image in the Images folder to select the image for index.html (Figure 3–27).

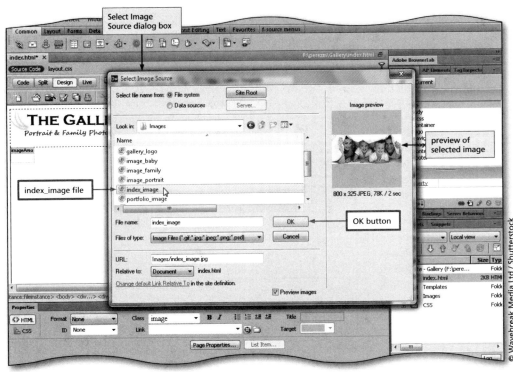

Figure 3–27

- Click the OK button in the Select Image Source dialog box to open the Image Tag Accessibility Attributes dialog box.

- In the Alternate text text box, type Home family portrait to add the alt tag necessary for accessibility.

- Click the OK button to display the index image in the imageArea region (Figure 3–28).

 **Experiment**

- Click the Brightness and Contrast button in the Property inspector, and then click the OK button. Use the slider to change the brightness of the image. When you are finished, click the Cancel button to return the image to its original state.

Figure 3–28

- Click the Save button on the Standard toolbar to save your work.

Creating Additional Pages for the Site

After creating the template and home page, the next step is to create the other pages of the Gallery site to which the home page links. The plan for the Gallery site specifies that the site should contain six pages: index.html (the home page), services.html, portfolio.html, pricing.html, session.html, and contract.html. You can design each of these additional pages using the Gallery Template to set the standard structure of the page. If any design change is necessary, you only need to change the template. Dreamweaver then updates all of the pages automatically. The common elements such as the logo, navigation, and footer remain unchangeable, while the editable regions can display different pictures and content on each page.

To Create the Services Web Page

The Gallery specializes in portrait and family photography in a variety of beautiful natural settings throughout the Florida area. These services will be detailed in a page named services.html. The following steps create the services page using the Gallery Template.

1

- Click the Close button on the index.html tab to close the home page and display the Welcome screen.

- Click More in the Create New list to display the New Document dialog box.

- Click Page from Template in the left pane of the New Document dialog box to create a page from the template.

- If necessary, click Gallery in the Site list to create a page for the Gallery site (Figure 3–29).

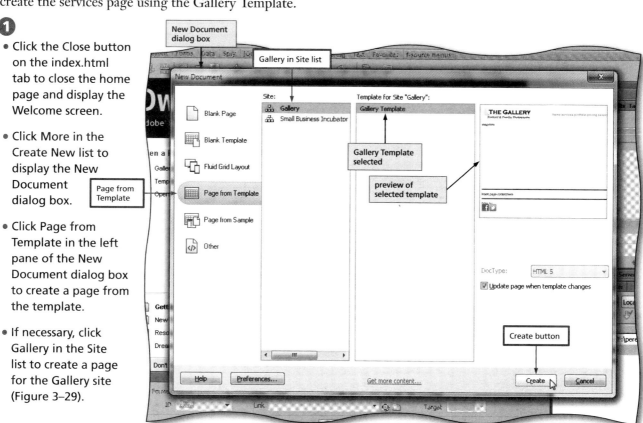

Figure 3–29

2

- Click the Create button in the New Document dialog box to create a new page based on the template for the Gallery site.

- Click File on the Application bar and then click Save As to display the Save As dialog box.

- If necessary, select the text in the File name text box and then type services.html to name the new page (Figure 3–30).

Q&A Which folder should I select when saving the services.html file?

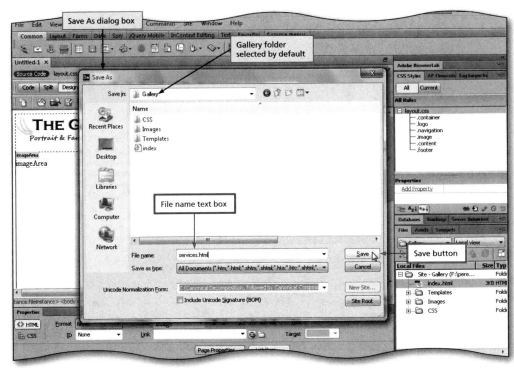

Figure 3–30

Save the services.html page in the root Gallery folder, which is the folder displayed by default in the Save As dialog box.

3

- Click the Save button in the Save As dialog box to save the document as services.html.

- Select the text, imageArea, in the imageArea region and then press the DELETE key to delete the text.

- Click Insert on the Application bar and then click Image on the Insert menu to display the Select Image Source dialog box.

- If necessary, scroll down and click services_image to select the services image (Figure 3–31).

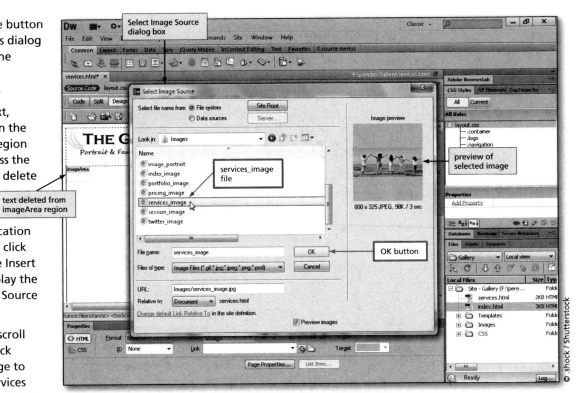

Figure 3–31

© shock / Shutterstock

4

- Click the OK button in the Select Image Source dialog box to display the Image Tag Accessibility Attributes dialog box.

- In the Alternate text text box, type `Services family portrait` to add the alt tag necessary for accessibility (Figure 3–32).

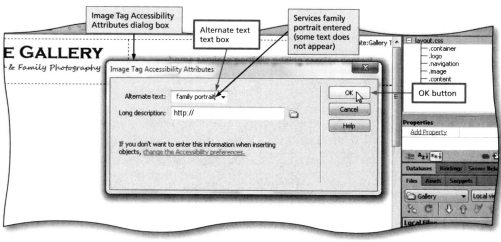

Figure 3–32

5

- Click the OK button to display the services image in the imageArea region (Figure 3–33).

6

- Click the Save button on the Standard toolbar to save your work.

Figure 3–33

© .shock / Shutterstock

To Create the Portfolio Web Page

The portfolio page showcases portraits and family photos in which the personality of the subjects shines through. The following steps create the portfolio page using the Gallery Template.

• Click the Close button on the services.html tab to close the services page and display the Welcome screen.

• Click More in the Create New list to display the New Document dialog box.

• Click Page from Template in the left pane of the New Document dialog box to create a page from the template.

• If necessary, click Gallery in the Site list to create a page for the Gallery site (Figure 3–34).

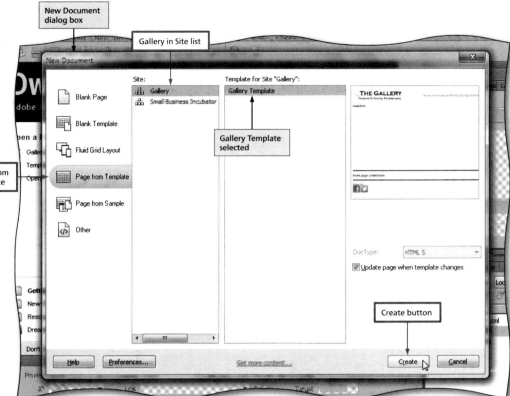

Figure 3–34

• Click the Create button in the New Document dialog box to create a new page based on the template for the Gallery site.

• Click File on the Application bar and then click Save As to display the Save As dialog box.

• If necessary, select the text in the File name text box and then type `portfolio.html` to name the new page (Figure 3–35).

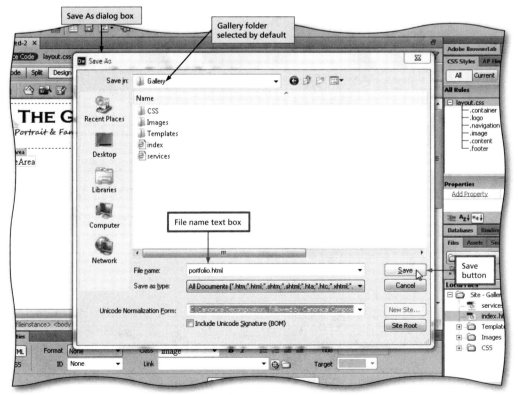

Figure 3–35

- Click the Save button in the Save As dialog box to save the document as portfolio.html.

- Select the text, imageArea, in the imageArea region and then press the DELETE key to delete the text.

- Click Insert on the Application bar and then click Image on the Insert menu to display the Select Image Source dialog box.

- Click portfolio_image to select the portfolio image in the Images folder (Figure 3–36).

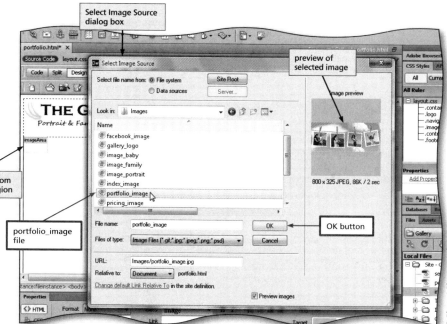

Figure 3–36

- Click the OK button in the Select Image Source dialog box to open the Image Tag Accessibility Attributes dialog box.

- In the Alternate text text box, type `Portfolio family portrait` to add the alt tag necessary for accessibility (Figure 3–37).

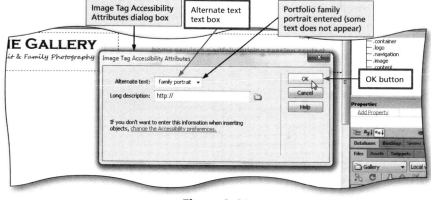

Figure 3–37

- Click the OK button to display the Gallery portfolio image in the imageArea region (Figure 3–38).

- Click the Save button on the Standard toolbar to save your work.

Figure 3–38

To Create the Pricing Web Page

The package pricing information for the Gallery will be displayed on the pricing.html page. The following steps create the pricing page using the Gallery Template.

1

- Click the Close button on the portfolio.html tab to close the portfolio page and display the Welcome screen.

- Click More in the Create New list to display the New Document dialog box.

- Click Page from Template in the left pane of the New Document dialog box to create a page from the template. If necessary, click Gallery in the Site list, and then click the Create button to create a new page based on the template for the Gallery site.

- Click File on the Application bar and then click Save As to display the Save As dialog box.

- If necessary, select the text in the File name text box and then type `pricing.html` to name the new pricing page created from the template (Figure 3–39).

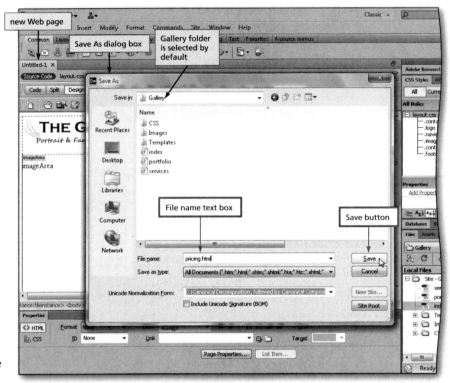

Figure 3–39

2

- Click the Save button in the Save As dialog box to save the document as pricing.html.

- Select the text, imageArea, and then press the DELETE key to delete the text.

- Click Insert on the Application bar and then click Image on the Insert menu to display the Select Image Source dialog box.

- Click pricing_image to select the pricing image in the Images folder (Figure 3–40).

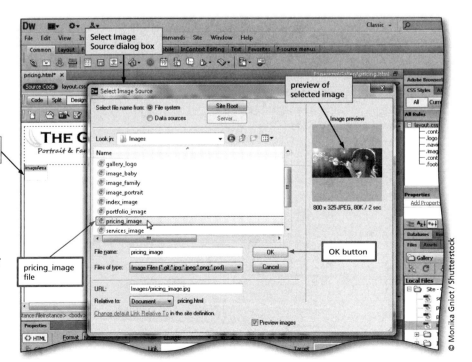

Figure 3–40

- Click the OK button in the Select Image Source dialog box to open the Image Tag Accessibility Attributes dialog box.

- In the Alternate text text box, type `Pricing family portrait` to add the alt tag necessary for accessibility.

- Click the OK button in the Image Tag Accessibility Attributes dialog box to display the pricing image in the imageArea region (Figure 3–41).

- Click the Save button on the Standard toolbar to save your work.

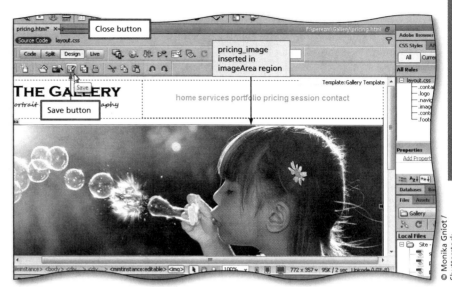

Figure 3–41

© Monika Gniot / Shutterstock

To Create the Session Web Page

Each photo shoot at the Gallery is a memorable experience. To prepare for a one-hour photography session at a selected venue, each client must decide what to wear and what to bring. The session.html page prepares each family for their special photo shoot. The following steps create the session page using the Gallery Template.

- Click the Close button on the pricing.html tab to close the pricing page and display the Welcome screen.

- Click More in the Create New list to display the New Document dialog box.

- Click Page from Template in the left pane of the New Document dialog box to create a page from the template. If necessary, click Gallery in the Site list, and then click the Create button to create a new page based on the template for the Gallery site.

- Click File on the Application bar and then click Save As to display the Save As dialog box. If necessary, select the text in the File name text box and then type `session.html` to name the session page (Figure 3–42).

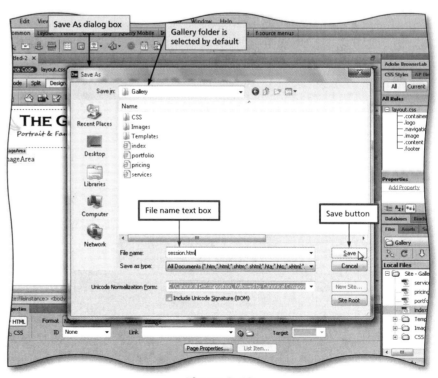

Figure 3–42

- Click the Save button in the Save As dialog box to save the document as session.html.

- Select the text, imageArea, and press the DELETE key to delete the text.

- Click Insert on the Application bar and then click Image on the Insert menu to display the Select Image Source dialog box.

- Click session_image to select the session image in the Images folder (Figure 3–43).

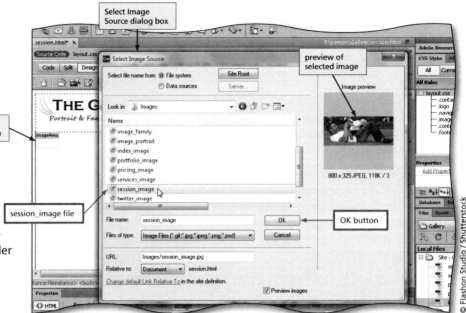

Figure 3–43

- Click the OK button to open the Image Tag Accessibility Attributes dialog box.

- In the Alternate text text box, type `Session family portrait` to add the alt tag necessary for accessibility.

- Click the OK button to display the session image in the imageArea region (Figure 3–44).

④

- Click the Save button on the Standard toolbar to save your work.

Figure 3–44

To Create the Contact Web Page

Every business site should provide contact details such as location, phone numbers, and hours; and for a photography studio, possible session times also should be included. The following steps create the contact page using the Gallery Template.

1

- Click the Close button on the session.html tab to close the session page and display the Welcome screen.

- Click More in the Create New list to display the New Document dialog box.

- Click Page from Template in the left pane of the New Document dialog box to create a page from the template. If necessary, click Gallery in the Site list, and then click the Create button to create a new page based on the template for the Gallery site.

- Click File on the Application bar and then click Save As to display the Save As dialog box. If necessary, select the text in the File name text box and then type `contact.html` to name the contact page (Figure 3–45).

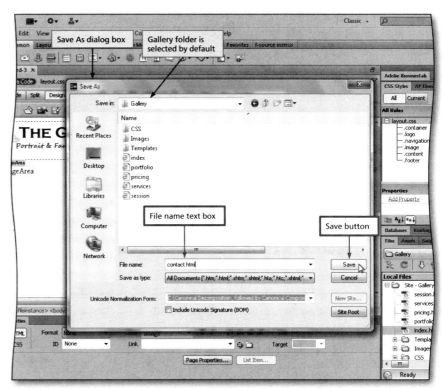

Figure 3–45

2

- Click the Save button in the Save As dialog box to save the document as contact.html.

- Select the text, imageArea, and then press the DELETE key to delete the text.

- Click Insert on the Application bar, click Image on the Insert menu to display the Select Image Source dialog box, and then click contact_image to select the contact image in the Images folder (Figure 3–46).

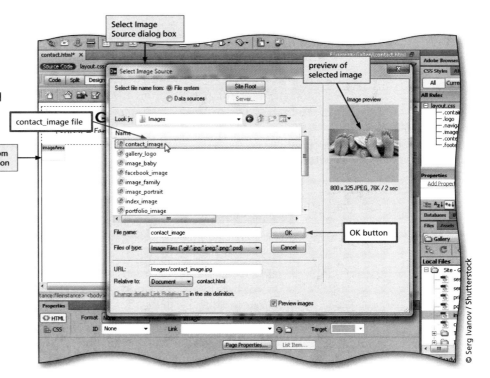

Figure 3–46

© Serg Ivanov / Shutterstock

3

- Click the OK button to display the Image Tag Accessibility Attributes dialog box.

- In the Alternate text text box, type `Contact family portrait` to add the alt tag necessary for accessibility.

- Click the OK button to display the contact image in the imageArea region (Figure 3–47).

4

- Click the Save button on the Standard toolbar to save your work.

- Click the Close button on the contact.html tab to close the contact page and display the Welcome screen.

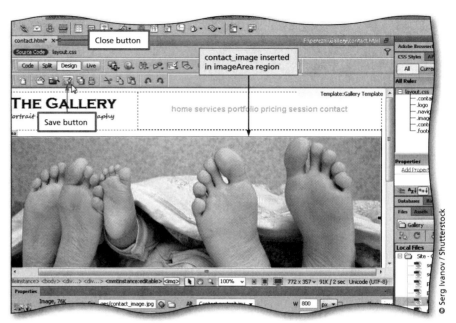

Figure 3–47

Break Point: If you wish to take a break, this is a good place to do so. To resume at a later time, start Dreamweaver, and continue following the steps from this location forward.

Adding Links to the Gallery Site

Web site navigation is the pathway people take to visit the pages in a site. Web site navigation must be well constructed, easy to use, and intuitive. Thoughtful and effective navigation tools guide users to other pages on the site and contribute to the accessibility of each page. The fundamental tool for Web navigation is the link, which connects a Web page to another page or file. If you place the mouse pointer over a link in a browser, the Web address of the link appears in the status bar. This location is the Web page or file that opens when you click the link.

Plan
Ahead

Identify the navigation of the site

Before you use links to create connections from one document to another on your Web site or within a document, keep the following guidelines in mind:

- **Prepare for links.** Some Web designers create links before creating the associated pages. Others prefer to create all of the files and pages first, and then create links. Choose a method that suits your work style, but be sure to test all of your links before publishing your Web site.

- **Link to text or images.** You can select any text or image on a page to create a link. When you do, visitors to your Web site can click the text or image to open another document or move to another place on the page.

- **Know the path or address.** To create relative links to pages in your site, the text files need to be stored in the same root folder or a subfolder in the root folder. To create absolute links, you need to know the URL of the Web page. To create e-mail links, you need to know the e-mail address.

- **Test the links.** Test all of the links on a Web page when you preview the page in a browser. Fix any broken links before publishing the page.

You can connect to another page using relative links or absolute links. A **relative link** connects Web pages within the site. For example, if a visitor begins on the home page of the Gallery site and clicks the link to the Contact page, the visitor is using a relative link. When you link text or an image to any file listed in the Files panel for the current site, you are creating a relative link. An **absolute link** means that the linked resource resides on another Web site outside of the current one, such as Facebook or Twitter. To create an absolute link, you provide the complete Web site address of the linked resource. For example, to include a link to the home page of the Professional Photographers of America (PPA) Web site, provide *http://www.ppa.com* as the complete Web address.

Another type of link in Dreamweaver is an **e-mail link**, which connects to a particular e-mail address. Clicking an e-mail link starts the user's default e-mail application and then opens a blank e-mail message containing the recipient's e-mail address.

BTW

Creating Links
To create links in Dreamweaver, you can use the Link box, the Browse For File button, or the Point to File button in the Property inspector. You also can use the Hyperlink button on the Insert bar.

To Open the Gallery Template Again

Add links to the Gallery Template so that any documents you create from the template will already contain the links to the other Web pages in the site. When you save the template with the links, Dreamweaver also updates all of the pages based on that template, which is a significant time-saver. The following steps reopen the Gallery Template.

1 On the Welcome screen, click Open to display the Open dialog box with the Gallery folder open.

2 Double-click the Templates folder in the Open dialog box and then click Gallery Template to select the Gallery Template (Figure 3–48).

3 Click the Open button to display the Gallery Template.

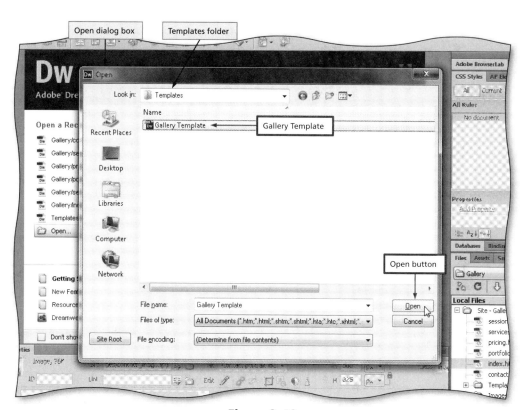

Figure 3–48

To Add Relative Links to the Gallery Template

Visitors can enter the Gallery site through any page within the site, not just the home page. Search engines, links from other Web sites, and bookmarks allow other pages to be used as entry points. Users must find their way around a Web site easily using relative links. You already have used the Point to File button to create links to other pages on the Gallery site. In fact, using the Point to File button is the easiest way to create a relative link. The following steps create relative links to each page within the site.

1

- Select the text, home, in the navigation region of the Gallery Template to select the link text (Figure 3–49).

Q&A

Why am I creating a relative link?

You use relative links when the linked documents are in the same site, such as those in your Gallery site.

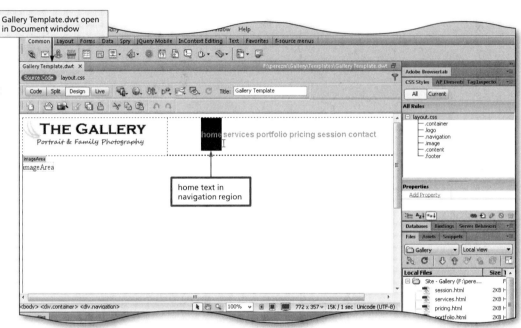

Figure 3–49

2

- Drag the Point to File button in the Property inspector to the index.html file in the Files panel to display a link line (Figure 3–50).

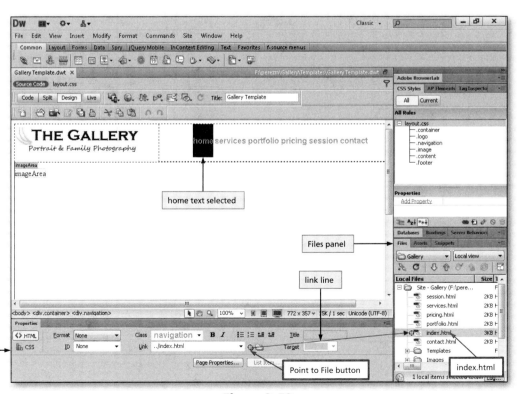

Figure 3–50

❸

- Release the mouse button to create the link to index.html.

- Select the text, services, in the navigation region to select the link text.

- Drag the Point to File button in the Property inspector to the services.html file in the Files panel to prepare to create a relative link to services.html (Figure 3–51).

Q&A Why did the text link for home in the navigation region change to blue underlined text?

Dreamweaver changes all text links to blue and underlines them by default. Later in this chapter, you add a new CSS rule to change the color to yellow and prevent the underlining.

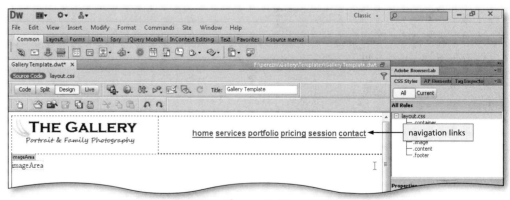

Figure 3–51

❹

- Release the mouse button to create the link to services.html.

- Select the text, portfolio, in the navigation region.

- Drag the Point to File button to the portfolio.html file to create a link.

- Select the text, pricing, in the navigation region, and then drag the Point to File button to the pricing.html file to create a link.

- Select the text, session, in the navigation region, and then drag the Point to File button to the session.html file to create a link.

- Select the text, contact, in the navigation region, and then drag the Point to File button to the contact.html file to create a link.

- Click a blank area of the page to deselect the text (Figure 3–52).

Q&A Why does an icon of a ship's wheel appear from time to time?

That icon is the Code Navigator icon. It often appears when you select text or objects on a page. You can click it to display a list of code sources related to the selection. You don't need to use it in these steps, so you can ignore it for now.

Figure 3–52

Other Ways
1. Type file name in Link box
2. Click Browse For File button in Property inspector
3. On Insert menu, click Hyperlink
4. On Insert bar, click Hyperlink
5. Select text for linking, right-click selected text, click Make Link
6. SHIFT+drag to file

To Add Absolute Links to the Gallery Template

The Gallery site has a presence on Facebook and Twitter, which means the company has set up pages on Facebook and Twitter to promote its photography business. The Facebook and Twitter logos in the footer of the Gallery template each use an absolute link to open the Gallery's pages at Facebook and Twitter. These social networking sites are not part of the Gallery site, so each logo image uses an absolute link to connect to these outside sites. To create an image link, select the image, and then type the Web address in the Link text box. The following steps create absolute links to the Gallery's Facebook and Twitter pages.

1

- If necessary, scroll down in the Document window and then click the Facebook logo in the footer region to select the image (Figure 3–53).

Q&A

Can I create links on a new page that doesn't contain any text or images yet?

No. You must select something on a page that becomes the link to another location, so you need to add text or images before creating links. If you want to create links on a new page, it's a good idea to save the page before making the links.

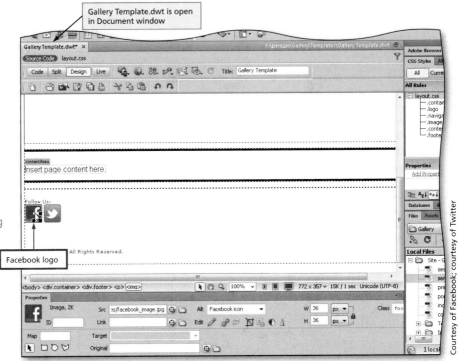

Figure 3–53

2

- Click the Link text box in the Property inspector and then type `https://www.facebook.com/TheGalleryPortraitAndFamilyPhotography` to add an absolute link to the Gallery's Facebook page (Figure 3–54).

Q&A

Why does the Facebook address in this step include https:// instead of http://?

A URL that begins with https:// identifies a secure Web site (Hypertext Transfer Protocol Secure). When a user connects to a Web site via HTTPS, the Web site encrypts the session with a digital certificate, which verifies the security of the connection.

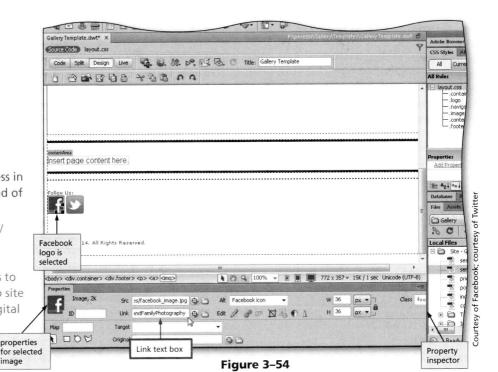

Figure 3–54

3

- Click the Twitter logo in the footer region of the Gallery Template file to select the image.

- Click the Link text box and then type `https://twitter.com/TheGalleryPFP` to add an absolute link to the Gallery's Twitter page (Figure 3–55).

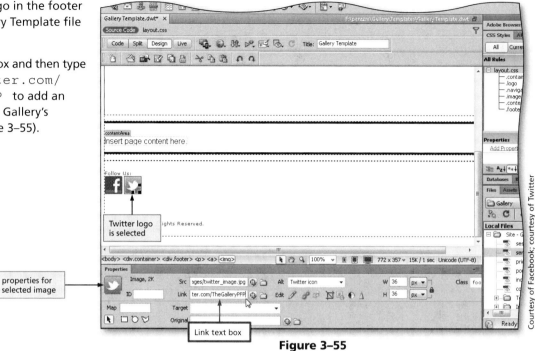

Twitter logo is selected

properties for selected image

Link text box

Figure 3–55

4

- Click the Save All button on the Standard toolbar to display the Update Template Files dialog box with six files listed for updating (Figure 3–56).

Q&A Why does the Update Template Files dialog box appear at this point?

After changing the template (in this case, by adding six text links and two image links), Dreamweaver allows you to make the same changes to the documents based on the template.

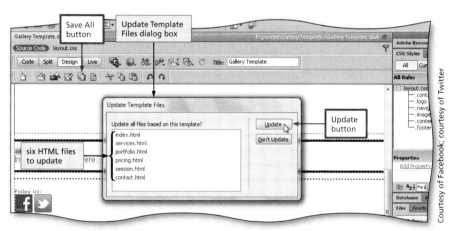

Save All button

Update Template Files dialog box

six HTML files to update

Update button

Figure 3–56

5

- Click the Update button to update all of the files based on this template and to open the Update Pages dialog box (Figure 3–57).

Q&A Are all of the links ready to be tested?

The template now includes all of the necessary links to all pages within the Web site and is ready for testing.

6

- Click the Close button to update the HTML pages within the Gallery site.

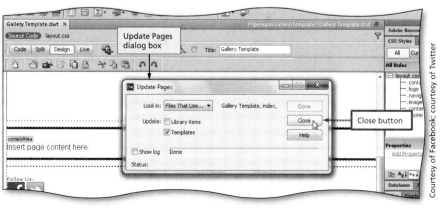

Update Pages dialog box

Close button

Figure 3–57

To Add an E-mail Link to the Gallery Template

The Contact page of the Gallery site provides a contact phone number and e-mail address. When visitors click an e-mail link, the default e-mail program installed on their computer opens a new e-mail message. The e-mail address you specify is inserted in the To box of the e-mail message header. The following steps show how to use the Insert menu to create an e-mail link on the home page.

1

- Double-click contact.html in the Files panel to open the contact page.

- Select the text, Insert page content here, in the contentArea region, type `Take the first step and contact us today to schedule your photography session!`, and then press the ENTER key to add the contact text.

- Type `(643) 555-0324` and then press the SHIFT+ENTER keys to add the contact phone number and a line break.

- Type `TheGalleryPFP@ thegallery.net` to add the e-mail address (Figure 3–58).

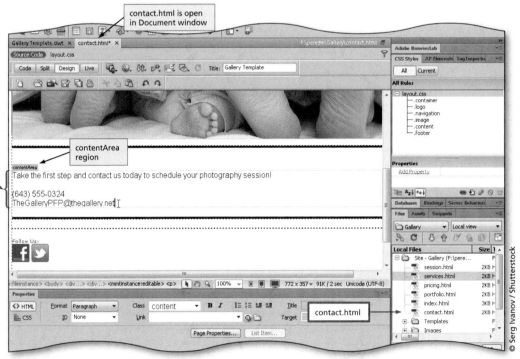

Figure 3–58

2

- Select the text, TheGalleryPFP@ thegallery.net, to select the e-mail address.

- Click Insert on the Application bar to display the Insert menu (Figure 3–59).

Q&A

Will clicking the e-mail link open my Internet e-mail such as Gmail or Hotmail?

No. You can copy and paste an e-mail address from a Web site into your Internet e-mail. An e-mail link opens automatically only in a local e-mail program such as Outlook.

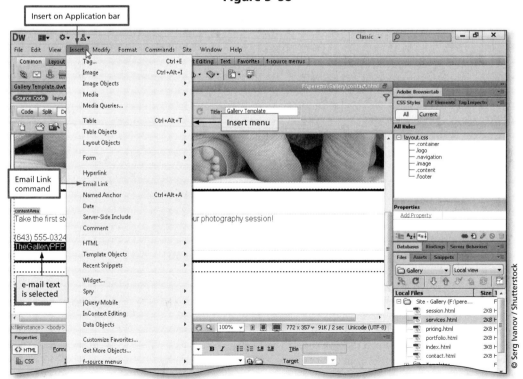

Figure 3–59

- Click Email Link on the Insert menu to display the Email Link dialog box (Figure 3–60).

Q&A What information does the Email Link dialog box already contain?

The Text and Email text boxes in the Email Link dialog box already contain the display text (the text displayed on the Web page) and the e-mail address (the recipient of the e-mail the user creates).

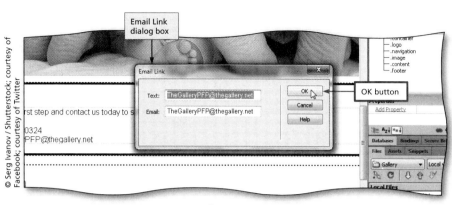

Figure 3–60

- Click the OK button in the Email Link dialog box to create an e-mail link.

- Click a blank area of the page to deselect the text (Figure 3–61).

- Click the Save button on the Standard toolbar to save the contact.html page.

- Click the Close button on the contact.html tab to close the document.

- Click the Close button on the Gallery Template.dwt tab to close the template and display the Welcome screen.

Figure 3–61

Other Ways

1. Click Email Link button on Insert bar
2. In Link box, type mailto: followed by e-mail address

Break Point: If you wish to take a break, this is a good place to do so. To resume at a later time, start Dreamweaver, and continue following the steps from this location forward.

Formatting Links

Dreamweaver refers to link text and its colors using the same terms that CSS uses. The color for link text is called the link color. The color of a link after it has been clicked is called the visited color. In the Gallery site, the link color of the text in the navigation region is blue. By default, link text is also underlined in the same color as the text. Adding relative links for the text in the navigation region made the links fully functional. Visitors can click each link to open the corresponding Web page.

To provide more interaction on the site, you can use CSS styles to format the links as rollover text instead of displaying the links in blue with underlining. A **rollover link** changes color when the mouse rolls over it. Rollover links in a Web site design allow you to change or highlight an image or text when the mouse points to it. This change in formatting provides an additional cue indicating that users can interact with the object by clicking it. When a mouse points to the orange links in the navigation area of the Gallery site, the rollover style can change the color to yellow, creating an interesting focal point to draw attention to the navigation links. CSS link styles are classified as page properties, which you can access using the Page Properties button in the Property inspector.

BTW

Rollover Images
Similar to rollover links, you can include rollover images that change when a user points to them on a Web page. To do so, you need two images: the original image, such as a button, and the rollover image, such as the button highlighted. Click Insert on the Application bar, point to Image Objects, and then click Rollover Image.

To Format a Link as Rollover Text

The following steps remove the blue underlining from text in the navigation region and add a rollover style to format the links.

1

- Open the Gallery Template file.

- If necessary, click a blank area of the Gallery Template to deselect any objects.

- Click the Page Properties button in the Property inspector to display the Page Properties dialog box (Figure 3–62).

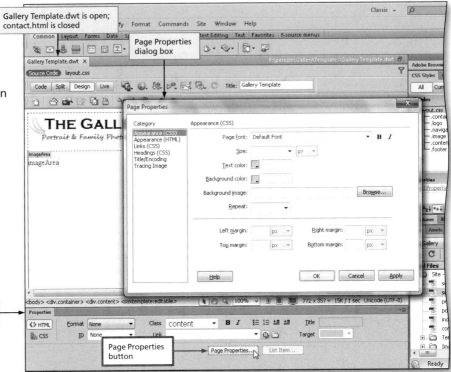

Figure 3–62

2

- Click the Links (CSS) category to display the Links options.

- Click the Link color text box and then type #FF9900 to change the link color to orange.

- Click the Rollover links text box and then type #FFCC00 to change the rollover link text color to yellow.

- Click the Visited links text box and then type #FF9900 to change the visited link text color to orange.

- Click the Underline style button and then click Never underline on the Underline style list to remove the default underline for the link (Figure 3–63).

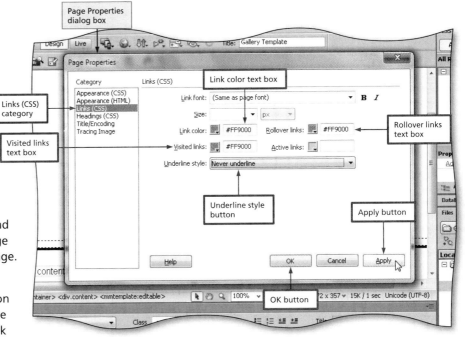

Figure 3–63

What is a visited link?

Before you visit a page, the link is displayed in a certain color by default. After you visit it, the link changes color. With Dreamweaver, you can set the visited link color to your preference.

3

- Click the Apply button in the Page Properties dialog box to change the link colors of the navigational controls.

- Click the OK button to close the Page Properties dialog box.

- Click the Save All button on the Standard toolbar to display the Update Template Files dialog box.

- Click the Update button to update all six HTML files based on this template and to open the Update Pages dialog box.

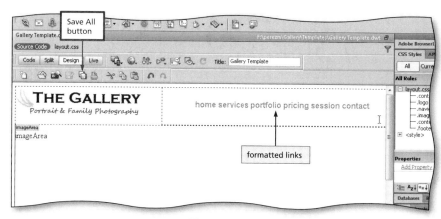

Figure 3–64

- Click the Close button in the Update Pages dialog box to close the Update Pages dialog box (Figure 3–64).

To Test the Rollover Links in a Browser

The following steps open and then preview the index.html page in Internet Explorer to test the rollover links.

1 Close the Gallery Template file.

2 Double-click index.html in the Files panel to open the index.html document.

3 Click the 'Preview/Debug in browser' button on the Document toolbar.

4 Click Preview in IExplore in the Preview/Debug in browser list to display the Gallery Web site in Internet Explorer (Figure 3–65).

© Wavebreak Media Ltd / Shutterstock; used with permission from Microsoft Corporation

Figure 3–65

5 Point to each link to view the rollover effect and then click each link to view the pages in the browser.

6 Click the Internet Explorer Close button to exit the browser.

7 Click the Close button on the index.html tab to close the index page and display the Welcome screen.

Modifying the CSS Style Sheet

When testing the links in the Gallery Web pages, you may have noticed an orange border around the Facebook logo and the Twitter logo. When you changed the link color to orange, the two image links on the Gallery template also were set to include an orange border. As you design a Web site, you may create new CSS rules or change the initial CSS rules in the style sheet for the site. In this case, you can modify a CSS rule to remove the orange border from the image links.

Creating Compound Styles

A compound style applies to two or more tags, classes, or IDs. In Chapter 2, you added a class selector style to the style sheet to identify a region by providing a name that begins with a period such as .logo or .footer. Other selector styles include a tag selector, which redefines an HTML tag such as an h1 heading, and an ID selector, which begins with a # symbol to define a block element such as a paragraph. A **compound selector** is not a different type of selector, but is actually a combination of the different types of selector styles. For example, in the Gallery site, a compound selector style applies to an image element when it is used as a link because you are combining the styles for an image tag selector and for a link.

An **anchor tag** creates a link to another page or document, or to a location within the same page. The anchor tag is <a> and refers to a clickable hyperlink element. The most common use of the anchor tag is to make links to other pages. In the Gallery site, the selector name of the style assigned to the Facebook and Twitter image links includes references to an anchor and an image: a img. The *img* stands for image, and the *a* stands for anchor. The selector *a img* is a compound selector because it combines the anchor tag (a) and the image tag (img) to place (or anchor) an image link at a desired location. To remove the border from the image links, you can use the New CSS Rule button on the CSS Styles panel to add a new CSS rule for the *a img* compound selector.

To Add New CSS Rules with a Compound Selector

To remove the orange border from the Facebook and Twitter image links, you can add a new CSS rule using a compound selector within the layout.css file. The following steps modify the layout.css style sheet to add a new CSS rule for the Gallery Web site.

1

- On the Welcome screen, click Open to display the Open dialog box, double-click the Templates folder, and then double-click Gallery Template.dwt to display the Gallery Template (Figure 3–66).

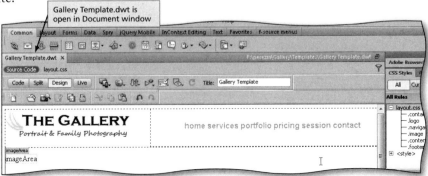

Figure 3–66

2

- Click the New CSS Rule button at the bottom of the CSS Styles panel to display the New CSS Rule dialog box (Figure 3–67).

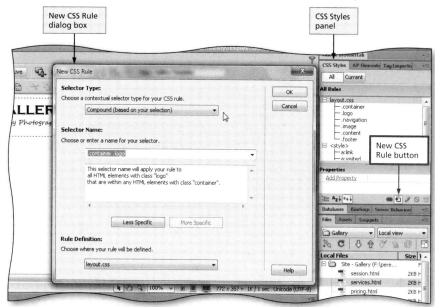

Figure 3–67

3

- If necessary, click the Selector Type button and then click 'Compound (based on your selection)' to create a compound selector type.

- In the Selector Name text box, type `a img` to create a CSS rule that applies to anchor image elements displayed as links.

- If necessary, click the Rule Definition button and then click layout.css to add the new rule to the layout.css file (Figure 3–68).

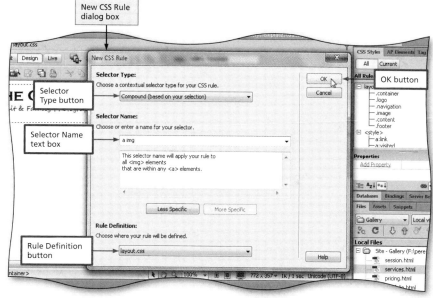

Figure 3–68

4

- Click the OK button in the New CSS Rule dialog box to display the 'CSS Rule Definition for a img in layout.css' dialog box.

- Click Border in the Category list to display the Border options.

- Click the Top box arrow in the Style section, and then click none to remove the default border around the image links (Figure 3–69).

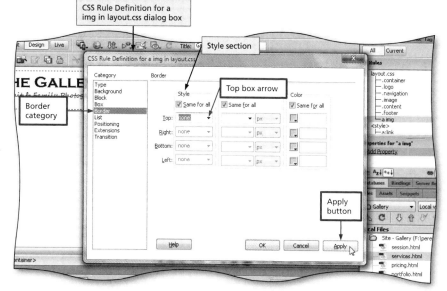

Figure 3–69

5

- Click the Apply button in the 'CSS Rule Definition for a img in layout.css' dialog box to apply the CSS rules for a img within the layout.css style sheet.

- Click the OK button to define the new CSS rule for the image border.

- Click the Save All button on the Standard toolbar to save your work.

- Double-click index.html in the Files panel and then scroll down the page to view the Facebook and Twitter images (Figure 3–70).

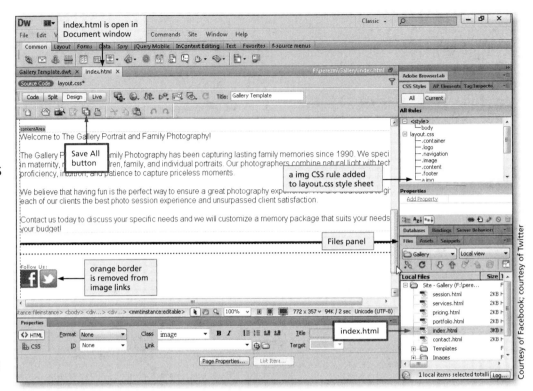

Figure 3–70

Q&A

Why does the orange border no longer appear around the Facebook and Twitter images?

The new CSS rule removed the border from all image links in files attached to the layout.css style sheet.

6

- Close the index page.

BTW

Image Placeholders and Adobe Fireworks
If Adobe Fireworks is installed on the same computer as Dreamweaver, you can click an image placeholder in Dreamweaver to open a new image in Fireworks that has the same dimensions as the placeholder.

Adding an Image Placeholder

After designing an initial Web site for a customer, you may not have access to the final images to complete the pages. For example, the customer might need to take photos of a product still being developed. In this case, you still can complete the design of the Web site. Dreamweaver contains a feature called an **image placeholder**, which reserves space on a Web page for an image during the design process by inserting a temporary photo in its place. A prototype of a Web site may contain image placeholders so the business owner can approve the prototype and the designers can continue working and complete the site. An image placeholder allows you to define the size and location of an image in your design without inserting the actual image. When you replace an image placeholder with the final graphic, the graphic uses the properties of the image placeholder, including the size, which saves you time and preserves the design of the page.

To Define an Image Placeholder

The portfolio page of the Gallery site contains portrait images taken by the photography studio to provide ideas for families to consider when planning their own photo sessions. To provide consistency in the photos, the portrait page can contain three image placeholders, each with a preset size of 150 pixels by 200 pixels. The following steps define an image placeholder on the portfolio page.

1

- In the Files panel, double-click portfolio.html to open the portfolio page.

- Scroll down and select the text, Insert page content here, in the contentArea region, type `Portraits`, and then press the SHIFT+ENTER keys to insert the text and a line break (Figure 3–71).

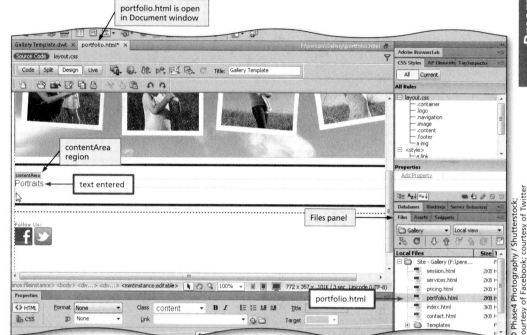

Figure 3–71

2

- Click Insert on the Application bar and then point to Image Objects on the Insert menu to display the Image Objects submenu (Figure 3–72).

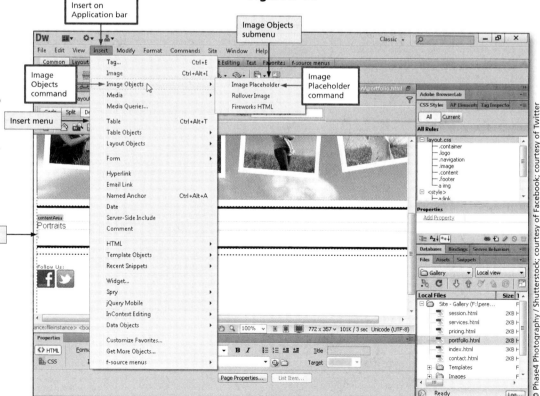

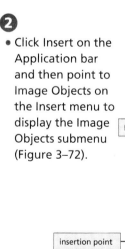

Figure 3–72

3

- Click Image Placeholder on the Image Objects submenu to display the Image Placeholder dialog box.

- In the Name text box, type `Portrait` to name the image placeholder.

- Select the value in the Width text box and then type `150` to set the width of the image placeholder.

- Select the value in the Height text box and then type `200` to set the height of the image placeholder.

- Click the Alternate text text box and then type `Portrait picture` to set the alternate text of the image placeholder (Figure 3–73).

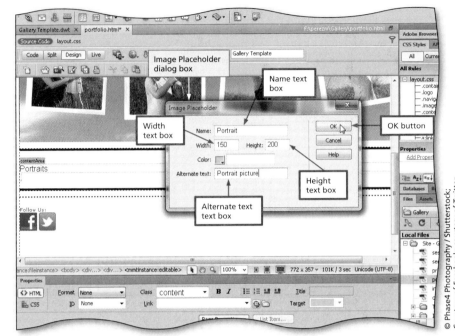

Figure 3–73

4

- Click the OK button in the Image Placeholder dialog box to create an image placeholder named Portrait (Figure 3–74).

Q&A

Does the image placeholder have to be gray?

No. You can set the color of the image placeholder using the Color button in the Image Placeholder dialog box.

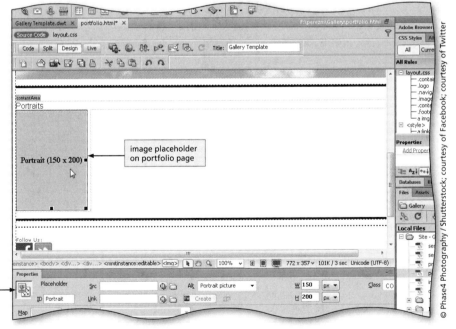

Figure 3–74

Other Ways

1. On Insert bar, click Images button, click Image Placeholder

To Replace an Image Placeholder

The following steps replace the image placeholder with the actual image for the portfolio page.

- Double-click the Portrait image placeholder to display the Select Image Source dialog box.

- Click the image_portrait file in the Select Image Source dialog box to select the replacement for the image placeholder (Figure 3–75).

Q&A Why did I insert an image placeholder if I immediately replace it with the actual image?

An image placeholder lets you set properties for an image so you can preserve the page design when you insert a photo or other image. When producing a Web site, you might not have the images as you design the pages. You immediately replace the image placeholder in these steps to practice the technique.

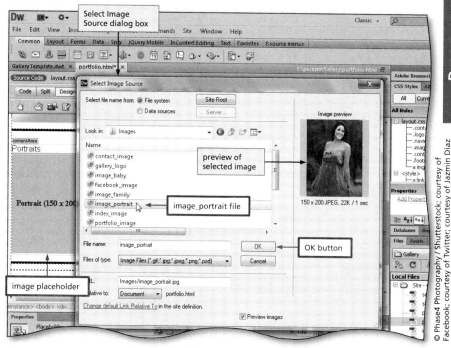

Figure 3–75

- Click the OK button in the Select Image Source dialog box to replace the image placeholder (Figure 3–76).

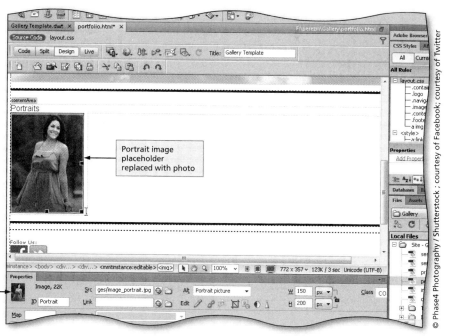

Figure 3–76

Other Ways

1. Select image placeholder, click Browse for File button on Property inspector

To Add Image Placeholders

The following steps add two other image placeholders and replacement images.

- Click to the right of the Portrait image to place the insertion point after the image.

- Press the ENTER key, type Family, and then press the SHIFT+ENTER keys to insert the text and a line break.

- Click Insert on the Application bar, point to Image Objects on the Insert menu, and then click Image Placeholder on the Image Objects submenu to display the Image Placeholder dialog box.

- In the Name text box, type Family to name the image placeholder.

- Select the value in the Width text box and then type 150 to set the width of the image placeholder.

- Select the value in the Height text box and then type 200 to set the height of the image placeholder.

- Click the Alternate text text box and then type Family picture to set the alternate text of the image placeholder (Figure 3–77).

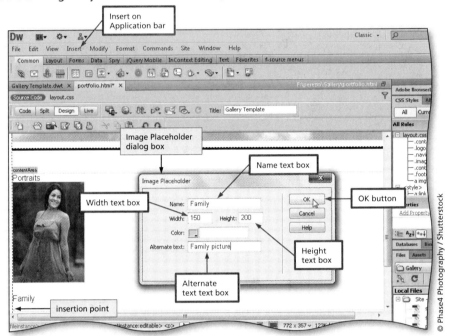

Figure 3–77

- Click the OK button in the Image Placeholder dialog box to create an image placeholder named Family.

- Double-click the Family image placeholder to display the Select Image Source dialog box, and then click the image_family file to select the replacement for the image placeholder.

- Click the OK button in the Select Image Source dialog box to replace the image placeholder (Figure 3–78).

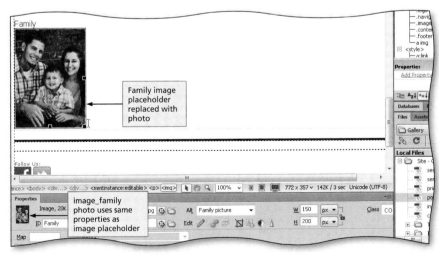

Figure 3–78

- Click to the right of the Family image to place the insertion point after the image.

- Press the ENTER key, type Baby, and then press the SHIFT+ENTER keys to insert the text and a line break.

- Click Insert on the Application bar, point to Image Objects on the Insert menu, and then click Image Placeholder on the Image Objects submenu to display the Image Placeholder dialog box.

- In the Name text box, type `Baby` to name the image placeholder.

- Select the value in the Width text box and then type `150` to set the width of the image placeholder.

- Select the value in the Height text box and then type `200` to set the height of the image placeholder.

- Click the Alternate text text box and then type `Baby picture` to set the alternate text of the image placeholder (Figure 3–79).

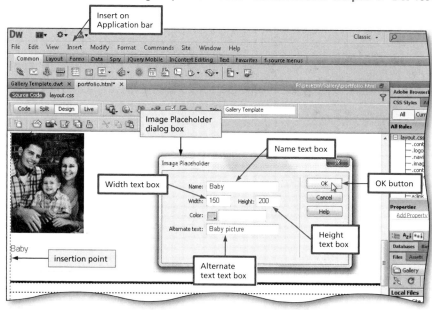

Figure 3–79

 4

- Click the OK button in the Image Placeholder dialog box to create an image placeholder named Baby.

- Double-click the Baby image placeholder to open the Select Image Source dialog box, and then click the image_baby file to select the replacement for the image placeholder.

- Click the OK button to replace the image placeholder (Figure 3–80).

 5

- Click the Save All button on the Standard toolbar to save the Gallery Web site.

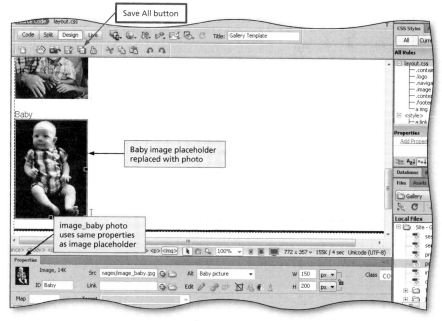

Figure 3–80

To View the Site in the Browser

The following steps preview the home page of the Gallery Web site using Internet Explorer.

1 Click the 'Preview/Debug in browser' button on the Document toolbar to display a list of browsers.

2 Click Preview in IExplore in the browser list to display the Gallery Web site in Internet Explorer.

3 Click each link on the page to test it. Scroll down as necessary to click the Facebook and Twitter image links. Click the Back button in the browser to return to the home page.

4 Click the Internet Explorer Close button to close the browser.

To Quit Dreamweaver

The following steps quit Dreamweaver and return control to the operating system.

1 Click the Close button on the right side of the Application bar to close the window.

2 If Dreamweaver displays a dialog box asking you to save changes, click the No button.

Chapter Summary

In this chapter, you were introduced to images and links and learned how to use placeholders. You began the chapter by modifying the Gallery template to add an editable region for images. Next, you added five new pages to the Gallery site and inserted graphics with alternate text on each page. You also used relative links to link the pages within the site, and you used absolute links connecting to Facebook and Twitter to provide a social networking presence for the site. In addition, you included an e-mail link that visitors can click to contact the owner of the Gallery photography studio. You modified a CSS rule to format all of the links as rollover links. Finally, you added image placeholders to a Web page and then replaced them with photos. The following tasks are all the new Dreamweaver skills you learned in this chapter:

1. Modify a Dreamweaver Template by Editing a CSS Rule (DW 144)
2. Modify a Dreamweaver Template by Adding an Editable Region (DW 146)
3. Copy Files into the Images Folder (DW 150)
4. Insert a Logo Image in the Template (DW 152)
5. Insert Social Networking Icons in the Template (DW 155)
6. Insert an Image on the Home Page (DW 157)
7. Create the Services Web Page (DW 159)
8. Create the Portfolio Web Page (DW 161)
9. Create the Pricing Web Page (DW 164)
10. Create the Session Web Page (DW 165)
11. Create the Contact Web Page (DW 166)
12. Add Relative Links to the Gallery Template (DW 170)
13. Add Absolute Links to the Gallery Template (DW 172)
14. Add an E-mail Link to the Gallery Template (DW 174)
15. Format a Link as Rollover Text (DW 176)
16. Add New CSS Rules with a Compound Selector (DW 178)
17. Define an Image Placeholder (DW 181)
18. Replace an Image Placeholder (DW 183)
19. Add Image Placeholders (DW 184)

Apply Your Knowledge

Reinforce the skills and apply the concepts you learned in this chapter.

Adding Images and a Link to a Web Page

Note: To complete this assignment, you will be required to use the Data Files for Students. Visit www.cengage.com/ct/studentdownload for detailed instructions on downloading the Data Files for Students or contact your instructor for information about accessing the required files.

Instructions: In this activity, you complete a Web page about the Mayan ruins of Tulum located on the Yucatán Peninsula in Mexico. To do so, you add images and a link to an existing Web page. The completed Web page is displayed in Figure 3–81.

Perform the following tasks:
1. Use Windows Explorer to copy the apply3.html file and the Images folder from the Chapter 03\Apply folder into the *your last name and first initial*\Apply folder.
2. Start Dreamweaver. Use the Sites button on the Files panel to display the Web sites created with Dreamweaver and the drives on your computer. Select the Apply site.
3. Open apply3.html. Select the word, Tulum, in the first sentence (not the main heading). Use the Link box on the Property inspector to insert a link to `http://en.wikipedia.org/wiki/Tulum`.

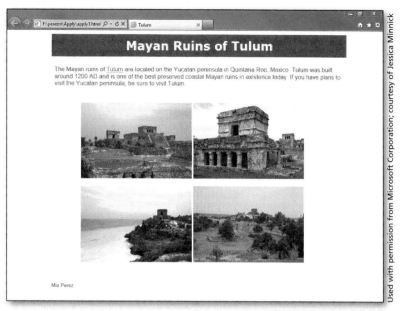

Figure 3–81

4. Double-click the image placeholder Tulum1 to display the Select Image Source dialog box, double-click the Images folder, and then click the tulum_image_1 file to select the replacement for the image placeholder. Click the OK button in the Select Image Source dialog box to replace the image placeholder.

5. Double-click the image placeholder Tulum2 to display the Select Image Source dialog box, and then click the tulum_image_2 file to select the replacement for the image placeholder. Click the OK button in the Select Image Source dialog box to replace the image placeholder.

6. Click to the right of Tulum_image_2, and then press the ENTER key. Use the Image command on the Insert menu to insert a new image, tulum_image_3.jpg. Enter `Tulum Picture 3` as the alternate text.

7. Place the insertion point after tulum_image_3.jpg and press the SPACEBAR. Use the Image command on the Insert menu to insert a new image, tulum_image_4.jpg. Enter `Tulum Picture 4` as the alternate text.

8. Replace the text, Your name here, with your first and last names.

9. Save your changes and then view your document in your browser. Compare your document to Figure 3–81. Make any necessary changes and then save your changes.

10. Submit the document in the format specified by your instructor.

Extend Your Knowledge

Extend the skills you learned in this chapter and experiment with new skills. You may need to use Help to complete the assignment.

Modifying Page Properties

Note: To complete this assignment, you will be required to use the Data Files for Students. Visit www.cengage.com/ct/studentdownload for detailed instructions on downloading the Data Files for Students or contact your instructor for information about accessing the required files.

Instructions: In this activity, you modify a Web page describing sites to visit on Maui. First you modify the page properties by selecting an image to use as a background. Then you establish link colors on the Web page. The page property changes are provided in Table 3–1. The completed Web page is displayed in Figure 3–82.

Continued >

Extend Your Knowledge *continued*

Table 3–1 Page Properties for Extend3.html		
Category	**Property**	**Value**
Appearance (CSS)	Background Image	Use the Browse button to navigate to Images/page_background
Links (CSS)	Link color	#60
	Rollover links	#F60
	Visited links	#009

Perform the following tasks:

1. Use Windows Explorer to copy the extend3.html file and the Images folder from the Chapter 03\ Extend folder into the *your last name and first initial*\Extend folder (the F:\perezm folder, for example).

2. Start Dreamweaver. Use the Sites button on the Files panel to select the Extend site.

3. Open extend3.html. Click the Page Properties button on the Property inspector, and then enter the page properties shown in Table 3–1.

4. Replace the text, Your name here, with your first and last names.

5. Save your changes and then view your document in your browser.

6. Point to the words, Maui and Haleakala National Park, to view the link changes. Click each link and then use your browser's Back button to return to the page to view the link color change.

7. Compare your document to Figure 3–82. Make any necessary changes and then save your changes.

8. Submit the document in the format specified by your instructor.

Figure 3–82

Make It Right

Analyze a Web site and suggest how to improve its design.

Editing CSS Rule Definitions on a Web Page

Note: To complete this assignment, you will be required to use the Data Files for Students. Visit www.cengage.com/ct/studentdownload for detailed instructions on downloading the Data Files for Students or contact your instructor for information about accessing the required files.

Instructions: The Learn HTML Web page provides tips for using HTML5. In this activity, you edit CSS rule definitions for four class selectors in the Learn HTML Web page. The CSS rule definitions for the page are provided in Table 3–2. The completed Web page is shown in Figure 3–83 on the next page.

Table 3–2 CSS Rule Definitions for right3.html		
CSS Rule Definition for .container		
Category	**Property**	**Value**
Box	Width	800px
	Right Margin	auto
	Left Margin	auto
CSS Rule Definition for .header		
Category	**Property**	**Value**
Type	Font-family	Verdana, Geneva, sans-serif
	Font-size	24pt
	Font-weight	bold
	Color	#C60
Background	Background-color	#FF9
Box	Bottom Margin	5px
CSS Rule Definition for .sidebar		
Category	**Property**	**Value**
Type	Font-family	Georgia, Times New Roman, Times, serif
	Font-size	14pt
	Font-style	italic
Background	Background-color	#F96
Box	Width	150px
	Height	500px
	Padding	15px, same for all
	Right Margin	10px
CSS Rule Definition for .content		
Category	**Property**	**Value**
Type	Font-family	Arial, Helvetica, sans-serif
Background	Background-color	#FFF
Box	Width	550px
	Height	500px
	Float	left

Continued >

Make It Right *continued*

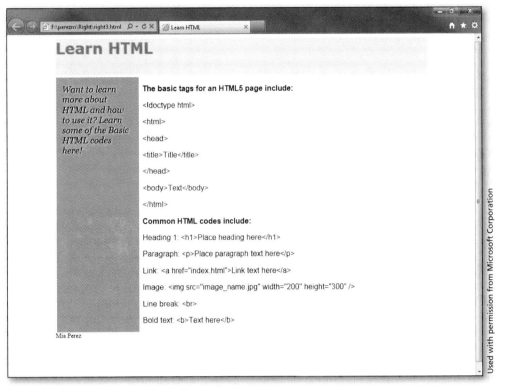

Figure 3–83

Perform the following tasks:

1. Use Windows Explorer to copy the right3.html file from the Chapter 03\Right folder into the *your last name and first initial*\Right folder (the F:\perezm folder, for example).

2. View right3.html in your browser to see its current design, and then close the browser.

3. Start Dreamweaver and open right3.html.

4. Edit the CSS rules for container, header, sidebar, and content. Refer to Table 3–2 to edit the CSS rules for each region. Apply and accept your changes.

5. Replace the text, Your name here, with your first and last names.

6. Save your changes and then view the Web page in your browser. Compare your page to Figure 3–83. Make any necessary changes and then save your changes.

7. Submit the document in the format specified by your instructor.

In the Lab

Design and/or create a document using the guidelines, concepts, and skills presented in this chapter. Labs are listed in order of increasing difficulty.

Lab 1: Adding Images and Links to the Healthy Lifestyle Web Site

Note: To complete this assignment, you will be required to use the Data Files for Students. Visit www.cengage.com/ct/studentdownload for detailed instructions on downloading the Data Files for Students or contact your instructor for information about accessing the required files.

Problem: You are creating an internal Web site for your company that features information about how to live a healthy lifestyle. Employees at your company will use this Web site as a resource for nutrition, exercise, and other health-related tips. In Chapter 2, you developed the template and home page for this Healthy Lifestyle Web site. Now you need to create other Web pages for the site and update the template with links to each page. You also need to modify CSS rules for the template header and navigation, and then adjust link colors and enhance the site with images.

First, use the Lifestyle Template to create Web pages for Nutrition, Exercise, Habits, and Sign Up. Update the Lifestyle Template with links to each page. Next, modify the Lifestyle CSS rule definitions to improve the formatting and design of the pages. Finally, add an image to the Home Web page, and then add images and text to the Nutrition Web page. The revised CSS rule definitions for the header and navigation are provided in Table 3–3. The page property values are provided in Table 3–4. The updated home page is shown in Figure 3–84, and the Nutrition page is shown in Figure 3–85.

Table 3–3 Modified CSS Rule Definitions for Lifestyle.css

CSS Rule Definition Updates for .header in Lifestyle.css

Category	Property	Value
Type	Color	#C30
Background	Background-color	Remove value
Block	Text-align	center
Box	Height	Remove value
	Top Padding	20px
	Bottom Padding	20px

CSS Rule Definition Updates for .navigation in Lifestyle.css

Category	Property	Value
Background	Background-image	Use the Browse button to select navigation_background in the Images folder
	Background color	Remove value
Box	Height	Remove value
Border	Style	Uncheck Same for all
	Top Style	dotted
	Right Style	Remove value
	Bottom Style	dotted
	Left Style	Remove value
	Width	Uncheck Same for all
	Top Width	medium
	Right Width	Remove value
	Bottom Width	medium
	Left Width	Remove value
	Color	Uncheck Same for all
	Top Color	#630
	Right Color	Remove value
	Bottom Color	#630

CSS Rule Definition Updates for .footer in Lifestyle.css

Category	Property	Value
Border	Top Width	thin
	Top Style	solid
	Top Color	#333

Continued >

In the Lab continued

Table 3–4 Page Properties for Healthy Lifestyle Pages		
Category	Property	Value
Links (CSS)	Link color	#C30
	Rollover links	#690
	Visited links	#C30
	Underline style	Show underline only on rollover

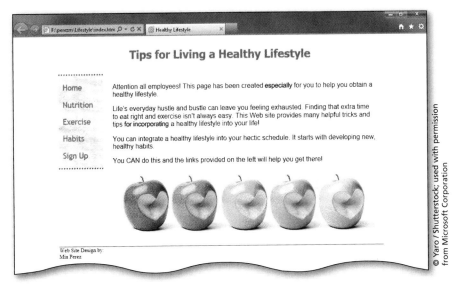

Figure 3–84

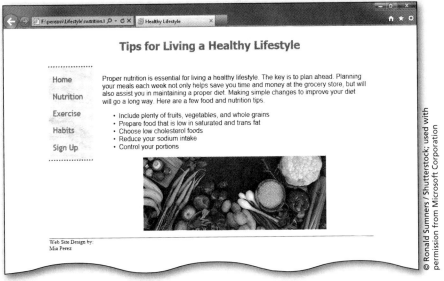

Figure 3–85

Perform the following tasks:

1. Use Windows Explorer to copy the Images folder and the nutrition.txt file from the Chapter 03\ Lab1 folder into the *your last name and first initial*\Lifestyle folder (the F:\perezm folder, for example).

2. Start Dreamweaver. Use the Sites button on the Files panel to select the Healthy Lifestyle site.

3. On the Dreamweaver Welcome screen, click More in the Create New list. In the New Document dialog box, select Page from Template, Site: Healthy Lifestyle, and Template for Site "Healthy Lifestyle": Lifestyle Template. Save the new Web page using `nutrition.html` as the file name. Close the file.

4. On the Dreamweaver Welcome screen, click More in the Create New list. In the New Document dialog box, select Page from Template, Site: Healthy Lifestyle, and Template for Site "Healthy Lifestyle": Lifestyle Template. Save the new Web page using `exercise.html` as the file name. Close the file.

5. Use the same method as in Steps 3 and 4 to create two more Web pages, using `habits.html` and `signup.html` as the file names. Close the files.

6. Open the Lifestyle Template.dwt file.

7. Select the word, Home, in the navigation bar. Use the Point to File button on the Property inspector to create a relative link to index.html. (*Hint*: You may need to use the scroll bar on the Files panel to scroll down and view the index.html file.)

8. Select the word, Nutrition, in the navigation bar. Use the Point to File button on the Property inspector to create a relative link to nutrition.html.

9. Select the word, Exercise, in the navigation bar. Use the Point to File button on the Property inspector to create a relative link to exercise.html.

10. Use the same method as in Steps 8 and 9 to create a link from the Habits text in the navigation bar to the habits.html file, and from the Sign Up text in the navigation bar to the signup.html file.

11. Use the Edit Rule button on the CSS Styles panel to edit the CSS rules for the header, navigation, and footer in Lifestyle.css. Refer to Table 3–3 for the updated values. Only update the values listed in the table; keep the other values the same. Apply and accept your changes.

12. Click the Page Properties button on the Property inspector and refer to Table 3–4 to change the page property values. Apply and accept your changes.

13. Click the Save All button on the Standard toolbar to save your changes. Click the Update button in the Update Template Files dialog box. Click the Close button in the Update Pages dialog box. Close the template.

14. Open index.html.

15. Place your insertion point after the last sentence in the content area and press the ENTER key. Use the Image command on the Insert menu to insert home_image. Use `Home image` as the alternate text.

16. Save your changes and view the document in your browser. Compare your document to Figure 3–84. Make any necessary changes, save your changes, and then close index.html.

17. Open nutrition.html.

18. Replace the text, Insert content here, with the text in the nutrition.txt file. Use the Unordered List button on the Property inspector to create an unordered list for the five lines of text below the paragraph, beginning with "Include plenty of…" and ending with "…your portions".

19. Place your insertion point after the last list item and press the ENTER key. Use the Unordered List button to remove the bullet. Use the Insert Image command on the Insert menu to insert the nutrition_image picture. Use `Nutrition image` as the alternate text.

20. Use the Format menu on the Application bar to center-align the picture.

21. Save your changes and view the document in your browser. Compare your document to Figure 3–85. Make any necessary changes and save your changes.

22. Click the links on the navigation bar to view the other pages, and to confirm that each item on the navigation bar is linked to the correct page. Make any necessary changes and then save your changes.

23. Submit the documents in the format specified by your instructor.

In the Lab

Lab 2: Adding Images and Links to the Designs by Dolores Web Site

Note: To complete this assignment, you will be required to use the Data Files for Students. Visit www.cengage.com/ct/studentdownload for detailed instructions on downloading the Data Files for Students or contact your instructor for information about accessing the required files.

Problem: You are creating a Web site for Designs by Dolores, a Web site design company. The site provides information about the company and its services. In Chapter 2, you developed the template and home page for the Designs by Dolores Web site. Now you need to create the Web pages for the site, insert a logo, and update the template with links to each page. You also need to modify the link colors and enhance the site with images.

First, use the Designs Template to create Web pages called About Us, Services, Pricing, Web Hosting, and Contact Us. Update the Design Template with links to each page. Next, modify the Lifestyle CSS rule definitions to improve the formatting and design of the pages. Finally, add text and an image to the About Us Web page. The revised CSS rule definitions for the header, navigation, content, and footer are provided in Table 3–5. The page property values are provided in Table 3–6. The updated home page is shown in Figure 3–86, and the About Us page is shown in Figure 3–87.

Table 3–5 CSS Rule Definition Updates for Designs by Dolores

CSS Rule Definition Updates for .header in Designs.css

Category	Property	Value
Type	Font-family	Remove value
	Font-size	Remove value
	Color	Remove value
Box	Top Padding	Remove value
	Bottom Padding	Remove value
	Left Padding	Remove value

CSS Rule Definition Updates for .navigation in Designs.css

Category	Property	Value
Background	Background-color	#036

CSS Rule Definition Updates for .content in Designs.css

Category	Property	Value
Box	Left Padding	Remove value
Border	Bottom Color	#036

CSS Rule Definition Updates for .footer in Designs.css

Category	Property	Value
Type	Color	#333
Background	Background-color	#FC0

Table 3–6 Page Properties		
Category	Property	Value
Links (CSS)	Link color	#FFF
	Rollover links	#FF0
	Visited links	#FFF
	Underline style	Never underline

Figure 3–86

Figure 3–87

Continued >

In the Lab *continued*

Perform the following tasks:

1. Copy the Images folder and the about.txt file from the Chapter 03\Lab2 folder into the *your last name and first initial*\Designs folder (the F:\perezm folder, for example).

2. Start Dreamweaver. Use the Sites button on the Files panel to select the Designs by Dolores site.

3. Create a new Web page using the Designs Template. Save the new page in the root folder for the Designs site using `about.html` as the file name. Close the file.

4. Use the same method as in Step 3 to create four more Web pages and use `services.html`, `pricing.html`, `hosting.html`, and `contact.html` as the file names.

5. Close all open documents, and then open Designs Template.dwt.

6. Select the word, Home, in the navigation bar. Use the Point to File button on the Property inspector to create a relative link to index.html.

7. Select the words, About Us, in the navigation bar. Use the Point to File button to create a relative link to about.html.

8. Use the same method as in Steps 6 and 7 to create links for Services, Pricing, Web Hosting, and Contact Us.

9. Use the Edit Rule button on the CSS Styles panel to edit the CSS rules for the header, navigation, content, and footer in Lifestyle.css. Refer to Table 3–5 for the updated values. Only update the values listed in the table; keep the other values the same. Apply and accept your changes.

10. Delete the text, Designs by Dolores, in the header. Insert the designs_logo image in the header. Use `Business logo` as the alternate text.

11. Click a blank area of the page, and then click the Page Properties button on the Property inspector and refer to Table 3–6 to change the page property values. Apply and accept your changes.

12. Save your changes to the Designs Template and the Designs.css file. Update the files that use the template and then close the template.

13. Open index.html.

14. Place your insertion point to the left of the word, Welcome, in the content area. Insert the home_image image. Use `Home image` as the alternate text.

15. Right-click the image, point to Align on the shortcut menu, and then click Left to left-align the image.

16. Place your insertion point after the last sentence, Please contact us today!, and then press the ENTER key.

17. Save your changes and view the document in your browser. Compare your document to Figure 3–86. Make any necessary changes, save your changes, and then close index.html.

18. Open about.html.

19. Replace the text, Insert content here, with the text in about.txt.

20. Insert the about_image picture after the last paragraph, use `About image` as the alternate text, and then center-align the picture on the page.

21. Save your changes and view the document in your browser. Compare your document to Figure 3–87. Make any necessary changes and then save your changes.

22. Click the links on the navigation bar to view the other pages, and to confirm that each item on the navigation bar is linked to the correct page. Make any necessary changes and then save your changes.

23. Submit the documents in the format specified by your instructor.

In the Lab

Lab 3: Adding Images and Links to the Justin's Lawn Care Service Web Site

Note: To complete this assignment, you will be required to use the Data Files for Students. Visit www.cengage.com/ct/studentdownload for detailed instructions on downloading the Data Files for Students or contact your instructor for information about accessing the required files.

Problem: You are creating a Web site for a new lawn care company, Justin's Lawn Care Service. The Web site will provide information to customers, including descriptions of services and pricing. In Chapter 2, you developed the template and home page for the Justin's Lawn Care Service site. Now you need to create the Web pages for the site, add a logo to the template, update the template with links to each page, modify the link colors, and enhance the site with images. You also will modify the CSS rule definitions.

First, use the Lawn Template to create Web pages for Services, Landscape, Prices, Quote, and Contact. Update the Design Template with links to each page. Next, modify the Lawn CSS rule definitions to improve the format and design of the pages. Add a logo and an image to the template, and then set an image as the page background. Finally, add images to the Landscape Web page. The revised CSS rule definitions for the header, navigation, content, and footer are provided in Table 3–7. The page property values are provided in Table 3–8. The Landscape page is shown in Figure 3–88, and the Contact page is shown in Figure 3–89.

Table 3–7 Updated CSS Rule Definitions for Justin's Lawn Care Service

CSS Rule Definition for .header in Lawn.css

Category	Property	Value
Type	Font-family	Remove value
	Font-size	Remove value
	Font-weight	Remove value
	Font-style	Remove value
	Color	Remove value
Background	Background-color	Remove value
Box	Top Padding	Remove value

CSS Rule Definition for .navigation in Lawn.css

Category	Property	Value
Type	Font-weight	bold
Background	Background-color	Remove value
Border	Style	Remove value (keep Same for all box checked)
	Width	Remove value (keep Same for all box checked)
	Color	Remove value (keep Same for all box checked)

Table 3–8 Page Properties

Category	Property	Value
Appearance (CSS)	Background image	Use the Browse button to select background_image in the Images folder
Links (CSS)	Link color	#030
	Rollover links	#090
	Visited links	#030
	Underline style	Show underline only on rollover

Continued >

In the Lab *continued*

Figure 3–88

Figure 3–89

Perform the following tasks:

1. Start Dreamweaver. Use the Sites button on the Files panel to select the Justin's Lawn Care Service site.

2. Use the Lawn_Template to create the following Web pages: `services.html`, `landscape.html`, `prices.html`, `quote.html`, and `contact.html`. Save and close each new Web page.

3. Open Lawn_Template.dwt. Add the appropriate relative links to Home, Services, Landscape, Prices, Quote, and Contact in the navigation bar.

4. Edit the CSS rules for the header and navigation in Lawn.css. Refer to Table 3–7 to modify the values. Apply and accept your changes.

5. Replace the text in the header with the lawn_logo image. Use `Business logo` as the alternate text.

6. Replace the text in the sidebar with the sidebar_image. image. Use `Sidebar image` as the alternate text.

7. Add page properties as specified in Table 3–8. Apply and accept your changes.

8. Save your changes to the Lawn Template and the Lawn.css file. Update files and then close the template file.

9. Open landscape.html.

10. Replace the text, Insert content here, with `Landscaping Ideas.`

11. Press the ENTER key and then insert the landscape1 and landscape2 images. Press the SHIFT+ENTER keys and then insert the landscape3 and landscape4 images. Press the SPACEBAR to separate the images that appear on one line. Use `Landscape image 1`, `Landscape image 2`, `Landscape image 3`, and `Landscape image 4` for the alternate text, respectively.

12. Click to the right of Landscape image 4, press the ENTER key, and then type `Plants and Flowers` below the landscape images.

13. Press the ENTER key and then insert the plant1, plant2, plant3, plant4, and plant5 images. Press the SPACEBAR to separate the images. Use `Plant image 1`, `Plant image 2`, `Plant image 3`, `Plant image 4`, and `Plant image 5` for the alternate text, respectively.

14. Save your changes and view the document in your browser. Compare your document to Figure 3–88. Make any necessary changes and save your changes.

15. Open contact.html.

16. Replace the text, Insert content here, with `For additional information about our services, please contact us at (777) 555-2629 or e-mail us at JustinsLCS@justin.net.`

17. Add an e-mail link, `JustinsLCS@justin.net`, to the e-mail text.

18. Save your changes and view the document in your browser. Compare your document to Figure 3–89. Make any necessary changes and save your changes.

19. Click the links on the navigation bar to view the other pages, and to confirm that each item on the navigation bar is linked to the correct page. Make any necessary changes and save your changes.

20. Submit the documents in the format specified by your instructor.

Cases and Places

Apply your creative thinking and problem solving skills to design and implement a solution.

1: Adding Web Pages and Links for Moving Venture Tips

Personal

You have created the home page for Moving Venture Tips and now need to create the other Web pages for the site. After creating the pages, you will add links to the pages from the navigation bar on the Move Template. You have decided to modify the font color and background color for the

Continued >

Cases and Places *continued*

header and navigation. You also want to modify the page properties by adding link colors. You also will add an image to the Rentals page. Use the Move Template to create Web pages for Budget, Rentals, Tips, and Contact. After creating these pages, link the text in the navigation bar to each page in the site. Modify the CSS rules for the header by changing the type color and background-color. Modify the CSS rules for the navigation by changing the background-color, border style, and box width. Add page properties for Links (CSS) by defining a color for the link color, rollover links, and visited links. Add text about rental information to the Rentals page. Create an Images folder and save it within the Move root folder. Add an image to the Rentals page. Check the spelling using the Commands menu and correct all misspelled words. Submit the document in the format specified by your instructor.

2: Adding Web Pages, Images, and Links for Student Campus Resources

Academic

You have created the home page for Student Campus Resources and now need to create the other Web pages for the site. After creating the pages, you will add links to the pages from the navigation bar on the Campus Template. You have decided to modify the font color and background color for the navigation. You also want to modify page properties by adding a background color and link colors. You also will add text and an image to the Activities page. Use the Campus Template to create Web pages for Activities, Committees, Events, and Contact. After creating these pages, link the text in the navigation bar to each page in the site. Modify the CSS rules for the navigation by changing the type color, background-color, and border colors (if you have borders). Modify the CSS rules for the content by changing or removing the box height. Modify the CSS rules for the footer by changing the font-family and color (in the Type category). Add page properties for Appearance (CSS) by defining a background color. Add page properties for Links (CSS) by defining a color for the link color, rollover links, and visited links. Add text describing student activities information to the Activities page. Create an Images folder and save it within the Campus root folder. Add an image to the Activities page and center-align the image on the page. Check the spelling using the Commands menu and correct all misspelled words. Submit the document in the format specified by your instructor.

3: Adding Web Pages, Images, and Links for French Villa Roast Café

Professional

You have created the home page for French Villa Roast Café and now need to create the other Web pages for the site. After creating the pages, you will add links to the pages from the navigation on the Cafe Template. You have decided to modify the font color for the navigation, and the background color and padding for the content. You also want to modify the page properties by adding a background image and link colors. You also will add an image to the Home and About pages. Use the Cafe Template to create Web pages for About, Menu, Rewards, and Contact. After creating these pages, link the text in the navigation bar to each page in the site. Modify the CSS rules for the navigation by changing the type color. Modify the CSS rules for the content by changing the background color and adjusting the box padding. Add page properties for Appearance (CSS) by defining a background image. Add page properties for Links (CSS) by defining a color for the link color, rollover links, and visited links. Add text to the About page. Add text to the Contact page and include an e-mail link. Create an Images folder and save it within the Cafe root folder. Add an image to the Home page and right-align the image on the page. Add an image to the About page and center-align the image on the page. Check the spelling using the Commands menu and correct all misspelled words. Submit the document in the format specified by your instructor.

4 | Exploring Tables and Forms

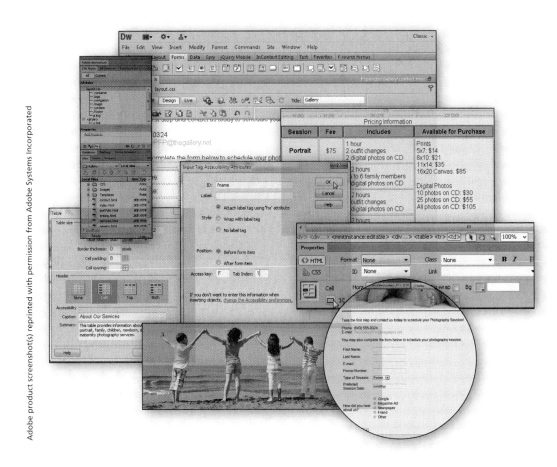

Objectives

You will have mastered the material in this chapter when you can:

- Design a Web page using tables
- Insert a table
- Format text in a table
- Modify a table structure
- Design a Web page using forms
- Insert a table into a form

- Describe and add text fields
- Add input tag accessibility attributes
- Describe and add a menu and a radio group
- Describe and add buttons
- View and test a form

4 | Exploring Tables and Forms

Introduction

To structure text, images, links, and forms into a grid pattern that contains rows and columns, Web designers employ tables. Web pages use tables to organize information. Dreamweaver's tables are very similar to the table feature in word processing programs such as Microsoft Word. A table allows you to add vertical and horizontal structure to a Web page by creating columns of text and laying out tabular data. You can delete, split, and merge rows and columns; modify table, row, or cell properties to add color and adjust alignment; and copy, paste, and delete cells in the table structure.

Tables also enable the Web site designer to lay out and create forms in a structured format. Web page forms can provide visitors with dynamic information as well as request and process information and feedback they provide. Web forms are highly versatile tools for conducting surveys, requesting job applications, and providing order forms, tests, automated responses, user questions, and online reservations.

Project — Formatted Tables and Forms

Continuing the design of the Gallery site, a table in the Services page provides information about the different types of photo services, and a table in the Pricing page itemizes session fees. The Services table, shown in Figure 4–1a, highlights the services the Gallery offers, including those for portrait, family, children, newborn, and maternity photography in a natural, relaxed setting. The Pricing Information table, shown in Figure 4–1b, is intended to answer customer questions about the costs for a photography portrait session as well as product offerings.

In addition to structuring information, a table also can be used to create a form. In this chapter, you add a form to the Contact page of the Gallery site. The Contact form, shown in Figure 4–1c, contains a request to schedule or inquire about a photography session. The form provides descriptive labels and Form objects, such as a text box, list menu, and radio buttons, allowing users to submit their information to the Gallery photography studio. As you complete the activities in this chapter, you will find that forms are one of the more important sources of interactivity on the Web and a standard tool for the Web page designer.

© .shock / Shutterstock

(a) Table on Services page

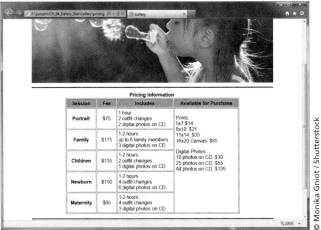

© Monika Gniot / Shutterstock

(b) Table on Pricing page

© Serg Ivanov / Shutterstock

(c) Table and form on Contact page

Figure 4–1

Overview

As you read this chapter, you will learn how to create the Web page project shown in Figure 4–1 by performing these general tasks:

• Insert a table into a Dreamweaver Web page.

• Format the table.

• Merge cells.

• Add accessibility attributes.

• Insert a form into a Dreamweaver Web page.

• Add labels, text fields, menus, and radio buttons to a form.

• Add a Submit button and a Reset button.

General Project Guidelines

When adding content to pages within a Web site, consider if the information would best be organized logically within a table format using rows and columns. If you intend to collect information from your site's visitors, a form can provide feedback. As you add two tables as shown in Figure 4–1a and Figure 4–1b, and a form in a Web page as shown in Figure 4–1c, you should follow these general guidelines:

1. **Determine what to insert into tables.** Create and organize the new content for the Web pages on your site. Consider whether you can organize and format some content in tables.

2. **Consider table layout.** Define the caption for the table, number of rows and columns, table width, border size, and cell padding. Determine whether cells within the table need to be merged to provide a better layout. If so, determine which cells need to be merged.

3. **Plan the format of the form pages.** Determine the purpose of the form. Consider the layout of the form.

4. **Consider the data you will collect.** Use the data to determine rows and columns that will be included as part of the form. Decide what the labels should state within the form. Consider the audience that will use the form and what is the best way to collect the data.

5. **Determine the types of controls you will add to the form.** The controls you add will determine and affect the type of data you can collect.

More specific details about these guidelines are presented at appropriate points throughout the chapter as necessary. The chapter also identifies the actions performed and decisions made regarding these guidelines during the development of the pages within the site shown in Figure 4–1.

To Start Dreamweaver and Open the Gallery Site

Each time you start Dreamweaver, it opens to the last site displayed when the program closed. The following steps start Dreamweaver and open the Gallery Web site.

1 Click the Start button on the Windows 7 taskbar to display the Start menu, and then type `Dreamweaver CS6` in the 'Search programs and files' box.

2 Click Adobe Dreamweaver CS6 in the list to start Dreamweaver.

3 If the Dreamweaver window is not maximized, click the Maximize button next to the Close button on the Application bar to maximize the window.

4 If the Gallery site is not displayed in the Files panel, click the Sites button on the Files panel toolbar and then click Gallery to display the files and folders in the Gallery site.

Understanding Tables

Web designers use tables to present large amounts of detailed information that can be uniformly structured into rows and columns. If a paragraph becomes cumbersome and repetitive within a Web page, consider using a table instead to format the information in an easy-to-follow layout. A table consists of **rows**, which extend horizontally from left to right; **columns**, which extend vertically from top to bottom; and **cells,** which define the area where a row and a column intersect. Each cell in the table can contain any standard element you use on a Web page, including text, images, and other objects.

Plan
Ahead

Determine what to insert into tables.
Readability is a key factor on any Web site because it enhances the user experience. Visitors prefer not to read lengthy paragraphs online. By taking advantage of simple tables to structure content, you can allow users to find the information they need in a few seconds.

Designing Table Layouts

Creating a table in Dreamweaver is a simple three-step process that begins with inserting a table, configuring the table layout, and finally adding the content, such as the text or images. After you insert a table, a Table dialog box opens that requests the number of rows and columns, table width, and other property settings as shown in Table 4–1.

Table 4–1 Table Dialog Box Settings		
Attribute	**Default**	**Description**
Rows	3	Determines the number of rows in the table
Columns	3	Determines the number of columns in the table
Table width	200 pixels	Specifies the width of the table in pixels or as a percentage of the browser window's width
Border thickness	1	Specifies the border width in pixels
Cell padding	None	Specifies the number of pixels between a cell's border and its contents
Cell spacing	None	Specifies the number of pixels between adjacent table cells
Header	None	Specifies whether the top row or first column is designated as a header row or column
Caption	None	Provides a table heading
Summary	None	Provides a table description; used by screen readers

Plan
Ahead

Consider table layout.

Choose a caption for the table. A caption provides a table heading. For a simple table, the caption should act as an adequate summary of the table's contents.

Determine the number of rows and columns. Analyze the information that will be placed in the table structure to determine how many row and column elements are necessary.

Format the table. Format a Web page table by setting various properties within the table. In Dreamweaver, you set the format properties entering text into the table structure. You set table properties — such as the width of the table, the width of the table's border, and the cell padding that determines the number of pixels between a cell's content and the cell boundaries — in the Table dialog box or Property inspector. Before formatting the table, determine if you need to merge cells together to provide a larger single cell or if you need to split a cell into two or more cells. You also can change the properties of a certain cell or selection of cells instead of the whole table.

The Services Web page highlights the Gallery's specialization of photography services in portrait, family, children, newborn, and maternity photos to capture memorable personal moments. By adding a table to display these services, as shown in Figure 4–2, customers can quickly evaluate if the Gallery's services match their photography needs. The Services table contains an opening caption, five horizontal rows and two vertical columns, a table width of 800 pixels, a 0-pixel border around the contents of the table, a cell padding of 8 pixels, and a left header. The left header appears centered and in a bold text. By default, the contents of the column or row that you designate as a header are displayed in a bold font and centered. The default is to create a table with no headers — that is, all rows and columns look the same. The Caption option displays a title outside of the table. If you specify a caption, the Align Caption option indicates where the table caption appears in relation to the table.

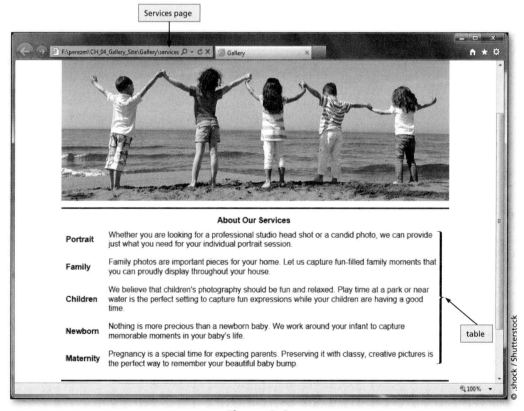

Figure 4–2

BTW

Providing a Table Summary
Summaries are especially useful in tables of data, such as a bus schedule or price list, where the content of a cell is meaningless without a column or row heading. For data tables, summaries should describe the layout of the table.

Adding a Summary to Provide Accessibility

When you create a table in Dreamweaver, the Summary option provides a table description for use in accessible content. Accessibility refers to making Web tables usable for people with visual, auditory, motor, and other disabilities. Summary text is similar to the alt text you added for images in Chapter 3. You add Summary text to the tables you create in this chapter. Screen readers recite the summary text to indicate the primary purpose of the table, but the text does not appear in the user's browser. The Summary text box is optional, and serves to improve the accessibility of your table.

To Insert a Table into the Services Page

The following steps insert a table with five rows and two columns into the Services Web page.

1

- Double-click the services.html file in the Files panel to display the contents of the Services page.

- Scroll down the page and select the placeholder text, Insert page content here, in the contentArea region, and then press the DELETE key to delete the placeholder text.

- Click Insert on the Application bar to display the Insert menu (Figure 4–3).

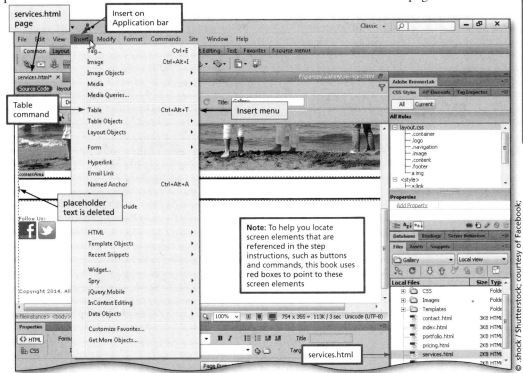

Figure 4–3

2

- Click Table on the Insert menu to display the Table dialog box (Figure 4–4).

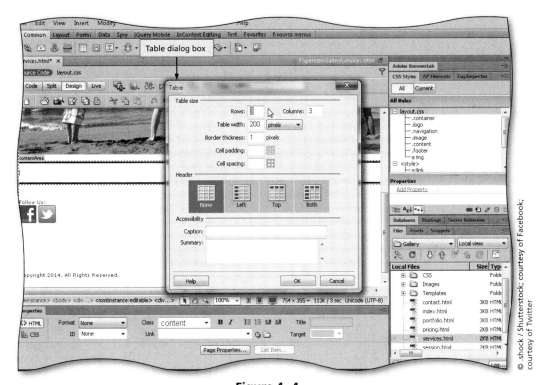

Figure 4–4

3

- In the Rows text box, type 5 to change the number of rows.

- Double-click the Columns text box and then type 2 to change the number of columns.

- Double-click the Table width text box and then type 800 to set the width to 800 pixels.

- Double-click the Border thickness text box and then type 0 to set the border thickness.

- Click the Cell padding text box and then type 8 to set the number of cell padding pixels (Figure 4–5).

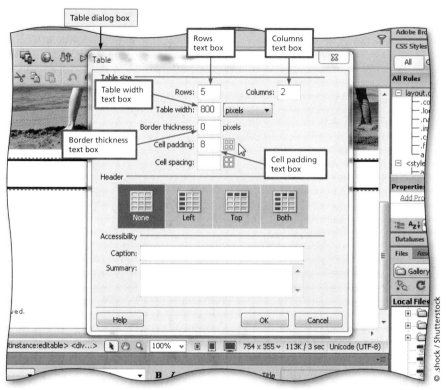

Figure 4–5

4

- Click Left in the Header area to create a left header in the table.

- Click the Caption text box and then type About Our Services to display a caption at the top of the table.

- Click the Summary text box and then type This table provides information about our portrait, family, children, newborn, and maternity photography services. to provide accessibility summary text (Figure 4–6).

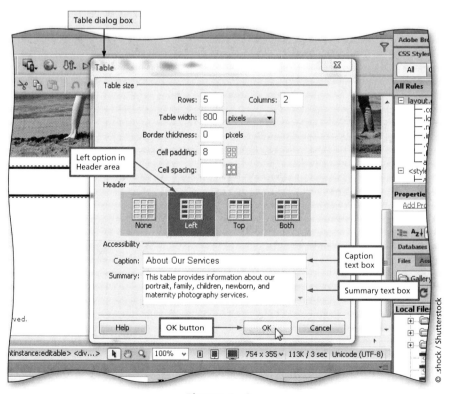

Figure 4–6

⑤
- Click the OK button in the Table dialog box to create a table in the Services page (Figure 4–7).

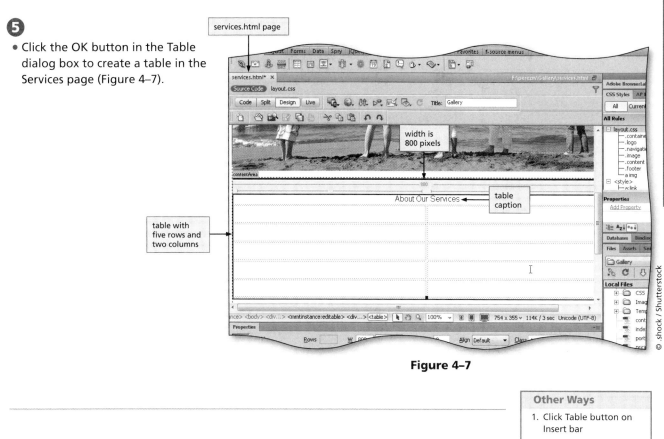

Figure 4–7

Other Ways

1. Click Table button on Insert bar

To Add Text to the Table

After you use the Table dialog box to create the structure of the table, the entire table is selected automatically. Therefore, be careful not to type anything when the table is selected or that text will replace the whole table. Instead, click a cell in the table before starting to enter content. In the table you just inserted, all of the cells are equal in width, but they expand as you add the text shown in Table 4–2 to describe the services provided by the Gallery photography studio.

Table 4–2 About Our Services

Service	Description
Portrait	Whether you are looking for a professional studio head shot or a candid photo, we can provide just what you need for your individual portrait session.
Family	Family photos are important pieces for your home. Let us capture fun-filled family moments that you can proudly display throughout your house.
Children	We believe that children's photography should be fun and relaxed. Play time at a park or near water is the perfect setting to capture fun expressions while your children are having a good time.
Newborn	Nothing is more precious than a newborn baby. We work around your infant to capture memorable moments in your baby's life.
Maternity	Pregnancy is a special time for expecting parents. Preserving it with classy, creative pictures is the perfect way to remember your beautiful baby bump.

The following steps add text to the table displayed on the Services page.

1

- Click in row 1, column 1 and then type `Portrait` to add text to the first row of the table.

- Click in row 2, column 1 and then type `Family` to add text to the second row.

- Click in row 3, column 1 and then type `Children` to add text to the third row.

- Click in row 4, column 1 and then type `Newborn` to add text to the fourth row.

- Click in row 5, column 1 and then type `Maternity` to add text to the fifth row (Figure 4–8).

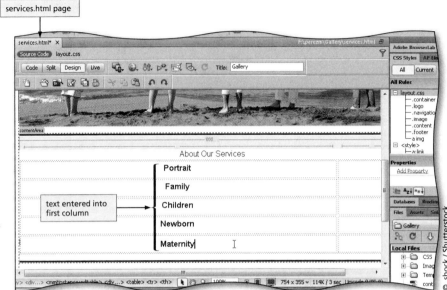

Figure 4–8

Q&A Why is the text centered within the cells?

When you enter text in a cell, Dreamweaver aligns the text in the center of each cell by default.

 2

- Click in row 1, column 2 and then type `Whether you are looking for a professional studio head shot or a candid photo, we can provide just what you need for your individual portrait session` to add a description to the first row of the table.

- Enter the remaining descriptions for the services shown in Table 4–2 in rows 2, 3, 4, and 5 to finish entering the text descriptions in the table (Figure 4–9).

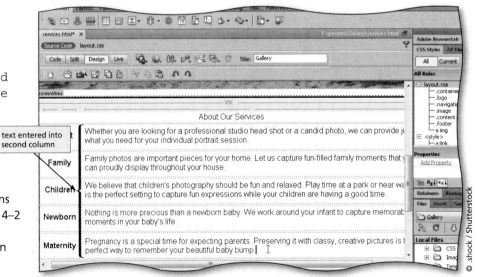

Figure 4–9

Formatting Tables

Each table within a Web page can be formatted to meet the design needs of the site. You use the Property inspector to format table attributes such as merging cells or setting the width of a table column. The Property inspector options change depending on the selected object within the table. When a table is selected, the Property inspector displays table properties, as shown in Figure 4–10. When another table element — a row, column, or cell — is selected, the Property inspector displays properties for the selected element. Table 4–3 describes the table elements in the Property inspector.

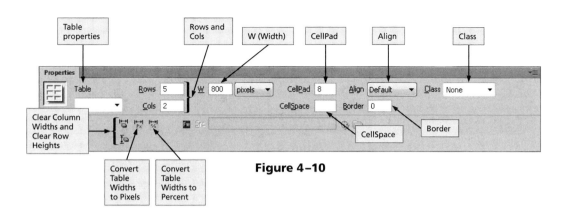

Figure 4–10

Table 4–3 Table Properties

Table Element	Description
Table	Specifies the table ID, an identifier used for Cascading Style Sheets, scripting, and accessibility. A table ID is not required; however, it always is a good idea to add this identifier.
Rows and Cols	The number of rows and columns in the table.
W	Specifies the minimum width of the table in either pixels or percent. If a size is not specified, the size can vary depending on the monitor and browser settings. A table width specified in pixels is displayed as the same size in all browsers. A table width specified in percent is altered in appearance based on the user's monitor resolution and browser window size.
CellPad	The number of pixels between the cell border and the cell content.
CellSpace	The number of pixels between adjacent table cells.
Align	Determines where the table appears, relative to other elements in the same paragraph such as text or images. The default alignment is to the left.
Class	An attribute used with Cascading Style Sheets.
Border	Specifies the border width in pixels.
Clear Column Widths and Clear Row Heights	Delete all specified row height or column width values from the table.
Convert Table Widths to Pixels	Sets the width of each column in the table to its current width expressed as pixels.
Convert Table Widths to Percent	Sets the width of the table and each column in the table to their current widths expressed as percentages of the Document window's width.

Setting Cell, Row, and Column Properties

When you select a cell, row, or column, the properties in the upper pane of the Property inspector are the same as the standard properties for text. You can use these properties to include standard HTML formatting tags within a cell, row, or column. The part of the table you select determines which properties are displayed in the lower pane of the Property inspector. The properties for cells, rows, and columns are the same except for one element — the icon displayed in the upper-left corner of the Property inspector. On the next page, Figure 4–11 shows the cell properties, Figure 4–12 shows the row properties, and Figure 4–13 shows the column properties of the Property inspector. Table 4–4 on the next page describes the Property inspector elements for the cell, row, and column properties.

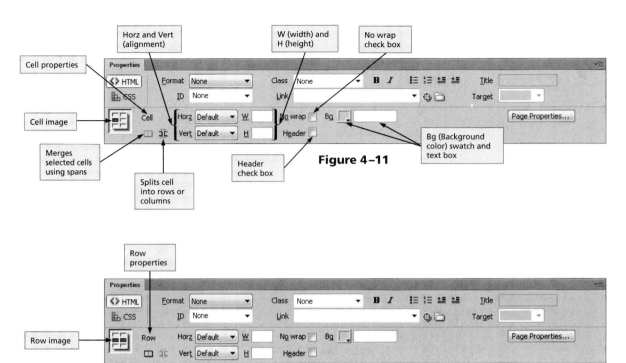

Figure 4–11

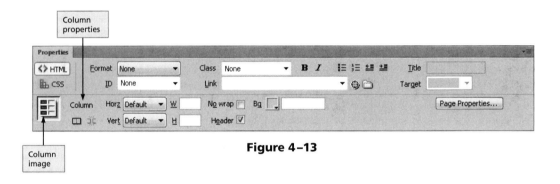

Figure 4–12

Figure 4–13

Table 4–4 Cell, Row, and Column Properties	
Table Element	**Description**
Horz	Specifies the horizontal alignment of the contents of a cell, row, or column. The contents can be aligned to the left, right, or center of the cells.
Vert	Specifies the vertical alignment of the contents of a cell, row, or column. The contents can be aligned to the top, middle, bottom, or baseline of the cells.
W and H	Specifies the width and height of selected cells in pixels or as a percentage of the entire table's width or height.
No wrap	Prevents line wrapping, keeping all text in a given cell on a single line. If No wrap is enabled, cells widen to accommodate all data as it is typed or pasted into a cell.
Bg (Background color)	Sets the background color of a cell, row, or column selected from the color picker (use the Bg button) or specified as a hexadecimal number (use the Bg text box).
Header	Formats the selected cells as table header cells. The contents of table header cells are bold and centered by default.
Merges selected cells using spans	Combines selected cells, rows, or columns into one cell (available when two or more cells, rows, or columns are selected).
Splits cell into rows or columns	Divides a cell, creating two or more cells (available when a single cell is selected).

Selecting Table Elements

Dreamweaver can change the format of the text of an entire table, a specific row or column, or an individual cell within a table. You can select the entire table by clicking the outermost edge of the table and then apply the format changes to all of the text within the table, such as selecting a new font or font size. The pointer changes to an arrow with a table grid icon when you select the table. When a table is selected, sizing handles appear on the table's lower and right edges, as shown in Figure 4–14. You can resize the table by dragging its sizing handles.

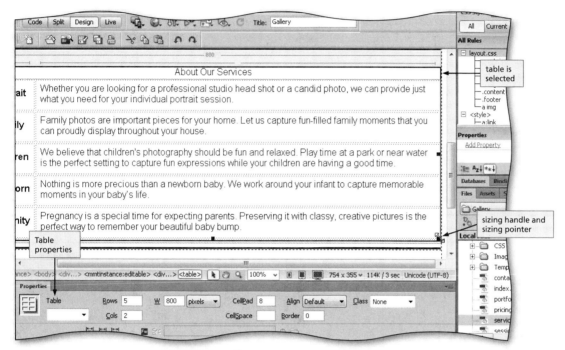

Figure 4–14

To select a single row or column to format the contents of the cells, position the insertion point to the left edge of a row or the top edge of a column. A small black selection arrow appears, as shown in Figure 4–15, indicating the row or column that you are about to select. In addition, the cells in the row or column are outlined in red. Click to select the row or column of cells. You then can apply formatting options to customize the table so it matches the design and layout of your site.

Figure 4–15

To Format Table Text

You can modify the appearance of text within the table, such as changing text alignment, by first selecting the entire table, a row or a column, or an individual cell and then applying the format. The following steps change the alignment of text within a column of the Services table.

- Point to the top of the left column in the table until a selection arrow appears (Figure 4–16).

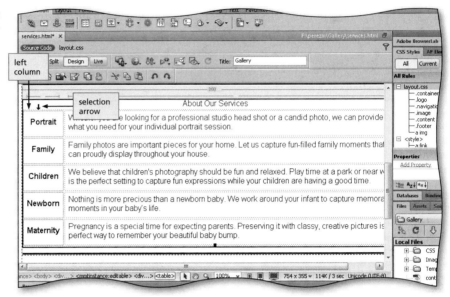

Figure 4–16

2

- Click to select the first column of cells.

- Click Format on the Application bar and then point to Align to display the Align submenu (Figure 4–17).

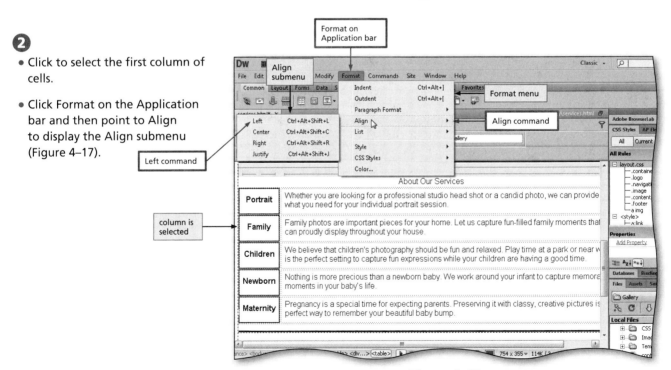

Figure 4–17

- Click Left on the Align submenu to left-align the text in the first column of the table (Figure 4–18).

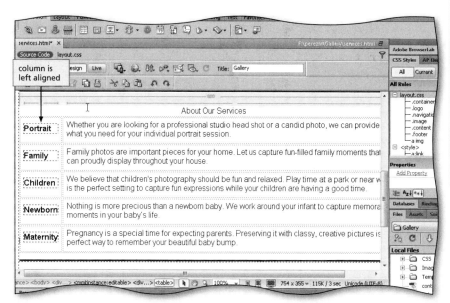

Figure 4–18

Other Ways
1. Click panel options button, click Close or click Close Tab Group 2. To align, right-click text, point to Align, click alignment command

To Format the Table Caption

The caption, which displays a description of the contents of the table, appears outside the table. After selecting the caption, you can use the Property inspector to apply a format such as Bold to the selected text. The following steps bold the text caption and view the finished table in a browser.

- Select the text, About Our Services, to select the caption for formatting (Figure 4–19).

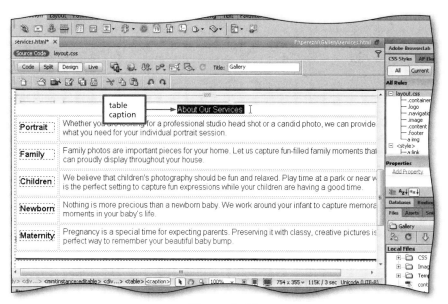

Figure 4–19

2

- In the Property inspector, click the Bold button to bold the selected text, and then click outside the caption to view the bold text (Figure 4–20).

Can I use the Property inspector to format text within a table?

Yes. You can use the Property inspector to format every element within the table structure.

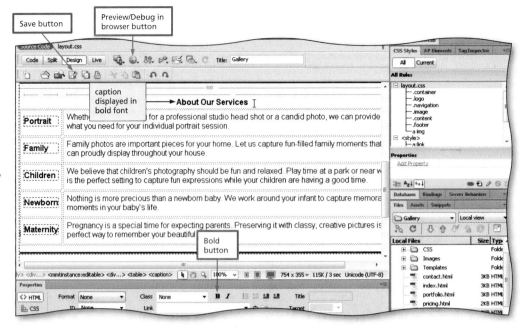

Figure 4–20

3

- Click the Save button on the Standard toolbar to save your work.

- Click the 'Preview/ Debug in browser' button on the Document toolbar to display a list of browsers.

- Click Preview in IExplore in the list to display the Services page in Internet Explorer (Figure 4–21).

4

- Click the Internet Explorer Close button to close the browser.

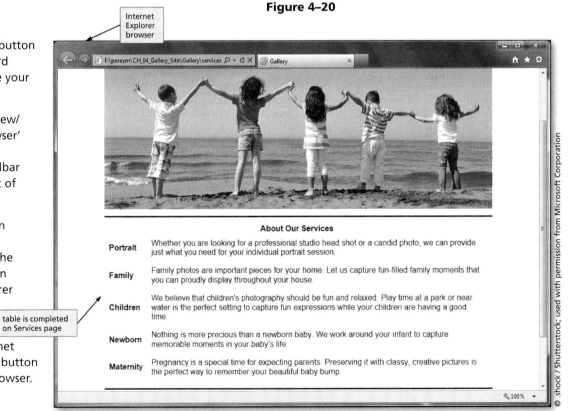

Figure 4–21

Understanding the HTML Structure of a Table

As you work with and become more familiar with tables, it is helpful to have a basic understanding of the HTML structure of a table. For example, suppose you have a table with two rows and two columns, displaying a total of four cells, such as the following:

First cell	Second cell
Third cell	Fourth cell

The general syntax of the table is:

```
<table>
  <tr>
    <td> First cell </td>
    <td> Second cell </td>
  </tr>
  <tr>
    <td> Third cell </td>
    <td> Fourth cell </td>
  </tr>
</table>
```

When you click in a table in Dreamweaver, the tag selector displays the <table>, <td>, and <tr> tags. The <table> tag indicates the whole table. Clicking the <table> tag in the tag selector selects the whole table. The <td> tag indicates **table data**. Clicking the <td> tag in the tag selector selects the cell containing the insertion point. The <tr> tag indicates **table row**. Clicking the <tr> tag in the tag selector selects the row containing the insertion point.

To Insert a Table into the Pricing Page

A primary reason for visiting the Gallery site is to identify the pricing for a photography session and photo costs. Pricing information tables play an important role for every company that offers products or services. A pricing table must be simple, but at the same time clearly differentiate between features and prices of different products and services. The Gallery studio carefully examined its product portfolio and selected the most important features to present in its pricing plan. The following steps insert a table with six rows and four columns into the Pricing Web page.

1

- Click the Close button on the services.html tab to close the Services page.

- In the Files panel, double-click pricing.html to display the Pricing page.

- Select the text, Insert page content here, in the contentArea region and then press the DELETE key to delete the placeholder text.

- Click Insert on the Application bar to display the Insert menu (Figure 4–22).

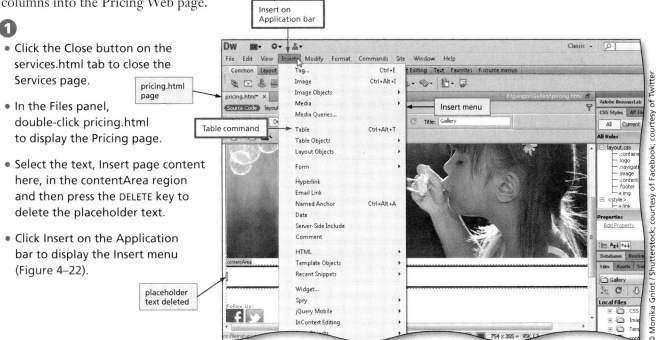

Figure 4–22

2

● Click Table on the Insert menu to display the Table dialog box (Figure 4–23).

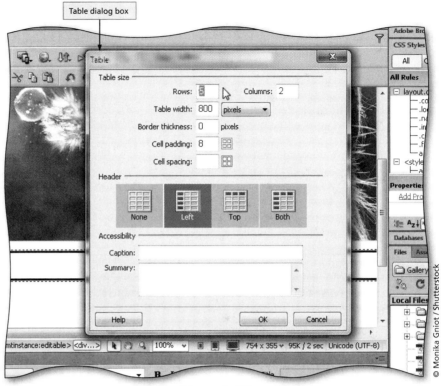

Figure 4–23

3

● In the Rows text box, type 6 to change the number of rows.

● Double-click the Columns text box and then type 4 to change the number of columns.

● Double-click the Table width text box and then type 650 to set the width.

● Double-click the Border thickness text box and then type 1 to set the border thickness.

● Double-click the Cell padding text box and then type 4 to set the cell padding.

● Click Both in the Header area to create top and left headers in the table.

● Click the Caption text box and then type Pricing Information to display a caption at the top of the table.

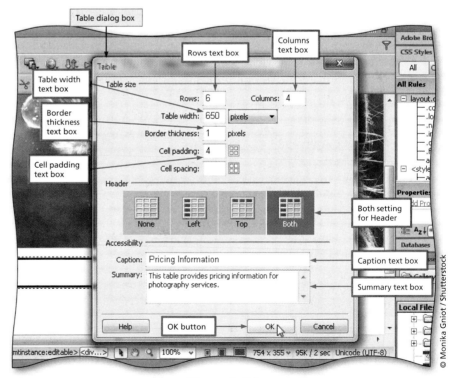

Figure 4–24

● Click the Summary text box and type This table provides pricing information for photography services. to provide accessibility summary text (Figure 4–24).

④
- Click the OK button in the Table dialog box and scroll down to view the table in the Services page (Figure 4–25).

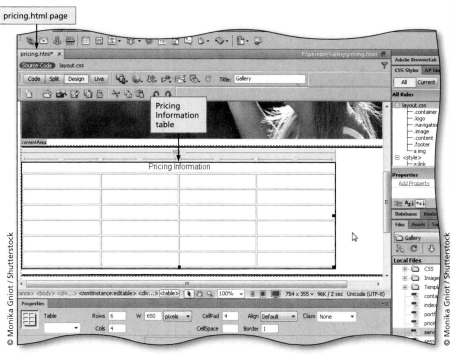

Figure 4–25

Other Ways

1. Click Table button on Insert bar

Merging and Splitting Cells in a Table

You can create a cell that spans two or more other cells by **merging** cells. You should be familiar with the concept of merging cells if you have worked with spreadsheets or word processing tables. Using HTML, merging cells is a more complicated process. Dreamweaver, however, simplifies the process of merging cells by hiding some complex HTML table restructuring code behind an easy-to-use interface in the Property inspector. By merging and splitting cells, you can set alignments that are more complex than straight rows and columns.

To merge two or more cells, select the cells and then click the 'Merges selected cells using spans' button in the Property inspector. The selected cells must be contiguous and in the shape of a line or a rectangle. You can merge any number of adjacent cells as long as the entire selection is a line or a rectangle. To split a cell, click the cell and then click the 'Splits cell into rows or columns' button in the Property inspector to display the Split Cell dialog box. In the Split Cell dialog box, specify how to split the cell and then click the OK button. You can split a cell into any number of rows or columns, regardless of whether it was merged previously. When you split a cell into two rows, the other cells in the same row as the split cell are not split. If you split a cell into two or more columns, the other cells in the same column are not split. To select a cell quickly, click in the cell and then click the <td> tag on the tag selector.

BTW

Deleting a Row or Column
Select a row or column and then press the DELETE key. You also can delete a row or column by clicking a cell, clicking Modify on the Application bar, pointing to Table, and then clicking Delete Row or Delete Column on the submenu. Or click a cell within a row or column, right-click to display the context menu, point to Table, and then click Delete Row or Delete Column.

To Merge Cells in a Table

You can use the Property inspector to merge cells before or after text is added to the cell. The last column in the Pricing Information table should have a header with the text, Available for Purchase. The five current cells should be merged into one cell to list the prices for the photo prints and digital photos. The following steps merge multiple cells in a table.

- Click in row 2, column 4 of the Pricing Information table and then drag to select the next four cells in the last column to select a total of five cells (Figure 4–26).

Q&A

When you select multiple cells to merge, must the selection be shaped as a rectangle?

Yes. A merged cell selection must be in a contiguous line and in the shape of a rectangle.

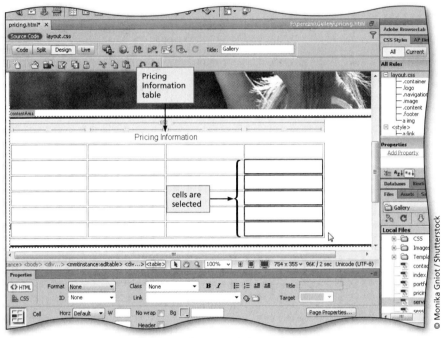

Figure 4–26

- In the Property inspector, click the 'Merges selected cells using spans' button to merge the five selected cells in the last column into one cell (Figure 4–27).

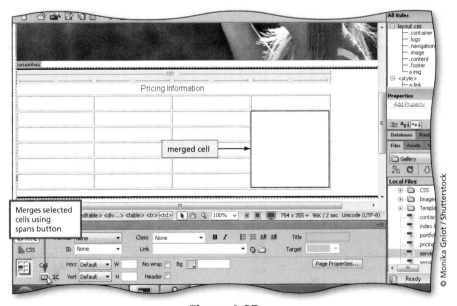

Figure 4–27

Other Ways

1. Right-click selected cells, click Merge Cells
2. On Modify menu, point to Table, click Merge Cells

To Add Text to the Pricing Table

After inserting the table in the Pricing page, you are ready to enter and format the pricing text. Table 4–5 lists the Gallery's pricing options. Notice that several cells in Table 4–5 contain more than one line of text. To add multiple lines of text within the same cell, press the SHIFT+ENTER keys at the end of a line to insert a line break.

Table 4–5 Pricing Information Table			
Session	**Fee**	**Includes**	**Available for Purchase**
Portrait	$75	1 hour 2 outfit changes 2 digital photos on CD	Prints 5x7: $14 8x10: $21
Family	$175	1-2 hours up to 6 family members 3 digital photos on CD	11x14: $35 16x20 Canvas: $85
Children	$135	1-2 hours 2 outfit changes 5 digital photos on CD	Digital Photos 10 photos on CD: $30 25 photos on CD: $55 All photos on CD: $105
Newborn	$150	1-2 hours 4 outfit changes 6 digital photos on CD	
Maternity	$95	1-2 hours 4 outfit changes 3 digital photos on CD	

The following steps add text to the table displayed on the Pricing page.

1

- Click in row 1, column 1 and then type `Session` to add text to the first row of the table.

- Click in row 2, column 1 and then type `Portrait` to add text to the second row.

- Enter the remaining text for the first column shown in Table 4–5 to complete the text for the first column (Figure 4–28).

Q&A

Why was the text in the first column in the table automatically bolded?

In the Table dialog box, the Header was set to Both. The first column and first row of this table are each headers, which are displayed in bold automatically.

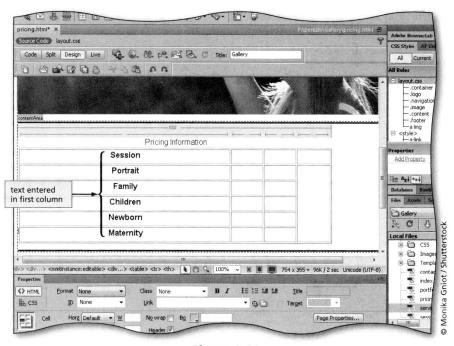

Figure 4–28

- Click in row 1, column 2 and then type `Fee` to add text to the second column.

- Enter the remaining descriptions for the fees shown in Table 4–5 to finish entering the text for the second column (Figure 4–29).

Q&A

Why were the fees not displayed in bold text in the table?

The Fee column is not a header. When you select Both as the header option, Dreamweaver bolds the text in the first row and first column.

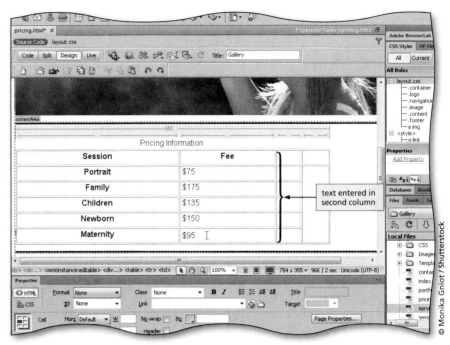

Figure 4–29

- Click in row 1, column 3 and then type `Includes` to add text to the third column.

- Click in row 2, column 3, type `1 hour`, and then press the SHIFT+ENTER keys to create a line break.

- On the next line in the same cell, type `2 outfit changes` and then press the SHIFT+ENTER keys to create a line break.

- On the next line in the same cell, type `2 digital photos on CD` to complete the cell text.

- Enter the remaining text for the third column shown in Table 4–5 to complete the text for the third column, pressing the SHIFT+ENTER keys to create line breaks between each line of text within a cell (Figure 4–30).

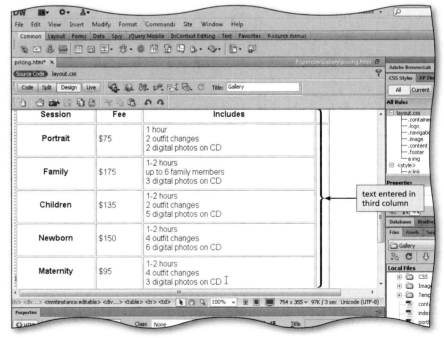

Figure 4–30

- Click in row 1, column 4 and then type `Available for Purchase` to add text to the last column.

- In the merged cell, type the text shown in Table 4–5 to complete the table, pressing the SHIFT+ENTER keys to create line breaks between each line of text within the cell except before the text, Digital Photos, where you insert two line breaks (Figure 4–31).

- Click the Save All button on the Standard toolbar to save your work.

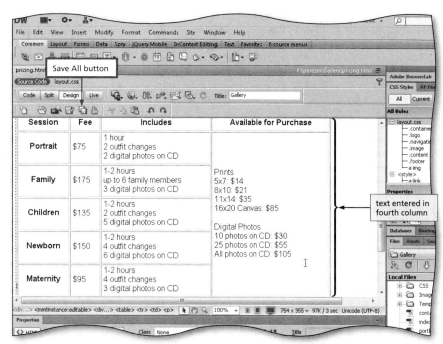

Figure 4–31

To Align the Table

When you insert a table, Dreamweaver adds the table to the page with the default left alignment. You can set the alignment of a table to Left, Center, or Right. With the table selected, you can use the Align property in the Property inspector to change the alignment of the Pricing Information table to Center. The following steps align a table in the center of the page.

1

- Click the outermost edge of the Pricing Information table to select the entire table (Figure 4–32).

Q&A

Why does the table have a red border and black handles?

The border and handles indicate that the table is selected.

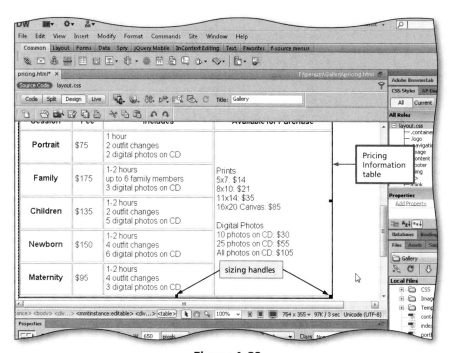

Figure 4–32

● Click the Align button in the
Property inspector and then click
Center to center the table on the
page (Figure 4–33).

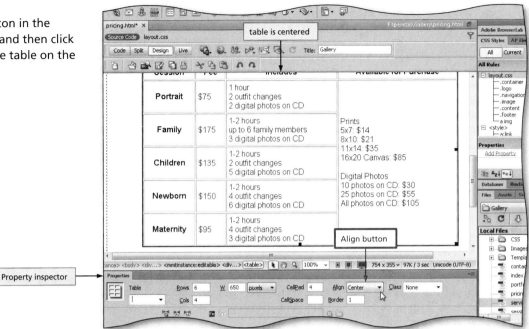

Figure 4–33

To Adjust the Table Width

If you overestimate or underestimate the table width when first inserting a table into the Document
window, you can adjust the table width using the Property inspector.

The Pricing Information table is too wide for the text it contains and needs to be adjusted. You change the
table width by selecting the table and then changing the width in the Property inspector. The following steps
adjust the width of the table.

● With the Pricing Information table
selected, double-click the W box in
the Property inspector to select the
width value (Figure 4–34).

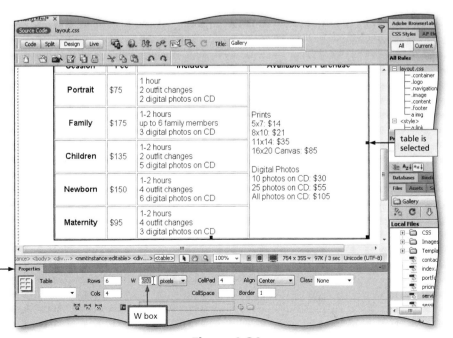

Figure 4–34

2

- Type 575 in the W box to adjust the width of the table and then scroll up to display the size (Figure 4–35).

Q&A What is the visual feedback displayed above the table?

Visual feedback is the green ruler above the table. It displays the number of pixels for the table width.

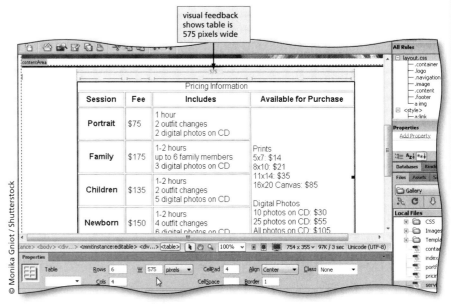

Figure 4–35

To Adjust the Column Width

In addition to changing the width of the entire table, Dreamweaver assists you in setting properties for column width or row height. When you change the width of a single cell in Dreamweaver, the remaining cells in that column are adjusted automatically to match the newly defined cell width. You also can change the row height of a single cell to define the height of the entire row. To change the width of a column row, you can adjust the width of the column by dragging the divider between the columns to the specified width or by entering the value in the W box (the Column Width property) in the Property inspector.

To create a custom fit for the text within each column in the Pricing Information table, you can set the four column widths to 90, 60, 200, and 225 pixels, respectively. Set the first two columns by dragging the right border of the column you want to change to the correct width measurement. As you drag the right border of the column, visual feedback above the table displays the exact number of pixels in each table column. Use the W box in the Property inspector to set the appropriate pixel measurements in the third and fourth columns. The following steps adjust the column width of multiple columns.

1

- Click the right border of the first column in the table and drag to the right until the pixel size is displayed in the visual feedback as 90 pixels (Figure 4–36).

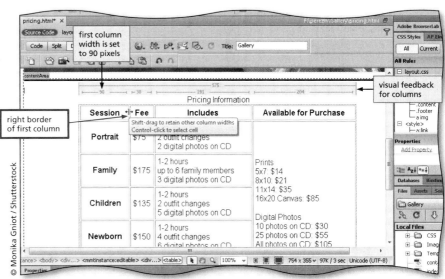

Figure 4–36

2

- Click the right border of the second column and drag to the right until the pixel size is displayed in the visual feedback as 60 pixels (Figure 4–37).

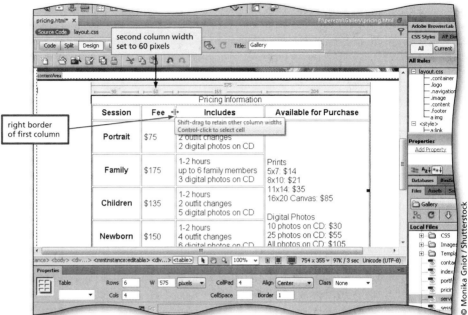

Figure 4–37

3

- Point to the top edge of the third column to display a selection arrow and then click to select the column (Figure 4–38).

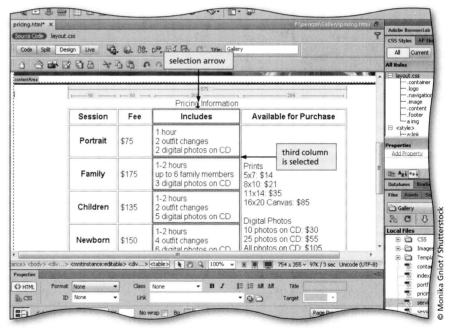

Figure 4–38

4

- Click the W box in the Property inspector and then type 200 to enter a width value.

- Press the ENTER key to resize the third column (Figure 4–39).

Q&A

Two numbers appear for each column width. What do both of these values mean?

The first number in the column width specifies the value placed in the code. The second value in parentheses represents the width of the column in the window size indicated on the status bar.

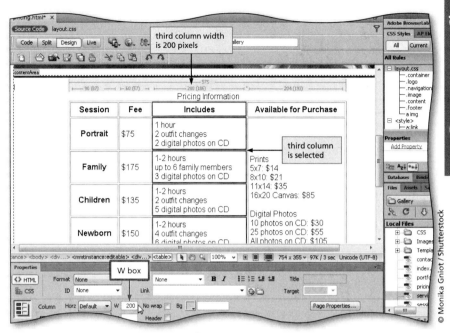

Figure 4–39

5

- Point to the top edge of the fourth column to display a selection arrow, and then click to select the column.

- Click the W box in the Property inspector, type 225, and then press the ENTER key to resize the fourth column (Figure 4–40).

6

- Click the Save button on the Standard toolbar to save your work.

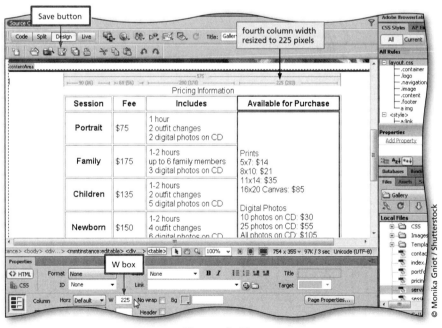

Figure 4–40

Changing the Default Cell Alignment

The default horizontal cell alignment for text is left. When you enter text in a cell, it appears along the left margin of the cell. You can change the horizontal alignment using the Horz button in the Property inspector by clicking the cell and then changing the Horz setting to Center or Right. The default vertical cell alignment is Middle, which aligns the cell content in the middle of the cell. You change the vertical alignment using the Vert button in the Property inspector. Table 4–6 describes the cell alignment options.

Table Formatting Conflicts
When a property, such as alignment, is set to one value for the whole table and another value for individual cells, cell formatting takes precedence over row formatting; this in turn takes precedence over table formatting. The order of precedence for table formatting is cells, rows, and table.

Table 4–6 Cell Alignment Options

Alignment	Description
Default	Specifies a baseline alignment; default may vary depending on the user's browser
Baseline	Aligns the cell content at the bottom of the cell (same as Bottom)
Top	Aligns the cell content at the top of the cell
Bottom	Aligns the cell content at the bottom of the cell
TextTop	Aligns the top of the image with the top of the tallest character in the text line
Absolute Middle	Aligns the middle of the image with the middle of the text in the current line
Absolute Bottom	Aligns the bottom of the image with the bottom of the line of text
Left	Places the selected image on the left margin, wrapping text around it to the right; if left-aligned text precedes the object on the line, it generally forces left-aligned objects to wrap to a new line
Middle	Aligns the middle of the image with the baseline of the current line
Right	Places the image on the right margin, wrapping text around the object to the left; if right-aligned text precedes the object on the line, it generally forces right-aligned objects to wrap to a new line

To Align Text within a Table Cell

Alignment properties can be applied to a single cell, to multiple cells, or to the entire table. The following steps align the text within the table cells.

- Point to the top edge of the second column to display the selection arrow and then click to select the column.

- Click the Horz button in the Property inspector and then click Center to center the text horizontally in the second column (Figure 4–41).

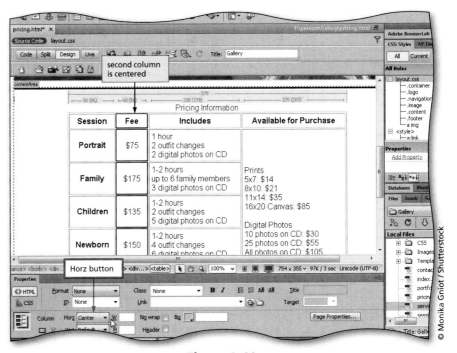

Figure 4–41

© Monika Gniot / Shutterstock

2

- Click the merged cell in the fourth column to select the cell.

- Click the Vert button in the Property inspector and then click Top to top-align the text in the cell (Figure 4–42).

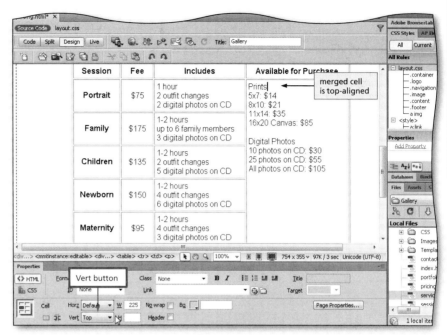

Figure 4–42

To Set the Background Color of Cells in a Table

You can apply background color to an entire table or to individual cells. If a background color is not specified, the table uses the same background color as the Web page. To set the background color of a table or its cells, you click the Bg (Background color) button in the Property inspector to display a color palette, also called a color picker. You can enter a hexadecimal color preceded by a number sign (#), such as #0000FF, in the Bg text box if you know the hexadecimal code.

To follow the color design of the page and to emphasize the header row, you can apply an orange background color to the first row of the Pricing Information table. The following steps set the background color of the cells in a table.

- Point to the left edge of the first row to display a selection arrow and then click to select the row.

- Click the Bg button in the Property inspector to display the color picker, and then point to the #FF9900 orange box to display its hexadecimal number (Figure 4–43).

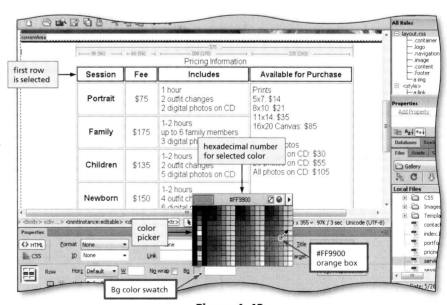

Figure 4–43

2

- Click the #FF9900 orange box to change the background color of the selected row (Figure 4–44).

Q&A

Can I type the hexadecimal code for the background color instead of using the color picker?

Yes. You can type the hexadecimal color in the Bg text box to identify the background color.

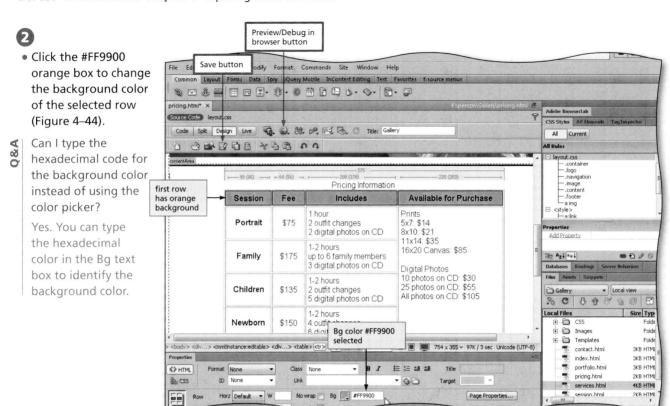

Figure 4–44

 3

- Click the Save button on the Standard toolbar to save your work.

- Click the 'Preview/Debug in browser' button on the Document toolbar to display a list of browsers.

- Click Preview in IExplore in the list to display the Pricing page in Internet Explorer. Scroll down the page to view the table (Figure 4–45).

4

- Click the Internet Explorer Close button to close the browser.

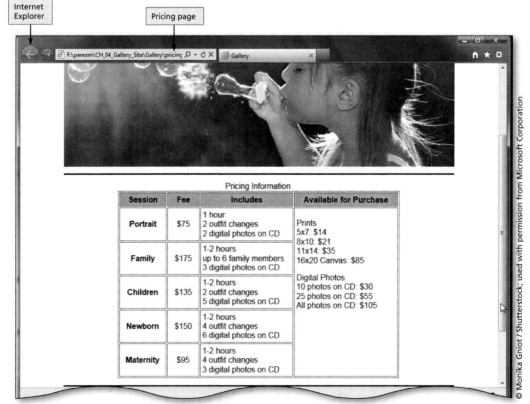

Figure 4–45

Break Point: If you wish to take a break, this is a good place to do so. To resume at a later time, start Dreamweaver and continue following the steps from this location forward.

Understanding How Forms Work

Forms are interactive elements that provide a way for a Web site visitor to interact with the site. Users complete forms to give feedback, submit an order for merchandise or services, request information, make reservations, and so on. The Gallery photography studio has requested that its Web site use a form to collect information from customers who are interested in scheduling a photography session. Dreamweaver provides a number of Form objects to collect input including text boxes, list boxes, menus, radio buttons, and buttons. After a user enters information into the form elements, Dreamweaver's validation ensures that the information is filled in and properly formatted.

Plan the format of the form pages.
To integrate a form page into your Web site, format it the same way you format other pages on the site. Use the same background image, for example, and the same font and styles for the text. To minimize the amount of scrolling users must do to complete the form, sketch the form and plan where to insert the controls such as text boxes and labels. If you are converting a paper-based form into a Web form, identify how you can condense the form to fit the screen and what types of electronic controls you will use where users enter information on the printed form.

Plan
Ahead

Creating a Form

To insert a form on a Web page, you can use the Form button on the Forms tab of the Insert bar. Dreamweaver creates the JavaScript necessary for processing the form, inserts the beginning <form> and ending </form> tags into the source code, and then displays a dotted red outline to represent the form in Design view. Unlike a table, you cannot resize a form by dragging the borders. Rather, the form expands as you insert objects. When viewed in a browser, the form outline is not displayed; therefore, no visible border exists to turn on or off.

Consider the data you will collect.
Use familiar labels that clearly describe the data users should enter. For example, if you are requesting a date, consider that your audience may be unsure which date format to use. Provide a sample or layout of the date such as mm/dd/yyyy. If necessary, include simple, clear instructions about what information users should provide. Consider providing radio button options instead of expecting the user to write a narrative answer. Finally, organize the input fields so the order is logical and predictable to users.

Plan
Ahead

Form Processing

A form provides an effective way to collect data from a Web site visitor. Forms, however, do not process data. Typically, forms contain objects that users interact with by entering or selecting information, for example. After providing the requested information, users click a button to submit the data to a server for processing or to e-mail it to a designated e-mail address. This chapter covers the insertion of tables and forms, but form processing is beyond the scope of this text.

Dreamweaver **behaviors** are ways to add scripting events to the elements on your Web page. Dreamweaver writes the JavaScript and the event attribute for you. You can create behaviors to validate form information. In Chapter 5, you add Spry form validation to the form created in this chapter to provide a Dreamweaver behavior that checks the user's entries into the form.

To submit a form, a Web site must run a script to process the form input data. A script is a text file that is executed within an application and usually is written in Perl, VBScript, JavaScript, Java, or C++. These scripts reside on a server. Therefore, they are called server-side scripts. Other server-side technologies include Adobe ColdFusion, ASP.NET, PHP, and JavaServer Pages (JSP). Some types of database applications support these technologies. A common way to process form data is through a Common Gateway Interface (CGI) script. When a browser collects data, it sends the data to a Hypertext Transfer Protocol (HTTP) server (a gateway) specified in the HTML form. The server starts a program (which also is specified in the HTML form) that can process the collected data. The gateway can process the data as you specify. It may return customized HTML based on the user's entries, log the data to a file, or e-mail the data to someone.

To Insert a Form into a Web Page

Using a form, the Contact page in the Gallery site collects the customer's name, e-mail address, and phone number, the type of session, the preferred session date, and how the customer heard about the photography studio. The following steps insert a form into the Contact page.

1

- Click the Close button on the pricing.html tab to close the Pricing page.

- In the Files panel, double-click contact.html to display the Contact page.

- Scroll down, click after the e-mail address, TheGalleryPFP@thegallery.net, and then press the ENTER key to move the insertion point to the next line.

- Type You may complete the form below to schedule your photography session. and then press the ENTER key to enter text that introduces the form (Figure 4–46).

Figure 4–46

2

- Click Insert on the Application bar to display the Insert menu, and then point to Form to display the Form submenu (Figure 4–47).

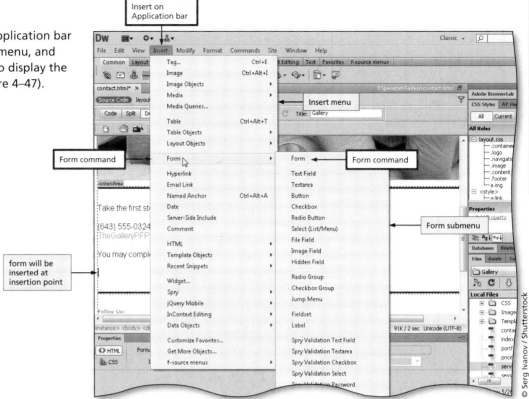

Figure 4–47

3

- Click Form on the Form submenu to insert a form outlined by a dotted red line (Figure 4–48).

Will the dotted red outline of the form appear on the Web page?

No. The form outline is a guide to help you select the form and insert text and form objects. When the form is displayed on a Web page, the outline does not appear.

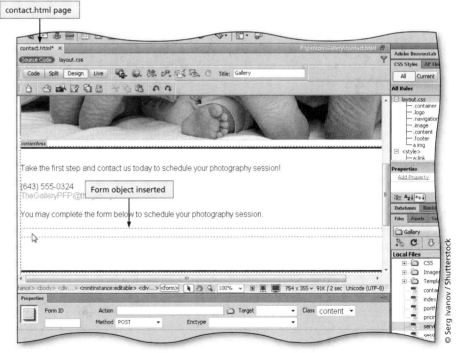

Figure 4–48

4

- Click the Form ID text box in the Property inspector and then type `contact` to name the form (Figure 4–49).

Q&A

Why should I name the form?

Although a form does not need a name to work, a name is helpful if you use Dreamweaver behavior or JavaScript to interact with the form.

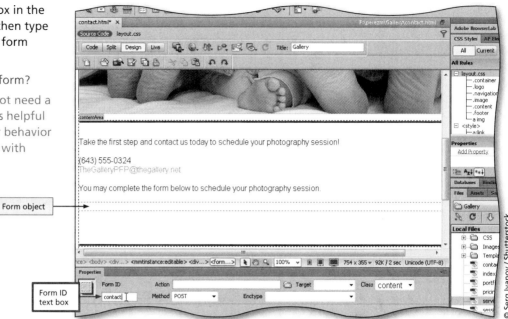

Figure 4–49

Other Ways

1. Click Form button on Forms tab of Insert bar
2. Click Form button on Insert bar

BTW

Using the Action Text Box

If the Gallery site was running on a server and connected to a database, the Action text box in the Property inspector would contain the server's Web address. When a visitor to the site completes the form and clicks the Submit button, the form information would be sent to a database on a Web server. Typically, a CGI script would run after the visitor clicks the Submit button, which sends the form data to the company's server.

Using the Form Property Inspector

As you have seen, the Property inspector options change depending on the selected object. Figure 4–50 shows the Property inspector for a selected form. Table 4–7 lists the form elements and descriptions.

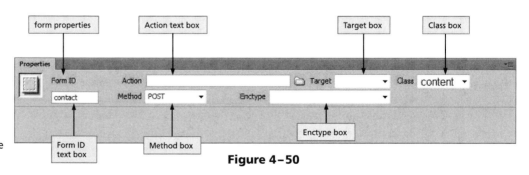

Figure 4–50

Table 4–7 Form Elements

Form Element	Description
Form ID	Naming a form makes it possible to reference or control the form with a behavior.
Action	Contains the mailto address where the form results can be e-mailed or specifies the URL to the dynamic page or script that will process the form.
Target	Specifies the window or frame in which to display the data after processing if a script specifies that a new page should be displayed. The four targets are _blank, _parent, _self, and _top. The _blank target opens the referenced link (or processed data) in a new browser window, leaving the current window untouched. The **_blank target** is the one most often used with a jump menu, which is discussed later in this chapter. The **_self-target** opens the destination document in the same window as the one in which the form was submitted. The two other targets mostly are used with frames.
Method	Indicates the method by which the form data is transferred to the server. The three options are **POST**, which embeds the form data in the HTTP request; **GET**, which appends the form data to the URL requesting the page; and **Default**, which uses the browser's default setting to send the form data to the server.
Enctype	Specifies a **MIME** (Multipurpose Internet Mail Extensions) type for the data being submitted to the server so the server software will know how to interpret the data.
Class	Sets style sheet attributes and/or attaches a style sheet to the current document.

To Insert a Table to Design the Form

The first step in building a Web form is to create the table structure to assist with the layout of the form elements. The table structure dictates how the form appears on the Web page. Although a table is not required to build a form, using one helps to organize the layout of a form.

The form on the Contact page of the Gallery site should contain seven rows and two columns. The following steps insert a table to design the form.

- If necessary, click within the dotted red form area to place the insertion point within the Form object.

- Click Insert on the Application bar and then click Table on the Insert menu to display the Table dialog box.

- In the Rows text box, type 7 to change the number of rows.

- Double-click the Columns text box and then type 2 to change the number of columns.

- Double-click the Table width text box and then type 325 to set the width to 325 pixels.

- Double-click the Border thickness text box and then type 0 to set the border thickness.

- Double-click the Cell padding text box and then type 4 to set the number of cell padding pixels.

- Click None in the Header area to create a table without a header (Figure 4–51).

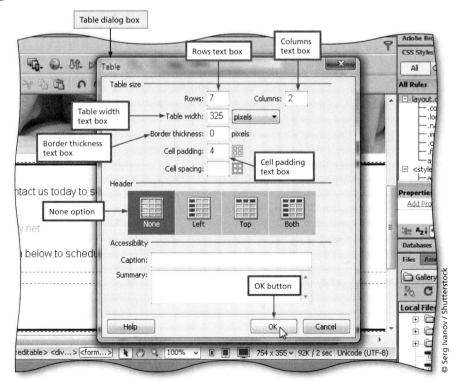

Figure 4–51

© Serg Ivanov / Shutterstock

- Click the OK button in the Table dialog box to create a table in the Contact page (Figure 4–52).

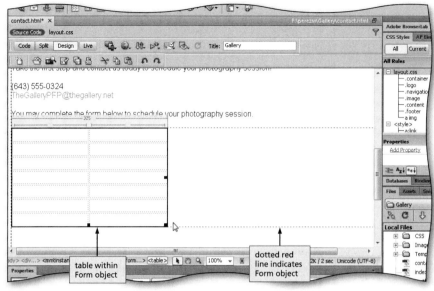

Figure 4–52

Other Ways
1. Click Table button on Insert bar

To Add Labels to the Table

Labels identify the type of data users should enter into a form. The first column in the table displays the labels of the requested information for the form. The second column displays the form elements such as text boxes and radio buttons to collect the user's information. The following steps add labels to the table structure displayed in the form.

- Click in row 1, column 1 and then type `First Name:` to add a label to the first row of the table.

- Click in row 2, column 1 and then type `Last Name:` to add a label to the second row.

- Click in row 3, column 1 and then type `E-mail:` to add a label to the third row.

- Click in row 4, column 1 and then type `Phone Number:` to add a label to the fourth row.

- Click in row 5, column 1 and then type `Type of Session:` to add a label to the fifth row.

- Click in row 6, column 1 and then type `Preferred Session Date:` to add a label to the sixth row.

- Click in row 7, column 1 and then type `How did you hear about us?` to add a label to the last row (Figure 4–53).

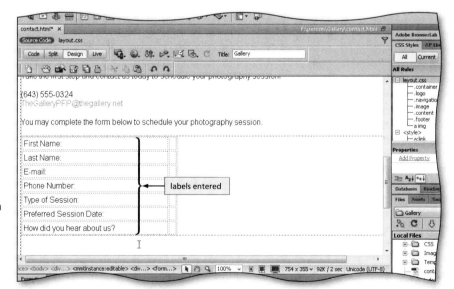

Figure 4–53

2

- Click the W box in the Property inspector and then type `125` to enter a column width.

- Press the ENTER key to adjust the column width of the first column (Figure 4–54).

🔍 **Experiment**

- Type 200 in the H box in the Property inspector to change the row height. When you are finished, click the Edit menu and then click Undo to return the row to its original state.

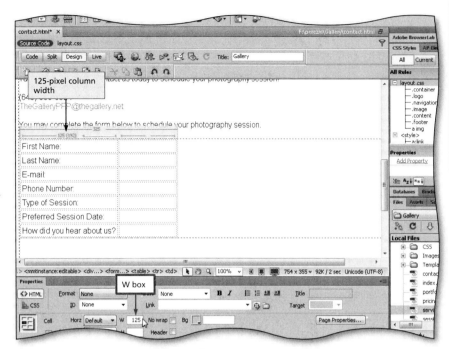

Figure 4–54

Adding Form Objects

In Dreamweaver, users enter data in forms using form objects such as text fields, check boxes, and radio buttons. Each form object should have a unique name — except radio buttons within the same group, which should share the same name. To insert a form field in a Web page, first add the field and any descriptive labels, and then modify the properties of the field. All Dreamweaver form objects are available on the Forms tab of the Insert bar (Figure 4–55). Table 4–8 on the next page lists the names and descriptions of buttons on the Forms tab.

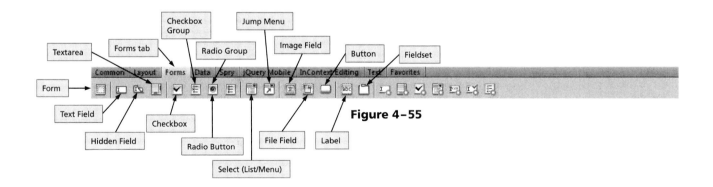

Figure 4–55

Table 4–8 Forms Tab on Insert Bar	
Button Name	**Description**
Form	Inserts a form into the Document window
Text Field	Accepts any type of alphanumeric text entry
Hidden Field	Stores information entered by a user and then uses that data within the site database
Textarea	Provides a multiline text entry field
Checkbox	Represents a selection
Checkbox Group	Allows multiple responses in a single group of options, letting the user select as many options as apply
Radio Button	Represents an exclusive choice; only one item in a group of buttons can be selected
Radio Group	Represents a group of radio buttons
Select (List/Menu)	List displays option values within a scrolling list that allows users to select multiple options; Menu displays the option values in a pop-up menu that allows users to select only a single item
Jump Menu	A special form of a pop-up menu that lets the viewer link to another document or file
Image Field	Creates a custom, graphical button
File Field	Allows users to browse to a file on their computers and upload the file as form data
Button	Performs actions when clicked; Submit and Reset buttons send data to the server and clear the form fields, respectively
Label	Provides a way to associate the text label for a field with the field structurally
Fieldset	Inserts a container tag for a logical group of form elements

Plan Ahead

Determine the types of controls you will add to the form.
After creating a form and inserting a table to align the form contents, you add objects that users select to enter information into the form. Select objects based on the type of information you want to collect: Text Fields and Textareas for text entries such as addresses, Checkboxes, Radio Buttons, and List/Menus for options users select, and Buttons for submitting and resetting the form.

BTW

Text Fields and Text Areas
When you insert a text field, you can use the Property inspector to set its type as Single-line, Multi-line, or Password. Inserting a multi-line text field is the same as inserting a Text Area object.

Adding Text Fields

A **Text Field** is a form object in which users enter a response. Forms support three types of Text Field objects: single-line, multiple-line, and password. Data entered into a Text Field object can consist of alphanumeric and punctuation characters. When you insert a text field into a form, the Input Tag Accessibility Attributes dialog box opens (Figure 4–56). Table 4–9 contains a description of the options in this dialog box. By setting accessibility attributes for each form object, visitors can complete the form even if they use a screen reader.

Figure 4–56

Table 4–9 Options in the Input Tag Accessibility Attributes Dialog Box	
Attribute Name	**Description**
ID	Assigns an ID value that can be used to refer to a field from JavaScript and also used as an attribute value.
Label	The descriptive text that identifies the form object.
Style	Provides three options for the <label> tag: Attach label tag using 'for' attribute allows the user to associate a label with a form element, even if the two are in different table cells. 'Wrap with label tag' wraps a label tag around the form item. 'No label tag' turns off the Accessibility option.
Position	Determines the placement of the label text in relation to the form object — before or after the form object.
Access key	Selects the form object in the browser using a keyboard shortcut (one letter). This key is used in combination with the CTRL key (Windows) to access the object.
Tab Index	Specifies the order in which the form objects are selected when pressing the TAB key. The tab order goes from the lowest to highest number.

To Add Text Fields to a Form

The user enters his or her information in the First Name, Last Name, E-mail, and Phone Number Text Field objects in the form on the Contact page. The following steps add several Text Fields to a form.

1

- Click the Forms tab on the Insert bar to display the Form buttons (Figure 4–57).

Figure 4–57

2

- In the table, click row 1, column 2 and then click the Text Field button on the Forms tab to insert a Text Field object and display the Input Tag Accessibility Attributes dialog box (Figure 4–58).

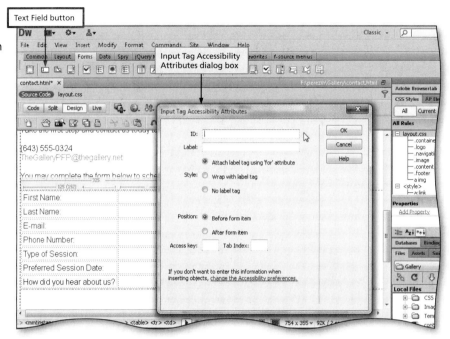

Figure 4–58

3

- In the ID text box, type fname to assign a name to the First Name text field.

- Click the Access key text box and then type F to set the access key.

- Click the Tab Index text box and then type 1 to set the tab position (Figure 4–59).

Q&A How does the Access key work?

When the page opens in the browser, you can press the ALT+F keys as a keyboard shortcut to access this field.

Q&A What is the purpose of the tab index?

When you press the TAB key, you progress from one form item to another in the order specified by the tab index.

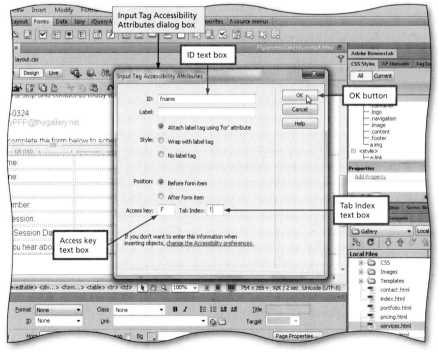

Figure 4–59

4

- Click the OK button in the Input Tag Accessibility Attributes dialog box to add the First Name text field (Figure 4–60).

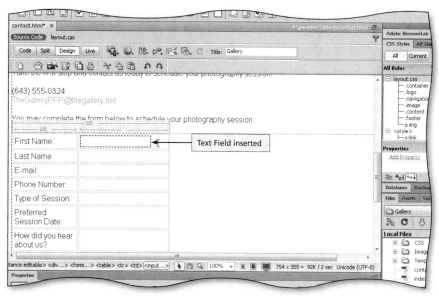

Figure 4–60

5

- Click row 2, column 2 and then click the Text Field button on the Forms tab to insert a Text Field object and display the Input Tag Accessibility Attributes dialog box.

- In the ID text box, type `lname` to assign a name to the Last Name text field.

- Click the Access key text box and then type `L` to set the access key.

- Click the Tab Index text box and then type `2` to set the tab position (Figure 4–61).

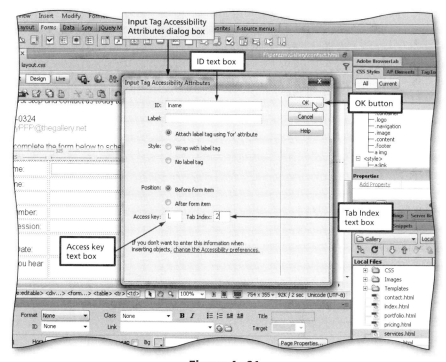

Figure 4–61

6

- Click the OK button in the Input Tag Accessibility Attributes dialog box to add the Last Name text field.

- Click row 3, column 2 and then click the Text Field button on the Forms tab to insert a Text Field object and display the Input Tag Accessibility Attributes dialog box.

- In the ID text box, type `email` to assign a name to the E-mail text field.

- Click the Access key text box and then type `E` to set the access key.

- Click the Tab Index text box and then type `3` to set the tab position (Figure 4–62).

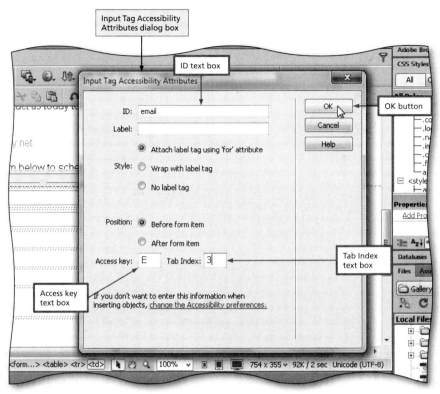

Figure 4–62

7

- Click the OK button in the Input Tag Accessibility Attributes dialog box to add the E-mail text field.

- Click row 4, column 2 and then click the Text Field button on the Forms tab to insert a Text Field object and display the Input Tag Accessibility Attributes dialog box.

- In the ID text box, type `phone` to assign a name to the Phone Number text field.

- Click the Access key text box and then type `P` to set the access key.

- Click the Tab Index text box and type `4` to set the tab position (Figure 4–63).

8

- Click the OK button in the Input Tag Accessibility Attributes dialog box to add the Phone Number text field.

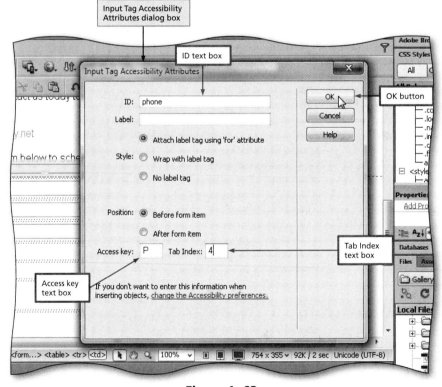

Figure 4–63

Other Ways

1. On Insert menu, point to Form, click Text Field

Adding a Menu

Another way to provide form field options for your Web site visitors is to use the Select (List/Menu) button on the Forms bar to include lists and menus. These objects let users select one or more options from many choices within a limited space. A **list** provides a scroll bar with up and down arrows enabling users to scroll the list, whereas a menu displays a pop-up list. Multiple selections can be made from a list, while users can select only one item from a menu.

You also can offer your Web site visitors a range of choices by using a pop-up menu. This type of menu (also called a drop-down menu) allows a user to select a single item from a list. A pop-up menu is useful when you have a limited amount of space because it occupies only a single line in the form. Only one option is visible when the menu is displayed in the browser. Clicking an arrow button displays the entire list. The user then clicks one of the menu items to make a selection.

To Add a Menu to a Form

As customers fill out the Contact form, they request a type of photography session: Portrait, Family, Children, Newborn, and Maternity. To allow this selection, you can insert a Select object in the form and then set the five session types as list values. By default, a Select object uses Menu as its Type property, which means users click the menu's arrow button to display all the list values. The following steps add a menu to the form on the Contact page.

1

- Click row 5, column 2 to select the cell (Figure 4–64).

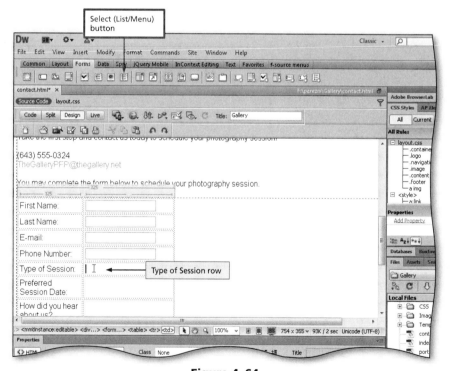

Figure 4–64

- Click the Select (List/Menu) button on the Forms tab to insert a Select object and display the Input Tag Accessibility Attributes dialog box.

- In the ID text box, type type to assign a name to the Select object.

- Click the Access key text box and type T to set the access key.

- Click the Tab Index text box and type 5 to set the tab position (Figure 4–65).

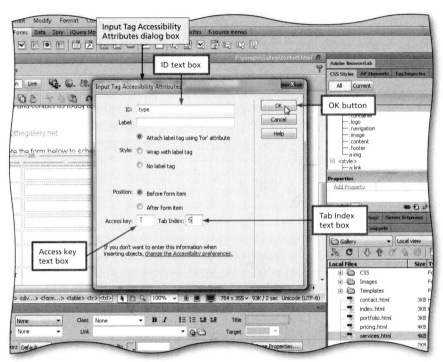

Figure 4–65

- Click the OK button in the Input Tag Accessibility Attributes dialog box to add the Select object to the form.

- If necessary, click the Select object to display its properties in the Property inspector (Figure 4–66).

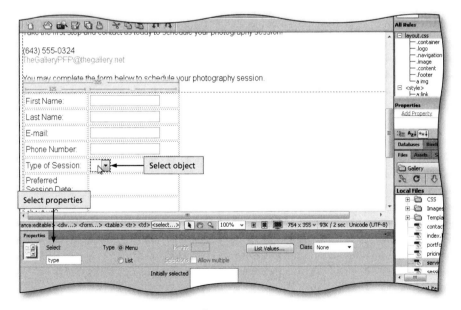

Figure 4–66

- Click the List Values button in the Property inspector to display the List Values dialog box (Figure 4–67).

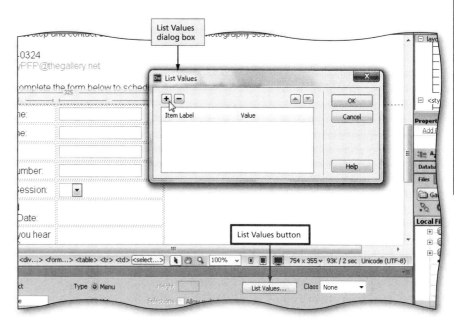

Figure 4–67

- In the text box for entering an Item Label, type Portrait to enter an Item Label for the first menu item (Figure 4–68).

Q&A What is an Item Label?

An Item Label is the text that appears on a menu. Users click this text to select the menu item.

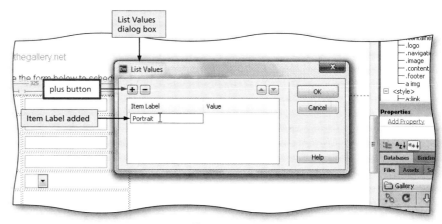

Figure 4–68

- Click the plus button and then type Family to enter an Item Label for the second menu item.

- Click the plus button and then type Children to enter an Item Label for the third menu item.

- Click the plus button and then type Newborn to enter an Item Label for the fourth menu item.

- Click the plus button and then type Maternity to enter an Item Label for the fifth menu item (Figure 4–69).

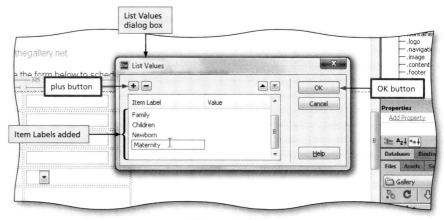

Figure 4–69

● Click the OK button in List Values dialog box to update the Select object in the form.

● In the Initially selected list box, click Portrait to select the initial value to display in the Select object (Figure 4–70).

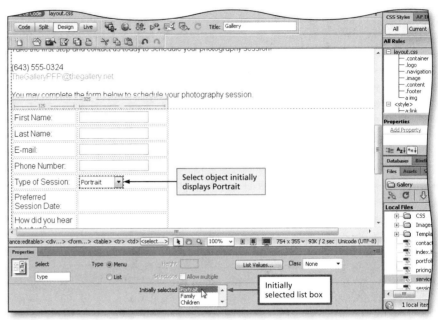

Figure 4–70

Text Field Properties

When you select a Text Field within a form, the Property inspector displays the properties of that field, as shown in Figure 4–71. Table 4–10 lists the attribute names and descriptions for a single-line text field.

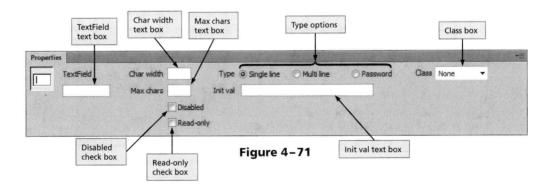

Figure 4–71

Table 4–10 Properties for a Single-Line Text Field

Attribute Name	Description
TextField	Assigns a unique name to the Form object
Char width	Specifies the width of the field in characters
Max chars	Specifies the maximum number of characters that can be entered into the field
Type	Designates the field as a single-line, multiple-line, or password field
Init val	Assigns the value that is displayed in the field when the form first loads
Class	Establishes an attribute used with Cascading Style Sheets
Disabled	Disables the text area
Read-only	Lets users read but not change the value in the field

To Add an Initial Value to a Text Field

When the Contact page opens in a browser, the Preferred Session Date Text Field in the form requests a date in a format such as 11/18/2014. However, the date format can differ depending on the locale of the user. To clarify the format users should follow when entering the date, specify an initial value such as mm/dd/yyyy to identify the month, day, and year order. To do so, use the Init val property in the Text Field Property inspector. An **Init val** displays an initial value such as a number, text, or a note to the Web site visitor to identify the correct format for entering a value. The following steps add a Text Field object with an initial value to the form on the Contact page.

- Click row 6, column 2 to place the insertion point in the cell.

- Click the Text Field button on the Forms tab to insert a Text Field object and display the Input Tag Accessibility Attributes dialog box.

- In the ID text box, type `date` to assign a name to the Text Field object.

- Click the Access key text box and type `D` to set the access key.

- Click the Tab Index text box and type `6` to set the tab position (Figure 4–72).

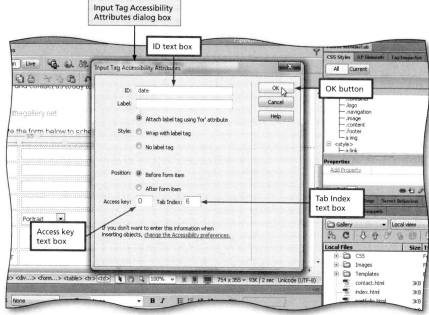

Figure 4–72

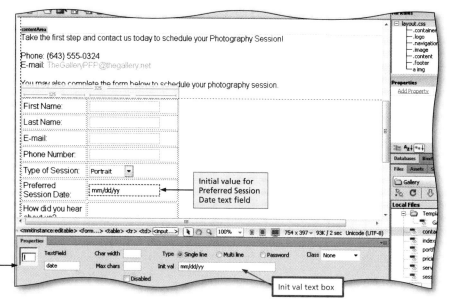

Figure 4–73

2

- Click the OK button in the Input Tag Accessibility Attributes dialog box to add the Text Field object to the form.

- Click the Text Field object you just added to display its properties in the Property inspector.

- Click the Init val text box in the Property inspector, type `mm/dd/yy` and then press the ENTER key to display an initial value for the Text Field object (Figure 4–73).

Other Ways

1. On Insert menu, point to Form, click Text Field

Adding Radio Buttons in a Radio Group

In addition to Text Field and Select objects, you can use radio buttons to collect user information on a form. A Radio Button object includes the button and a descriptive label. For example, you might insert a New Customer radio button on a form that users click if they are new customers. Use the Radio Button Property inspector shown in Figure 4–74 to specify the Checked value, which is the value you want to send to the server-side script or application when a user selects this radio button. For the Initial state, select Checked if you want an option to appear selected when the form first loads in the browser.

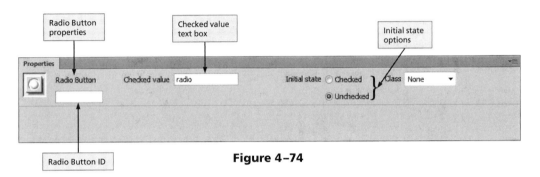

Figure 4–74

A **Radio Group** object includes two or more related radio buttons and allows a user to select only one radio button in the group. If one button is already selected in the group and the user clicks a different radio button, the previously selected button is deselected. When you insert a Radio Group object, you use the Radio Group dialog box to name the object and to provide labels and values for each radio button in the group, which streamlines the process of adding a set of radio buttons to a form.

To Add a Radio Group

The last request on the Gallery Contact form asks "How did you hear about us?" For reporting purposes, instead of asking an open-ended question that leads to a variety of answers, provide a predefined set of responses in a Radio Group object so the responses are easier to interpret. In addition, it is less time consuming for customers to choose one of five choices than typing a response. The following steps add a Radio Group object to the Contact form.

- Click row 7, column 2 to place the insertion point in the cell (Figure 4–75).

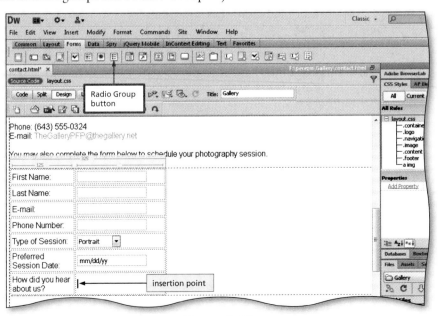

Figure 4–75

- Click the Radio Group button on the Forms tab to insert a Radio Group object and display the Radio Group dialog box.

- In the Name text box, type `referral` to name the Radio Group object.

- Click the first Radio label and then type `Google` to provide a descriptive label for the first radio button in the group (Figure 4–76).

Q&A

Why does the radio group already have two default buttons?

Every radio group includes at least two radio buttons.

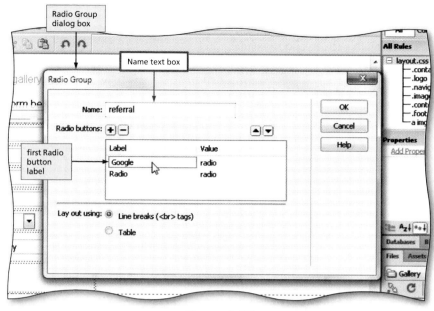

Figure 4–76

- Click the second Radio label and then type `Magazine Ad` to provide a descriptive label for the second radio button in the group.

- Click the plus button, click the third Radio label, and then type `Newspaper` to provide a descriptive label for the third radio button.

- Click the plus button, scroll down, click the fourth Radio label, and then type `Friend` to provide a descriptive label for the fourth radio button.

- Click the plus button, scroll down, click the fifth Radio label, and then type `Other` to provide a descriptive label for the fifth radio button (Figure 4–77).

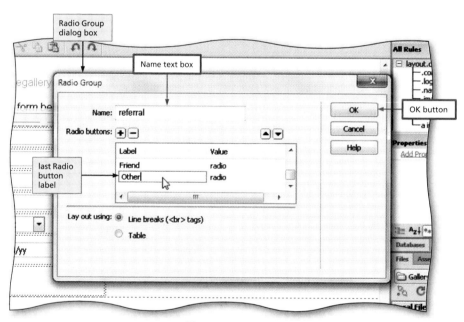

Figure 4–77

4

• Click the OK button in the Radio Group dialog box to add the Radio Group object to the form.

• Click outside the form to deselect the Radio Group object (Figure 4–78).

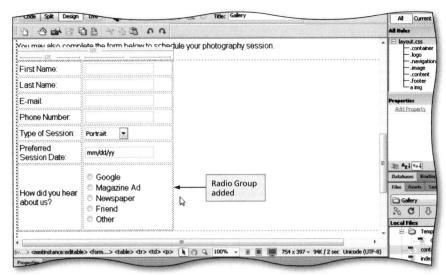

Figure 4–78

Other Ways
1. On Insert menu, point to Form, click Radio Group

To Add Submit and Reset Buttons to a Form

Each Web form needs at least one button for submitting the data to a server or application. The most common button on a Web form is a Submit button, which sends data from the completed form to the processing application. Most Web forms also include a Reset button, which clears all the fields in the form. When you add a button to a form, Dreamweaver automatically assigns it properties suitable for a Submit button.

The following steps add Submit and Reset buttons to a form.

1

• If necessary, click to the right of the table and then press the ENTER key to place the insertion point on the line below the table (Figure 4–79).

Q&A

Does it matter where the Submit button is placed on the Web page?

Yes. The Submit button must be placed on the form, which is indicated by the dotted red outline in Dreamweaver. Otherwise, the button is not considered part of the Form object.

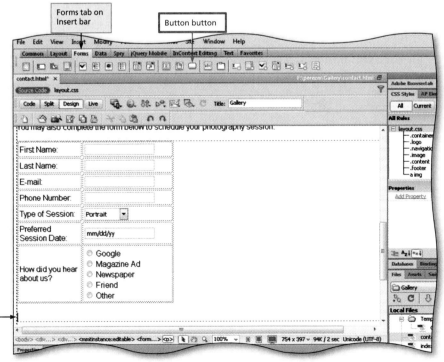

Figure 4–79

- Click the Button button on the Forms tab to insert a Button object and display the Input Tag Accessibility Attributes dialog box.

- In the ID text box, type `submit` to assign a name to the Button object.

- Click the Access key text box and type `S` to set the access key.

- Click the Tab Index text box and type `7` to set the tab position (Figure 4–80).

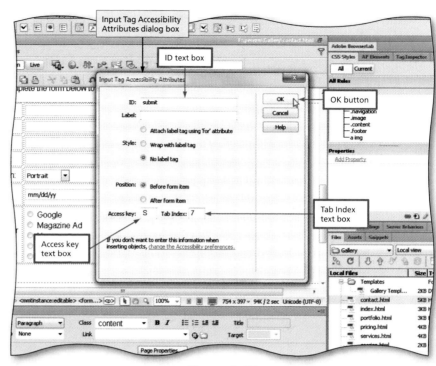

Figure 4–80

- Click the OK button in the Input Tag Accessibility Attributes dialog box to add the button to the form.

- If necessary, click the Submit form option button in the Property inspector to select the Submit form action (Figure 4–81).

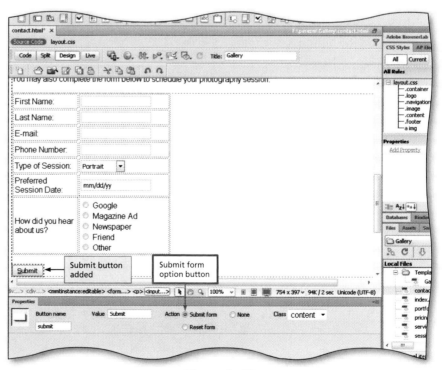

Figure 4–81

- Click to the right of the Submit button and press the ENTER key to place the insertion point on the line below the Button object.

- Click the Button button on the Forms tab to open the Input Tag Accessibility Attributes dialog box.

- In the ID text box, type reset to assign a name to the Button object.

- Click the Access key text box and type R to set the access key.

- Click the Tab Index text box and type 8 to set the tab position (Figure 4–82).

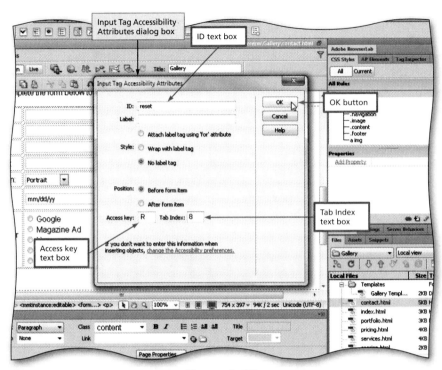

Figure 4–82

- Click the OK button in the Input Tag Accessibility Attributes dialog box to add the button to the form.

- Click the Reset form option button in the Property inspector to select the Reset form action and change the text from Submit to Reset on the second Button object (Figure 4–83).

Q&A

Does the Reset button clear the form in the browser?

Yes. Users click the Reset button to clear information from the form.

- Click the Save All button on the Standard toolbar to save your work.

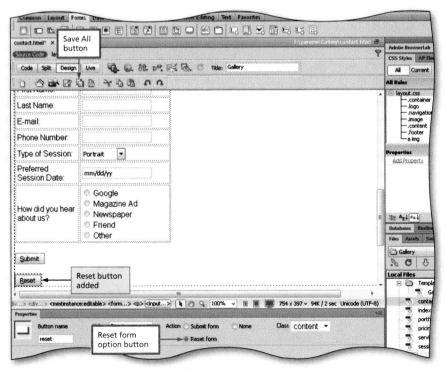

Figure 4–83

Other Ways

1. On Insert menu, point to Form, click Button

To View and Test the Site in the Browser

When you test the form on the Contact page, keep in mind that clicking the Submit button does not send the form data to a server. However, clicking the Reset button clears the data from the form. The following steps preview and test the table and Form objects on the Contact page using Internet Explorer.

1 Click the 'Preview/Debug in browser' button on the Document toolbar to display a list of browsers.

2 Click Preview in IExplore in the list to display the Gallery Web site in Internet Explorer.

3 Open the Services page to view the first table, and then open the Pricing page to view the second table.

4 Open the Contact page and enter your information into the Web form.

5 Click the Internet Explorer Close button to close the browser.

To Quit Dreamweaver

The following steps quit Dreamweaver and return control to the operating system.

1 Click the Close button on the right side of the Application bar to close the window.

2 If Dreamweaver displays a dialog box asking you to save changes, click the No button.

Chapter Summary

In this chapter, you learned how to add tables and a form to the Gallery site that provide detailed information to your customers about photography services. You formatted the table by changing alignment, setting column width, merging cells, applying bold, and specifying cell background color. You added a text field, a radio group, and Submit and Reset buttons to the Contact form. Finally, you viewed and tested the tables and the form in your browser. The following tasks are all the new Dreamweaver skills you learned in this chapter:

1. Insert a Table into the Services page (DW 207)
2. Add Text to the Table (DW 209)
3. Format Table Text (DW 214)
4. Format the Table Caption (DW 215)
5. Merge Cells in a Table (DW 220)
6. Align the Table (DW 223)
7. Adjust the Table Width (DW 224)
8. Adjust the Column Width (DW 225)
9. Align Text within a Table Cell (DW 228)
10. Set the Background Color of Cells in a Table (DW 229)
11. Insert a Form into a Web Page (DW 232)
12. Insert a Table to Design the Form (DW 235)
13. Add Text Fields to a Form (DW 239)
14. Add a Menu to a Form (DW 243)
15. Add an Initial Value to a Text Field (DW 247)
16. Add a Radio Group (DW 248)
17. Add Submit and Reset Buttons to a Form (DW 250)

Apply Your Knowledge

Reinforce the skills and apply the concepts you learned in this chapter.

Adding a Table to a Web Page

Instructions: In this activity, you create a table with six rows and four columns that contains information about a training schedule. The completed Web page is displayed in Figure 4–84.

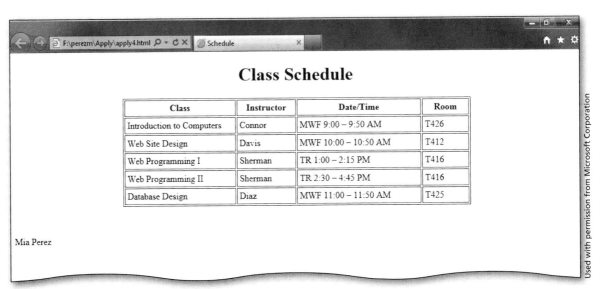

Used with permission from Microsoft Corporation

Figure 4–84

Perform the following tasks:

1. Start Dreamweaver. Use the Sites button on the Files panel to select the Apply site.

2. Create a new HTML document, and then save it as `apply4.html`.

3. Enter `Schedule` as the document title.

4. Use `Class Schedule` as the document heading. Select the document heading and use Heading 1 as the format in the Property inspector. Center the heading on the page.

5. After the Class Schedule heading, press the ENTER key. Insert a table, enter the following values in the Table dialog box, and then click the OK button:

   - Rows: `6`
   - Columns: `4`
   - Table width: `620`
   - Border thickness: `1` pixel
   - Cell padding: `4`
   - Header: `Top`
   - Summary: `This table provides class schedule information.`

6. Enter the text as shown in Table 4–11.

7. Change the width of the first column to `190`, the width of the second column to `95`, the width of the third column to `210`, and the width of the fourth column to `73`.

8. Center-align the table on the page.

9. Click to the right of the table. Press the ENTER key twice. Type your first and last names.

10. Save your changes and then view your document in your browser. Compare your document to Figure 4–84. Make any necessary changes, and then save your changes.

11. Submit the document in the format specified by your instructor.

Table 4–11 Class Schedule			
Column 1	**Column 2**	**Column 3**	**Column 4**
Class	Instructor	Date/Time	Room
Introduction to Computers	Connor	MWF 9:00 – 9:50 AM	T426
Web Site Design	Davis	MWF 10:00 – 10:50 AM	T412
Web Programming I	Sherman	TR 1:00 – 2:15 PM	T416
Web Programming II	Sherman	TR 2:30 – 4:45 PM	T416
Database Design	Diaz	MWF 11:00 – 11:50 AM	T425

Extend Your Knowledge

Extend the skills you learned in this chapter and experiment with new skills. You may need to use Help to complete the assignment.

Creating a Form

Note: To complete this assignment, you will be required to use the Data Files for Students. Visit www.cengage.com/ct/studentdownload for detailed instructions or contact your instructor for information about accessing the required files.

Instructions: In this activity, you create a form using a table layout and add two buttons. First, you insert four Text Fields. Next, you insert a Text Area and a Checkbox Group with three check box options. Finally, you insert the Submit and Reset buttons. The completed Web page is displayed in Figure 4–85.

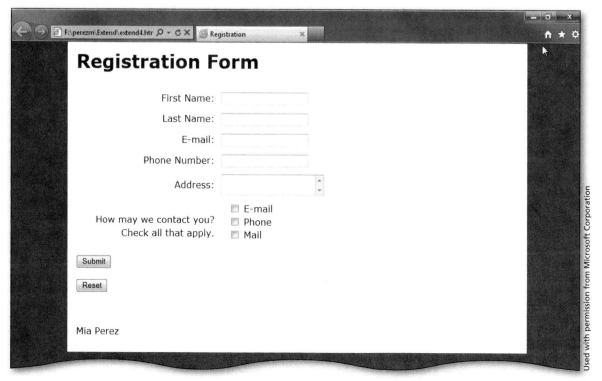

Used with permission from Microsoft Corporation

Figure 4–85

Continued >

Extend Your Knowledge *continued*

Perform the following tasks:

1. Use Windows Explorer to copy the extend4.html file from the Chapter 04\Extend folder into the *your last name and first initial*\Extend folder.

2. Start Dreamweaver. Use the Sites button on the Files panel to select the Extend site. Open extend4.html.

3. In the first row, second column of the table, insert a Text Field object for First Name. Use `fname` for the ID, use `F` for the Access key, and set the Tab Index to `1`.

4. In the second row, second column of the table, insert a Text Field object for Last Name. Use `lname` for the ID, use `L` for the Access key, and set the Tab Index to `2`.

5. In the third row, second column of the table, insert a Text Field object for E-mail. Use `email` for the ID, use `E` for the Access key, and set the Tab Index to `3`.

6. In the fourth row, second column of the table, insert a Text Field object for Phone Number. Use `phone` for the ID, use `P` for the Access key, and set the Tab Index to `4`.

7. In the fifth row, second column of the table, insert a Text Area object for Address. (*Hint*: Point to the buttons on the Form bar to identify the TextArea button.) Use `address` for the ID, use `A` for the Access key, and set the Tab Index to `5`.

8. In the second column, last row of the table, insert a Checkbox Group object. Use `contact` for the Checkbox Group name. Use `E-mail`, `Phone`, and `Mail` as the labels for the check boxes.

9. Insert a button below the table. Use `submit` for the ID, use `S` for the Access key, and set the Tab Index to `6`. Confirm that the Submit form option button is selected in the Property inspector.

10. Click after the Submit button and then press the ENTER key. Insert another button. Use `reset` for the ID, use `R` for the Access key, and set the Tab Index to `7`. In the Property inspector, select the Reset form option button.

11. Replace the text, Your name here, with your first and last names.

12. Save your changes and then view your document in your browser.

13. Compare your document to Figure 4–85. Make any necessary changes and then save your changes.

14. Submit the document in the format specified by your instructor.

Make It Right

Analyze a Web page and suggest how to improve its design.

Creating a Table to Lay Out Content

Note: To complete this assignment, you will be required to use the Data Files for Students. Visit www.cengage.com/ct/studentdownload for detailed instructions or contact your instructor for information about accessing the required files.

Instructions: The Local Library Events Web page provides a list of events at local libraries. In this activity, you modify the list by making it into a table. First, you insert a table with six rows and four columns. Next, you move the content into the table and modify the column widths. Finally, you center-align the table on the page. The modified Web page is shown in Figure 4–86.

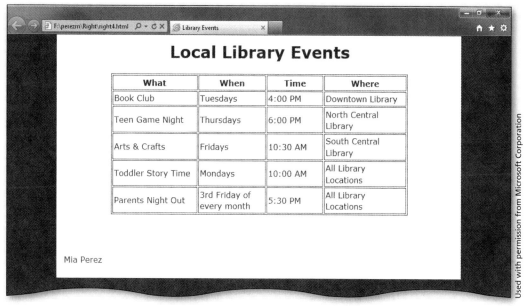

Figure 4–86

Perform the following tasks:

1. Use Windows Explorer to copy the right4.html file from the Chapter 04\Right folder into the *your last name and first initial*\Right folder.

2. Open right4.html in your browser to view its contents and then close the browser.

3. Start Dreamweaver. Use the Sites button on the Files panel to select the Right site. Open right4.html.

4. Insert a table below the page heading. Enter the following values in the Table dialog box and then click the OK button:

   - Rows: 6
   - Columns: 4
   - Table width: 600
   - Border thickness: 1 pixel
   - Cell padding: 2
   - Header: Top
   - Summary: This table contains library event information.

5. Use What, When, Time, and Where as the table headings in the first row.

6. Move the Book Club event information into the second row. Move the Teen Game Night event information into the third row. Move the Arts & Crafts event information into the fourth row. Move the Toddler Story Time event information into the fifth row. Move the Parents Night Out event information into the sixth row.

7. Change the width of the first column to 165, the width of the second column to 130, the width of the third column to 108, and the width of the fourth column to 161.

8. Center-align the table on the page.

9. Delete any remaining event text below the table.

10. Replace the text, Your name here, with your first and last names.

11. Save your changes and then view the Web page in your browser. Compare your page to Figure 4–86. Make any necessary changes and then save your changes.

12. Submit the document in the format specified by your instructor.

In the Lab

Design and/or create a document using the guidelines, concepts, and skills presented in this chapter. Labs are listed in order of increasing difficulty.

Lab 1: Adding Table and Form Elements to the Healthy Lifestyle Web Site

Problem: You are creating an internal Web site for your company that features information about how to live a healthy lifestyle. Employees at your company will use this Web site as a resource for nutrition, exercise, and other health-related tips. In Chapter 2, you created the Habits and Sign Up Web pages. Now you need to add content to these pages. You need to add a table with information about how to change a habit on the Habits Web page. You need to create a form on the Sign Up Web page where employees can sign up to receive weekly e-mail tips.

First, insert a table and add content to the table on the Habits page. Then, add a form, a table, and form elements to the Sign Up page. The Habits page is shown in Figure 4–87. The Sign Up page is shown in Figure 4–88.

Used with permission from Microsoft Corporation

Figure 4–87

Used with permission from Microsoft Corporation

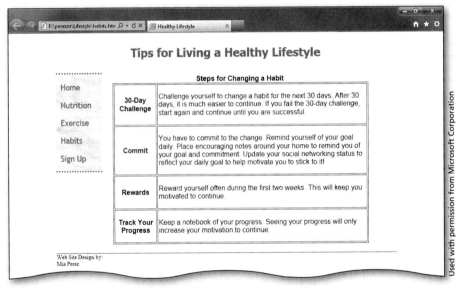

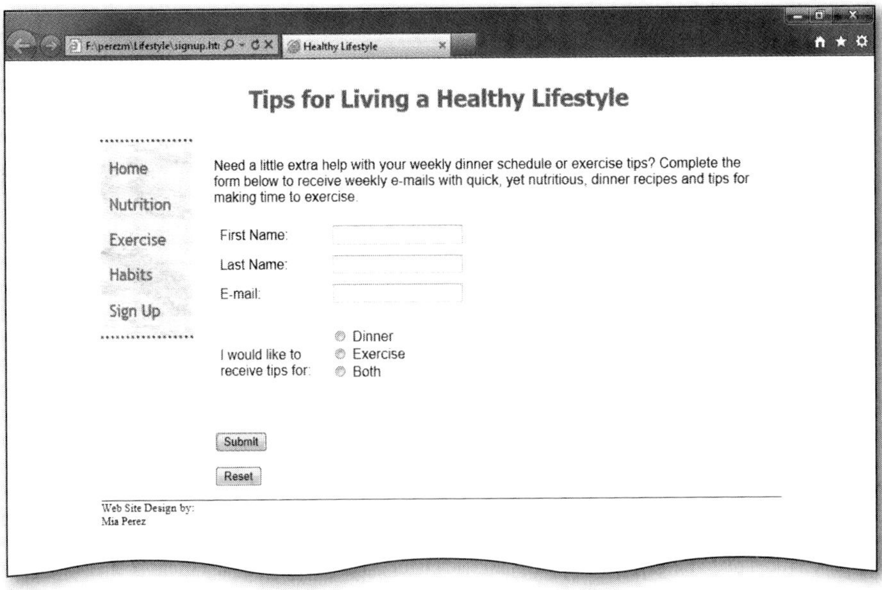

Figure 4–88

Perform the following tasks:

1. Start Dreamweaver. Use the Sites button on the Files panel to select the Healthy Lifestyle site. Open habits.html.

2. Delete the text, Insert content here, in the editable region. Insert a table. Enter the following values in the Table dialog box and then click the OK button:
   - Rows: `4`
   - Columns: `2`
   - Table width: `600`
   - Border thickness: `1` pixel
   - Cell padding: `2`
   - Header: `Left`
   - Caption: `Steps for Changing a Habit`
   - Summary: `This table provides information about how to change a habit.`

3. Enter the text as shown in Table 4–12 on the next page.

4. Bold the table caption, Steps for Changing a Habit.

5. Save your changes and view the document in your browser. Compare your document to Figure 4–87. Make any necessary changes and save your changes. Close habits.html.

6. Open signup.html.

7. Delete the text, Insert content here, in the editable region. Insert a table. Enter the following values in the Table dialog box and then click the OK button:
   - Rows: `4`
   - Columns: `2`
   - Table width: `500`
   - Border thickness: `0` pixels
   - Cell padding: `5`
   - Header: `None`
   - Summary: `This table is a form to enable employees to sign up to receive weekly e-mail tips.`

8. Type `First Name:` in the first row, first column. Insert a Text Field in the first row, second column. Use `fname` for the ID, `F` for the Access key, and `1` for the Tab Index.

9. Type `Last Name:` in the second row, first column. Insert a Text Field in the second row, second column. Use `lname` for the ID, `L` for the Access key, and `2` for the Tab Index.

10. Type `E-mail:` in the third row, first column. Insert a Text Field in the third row, second column. Use `email` for the ID, `E` for the Access key, and `3` for the Tab Index.

11. Type `I would like to receive tips for:` in the fourth row, first column. Insert a Radio Group in the fourth row, second column. Use `tips` for the name. Use `Dinner`, `Exercise`, and `Both` as the radio button labels.

12. Insert a button below the table. Use `submit` for the ID, use `S` for the Access key, and set the Tab Index to `4`. Confirm that the Submit form option button is selected in the Property inspector.

Continued >

In the Lab *continued*

13. Click after the Submit button and then press the ENTER key. Insert another button. Use `reset` for the ID, use `R` for the Access key, and set the Tab Index to `5`. In the Property inspector, select the Reset form option button.

14. Save your changes and view the document in your browser. Compare your document to Figure 4–88. Make any necessary changes and save your changes.

15. Submit the documents in the format specified by your instructor.

Table 4–12 Steps for Changing a Habit	
Column 1	**Column 2**
30-Day Challenge	Challenge yourself to change a habit for the next 30 days. After 30 days, it is much easier to continue. If you fail the 30-day challenge, start again and continue until you are successful.
Commit	You have to commit to the change. Remind yourself of your goal daily. Place encouraging notes around your home to remind you of your goal and commitment. Update your social networking status to reflect your daily goal to help motivate you to stick to it!
Rewards	Reward yourself often during the first two weeks. This will keep you motivated to continue.
Track Your Progress	Keep a notebook of your progress. Seeing your progress will only increase your motivation to continue.

In the Lab

Lab 2: Adding Table and Form Elements to the Designs by Dolores Web Site

Problem: You are creating a Web site for Designs by Dolores, a Web site design company. The site provides information about the company and its services. In Chapter 2, you created the Services, Pricing, and Contact Us pages. Now you need to add content to the Services and Pricing pages. Additionally, you need to create a form on the Contact Us page.

First, insert a table on the Services Web page and add content. Next, insert a table on the Pricing Web page and add content. Finally, create a form on the Contact Us Web page. The Services page is shown in Figure 4–89. The Pricing page is shown in Figure 4–90. The Contact Us page is shown in Figure 4–91.

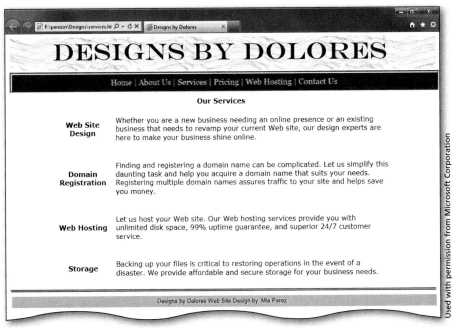

Figure 4–89

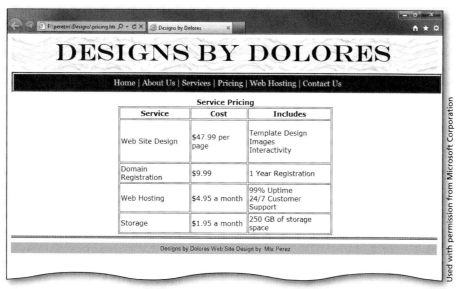

Figure 4–90

Figure 4–91

Perform the following tasks:

1. Start Dreamweaver. Use the Sites button on the Files panel to select the Designs by Dolores site. Open services.html.

2. Delete the text, Insert content here, in the editable region. Insert a table. Enter the following values in the Table dialog box and then click the OK button:

   - Rows: 4
   - Columns: 2
   - Table width: 800
   - Border thickness: 0   pixels
   - Cell padding: 5

Continued >

- Header: `Left`
- Caption: `Our Services`
- Summary: `This table contains information about the services offered by Designs by Dolores.`

3. Enter the text as shown in Table 4–13.

4. Bold the table caption, Our Services. Center-align the table on the page.

5. Save your changes and view the document in your browser. Compare your document to Figure 4–89. Make any necessary changes and save your changes. Close services.html.

6. Open pricing.html.

7. Delete the text, Insert content here, in the editable region. Insert a table. Enter the following values in the Table dialog box and then click the OK button:
   - Rows: `5`
   - Columns: `3`
   - Table width: `500`
   - Border thickness: `1` pixel
   - Cell padding: `2`
   - Header: `Top`
   - Caption: `Service Pricing`
   - Summary: `This table contains service pricing information.`

8. Enter the text as shown in Table 4–14.

9. Bold the table caption, Service Pricing. Center-align the table on the page.

10. Save your changes and view the document in your browser. Compare your document to Figure 4–90. Make any necessary changes and save your changes. Close pricing.html.

11. Open contact.html.

12. Replace the text, Insert content here, with `Need a Web site? Please complete the form below and one of our professionals will be in touch with you soon.`

13. Press the ENTER key. Insert a form and then insert a table within the form. Enter the following values in the Table dialog box and then click the OK button:
    - Rows: `5`
    - Columns: `2`
    - Table width: `600`
    - Border thickness: `0` pixels
    - Cell padding: `5`
    - Header: `None`
    - Summary: `This table contains the contact form.`

14. Type `First Name:` in the first row, first column. Insert a Text Field in the first row, second column. Use `fname` for the ID, `F` for the Access key, and `1` for the Tab Index.

15. Type `Last Name:` in the second row, first column. Insert a Text Field in the second row, second column. Use `lname` for the ID, `L` for the Access key, and `2` for the Tab Index.

16. Type `E-mail:` in the third row, first column. Insert a Text Field in the third row, second column. Use `email` for the ID, `E` for the Access key, and `3` for the Tab Index.

17. Type `Phone Number:` in the fourth row, first column. Insert a Text Field in the third row, second column. Use `phone` for the ID, `P` for the Access key, and `4` for the Tab Index.

18. Type `Business Industry:` in the fifth row, first column. Insert a Select (List/Menu) object in the fifth row, second column. Use `industry` for the name, `I` for the Access key, and `5` for the Tab Index.

19. Click the List Values button in the Property inspector and use `Financial, Law, Retail, Government, Engineering, Education,` and `Other` as the radio button labels. Use Other for the Initially selected value.

20. Insert a button below the table. Use `submit` for the ID, use `S` for the Access key, and set the Tab Index to `6`. Confirm that the Submit form option button is selected in the Property inspector.

21. Click after the Submit button and then press the ENTER key. Insert another button. Use `reset` for the ID, use `R` for the Access key, and set the Tab Index to `7`. In the Property inspector, select the Reset form option button.

22. Save your changes and view the document in your browser. Compare your document to Figure 4–91. Make any necessary changes and save your changes.

23. Submit the documents in the format specified by your instructor.

Table 4–13 Our Services	
Column 1	**Column 2**
Web Site Design	Whether you are a new business needing an online presence or an existing business that needs to revamp your current Web site, our design experts are here to make your business shine online.
Domain Registration	Finding and registering a domain name can be complicated. Let us simplify this daunting task and help you acquire a domain name that suits your needs. Registering multiple domain names assures traffic to your site and helps save you money.
Web Hosting	Let us host your Web site. Our Web hosting services provide you with unlimited disk space, 99% uptime guarantee, and superior 24/7 customer service.
Storage	Backing up your files is critical to restoring operations in the event of a disaster. We provide affordable and secure storage for your business needs.

Table 4–14 Service Pricing		
Column 1	**Column 2**	**Column 3**
Service	Cost	Includes
Web Site Design	$47.99 per page	Template Design Images Interactivity
Domain Registration	$9.99	1 Year Registration
Web Hosting	$4.95 a month	99% Uptime 24/7 Customer Support
Storage	$1.95 a month	250 GB of storage space

In the Lab

Lab 3: Adding Tables and Form Elements to the Justin's Lawn Care Service Web Site

Problem: You are creating a Web site for a new lawn care company, Justin's Lawn Care Service. The Web site will provide information to customers, including descriptions of services and pricing. In Chapter 2, you created the Services, Prices, and Quote Web pages. Now you need to add content to the Services and Prices page. You also need to create a form on the Quote page.

First, insert a table on the Services Web page and add content. Next, insert a table on the Prices Web page and add content. Finally, create a form on the Quote Web page. The Services page is shown in Figure 4–92. The Prices page is shown in Figure 4–93. The Quote page is shown in Figure 4–94.

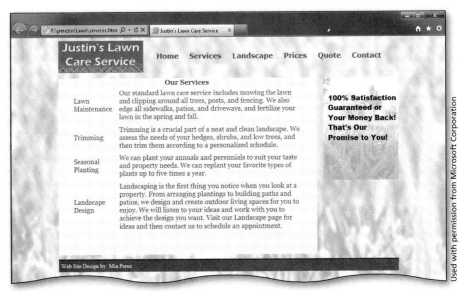

Figure 4–92

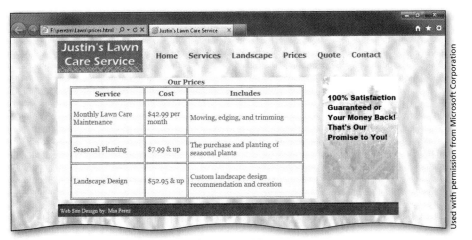

Figure 4–93

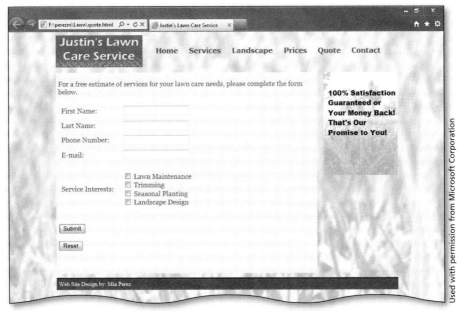

Figure 4–94

Perform the following tasks:

1. Start Dreamweaver. Use the Sites button on the Files panel to select the Justin's Lawn Care Service site. Open services.html.

2. Delete the text, Insert content here, in the editable region. Insert a table. Enter the following values in the Table dialog box and then click the OK button:
   - Rows: `4`
   - Columns: `2`
   - Table width: `550`
   - Border thickness: `0` pixels
   - Cell padding: `4`
   - Header: `None`
   - Caption: `Our Services`
   - Summary: `This table contains service pricing information.`

3. Enter the text as shown in Table 4–15 on page 267.

4. Bold the table caption, Our Services. Center-align the table on the page.

5. Save your changes and view the document in your browser. Compare your document to Figure 4–92. Make any necessary changes and save your changes. Close services.html.

6. Open prices.html.

7. Delete the text, Insert content here, in the editable region. Insert a table. Enter the following values in the Table dialog box and then click the OK button:
   - Rows: `4`
   - Columns: `3`
   - Table width: `550`
   - Border thickness: `1` pixel
   - Cell padding: `4`

Continued >

- Header: `Top`
- Caption: `Our Prices`
- Summary: `This table contains price information.`

8. Enter the text as shown in Table 4–16.

9. Bold the table caption, Our Prices. Center-align the table on the page.

10. Save your changes and view the document in your browser. Compare your document to Figure 4–93. Make any necessary changes and save your changes. Close prices.html.

11. Open quote.html.

12. Replace the text, Insert content here, with `For a free estimate of services for your lawn care needs, please complete the form below.`

13. Press the ENTER key. Insert a form, and then insert a table. Enter the following values in the Table dialog box and then click the OK button:

- Rows: `5`
- Columns: `2`
- Table width: `600`
- Border thickness: `0` pixels
- Cell padding: `5`
- Summary: `This table contains the quote form.`

14. Type `First Name:` in the first row, first column. Insert a Text Field in the first row, second column. Use `fname` for the ID, `F` for the Access key, and `1` for the Tab Index.

15. Type `Last Name:` in the second row, first column. Insert a Text Field in the second row, second column. Use `lname` for the ID, `L` for the Access key, and `2` for the Tab Index.

16. Type `Phone Number:` in the third row, first column. Insert a Text Field in the third row, second column. Use `phone` for the ID, `P` for the Access key, and `3` for the Tab Index.

17. Type `E-mail:` in the fourth row, first column. Insert a Text Field in the fourth row, second column. Use `email` for the ID, `E` for the Access key, and `4` for the Tab Index.

18. Type `Service Interests:` in the fifth row, first column. Insert a Checkbox Group in the fifth row, second column. Use `services` for the name. Use `Lawn Maintenance`, `Trimming`, `Seasonal Planting`, and `Landscape Design` as the check box labels.

19. Insert a button below the table. Use `submit` for the ID, use `S` for the Access key, and set the Tab Index to `5`. Confirm that the Submit form option button is selected in the Property inspector.

20. Click after the Submit button and then press the ENTER key. Insert another button. Use `reset` for the ID, use `R` for the Access key, and set the Tab Index to `6`. In the Property inspector, select the Reset form option button.

21. Save your changes and view the document in your browser. Compare your document to Figure 4–94. Make any necessary changes and save your changes.

22. Submit the documents in the format specified by your instructor.

Table 4–15 Our Services

Column 1	Column 2
Lawn Maintenance	Our standard lawn care service includes mowing the lawn and clipping around all trees, posts, and fencing. We also edge all sidewalks, patios, and driveways, and fertilize your lawn in the spring and fall.
Trimming	Trimming is a crucial part of a neat and clean landscape. We assess the needs of your hedges, shrubs, and low trees, and then trim them according to a personalized schedule.
Seasonal Planting	We can plant your annuals and perennials to suit your taste and property needs. We can replant your favorite types of plants up to five times a year.
Landscape Design	Landscaping is the first thing you notice when you look at a property. From arranging plantings to building paths and patios, we design and create outdoor living spaces for you to enjoy. We will listen to your ideas and work with you to achieve the design you want. Visit our Landscape page for ideas and then contact us to schedule an appointment.

Table 4–16 Our Prices

Column 1	Column 2	Column 3
Service	Cost	Includes
Monthly Lawn Care Maintenance	$42.99 per month	Mowing, edging, and trimming
Seasonal Planting	$7.99 & up	The purchase and planting of seasonal plants
Landscape Design	$52.95 & up	Custom landscape design recommendation and creation

Cases and Places

Apply your creative thinking and problem solving skills to design and implement a solution.

1: Adding Tables and a Form to the Moving Venture Tips Web Site

Personal

You have created all of the pages for Moving Venture Tips and now need to add a table to the Budget page, and a table and a form to the Contact page. You have decided to provide an example budget on the Budget page. Use a table with a minimum of five rows and two columns to create a sample budget on the Budget page. Use Expense, Budget, January, February, and March as the column headings and then include budget information for each row and column. Insert a form on the Contact page, and then insert a table with four rows and two columns. Use First Name, Last Name, E-mail, and I would like to receive more information about as labels in the first column. Insert text fields for First Name, Last Name, and E-mail in the second column. Insert a Radio Group in the last row, second column of the table. Use Moving Tips, Budgeting, and Rentals as labels. Insert a Submit button and a Reset button below the table. Check the spelling using the Commands menu and correct all misspelled words. Submit the document in the format specified by your instructor.

2: Adding Tables and a Form to Student Campus Resources

Academic

You have created all of the pages for Student Campus Resources and now need to add a table to the Events page, and a table and a form to the Contact page. Use a table with a minimum of six rows and four columns to list campus events on the Events page. Use Event Name, Date, Time, Location as your column headings and then complete the table by listing at least four different events. Insert a form on the Contact page, and then insert a table with four rows and two columns. Use First

Name, Last Name, E-mail, and I would like to receive weekly e-mails about as labels in the first column. Insert text fields for First Name, Last Name, and E-mail in the second column, and insert a Select (List/Menu) in the last row, second column. Use Activities, Committees, Events, and All of the above as the list values. Insert a Submit button and a Reset button below the table. Check the spelling using the Commands menu and correct all misspelled words. Submit the document in the format specified by your instructor.

3: Adding Web Pages, Images, and Links for French Villa Roast Café

Professional

You have created all of the pages for French Villa Roast Café and now need to add a table to the Menu page, and a table and a form to the Contact page. Use a table with a minimum of six rows and four columns to create a menu on the Menu page. Create column headings for the menu and then fill out the table by listing at least five different menu items. Insert a form on the Rewards page, and then insert a table with five rows and two columns. Use First Name, Last Name, E-mail, Phone Number, and How many times a month do you purchase coffee from us? as labels in the first column. Insert text fields for First Name, Last Name, and E-mail in the second column. Insert a Radio Group in the last row, second column. Use 1-3, 4-9, and 10+ as labels. Insert a Submit button and a Reset button below the table. Check the spelling using the Commands menu and correct all misspelled words. Submit the document in the format specified by your instructor.

5 | Creating Interactive Web Pages with Spry and Adobe Widgets

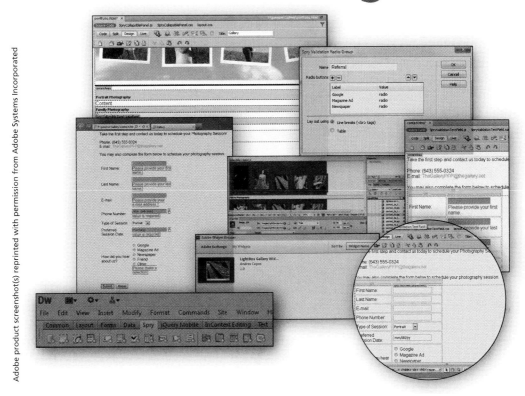

Objectives

You will have mastered the material in this chapter when you can:

- Describe the Spry framework
- Identify Spry widgets
- Apply Spry validation to a text field
- Edit a Web page form with Spry validation
- Test a Web page form with Spry validation
- Add the Spry Collapsible Panel to a Web page
- Install Adobe AIR

- Create an Adobe account
- Use the Adobe Dreamweaver Widget Browser
- Review widgets in Adobe Exchange and save a widget to My Widget
- Add a LightBox Gallery widget to a collapsible panel within a Web page
- Review the LightBox Gallery widget in a browser

5 | Creating Interactive Web Pages with Spry and Adobe Widgets

Introduction

Interactive Web pages enhance a visitor's experience by providing a more memorable Web site experience. This helps attract new customers and retain current customers. A Web site can offer interactivity in many ways, such as an interactive navigational menu bar, validation for form fields, an image slide show viewer, or collapsible panels that organize large amounts of information on a page. Providing interactivity normally would require coding knowledge and time to create the interactive objects. Fortunately, Adobe Dreamweaver CS6 tools can provide interactivity without requiring coding knowledge.

Interactive objects make it easier to use a Web site and can provide stunning visual appeal for visitors to your Web site. For example, a Web form allows you to enter personal information. But if you leave several contact fields blank, the correct information will not be forwarded to the site owner. A feature within Dreamweaver, called **Spry validation**, allows you to check for valid entries in Web forms. The form can verify each entry using conditions that you specify, such as requiring users to fill in a form text field or enter an e-mail address in a valid format.

As a Web designer, you can use Dreamweaver to add powerful Spry elements, such as a collapsible panel, directly into a Web page. Spry Collapsible Panels have at least one tab and a content area. The content area is displayed or hidden when you click the corresponding tab. Other Adobe widgets beyond the installed Spry widgets are also available for download, such as the LightBox Gallery, which allows a visitor to click a thumbnail image from a gallery to display a larger version, dimming the page content behind it, so the gallery acts as a slide show. Interactive widgets introduce a realm of engagement and interactivity well beyond a static page.

Project — Interactive Web Pages

In Chapter 4, you created a form on the Contact Web page to acquire potential customer information for the Gallery Portrait and Family Photography studio. The form contained fields for customers to enter their contact information. Currently, the form fields do not contain any type of validation, so a visitor does not have to complete each field properly. Adding Spry validation widgets, as shown in Figure 5–1a, to the Gallery's Contact page verifies that the information entered into the form meets specified conditions. Additionally, the Gallery owners are interested in showcasing more of their work on the Web site. In Chapter 5, you create a visually appealing, streamlined Portfolio Web page (Figures 5–1b, 5–1c, 5–1d) with an animated slide show using a LightBox Gallery widget within a Spry Collapsible Panel.

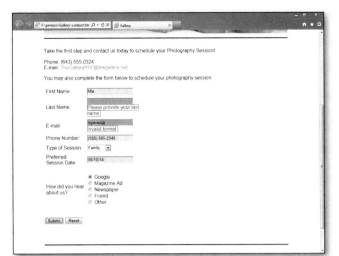

(a) Spry validation on the Contact page

(b) Portfolio page with Spry Collapsible Panels

© Phase4 Photography / Shutterstock; courtesy of Facebook; courtesy of Twitter

(c) Lightbox Gallery widget in an expanded panel

© Phase4 Photography / Shutterstock; source Jazmin Diaz

(d) Zoom effect on Portfolio page

© Phase4 Photography / Shutterstock; source Jazmin Diaz

Figure 5–1

Overview

As you read this chapter, you will learn how to create interactive Web pages by performing these general tasks:

- Explore Spry options and how they work.
- Modify a form with Spry validation.
- Test a Spry validation Web page.
- Add the Spry Collapsible Panel widget to a Web page.
- Add headings to a collapsible panel.
- Create an Adobe account to access Adobe Exchange.
- Explore the Adobe Widget Browser.
- Add the LightBox Gallery widget within the Spry Collapsible Panel.

Plan
Ahead

General Project Guidelines

When adding Spry widgets to a Web site, consider the appearance and the characteristics of the completed site. As you add Spry widgets, you should follow these general guidelines:

1. **Review what you need to publish a Web site that uses Spry widgets.** A Web page containing a Spry widget needs support files that specify the formatting and behavior of the widget. Dreamweaver creates the support files for you automatically. You need to upload these files to the server when you publish the site.

2. **Select Spry widgets.** Determine which Spry widgets you will use, the page(s) in which they will be used, and where they will be placed on a page.

3. **Edit and format Spry widgets.** Change the content and format of the widgets to match the design of your Web site.

4. **Add custom content to a widget.** After adding a widget to a Web page and then customizing it to fit the page, add text or images as necessary to suit the purpose of the widget.

   More specific details about these guidelines are presented at appropriate points throughout the chapter.

To Start Dreamweaver and Open the Gallery Site

Each time you start Dreamweaver, it opens to the last site displayed when you closed the program. The following steps start Dreamweaver and open the Gallery Web site.

1 Click the Start button on the Windows 7 taskbar to display the Start menu, and then type `Dreamweaver CS6` in the 'Search programs and files' box.

2 Click Adobe Dreamweaver CS6 in the list to start Dreamweaver.

3 If the Dreamweaver window is not maximized, click the Maximize button next to the Close button on the Application bar to maximize the window.

4 If the Gallery site is not displayed in the Files panel, click the Sites button on the Files panel toolbar and then click Gallery to display the files and folders in the Gallery site.

Understanding the Spry Framework

Dreamweaver's Spry tools provide additional Web page customization options to create a rich, interactive user experience. The **Spry framework for AJAX** is a JavaScript library that supports a set of reusable widgets. **AJAX (Asynchronous JavaScript and XML)** combines several Web technologies to develop a Web application that efficiently works with a Web browser to process and transfer customer data to a Web server. A **Spry widget** is an element containing HTML, CSS, and JavaScript code to facilitate interaction with a user. Examples of a Spry widget include form validation, collapsible panels, and a menu bar.

A Spry widget consists of three parts: structure, behavior, and styling. HTML code defines the **widget structure**. The **widget behavior** is determined by the JavaScript code and controls how the widget responds to a user when an event begins. CSS code controls the **widget styling**, which defines the appearance of the widget.

The use of Dreamweaver's Spry widgets allows you to integrate interactivity into your Web pages without needing to know how to code in HTML, CSS, or JavaScript. Additionally, you can edit a widget's style to suit your needs.

Spry Widget Options

You can add many Spry widget options to a Web page. To locate the Spry Widget category on the Insert bar, click the Spry tab. The Spry category on the Insert bar is shown in Figure 5–2. Table 5–1 describes the Spry widgets available through Adobe Dreamweaver CS6.

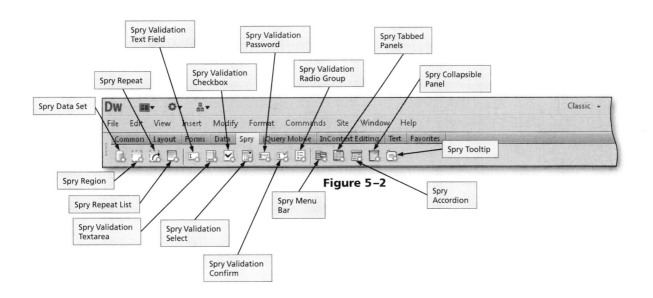

Figure 5–2

Table 5–1 Spry Widgets	
Command	**Description**
Spry Data Set	JavaScript object containing pointers to XML data sources
Spry Region	JavaScript object used to enclose data objects, which includes two types of regions: the Spry Region wraps around data objects, such as tables; and the Spry Detail Region works with a master table object to allow dynamic updating of data
Spry Repeat	Data structure that displays repeating data such as a set of photograph thumbnails
Spry Repeat List	Data structure that displays data as an ordered list, an unordered or a bulleted list, a definition list, or a drop-down list
Spry Validation Text Field	Data structure that displays valid or invalid data when a site visitor enters text
Spry Validation Textarea	Widget that validates the text entered into a Textarea field
Spry Validation Checkbox	Widget that displays valid or invalid states when the user selects or fails to select a check box
Spry Validation Select	Drop-down menu that displays valid or invalid states when the user makes a selection
Spry Validation Password	Widget that validates password input
Spry Validation Confirm	Widget that confirms valid data was entered
Spry Validation Radio Group	Widget that validates an option is selected within the radio group
Spry Menu Bar	Widget that provides a set of navigational menu items that display submenus or links when a site visitor points to the item
Spry Tabbed Panels	Widget used to organize content into panels; users can access panels by clicking a tab at the top of the panel
Spry Accordion	Widget that resembles a standard vertical menu bar; the panels collapse and expand to display or hide content panels
Spry Collapsible Panel	Widget that stores content in a compact space; users can hide or reveal the content by clicking a tab
Spry Tooltip	Widget that displays an on-screen tip when users point to forms and other data

BTW

Getting More Widgets
The Adobe Exchange offers more widgets than the built-in Spry widgets included with Dreamweaver. Click the Extend Dreamweaver button on the Application bar and then click Browse for Web Widgets.

Adding Spry to a Web Page

Dreamweaver provides various Spry options, as described in Table 5–1, including Spry validation widgets, which you will integrate into the form on the Contact Web page. On the Portfolio Web page, you will add a Spry Collapsible Panel widget to display many sample photos in a compact space. Each page should also contain a Submit button so users can submit their form information to the server for processing, and a Reset button so users can clear the form and re-enter data as necessary.

Plan Ahead

> **Select Spry widgets.**
> Although Dreamweaver provides 15 Spry widgets for adding interactivity to Web sites, this chapter uses the Spry Validation Text Field, Spry Validation Radio Group, and Spry Collapsible Panel widgets. Consider incorporating these widgets in the following circumstances:
>
> • **Spry Validation Text Field.** Determine which fields require validation in the Web form such as required fields or e-mail format fields. Use this type of Spry widget when your Web site includes a form and you want to make sure you capture data from the user.
>
> • **Spry Validation Radio Group.** Use this type of Spry widget when you want to require users to make a selection from a group.
>
> • **Spry Collapsible Panel.** Use this type of Spry widget when you want to organize large amounts of information on a page in a compact space.

Using Spry Validation Widgets

The form created on the Contact page in Chapter 4 contains fields for user entry. Currently, these fields do not contain rules to require or validate that users complete the form. For example, all e-mail addresses have a prescribed format starting with a unique name followed by an @ symbol and a domain name. An e-mail address should not contain a blank space. Dreamweaver includes a built-in Spry validation widget that tests the e-mail address a user enters to determine whether the address matches the format.

After the user completes the form and clicks the Submit button, Spry validation confirms that users filled in the required text fields in the specified format. The form displays an error message if a required text field is blank or the text field is not in the prescribed format. In Dreamweaver, you use the Preview states property of a widget to view its message options. These options vary depending on the widget. The following states are some of the general options available.

Initial state. Determines how the text field is displayed in a browser before the user clicks the Submit button; does not include a message

Required state. Specifies the error message to display if the user does not enter text into the text field

Valid state. Determines that the user entered the text in the correct format; does not include a message

Invalid state. Specifies the error message to display if the user does not enter the text in the specified format

To Open the Contact Web Page and Select the Spry Category on the Insert Bar

Before adding Spry widgets to the Contact Web page, you must open the page in Dreamweaver and display the Spry category on the Insert bar. The following steps open the Contact Web page and display the Spry category on the Insert bar.

1

- Double-click contact.html in the Files panel to display the Contact page in the Document window (Figure 5–3).

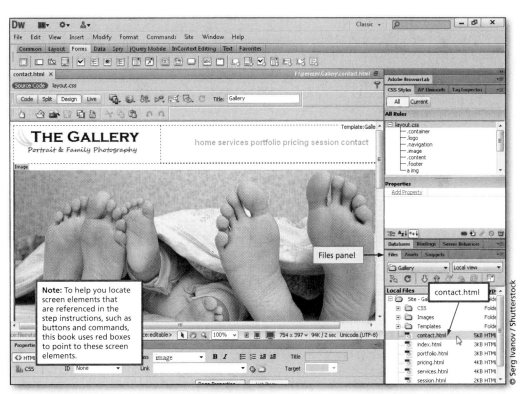

Note: To help you locate screen elements that are referenced in the step instructions, such as buttons and commands, this book uses red boxes to point to these screen elements.

Figure 5–3

2

- Click the Spry tab on the Insert bar to display the Spry options (Figure 5–4).

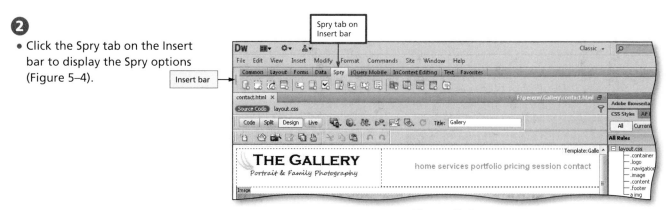

Figure 5–4

To Add Spry Validation to a Required Text Field

If a visitor submits a partially completed form, the Gallery might not have enough information to contact this new customer. Add Spry validation to the form to make certain that visitors complete the required text fields on the form. To confirm that the user does not leave a form text field blank, use the Spry Validation Text Field widget to verify that the text box contains text. You can add the Spry Validation Text Field widget using the Insert menu on the Application bar or using the Spry Validation Text Field button on the Insert bar. After adding the Spry Validation Text Field, you use the Property inspector to modify the Spry TextField name and the Preview states setting. Recall that the Preview states property determines the error message to display if the user does not enter information or enters invalid information. The following steps apply Spry validation to the First Name text field.

1

- If necessary, scroll down the page and click the First Name text field to select the field.

- Click the Spry Validation Text Field button on the Insert bar to change the text field to a Spry Validation Text Field, and to display the properties of the First Name text field on the Property inspector (Figure 5–5).

Q&A Why do dashed lines appear around the text field?

The dashed lines indicate that the text field is selected.

Q&A Does clicking the Spry Validation Text Field button create a second text field for the First Name field?

No. Clicking this button applies Spry validation to the selected text field.

Q&A Why does the Property inspector change after I click the Spry Validation Text Field button?

Now that you have applied Spry validation to the text field, the Property inspector displays validation properties for the selected text field.

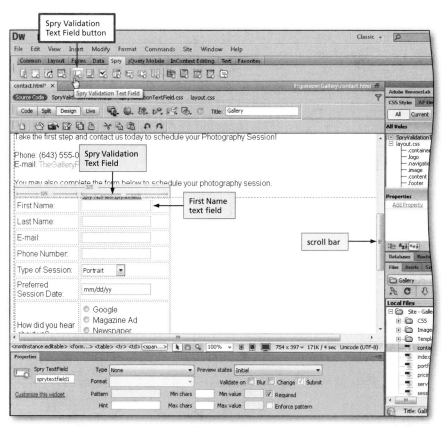

Figure 5–5

2

- Double-click the Spry TextField text box on the Property inspector and then type firstName to rename the field (Figure 5–6).

Q&A

The Required check box on the Property inspector is checked, though I did not click it. Is this box supposed to be checked?

Yes. By default, the Required check box is checked. Because Spry validation is designed to require text input from the user, Required is the default setting of a Spry Validation Text Field.

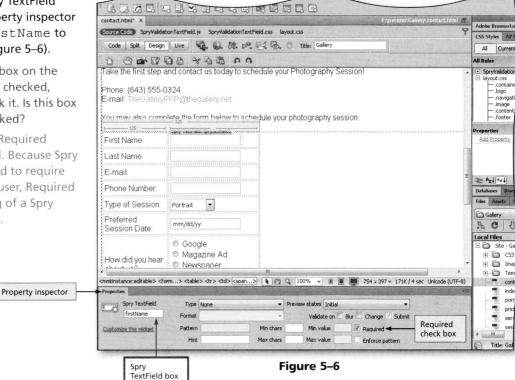

Property inspector

Spry TextField box

Figure 5–6

3

- Click the Preview states button to display the list of Preview states.

- Click Required to select the setting and to display the A value is required error message next to the firstName Spry Validation Text Field (Figure 5–7).

Q&A

What is the Preview states required error message?

If a user does not enter text in the firstName field, the Required Preview states sets an error message to appear after the text field, advising the user that the field is required.

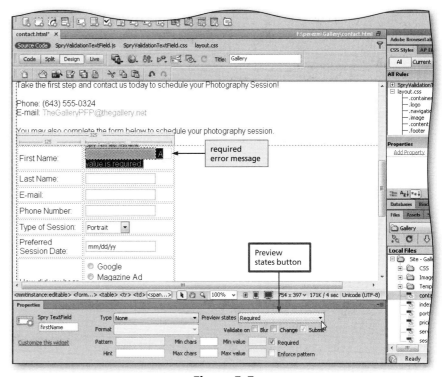

required error message

Preview states button

Figure 5–7

4

- Select the text, A value is required, next to the firstName Spry Validation Text Field to prepare to replace it (Figure 5–8).

Q&A Should I also select the period at the end of the message?

No. Delete all the text except for the period. A period should appear at the end of the replacement text.

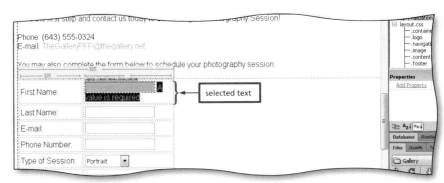

Figure 5–8

5

- Type Please provide your first name to change the required error message (Figure 5–9).

Q&A Does the error message appear on the Contact page when I view it in my browser?

No. This text appears only if the user clicks the Submit button without entering text into the firstName field.

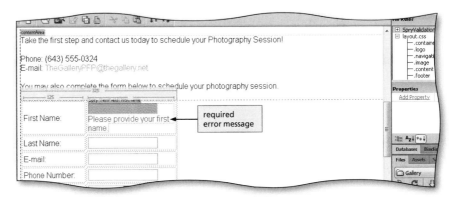

Figure 5–9

6

- Click the Save button on the Standard toolbar to save your work and display the Copy Dependent Files dialog box (Figure 5–10).

Q&A Why is the Copy Dependent Files dialog box displayed?

When you insert a Spry widget, Dreamweaver creates dependent asset files that you need to include the widget on the Web page. Be sure to copy the files to your site so the widget works properly.

Q&A On the contact.html page tab, what do SpryValidationTextField .js and SpryValidationTextField.css refer to?

SpryValidationTextField.js indicates that a JavaScript (.js) file named SpryValidationTextField.

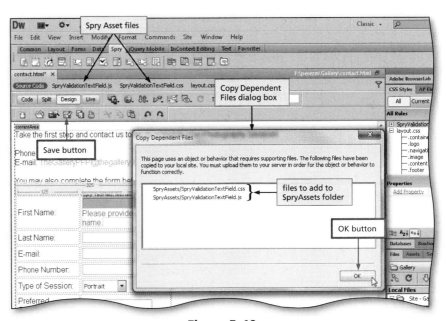

Figure 5–10

js is attached to the Contact page. This file is contained within the SpryAssets folder. SpryValidationTextField.css indicates that another CSS file is attached to the Contact page. The rules for SpryValidationTextField.css appear in the CSS Styles panel. This file is also contained within the SpryAssets folder.

7

- Click the OK button to save the Spry Asset files and to create the SpryAssets folder in the Files panel (Figure 5–11).

Q&A

What is the SpryAssets folder?

The SpryAssets folder contains the Spry widget files associated with the Web site.

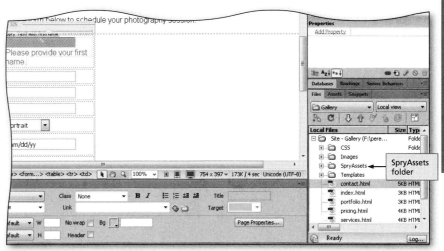

Figure 5–11

Other Ways

1. On Application bar, click Insert, point to Spry, click Spry Validation Text Field

To Add Spry Validation to Another Required Text Field

In addition to entering text in the First Name field, users also should be required to enter text in the Last Name field so the form contains complete contact information when users submit it. The following steps apply Spry Validation to the Last Name text field.

1

- Click the Last Name text field to select the field.

- Click the Spry Validation Text Field button on the Insert bar to change the text field to a Spry Validation Text Field (Figure 5–12).

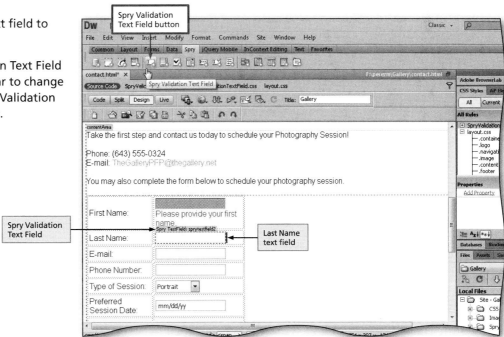

Figure 5–12

2

- Double-click the Spry TextField text box on the Property inspector and then type `lastName` to rename the Spry Validation Text Field.

- Click the Preview states button and then click Required to display the A value is required error message (Figure 5–13).

Why is the Spry TextField value entered as one word, as in lastName?

The name cannot include any spaces, so to use a descriptive name containing two words, you remove the space but capitalize the first letter of the second word to make the name easy to read.

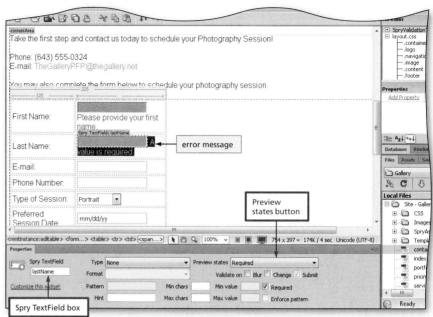

Figure 5–13

3

- Select the text, A value is required, next to the lastName Spry Validation Text Field, and then type `Please provide your last name` to change the required error message (Figure 5–14).

4

- Click the Save button on the Standard toolbar to save your work.

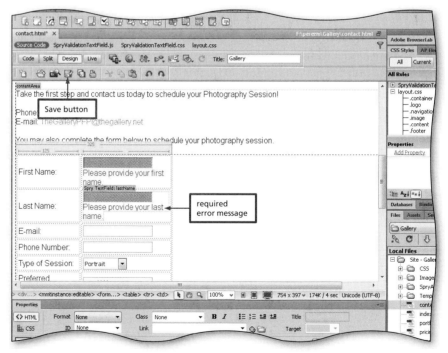

Figure 5–14

To Add Spry Validation to an E-mail Text Field

You want to make the E-mail text field a required text field and add validation to confirm that the text matches the format of an e-mail address, as in myname@email.com. If the e-mail address entered is missing the @ symbol or the domain, it does not pass the validation test. The following steps apply Spry validation to an e-mail text field.

- Click the E-mail text field to select the field.

- Click the Spry Validation Text Field button on the Insert bar to change the text field to a Spry Validation Text Field (Figure 5–15).

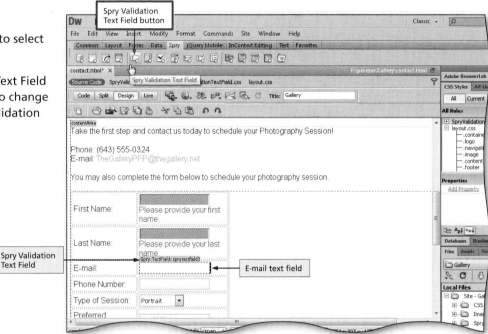

Figure 5–15

- Double-click the Spry TextField text box on the Property inspector and then type e-mail to rename the Spry Validation Text Field (Figure 5–16).

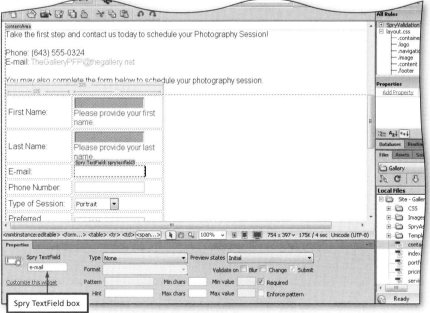

Figure 5–16

- Click the Type button and then click Email Address to select the type of Spry validation widget.

- Click the Preview states button, and then click Required to select the setting and to view the A value is required error message (Figure 5–17).

Q&A

How does the Email Address type validate the E-mail text field?

The Email Address type reviews the text to confirm that it is in the proper e-mail format, such as myname@email.com. If the e-mail address is not in the proper format, the form displays the Invalid format error message.

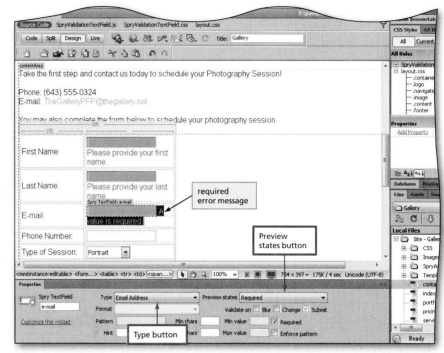

Figure 5–17

- Select the text, A value is required, next to the e-mail Spry Validation Text Field to prepare to replace it.

- Type `Please provide your`, press ENTER, and then type `e-mail address` to change the required error message for the e-mail Spry Validation Text Field (Figure 5–18).

- Click the Save button on the Standard toolbar to save your work.

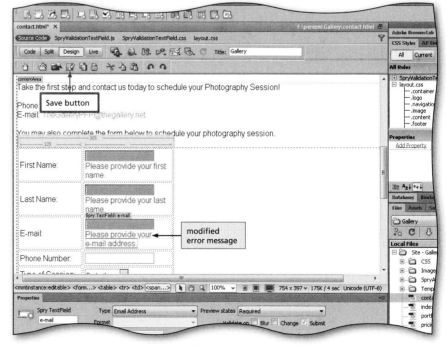

Figure 5–18

To Add Spry Validation to a Phone Number Text Field

You want to make the Phone Number text field a required text field and add validation to confirm that the text entered is a phone number. The validation test checks that the phone number entered is in the proper format: (999) 999-9999. If the phone number entered is missing a number or is not in the required format, it will not pass the validation test. The following steps apply Spry validation to a phone number text field.

- If necessary, scroll down the page, and then click the Phone Number text field to select the field.

- Click the Spry Validation Text Field button on the Insert bar to change the text field to a Spry Validation Text Field (Figure 5–19).

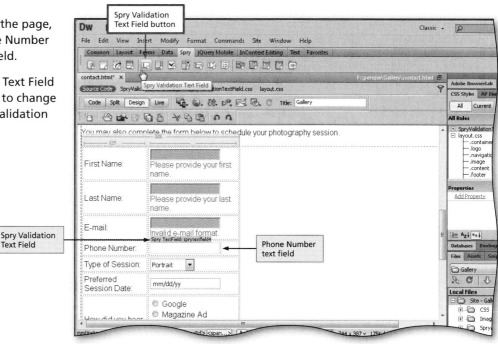

Figure 5–19

- Double-click the Spry TextField text box on the Property inspector and then type `phoneNumber` to rename the field.

- Click the Type button and then click Phone Number to select the type of Spry validation widget (Figure 5–20).

Q&A

How does the Phone Number type validate the Phone Number text field?

The Phone Number type only allows numbers and phone number formatting characters, such as parentheses ( ) and a hyphen (-). If the phone number entered is not within the specified pattern, such as (999) 999-9999, the Web page displays the Invalid format error message.

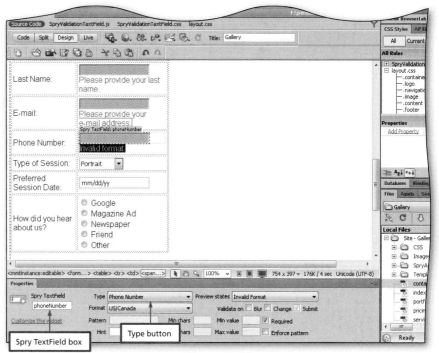

Figure 5–20

- If necessary, click the Format button on the Property inspector, and then click US/Canada to select the phone number format.

- Click the Hint text box and then Type `(999) 999-9999` to provide a reminder for users about the valid format.

- Click the Enforce pattern check box to select it (Figure 5–21).

Q&A What is the US/Canada format?

The US/Canada format is a specific phone number pattern that includes parentheses ( ) and a hyphen (-), such as (999) 999-9999.

Q&A What is the Hint property?

When the browser displays the form, (999) 999-9999 appears in the Phone Number field to show the desired format. When the user moves the insertion point to the Phone Number text field, the hint disappears, allowing the user to type a phone number.

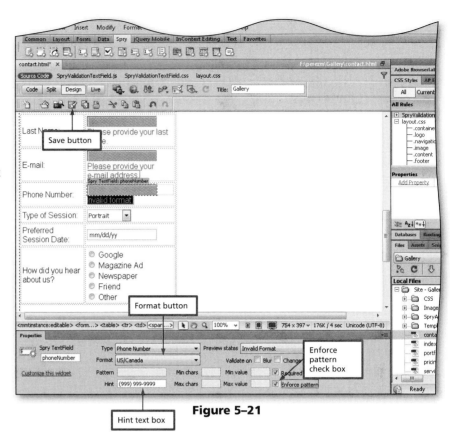

Figure 5–21

Q&A What happens to the Phone Number text field when the Enforce pattern check box is selected?

As a user enters a phone number into the field, the parentheses ( ) and hyphen (-) are inserted automatically.

- Click the Save button on the Standard toolbar to save your work.

To Add Spry Validation to Another Required Text Field

Next, you can make the Preferred Session Date text field a required text field and add validation to confirm that the text entered is a date. The validation test checks that the date entered is in the specified date format, such as mm/dd/yy. If the user enters a date in a different format, it will not pass the validation test. The following steps apply Spry validation to the Preferred Session Date text field.

1

- Click the Preferred Session Date text field to select the field.

- Click the Spry Validation Text Field button on the Insert bar to change the text field to a Spry Validation Text Field (Figure 5–22).

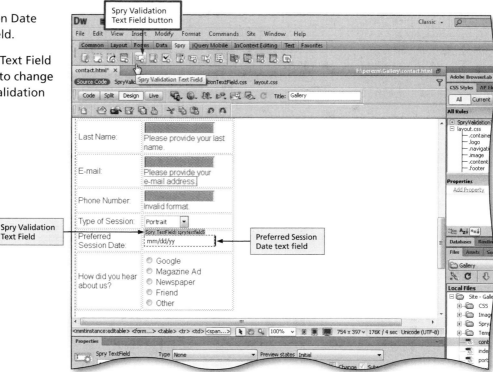

Figure 5–22

2

- Double-click the Spry TextField text box and then type `sessionDate` to rename the Spry Validation Text Field.

- Click the Type button and then click Date to select the type of Spry validation widget (Figure 5–23).

Q&A

How does the Date type validate the Date text field?

The Date type reviews the text the user enters to confirm it is a date in the specified date format, such as mm/dd/yy. If the user enters a date in a different format, the Web page displays the Invalid format error message.

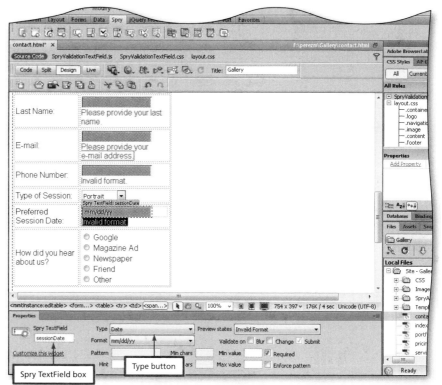

Figure 5–23

- If necessary, click the Format button on the Property inspector, and then click mm/dd/yy to select a date format.

- Click the Hint text box and then type `mm/dd/yy` to provide a reminder for users about the valid format.

- Click the Enforce pattern check box to select it (Figure 5–24).

Q&A

What happens to the Preferred Session Date text field when the Enforce pattern check box is selected?

As a user types a date into the field, the form inserts forward slashes (/) automatically where appropriate.

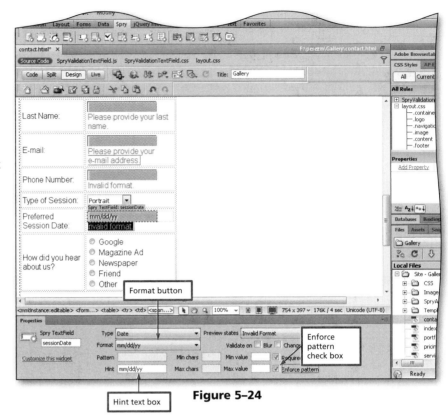

Figure 5–24

- Click the Preferred Session Date text field to select the field.

- Double-click the value in the Init val text box and then press the DELETE key to delete the value (Figure 5–25).

Q&A

Why do I need to delete the initial value?

Deleting the initial value means the user does not need to delete it before typing a date. The Hint property displays mm/dd/yy in the Preferred Session Date text field until the user clicks the text box so the user can enter a date.

- Click the Save button on the Standard toolbar to save your work.

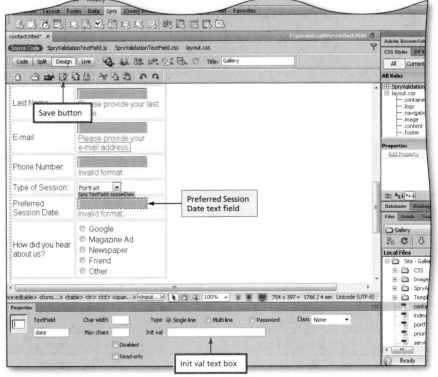

Figure 5–25

To Add Spry Validation to a Radio Group

You want to make the 'How did you hear about us?' Radio Group a required response. To complete this task, first delete the existing Radio Group and then insert the Spry Validation Radio Group widget. Inserting the Spry Validation Radio Group widget adds two files to the SpryAssets folder: SpryValidationRadioGroup.js and SpryValidationRadioGroup.css. The following steps delete the existing Radio Group and insert a Spry Validation Radio Group widget.

1

- Select the 'How did you hear about us?' Radio Group to select it and then press the DELETE key to delete it (Figure 5–26).

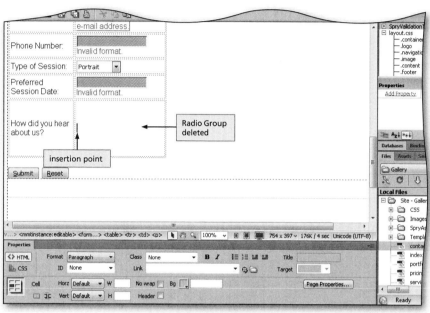

Figure 5–26

2

- Click the Spry Validation Radio Group button on the Insert bar to insert a Spry Validation Radio Group widget and to open the Spry Validation Radio Group dialog box.

- In the Name text box, replace the text RadioGroup1 with `Referral` to provide a unique name (Figure 5–27).

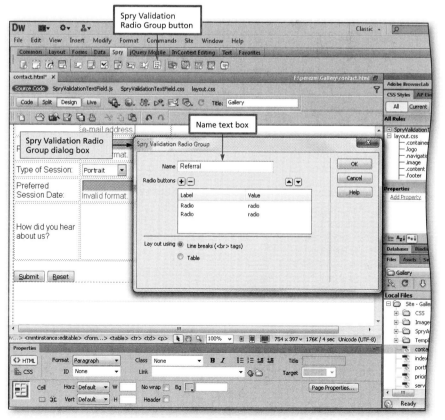

Figure 5–27

- Click the first Radio label in the Label list and then type `Google` to replace the text.

- Click the second Radio label in the Label list and then type `Magazine Ad` to replace the text (Figure 5–28).

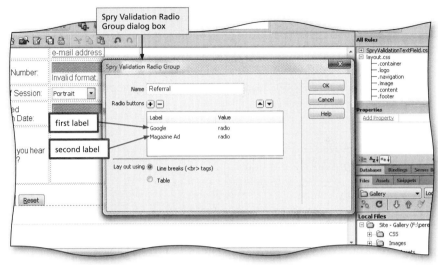

Figure 5–28

- Click the plus button to add another label and value to the list.

- Click the third Radio label in the Label list and then type `Newspaper` to replace the text (Figure 5–29).

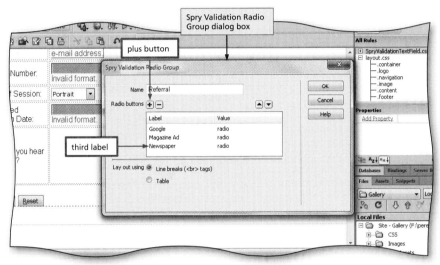

Figure 5–29

- Click the plus button to add another label and value to the list.

- If necessary, use the vertical scroll bar to scroll down and click the fourth Radio label in the Label list, and then type `Friend` to replace the text (Figure 5–30).

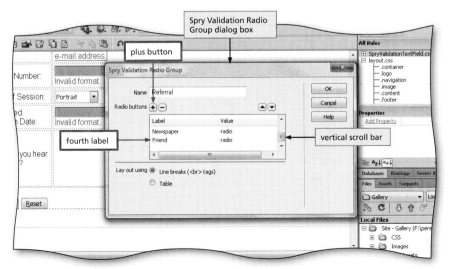

Figure 5–30

- Click the plus button to add another label and value to the list.

- Click the fifth Radio label in the Label list and then type `Other` to replace the text (Figure 5–31).

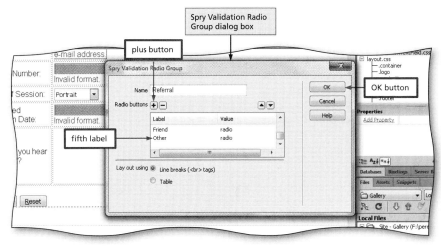

Figure 5–31

- Click the OK button to insert the Spry Validation Radio Group.

- Double-click the text in the Spry Radio Group text box on the Property inspector and then type `referral` to rename the Spry Radio Group (Figure 5–32).

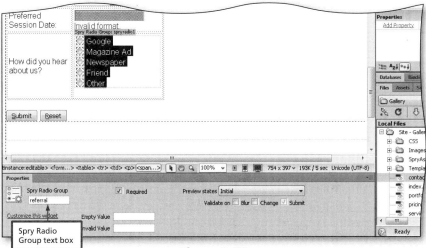

Figure 5–32

- Click the Save button on the Standard toolbar to save your work and display the Copy Dependent Files dialog box (Figure 5–33).

- Click the OK button to save the Spry Asset files to the SpryAssets folder.

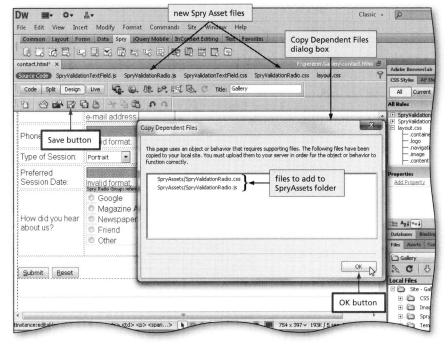

Figure 5–33

To Test the Spry Validation Form in a Browser

The following steps open and then preview the contact.html page in Internet Explorer to test the form.

1

- Click the 'Preview/ Debug in browser' button on the Document toolbar to display a list of browsers.

- Click Preview in IExplore in the 'Preview/Debug in browser' list to display the Gallery Web site in Internet Explorer (Figure 5–34).

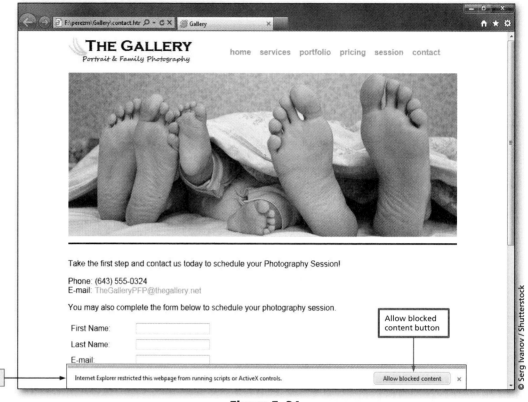

Figure 5–34

2

- If a security message appears regarding ActiveX controls, click the 'Allow blocked content' button to allow the Spry validation form to work properly.

- Click the Submit button to test the form (Figure 5–35).

 Q&A What should happen after I click the Submit button?

Each field should display a required validation message, except for the Type of Session field.

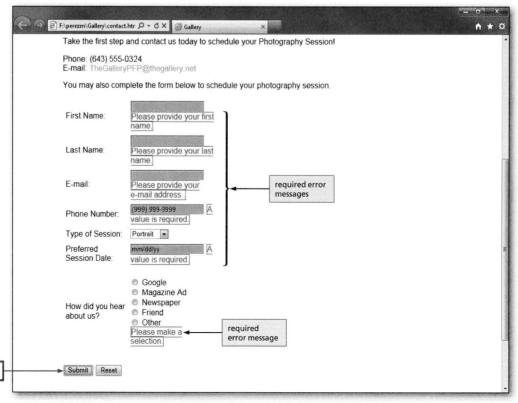

Figure 5–35

- Click the First Name text field, type your first name, click the Last Name text field, and then type your last name to test those fields.

- Click the E-mail text box, and then type the beginning part of your e-mail address, such as perezm, to test the field.

- Click the Submit button again to submit the form (Figure 5–36).

Q&A

Should I include the @ symbol or domain when I type my e-mail address?

No. You are testing the E-mail field to make sure it shows the Invalid format error message.

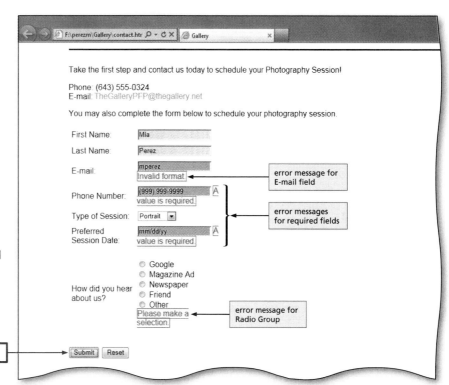

Figure 5–36

- Click the E-mail text box and then type your complete e-mail address to test the field again.

- Click the Phone Number text box and then type your phone number to enter a phone number.

- Click the Type of Session box arrow and then click Family to select a session type.

- Click the Preferred Session Date text box and then type today's date to enter a date.

- Click the Friend option button to select an option in the 'How did you hear about us?' Radio Group (Figure 5–37).

- Click the Submit button to submit the form, and then close Internet Explorer.

- Close the contact.html page.

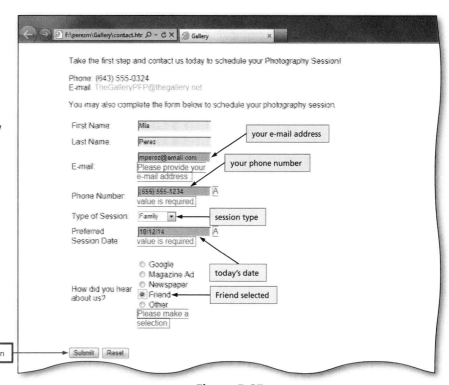

Figure 5–37

Other Ways

1. Press ALT+S

Break Point: If you wish to take a break, this is a good place to do so. To resume at a later time, start Dreamweaver and continue following the steps from this location forward.

Using Spry Collapsible Panels

The Portfolio Web page showcases the Gallery's photography work. Eventually, many images will be included on the page. To maximize space and minimize the length of the page, you can use Spry Collapsible Panel widgets to organize and store the images. You will add three Spry Collapsible Panels to the Portfolio Web page. The first Spry Collapsible Panel is for showcasing the Gallery's portrait photography. The second Spry Collapsible Panel is for family photography, and the third Spry Collapsible Panel is for baby photography.

To Add Spry Collapsible Panels to a Web Page

The following steps open the Portfolio Web page and insert Spry Collapsible Panel widgets.

1
- Double-click portfolio.html in the Files panel to display the Portfolio page in the Document window.

- Select the text and images within the editable region (Figure 5–38).

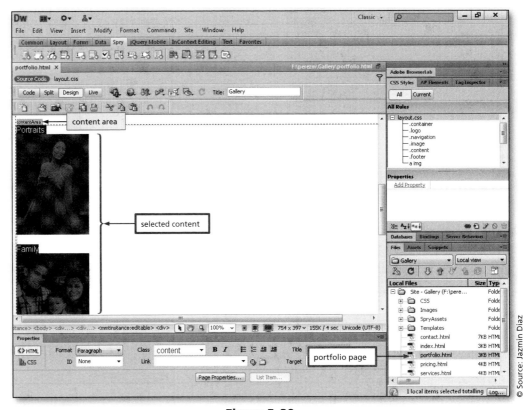

Figure 5–38

© Source: Jazmin Diaz

2

- Press the DELETE key to delete the selected text and images.

- Click the Spry Collapsible Panel button on the Insert bar to insert a Spry Collapsible Panel and to display the properties of the Spry Collapsible Panel on the Property inspector (Figure 5–39).

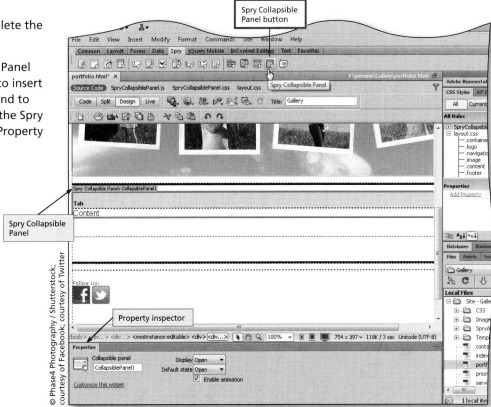

© Phase4 Photography / Shutterstock; courtesy of Facebook; courtesy of Twitter

Figure 5–39

3

- Double-click the Collapsible panel text box on the Property inspector and then type `portraitPanel` to rename the collapsible panel.

- Click the Default state button and then click Closed to set the Default state of the panel.

- Click the Display button and then click Closed to set the Display state of the panel (Figure 5–40).

 What does it mean to set the Default state to Closed for the Spry Collapsible Panel?

Selecting Closed as the Default state means the collapsible panel will be closed when the Portfolio Web page is displayed in a browser. Users can click the collapsible panel to view its content. Setting the Default state to Closed saves space on the Web page when it opens.

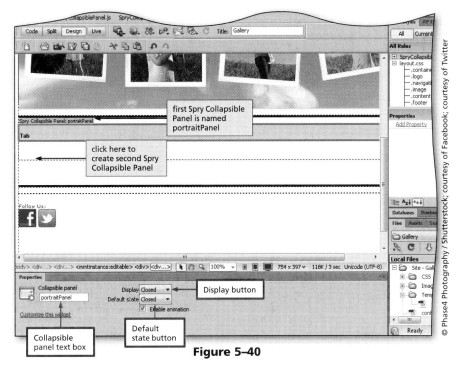

© Phase4 Photography / Shutterstock; courtesy of Facebook; courtesy of Twitter

Figure 5–40

• Click below the Content line in the portraitPanel to deselect the panel.

• Click the Spry Collapsible Panel button on the Insert bar to insert a second Spry Collapsible Panel (Figure 5–41).

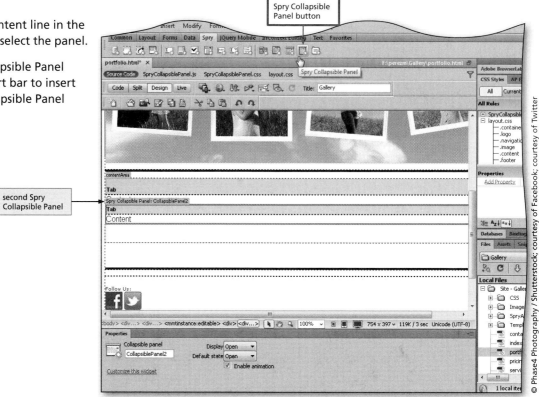

Figure 5–41

• Double-click the Collapsible panel text box on the Property inspector and then type `familyPanel` to rename the collapsible panel.

• Click the Default state button and then click Closed to set the Default state of the panel.

• Click the Display button and then click Closed to set the Display state of the panel (Figure 5–42).

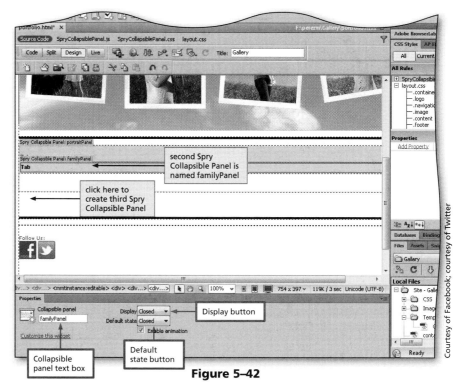

Figure 5–42

6

- Click below the Content line in the familyPanel to place the insertion point.

- Click the Spry Collapsible Panel button on the Insert bar to insert a third Spry Collapsible Panel (Figure 5–43).

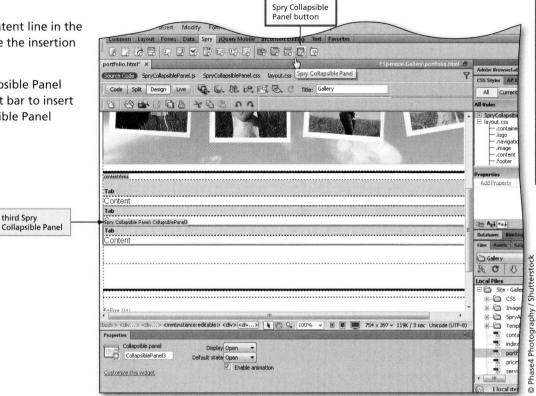

third Spry Collapsible Panel

Figure 5–43

7

- Double-click the Collapsible panel text box on the Property inspector and then type `babyPanel` to rename the collapsible panel.

- Click the Default state button and then click Closed to set the Default state of the panel.

- Click the Display button and then click Closed to set the Display state of the panel (Figure 5–44).

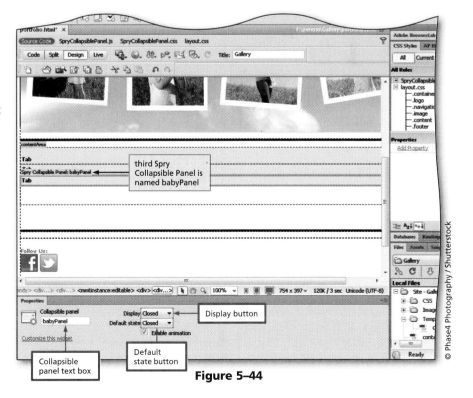

third Spry Collapsible Panel is named babyPanel

Display button

Collapsible panel text box

Default state button

Figure 5–44

- Click below the babyPanel to deselect it.

- Click the Save button on the Standard toolbar to save your work and display the Copy Dependent Files dialog box (Figure 5–45).

Q&A

Why does the text, SpryCollapsiblePanel.js and SpryCollapsiblePanel.css, appear on the portfolio.html page tab?

These are the files the browser needs to display the Spry Collapsible Panel widgets. The JavaScript and CSS files are contained within the SpryAssets folder in the Files panel.

- Click the OK button to save the Spry Asset files in the SpryAssets folder.

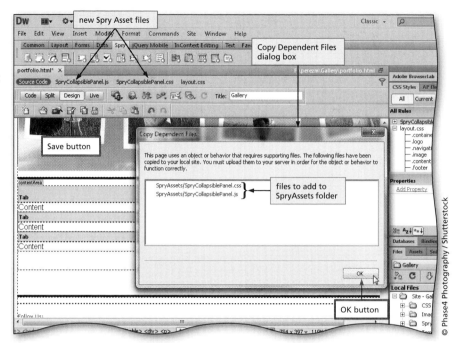

Figure 5–45

To Add Panel Headings

Tab is the default name for each Collapsible Panel widget. The following steps rename each tab to identify the type of images each tab will contain.

①

- Delete the text, Tab, in the portraitPanel Collapsible Panel widget and then type Portrait Photography to rename the tab (Figure 5–46).

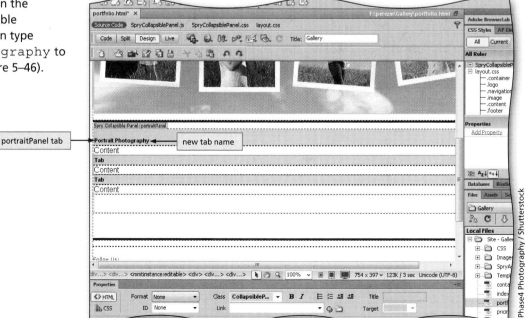

Figure 5–46

②

- Delete the text, Tab, in the familyPanel Collapsible Panel widget and then type `Family Photography` to rename the tab.

- Delete the text, Tab, in the babyPanel Collapsible Panel widget and then type `Baby Photography` to rename the tab (Figure 5–47).

 Q&A Why is it important to name each tab?

The tab title is always displayed on the Web page, even if the panel is closed. Naming each tab identifies the contents within each panel.

③

- Click the Save button on the Standard toolbar to save your work.

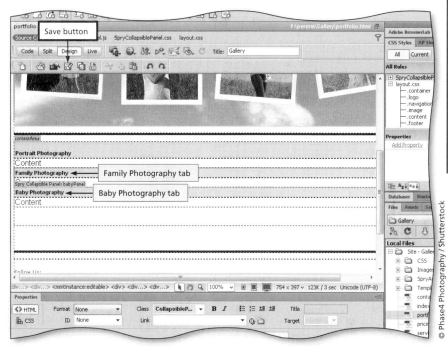

Figure 5–47

Using the Adobe Widget Browser

In addition to Spry widgets, Adobe provides a resource for other widgets through the Adobe Widget Browser. The **Adobe Widget Browser** is an Adobe AIR application that allows Web site designers to access a library of online widgets in Adobe Exchange. **Adobe Integrated Runtime (AIR)** applications allow Web site developers access to Web resources. **Adobe Exchange** provides many types of Web tags, scripts, and other custom Web items available for download. Widget designers can upload their custom-built widgets to Adobe Exchange to share widgets that they create.

In Adobe Exchange, you can select widgets to use in your Dreamweaver Web site and add them to a personal list of available widgets called My Widgets in the Adobe Widget Browser. You can add the widgets stored in My Widgets to your Web pages. You must accept terms and conditions before using some widgets.

One widget available in the Adobe Widget Browser is called the LightBox Gallery widget. You can use this widget to enhance the Gallery Web site. The LightBox Gallery widget can display sample portrait, family, and baby photography, and let visitors select and view a larger version of a photo. When a visitor clicks a photo, the widget animates the photo so it grows larger on the Web page to create an engaging effect.

Before you can access the widgets in the Adobe Widget Browser, you must install Adobe AIR and create an Adobe account. You do not need to pay a fee to download and install Adobe AIR or create an Adobe account. In addition, you do not need an Adobe account to create a My Widgets folder and store widgets in it.

BTW

Adobe Widget Browser Help
To find help when using the Adobe Widget Browser, open the Dreamweaver Help file and then search for Widget Browser.

To Check Your Computer for the Adobe AIR Application

Before you can use the Adobe Widget Browser, make sure Adobe AIR is installed on your computer. The following steps check to see if Adobe AIR is installed on your computer.

1

- Click the Extend Dreamweaver button on the title bar to display the Extend Dreamweaver options (Figure 5–48).

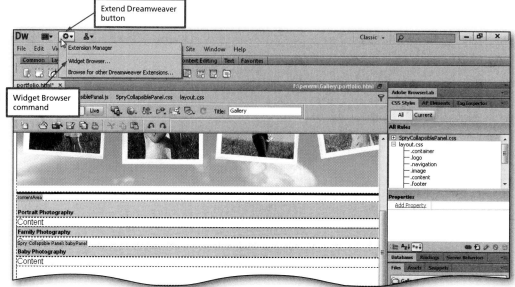

Figure 5–48

2

- Click Widget Browser to select it, and then click the Accept button in the End User License Agreement dialog box to display the Adobe Widget Browser window, which confirms Adobe AIR is installed on your computer (Figure 5–49).

 Q&A

What does it mean if the Adobe Widget Browser window is not displayed?

That means Adobe AIR is not installed on your computer. Complete the steps in the "To Download and Install Adobe AIR" section.

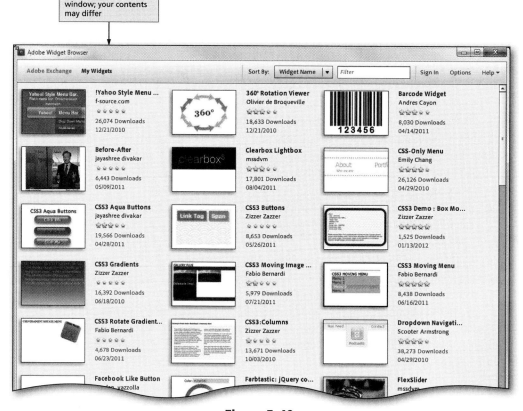

Figure 5–49

To Download and Install Adobe AIR

If Adobe AIR is not installed on your computer, you can download the program and install it free of charge. The following steps download and install the Adobe AIR application.

1

- Open your browser and type `http://get.adobe.com/air/` in the Address bar to open the Adobe AIR page on the Adobe Web site (Figure 5–50).

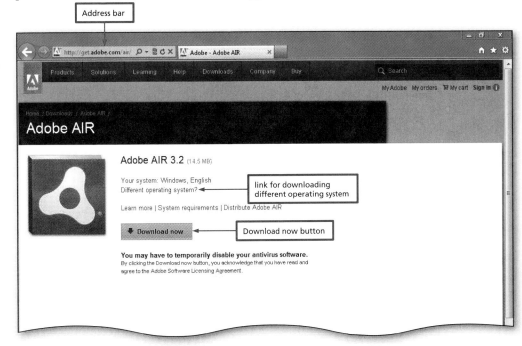

Figure 5–50

2

- If the Adobe AIR page shows your correct operating system, click the Download now button to begin downloading Adobe AIR.

- If the Adobe AIR page shows the incorrect operating system, click the 'Different operating system?' link, use the arrow buttons for Steps 1 and 2 to select your operating system and version, and then download the application (Figure 5–51).

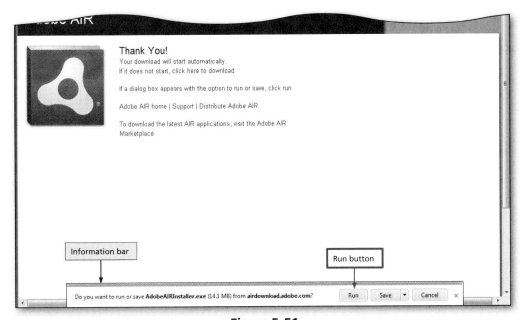

Figure 5–51

 Why does this Web site provide a download for a different version of Adobe AIR?

Adobe may have released a newer version of Adobe AIR. Continue with the installation.

3

- On the Information bar, click the Run button to start installing Adobe AIR and display the Adobe AIR Setup dialog box (Figure 5–52).

Q&A

What should I do if a User Account Control dialog box appears asking if I want to allow this program to make changes to my computer?

Click the OK or Allow button to install the software.

Figure 5–52

4

- Click the I Agree button in the Adobe AIR Setup dialog box to accept the license terms and to display the Installation completed message in the Adobe AIR Setup dialog box (Figure 5–53).

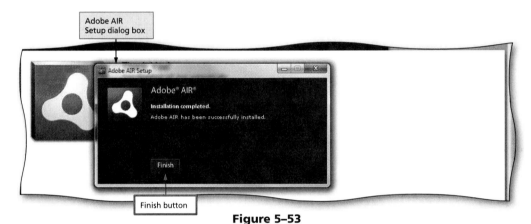

Figure 5–53

5

- Click the Finish button in the Adobe AIR Setup dialog box to finish installing the software.

- Close your browser.

To Create an Adobe Account

If you do not have an Adobe account, you must create one so you can view and select widgets to use in your Web sites. The following steps create an Adobe account if you do not already have one.

1

- If necessary, click the Extend Dreamweaver button on the title bar, and then click Widget Browser to display the Adobe Widget Browser window (Figure 5–54).

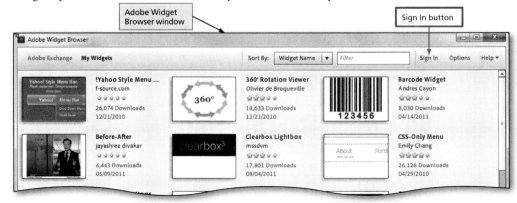

Figure 5–54

2

- Click the Sign In button to display the Sign In dialog box (Figure 5–55).

Q&A I have a Sign Out button, not a Sign In button. What should I do?

You are already signed in to your Adobe account. Skip ahead to the "To View and Add Widgets from Adobe Exchange to My Widgets" section.

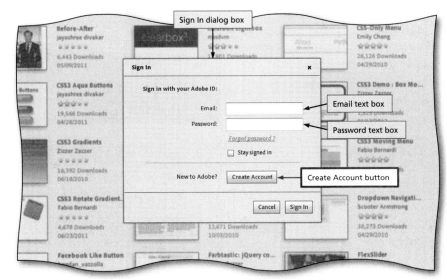

Figure 5–55

3

- If you need to create an account, click the Create Account button to visit the Adobe Web site (Figure 5–56).

Q&A I already have an Adobe account. Do I need to create another one?

No. If you already have an Adobe account, enter your e-mail address and password in the Sign In dialog box or the 'Returning members sign in' section of the Adobe Sign In page.

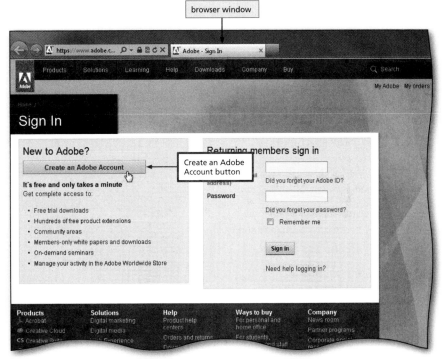

Figure 5–56

- If you need to create an account, click the Create an Adobe Account button to display the Join Adobe page (Figure 5–57).

Q&A

What should I do if I do not need to create an account?

Skip ahead to Step 2 in the next section, "To Sign in to the Adobe Widget Browser."

Account Details form

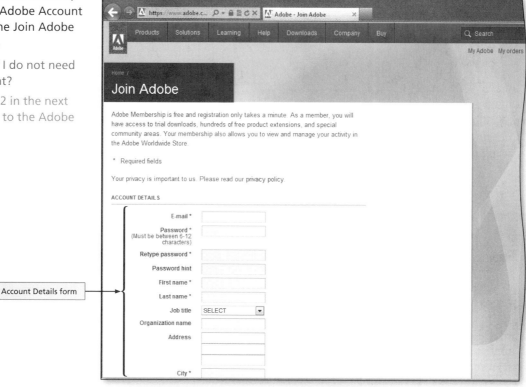

Figure 5–57

- Complete the Account Details form to create an Adobe account.

- Click the Continue button at the bottom of the form to display the Thank You page (Figure 5–58).

- Click the Continue button and then close your browser to finish creating an Adobe account.

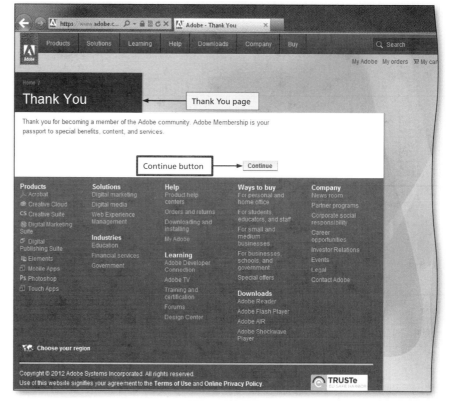

Figure 5–58

To Sign in to the Adobe Widget Browser

Once you have established an Adobe account, you need to sign in to access Adobe Exchange using the Adobe Widget Browser window so you can review the widgets available in Adobe Exchange. The following steps sign you in to your Adobe account through the Adobe Widget Browser.

1 If necessary, open the Adobe Widget Browser.

2 Click the Sign In button to display the Sign In dialog box.

3 Enter the e-mail address and password used to create your Adobe account to sign in.

4 Click the Sign In button to finish signing in to your Adobe account and display the Adobe Widget Browser (Figure 5–59).

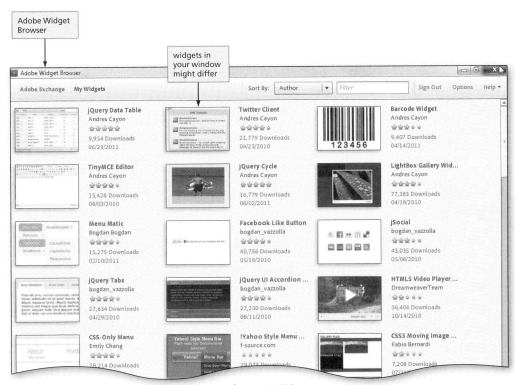

Figure 5–59

To View and Add Widgets from Adobe Exchange to My Widgets

Once you have signed in to the Adobe Widget Browser, you can review the widgets available in Adobe Exchange and add them to My Widgets. My Widgets is the location where you store saved widgets from Adobe Exchange. The following steps view widgets and add them to My Widgets.

● Scroll down to view all of the widgets available through Adobe Exchange (Figure 5–60).

Figure 5–60

2

● Scroll to the top of the window, click the Filter text box, and then type `light` to search for the LightBox Gallery Widget (Figure 5–61).

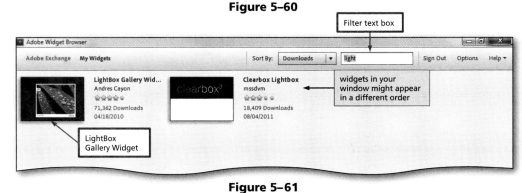

Figure 5–61

3

● Click the LightBox Gallery Widget to select it.

● Review the information provided in the window to learn more about this widget (Figure 5–62).

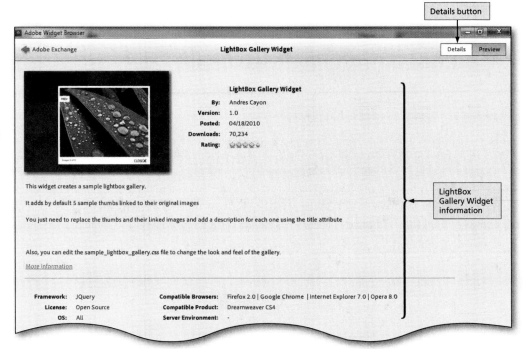

Figure 5–62

4

- Click the Preview button to see a live view of the widget.

- If necessary, click the <default> option in the Widget Presets list to select the default developer preset.

- On the Live View tab, click the first sample image to view it (Figure 5–63).

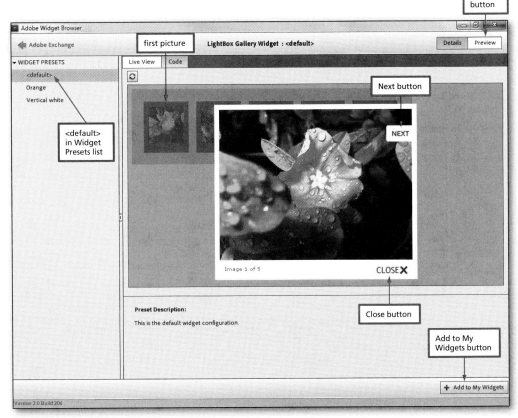

Figure 5–63

5

- Point to an image and then click the Next button to view all of the images.

- Click the Close button when you are finished viewing the images.

- Click the Add to My Widgets button to add the LightBox Gallery widget to My Widgets and to display the Widget Added dialog box (Figure 5–64).

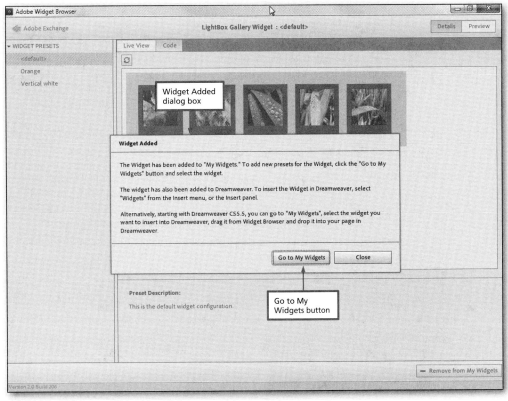

Figure 5–64

6

- Click the Go to My Widgets button in the Widget Added dialog box to view the widget in My Widgets (Figure 5–65).

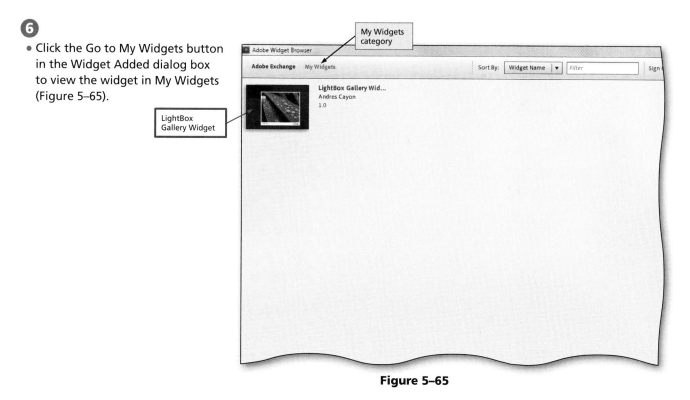

Figure 5–65

Editing and Formatting Widgets

After you add the LightBox Gallery widget to My Widgets, you can create a preset for the widget to modify the widget's formatting and properties. A **preset** is a set of predefined properties for the widget. You create a preset to specify the background color, size, text color, margins, padding, and other properties of the widget. The preset you create provides custom properties specifically designed for the Gallery Web site.

Plan
Ahead

Edit and format Spry widgets.
After you add Spry widgets and widgets from other sources to a Web page, typically you need to edit and format the widgets. Consider what you must change to have the widget fit the design of your Web page. For example, you might need to change properties such as background color to match the style of the Web page. You also might need to modify widget controls so they work for your Web site.

To Create a Widget Preset for the LightBox Gallery Widget

The LightBox Gallery widget gives a rich, interactive look and feel to the Gallery Web site. The widget includes buttons for navigation. Before anyone can use the buttons, you need to update the file path for the widget's buttons to associate them with the Gallery site. Otherwise, the widget will not display the buttons. In addition, you can format the widget to match the design of the Gallery site. To do so, you create presets using the Adobe Widget Browser. The following steps create a preset for the LightBox Gallery widget.

1

- In My Widgets, click the LightBox Gallery widget to select it and display options for modifying the widget (Figure 5–66).

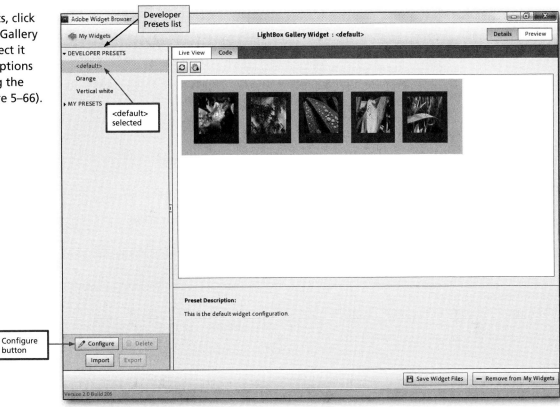

Figure 5–66

2

- If necessary, click <default> in the Developer Presets list to select it.

- Click the Configure button to create a new preset (Figure 5–67).

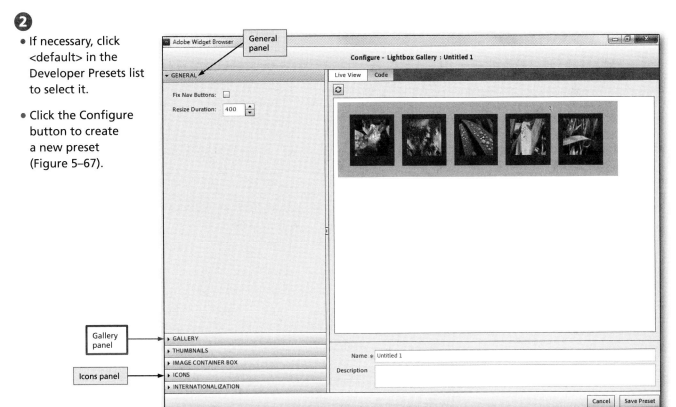

Figure 5–67

- Click the Gallery panel to display the Gallery preset options.

- Double-click the Width text box and then type 800 to modify the widget width (Figure 5–68).

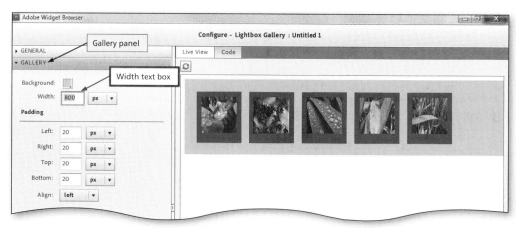

Figure 5–68

- Click the Icons panel to display the file paths for the loading icon and navigation buttons.

- Click in the Loading text box before the text, /images/ lightbox/lightbox-ico-loading.gif, and then type /your last name and first initial/ Gallery to define the location of the Loading icon within the Gallery Web site.

- Click in the Previous Button text box before the text, /images/lightbox/ lightbox-btn-prev.gif, and then type /your last name and first initial/Gallery to define the location of the Previous Button within the Gallery Web site.

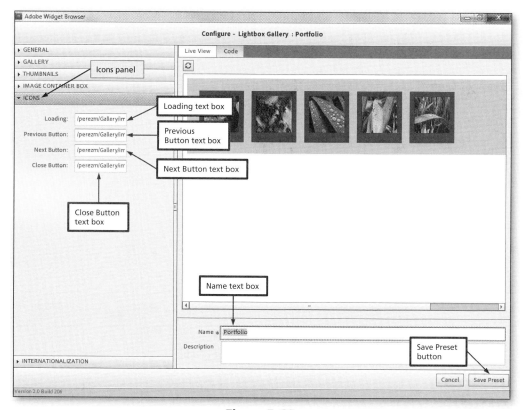

Figure 5–69

- Click in the Next Button text box before the text, /images/lightbox/lightbox-btn-next.gif, and then type /your last name and first initial/Gallery to define the location of the Next Button within the Gallery Web site.

- Click in the Close Button text box before the text, /images/lightbox/lightbox-btn-close.gif, and then type /your last name and first initial/Gallery to define the location of the Close Button within the Gallery Web site.

- Select the text in the Name text box and then type Portfolio to name the preset (Figure 5–69).

Q&A

What is the purpose of changing the file paths on the Icons panel?

The Icons panel lists properties for the buttons in the LightBox Gallery widget. The file path of the buttons must be updated to include the file path of the Gallery Web site otherwise they will not be displayed.

5

- Click the Save Preset button to save the preset and display the Portfolio preset in the My Presets list (Figure 5–70).

Q&A

What does it mean when the Portfolio preset is displayed in the My Presets list?

Now that you have defined the Portfolio preset, the widget width will be adjusted as you specified when you add the widget to the Gallery site. Additionally, the widget will include loading icon and navigation buttons.

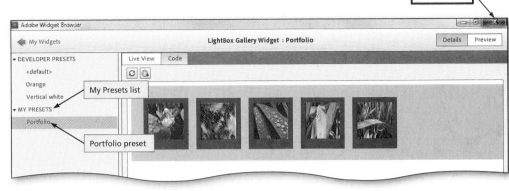

Close button

Figure 5–70

6

- Click the Close button to close the Adobe Widget Browser.

Adding an Adobe Widget to a Web Page

To enhance the appearance of the Gallery Web site, you can insert the LightBox Gallery widget within each collapsible panel on the Portfolio page. You then can replace the default images in the widget with images from your data files.

Add custom content to a widget.
You add a widget to a Web page to enhance the page. Spry widgets and widgets you add from other sources might contain default content that needs to be replaced to fit the purpose of your Web page. For example, you can add images to each tab in a Spry Collapsible Panel widget. When you plan the Web page, determine what content should appear in widgets and what purpose it serves to include them in widgets. If inserting content in widgets does not enhance the page, add the content to the page as you usually do.

Plan
Ahead

The LightBox Gallery widget is an interactive way to showcase images on a Web site. Two image files are linked to each image within the widget; one image is the thumbnail image and the other is the full-size image. A **thumbnail image** is a smaller version of an image. Web pages typically display thumbnail images to conserve file size. The Portfolio page will use the LightBox Gallery widget to display thumbnails of sample portrait, family, and baby photography. Visitors can click a thumbnail image to display a full-size version of the photo.

To Copy Files to a New Lightbox Folder

Before adding images to the LightBox Gallery widget, you must create a Lightbox folder in the Images folder for the Gallery site. To complete this assignment, you will be required to use the Data Files for Students. Visit www.cengage.com/ct/studentdownload for detailed instructions or contact your instructor for information about accessing the required files. The following steps create the Lightbox folder and copy 30 files from the Data Files for Students to the Gallery site.

1 If necessary, insert the USB flash drive containing your student data files, and then use Windows Explorer to open a window displaying the contents of the drive containing the Data Files for Students.

2 Double-click the Chapter 05 folder and then double-click the Gallery folder to open it.

3 Click the baby_1a_thumb file, or the first file in the list, hold down the SHIFT key, and then click the portrait_5b file, or the last file in the list, to select the images needed for the site.

4 Right-click the selected files, click Copy on the context menu, and then navigate to the *your last name and first initial* folder on Removable Disk F: to prepare to copy the files.

5 Double-click the Gallery folder, and then double-click the Images folder to open the Images folder.

6 Create a new folder named Lightbox, and then double-click the Lightbox folder to open it.

7 Right-click a blank spot in the open window, and then click Paste on the context menu to copy the 30 image files into the Lightbox folder (Figure 5–71).

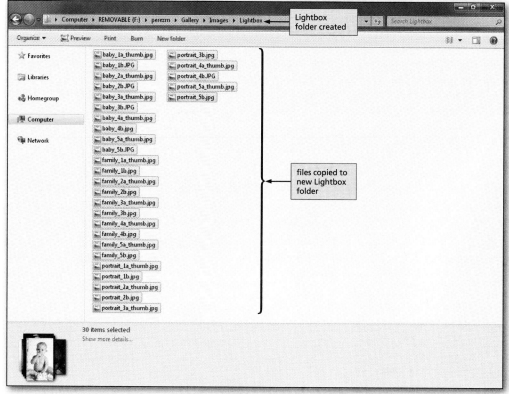

Figure 5–71

To Add the LightBox Gallery Widget to a Web Page

The image files you copied into the Lightbox folder include thumbnail images and full-size images of the photos to display on the Portfolio page. First, you insert a LightBox Gallery widget within each collapsible panel on the Portfolio page. Then you can link each image to a thumbnail image file and a full-size image file. Each collapsible panel will use a LightBox Gallery widget to showcase the Gallery's photography. The following steps add the LightBox Gallery widget within a Collapsible Panel widget on the Portfolio Web page.

- In portfolio.html, scroll to view the right edge of the portraitPanel (Figure 5–72).

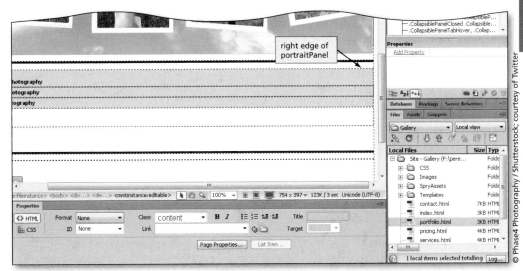

right edge of portraitPanel

Figure 5–72

2

- Point to the panel and then click the 'Click to show panel content' button to display the portraitPanel content.

- Delete the text, Content, below the Portrait Photography panel name to create space for the LightBox Gallery widget.

- Click Insert on the Application bar to display the Insert menu (Figure 5–73).

Q&A

Why don't I see the panel below the Portrait Photography panel name now?

The panel is now empty, so it looks like it has been removed. However, it is still part of the page.

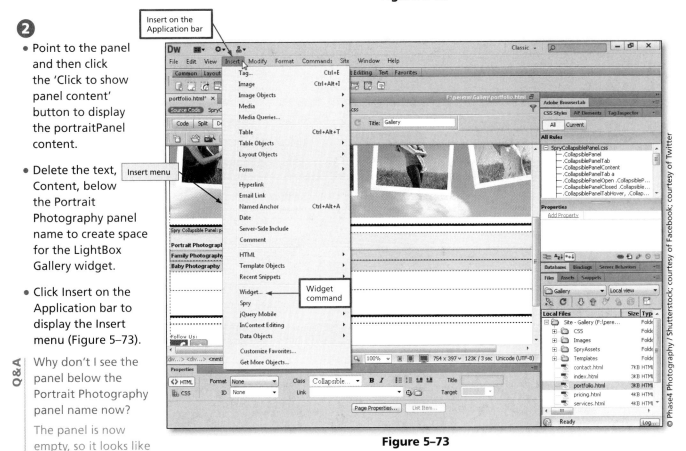

Insert on the Application bar

Insert menu

Widget command

Figure 5–73

3

- Click Widget on the Insert menu to display the Widget dialog box.

- If necessary, click the Widget button and then click LightBox Gallery Widget (1.0) to select the widget.

- Click the Preset button and then click Portfolio to select the page for the widget (Figure 5–74).

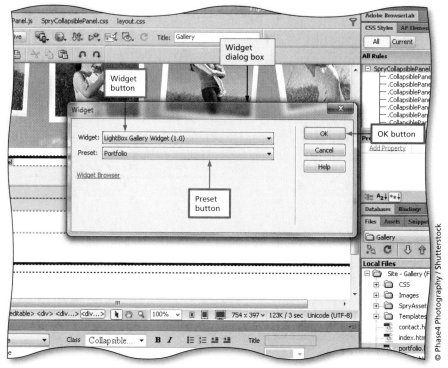

Figure 5–74

4

- Click the OK button in the Widget dialog box to insert the LightBox Gallery widget (Figure 5–75).

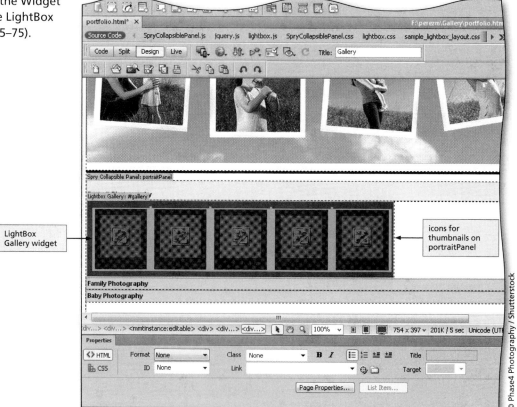

Figure 5–75

5

- If necessary, scroll to the right and then click the 'Click to show panel content' button on the familyPanel to display the familyPanel content.

- Delete the text, Content, below the Family Photography panel name.

- Click Insert on the Application bar, and then click Widget to display the Widget dialog box.

- Confirm that the Widget button displays LightBox Gallery Widget (1.0) and the Preset button displays Portfolio to select the widget and the preset.

- Click the OK button in the Widget dialog box to insert the LightBox Gallery widget (Figure 5–76).

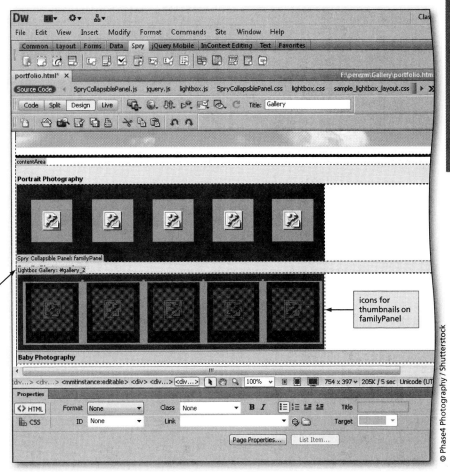

Figure 5–76

6

- If necessary, scroll to the right and click the 'Click to show panel content' button on the babyPanel to display the babyPanel content.

- Delete the text, Content, below the Baby Photography panel name.

- Click Insert on the Application bar to display the Insert menu and then click Widget to display the Widget dialog box.

- Confirm that the Widget button displays LightBox Gallery Widget (1.0) and the Preset button displays Portfolio to select the widget and the preset.

- Click the OK button in the Widget dialog box to insert the LightBox Gallery widget (Figure 5–77).

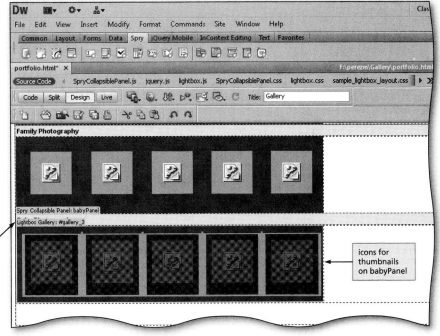

Figure 5–77

- Click below the widget to deselect it, and then click the Save button on the Standard toolbar to save your work and display the Copy Dependent Files dialog box (Figure 5–78).

- Click the OK button in the Copy Dependent Files dialog box to save the Spry Asset and image files.

Q&A

I noticed that image files are being saved to the Images folder. What are these image files?

The LightBox Gallery widget contains default image files. You will replace them shortly.

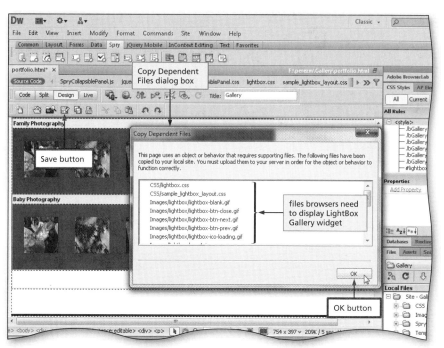

Figure 5–78

To Add Images to the Portrait Photography LightBox Gallery Widget

The Portrait Photography LightBox Gallery widget is now ready for the images to display on the Portfolio page. The following steps replace the default images in the LightBox Gallery widget with images to showcase the Gallery's portrait photography.

- In the Files panel, expand the Images folder and then expand the Lightbox folder to view the image files contained in the folder (Figure 5–79).

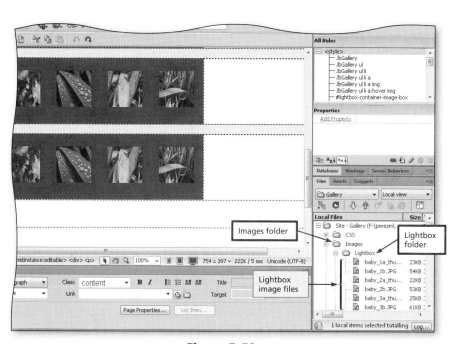

Figure 5–79

2

- If necessary, scroll to display the portrait image files.

- Click the first image on the Portrait Photography tab to select the image.

- Click the Point to File button next to the Src text box, and then drag the pointer to portrait_1a_thumb .jpg in the Files panel to select the image.

- Click the Point to File button next to the Link text box, and then drag the pointer to portrait_1b.jpg in the Files panel to link to the image (Figure 5–80).

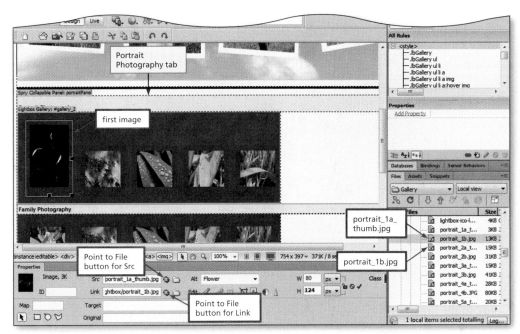

Figure 5–80

Q&A

Why do I need to select one file in the Src text box and a different file in the Link text box?

Because the LightBox Gallery widget accommodates a thumbnail image and a full-size image, you must select a source file for the thumbnail. You also link to the file containing the full-size image so that when the thumbnail is clicked, the full-size image opens.

3

- Click the second image on the Portrait Photography tab to select the image.

- Click the Point to File button next to the Src text box, and then drag the pointer to portrait_2a_thumb. jpg in the Files panel to select the image.

- Click the Point to File button next to the Link text box, and then drag the pointer to portrait_2b.jpg in the Files panel to link to the image (Figure 5–81).

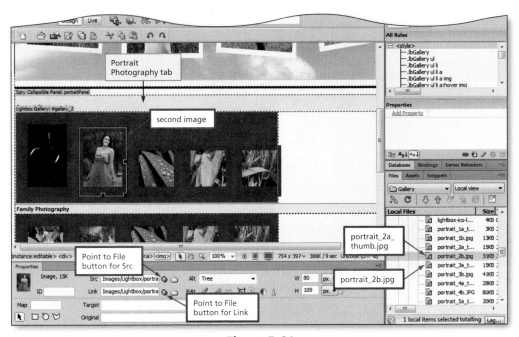

Figure 5–81

- Click the third image on the Portrait Photography tab to select the image.

- Click the Point to File button next to the Src text box, and then drag the pointer to portrait_3a_thumb. jpg in the Files panel to select the image.

- Click the Point to File button next to the Link text box, and then drag the pointer to portrait_3b.jpg in the Files panel to link to the image (Figure 5–82).

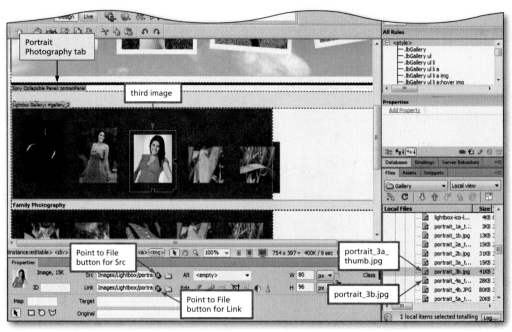

Figure 5–82

- Click the fourth image on the Portrait Photography tab to select the image.

- Click the Point to File button next to the Src text box, and then drag the pointer to portrait_4a_thumb. jpg in the Files panel to select the image.

- Click the Point to File button next to the Link text box, and then drag the pointer to portrait_4b.jpg in the Files panel to link to the image.

- Click the fifth image on the Portrait Photography tab to select the image.

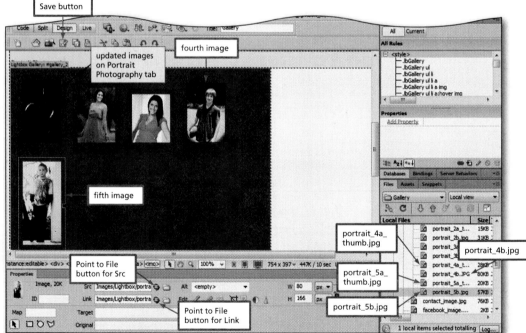

Figure 5–83

- Click the Point to File button next to the Src text box, and then drag the pointer to portrait_5a_thumb.jpg in the Files panel to select the image.

- Click the Point to File button next to the Link text box, and then drag the pointer to portrait_5b.jpg in the Files panel to link to the image (Figure 5–83).

- Click outside of the widget to deselect it and then click the Save button on the Standard toolbar to save your work.

To Add Images to the Family Photography LightBox Gallery Widget

The Family Photography LightBox Gallery widget is ready for images. The following steps show you how to replace the current images in the LightBox Gallery widget with images to showcase the Gallery's family photography.

1

- If necessary, scroll to display the family files in the Files panel.

- Click the first image on the Family Photography tab to select it.

- Click the Point to File button next to the Src text box, and then drag the pointer to family_1a_thumb.jpg in the Files panel to select the image.

- Click the Point to File button next to the Link text box, and then drag the pointer to family_1b.jpg in the Files panel to link to the image (Figure 5–84).

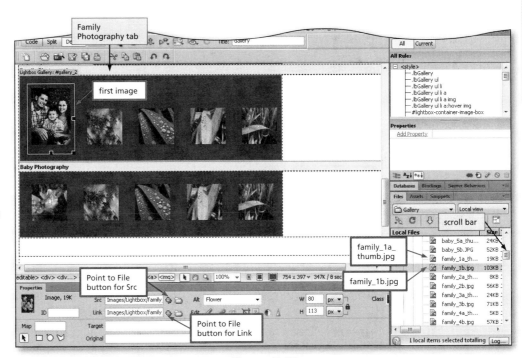

Figure 5–84

2

- Click the second image on the Family Photography tab to select the image.

- Click the Point to File button next to the Src text box, and then drag the pointer to family_2a_thumb.jpg in the Files panel to select the image.

- Click the Point to File button next to the Link text box, and then drag the pointer to family_2b.jpg in the Files panel to link to the image (Figure 5–85).

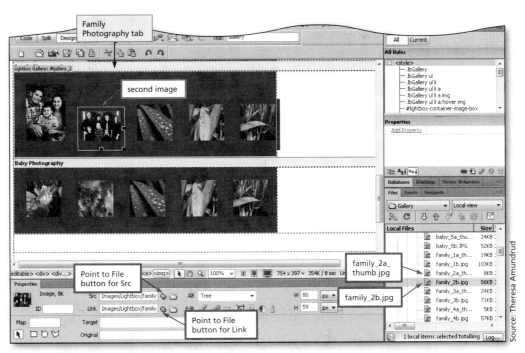

Figure 5–85

Source: Theresa Amundrud

3

- Click the third image on the Family Photography tab to select the image.

- Click the Point to File button next to the Src text box, and then drag the pointer to family_3a_thumb.jpg in the Files panel to select the image.

- Click the Point to File button next to the Link text box, and then drag the pointer to family_3b.jpg in the Files panel to link to the image (Figure 5–86).

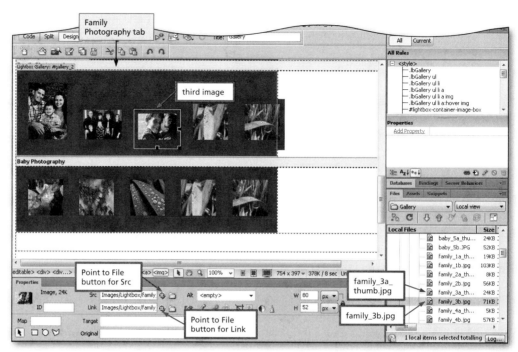

Figure 5–86

4

- Click the fourth image on the Family Photography tab to select the image.

- Click the Point to File button next to the Src text box, and then drag the pointer to family_4a_thumb.jpg in the Files panel to select the image.

- Click the Point to File button next to the Link text box, and then drag the pointer to family_4b.jpg in the Files panel to link to the image.

- Click the fifth image on the Family Photography tab to select the image.

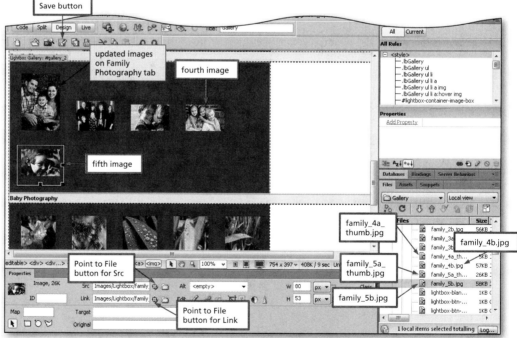

Figure 5–87

- Click the Point to File button next to the Src text box, and then drag the pointer to family_5a_thumb.jpg in the Files panel to select the image.

- Click the Point to File button next to the Link text box, and then drag the pointer to family_5b.jpg in the Files panel to link to the image (Figure 5–87).

5

- Click outside of the widget to deselect the widget and then click the Save button on the Standard toolbar to save your work.

To Add Images to the Baby Photography LightBox Gallery Widget

The Baby Photography LightBox Gallery widget is ready for images. The following steps show you how to replace the current images in the LightBox Gallery widget with images to showcase the Gallery's baby photography.

1

- If necessary, scroll to display the baby files in the Files panel.

- Click the first image on the Baby Photography tab to select it.

- Click the Point to File button next to the Src text box, and then drag the pointer to baby_1a_thumb.jpg in the Files panel to select the image.

- Click the Point to File button next to the Link text box, and then drag the pointer to baby_1b.jpg in the Files panel to link to the image (Figure 5–88).

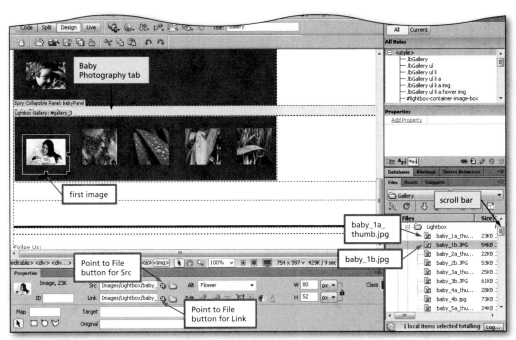

Figure 5–88

2

- Click the second image on the Baby Photography tab to select the image.

- Click the Point to File button next to the Src text box, and then drag the pointer to baby_2a_thumb.jpg in the Files panel to select the image.

- Click the Point to File button next to the Link text box, and then drag the pointer to baby_2b.jpg in the Files panel to link to the image (Figure 5–89).

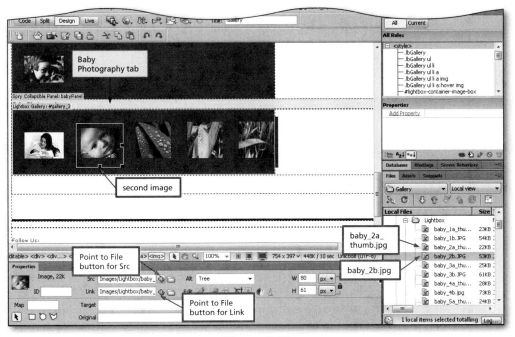

Figure 5–89

- Click the third image on the Baby Photography tab to select the image.

- Click the Point to File button next to the Src text box, and then drag the pointer to baby_3a_thumb.jpg in the Files panel to select the image.

- Click the Point to File button next to the Link text box, and then drag the pointer to baby_3b.jpg in the Files panel to link to the image.

- Click the fourth image on the Baby Photography tab to select the image.

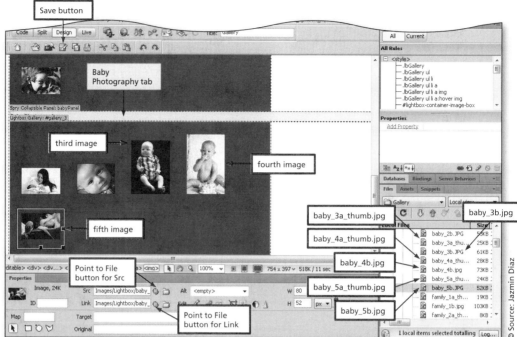

Figure 5–90

- Click the Point to File button next to the Src text box, and then drag the pointer to baby_4a_thumb.jpg in the Files panel to select the image.

- Click the Point to File button next to the Link text box, and then drag the pointer to baby_4b.jpg in the Files panel to link to the image.

- Click the fifth image on the Baby Photography tab to select the image.

- Click the Point to File button next to the Src text box, and then drag the pointer to baby_5a_thumb.jpg in the Files panel to select the image.

- Click the Point to File button next to the Link text box, and then drag the pointer to baby_5b.jpg in the Files panel to link to the image (Figure 5–90).

- Click outside of the widget to deselect it and then click the Save button on the Standard toolbar to save your work.

To View the Collapsible Panels and LightBox Gallery Widget in a Browser

The following steps open the portfolio.html page in Internet Explorer to view the collapsible panels and LightBox Gallery widgets.

1 Click the 'Preview/Debug in browser' button on the Document toolbar to display a list of browsers.

2 Click Preview in IExplore in the 'Preview/Debug in browser' list to display the Gallery Web site in Internet Explorer.

3 If necessary, click the 'Allow blocked content' button to allow the widgets to work properly.

4 Click the Portrait Photography tab to display the first LightBox Gallery widget.

5 Click the first thumbnail on the Portrait Photography tab to display the full-size image (Figure 5–91).

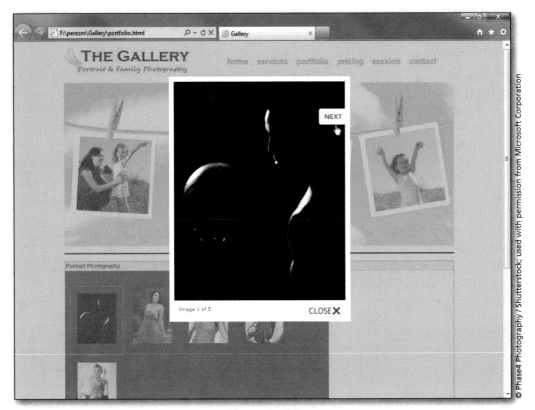

Figure 5–91

6 Click each image to advance to the next image, and then click the last image to close it.

7 Click the Portrait Photography tab to close the Portrait Photography collapsible panel.

8 Click the Family Photography tab to display the second LightBox Gallery widget.

9 Click the first thumbnail on the Family Photography tab to display the full-size image, and then click each image to display it.

10 Click the Family Photography tab to close the Family Photography collapsible panel.

11 Click the Baby Photography tab to display the third LightBox Gallery widget.

12 Click the first thumbnail on the Baby Photography tab to display the full-size image, and then click each image to display it.

13 Close your browser, and then close Dreamweaver.

Chapter Summary

In this chapter, you applied Spry validation to a form on the Contact page using Dreamweaver's Spry validation widgets. You also used Spry to insert Collapsible Panel widgets. You used the Adobe Widget Browser to view widgets within Adobe Exchange and then saved a widget to My Widgets. You added LightBox Gallery widgets to the Spry Collapsible Panels. You also learned how to replace images in the LightBox Gallery widget. You displayed the updated Web pages in a browser and tested the form validation, the collapsible panels, and the LightBox Gallery widget. The following tasks are all the new Dreamweaver skills you learned in this chapter:

1. Open the Contact Web Page and Select the Spry Category on the Insert Bar (DW 275)
2. Add Spry Validation to a Required Text Field (DW 276)
3. Add Spry Validation to Another Required Text Field (DW 279)
4. Add Spry Validation to an E-mail Text Field (DW 281)
5. Add Spry Validation to a Phone Number Text Field (DW 283)
6. Add Spry Validation to Another Required Text Field (DW 284)
7. Add Spry Validation to a Radio Group (DW 287)
8. Test the Spry Validation Form in a Browser (DW 290)
9. Add Spry Collapsible Panels to a Web Page (DW 292)
10. Add Panel Headings (DW 296)
11. Check Your Computer for the Adobe AIR Application (DW 298)
12. Download and Install Adobe AIR (DW 299)
13. Create an Adobe Account (DW 300)
14. View and Add Widgets from Adobe Exchange to My Widgets (DW 303)
15. Create a Widget Preset for the LightBox Gallery Widget (DW 306)
16. Add the LightBox Gallery Widget to a Web Page (DW 311)
17. Add Images to the Portrait Photography LightBox Gallery Widget (DW 314)
18. Add Images to the Family Photography LightBox Gallery Widget (DW 317)
19. Add Images to the Baby Photography LightBox Gallery Widget (DW 319)

Apply Your Knowledge

Reinforce the skills and apply the concepts you learned in this chapter.

Adding Spry Collapsible Panels to a Web Page

Instructions: In this activity, you create a new Web page that contains information about Spry widgets. You add three Spry Collapsible Panels, change the tab headings, and add content to each tab. The completed Web page is shown in Figure 5–92.

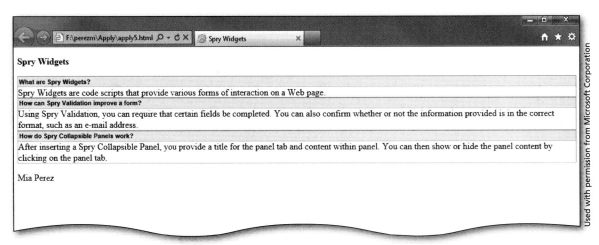

Figure 5–92

Perform the following tasks:

1. Start Dreamweaver. Use the Sites button on the Files panel to display the Web sites created with Dreamweaver and the drives on your computer. Select the Apply site.

2. Create a new HTML file and name it `apply5.html`.

3. Title the document `Spry Widgets`.

4. Type `Spry Widgets` at the top of the page and bold the heading.

5. Press the ENTER key and insert a Spry Collapsible Panel widget.

6. Use `spryPanel1` as the Collapsible panel name on the Property inspector, set the Default state to Closed and set the Display to Closed.

7. Replace the Tab name, Tab, with `What are Spry Widgets?`. Answer this question, using your own words, in the content area of spryPanel1.

8. Click below spryPanel1 and insert another Spry Collapsible Panel widget.

9. Use `spryPanel2` as the Collapsible panel name and set the Default state to Closed.

10. Replace the Tab name, Tab, with `How can Spry Validation improve a form?`. Answer this question, using your own words, in the content area of spryPanel2.

11. Click below spryPanel2 and insert another Spry Collapsible Panel widget.

12. Use `spryPanel3` as the Collapsible panel name and set the Default state to Closed.

13. Replace the Tab name, Tab, with `How do Spry Collapsible Panels work?`. Answer this question, using your own words, in the content area of spryPanel3.

14. Include your name below the third panel.

15. Save your changes and copy dependent files. View your document in your browser and click each panel to view its content.

16. Compare your document to Figure 5–92. Make any necessary changes and then save your changes.

17. Submit the document in the format specified by your instructor.

Extend Your Knowledge

Extend the skills you learned in this chapter and experiment with new skills. You may need to use Help to complete the assignment.

Adding Spry Tabbed Panels

Instructions: In this activity, you create a new Web page with information about Spry Tabbed Panels. You add Spry Tabbed Panels, change the tab headings, and add content to each tab. The completed Web page is shown in Figure 5–93.

Figure 5–93

Continued >

Extend Your Knowledge *continued*

Perform the following tasks:

1. Start Dreamweaver. Use the Sites button on the Files panel to select the Extend site.

2. Create a new HTML file and name it `extend5.html`.

3. Title the document `Spry Tabbed Panels`.

4. Type `Using Spry Tabbed Panels` at the top of the page and bold the heading.

5. Press the ENTER key and use the Insert bar to insert a Spry Tabbed Panels widget.

6. Replace the Tab name, Tab 1, with `Step 1`. In the content area of the Step 1 tab, replace the text, Content 1, with `Place your insertion point where you want to insert the Spry Tabbed Panels and then click the Spry Tabbed Panels button on the Insert bar` to describe how to insert the widget.

7. Replace the Tab name, Tab 2, with `Step 2`.

8. Click the eye-open button to show panel content on the Step 2 tab. In the content area of the Step 2 tab, replace the text, Content 2, with `Rename each tab and insert content within the content area of each tab. If necessary, use the Property inspector to add additional tabbed panels` to insert the Step 2 text.

9. Include your name below the panels.

10. Save your changes and copy dependent files. View your document in your browser and click each panel to view its content.

11. Compare your document to Figure 5–93. Make any necessary changes and then save your changes.

12. Submit the document in the format specified by your instructor.

Make It Right

Analyze a Web page and suggest how to improve its design.

Adding Spry Validation to a Web Page Form

Note: To complete this assignment, you will be required to use the Data Files for Students. Visit www.cengage.com/ct/studentdownload for detailed instructions or contact your instructor for information about accessing the required files.

Instructions: In this activity, you modify a Web page form using Spry Validation widgets. You apply Spry validation to four text fields and change the validation properties for each text field. The completed Web page is shown in Figure 5–94.

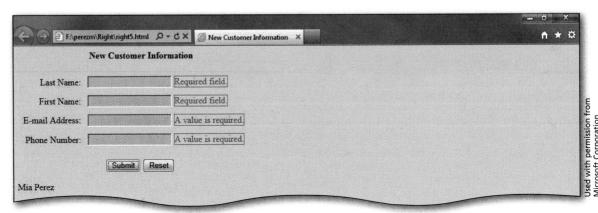

Figure 5–94

Perform the following tasks:

1. Use Windows Explorer to copy the right5.html file from the Chapter 05\Right folder provided with your data files into the *your last name and first initial*\Right folder.

2. Start Dreamweaver and open right3.html. If necessary, click the Spry tab on the Insert bar.

3. Click the Last Name text field and use the Spry Validation Text Field button to add validation to this field.

4. In the Property inspector, identify the Spry TextField as `lastName`. Set the Preview states as Required.

5. Select the text, A value is required, that was inserted after the lastName Spry Validation Text Field and then type `Required field`.

6. Click the First Name text field and use the Spry Validation Text Field button to add validation to this field.

7. In the Property inspector, identify the Spry TextField as `firstName`. Set the Preview states as Required.

8. Select the text, A value is required, that was inserted after the firstName Spry Validation Text Field and delete it. Type `Required field`.

9. Click the E-mail Address text field and use the Spry Validation Text Field button to add validation to this field.

10. In the Property inspector, identify the Spry TextField as `Email`. Set the Type to Email Address.

11. Click the Phone Number text field and use the Spry Validation Text Field button to add validation to this field.

12. In the Property inspector, identify the Spry TextField as `phoneNumber`. Set the Type to Phone Number. If necessary, set the Format as US/Canada and then click the Enforce pattern check box.

13. Include your name below the form.

14. Save your changes and copy dependent files. View the document in your browser and click the Submit button to test the Spry validation.

15. Compare your document to Figure 5–94. Make any necessary changes, and then save your changes.

16. Submit the document in the format specified by your instructor.

In the Lab

Design and/or create a document using the guidelines, concepts, and skills presented in this chapter. Labs are listed in order of increasing difficulty.

Lab 1: Modifying the Sign Up Form with Spry Validation

Problem: You are creating an internal Web site for your company that features information about how to live a healthy lifestyle. Employees at your company will use this Web site as a resource for nutrition, exercise, and other health-related tips. You have created a form on the Sign Up Web page where employees will enter information to sign up for weekly health tips. Now you need to modify the form with Spry validation.

First, modify the First Name, Last Name, and E-mail text fields with Spry Validation Text Fields. Next, format the properties for each field. Finally, remove the current Radio Group and replace it with a Spry Validation Radio Group. The updated Sign Up page is shown in Figure 5–95.

Continued >

In the Lab *continued*

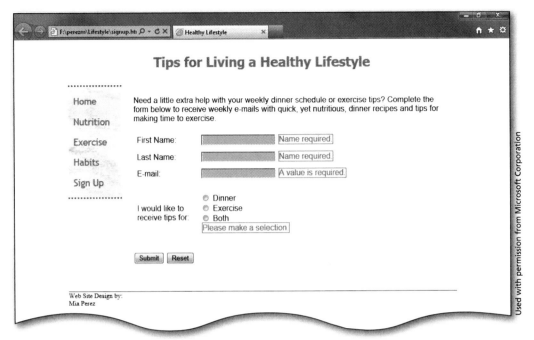

Figure 5–95

Perform the following tasks:

1. Start Dreamweaver. Use the Sites button on the Files panel to select the Healthy Lifestyle site. Open signup.html.

2. Click the First Name text field and use the Spry Validation Text Field button to add validation to this field.

3. In the Property inspector, identify the Spry TextField as `firstName`.

4. Click the Preview states box arrow and then select Required. Select the text, A value is required, that was inserted after the firstName Spry Validation Text Field and delete it. Type `Name required`.

5. Click the Last Name text field and use the Spry Validation Text Field button to add validation to this field.

6. In the Property inspector, identify the Spry TextField as `lastName`.

7. Click the Preview states box arrow and then select Required. Select the text, A value is required, that was inserted after the lastName Spry Validation Text Field and delete it. Type `Name required`.

8. Click the E-mail Address text field and use the Spry Validation Text Field button to add validation to this field.

9. In the Property inspector, identify the Spry TextField as `Email`. Click the Type box arrow and then select Email Address.

10. Delete the current Radio Group and insert a Spry Validation Radio Group.

11. Use `Tips` as the Radio Group name. Use `Dinner` and `Exercise` as the first two Labels. Add one more radio button and use `Both` for the Label name.

12. Save your changes and copy dependent files. View the document in your browser and click the Submit button to test the Spry validation.

13. Compare your document to Figure 5–95. Make any necessary changes, and then save your changes.

14. Submit the document in the format specified by your instructor.

In the Lab

Lab 2: Modifying the Contact Us Form with Spry Validation

Problem: You are creating a Web site for Designs by Dolores, a Web site design company. The site provides information about the company and its services. In Chapter 4, you created a form on the Contact Us page for the Designs by Dolores Web site. Now you need to modify the form by adding Spry validation.

First, modify the First Name, Last Name, E-mail, and Phone Number text fields by applying Spry Validation Text Fields. Next, format the properties for each field. The updated Contact Us page is shown in Figure 5–96.

Figure 5–96

Perform the following tasks:

1. Start Dreamweaver. Use the Sites button on the Files panel to select the Designs by Dolores site. Open contact.html.

2. Modify the First Name text field by applying a Spry Validation Text Field.

3. In the Property inspector, identify the Spry TextField as `firstName`.

4. Select Required for the Preview states. Select the text, A value is required, that was inserted after the firstName Spry Validation Text Field and delete it. Type `Please provide your first name.`

5. Modify the Last Name text field by applying a Spry Validation Text Field.

6. In the Property inspector, identify the Spry TextField as `lastName`.

7. Select Required for the Preview states. Select the text, A value is required, that was inserted after the lastName Spry Validation Text Field and delete it. Type `Please provide your last name.`

8. Modify the E-mail text field by applying a Spry Validation Text Field.

9. In the Property inspector, identify the Spry TextField as `Email`. Use Email Address for the Type.

Continued >

In the Lab *continued*

10. Modify the Phone Number text field by applying a Spry Validation Text Field.

11. In the Property inspector, identify the Spry TextField as `phoneNumber`. Use Phone Number for the Type. Click the Enforce pattern check box.

12. Save your changes and copy dependent files. View the document in your browser and click the Submit button to test the Spry validation.

13. Compare your document to Figure 5–96. Make any necessary changes, and then save your changes.

14. Submit the document in the format specified by your instructor.

In the Lab

Lab 3: Modifying the Quote Form with Spry Validation and Adding a Calendar Widget

Problem: You are creating a Web site for a new lawn care company, Justin's Lawn Care Service. The Web site will provide information to customers, including descriptions of services and pricing. In Chapter 4, you developed a form for the Quote page. Now you need to modify the form by adding Spry Validation.

First, modify the First Name, Last Name, Phone Number, and E-mail text fields by applying Spry Validation Text Fields. Next, format the properties for each field. The updated Quote page is shown in Figure 5–97.

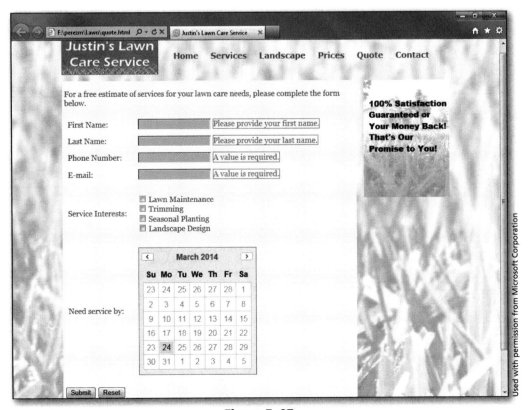

Figure 5–97

Perform the following tasks:

1. Start Dreamweaver. Use the Sites button on the Files panel to select the Justin's Lawn Care Service site. Open quote.html.

2. Modify the First Name text field by applying a Spry Validation Text Field. Use `firstName` as the Spry TextField.

3. Select Required for the Preview states and use `Please provide your first name` as the Required text.

4. Modify the Last Name text field by applying a Spry Validation Text Field. Use `lastName` as the Spry TextField.

5. Select Required for the Preview states and use `Please provide your last name` as the Required text.

6. Modify the Phone Number text field by applying a Spry Validation Text Field. Use `phoneNumber` as the Spry TextField.

7. Use Phone Number for the Type and enforce the pattern.

8. Modify the E-mail text field by applying a Spry Validation Text Field. Use `Email` as the Spry TextField.

9. Select Email Address for the Type.

10. Add a row below Service Interests. (*Hint*: On the Insert menu, select Table Objects, and then select Insert Row Below.)

11. Type `Need Service by:` in the first column of the new row.

12. Open the Adobe Widget Browser. Browse Adobe Exchange for the YUI Calendar for Widget Browser. If this widget is not available on Adobe Exchange, select a different calendar widget. Add the widget to My Widgets.

13. Add the YUI Calendar for Widget Browser in the second column of the last row.

14. Save your changes and copy dependent files. View the document in your browser and click the Submit button to test the Spry validation. Click a date on the calendar to test the widget.

15. Compare your document to Figure 5–97. Make any necessary changes, and then save your changes.

16. Submit the document in the format specified by your instructor.

Cases and Places

Apply your creative thinking and problem solving skills to design and implement a solution.

1: Adding Spry Validation to the Contact Page for Moving Venture Tips

Personal

You have created a form on the Contact page for Moving Venture Tips and now need to add Spry validation to the form fields. Review the form elements added to the Contact page and determine which Spry validation widgets should be integrated. Update text fields to Spry Validation Text Fields and update the property type for e-mail to Email Address. Remove the existing Radio Group and add a Spry Validation Radio Group. Check the spelling using the Commands menu, and correct all misspelled words. Save and open the Contact page in your Web browser to test the Spry validation fields. Submit the document in the format specified by your instructor.

Continued >

Cases and Places *continued*

2: Adding Spry Validation to the Contact Page for Student Campus Resources

Academic

You have created a form on the Contact page for Student Campus Resources and now need to add Spry validation to the form fields. You also have created the Committees page and now need to add Spry Collapsible Panels and content. Review the form elements added to the Contact page and determine which Spry validation widgets should be integrated. Update text fields to Spry Validation Text Fields and update the property type for e-mail to Email Address. Save and open the Contact page in your Web browser to test the Spry validation fields. Add four Spry Collapsible Panels to the Committees page. Give each panel heading a committee name and provide content about the committee within each panel content area. Name each Spry Collapsible Panel on the Property inspector and set the Default state to Closed. Save and open the Committees page in your Web browser to review the panels. Check the spelling using the Commands menu, and correct all misspelled words. Submit the document in the format specified by your instructor.

3: Adding Web Pages, Images, and Links for French Villa Roast Café

Professional

You have created a form on the Rewards page for French Villa Roast Café and now need to add Spry validation to the form fields. You also have created the Contact page and now need to add a widget to the page. Review the form elements added to the Rewards page and determine which Spry validation widgets should be integrated. Update text fields to Spry Validation Text Fields. Update the property type for e-mail to Email Address. Update the property type for phone number to Phone Number and enforce the pattern. Remove the existing Radio Group and add a Spry Validation Radio Group. Save and open the Rewards page in your Web browser to test the Spry validation fields. Add the Facebook Like Button widget to the Contact page below the contact information. Save and open the Contact page in your Web browser to view the widget. Check the spelling using the Commands menu, and correct all misspelled words. Submit the document in the format specified by your instructor.

6 | Enhancing Web Pages with Audio and Video

Objectives

You will have mastered the material in this chapter when you can:

- Describe how to add multimedia using plug-ins and HTML5

- Explain audio codecs

- Embed a sound file in a Web page

- Set the parameters of an audio file

- Add HTML5 code to play an audio file

- Describe video formats

- Embed a video file in a Web page

- Set the parameters of a video file

- Add HTML5 code to play a video file

6 | Enhancing Web Pages with Audio and Video

Introduction

A captivating Web site is about more than text and information; it is also a medium for expressing artistic creativity and optimizing the presentation of information. Web pages can include media objects, also called multimedia and interactive media, to enhance the experience for users. On multimedia Web pages, users can play audio files such as music, sound effects, or speech and video files to augment the content on the page. HTML5 adds multimedia capabilities such as audio and video to basic HTML pages.

Listening to music on the Web is a popular pastime. Audio files can include a podcast of a breaking news story posted to a Web site, a broadcast of an Internet radio station, background music that sets a mood for the Web page, or a song downloaded from the latest album of a hot new music group. The Web also includes video files that showcase product demos, professional broadcasts, music videos, prerecorded television clips, and marketing videos, such as one that provides a tour of a scenic vacation destination.

In this chapter, you learn how to add interactive media to a Web page. Interactive media is not suitable for all Web sites. You most likely have visited Web sites that contained so many bits and pieces of multimedia that, instead of being a constructive part of the site, distracted you from the experience the Web designers intended. Therefore, when adding media elements to your Web page, you need to consider the value of each element and how it will enhance or detract from your site.

Project — Web Pages with Multimedia

The Gallery photography studio has decided to follow the current trend in digital marketing by adding multimedia elements to enhance the visual material on its site. Adding audio and video to a Web page can be accomplished in two ways in most modern browsers. Older browsers use plug-ins to play audio and video files. Newer browsers use HTML5, which supports video and audio tags that allow users to play video and audio files without an external plug-in or player. The chapter project adds audio and video files using both the plug-in and HTML5 methods.

The Portfolio page of the Gallery site shown in Figure 6–1a plays an audio file with soft background music to create a mood that can enhance a visitor's browsing experience. To bring the Gallery's photos to life, a video slide show plays in the Session page, as shown in Figure 6–1b. Professional, polished multimedia in small doses adds entertainment value to the Gallery site while showcasing additional photos in the video.

(a) The Portfolio page plays an audio file

(b) The Session page plays a video file

Figure 6–1

Overview

As you read this chapter, you will learn how to create the Web page project shown in Figure 6–1 by performing these general tasks:

- Add sound to a Web page using a plug-in.
- Set parameters for the audio plug-in.
- Add sound to a Web page using HTML5 code.
- Add video to a Web page using a plug-in.
- Set parameters for the video plug-in.
- Add video to a Web page using HTML5 code.

General Project Guidelines

When adding interactive multimedia to a Web site, consider the purpose and audience of the site. As you create the Web pages shown in Figure 6–1, you should follow these general guidelines:

1. **Consider adding multimedia files to a Web page.** Web pages can be more engaging and useful with multimedia elements such as audio and video.

2. **Add sound to a Web page.** If spoken language or music will add an entertainment or instructional value, consider adding sound to a Web page, such as the splash page, which is the page that opens when most visitors enter your site. You can use sound elements to guide visitors through your site, especially if you use interactive media controls.

3. **Determine how to embed multimedia files.** Dreamweaver allows you to embed multimedia files using plug-ins or HTML5 code. Select the method that is most appropriate for the purpose and audience of your Web page.

4. **Play video on a Web page.** Video is especially effective for training visitors, displaying demonstrations, or marketing a service.

More specific details about these guidelines are presented at appropriate points throughout the chapter. The chapter also identifies the actions performed and decisions made regarding these guidelines during the development of the pages within the site shown in Figure 6–1.

To Start Dreamweaver and Open the Gallery Site

Each time you start Dreamweaver, it opens to the last site displayed when you closed the program. The following steps start Dreamweaver and open the Gallery Web site.

1 Click the Start button on the Windows 7 taskbar to display the Start menu, and then type `Dreamweaver CS6` in the 'Search programs and files' box.

2 Click Adobe Dreamweaver CS6 in the list to start Dreamweaver.

3 If the Dreamweaver window is not maximized, click the Maximize button next to the Close button on the Application bar to maximize the window.

4 If the Gallery site is not displayed in the Files panel, click the Sites button on the Files panel toolbar and then click Gallery to display the files and folders in the Gallery site.

Adding Multimedia with Plug-ins or HTML5

Recently, the computing world has changed from one in which people use a single device — primarily a PC — to one in which they use several platforms such as Web-enabled tablets and smartphones. Now that the user experience spans many devices, Web standards play a more important role than ever, for both users and developers.

HTML5 Web standards within the newest browsers are replacing older technologies that did not support mobile devices with limited resources. For example, an iPhone or iPad does not support Web pages that display Flash animation, which requires a plug-in. A **plug-in** is a piece of software that acts as an add-on to a Web browser and provides additional functionality to the browser. Plug-ins can allow a browser to display additional content it was not originally designed to display. Recent browsers that run on traditional computers and mobile devices do not support legacy plug-ins, but instead rely on the new HTML5 Web technology to play audio and video elements. Plug-ins provide support for particular data types, which allows the site visitor to view specific documents, watch animation, or play multimedia files within a Web browser without opening an outside program.

Consider adding multimedia files to a Web page.
Multimedia files can enhance Web pages by providing engaging experiences. Visitors can listen to music that highlights the mood of the page or play a speech related to the Web page content. In addition to audio content, visitors can watch videos that inform, instruct, or entertain them.

    To add multimedia files using Dreamweaver, you can insert a link to an audio or video file, or you can embed an audio or video file in the Web page. If you link to a multimedia file, you reduce the bandwidth your Web site must manage. (Video files in particular use more bandwidth than other types of files.) Embedding a multimedia file offers the advantage of letting visitors play the file without leaving the Web page. Web designers also can specify how the embedded multimedia file plays in the browser: automatically when the page opens, or only when the user clicks the Play button in the playback controls.

Plan Ahead

    Older browsers require users to install a plug-in before the browsers can display media objects. For instance, the Flash Player is required to view Flash files. The Flash Player is available as an add-on for Firefox and as an ActiveX control for Microsoft Internet Explorer. Adobe provides this player as a free download at the Adobe Web site. Other popular plug-ins include QuickTime Player to play movie files and Acrobat Reader to display PDF files.

    In this chapter, you add an audio file and a video file as plug-ins for legacy browsers, and then use HTML5 code to play multimedia files on newer browsers for cross-platform functionality. Developers face the possibility that many users are still using legacy browsers while others are using the most recent browsers. Quickly, HTML5 is becoming the prominent Web standard as the next wave of innovation pushes the boundary toward a world without plug-ins. This chapter provides instruction for adding media objects using plug-ins and HTML5 to meet the needs of your varied audience.

Adding Sound

Adding sound to your Web page can add interest and set a mood. Background music can enhance your theme. Attaching audio to objects, such as the sound of clicking buttons, can provide valuable feedback to the Web site visitor. Sound, however, can be expensive in terms of the Web site size and download time, particularly for those visitors with slower access. Another consideration is that adding sound to a Web page can be a challenging and confusing task. Most computers today have some type of sound card and some type of sound-capable plug-in that works with the browser. The difficult part, though, is generating desirable sounds.

    As you work with sound, you will discover that the Web supports several popular sound file formats, along with their corresponding plug-ins, but has not settled on a standard audio file format. The trick is to find a widely used audio file format that most browsers handle in the same way. Some factors to consider before deciding on a format and method for adding sound are the sound's purpose, your audience, the file size, the sound quality you want, and the differences between browsers.

Plan
Ahead

Add sound to a Web page.
As with any other technology, you first determine what you want to accomplish by adding a sound. Your sound should serve a clear and useful purpose or have an entertaining or instructional value. In most instances, when you add a background sound, keep the sound short, do not loop the sound, and provide your visitor with an option to stop the sound. Based on which browser you are using, a sound file is placed within a Web page using two methods:

- **Add sound using a plug-in.** Sound embedded into a Web page that supports plug-ins can be set to play automatically as the page loads, or a user can click on audio controls to start the audio file on command.

- **Add sound using HTML5 code.** HTML5 defines a new element for modern and mobile browsers that specifies a standard way to embed an audio file in a Web page: the <audio> element. The audio HTML5 element allows you to deliver audio files directly through the browser, without the need for a plug-in.

Understanding Audio Codecs

A **codec** is a program that allows you to read and save media files, often compressing them to reduce file size. For example, a codec can compress large audio files such as WAV files into a much smaller MP3 audio file type. All codecs involve a trade-off between the amount of compression and the resulting quality. If you compress a media file too much, the quality of the sound or video becomes poor and unusable. If you do not compress a media file enough, the file may take too long to download into a browser. To play an MP3 file that has been compressed with the MP3 codec, a computer must use the codec to also decompress the file locally to play the audio file.

Table 6–1 lists and describes the commonly used sound file formats.

Table 6–1 Common Sound File Formats

File Name Extension	Description
.aif (Audio Interchange File Format, or AIFF)	The AIFF format can be played by most browsers and does not require a plug-in; you also can record AIFF files from a CD, tape, microphone, and so on. The large file size, however, limits the length of sound clips that you can use on your Web pages.
.midi (Musical Instrument Digital Interface, or MIDI)	The MIDI format is for instrumental music. MIDI files are supported by many browsers and do not require a plug-in. The sound quality is good, but it can vary depending on a visitor's sound card. A small MIDI file can provide a long sound clip. MIDI files cannot be recorded and must be synthesized on a computer with special hardware and software.
.mp3 (Moving Picture Experts Group Audio, or MPEG Audio Layer-3, or MP3)	The MP3 format is a compressed format that allows for small sound files with very good sound quality. You can stream an MP3 file so that a visitor does not have to wait for the entire file to download before hearing it. To play MP3 files, visitors must download and install a plug-in such as QuickTime, Windows Media Player, or RealPlayer.
.ogg, .ogv, .oga, .ogx, .spx (Ogg container format)	The Ogg multimedia framework supports audio, video, text, and metadata. The Ogg file type is primarily associated with the Ogg Vorbis software by the Xiph.Org Foundation. Ogg is an open and standardized format designed for streaming and manipulating media files. Ogg's various codecs have been incorporated into many free and proprietary media players.
.qt, .qtm, .mov (QuickTime)	These formats are designed for audio and video files by Apple. QuickTime is included with the Mac operating system, and most Mac applications can use files in these formats. PCs also can play media in the QuickTime format, though they need a QuickTime driver to do so.
.ra, .ram, .rpm (RealAudio)	The RealAudio format has a very high degree of compression, with smaller file sizes than the MP3 format. Whole song files can be downloaded in a reasonable amount of time. The files also can be streamed, so visitors can begin listening to the sound before the file has been downloaded completely. The sound quality is poorer than that of MP3 files. Visitors must download and install the RealPlayer helper application or plug-in to play these files.
.wav (Waveform Audio File Format, or WAV)	WAV-formatted files have good sound quality, are supported by many browsers, and do not require a plug-in. The large file size, however, severely limits the length of sound clips that you can use on your Web pages.

To Copy the Audio and Video Folders

Before adding an audio or video file to a site, you must add the multimedia files to the file structure of the site. To complete the following steps, you will be required to use the Data Files for Students. Visit www.cengage.com/ct/studentdownload for detailed instructions or contact your instructor for information about accessing the required files. The following steps show how to copy files from the Data Files for Students to the Gallery site.

1 Use Windows Explorer to navigate to the storage location of the Data Files for Students.

2 Double-click the Chapter 06 folder, and then double-click the Gallery folder to open the folders.

3 Click the Audio folder, hold down the shift key, and then click the Video folder to select both folders (Figure 6–2).

4 Right-click the selected folders, click Copy on the context menu, and then navigate to the *your last name and first initial* folder on Removable Disk F: to prepare to copy the files.

5 Double-click the Gallery folder, right-click anywhere in the open window, and then click Paste on the context menu to copy the folders into the Gallery site.

6 Click the background of the window to deselect the folders. Verify that the Gallery site now contains an Audio folder and a Video folder (Figure 6–3).

Figure 6–2

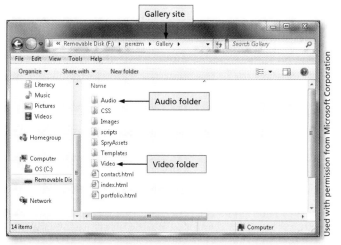

Figure 6–3

Embedding a Sound File Using a Plug-In

To incorporate a background sound file directly into a Web page, you can embed the audio file using Dreamweaver. However, the sound plays only if visitors to your site have the appropriate plug-in. **Embedding** a sound file means that the music plays in a browser with an installed plug-in that supports the codec of the audio file. Audio files can be played on legacy browsers and some present-day browsers using third-party browser plug-ins such as Flash, QuickTime, and Silverlight.

To Embed a Background Sound Using a Plug-In

When the Portfolio page opens in the Gallery site, the photographer has requested that a soft background song set the mood as the visitor views the photographs within the collapsible panels. The portfolio_audio.mp3 song can begin automatically when the page is launched and play one time for 1 minute and 37 seconds (the length of the music file). The following steps embed a song that plays using a browser plug-in.

1

• In the Files panel, double-click portfolio.html to open the Portfolio page.

• Click below the Baby Photography Collapsible Panel to place the insertion point below the Collapsible Panel (Figure 6–4).

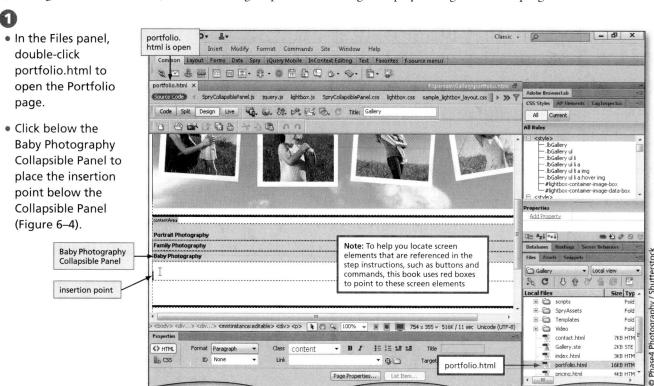

Figure 6–4

2

- If necessary, click the Common tab on the Insert bar to display the Common category of buttons.

- Click the Media button arrow on the Insert bar to display the Media menu (Figure 6–5).

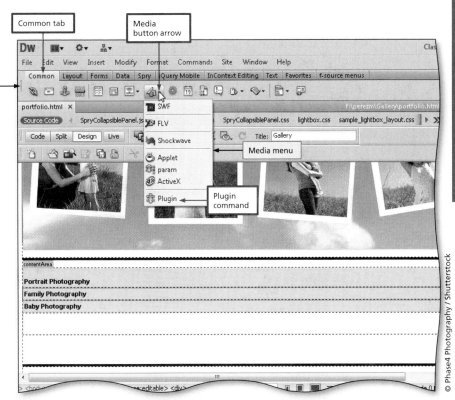

Figure 6–5

3

- Click Plugin to display the Select File dialog box.

- Double-click the Audio folder to display the audio files.

- Click portfolio_audio.mp3 to select the sound file for the Portfolio page (Figure 6–6).

Q&A

Why does the appearance of the Media button change after I select Plugin?

The Media button displays an icon for the last media object you selected. For example, after you select Plugin, the Media button displays a Plugin icon. If you want to insert another Plugin object, you can click the Media button instead of the Media button arrow.

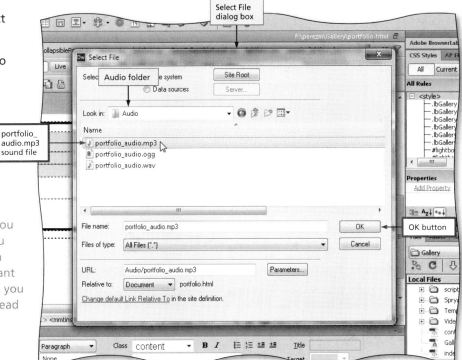

Figure 6–6

- Click the OK button to insert a placeholder for the background sound file (Figure 6–7).

Q&A

Why does Dreamweaver insert a placeholder for the sound file?

When you add a background sound to a page, Dreamweaver inserts a placeholder as a cue that the page includes a media object.

placeholder for background sound

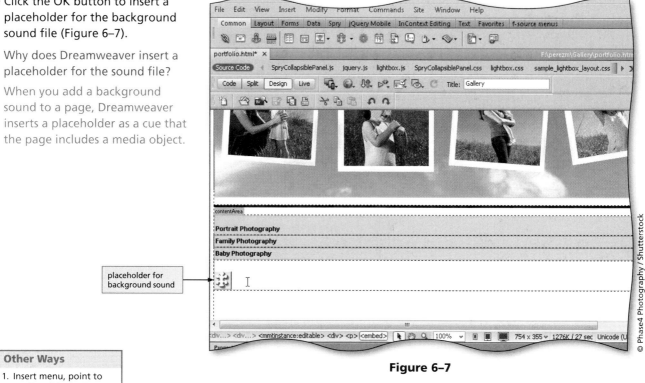

© Phase4 Photography / Shutterstock

Figure 6–7

Other Ways

1. Insert menu, point to Media, click Plugin

Setting the Parameters of a Sound File

After embedding an audio file as a plug-in file, you can define parameters to play the file automatically when the site loads, display a controller that provides a stop button, or loop the song continuously. (A parameter is similar to a property; it defines the characteristics of object.) You also can define the default volume of the sound or the music so that it plays softly in the background, and determine how long to play the song. You set the parameters of embedded sound files in the Property inspector. Table 6–2 lists frequently used parameters for the audio plug-in.

Table 6–2 Audio Plug-In Parameters	
Parameter	**Description**
autoplay	Defines whether the player should start automatically
hidden	Informs the browser not to display the audio controls
loop	Informs the browser to play the audio file continuously

To Set the Parameters of a Sound File

When sound is integrated into the Portfolio page, you can determine how the song plays within the browser. Set the parameters of the plug-in in the Property inspector to specify the source file, set the background sound to play automatically, and hide the audio controls. The following steps set the parameters of the audio plug-in object.

1

• If necessary, select the sound placeholder.

• Click the Parameters button in the Property inspector to display the Parameters dialog box (Figure 6–8).

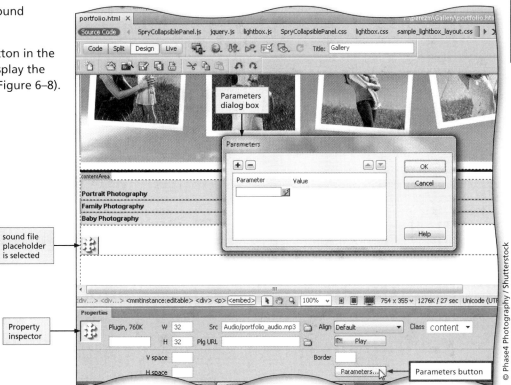

Figure 6–8

2

• Type `id` to enter the name of the first audio parameter.

• Press the TAB key twice to move the insertion point to the Value column.

• Type `portfolio_audio` in the Value column to specify the file name of the embedded sound object (Figure 6–9).

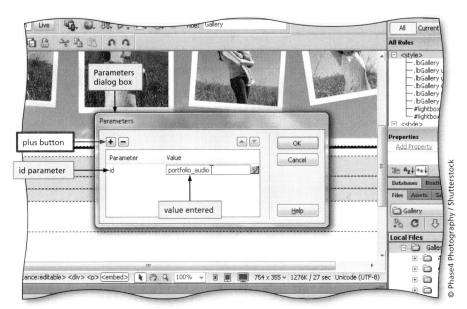

Figure 6–9

3

- Click the plus button and then type `hidden` to enter the name of the second parameter.

- Press the TAB key twice to move the insertion point to the Value column.

- Type `true` in the Value column to specify that the audio controls are hidden in the Web page (Figure 6–10).

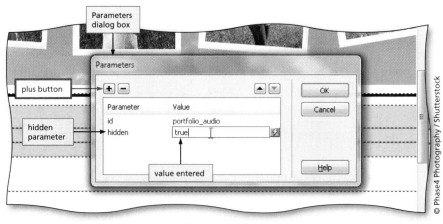

Figure 6–10

© Phase4 Photography / Shutterstock

4

- Click the plus button and then type `autoplay` to enter the name of the third parameter.

- Press the TAB key twice to move the insertion point to the Value column.

- Type `true` in the Value column to specify that the audio automatically plays when the Web page opens (Figure 6–11).

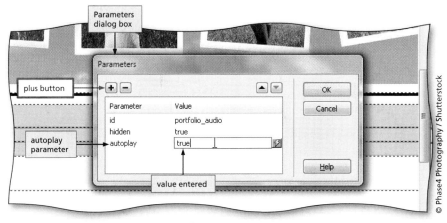

Figure 6–11

© Phase4 Photography / Shutterstock

5

- Click the plus button and then type `loop` to enter the name of the fourth parameter.

- Press the TAB key twice to move the insertion point to the Value column.

- Type `false` in the Value column to specify that the sound file plays one time only (Figure 6–12).

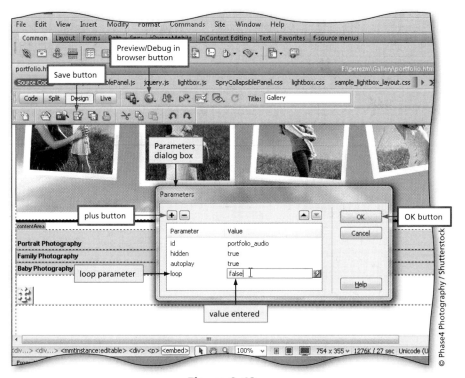

Figure 6–12

© Phase4 Photography / Shutterstock

6

- Click the OK button to apply the parameters to the background sound file.

- Click the Save button on the Standard toolbar to save your changes.

- Click the 'Preview/ Debug in browser' button on the Document toolbar and then click Preview in IExplore in the browser list to display the Portfolio page in Internet Explorer.

- If necessary, click the 'Allow blocked content' button to play the audio file (Figure 6–13).

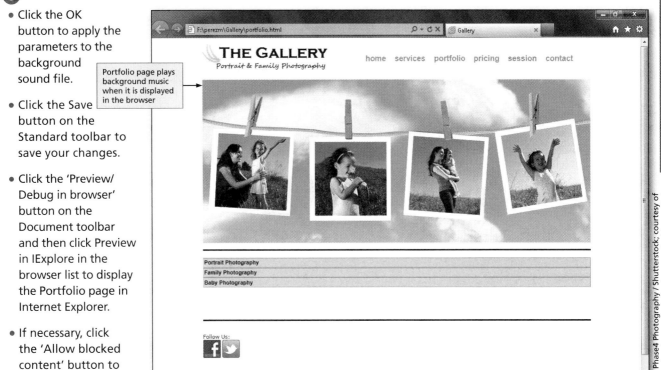

Portfolio page plays background music when it is displayed in the browser

Figure 6–13

© Phase4 Photography / Shutterstock; courtesy of Facebook; courtesy of Twitter; used with permission from Microsoft Corporation

Q&A What should happen when I open the Portfolio page in a browser?

The music file should automatically play one time.

Q&A The audio file seems to play, but I don't hear any sound. What is the problem?

Check to see if the volume on your speakers is muted in the notification area on a Windows computer. On a Mac, click the Volume icon on the menu bar and then drag the slider.

7

- Click the Internet Explorer Close button to close the browser.

Adding an Audio File with HTML5 Code

When you design a Web site, you cannot control which browser visitors use to view your site. Browsers translate the HTML5 code, allowing you to read text, view images, play videos, and listen to audio clips on Web sites, though each browser renders the Web page content in a different way.

A typical browser supports multiple audio file types, regardless of the operating system running it, but no universal audio or video file type is supported by all browsers. As an HTML5 developer, you need to make sure that audio content is created in different formats that each browser supports. Table 6–3 displays the audio formats supported by the major browsers using HTML5 code.

Table 6–3 HTML5 Audio Formats	
Browser	**Supported Audio Format**
Apple Safari	MP3, WAV
Google Chrome	MP3, OGG, WAV
Microsoft Internet Explorer	MP3
Mozilla Firefox	MP3, WAV
Opera	OGG, WAV

Because one audio file type is not supported by all browsers, you should develop your audio content in at least two formats such as OGG and MP3 in order to cover the major browser types. The browser listing shown in Table 6–3 may change as HTML5 gains popularity and more software development companies use it.

Coding HTML5 Audio

BTW

Audio and Video Controls
Each browser renders audio and video controls with a different visual layout, but the usual controls, such as play/pause button, position slider, current track position indicator, and volume control, are all present.

HTML5 media tags for audio and video media elements aim to standardize elements on a page and to reduce the use of plug-ins. To embed sounds, music, and other audio on a Web page, Web developers can use the HTML5 audio element, which is supported by Internet Explorer 9 and 10, Firefox, Opera, Chrome, and Safari. In the audio element, you can include the *controls* attribute to display a control bar that viewers use to play the sound or the music file. Following the audio element, use the HTML5 source element to list more than one format for the audio file so it plays in multiple browsers. Three audio formats are compatible with the audio element: .mp3, .wav, and .ogg. To play one audio file, include three source elements: one for each audio format. The first source element typically specifies the file in the MP3 format.

To add HTML5 audio code to a Web page, you can work in Code view. By typing the following HTML5 code in Code view, a modern browser can play the portfolio_audio.mp3, portfolio_audio.wav, or portfolio_audio.ogg audio file stored in the Audio folder without using plug-ins or without a third-party service hosting the audio files.

```
<audio controls autoplay="autoplay" loop>
<source src="Audio/portfolio_audio.mp3" type="audio/mp3"/>
<source src="Audio/portfolio_audio.wav" type="audio/wav"/>
<source src="Audio/portfolio_audio.ogg" type="audio/ogg"/>
</audio>
```

The five lines of HTML5 code play the portfolio_audio sound files as follows:

Line 1: Audio controls will be displayed with a Play button and a Pause button. The autoplay attribute sets the audio to start playing as soon as the file is available. Because "autoplay" is the default value for the autoplay attribute, you can omit it and still have the browser play the audio file automatically. The loop attribute specifies that the audio starts playing again immediately after it finishes.

Line 2: The source element specifies the name, location, and type of the MP3 file to play. The src attribute identifies the path and name of the audio file.

Line 3: The source element specifies the name, location, and type of the WAV file to play. The src attribute identifies the path and name of the audio file.

Line 4: The source element specifies the name, location, and type of the OGG file to play. The src attribute identifies the path and name of the audio file.

Line 5: This statement identifies the closing tag of the HTML5 audio element.

To Add an Audio File Using HTML5 Code

Insert the HTML5 audio element between the opening <body> and closing </body> tags in the HTML code for the Portfolio page. The following steps add HTML5 code to the Web page to include an audio file with controls on the page.

- Click the placeholder for the audio file plug-in on the Portfolio page and then press the DELETE key to delete it (Figure 6–14).

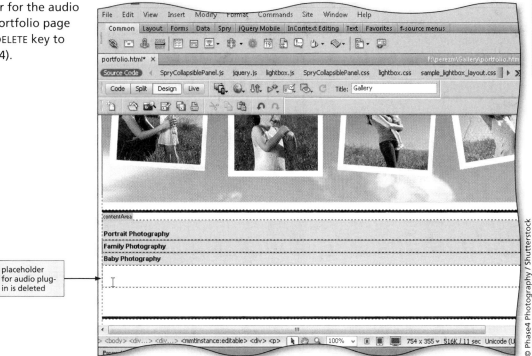

placeholder for audio plug-in is deleted

© Phase4 Photography / Shutterstock

Figure 6–14

- Click the Code button on the Standard toolbar to display the HTML code for the Portfolio page (Figure 6–15).

Q&A

The insertion point is in a different place in my code than it is in Figure 6–15. What should I do?

Move the insertion point so it appears where shown in Figure 6–15.

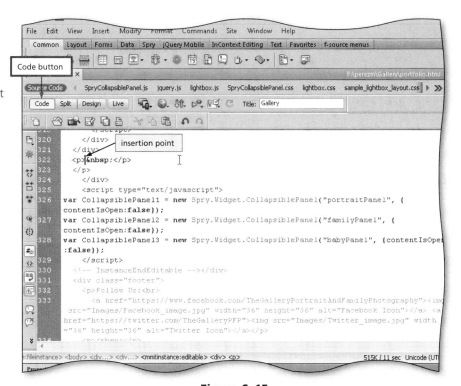

Figure 6–15

3

- Press the ENTER key twice and then click the blank line to begin a new line of HTML5 code.

- Type `<audio controls autoplay="autoplay" loop>` to display the audio controls and to play the song continuously when the browser loads the page in an HTML5-supported browser (Figure 6–16).

code for audio controls

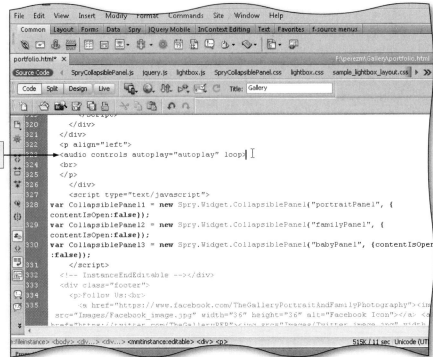

Figure 6–16

4

- Press the ENTER key to begin a new line.

- Type `<source src= "Audio/portfolio_audio. mp3" type="audio/mp3" />` to play the audio file in a browser that supports .mp3 files (Figure 6–17).

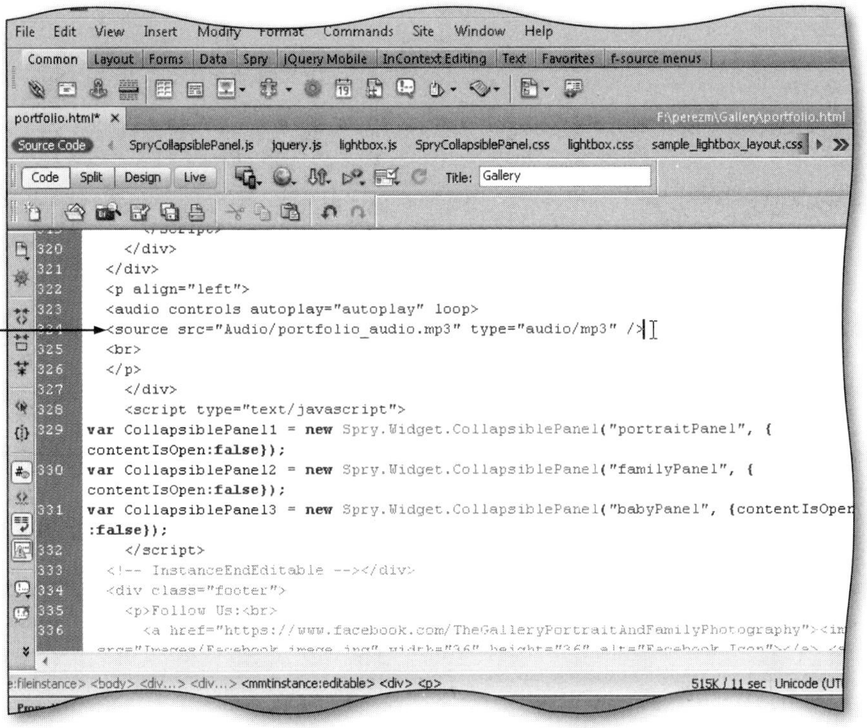

code for .mp3 file

Figure 6–17

5

- Press the ENTER key to begin a new line.

- Type `<source src= "Audio/portfolio_audio. wav" type="audio/wav" />` to play the audio file in a browser that supports .wav files (Figure 6–18).

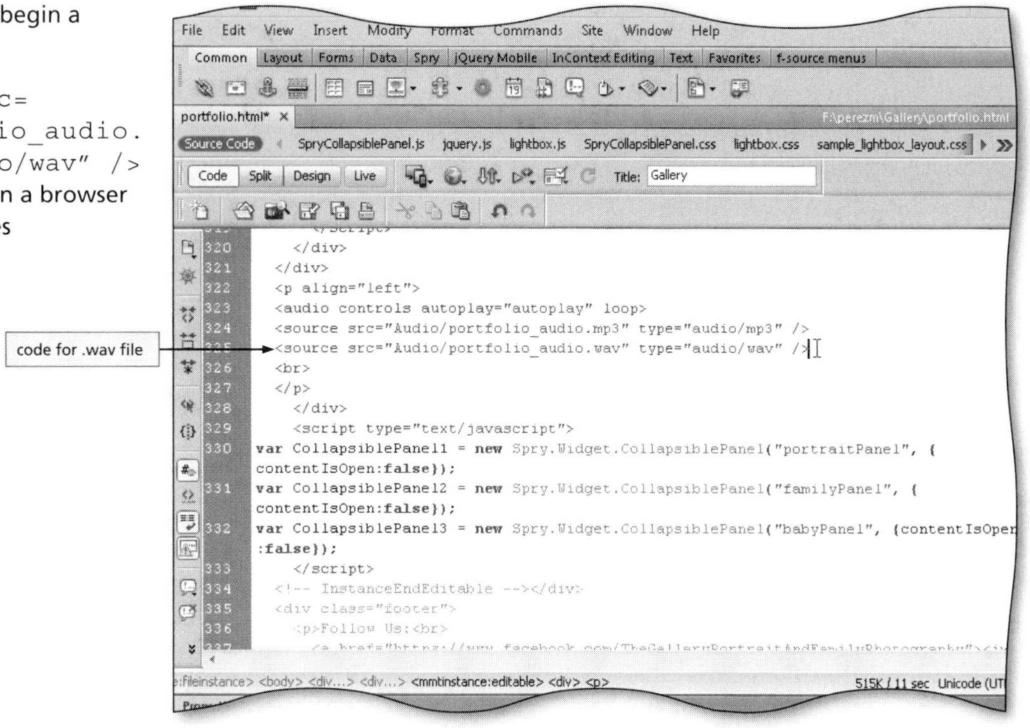

code for .wav file →

Figure 6–18

6

- Press the ENTER key to begin a new line.

- Type `<source src= "Audio/portfolio_audio. ogg" type="audio/ogg" />` to play the audio file in a browser that supports .ogg files (Figure 6–19).

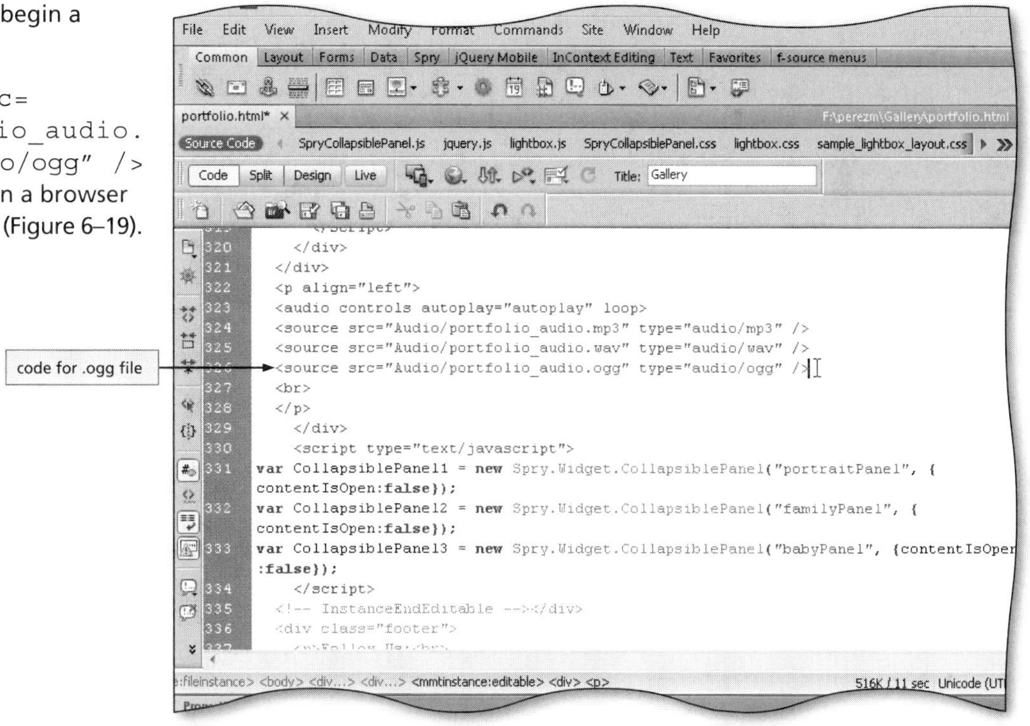

code for .ogg file →

Figure 6–19

7

- Press the ENTER key to begin a new line.

- Type `</audio>` to close the HTML5 audio tag (Figure 6–20).

Q&A

As I began to type </audio> to close the audio HTML5 code tag, why did Dreamweaver complete the tag for me?

When you establish an opening tag and begin to insert a closing tag, Dreamweaver automatically provides the text of the closing tag.

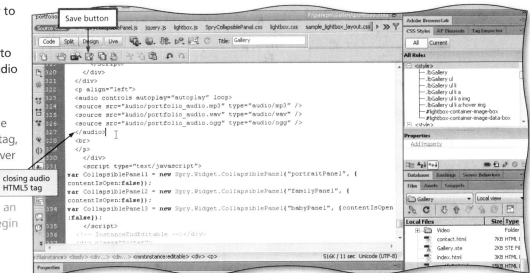

Figure 6–20

8

- Click the Save button on the Standard toolbar to save your changes.

To Display an HTML5 Audio Control in Different Browsers

When you include the HTML5 audio element in a Web page, each browser displays the audio playback control including a Play/Pause button, the time, a Mute button, and volume controls. In most situations, it is a best practice to display audio controls so users can mute the sound. The audio controls look different in various browsers. The following steps display the audio control in the Portfolio page in various browsers.

1

- Click the 'Preview/ Debug in browser' button on the Document toolbar and then click Preview in IExplore in the browser list to display the Portfolio page.

- If necessary, click the 'Allow blocked content' button to view the audio controls in the Internet Explorer browser (Figure 6–21).

Figure 6–21

2

- Click the Internet Explorer Close button to close the browser.

- Click the 'Preview/ Debug in browser' button on the Document toolbar, and then click Preview in Firefox in the browser list to display the Portfolio page and view the audio controls in the Firefox browser (Figure 6–22).

Q&A When I click the 'Preview/Debug in browser' button, Firefox is not included in my list of browsers. What should I do?

Figure 6–22

Portfolio page in Firefox

audio control

If Firefox is installed on your computer, click Edit Browser List on the 'Preview/Debug in browser' list, click the plus button, enter the requested information in the Add Browser dialog box, click the OK button to close the dialog box, and then click the OK button to add Firefox to the list of browsers. If you do not have the Firefox browser installed on your computer, however, you cannot test the page using Firefox.

3

- Click the Firefox Close button to close the browser.

- Click the 'Preview/ Debug in browser' button on the Document toolbar, and then click Preview in chrome in the browser list to display the Portfolio page and view the audio controls in the Chrome browser (Figure 6–23).

Q&A When I click the 'Preview/Debug in browser' button, Chrome is not included in my list of browsers. What should I do?

Figure 6–23

Portfolio page in Chrome

audio control

If Chrome is installed on your computer, follow the instructions provided in Step 2, substituting Chrome for Firefox. If you do not have the Chrome browser installed on your computer, however, you cannot test the page using Chrome.

Should I install multiple browsers on my computer for testing purposes?

Yes. Each browser behaves differently, so you should test your projects in multiple browsers.

Experiment

- Allow the song to play all the way to the end. The song begins again because of the loop statement. Remove the loop statement from the HTML5 code. Save the code and refresh the page. The song only plays one time. Reinsert the loop statement and then save the document.

4

- Click the Chrome Close button to close the browser.

To Remove an HTML5 Audio Control and Loop

Audio on a Web page that plays repeatedly or too often can annoy visitors and people nearby them. With this in mind, the Gallery owners requested that you hide the audio controls on the Portfolio page and set the music to play only once. The following steps remove the audio control and the loop from the Portfolio page.

1

- If necessary, click the Code button on the Standard toolbar to return to Code view.

- Scroll down to find the audio tag, and then delete the word, controls, from the audio tag to remove the audio controls.

- Delete the word, loop, from the audio tag to remove the audio loop (Figure 6–24).

2

- Click the Close button on the portfolio.html tab to close the Portfolio page.

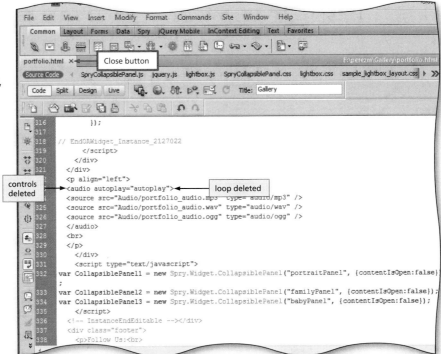

Figure 6–24

Break Point: If you wish to take a break, this is a good place to do so. To resume at a later time, start Dreamweaver and continue following the steps from this location forward.

Adding Video

By adding an appropriate amount of video to a Web page, you can entice your Web site visitors to learn more about your product and enrich their experience. Video clips should be only long enough to convey your meaning, and short enough so the video content loads quickly and creates a positive reaction from your customers. Each multimedia object should showcase your creativity and make your site stand apart from the average Web site.

Plan Ahead

Determine how to embed multimedia files.
As with audio files, you can embed video files by using a plug-in or using HTML5 code. The advantage of using a plug-in is that Dreamweaver CS6 provides tools for inserting plug-in objects without requiring you to enter HTML code. However, browsers are moving away from supporting plug-ins. So if you include plug-in objects, eventually your Web page will not be able to play the multimedia in the new browsers.

Instead of requiring plug-ins, browsers support HTML5 for multimedia objects. Embedding audio and video files by using HTML5 code means users can play the files in current and future browsers. However, embedding multimedia files with this method requires you to enter HTML5 code.

Understanding Video Formats

Many types of video formats such as MPEG-4 (.mp4), Windows Media (.wmv), Flash (.swf and .flv), RealVideo (.rm and .ram), and QuickTime (.mov) display full-motion video, which you can use to promote the message of your Web site. The MP4 format is the new and upcoming format for Web-based video. It is supported by YouTube, Flash players, and HTML5. Table 6–4 describes the various video formats.

BTW

Use Design Notes with Media Objects
If you want to store extra information with a media object, such as the name or location of the source file, you can add Design Notes to a media object. Right-click the media object in the document window, click Design Notes for Page on the context menu, and then enter the note.

Table 6–4 Video File Formats

File Name Extension	Description
.avi (Audio Video Interleave, or AVI)	AVI is supported by all computers running Windows.
.mov (QuickTime, developed by Apple)	QuickTime is a common format on the Internet for both Windows and Macintosh computers. The QuickTime Player is required to play QuickTime movies on a Windows computer.
.mpg or .mpeg (Moving Pictures Expert Group, or MPEG)	Files in the MPEG format run on Windows and Macintosh computers.
.mp4 (MPEG-4)	MPEG-4 (with H.264 video compression) is the new format for the Web. Online video publishers are moving to MP4 as the Internet sharing format for both Flash players and HTML5.
.rm or .ram (RealVideo, developed by RealMedia)	The RealVideo format allows the streaming of video, which reduces downloading time.
.swf (Flash SWF)	Files in the SWF format require the Adobe Flash Player.
.wmv, .wvx (Windows Media Format, or WMF)	The WMF format allows the streaming of video, uses the Windows Media Player, and plays on both Windows and Macintosh computers.

Embedding a Video File Using a Plug-In

To add rich media to a Web page, you can insert video files by using a plug-in or by entering HTML5 code. The most popular video plug-in is for the Flash player, which supports Flash videos such as animated text, business presentations, and visual effects and MP4 video files. The MP4 file type is the preferred modern video type that transmits over a narrower bandwidth compared to other video file types. The best format for smartphones and slate devices is MP4.

To Embed a Video Using a Plug-In

The Session page should open with an MP4 video named session_video.mp4, which features a guide to a Gallery photography session. The video highlights samples of the Gallery's photography. The following steps embed a video that plays using a browser plug-in.

1

- In the Files panel, double-click session. html to open the Session page.

- If necessary, click the Design button on the Document toolbar to display the page in Design view.

- Select the text, Insert page content here., and then press the DELETE key to delete the text (Figure 6–25).

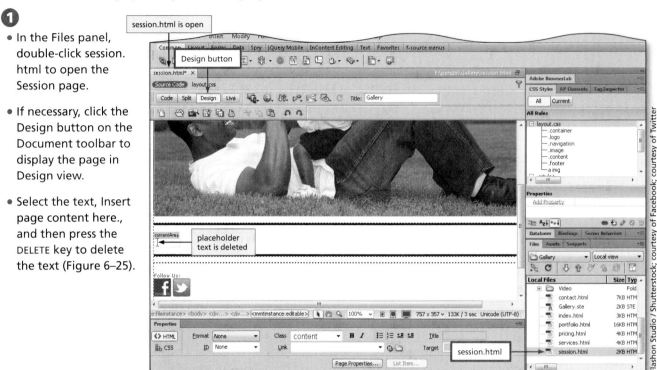

Figure 6–25

2

- Click the Media button arrow on the Insert bar to display the Media menu (Figure 6–26).

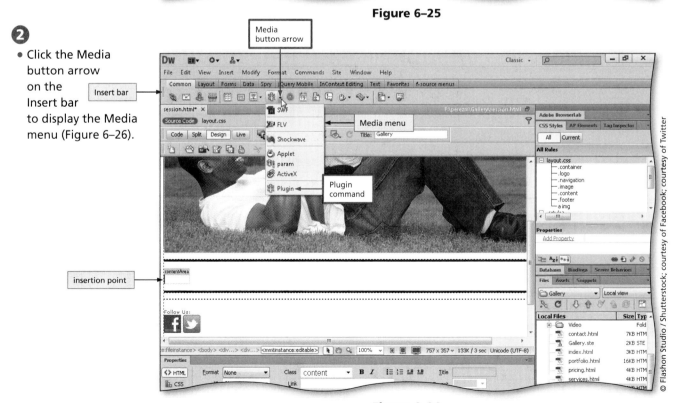

Figure 6–26

- Click Plugin to display the Select File dialog box.

- Double-click the Video folder to display the video files.

- Click the Up One Level button to navigate to the Gallery folder.

- Click session_video.mp4 to select the video file for the Session page (Figure 6–27).

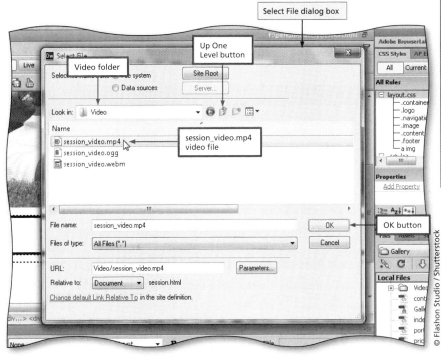

Figure 6–27

④

- Click the OK button to insert a 32 × 32 pixel placeholder for the video (Figure 6–28).

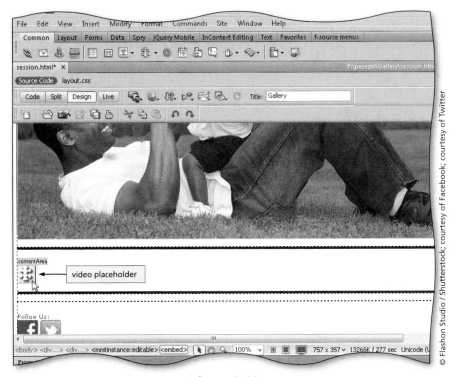

Figure 6–28

• Double-click the W text box in the Property inspector and then type 4 8 0 to set the width of the video placeholder.

• Double-click the H text box in the Property inspector and then type 3 7 5 to set the height of the video placeholder.

• Press the ENTER key to resize the video placeholder (Figure 6–29).

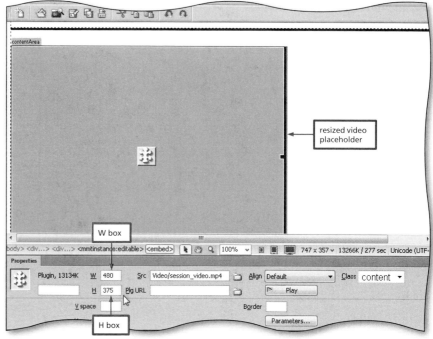

Figure 6–29

Other Ways

1. Insert menu, point to Media, click Plugin

To Set the Parameters of a Video File

The title of the session_video is Gallery Photography Session Guide. It should not start playing on the Session page until the visitor clicks the Play button in the video controls. The default setting for video plug-ins is to play the video automatically when the page opens. To allow the user to control the playback, you set the autoplay parameter to false. The following steps set the parameter of the video plug-in object.

• If necessary, select the video placeholder on the Session page.

• Click the Parameters button in the Property inspector to open the Parameters dialog box (Figure 6–30).

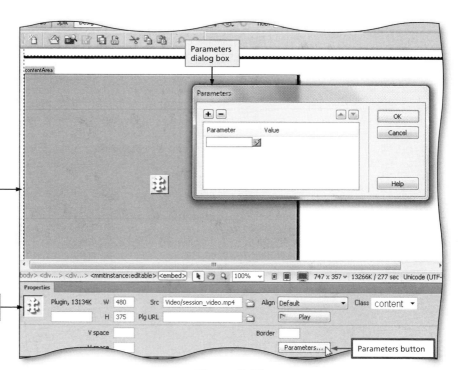

Figure 6–30

2

- Type `autoplay` to enter the name of the first parameter.

- Press the TAB key twice to move the insertion point to the Value column.

- Type `false` in the Value column to specify that the video must be started by the site visitor (Figure 6–31).

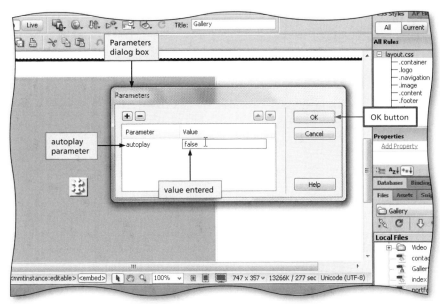

Figure 6–31

3

- Click the OK button to apply the parameter to the video file.

- With the placeholder selected, click Format on the Application bar to display the Format menu.

- Point to Align on the Format menu to display the Align submenu (Figure 6–32).

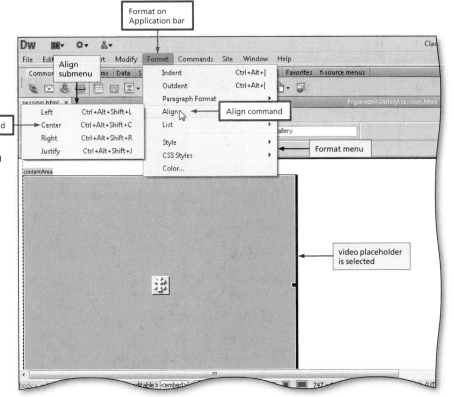

Figure 6–32

- Click Center on the Align submenu to center the video placeholder (Figure 6–33).

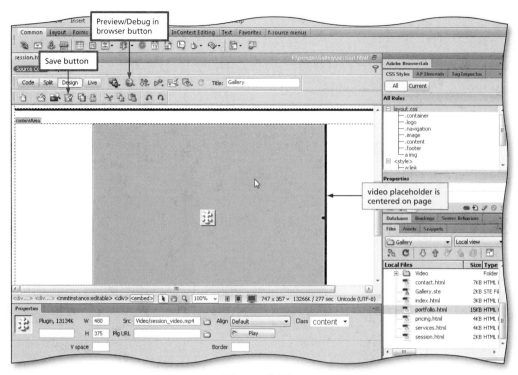

Figure 6–33

- Click the Save button on the Standard toolbar to save your changes.

- Click the 'Preview/ Debug in browser' button on the Document toolbar, and then click Preview in IExplore in the browser list to display the Session page in Internet Explorer.

- If necessary, click the 'Allow blocked content' button to give permission for playing the video file.

- Scroll down and click the Play button in the video controls to play the video (Figure 6–34).

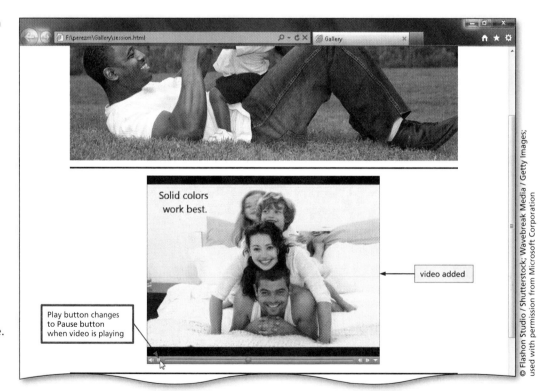

Figure 6–34

- Click the Internet Explorer Close button to close the browser.

Adding a Video File with HTML5 Code

In the same fashion as the audio file, modern browsers may not support the use of video plug-ins, but instead rely on the HTML5 video element. Table 6–5 lists the HTML5 video formats supported by the major browsers.

Table 6–5 HTML5 Video Formats	
Browser	**Supported Video Format**
Apple Safari 5	MP4
Google Chrome 6	MP4, WebM, Ogg
Microsoft Internet Explorer 9	MP4
Mozilla Firefox 4	WebM, Ogg
Opera 10.6	WebM, Ogg

Play video on the Web page.
You can add video to your Web page by having the browser download the video file to the user or by streaming the video content so that it plays while the file is downloading. Older browsers require their Web site visitors to download a plug-in to view common streaming formats such as Windows Media, RealMedia, and QuickTime. You then can use parameters to control the video object and how it is played on the Web page. Newer browsers support an HTML tag named <video> to play most video file types.

Plan
Ahead

Coding HTML5 Video

As with the HTML5 audio element, you can embed a video on a Web page using the HTML5 video element, which is supported by Internet Explorer 9 and 10, Firefox, Opera, Chrome, and Safari. In the video element, you can include the *controls* attribute to display a control bar that viewers use to control the playback of the video file. Following the video element, use the HTML5 source element to list more than one format for the video file so it plays in multiple browsers. Three video formats are compatible with the video element: .mp4, .webm, and .ogg. To play one video file, include three source elements — one for each video format. The first source element typically specifies the file in the MP4 format.

The following code supports the three video file formats used in various browsers.

BTW

Video Size
If the file size of a video is too large, it can take a long time to download and play when it is displayed in a Web page. You should therefore limit the size of your videos without sacrificing their quality. By optimizing the video's file size, the video can play quickly and smoothly.

```
<video width="480" height="360" controls autoplay="autoplay">
<source src="Video/session_video.mp4" type="video/mp4"/>
<source src="Video/session_video.webm" type="video/webm"/>
<source src="Video/session_video.ogg" type="video/ogg"/>
</video>
```

The five lines of HTML5 code play the session_video video files as follows:
Line 1: The video appears in a video player with a width of 480 pixels and a height of 360 pixels. Video controls will be displayed with a Play button and a Pause button. The autoplay attribute sets the video to start playing as soon as the file is available. As with audio files, "autoplay" is the default value for this attribute, so you can omit the value to still have the video play when the page is opened.

Displaying a Video in Live View
If QuickTime is installed on your computer, you can preview an embedded video in Live view.

Line 2: The source element specifies the name, location, and type of the MP4 file to play. The src attribute identifies the path and name of the audio file.

Line 3: The source element specifies the name, location, and type of the WebM file to play. The src attribute identifies the path and name of the audio file.

Line 4: The source element specifies the name, location, and type of the OGG file to play. The src attribute identifies the path and name of the audio file.

Line 5: This statement identifies the closing tag of the HTML5 video element.

To Add a Video File Using HTML5 Code

The Session page can use HTML5 code to embed a video file that any modern browser can play. The following steps add HTML5 code to the Web page to play a video file in a player with playback controls.

1

• Select the video file placeholder in the Session page and then press the DELETE key to delete the plug-in (Figure 6–35).

placeholder for video plug-in is deleted

© Flashon Studio / Shutterstock

Figure 6–35

2

• Click the Code button on the Standard toolbar to display the HTML code for the Session page (Figure 6–36).

Figure 6–36

3

- Press the ENTER key twice and then click the blank line to begin a new line of code.

- Type `<video width="480" height="360" controls autoplay="autoplay">` to display the video controls and to play the video automatically when the browser loads the page in an HTML5-supported browser (Figure 6–37).

Q&A Why do I need to specify width and height values for the video?

As the video loads in a browser, the width and height values determine how much space to reserve for the video in the player.

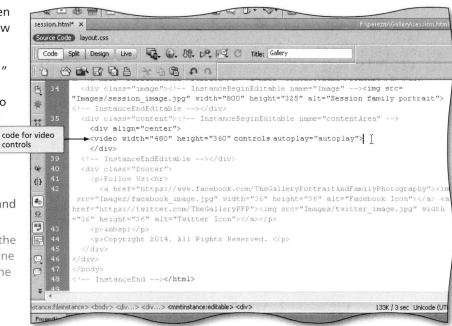

code for video controls

Figure 6–37

4

- Press the ENTER key to begin a new line.

- Type `<source src="Video/session_ video.mp4" type="video/ mp4" />` to play the video file in a browser that supports .mp4 files (Figure 6–38).

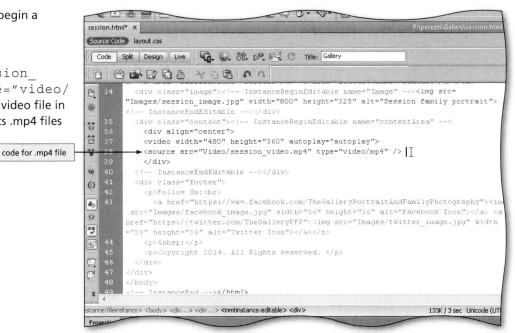

code for .mp4 file

Figure 6–38

5

- Press the ENTER key to begin a new line.

- Type `<source src="Video/ session_video.webm" type="video/webm" />` to play the video file in a browser that supports .webm files (Figure 6–39).

code for .webm file

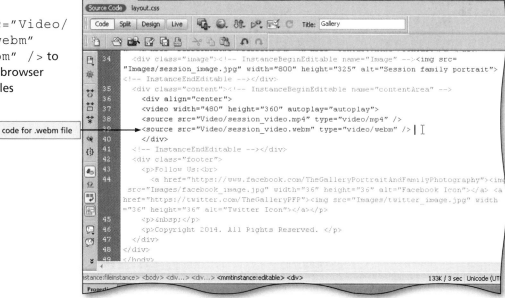

Figure 6–39

6

- Press the ENTER key to begin a new line.

- Type `<source src= "Video/session_video. ogg" type="video/ogg" />` to play the video file in a browser that supports .ogg files.

- Press the ENTER key to begin a new line.

- Type `</video>` to close the video HTML5 code tag (Figure 6–40).

code for .ogg file

closing video tag

Save button

Preview/Debug in browser button

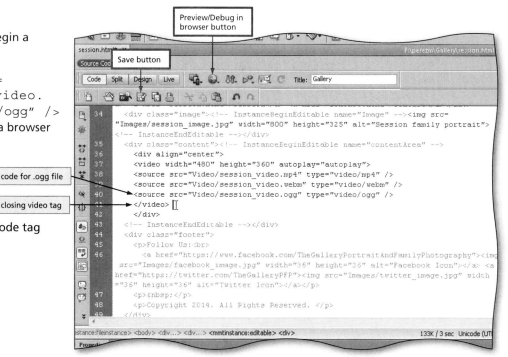

Figure 6–40

7

- Click the Save button on the Standard toolbar to save your changes.

- Click the 'Preview/ Debug in browser' button on the Document toolbar, and then click Preview in IExplore in the browser list to display the Session page in the Internet Explorer browser.

- If necessary, click the 'Allow blocked content' button to display the video controls and play the video (Figure 6–41).

 Why did the video play automatically?

In the HTML5 code, the video was set to autoplay.

© Flashon Studio / Shutterstock; Wavebreak Media / Getty Images; courtesy of Jazmin Diaz; used with permission from Microsoft Corporation

Figure 6–41

8

- Click the Internet Explorer Close button to close the browser.

- Click the Design button on the Document toolbar to display the page in Design view.

To Quit Dreamweaver

The following steps quit Dreamweaver and return control to the operating system.

1 Click the Close button on the right side of the Application bar to close the Dreamweaver window.

2 If Dreamweaver displays a dialog box asking you to save changes, click the No button.

Chapter Summary

In this chapter, you learned about Dreamweaver's audio and video multimedia objects. You added an audio file to the Portfolio page using both plug-ins and HTML5 code. You also customized the audio parameters of the background music. Next, you added a video to the Session page by first using a plug-in and then entering HTML5 code. The following tasks are all the new Dreamweaver skills you learned in this chapter:

1. Copy the Audio and Video folders (DW 337)
2. Embed a Background Sound Using a Plug-in (DW 338)
3. Set the Parameters of a Sound File (DW 341)
4. Add an Audio File Using HTML5 Code (DW 345)
5. Display an HTML5 Audio Control in Different Browsers (DW 348)
6. Embed a Video Using a Plug-in (DW 352)
7. Set the Parameters of a Video File (DW 354)
8. Add a Video File Using HTML5 Code (DW 358)

Apply Your Knowledge

Reinforce the skills and apply the concepts you learned in this chapter.

Adding a Video as a Plug-In

Note: To complete this assignment, you will be required to use the Data Files for Students. Visit www.cengage.com/ct/studentdownload for detailed instructions or contact your instructor for information about accessing the required files.

Instructions: In this activity, you insert a video as a plug-in and use the Property inspector to modify the width and height of the video. The completed Web page is displayed in Figure 6–42.

Figure 6–42

Perform the following tasks:

1. Copy the apply6.html file and the Video folder provided with your Data Files from the Chapter 06\Apply folder into the *your last name and first initial*\Apply folder.

2. Start Dreamweaver. Use the Sites button on the Files panel toolbar to select the Apply site. Open apply6.html.

3. Insert a blank line below the page heading, Vacation in Paradise, and then insert a plug-in on the blank line.

4. Navigate to the Video folder and select the sunset video file. Use the Property inspector to set the width to 720 and the height to 500.

5. Replace the text, Your name here, with your first and last names.

6. Save your changes and then view your document in your browser. If necessary, allow blocked content. Compare your document to Figure 6–42. Make any necessary changes, and then save your changes.

7. Submit the document in the format specified by your instructor.

Extend Your Knowledge

Extend the skills you learned in this chapter and experiment with new skills. You may need to use Help to complete the assignment.

Researching Audio and Video Converter Software

Instructions: In this activity, you research various audio and video converter software and write a summary of available converters.

Perform the following tasks:

1. Using your browser and favorite search engine, conduct some research to find software for at least two audio converters and two video converters.

2. Use a word processing program such as Microsoft Word to list each converter software you found. Include the name of the software, the cost of the software, the types of file format conversions that the software will perform, and the URL for each converter.

3. Save the document with the file name Extend6_*lastname_firstname*, using your first and last names as indicated. Submit the document in the format specified by your instructor.

Make It Right

Analyze a Web page and suggest how to improve its design.

Adding Video Using the HTML5 Video Tag

Note: To complete this assignment, you will be required to use the Data Files for Students. Visit www.cengage.com/ct/studentdownload for detailed instructions or contact your instructor for information about accessing the required files.

Instructions: In this activity, you insert the HTML5 video tag and three source file formats to accommodate various browsers. You also set the video to play automatically and make the controls visible on the Web page. The completed Web page is displayed in Figure 6–43.

Continued >

Make It Right *continued*

Figure 6–43

Perform the following tasks:

1. Copy the right6.html file and the Video folder provided with your Data Files from the Chapter 06\Right folder into the *your last name and first initial*\Right folder.

2. Start Dreamweaver. Use the Sites button on the Files panel toolbar to select the Right site. Open right6.html.

3. Place the insertion point at the end of the Web page heading, Tips for Slicing Carrots, and then press the ENTER key to insert a new line.

4. Change to Code view.

5. Delete the text, , and then press the ENTER key twice.

6. In the blank line, insert the video tag. Use 480 for the width and 320 for the height. Include the controls and autoplay attributes within the tag to show the controls and to automatically play the video on the Web page.

7. Include three source tags — one for each type of video file format (.mp4, .webm, and .ogg) — to play the right6 video file stored in the Video folder. Identify each type of video format used within the source tag.

8. Add the closing video tag after the last source tag. Change to Design view.

9. Replace the text, Your name here, with your first and last names.

10. Save your changes and then view your document in your browser. Compare your document to Figure 6–43. Make any necessary changes, and then save your changes.

11. Submit the document in the format specified by your instructor.

In the Lab

Design and/or create a document using the guidelines, concepts, and skills presented in this chapter. Labs are listed in order of increasing difficulty.

Lab 1: Adding a Video as a Plug-In

Note: To complete this assignment, you will be required to use the Data Files for Students. Visit www.cengage.com/ct/studentdownload for detailed instructions or contact your instructor for information about accessing the required files.

Problem: You are creating an internal Web site for your company that features information about how to live a healthy lifestyle. Employees at your company will use this Web site as a resource for nutrition, exercise, and other health-related tips. In Chapter 2, you created the Exercise Web page. Now you need to add a video to the page. You need to insert a video as a plug-in and then define its width and height. The completed Exercise page is shown in Figure 6–44.

Spotmatik / Getty Images; used with permission from Microsoft Corporation

Figure 6–44

Perform the following tasks:

1. Copy the Video folder provided with your Data Files from the Chapter 06\Lifestyle folder into the *your last name and first initial*\Lifestyle folder.

2. Start Dreamweaver. Use the Sites button on the Files panel toolbar to select the Healthy Lifestyle site. Open exercise.html.

3. Delete the text within the editable region, Insert content here. Insert a plug-in.

4. Navigate to the Video folder and select the lifestyle video file. Use the Property inspector to set the width to 480 and the height to 340. Center the video in the editable region.

5. Save your changes and then view your document in your browser. Compare your document to Figure 6–44. Make any necessary changes, and then save your changes.

6. Submit the document in the format specified by your instructor.

In the Lab

Lab 2: Adding Video Using the HTML5 Video Tag

Note: To complete this assignment, you will be required to use the Data Files for Students. Visit www.cengage.com/ct/studentdownload for detailed instructions or contact your instructor for information about accessing the required files.

Problem: You are creating a Web site for Designs by Dolores, a Web site design company. The site provides information about the company and its services. In Chapter 2, you created the Web Hosting page. Now you need to add a video to this page.

In this activity, you insert the HTML5 video tag and specify three source file formats to accommodate various browsers. You also make the video play automatically and make the controls visible on the Web page. The completed Web page is displayed in Figure 6–45.

Perform the following tasks:

1. Copy the Video folder provided with your Data Files from the Chapter 06\Designs folder into the *your last name and first initial*\Designs folder.

2. Start Dreamweaver. Use the Sites button on the Files panel toolbar to select the Designs by Dolores site. Open hosting.html.

3. Delete the text within the editable region, Insert content here. Use the Format menu to center-align the contents of the editable region.

4. Change to Code view and press the ENTER key twice.

5. In the blank line, insert the video tag. Use 480 for the width and 320 for the height. Include the controls and autoplay attributes within the tag to show the controls and to play the video automatically on the Web page.

6. Include three source tags — one for each type of video file format (.mp4, .webm, and .ogg) — for the designs video file in the Video folder. Identify each type of video format used within the source tag.

7. Add the closing video tag after the last source tag. Change to Design view.

8. Save your changes and then view your document in your browser. Compare your document to Figure 6–45. Make any necessary changes, and then save your changes.

9. Submit the document in the format specified by your instructor.

Figure 6–45

In the Lab

Lab 3: Adding Video Using the HTML5 Video Tag

Note: To complete this assignment, you will be required to use the Data Files for Students. Visit www.cengage.com/ct/studentdownload for detailed instructions or contact your instructor for information about accessing the required files.

Problem: You are creating a Web site for Justin's Lawn Care Service, a lawn care company. The site provides information about the company and its services. In Chapter 3, you added content to the Contact page. Now you need to add a video to this page.

In this activity, you insert the HTML5 video tag and specify three source file formats to accommodate various browsers. You also make the video play automatically and make the controls visible on the Web page. The completed Web page is displayed in Figure 6–46.

Anna Solovei / Getty Images; used with permission from Microsoft Corporation

Figure 6–46

Perform the following tasks:

1. Copy the Video folder provided with your Data Files from the Chapter 06\Lawn folder into the *your last name and first initial*\Lawn folder.

2. Start Dreamweaver. Use the Sites button on the Files panel to select the Justin's Lawn Care Service site. Open contact.html.

3. Place the insertion point at the end of the sentence in the editable region and then press the ENTER key. Center-align the new line.

4. Change to Code view, delete the text, , and then press the ENTER key twice.

5. In the blank line, insert the video tag. Use 480 for the width and 320 for the height. Include the controls and autoplay attributes within the tag.

6. Include three source tags — one for each type of video file format (.mp4, .webm, and .ogg) — for the lawn video file in the Video folder. Identify each type of video format used within the source tag.

7. Add the closing video tag after the last source tag. Change to Design view.

8. Save your changes and then view your document in your browser. Compare your document to Figure 6–46. Make any necessary changes, and then save your changes.

9. Submit the document in the format specified by your instructor.

Cases and Places

Apply your creative thinking and problem solving skills to design and implement a solution.

Note: To complete these assignments, you will be required to use the Data Files for Students. Visit www.cengage.com/ct/studentdownload for detailed instructions or contact your instructor for information about accessing the required files.

1: Adding Music as a Plug-In to the Moving Venture Tips Web Site

Personal

You have created all of the pages for Moving Venture Tips and now need to add content and music to the Tips page. Save the Audio folder provided with your Data Files into the *your last name and first initial*\Move folder. Create a page heading and then use an unordered list to make a list of moving tips on the Tips page. Insert an audio plug-in below the list and select the file in the Audio folder provided with your Data Files. Make the plug-in invisible on the page. Check the spelling using the Commands menu, and correct all misspelled words. Submit the document in the format specified by your instructor.

2: Adding Video Using HTML5 to Student Campus Resources

Academic

You have created all of the pages for Student Campus Resources and now can enhance the Home page with a video. You will embed the video file named campus using the HTML5 video tag. Save the Audio folder provided with your Data Files into the *your last name and first initial*\Campus folder. Use 480 for the width and 320 for the height. Show the controls on the page and set the video to autoplay. In the source tags, specify the three file formats of the video files in the Video folder provided with your Data Files: .mp4, .webm, and .ogg. Submit the document in the format specified by your instructor.

3: Adding Audio Using HTML5 to French Villa Roast Café

Professional

You have created all of the pages for French Villa Roast Café and now can enhance the home page with audio. You will embed the audio file named cafe_music using the HTML5 audio tag. Save the Audio folder provided with your Data Files into the *your last name and first initial*\Cafe folder. Do not show the controls. Set the audio to autoplay and loop. In the <source> tags, specify the .mp3 and .ogg files in the Audio folder provided with your Data Files. Submit the document in the format specified by your instructor.

7 Media Objects, Behaviors, and CSS3 Styling

Adobe product screenshot(s) reprinted with permission from Adobe Systems Incorporated

Objectives

You will have mastered the material in this chapter when you can:

- Describe media objects
- Describe Java applets and insert an applet into a Web page
- Describe a Flash movie
- Explain how to insert a Flash movie into a Web page
- Describe a Flash video
- Explain how to insert a Flash video into a Web page
- Describe Shockwave

- Explain how to insert a Shockwave movie into a Web page
- Describe plug-ins and ActiveX controls
- Understand and add behaviors
- Describe CSS3 style properties
- Add CSS3 properties to a Web page
- Add a transition using the CSS Transitions panel

7 | Media Objects, Behaviors, and CSS3 Styling

Introduction

Enhancing the user experience on a Web site through media objects, behaviors, and CSS3 styling attracts new customers and new business. It also provides your Web site with a media-rich look and feel. Media objects, also called multimedia, are files that can be displayed or executed within Web pages and add visual appeal. Examples of media objects include audio, video, Flash objects (SWF), Shockwave objects (DIR or DCR), Adobe Portable Document Format (PDF) files, Java applets, and other objects that can be viewed through a browser.

To include these media objects on Web pages, users of older browsers usually need to install a plug-in, ActiveX controls, or other helper applications. A plug-in is a small program installed on your computer that you use to view plug-in objects through a Web browser. After you install a plug-in, browsers usually recognize the plug-in application and display the media objects correctly. An **ActiveX control** is a Microsoft program for viewing ActiveX objects such as animation through a Web browser. A **helper application** is a separate program, such as Adobe Flash Player or Adobe Shockwave Player. As with plug-ins, users of older browsers can install an ActiveX control or a helper application to view and interact with media objects on Web pages.

You can add interactive enhancements in Dreamweaver by using behaviors. **Behaviors** are a combination of an action and an event. An **action** is what an object does, such as a photo growing in size on a Web page. An **event** is what the user does to trigger the action, such as clicking the photo. The resulting behavior is a mini JavaScript, or program, that runs in a Web browser when a user triggers the event. (**JavaScript** is a scripting language written as a text file.) For example, when a browser opens a new Web page, a behavior can make an image appear or fade on the page. Behaviors can enhance a Web site by adding interactivity so the Web page provides visual feedback, for example, when a user points to navigation text.

As discussed in Chapter 1, CSS3 is the new modern standard for Web site presentation. Using CSS styles, you can control font size, font color, background, and many other attributes of a Web page. The CSS file, layout.css, contains the styling definitions for the Gallery Web site. CSS3 provides new styling options, such as rounded corners, text effects, 2D and 3D transformations, and animations.

In this chapter, you learn about media objects, behaviors, and CSS3 styles. You learn how to add an applet media object to a Web page and define its parameters. You also learn how to add behaviors and new CSS3 styles to a Web page. You should use only suitable media objects and behaviors on a Web site. You most likely have visited Web sites that contained so many bits and pieces of multimedia and interaction that they were distracting rather than serving as a constructive part of the site. Therefore, when adding these elements to your Web site, you need to consider the value of each element and how it will enhance or detract from your site.

Project — Designing with Media and Behaviors

Continuing the design of the Gallery site, a Java applet in the Portfolio page can scroll text that invites visitors to view the portfolio of Gallery photographs. The Java applet, as shown in Figure 7–1a, appears below the Spry Collapsible Panels.

A behavior on the home page makes the main image appear faint at first and then gradually become more distinct. This type of basic animation is called a **transition**. You can use the CSS Styles panel or the CSS Transitions panel to add transitions to a Web page. Figure 7–1b shows the image transition for the home page. Another behavior on the Contact page displays a message in a dialog box. As shown in Figure 7–1c, when the Contact page opens, the pop-up message is displayed to offer a special discount for scheduling a photography session.

CSS3 properties on the Portfolio page format the Spry Collapsible Panels with rounded corners, also shown in Figure 7–1a. Additional CSS properties align the tab text and modify the font size. Using the CSS Transitions panel, you change the look of the Spry Collapsible Panel tabs when you hover over them with the mouse pointer. Finally, you apply a box-shadow property to the image editable region on the Gallery Template to create a drop-shadow effect for each image in the Image area, shown in Figure 7–1b. As you complete the activities in this chapter, you will find that using media objects and behaviors enhances the user experience.

(a) Java applet and CSS3 formatting on the Portfolio page

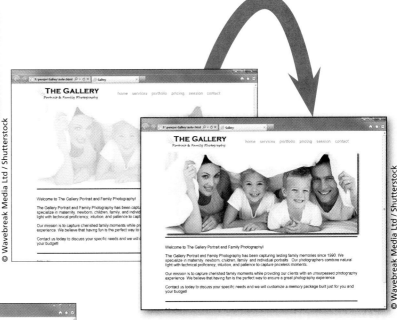

(b) Dreamweaver behavior animates a transition when the home page opens

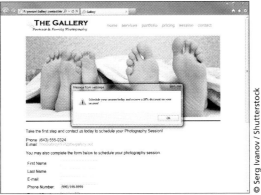

(c) Dreamweaver behavior displays a message box on the Contact page

Figure 7–1

Overview

As you read this chapter, you will learn how to add media objects, behaviors, and CSS properties to the Web pages shown in Figure 7–1 by performing these general tasks:

- Insert a Java applet on a Web page.
- Define Java applet properties and parameters.
- Add an action and event to create a behavior.
- Add the CSS3 border-radius property.
- Add the CSS font-size property.
- Add the CSS text-align property.
- Add CSS transitions.
- Add the CSS3 box-shadow property.

Plan Ahead

General Project Guidelines

When adding media to a Web site, consider the purpose and audience of the site. As you create the Web pages shown in Figure 7–1, you should follow these general guidelines:

1. **Design media objects for both older and new browsers.** Older browsers can support media objects with the right plug-ins and ActiveX controls, when required. Newer browsers support HTML5 and CSS3 properties, which negate the need for media plug-ins. Consider whether you will design for the latest browsers or older browsers. You may decide to design different sites to accommodate both.

2. **Enable JavaScript and determine what type of applet to use in your Web site.** Your audience may include users with older browsers that do not support behaviors, which use JavaScript. Consider including a link to an alternate Web page or site as a solution. Additionally, some users may disable JavaScript from running on their browser. Ask users to enable JavaScript to view the Web page properly.

3. **Determine what types of behaviors to use in your Web site.** You can include many types of behaviors within your Web site. Review the different behaviors and determine which ones are suitable for your Web site needs.

4. **Determine which CSS3 style properties to use in your Web site.** When designing a Web site, consider the different CSS3 style properties and how to best use them.

   More specific details about these guidelines are presented at appropriate points throughout the chapter as necessary. The chapter also identifies the actions performed and decisions made regarding these guidelines during the development of the pages within the site shown in Figure 7–1.

To Start Dreamweaver and Open the Gallery Site

Each time you start Dreamweaver, it opens to the last site displayed when you closed the program. The following steps start Dreamweaver and open the Gallery Web site.

1 Click the Start button on the Windows 7 taskbar to display the Start menu, and then type `Dreamweaver CS6` in the 'Search programs and files' box.

2 Click Adobe Dreamweaver CS6 in the list to start Dreamweaver.

3 If the Dreamweaver window is not maximized, click the Maximize button next to the Close button on the Application bar to maximize the window.

4 If the Gallery site is not displayed in the Files panel, click the Sites button on the Files panel toolbar and then click Gallery to display the files and folders in the Gallery site.

Adding Media Objects

Web designers use various types of multimedia, depending on the Web site content and audience. In Dreamweaver, you add media objects to a Web page using the Insert menu on the Application bar. The available media objects you can insert are described in Table 7–1.

Table 7–1 Media Objects	
Command	**Description**
SWF	Places a Flash movie at the insertion point using the <object> tag; when you insert a SWF object, a dialog box is displayed in which you browse to a SWF file; a SWF file is a compressed version of a Flash (.fla) file
FLV	Inserts a Flash video into a Web page
Shockwave	Places an Adobe Shockwave movie at the insertion point using the <object> and <embed> tags
Applet	Places a Java applet at the insertion point; when you insert a Java applet, a dialog box is displayed in which you specify the file that contains the applet's code, or click Cancel to leave the source unspecified; the Java applet is displayed only when the document is viewed in a browser
ActiveX	Places an ActiveX control at the insertion point; use the Property inspector to specify a source file and other properties for the ActiveX control
Plugin	Use when you want to insert something other than a Flash, Shockwave, Applet, or ActiveX object; displays a dialog box in which you specify the source file and parameters for the plug-in

Plan Ahead

Design media objects for both older and newer browsers.
As discussed in Chapter 6, Web standards are evolving and new browsers with HTML5 capabilities negate the need for many media objects. However, only the newest browsers support HTML5, thus requiring you to continue to integrate media objects for older browsers. Understand your target audience and design your multimedia elements to best reach them.

Using Adobe Flash

Adobe Flash is a program used to create animation files. Flash includes a collection of tools for animating and drawing graphics, adding sound, creating interactive elements, and playing movies. You create Flash objects using vector graphics. Because vector images are small files, browsers can download and display them on a Web page faster than bitmap graphics. This means Flash objects are an efficient way to deliver graphics over the Internet.

You also use Adobe Flash to create **Flash animation files** (also called Flash movies). Learning to use Flash and develop movies can be time-consuming. However, many Web sites provide free or nominally priced Flash animations that you can download and add to your Web site.

Most media objects require a helper program or plug-in before they can be displayed in the browser. For instance, the **Flash Player** is required to view Flash files. Most current Web browsers have incorporated the Flash Player, so you probably do not need to download and install it separately. Adobe also provides the player as a free download at the Adobe Web site. Plug-ins are discussed in more detail later in this chapter.

BTW

Indicating Copyright
If you use Flash movies or other content on your Web site that is copyrighted by someone else, make sure you have permission to use the content. Indicate that the material is copyrighted by including the word "copyright," displaying the © copyright symbol, or referencing the date of publication and the name of the creator.

As noted in Chapter 6, while current Web browsers support Flash, new and future Web browsers may not. For example, the Apple Safari Web browser does not support Flash nor does Microsoft Internet Explorer 10. Instead, these browsers are relying on the multimedia features of HTML5 to play audio and video and create special effects on Web pages. Web browser developers are favoring the HTML5 framework over the need for Flash, plug-ins, and ActiveX controls.

Animated media objects add visual and sensory appeal to a Web page. Besides entertainment, media objects can provide instructions. Multimedia elements also can guide visitors through your site. You can create these effects using the following types of Flash files:

- **Flash file (.fla):** This is the source file for any Flash movie and it is created in the Flash program. Opening an .fla file requires the Flash program. The file must be exported from Flash as an .swf to be viewed in a browser.

- **Flash movie file (.swf):** This is a compiled version of the Flash (.fla) file and it is optimized for viewing on the Web. This file can be played back in browsers and previewed in Dreamweaver, but cannot be edited in Flash.

- **Flash movie file (.swt):** This is a Flash template file. This format allows you to customize content in the .swf file using the Flash button in Dreamweaver.

- **Flash movie file (.swc):** This is a compressed file created in Flash. You can use it to customize media objects. Like the .fla file, an .swc file must be exported from Flash as an .swf to be viewed in a browser.

- **Flash video file (.flv):** This file type is used to deliver encoded audio and video over the Internet using the Adobe Flash Player. To view .flv files, users must have Flash Player 9 or later installed on their computers. If they do not, a dialog box appears that lets users install the latest version of the Flash Player.

To Insert a Flash Movie into Dreamweaver

If you choose to insert a Flash movie (.swf) into the Dreamweaver document window, you would perform the following steps.

1. Create a Media folder within the site root folder and copy the .swf file to the Media folder.

2. Click the document window where you want to insert the movie.

3. Click Insert on the Application bar, point to Media, and then click SWF.

4. In the Select SWF dialog box, select the .swf file you want to insert.

5. Type the title, access key, and tab index in the Object Tag Accessibility Attributes dialog box and then click OK.

6. Enter the width and height values of the movie in the W and H text boxes, respectively, in the Property inspector.

7. Add any other parameters in the Property inspector.

To Insert a Flash Video into Dreamweaver

If you choose to insert a Flash video (.flv) into a Dreamweaver document, you would perform the following steps.

1. Copy the .flv file to the Media folder.

2. Click the document window where you want to insert the movie.

3. Click Insert on the Application bar, point to Media, and then click FLV.

4. In the Insert FLV dialog box, select the Video type.

5. Click the Browse button and select the .flv file you want to insert.

6. Select the Skin and enter the W and H values.

7. Select Auto play and Auto rewind if desired, and then click OK.

8. Edit any other parameters in the Property inspector.

Using Shockwave

Another media type is **Shockwave,** which is designed for online movies and animation and often is used in game development. You use Adobe Director to create Shockwave files. Director often is referred to as an authoring tool or a development platform. In Director, a developer can use a combination of graphics, video, sound, text, animation, and other elements to create interactive multimedia. After the developer creates the multimedia, it can be compressed into a Shockwave file and then added to a Web page. To play Shockwave content within your browser, the Shockwave Player must be installed on your computer. You can download the latest version of the player from the Adobe Web site. According to the Adobe Web site, more than 280 million users have installed the Shockwave Player.

Following are the file formats associated with Director and Shockwave.

- **Director compressed resource file (.dcr):** The .dcr file is a compressed Shockwave file that can be previewed in Dreamweaver and viewed in your browser.

- **Director project file (.dir):** The .dir file is the source file created by Director. These files usually are exported through Director as .dcr files.

- **Director file (.dxr):** The .dxr file is a locked Director file.

- **Director cast file (.cst):** The .cst file contains additional information that is used in a .dxr or .dir file.

- **Director cast file (.cxt):** The .cxt file is a file locked for distribution purposes.

When you insert a Shockwave movie into a Web page, Dreamweaver uses both the <object> tag (for the ActiveX control) and the <embed> tag (for the plug-in) to get the best results in all browsers. When you make changes to the movie in the Property inspector, Dreamweaver maps your entries to the appropriate parameters for both the <object> and <embed> tags.

TO INSERT A SHOCKWAVE FILE INTO DREAMWEAVER

Inserting a Shockwave file is similar to inserting a Flash movie. If you choose to insert a Shockwave file into Dreamweaver, you would perform the following steps.

1. Click the document window where you want to insert the movie.

2. Click the Media button arrow in the Common category on the Insert bar, and then click Shockwave.

3. In the Select File dialog box, select the movie file you want to insert.

4. Enter the width and height values of the movie in the W and H text boxes in the Property inspector.

5. Add any necessary parameters in the Property inspector.

Using Java Applets

An **applet** is a small program written in **Java**, which is a programming language for the Web. An applet can be downloaded by any computer. An applet is embedded with the applet tag, <applet>, in the HTML document and runs in a browser. Hundreds of applets are available on the Internet — many of them for free. Some general categories of applets include text effects, audio effects, visual effects, navigation, games, and utilities.

The Gallery site can use an applet to scroll text on the Portfolio page. The scrolling text will catch the eye of a Web site visitor and encourage him or her to schedule a photography session.

Understanding the Structure of an Applet

As you work with and become more familiar with applets, it is helpful to have a basic understanding of the applet structure within an HTML document. Suppose, for example, you insert an applet that displays a scrolling message, such as the following:

Good morning!

The general syntax of the applet is:

```
<applet code="fileName.class" codebase = "codeBaseURL"
name="appletName"
    alt="alternateText" width="32" height="32">
    <param name="attributeName0" value="attributeValue">
    <param name="attributeName1" value="attributeValue">
    <param name="attributeName2" value="attributeValue">
</applet>
```

Following are the attributes contained within the applet tag.

- **Code:** The code tag is a required element that provides the name of the applet file. The applet file is created in a Java program and is then compiled into the .class file format.
- **Codebase:** The codebase tag defines the URL or the location of the .class file.
- **Name:** The name tag identifies a particular instance of the use of an applet.
- **Alt:** The alt tag is alternate text and informs users that the applet cannot be displayed if the user has a browser that does not support Java.
- **Width and Height:** The width and height tags define the size of the applet in pixels.
- **Param Name and Value:** The param name and value tags pass information to the applet. The information can include text to be displayed or define applet properties.

Plan Ahead

Enable JavaScript and determine what type of applet to use in your Web site.
You can find many types of applets to include in your Web site, such as those that scroll text, animate a navigation bar, or display a count-down timer. Search the Internet for Java applets and find one that is suitable for your multimedia Web site needs.

To Copy the Applets Folder to the Gallery Site Folder

Before adding an applet to a site, you must add the Applets folder to the file structure of the Gallery site. To complete this assignment, you will be required to use the Data Files for Students. Visit www.cengage.com/ct/studentdownload for detailed instructions or contact your instructor for information about accessing the required files. The following steps copy the Applets folder from the Data Files for Students to the Gallery site.

- Use Windows Explorer to navigate to the storage location of the Data Files for Students.

- Double-click the Chapter 07 folder and then double-click the Gallery folder to open the folders.

- Right-click the Applets folder, click Copy on the context menu, and then navigate to the *your last name* and *first initial folder* on Removable Disk F: to prepare to copy the files.

- Double-click the Gallery folder and then right-click anywhere in the open window to display the context menu.

- Click Paste on the context menu to paste the Applets folder into the Gallery folder (Figure 7–2).

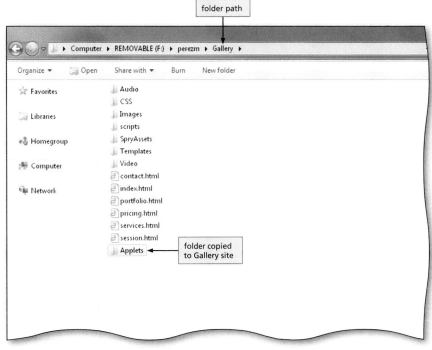

Figure 7–2

- Double-click the Applets folder to view its contents, which includes the Fader.class file (Figure 7–3).

Q&A

What is the Fader.class file?

The Fader.class file is a Java applet that scrolls text.

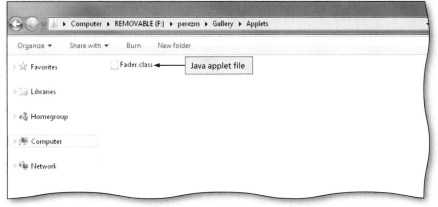

Figure 7–3

To Insert an Applet on a Web Page

Applets can increase the visual appeal of a Web page by animating text and other objects. Add the Fader applet to the Portfolio page to provide additional information about the Gallery photography experience. The applet vertically scrolls four lines of descriptive text below the collapsible panels. The following steps insert an applet on the Portfolio Web page.

● In the Files panel,
double-click the
portfolio.html file to
display the contents
of the Portfolio page
(Figure 7–4).

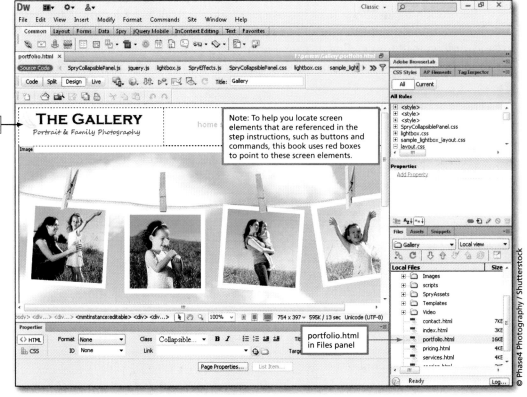

Figure 7–4

❷

● Scroll down the page
and click within
the contentArea
below the Baby
Photography
Collapsible Panel.

● Click Insert on the
Application bar
and then point to
Media to display
the Media submenu
(Figure 7–5).

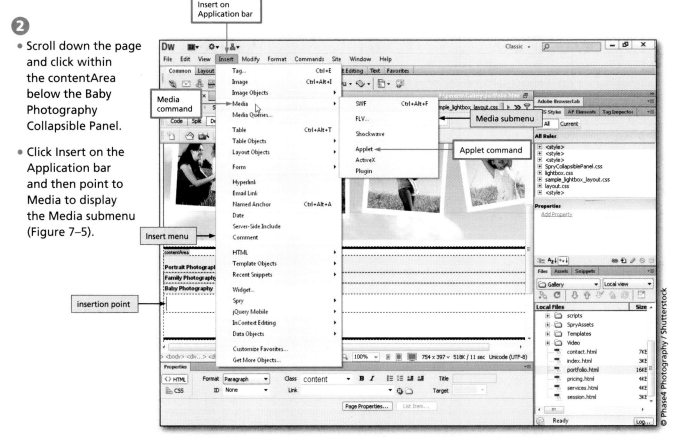

Figure 7–5

3

• Click Applet on the Media submenu to display the Select File dialog box (Figure 7–6).

Figure 7–6

4

• If necessary, navigate to the Gallery folder.

• Double-click the Applets folder to open it, and then click the Fader applet file to select it (Figure 7–7).

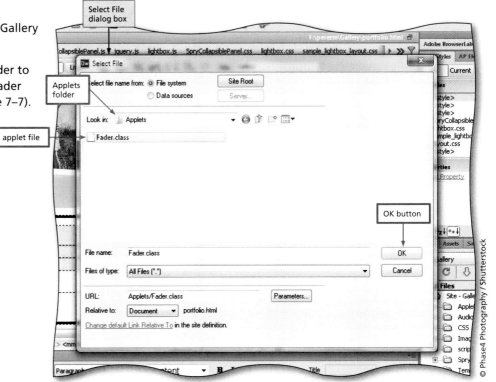

Figure 7–7

5

- Click the OK button to display the Applet Tag Accessibility Attributes dialog box (Figure 7–8).

Figure 7–8

6

- Type `Your browser does not support Java` in the Alternate text text box.

- Click the Title text box and then type `Photography Experience` (Figure 7–9).

Q&A

When is the alternate text displayed?

The alternate text is displayed when the user's browser does not support Java.

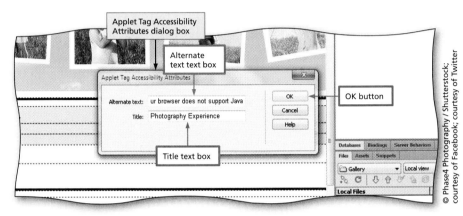

Figure 7–9

7

- Click the OK button to insert the applet (Figure 7–10).

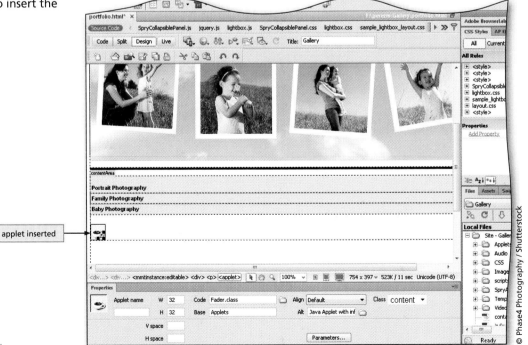

applet inserted

Figure 7–10

Other Ways

1. Click Media button arrow on Insert bar, click Applet

To Modify Applet Properties and Define Applet Parameters

Once the applet is inserted on the page, you need to modify applet properties and define parameters to run the applet properly. You edit applet properties and add parameters on the Property inspector. The following steps modify the applet properties and define the applet parameters.

- If necessary, click the applet to select it.

- In the Property inspector, click the Applet name text box and then type `textScroll` to name the applet.

- Double-click the W text box and then type `500` to change the width of the applet.

- Double-click the H text box and then type `90` to change the height of the applet.

- Click the Base text box and then insert a forward slash, `/`, after the word, Applets, to complete the file path (Figure 7–11).

Q&A Why do I need to include a forward slash after the word, Applets, in the Base text box?

The forward slash further defines the location of the applet file, Fader.class. Without the forward slash, the applet tag will not be able to locate the file.

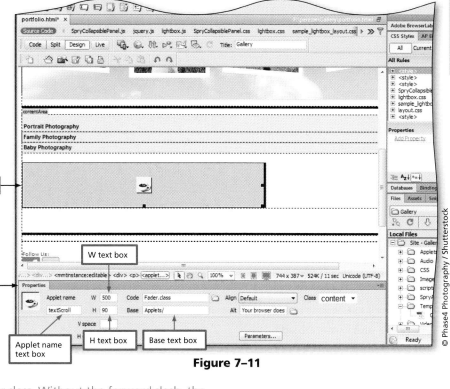

Figure 7–11

- Click the Parameters button to display the Parameters dialog box (Figure 7–12).

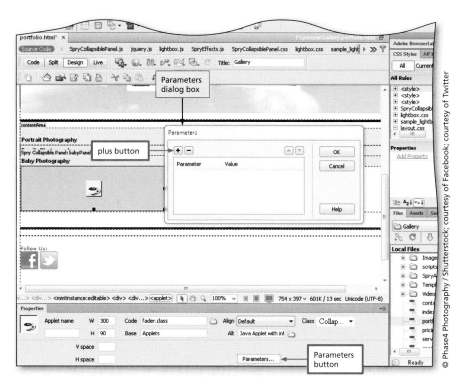

Figure 7–12

3

- Type `delay` to identify the first parameter.

- Click in the Value column and then type `20` to add the delay value (Figure 7–13).

What does the delay parameter do?

The delay parameter defines the amount of time it takes for the vertically scrolling line of text to appear on the screen. The value is the time in milliseconds.

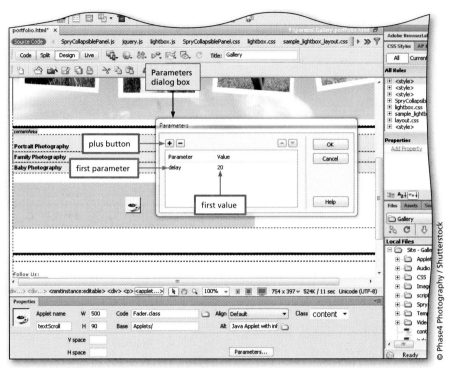

Figure 7–13

4

- Click the plus button to add another new parameter.

- Type `stop` to identify the second parameter.

- Click in the Value column next to the stop parameter and then type `4000` to add the stop value (Figure 7–14).

What does the stop parameter do?

The stop parameter designates the amount of time that each line of text will be displayed on the screen before advancing to the next line of text. The value is the time in milliseconds.

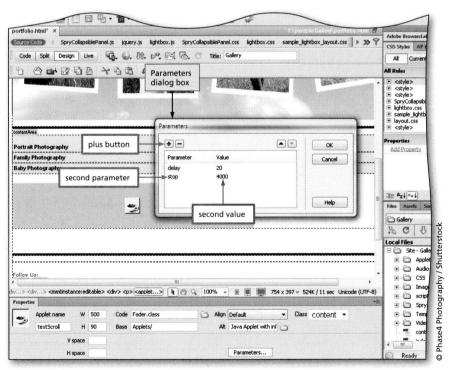

Figure 7–14

5

- Click the plus button to add another new parameter.

- Type `fontSize` to identify the third parameter.

- Click in the Value column next to the fontSize parameter and then type `16` to add the fontSize value (Figure 7–15).

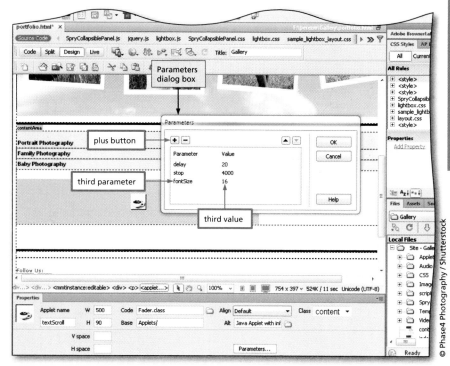

Figure 7–15

6

- Click the plus button to add another new parameter.

- Type `fontStyle` to identify the fourth parameter.

- Click in the Value column next to the fontStyle parameter and then type `1` to add the fontStyle value.

- Click the plus button to add another new parameter.

- Type `fontName` to identify the fifth parameter.

- Click in the Value column next to the fontName parameter and then type `Arial` to add the fontName value.

- Click the plus button to add another new parameter.

- Type `text0` to identify the sixth parameter.

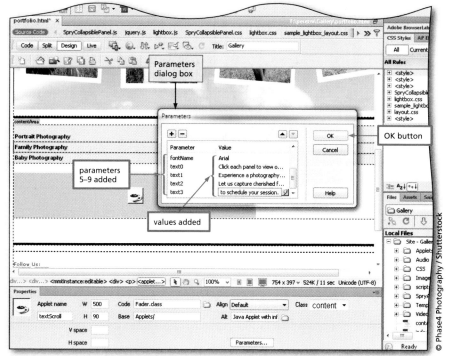

Figure 7–16

- Click in the Value column next to the text0 parameter and then type `Click each panel to view our portfolio.` to add the text0 value.

- Click the plus button to add another new parameter.

- Type `text1` to identify the seventh parameter.

● Click in the Value column next to the text1 parameter and then type `Experience a photography session centered around you.` to add the text1 value.

● Click the plus button to add another new parameter.

● Type `text2` to identify the eighth parameter.

● Click in the Value column next to the text2 parameter and then type `Let us capture cherished family moments.` to add the text2 value.

● Click the plus button to add another new parameter.

● Type `text3` to identify the ninth parameter.

● Click in the Value column next to the text3 parameter and then type `Contact us today to schedule your session.` to add the text3 value (Figure 7–16).

Q&A What are the text0 through text3 parameters?

The text0 parameter is the first line of text that appears in the applet, text1 is the second line, and so on.

7

● Click the OK button to close the Parameters dialog box.

● Click Format on the Application bar, point to Align, and then click Center to center-align the applet in the contentArea (Figure 7–17).

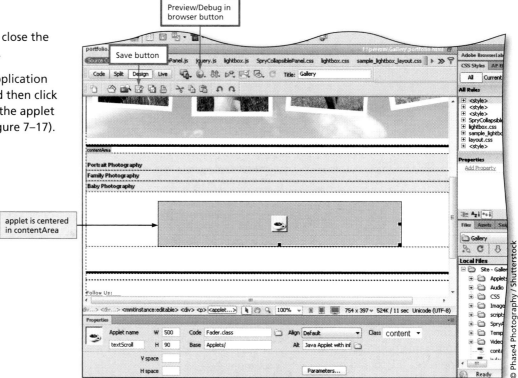

Figure 7–17

- Click the Save button on the Standard toolbar to save your work.

- Click the 'Preview/ Debug in browser' button on the Document toolbar to display a list of browsers, and then click Preview in IExplore in the browser list to display the applet in the Portfolio page in Internet Explorer.

- If necessary, click the 'Allow blocked content' button to run the applet (Figure 7–18).

- Click the Internet Explorer Close button to close the browser.

- Close the Portfolio page.

Figure 7–18

© Phase4 Photography / Shutterstock; used with permission from Microsoft Corporation

Using Plug-Ins and ActiveX

In addition to Flash files, you can insert other media objects — though you need special technology to do so. A Web browser is basically an HTML/scripting decoder. Recall that a plug-in is software that acts as an add-on to a Web browser and provides the browser with additional functionality. To work correctly, plug-in programs must be stored in the browser's Plugins folder. Many browsers include a Plugins folder within their own application folders. When the browser encounters a file with an extension of .swf, for example, it looks in the Plugins folder and attempts to start the Flash Player plug-in.

Some of the more popular plug-ins are Adobe Reader, Flash, and Shockwave; Microsoft Silverlight; Apple QuickTime; Java Virtual Machine; and RealPlayer. With these plug-ins, Web site visitors can use their browsers to view specialized content, such as animation, streaming audio and video, and formatted content. These plug-in programs and information are available for download on their related Web sites.

In contrast, ActiveX is a set of technologies developed by Microsoft and is an outgrowth of OLE (object linking and embedding). An ActiveX control uses ActiveX technologies and is limited to Windows environments. ActiveX objects use the <object> tag instead of the <embed> tag. When ActiveX is inserted, a ClassID or ActiveX control is specified; in many instances, parameters also are added.

To Add the Check Plugin Behavior to a Dreamweaver Document

If you choose to add the Check Plugin behavior to a Dreamweaver document, you would complete the following steps.

1. Select an object in the document window and then open the Behaviors panel.

2. Click the plus button in the Behaviors panel and then click Check Plugin on the Actions pop-up menu.

3. Select a plug-in in the Plugin list, or click the Enter option button and then type the exact name of the plug-in in the adjacent text box.

4. In the 'If found, go to URL' text box, specify a URL for visitors who have the plug-in.

5. In the 'Otherwise, go to URL' text box, specify an alternative URL for visitors who do not have the plug-in.

6. Click the 'Always go to first URL if detection is not possible' check box to select it, and then click the OK button.

Inserting Adobe Smart Objects

Another type of media object you can insert on a Web page is a Photoshop image file. Photoshop files have a PSD file name extension. When you insert Photoshop images in a Web page, Dreamweaver optimizes them as Web-ready images in a GIF, JPEG, or PNG format. To do so, Dreamweaver inserts the Photoshop image as a Smart Object, which maintains a connection to the original Photoshop (PSD) file. That means if the original Photoshop file is modified, the Smart Object in Dreamweaver is also modified. If the Smart Object in Dreamweaver is not in sync with the original Photoshop file, Dreamweaver displays an icon so you can update the image without accessing Photoshop or the original file.

To Insert a Smart Object into a Dreamweaver Document

If you choose to insert a Smart Object into the Dreamweaver document window, you would perform the following steps.

1. Click the document window where you want to insert the Smart Object.

2. Drag the Photoshop file from the Files panel to the insertion point.

3. In the Image Preview dialog box, set optimization settings as necessary and then click OK.

4. To update a Smart Object, select the Smart Object and then click the Update from Original button in the Property inspector.

Integrating Adobe Fireworks

Adobe Fireworks allows you to create static images (usually as GIF or JPEG files), animated GIFs, large images that have been divided (or sliced) into smaller components, rollover buttons, navigation bars, and partial Web pages. If you use Fireworks, you can create and optimize these objects in Fireworks, and then export them into a Dreamweaver site. To do so, you should set Fireworks as the primary external image editor.

BTW

Using Fireworks to Insert Navigation Bars, Rollover Images, and Buttons
If you use Fireworks as your primary image editor, you can create images in Fireworks and then use them as interactive content such as navigation bars, rollover images, and buttons in Dreamweaver. To insert a rollover image, for example, click Insert on the Application bar, point to Image Objects, click Rollover Image, and then select the Fireworks images you want to use.

To Set Fireworks as the Primary Image Editor in Dreamweaver

If you choose to set Fireworks as the primary image editor in Dreamweaver, you would perform the following steps.

1. In Dreamweaver, click Edit on the Application bar and then click Preferences.

2. Click the File Types/Editors category.

3. In the Extensions list, select a file extension (.gif, .jpg, or .png).

4. In the Editors list, select Fireworks, and then click Make Primary.

Adding Behaviors to Web Pages

A **behavior** is a combination of an event and an action, where the action is invoked as a result of the event. Behaviors are attached to specific elements on a Web page. The element can be a table, an image, a link, or another page element. You can think of an action as a task performed using JavaScript code. Some of the actions you can attach to page elements include playing various effects, displaying pop-up messages, calling JavaScript code, swapping images, and showing or hiding elements. When a behavior is initiated, Dreamweaver uses JavaScript to write the code. Recall that JavaScript is a scripting language written as a text file. After a behavior is attached to a page element, and when the event specified occurs for that element, the browser calls the action (the JavaScript code) that is associated with that event. A scripting language such as JavaScript gives Web site designers flexibility and control when creating a Web site, especially one that invites users to interact with page elements. Actions are described in Table 7–2.

BTW

More Behaviors
Additional behaviors are available through the Adobe Marketplace and Exchange. To access the behaviors, click the behavior plus (+) sign button and select Get More Behaviors. Your browser will launch and display the Adobe Marketplace and Exchange home page. You can then browse the site for additional behaviors and install the desired extension package.

Table 7–2 Actions	
Actions	**Description**
Call JavaScript	Executes a JavaScript when the specified event occurs
Change Property	Modifies an element's properties when the specified event occurs
Check Plugin	Confirms that required browser plug-ins are installed when the specified event occurs
Drag AP Element	Allows the user to drag an AP Element when the specified event occurs
Effects	Performs a selected effect when the event occurs
Go To URL	Opens a URL when the specified event occurs
Jump Menu	Generates a jump menu when the specified event occurs
Jump Menu Go	Associates a Go button with the jump menu when the specified event occurs
Open Browser Window	Opens a new browser window when the specified event occurs
Popup Message	Displays a pop-up message when the specified event occurs
Preload Images	Preloads script images when the specified event occurs
Set Text	Sets the text for a selected page element when the specified event occurs
Show-Hide Elements	Shows or hides page elements when the specified event occurs
Swap Image	Swaps an image with another image when the specified event occurs
Swap Image Restore	Resets the swapped image with the original image when the specified event occurs
Validate Form	Validates form data submitted when the specified event occurs

BTW

Editing Behaviors
A behavior can be edited after it has been added. To edit a behavior, double-click the action in the Behaviors panel to display the action's dialog box, and then edit the behavior as desired. Click the OK button when finished.

You add behaviors using the Tag Inspector panel with the Behaviors button selected, also called the Behaviors panel. Click the plus (+) button to display a menu of actions, and then select an action. After you select an action, you must select an event to trigger the action. Dreamweaver provides many types of events to invoke the behavior action. The default event for most actions is onClick, which means that the action occurs when the associated object is clicked. You select events using the Behaviors panel. To change the event for an action, click the event in the Behaviors panel to display a box arrow to the right of the event, which you click to view a menu of available events. (Note that the arrow for the pop-up menu is not displayed until you click the event.) Select the event you want to trigger the specified action. Common events are described in Table 7–3.

Table 7–3 Events	
Event	**Description**
onBlur	Action occurs when the element loses focus
onClick	Action occurs when the element is clicked
onDblClick	Action occurs when the element is double-clicked
onError	Action occurs when there is an error
onFocus	Action occurs when the element is selected
onKeyDown	Action occurs when the down arrow key is pressed
onKeyPress	Action occurs when a key is pressed
onKeyUp	Action occurs when the up arrow key is pressed
onLoad	Action occurs when the element loads on the Web page
onMouseDown	Action occurs when the mouse moves down
onMouseMove	Action occurs when the mouse moves
onMouseOut	Action occurs when the mouse is moved away from the element
onMouseOver	Action occurs when the mouse moves over the element
onMouseUp	Action occurs when the mouse moves up

Plan
Ahead

Determine what types of behaviors to use in your Web site.
Similar to applets, you can use behaviors to add animation and interactivity on your Web site. A behavior is an action and an event. For example, the Appear/Fade action makes a page element appear or fade within a specified time frame, which you specify. You also set the event that triggers the action, such as onClick. In this case, the behavior involves a user clicking a page element to trigger the Appear/Fade action. Review the behaviors Dreamweaver provides and determine which ones are suitable for your Web site needs.

To Insert an Effect Behavior on the Home Page

You can add the Appear/Fade effect to an image on the home page to gain the attention of visitors when they open the page. The effect makes the image slowly appear on the page. To trigger the Appear/Fade action, you can use the onLoad event, which means the action occurs when the page is loaded, or opens, in the browser. The following steps insert the Appear/Fade effect on the home page.

1

- In the Files panel, double-click the index.html file to display the contents of the home page (Figure 7–19).

Figure 7–19

2

- Click the Tag Inspector tab to display the Tag Inspector panel (Figure 7–20).

Figure 7–20

© Wavebreak Media Ltd / Shutterstock

• Click the Behaviors button to display the Behaviors panel (Figure 7–21).

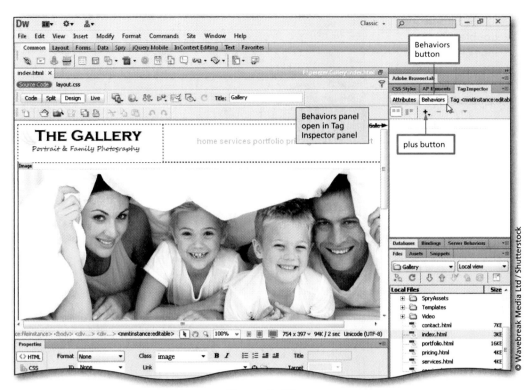

Figure 7–21

• Click the image on the home page to select it.

• Click the plus button on the Behaviors panel to display the action menu, and then point to Effects to display the Effects submenu (Figure 7–22).

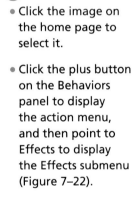

Figure 7–22

5

- Click the Appear/Fade effect command to display the Appear/Fade dialog box (Figure 7–23).

Figure 7–23

6

- Double-click the Effect duration text box and then type   4 0 0 0 to change the time the Appear/Fade effect should be displayed.

- Click the Effect button and then click Appear to change the effect type (Figure 7–24).

Q&A What is the Effect duration?

The Effect duration defines the amount of time it takes for an element to appear or fade on the page. The time is in milliseconds.

Q&A What is the difference between Fade and Appear?

The Fade effect makes the image slowly fade from the page. The Appear effect makes the image slowly appear on the page.

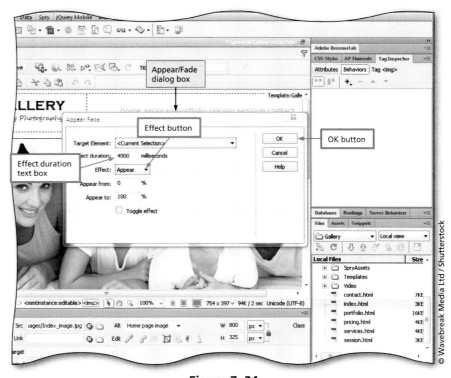

Figure 7–24

7

• Click the OK button in the Appear/Fade dialog box to add the action to the Behaviors panel.

• If necessary, click the onClick event in the Behaviors panel to display the box arrow (Figure 7–25).

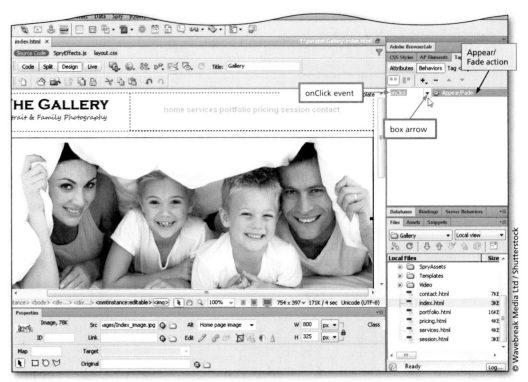

Figure 7–25

8

• Click the box arrow to display a list of events and then click onLoad to select an event (Figure 7–26).

Q&A

I selected the onLoad event, but it still appears as onClick. What should I do?

Select onLoad again.

Figure 7–26

- Click the Save button on the Standard toolbar to save your work.

- If a Copy Dependent Files dialog box is displayed, click the OK button.

- Click the 'Preview/ Debug in browser' button on the Document toolbar to display a list of browsers, and then click Preview in IExplore in the browser list to display the home page in Internet Explorer.

- If necessary, click the 'Allow blocked content' button to view the effect (Figure 7–27).

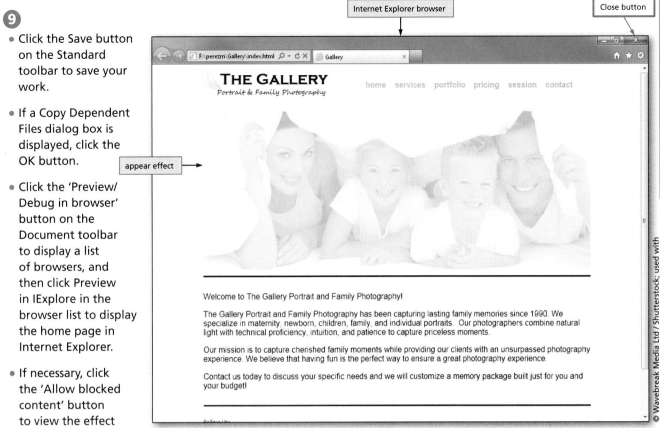

Figure 7–27

- Click the Internet Explorer Close button to close the browser.

- Close the home page.

To Insert a Pop-up Message on the Contact Page

When a potential customer visits the Contact page and starts to complete the form to inquire about scheduling a photography session, a dialog box is displayed with a special offer message. This type of dialog box is called a pop-up message, and is an interactive way to capture the customer's attention and gain new business. The following steps show how to insert a pop-up message on the Contact Web page.

1

- In the Files panel, double-click the contact.html file to display the Contact page (Figure 7–28).

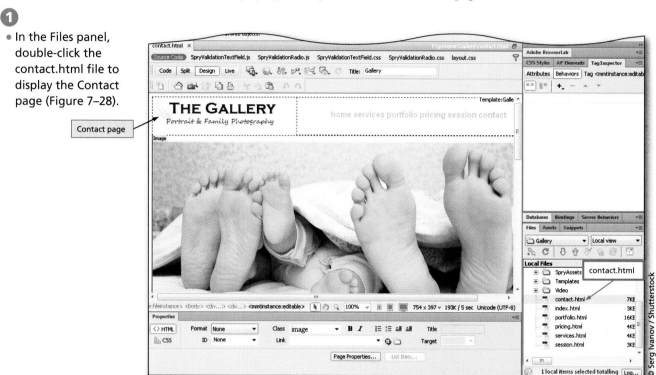

Figure 7–28

2

- If necessary, use the scroll bar to scroll down and then click the First Name Spry Text Field to select the field.

- Click the plus button on the Behaviors panel to display the action menu (Figure 7–29).

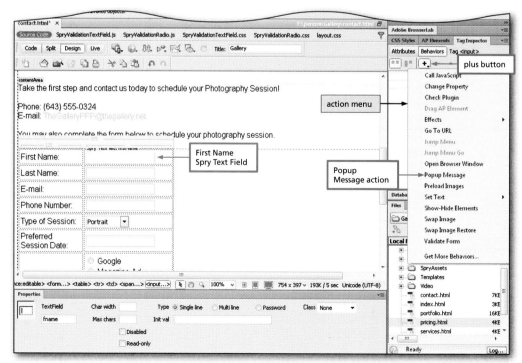

Figure 7–29

- Click the Popup Message action to display the Popup Message dialog box (Figure 7–30).

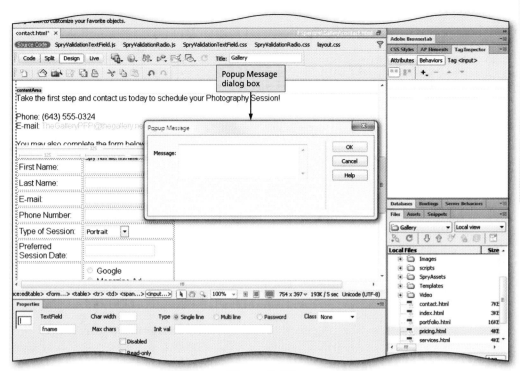

Figure 7–30

- Type Schedule your session today and receive a 20% discount on your session! in the Message text box to add the pop-up message text (Figure 7–31).

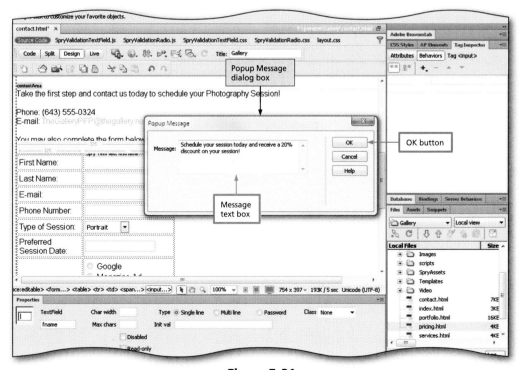

Figure 7–31

5

- Click the OK button in the Popup Message dialog box to add the action.

- Click the onBlur event in the Behaviors panel to display the box arrow (Figure 7–32).

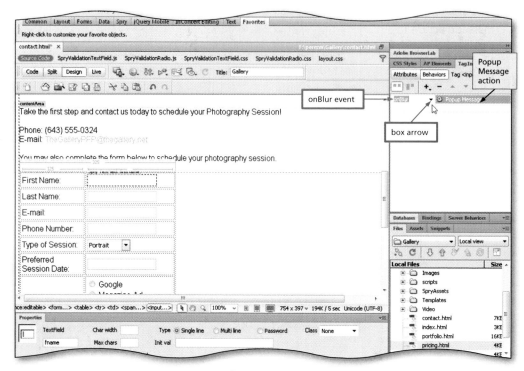

Figure 7–32

6

- Click the box arrow and then click onFocus to select the event (Figure 7–33).

Q&A

I selected the onFocus event, but it still appears as onBlur. What should I do?

Select onFocus again.

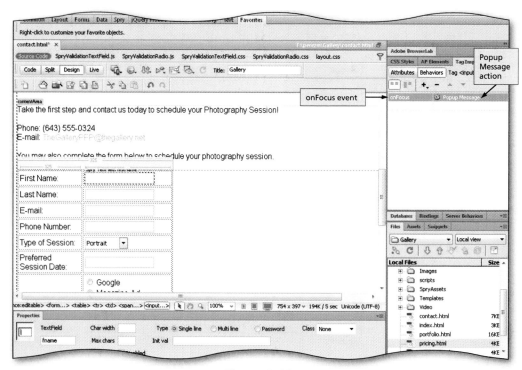

Figure 7–33

7

- Click the Save button on the Standard toolbar to save your work.

- Click the 'Preview/ Debug in browser' button on the Document toolbar to display a list of browsers, and then click Preview in IExplore in the browser list to display the Contact page in Internet Explorer.

- If necessary, click the 'Allow blocked content' button and then click the First Name text box to display the pop-up message (Figure 7–34).

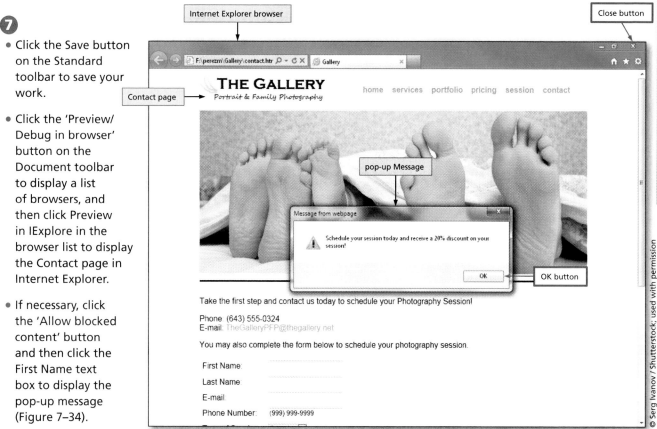

Internet Explorer browser

Close button

Contact page

pop-up Message

OK button

© Serg Ivanov / Shutterstock; used with permission from Microsoft Corporation

Figure 7–34

8

- Click the OK button to close the pop-up message.

- Click the Internet Explorer Close button to close the browser.

- Close the Contact page.

Styling Elements with CSS3

New CSS3 styling properties enrich a Web page with more visual appeal. Borders can be rounded, providing a sleek, smooth look. Multiple images can work together as a unified background. Shadow effects can add depth to text and objects. Graphics can be rotated or appear with a 3D effect on a Web page.

While these new CSS3 styling properties are desirable for design, not all browsers currently support all of the new CSS3 properties. As CSS3 standards become more widely used, more browsers will be optimized to accommodate the CSS3 standards. Some browsers require a tag prefix in the CSS code. For example, Google Chrome and Apple Safari require the prefix, -webkit-, in the CSS code when adding the border-image property. Fortunately, when adding a new CSS3 property that requires a special browser prefix, Dreamweaver CS6 knows which browser supports the CSS3 property and adds the required prefix code for you. Table 7–4 lists some of the new CSS3 properties and browser support information.

Table 7–4 CSS3 Style Properties

| | | Browser Support | | | | |
Property	Description	Internet Explorer 9	Mozilla Firefox	Google Chrome	Apple Safari	Opera
border-radius	Rounds border corners	X	X	X	X	X
box-shadow	Applies a drop-shadow effect	X	X	X	X	X
border-image	Creates a border using an image		X	X	X	X
background-size	Defines the size of a background image	X	X	X	X	X
background-origin	Defines the background image location	X	X	X	X	X
transform	Applies a 2D or 3D transformation to an element	X	X	X	X	X
transition	Changes an element from one style to another		X	X	X	X
resize	Defines whether a user can resize an element		X	X	X	
box-sizing	Defines how elements in a specific area are arranged	X	X	X	X	X
outline-offset	Draws an outline beyond the border edge		X	X	X	X

Plan Ahead

Determine which CSS3 style properties to use in your Web site.
When designing a Web site, consider the different CSS3 style properties and how to best use them. Follow these general guidelines:

• **Browser considerations.** Not all browsers support the new CSS3 standards. Review the properties you want to include and test them in various browsers.

• **Graphics.** If your Web site contains graphics and you would like the user to be able to interact with them, consider using 2D and 3D transforms.

• **Borders.** Rounded borders give a nice, clean look. If your Web site uses a lot of borders, consider rounding the edges.

To Add Rounded Borders to Spry Collapsible Panels

Rounded borders provide a smooth look to a Web page. The border-radius is a new CSS3 property that accomplishes this task. The border-radius is added to Spry Collapsible Panels on the Portfolio page to give the panels a more polished look. The property rounds the edges of each collapsible panel. The following steps add the border-radius property on the Portfolio Web page.

1

- In the Files panel, double-click the portfolio.html file to display the contents of the Portfolio page (Figure 7–35).

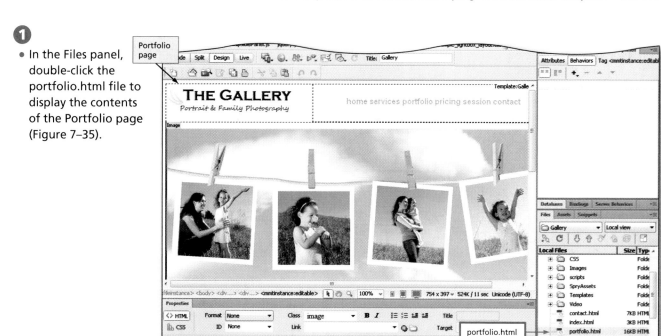

Figure 7–35

2

- Click the CSS Styles tab in the panel group to display the CSS Styles panel. If necessary, click the Current button to display the current style rules.

- If necessary, use the scroll bar in the document window to scroll down and display the Spry Collapsible Panels.

- Click the portraitPanel Spry Collapsible Panel to select it.

- Click .CollapsiblePanel in the Rules section of the CSS Styles panel to select the rule (Figure 7–36).

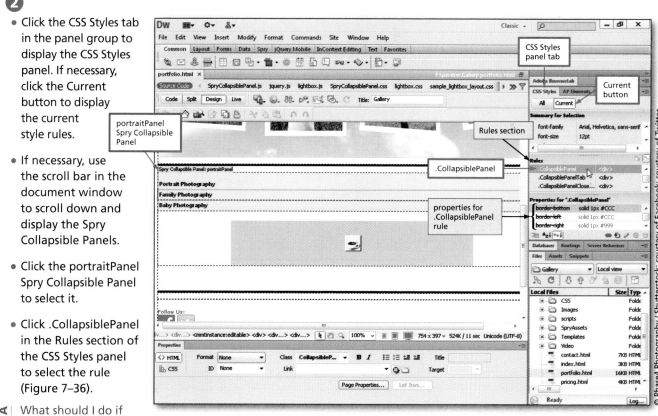

Figure 7–36

Q&A What should I do if the Rules section on the CSS Styles panel is displayed, but I cannot see anything to select below it?

Confirm that you selected the portraitPanel Spry Collapsible Panel. If so, use your mouse to adjust the size of the Rules section.

Q&A Why do I select the .CollapsiblePanel rule in the CSS Styles panel?

The .CollapsiblePanel rule contains the styles that apply to all the collapsible panels on the site.

3

- If necessary, use the scroll bar in the Properties for ".CollapsiblePanel" section to scroll down and display the Add Property link (Figure 7–37).

Q&A

I see the Properties for ".CollapsiblePanel" section on the CSS Styles panel, but I cannot see anything to select below it. What should I do?

Use your mouse to adjust the size of the Properties for ".CollapsiblePanel" section.

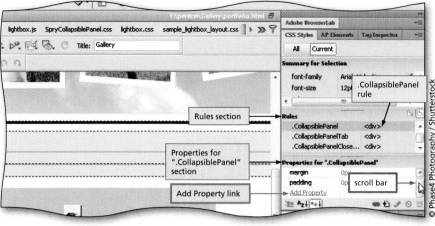

Figure 7–37

4

- Click the Add Property link to display a box arrow.

- Click the box arrow and then scroll down to border-radius to find the property (Figure 7–38).

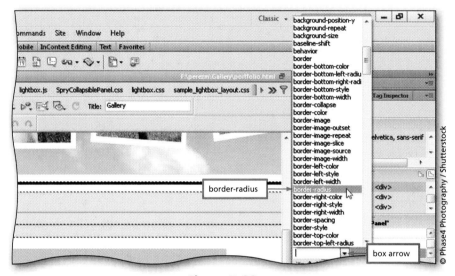

Figure 7–38

5

- Click border-radius to select it (Figure 7–39).

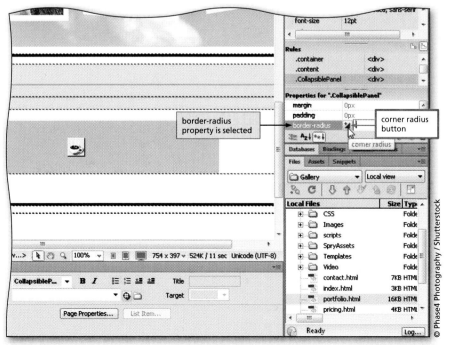

Figure 7–39

6

- Click the corner radius button to the right of the border-radius property to display controls for defining the property.

- If necessary, click the 'Same for all' check box to insert a check mark.

- In the Top Left text box, type 10 to change the width of the border, and then, if necessary, select px to set the measurement units to pixels (Figure 7–40).

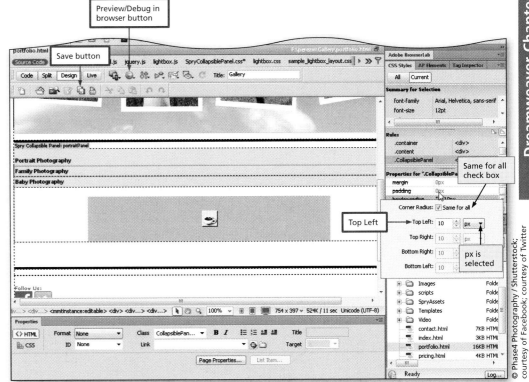

Figure 7–40

Q&A

What does 10px mean in relation to the border-radius?

10px is the amount of rounding that will be applied to each corner of each panel. The larger the number, the rounder the border.

7

- Click the Portfolio page to close the property box.

- Click the Save All button on the Standard toolbar to save your work.

- Click the 'Preview/ Debug in browser' button on the Document toolbar to display a list of browsers, and then click Preview in IExplore in the browser list to display the Portfolio page in Internet Explorer.

- If necessary, click the 'Allow blocked content' button to view the Spry Collapsible Panels (Figure 7–41).

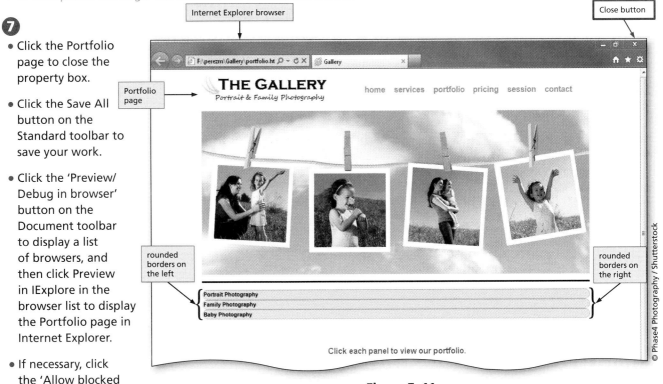

Figure 7–41

8

- Click the Internet Explorer Close button to close the browser.

© Phase4 Photography / Shutterstock; courtesy of Facebook; courtesy of Twitter

© Phase4 Photography / Shutterstock

To Add CSS Properties to Spry Collapsible Panels

To further improve the visual appeal of the Spry Collapsible Panels, you can increase the font size and center-align the panel tab text. You use the font-size property to define the size of a font, and you use the text-align property to align text. Add the font-size property to Spry Collapsible Panels on the Portfolio page to change the font size to 16 points. Add the text-align property to the Spry Collapsible Panels to center the tab text. The following steps add the font-size and text-align properties to the Spry Collapsible Panels on the Portfolio page.

- If necessary, click the portraitPanel Spry Collapsible Panel to select it, and then click .CollapsiblePanel in the Rules section of the CSS Styles panel to select the rule.

- If necessary, use the scroll bar in the Properties for ".CollapsiblePanel" section to scroll down and display the Add Property link (Figure 7–42).

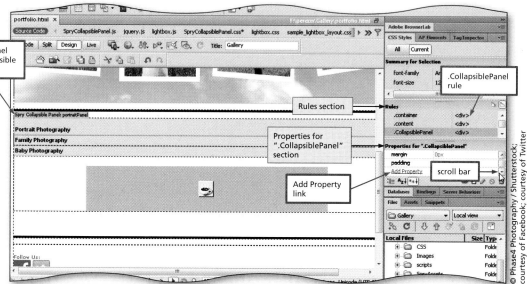

Figure 7–42

- Click the Add Property link to display a box arrow.

- Click the box arrow and then use the scroll bar to find the font-size property (Figure 7–43).

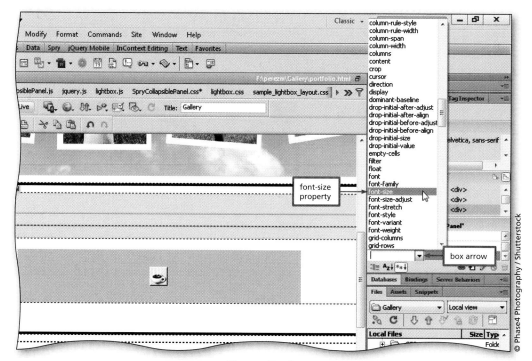

Figure 7–43

3

- Click the font-size property to select it.

- Type `16pt` in the text box to the right of font-size and then press the ENTER key to define the font size (Figure 7–44).

<div class="q-and-a">
Q&A

Why did the font size change for the text in all of the collapsible panels?

The .CollapsiblePanel rule applies to all of the collapsible panels in the site, so adding a property to the rule affects all of the collapsible panels.
</div>

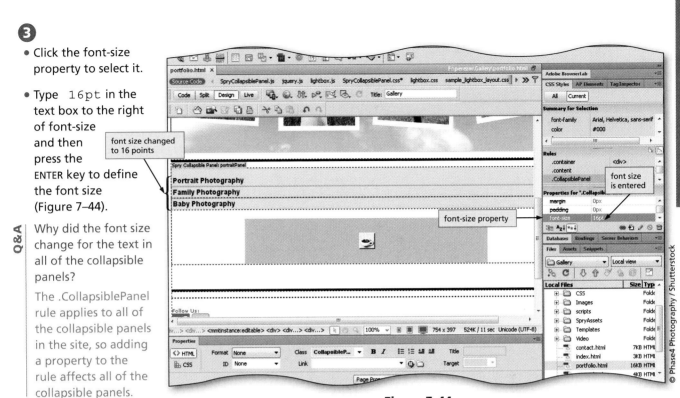

Figure 7–44

4

- Use the scroll bar in the Properties for ".CollapsiblePanel" section to scroll down and display the Add Property link.

- Click the Add Property link to display a box arrow.

- Click the box arrow and then use the scroll bar to find the text-align property (Figure 7–45).

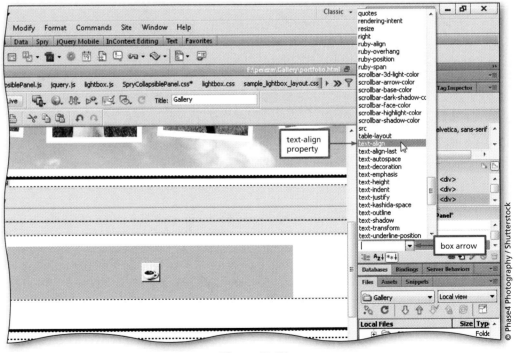

Figure 7–45

• Click the text-align property to select it.

• Type `center` in the text box to the right of text-align and then press the ENTER key to center-align the text (Figure 7–46).

• Click the Save All button on the Standard toolbar to save your work.

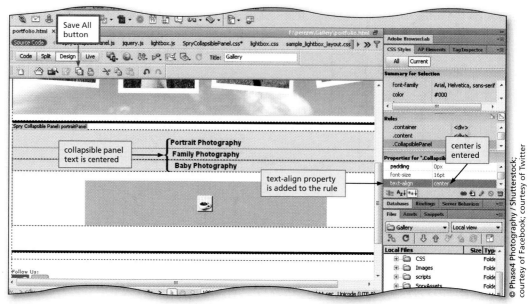

Figure 7–46

Using the CSS Transitions Panel

A **CSS3 transition** is an effect that sets an element to gradually change from one style to another. When viewed in a browser, an element such as a heading changes from dark blue to light blue. Before CSS3 was introduced, Web designers had to use Flash animations or a JavaScript file to create a similar effect.

The CSS Transitions panel is a new feature in Dreamweaver designed to let you add CSS3 transitions to your pages. The CSS Transitions panel allows you to apply element property changes triggered by an event. For example, navigation text can fade from one color to another when you hover over it, or a page background can change from one color to another when a page element is active.

To Add a CSS Transition Using the CSS Transitions Panel

You can make the Spry Collapsible Panels on the Portfolio page interactive by changing the background color of their tabs when your mouse hovers over each panel. You can use the CSS Transitions panel to add a new CSS transition to the Spry Collapsible Panels. The panels will change color when the mouse hovers over each one. The following steps add CSS transitions to the Spry Collapsible Panels on the Portfolio Web page.

1

- Click Window on the Application bar to display the Window menu and find the CSS Transitions command (Figure 7–47).

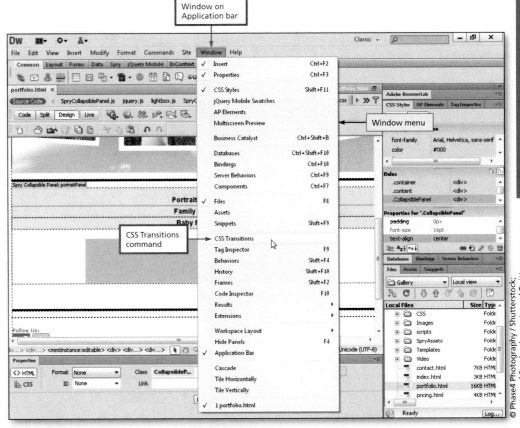

Figure 7–47

2

- Click CSS Transitions to display the CSS Transitions panel (Figure 7–48).

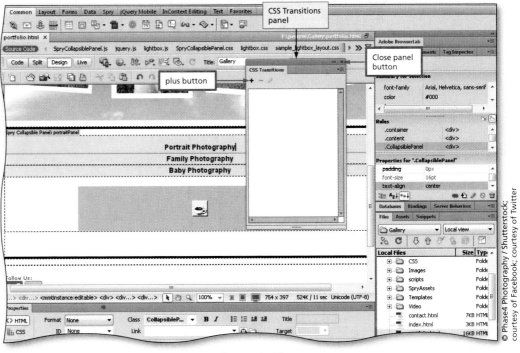

Figure 7–48

3

- Click the plus button to display the New Transition dialog box (Figure 7–49).

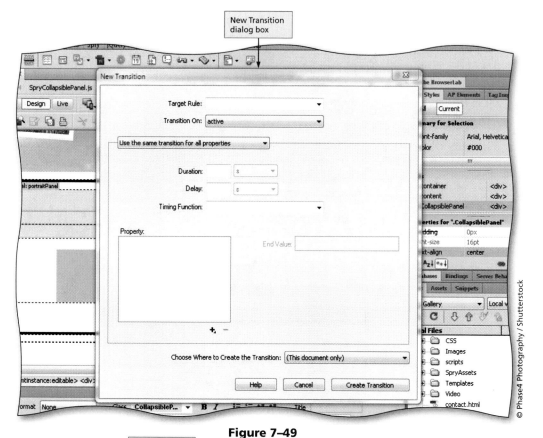

Figure 7–49

4

- Click the Target Rule box arrow and then click .Collapsible PanelTabHover to select the rule.

- Click the Transition On button and then click hover to select the trigger event (Figure 7–50).

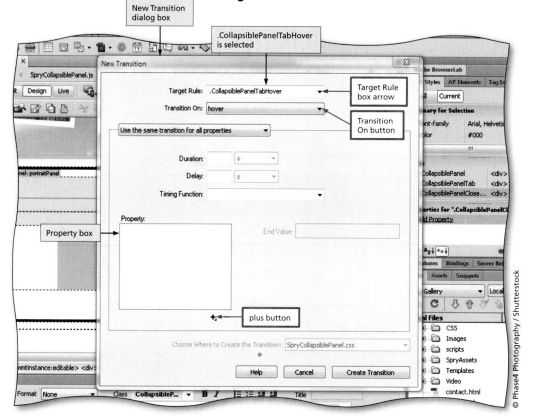

Figure 7–50

© Phase4 Photography / Shutterstock

5

- Click the plus button below the Property box to display a list of property options and find the background-color property (Figure 7–51).

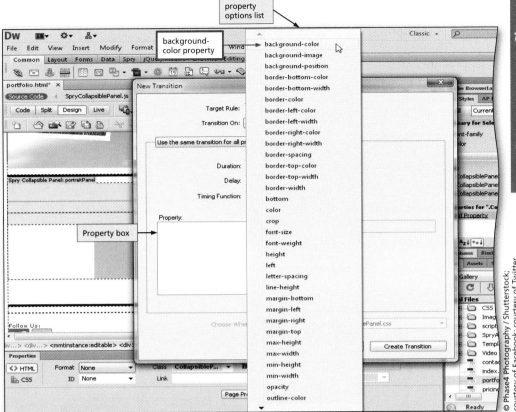

Figure 7–51

6

- Click background-color to select it and add it to the Property box.

- Type #003 in the End Value text box to define the background color (Figure 7–52).

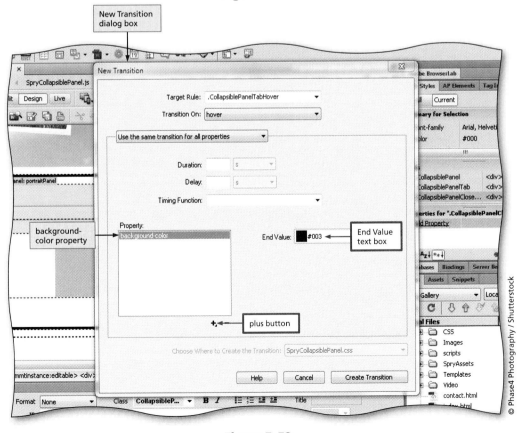

Figure 7–52

7

• Click the plus button below the Property box and then click color to select the property and add it to the Property box.

• Type #FFF in the End Value text box to define the text color (Figure 7–53).

Q&A

What does the color property do?

The color property defines the text color.

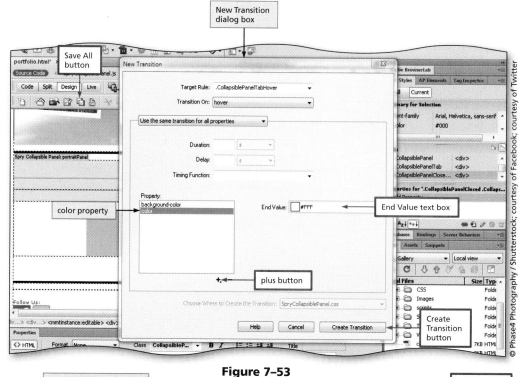

Figure 7–53

8

• Click the Create Transition button to add the transition to the CSS Transitions panel.

• Click the Save All button on the Standard toolbar to save your work.

• Click the 'Preview/ Debug in browser' button on the Document toolbar to display a list of browsers, and then click Preview in IExplore in the browser list to display the Portfolio page in Internet Explorer.

Figure 7–54

• If necessary, click the 'Allow blocked content' button to view the Spry Collapsible Panels.

• Point to the collapsible panels to view the CSS transitions (Figure 7–54).

9

• Click the Internet Explorer Close button to close the browser.

• Click the Close button on the CSS Transitions panel to close the panel.

• Close the Portfolio page.

To Add a Box Shadow to an Image

Web page images are enhanced with box shadows. The box-shadow property is a new CSS3 property that applies a drop-shadow effect to a page element. You can add a box shadow to the .image editable region on the Gallery template to apply a shadow to the main image on each page. The following steps add the box-shadow property to the Gallery template.

1

- Click Open on the Dreamweaver Welcome screen to display the Open dialog box. If necessary, navigate to the Gallery site on Removable Disk (F:).

- Double-click the Templates folder to display the template file, Gallery Template.dwt (Figure 7–55).

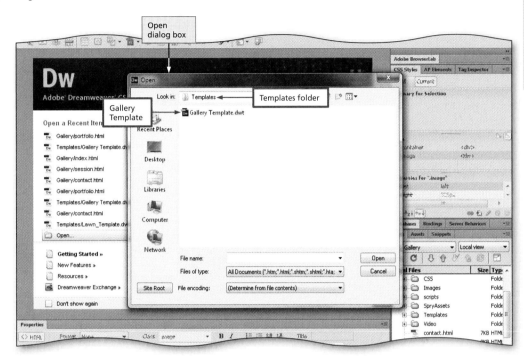

Figure 7–55

2

- Double-click the Gallery Template to open it.

- Click the word, Image, in the Image editable region to select the region.

- If necessary, use the scroll bar in the Properties section of the CSS Styles panel to scroll down and display the Add Property link (Figure 7–56).

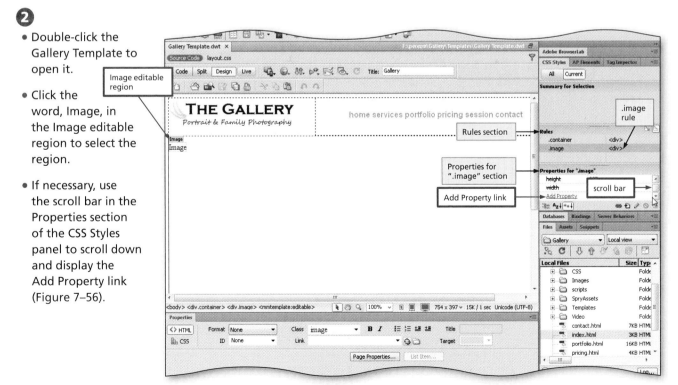

Figure 7–56

- Click the Add Property link to display a box arrow.

- Click the box arrow and then use the scroll bar to find the box-shadow property (Figure 7–57).

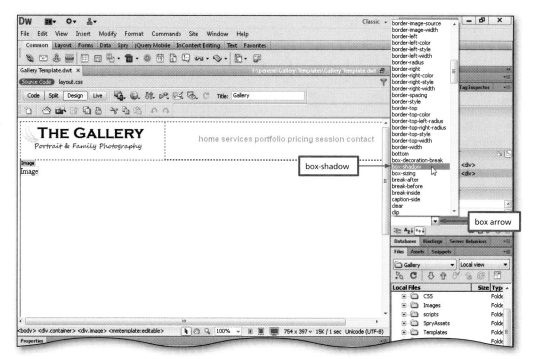

Figure 7–57

- Click the box-shadow property to select it (Figure 7–58).

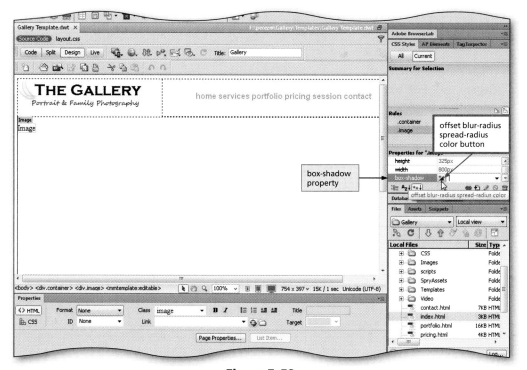

Figure 7–58

- Click the offset blur-radius spread-radius color button to display controls for defining the property.

- Replace the 0 with 7 in the X-Offset text box to change the value.

- Replace the 0 with 7 in the Y-Offset text box to change the value.

- Type 8 in the Blur radius text box to define the value.

- Type #333 in the Color text box to define the value (Figure 7–59).

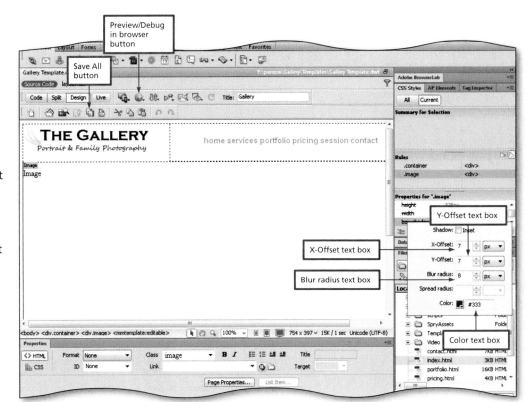

Figure 7–59

Q&A What does it mean to set the X-Offset to 7px?

The 7px X-Offset is the amount of shadow that appears along the horizontal edge, or bottom, of an element. The higher the number, the farther the shadow extends.

Q&A What does it mean to set the Y-Offset to 7px?

The 7px Y-Offset is the amount of shadow that appears along the vertical edge, or right side, of an element. The higher the number, the farther the shadow extends.

Q&A What does it mean to set the Blur radius to 8px?

The Blur radius gives a blurring effect to the shadow. The 8px Blur radius specifies the degree of a blur effect. The higher the number, the more the shadow is blurred.

- Click the document window to close the offset blur-radius spread-radius color property box.

- Click the Save All button on the Standard toolbar to save your work.

- Close the Gallery Template.

- In the Files panel, double-click index. html to open the home page.

- Click the 'Preview/ Debug in browser' button on the Document toolbar to display a list of browsers, and then click Preview in IExplore in the browser list to display the home page in Internet Explorer.

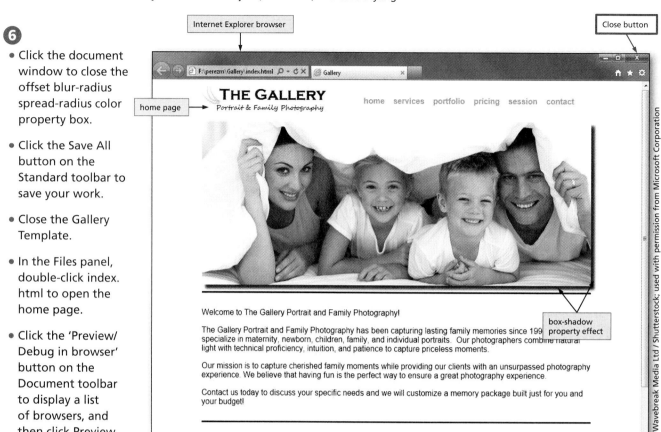

Figure 7–60

- If necessary, click the 'Allow blocked content' button to display the box shadows on the page (Figure 7–60).

7

- Click the Internet Explorer Close button to close the browser.

- Close the home page.

To Quit Dreamweaver

The following steps quit Dreamweaver and return control to the operating system.

1 Click the Close button on the right side of the Application bar to close the window.

2 If Dreamweaver displays a dialog box asking you to save changes, click the No button.

Chapter Summary

In this chapter, you added an applet media object to the Portfolio Web page to enhance the visual appeal. You modified properties and defined parameters for the applet. You also added a behavior to make the image slowly appear on the home page. You applied the Appear/Fade action and the onLoad event to an image to create a behavior. You added another behavior to show a pop-up message on the Contact page. To create this behavior, you added the Popup Message action and the onFocus event to the First Name Spry Text Field. You used the CSS3 border-radius property to round the corners of the Spry Collapsible Panels on the Portfolio page. You also used the CSS Transition panel to modify the look of the Spry Collapsible Panel tabs when the mouse hovers over each tab. Finally, you added the CSS3 box-shadow property to the Gallery Template to add a shadow to the Image editable area. You viewed the applet, behaviors, and CSS changes in your browser. The following tasks are all the new Dreamweaver skills you learned in this chapter:

1. Insert a Flash Movie into Dreamweaver (DW 374)
2. Insert a Flash Video into Dreamweaver (DW 374)
3. Insert a Shockwave File into Dreamweaver (DW 375)
4. Insert an Applet on a Web Page (DW 377)
5. Modify Applet Properties and Define Applet Parameters (DW 381)
6. Add the Check Plugin Behavior to a Dreamweaver Document (DW 386)
7. Insert a Smart Object into Dreamweaver (DW 386)
8. Set Fireworks as the Primary Image Editor in Dreamweaver (DW 387)
9. Insert an Effect Behavior on the Home Page (DW 388)
10. Insert a Pop-up Message on the Contact Page (DW 394)
11. Add Rounded Borders to Spry Collapsible Panels (DW 398)
12. Add CSS Properties to Spry Collapsible Panels (DW 402)
13. Add a CSS Transition Using the CSS Transitions Panel (DW 404)
14. Add a Box Shadow to an Image (DW 409)

Apply Your Knowledge

Reinforce the skills and apply the concepts you learned in this chapter.

Adding an Applet to a Web Page

Note: To complete this assignment, you will be required to use the Data Files for Students. Visit www.cengage.com/ct/studentdownload for detailed instructions or contact your instructor for information about accessing the required files.

Instructions: A company uses an applet to enhance its advertising on a Web page. In this activity, you modify a Web page by adding an applet. First, you insert the applet. Next, you modify the applet's properties, and then you define the applet's parameters. The modified Web page is shown in Figure 7–61.

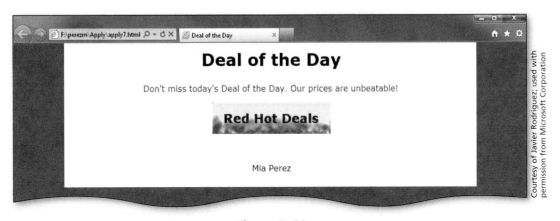

Courtesy of Javier Rodriguez; used with permission from Microsoft Corporation

Figure 7–61

Continued >

Apply Your Knowledge *continued*

Perform the following tasks:

1. Use Windows Explorer to copy the apply7.html file and the Applets folder from the Chapter 07\ Apply folder into the *your last name and first initial*\Apply folder.

2. Start Dreamweaver. Use the Sites button on the Files panel to select the Apply site. Open apply7.html.

3. Place the insertion point after the last line of text, Our prices are unbeatable!, and then press the ENTER key.

4. Insert an applet media object using Insert on the Application bar. Navigate to the Applets folder and select the fire applet file.

5. Enter Your browser does not support Java as the Alternate text and enter Hot Deal for the Title.

6. Change the width (W) to 234 and the height (H) to 60 in the Property inspector.

7. Insert a forward slash, /, after the word, Applets, in the Base text box.

8. Add the applet parameters and values provided in Table 7–5.

9. Click to the right of the applet. Press the ENTER key twice, and then type your first and last names.

10. Save your changes and then view your document in your browser. Compare your document to Figure 7–61. Make any necessary changes, and then save your changes.

11. Submit the document in the format specified by your instructor.

Table 7–5 Fire Applet Parameters and Values	
Parameter	**Value**
coolingfactor	2
coolingrows	50%
coolinglimit	50%
text	Red Hot Deals
textfont	Verdana
textsize	24
textcolor	003300

Extend Your Knowledge

Extend the skills you learned in this chapter and experiment with new skills. You may need to use Help to complete the assignment.

Adding Behaviors

Note: To complete this assignment, you will be required to use the Data Files for Students. Visit www.cengage.com/ct/studentdownload for detailed instructions or contact your instructor for information about accessing the required files.

Instructions: A Web page contains three images to demonstrate three behaviors. In this activity, you add the grow/shrink effect, the shake effect, and the appear/fade effect to images. First, you add the Grow/Shrink effect to the first image and define grow/shrink values in the Grow/Shrink dialog box. Next, you add the shake effect to the second image. Finally, you add the appear/fade effect to the third image and define appear/fade values in the Appear/Fade dialog box. The completed page is shown in Figure 7–62.

Figure 7–62

Perform the following tasks:

1. Use Windows Explorer to copy the extend7.html file from the Chapter 07\Extend folder into the *your last name and first initial*\Extend folder. Copy the three image files, lighthouse, sun, and sunset, into the *your last name and first initial*\Extend\Images folder.

2. Start Dreamweaver. Use the Sites button on the Files panel to select the Extend site. Open extend7.html.

3. Click the sun image to select it. Use the Tag Inspector Behaviors panel to add the Grow/Shrink effect to the selected picture. Use the following values for the Grow/Shrink dialog box:

    Target Element: `<Current Selection>`

    Effect duration: `3000` milliseconds

    Effect: `Shrink`

    Shrink from: `100%`

    Shrink to: `50%`

    Shrink to: `Top Left Corner`

    Click the Toggle effect check box to select it and then click the OK button to add the behavior.

4. Confirm that the event selected is onClick.

5. Click the lighthouse image to select it. Use the Tag Inspector Behaviors panel to add the Shake effect to the selected picture. Use <Current Selection> for the Target Element and click the OK button to add the behavior.

6. Confirm that the event selected is onClick.

Continued >

7. Click the sunset image to select it. Use the Tag Inspector Behaviors panel to add the Appear/Fade effect to the selected picture. Use the following values for the Appear/Fade dialog box:

Target Element: `<Current Selection>`

Effect duration: `5000` milliseconds

Effect: `Fade`

Fade from: `100%`

Fade to: `30%`

Click the Toggle effect check box to select it and then click the OK button to add the behavior.

8. Confirm that the event selected is onClick.

9. Save your changes and then view the Web page in your browser. Click each picture to view the behavior. Make any necessary changes, and then save your changes.

10. Submit the document in the format specified by your instructor.

Make It Right

Analyze a Web page and suggest how to improve its design.

Adding CSS3 Properties

Note: To complete this assignment, you will be required to use the Data Files for Students. Visit www.cengage.com/ct/studentdownload for detailed instructions or contact your instructor for information about accessing the required files.

Instructions: A Web page provides a list of new CSS3 properties in a box. In this activity, you round the borders of the box, add a shadow effect, and add a CSS transition. First, you apply the border-radius property and define its corner radius values. Next, you add the box-shadow property and define its X-Offset, Y-Offset, blur, and color values. Finally, you add a CSS transition that changes the font color when you hover over the text. The modified Web page is shown in Figure 7–63.

Figure 7–63

Perform the following tasks:

1. Use Windows Explorer to copy the right7.html file from the Chapter 07\Right folder into the *your last name and first initial*\Right folder.

2. Open right7.html in your browser to view its contents and design. Close the browser.

3. Start Dreamweaver. Use the Sites button on the Files panel to select the Right site. Open right7.html.

4. Use the CSS Styles panel to add the border-radius property for ".container". Enter `30px` as the corner radius value.

5. Add the box-shadow property for ".container". Enter the following property values:

   X-Offset: `20px`

   Y-Offset: `20px`

   Blur radius: `10px`

   Color: `#666`

6. Click the white border to select the div.container. Open the CSS Transitions panel and create a new transition with the following elements:

   Target Rule: `.container`

   Transition On: `hover`

   Property: `color`

   End Value: `#FF9`

7. Replace the text, Your name here, with your first and last names.

8. Save your changes and then view the Web page in your browser. Compare your page to Figure 7–63. Make any necessary changes, and then save your changes.

9. Submit the document in the format specified by your instructor.

In the Lab

Design and/or create a document using the guidelines, concepts, and skills presented in this chapter. Labs are listed in order of increasing difficulty.

Lab 1: Adding a Behavior to the Healthy Lifestyle Web Site

Problem: You are creating an internal Web site for your company that features information about how to live a healthy lifestyle. Employees at your company will use this Web site as a resource for nutrition, exercise, and other health-related tips. You have already added all of the Web site content. Now you need to add a behavior to the Sign Up page. You need to add a pop-up message that encourages employees to sign up to receive weekly e-mail tips.

Add the Popup Message behavior to the First Name text box. The pop-up message, as displayed in a browser, is shown in Figure 7–64.

Continued >

STUDENT ASSIGNMENTS

In the Lab continued

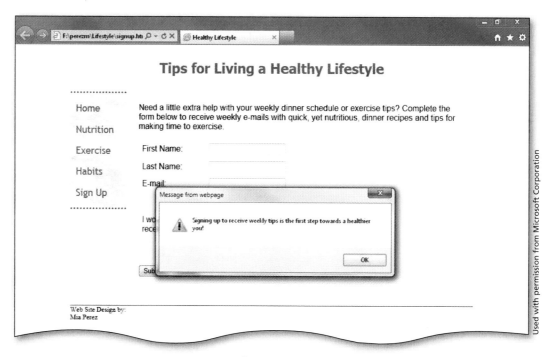

Used with permission from Microsoft Corporation

Figure 7–64

Perform the following tasks:

1. Start Dreamweaver. Use the Sites button on the Files panel to select the Healthy Lifestyle site. Open signup.html.

2. Click the First Name text box to select it. Use the Tag Inspector Behaviors panel to add the Popup Message.

3. Enter `Signing up to receive weekly tips is the first step towards a healthier you!` as the Message in the Popup Message dialog box. Click the OK button to add the behavior.

4. Change the event to onFocus.

5. Save your changes and then view the Web page in your browser. Click the First Name text box to view the pop-up message. Compare your page to Figure 7–64. Make any necessary changes, and then save your changes.

6. Submit the document in the format specified by your instructor.

In the Lab

Lab 2: Adding CSS3 Properties to the Designs by Dolores Web Site

Problem: You are creating a Web site for Designs by Dolores, a Web site design company. The site provides information about the company and its services. You have already added all of the Web site content. Now you need to add CSS3 properties to the template to further enhance the look of the Web site.

First, edit the rule for .header by modifying the bottom box margin to provide more space between the header area and navigation area. Next, add the box-shadow property to .header to give the logo a shadow effect. Finally, add the border-radius property to .footer to give it rounded corners. The updated home page is shown in Figure 7–65.

© Angela Waye / Shutterstock; used with permission from Microsoft Corporation

Figure 7–65

Perform the following tasks:

1. Start Dreamweaver. Use the Sites button on the Files panel to select the Designs by Dolores site. Open Designs_Template.dwt.

2. Click the header image to select the .header area of the template. Use the CSS Styles panel to edit the rule for .header. (*Hint*: Click the Edit Rule button, which displays a pencil icon.) Edit the box category by adding the value 10px to the bottom margin. Apply and accept your changes.

3. Use the CSS Styles panel to add the box-shadow property to .header. Assign the following property values:

   X-Offset: 6px

   Y-Offset: 6px

   Blur radius: 10px

   Color: #036

4. Click the document window to apply the changes.

5. Click the footer area to select .footer. Use the CSS Styles panel to add the border-radius property. Use 25px as the property value.

6. Use the Save All button on the Standard toolbar to save your work. Close Designs_Template.

7. Open index.html. View the document in your browser. Compare your document to Figure 7–65. Make any necessary changes, and then save your changes.

8. Submit the documents in the format specified by your instructor.

In the Lab

Lab 3: Adding CSS3 Properties to the Justin's Lawn Care Service Web Site

Problem: You are creating a Web site for a new lawn care company, Justin's Lawn Care Service. The Web site provides information to customers, including descriptions of services and pricing. In Chapter 2, you created the template for the Web site. Now you need to add CSS3 properties to the template to further enhance the site.

First, add the box-shadow property to .header to give the logo a shadow effect. Next, add the border-radius property to .content to give the content area rounded corners. Finally, add the border-radius property to .footer to give the footer area rounded corners. The updated home page is shown in Figure 7–66.

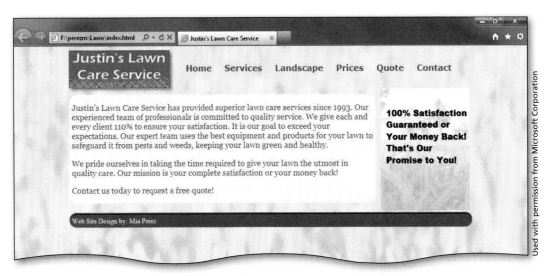

Figure 7–66

Perform the following tasks:

1. Start Dreamweaver. Use the Sites button on the Files panel to select the Justin's Lawn Care Service site. Open Lawn_Template.dwt.

2. Click the Justin's Lawn Care Service logo to select .header. Use the CSS Styles panel to add the box-shadow property to .header. Assign the following property values:

   X-Offset: 6px

   Y-Offset: 6px

   Blur radius: 10px

   Color: #036

3. Click the document window to apply the changes.

4. Click the content area to select .content. Use the CSS Styles panel to add the border-radius property. Use 15px as the property value.

5. Click the footer area to select .footer. Use the CSS Styles panel to add the border-radius property. Use 20px as the property value.

6. Use the Save All button on the Standard toolbar to save your work. Close Lawn_Template.

7. Open index.html and then view the document in your browser. Compare your document to Figure 7–66. Make any necessary changes, and then save your changes.

8. Submit the documents in the format specified by your instructor.

Cases and Places

Apply your creative thinking and problem solving skills to design and implement a solution.

1: Adding Behaviors to the Moving Venture Tips Web Site

Personal

You have added all of the content for the Moving Venture Tips Web site and now need to enhance the site with a behavior and CSS3 property. You have decided to add a pop-up message to the Contact page and modify the template navigation area by adding rounded borders. Add a pop-up message to the first name text box on the Contact page and use onFocus as the event to trigger the pop-up message action. The message should encourage the user to complete the form. Save your changes and close the Contact page. Then, open the template file and add the border-radius property to the navigation div. Use 25px as the border-radius value. Save your changes, open the home page, and view the Web site in your browser. Verify that the navigation area has rounded borders and the pop-up message is displayed on the Contact page when the insertion point is within the first name text box. Submit the document in the format specified by your instructor.

2: Adding CSS Transitions to Student Campus Resources

Academic

You have added all of the content for the Student Campus Resources Web site and now need to enhance the site with CSS transitions. You created a Spry Collapsible Panel on the Committees page and now need to add CSS properties and transitions to modify the look of the panels. On the Committees page, add the text-align property to the .CollapsiblePanelTab and use center as the value to center the panel tab text. Add the font-size property to the .CollapsiblePanelTab and use 12pt as the value to modify the size of the panel text. Use the CSS Transitions panel to add a transition for the target rule, .CollapsiblePanelTab. Use hover for the Transition On value. Add the background-color property and use a dark orange for the End Value color. Add the color property and use white as the End Value color. Save your changes, open the Committees page in your browser, and hover over the panels to view the CSS transitions. Submit the document in the format specified by your instructor.

3: Adding CSS3 Properties to French Villa Roast Café

Professional

You have added all of the content for French Villa Roast Café and now need to enhance the template with CSS3 properties. Open the template file and add the border-radius property to the header div. Use a minimum of 15px for the border-radius value. Next, add the box-shadow property to the header div and use 6px for the X- and Y-Offsets, 10px for the blur radius, and a dark red color value. Save your changes, open the home page, and view the Web site in your browser to see the CSS3 properties. Submit the document in the format specified by your instructor.

8 | Publishing a Web Site

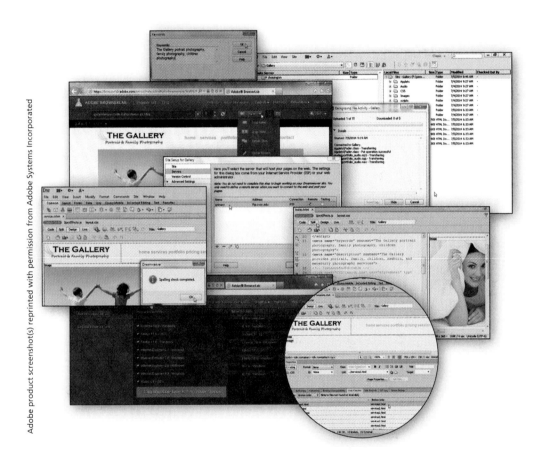

Objectives

You will have mastered the material in this chapter when you can:

- Understand copyright information
- Add meta tags
- Test a Web site using Adobe BrowserLab
- Test site links using the Link Checker
- Proofread a Web site

- Define a remote server
- Understand File Transfer Protocol (FTP)
- Define a remote server
- Publish a Web site to a remote server
- Manage Web site content

8 | Publishing a Web Site

Introduction

After designing an amazing Web site, it is now time to prepare, test, and publish the site to a Web server and share your Web presence. When preparing the final version of the site, make sure you have the proper rights to display all the images and you have identified the best keywords to describe your site for search engine purposes. It is vital that you test all aspects of your Web site before your efforts are displayed for the world to see. The testing phase, including viewing your layout and proofreading each page, ensures that all the hard work you put into your site is not undone by technical difficulties.

Publishing a Web site means transferring your files from your local host computer to your Web site host. Dreamweaver includes a File Transfer Protocol (FTP) interface that keeps your work flow seamless.

Project — Testing and Publishing the Gallery Site

The design phase of the Gallery site is complete, but the next phase of testing and publishing the site is critical to the success of the final project. Before the site is published to the Internet, each image and multimedia file should be checked to confirm that the correct copyright information is on file. Also consider how search engines will find your site. Place meta tags containing keywords and a site description in the head content of the site, which contains a variety of information, including the keywords that search engines use to identify your site. With the exception of the page title, all head content is invisible when viewed in the Dreamweaver document window or in a browser. Some head content is accessed by other programs, such as search engines, and some content is accessed by the browser. This chapter discusses the head content options and the importance of adding this content to Web pages. You must also proofread the entire site for typographical errors, check for broken links and typographical errors, and preview each page for proper layout design. Test all multimedia to verify it plays as intended.

The Gallery site shown in Figure 8–1a is being tested in Adobe BrowserLab, software from Adobe that ensures the site is displayed accurately across multiple browsers. In addition, the Link Checker can test the links in the site. After the site is thoroughly tested, it can then be published on a Web server. In this chapter, you publish a site to a remote server using an FTP built-in service, as shown in Figure 8–1b. As you complete the activities in this chapter, you will find that completing the detailed testing and publishing phase of development is crucial before customers begin accessing the Gallery site online.

(a) Testing the page in Adobe BrowserLab

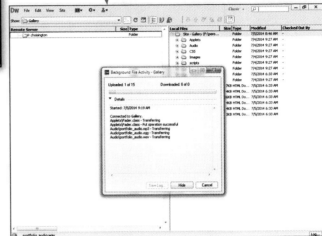

(b) Uploading the final page on a remote server

Figure 8–1

Overview

As you read this chapter, you will learn how to create the Web page project shown in Figure 8–1 by performing these general tasks:

- Determine if each image or multimedia file has the correct copyright information.
- Add keywords and a description to the head content.
- Test the site using Adobe BrowserLab.
- Test the links.
- Check spelling.
- Define a remote server.
- Publish a site to a remote server.

> **General Project Guidelines**
>
> As you prepare your Web site for publishing, consider the copyright information and how search engines can identify and access your site. Test the site thoroughly using the Adobe BrowserLab and Link Checker. Before publishing your site, identify the remote server to which you will add your files using the built-in FTP service in Dreamweaver. As you finalize and test your site using Adobe BrowserLab (Figure 8–1a) and publish your site using Dreamweaver's FTP built-in service (Figure 8–1b), you should follow these general guidelines:
>
> 1. **Determine copyright information.** Each image and multimedia object must be checked to ensure that the proper copyright permissions have been obtained.
>
> 2. **Plan head content.** Identify the keywords and descriptions to add to the head content for a page so that search engines and Web users can find your pages easily.
>
> 3. **Test a Web site for publication.** Before you publish a Web site, test it to verify it works as designed. Use Adobe BrowserLab to preview each page on a variety of browsers and use the Link Checker to validate links in the site. Proofread each page of the site carefully for grammatical and spelling errors.
>
> 4. **Determine where the site will be published.** A Web site can be published to a remote server of your choice using the FTP program within Dreamweaver.
>
> More specific details about these guidelines are presented at appropriate points throughout the chapter as necessary. The chapter also identifies the actions performed and decisions made regarding these guidelines during the development of the pages within the site shown in Figure 8–1a and Figure 8–1b.

To Start Dreamweaver and Open the Gallery Site

Each time you start Dreamweaver, it opens to the last site displayed when you closed the program. The following steps start Dreamweaver and open the Gallery Web site.

1 Click the Start button on the Windows 7 taskbar to display the Start menu, and then type `Dreamweaver CS6` in the 'Search programs and files' box.

2 Click Adobe Dreamweaver CS6 in the list to start Dreamweaver.

3 If the Dreamweaver window is not maximized, click the Maximize button next to the Close button on the Application bar to maximize the window.

4 If the Gallery site is not displayed in the Files panel, click the Sites button on the Files panel toolbar and then click Gallery to display the files and folders in the Gallery site.

Using Copyrighted Material

When you add images, widgets, multimedia, or any other element to a Web page, be aware of copyright issues. If you did not create the image or video yourself, you need permission from the owner before you can copy it and display it on your Web site. Even if an image or a multimedia file on the Web does not have a copyright notice, it is still protected by copyright laws.

Copyright protects content creators against unauthorized use of their work. If you improperly use copyrighted material, you could be violating the copyright law and be legally liable for damages to the content creator. To protect yourself, find and use materials that fall under the public domain, which means they are not protected by copyright and are freely available for use. The U.S. government, for example, releases images, videos, and other material to the public domain. If you are quoting an author's material or citing other works, you may be able to incorporate the material under fair use guidelines, which typically allow educational use. Otherwise, seek permission from the content creator to use his or her material.

In the case of the Gallery site, the photographer would not need permission to use his or her own photography, but would need the models or families used in the pictures to sign legal forms to release their images for publication on the Web.

At times, you may find an image that you want to use, but you may be unsure of where the image originated when you want to request digital rights to republish the image on your site. A Web site named TinEye (www.tineye.com) can help you determine the original source of an image. TinEye is a reverse image search engine that locates where an image came from on the Web, how it is being used, if modified versions of the image are available, or if you can access the image in a higher resolution.

Understanding Meta Tags

Before publishing a Web site to the Internet, consider how search engines will find your site. Which keywords should users enter to locate your site? HTML files consist of two main sections: the head section and the body section. The **head** section is one of the more important sections of a Web page because it contains vital information such as meta tags. Meta tags are used to specify the page description, keywords, author of the document, last modified date, and other information. **Meta tags** can help improve your Web site standings in search engines that use them. These special tags are hidden from the visible page but are read by most search engines, which use the tags to rank your site on results pages. Search engines use meta tags to understand the main topic or subject of a Web page. Although search engine algorithms keep getting smarter, they still depend on human direction in terms of content from meta tags.

BTW

Google and Meta Tags
The purpose of a meta tag is to provide a search engine with a description of the site. However, the Google search provider uses an actual text snippet from the beginning of the page for their description on the results page. Most other search engines use the meta description instead.

Plan head content.
Browsers and Web search tools refer to information contained in the head section of a Web page. Although this section is not displayed in the browser window, you can set the properties of the head elements to control how your pages are identified. At a minimum, you should set properties for the following head elements.

- **Keywords**: Enter keywords you anticipate users and search engines might use to find your page. Because some search engines limit the number of keywords or characters they track, enter only a few accurate, descriptive keywords.

- **Description**: Many search engines also read the contents of the Description text. Some search engines display the Description text in the search results, so be sure to enter a meaningful description.

Plan
Ahead

Head Content Elements

Dreamweaver makes it easy to add content to the head section through the Insert menu. To access the commands shown in Table 8–1, point to HTML on the Insert menu, and then point to the submenu of the command you want to select. (You also can display the Common category on the Insert bar and then click the Head button to display a list of head content commands.)

BTW

Viewing and Editing Head Content
You can view the elements in the head section of an HTML document by using the Head Content command on the View menu, displaying the document in Code view, or using the Code inspector.

Table 8–1 Head content elements	
Head Content Element	Description
Meta	A <meta> tag contains information about the current document. This information is used by servers, browsers, and search engines. HTML documents can have as many <meta> tags as needed. Each item uses a different set of tags.
Keywords	Keywords are a list of words that someone would type into a search engine's search field. Typically limit the keywords to about a dozen words for search engine optimization.
Description	The description contains a sentence or two that can be used in a search engine's results page to display information about the main topic of your page to the user.
Refresh	The <refresh> tag is processed by the browser to reload the page or load a new page after a specified amount of time has elapsed.
Base	The <base> tag sets the base URL to provide an absolute link and/or a link target that the browser can use to resolve link conflicts.
Link	The <link> element defines a relationship between the current document and another file. This is not the same as a link in the document window.

To Add Keywords and a Description Meta Tag

When your target audience searches for the Gallery Web site using a search engine such as Google or Bing, the search term or query may include phrases such as *The Gallery portrait photography*, *family photography*, or *children photography*. You enter these terms in the Keywords dialog box, and Dreamweaver inserts them in the meta tag for the page. Use a comma to separate search terms in the Keywords text box. Enter a description of the Gallery site using the Description dialog box. The text you enter appears in the results page of the search engine. Dreamweaver also enters this description in the meta tag for the page. The following steps add keywords and a description to the index.html page to provide meta tags for search engines.

1

- Open the index.html page to display the home page.

- Click Insert on the Application bar and then point to HTML to display the HTML submenu.

- Point to Head Tags on the HTML submenu to display the Head Tags submenu (Figure 8–2).

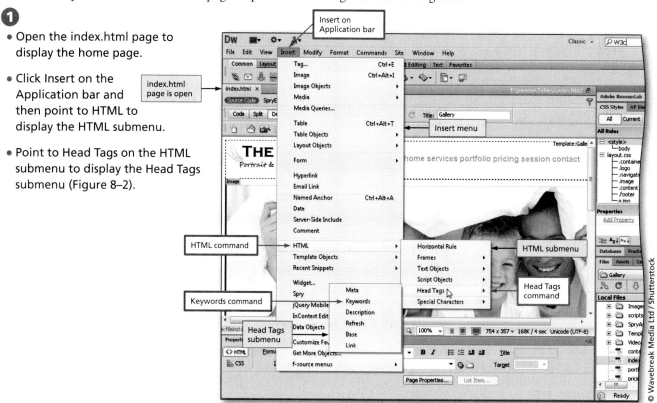

Figure 8–2

© Wavebreak Media Ltd / Shutterstock

2

- Click Keywords on the HTML submenu to display the Keywords dialog box (Figure 8–3).

Figure 8–3

3

- Type `The Gallery portrait photography, family photography, children photography` in the Keywords text box to add the keywords (Figure 8–4).

 Q&A

What does a search engine typically do with the keywords?

When a search engine begins a search for any of the keywords typed into a query, the matching Web pages are displayed with their hyperlinks and descriptions in the search results page.

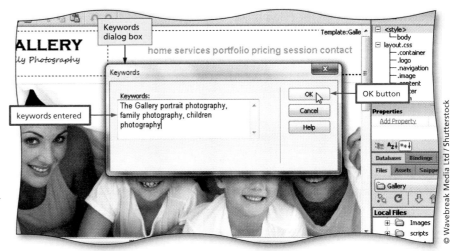

Figure 8–4

● Click the OK button to add the keywords to the head tag and close the Keywords dialog box.

● Click Insert on the Application bar and then point to HTML to display the HTML submenu.

● Point to Head Tags on the HTML submenu to display the Head Tags submenu (Figure 8–5).

Figure 8–5

● Click Description on the Head Tags submenu to open the Description dialog box (Figure 8–6).

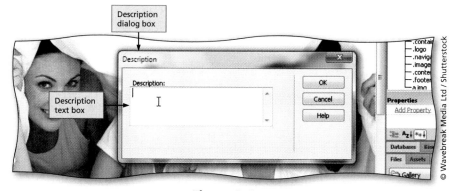

Figure 8–6

● Type The Gallery provides portrait, family, children, newborn, and maternity photography services in the Description text box to describe the Web page (Figure 8–7).

 What is the purpose of the description meta tag?

The description contains a sentence or two that can be displayed in a search engine's results page.

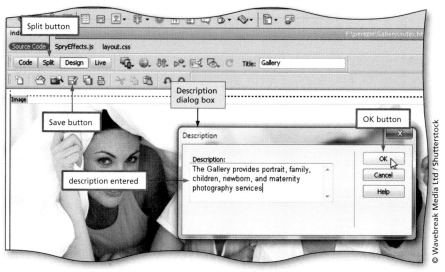

Figure 8–7

7
- Click the OK button to close the Description dialog box.

- Click the Save button on the Standard toolbar to save your work.

- Click the Split button on the Document toolbar and then scroll up as necessary to view the meta tags in the code (Figure 8–8).

8
- Click the Design button on the Document toolbar to return to Design view.

- Close the index.html page to display the Welcome screen.

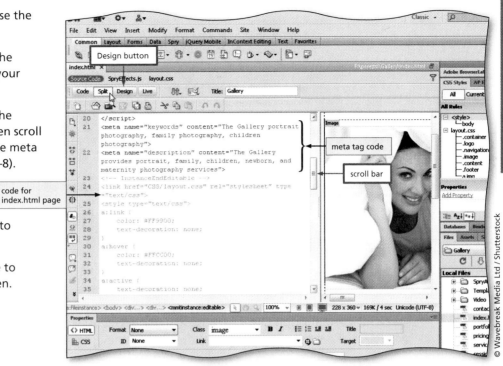

Figure 8–8

Other Ways
1. Click Head button on Common Insert bar

Testing a Web Site

Once all the pages are completed and linked, the site is ready for user testing. You can install multiple browsers on your local computer to test layout, links, and multimedia, but Adobe provides a testing environment to check your live code against all the major browsers. Each internal and external link within the site should be validated to confirm that it connects properly. Dreamweaver produces a report of all the broken links and displays it in the Results panel.

In addition to using Dreamweaver's built-in testing features, you should have users outside of the project team record errors and report on their experience of using the site. Before publishing your site, carefully proofread and check the site for typos and spelling errors. After collecting the results of user testing, you may need to return to the site design phase or begin the site development phase again to correct errors.

You can view each page of your site in Live view, which provides a realistic rendering of what your page looks like in a browser. Live view does not replace testing the page in multiple browsers, but rather provides another way of seeing what your page looks like "live" without having to leave the Dreamweaver workspace. While Live view reduces the need to jump back and forth to test an individual component in an actual browser, it is still crucial that you ensure the entire site renders consistently across multiple browsers and platforms. Because different Web browsers and operating systems can each display page layout differently, developers must test the site on all major browsers to make sure each aspect of the page is displayed properly.

Test a Web site for publication.

• **Test the site in multiple browsers.** Test the completed site in multiple browsers to determine consistency across different platforms. Adobe BrowserLab can take a screen shot of a Web site in different browsers to test each element used within the site.

• **Test the links of the site.** Test the links of the site to confirm that the site navigation works as intended. Before publishing a Web site, use the Dreamweaver Results panel to make sure your Web pages are professional and complete.

• **Proofread the site.** Read through all the text in the site to test for writing inconsistencies, spelling errors, and typographical mistakes.

Testing a Site using Adobe BrowserLab

Adobe BrowserLab, a feature of Dreamweaver CS6, is an online service that provides an easy solution for cross-browser testing. Adobe BrowserLab is part of the Creative Suite online services, which allow you to preview and compare screen shots of pages individually or in grouped Browser Sets. You can preview local Web content from within Dreamweaver, without having to post to a publicly accessible server first. For example, you can test the Gallery site in multiple versions of browsers including Chrome, Safari, Firefox, and Internet Explorer, and on Windows and Mac OS, eliminating the need to install each browser on your local computer. A Web page can look different in each browser. As a Web designer, you want each page to create a similar user experience regardless of the platform used. Adobe BrowserLab can accurately display potential browser compatibility issues and compare pages at a glance.

To Open Adobe BrowserLab

Before publishing the Gallery site to the Web, preview each page of the site in multiple browsers using Adobe BrowserLab. The following steps open Adobe BrowserLab to test the site. You need the Adobe ID and password that you created in Chapter 5 to use Adobe BrowserLab in these steps.

1

• Double-click the services.html file in the Files panel to display the contents of the Services page.

• Click the Adobe BrowserLab tab in the Panels group to display the Adobe BrowserLab panel (Figure 8–9).

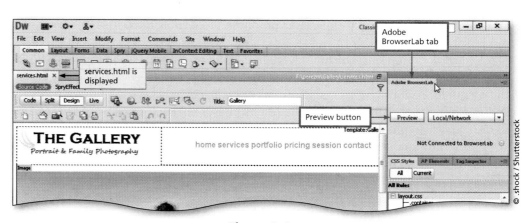

Figure 8–9

2

- Click the Preview button to launch Adobe BrowserLab in your default browser (Figure 8–10).

Q&A

What should I do if the Adobe BrowserLab Permission Settings dialog box is displayed?

With the Allow option button selected, click the OK button.

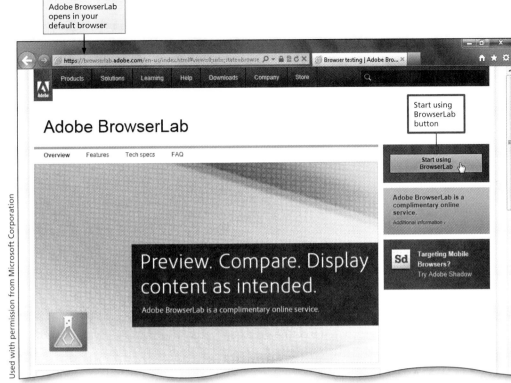

Figure 8–10

3

- If necessary, click the 'Start using BrowserLab' button to display the Sign In dialog box (Figure 8–11).

Q&A

What should I do if the Sign In dialog box is already displayed?

Skip ahead to Step 4.

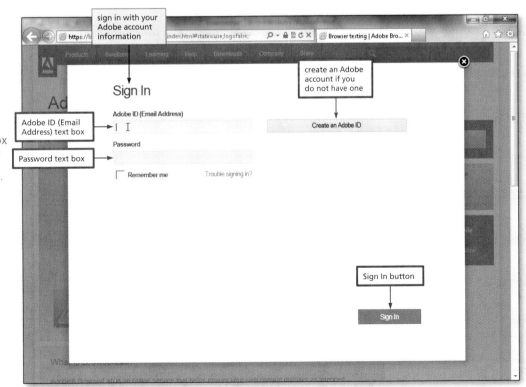

Figure 8–11

- In the Adobe ID (Email Address) and Password text boxes, type your Adobe ID and password.

- Click the Sign In button to send a verification e-mail to your personal e-mail address.

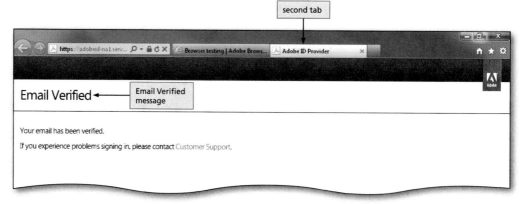

Figure 8–12

- If Adobe indicates it is verifying your e-mail address, open your personal e-mail program and locate the e-mail from Adobe with the subject line "Please verify your e-mail address" to find the verification e-mail sent by Adobe.

- Open the e-mail and click the link to verify your e-mail address and display the Email Verified message on the second tab of the Web page (Figure 8–12).

Q&A What should I enter for the Adobe ID and password?

Enter the Adobe ID and password you used when you accessed Adobe BrowserLab in Chapter 5.

Q&A What if Adobe does not need to verify my e-mail address?

Skip the rest of the steps in this section and compare your screen to Figure 8–14.

5

- Close the Adobe ID Provider tab in the browser to return to the Sign In Web page on the first tab of the browser.

- Read the Terms of Use and then click the check box to accept the Terms of Use (Figure 8–13).

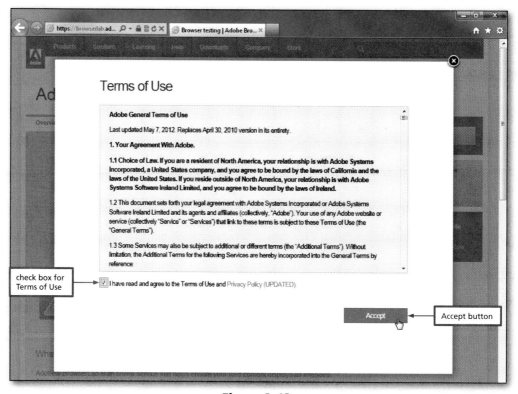

Figure 8–13

6

- Click the Accept button on the Terms of Use page to display Adobe BrowserLab with the Services page screen shot displayed in the Internet Explorer 6.0 browser (Figure 8–14).

Q&A What if I do not want to test my Web site with the Internet Explorer 6.0 browser?

Adobe BrowserLab is a cross-browser testing environment with multiple browser selections. You can select a different browser by clicking the browser button.

Q&A Is a live Web page being displayed in the browser?

No. Adobe BrowserLab takes a screen shot of the Web page. You can compare multiple browser screen shots at the same time.

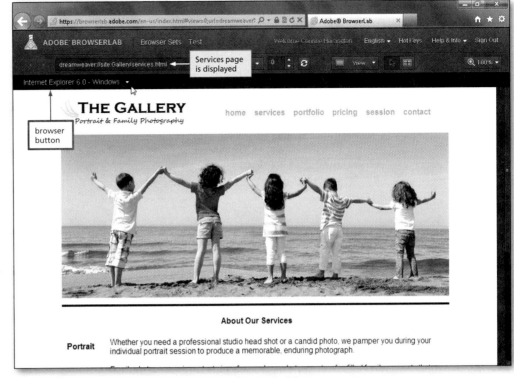

Figure 8–14

© .shock / Shutterstock

To Create a New Browser Set in Adobe BrowserLab

You can determine which browsers to use during the testing phase within Adobe BrowserLab. The Gallery's target audience may use older browsers on a PC or a Mac. Although BrowserLab tests the appearance of your page in a default browser set, you can select the browser you want to use for testing by creating your own browser set. The following steps create a custom browser set to test a site in multiple browsers.

1

- Click the Browser Sets button on the Adobe BrowserLab toolbar to add a new browser set (Figure 8–15).

Q&A Should I test the site in every browser ever made?

A good rule of thumb for testing a site within a browser is to test the current browsers as well as browsers that are at least two versions behind the most current browser version.

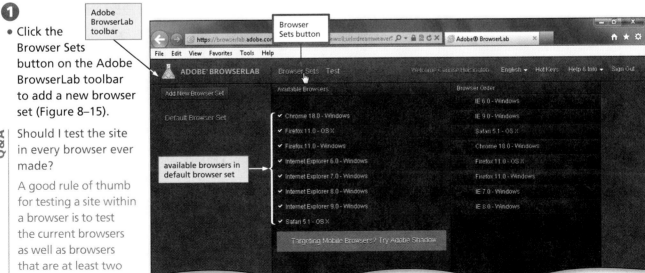

Figure 8–15

2

- Click the Add New Browser Set button to create a new browser set in Adobe BrowserLab (Figure 8–16).

Q&A Why is the new browser set named Custom Set #1?

Custom Set #1 is the default name for the first custom browser set you create.

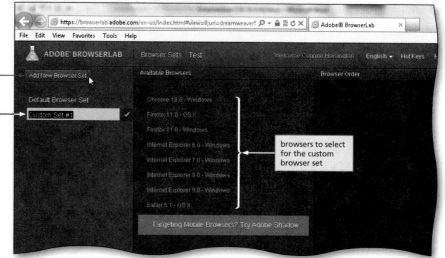

Add New Browser Set button

enter name for browser set

browsers to select for the custom browser set

Figure 8–16

3

- Type My Browser Set to rename the custom browser set (Figure 8–17).

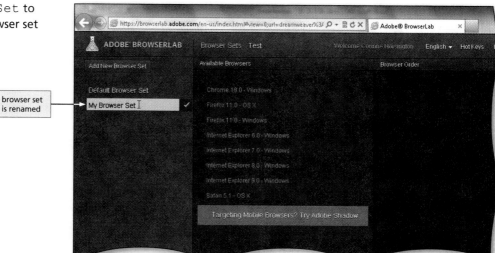

browser set is renamed

Figure 8–17

4

- Click the check box for each browser listed except Internet Explorer 6.0 to select seven browsers for the custom browser set (Figure 8–18).

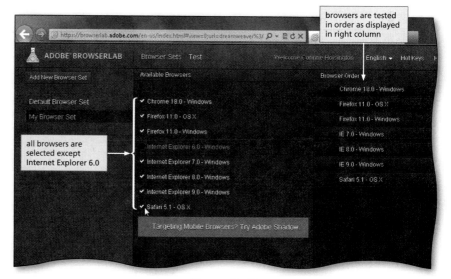

browsers are tested in order as displayed in right column

all browsers are selected except Internet Explorer 6.0

Figure 8–18

5

- Click the Test button on the Adobe BrowserLab toolbar to display the Gallery site in the first browser in the My Browser Set list (Figure 8–19).

Q&A

What if Chrome 18.0 is not the first browser listed?

Use the first browser listed. If necessary, you can change the order of the browsers in the right column. You may see different versions of browsers based on the most current versions.

Figure 8–19

© .shock / Shutterstock

6

- Click the browser button and then click Safari 5.1 – OS X to select the most recent version of the Safari browser (Figure 8–20).

Q&A

How do I test the other pages in the site?

In the Address bar of Adobe BrowserLab (not the address bar of the browser), replace the text, services.html, with the other page names such as index.html, portfolio.html, pricing .html, session.html, and contact.html.

Q&A

Why doesn't my video or music play when I test multimedia in Adobe BrowserLab?

Adobe BrowserLab takes screen shots of pages as they are displayed in the selected browser. Audio and video files cannot play within a screen shot. To test multimedia files, it is best to use an actual browser.

Figure 8–20

© .shock / Shutterstock

7

- Select each browser on the list to test the services.html page in multiple browsers.

To Change the View in Adobe BrowserLab

Instead of viewing each page in Adobe BrowserLab individually, you can change how a page is viewed. To compare a page in two browsers at the same time, you can change the view to 2-up View or Onion Skin view. The **2-up View** provides a side-by-side comparison so you can check the differences in appearance between two browsers. This makes it easier to focus on a specific area that might require certain modifications rather than switching from browser to browser. The **Onion Skin** view option exposes the minute differences by layering two copies of the Web page on top of each other. The following steps change the view in the Adobe BrowserLab to view the Services page in two browsers at the same time.

1

- Click the View button in the Adobe BrowserLab window to display the View list (Figure 8–21).

Figure 8–21

2

- Click 2-up View in the View list to display the services. html page in two browsers.

- If necessary, click the Change browser button on the left and then select the latest version of Safari.

- If necessary, click the Change browser button on the right and then select the latest version of Chrome (Figure 8–22).

Q&A Why does a shadow appear below the image in the Chrome browser?

Each browser establishes rules of how it displays HTML code. It is important to test each page in the most popular browsers to confirm how each page is displayed.

Figure 8–22

3

- Click the View button to display the View list.

- Click Onion Skin in the View list to display the services. html page in two browsers, with one page superimposed on top of the other, to compare how the page elements align (Figure 8–23).

Q&A

Why does this view look blurry?

The Services page looks blurry because the Safari and Chrome page views do not align perfectly.

Figure 8–23

 **Experiment**

- Click the View button and then click Show Rulers to display a ruler to gauge how far each element is apart on the page. When you are finished, click the View button and then click Hide Rulers to hide the rulers.

4

- Close the browser window to close Adobe BrowserLab.

Break Point: If you wish to take a break, this is a good place to do so. To resume at a later time, start Dreamweaver and continue following the steps from this location forward.

Testing using the Link Checker

Be sure to check and verify that the links work on your Web page. Links are not active within Dreamweaver; that is, you cannot open a linked document by clicking the link in the document window. You can check any type of link by displaying the page in a browser. Using a browser is the only available validation option for absolute or external links and e-mail links. For relative or internal links, Dreamweaver provides the Link Checker feature. Use the Link Checker to check internal links in a document, a folder, or an entire site.

A large Web site can contain thousands of links that can change over time. Dreamweaver's Link Checker searches for broken links and unreferenced files in a portion of a local site or throughout an entire local site. This feature is limited, however, because it verifies internal links only. A list of external links is compiled, but not verified. Keep in mind that you must check external links through a browser.

The Link Checker does have advantages, however. When you use this feature, the Link Checker displays a statistical report that includes information about broken

BTW

Site Testing
Before publishing your site, make sure that your pages look and work as expected in your targeted browsers, contain no broken links, and don't take too long to download.

links, orphaned files, and external links. An **orphaned file** is a file that is not connected to any page within the Web site. Although the orphaned file option is for informational purposes only, the orphaned file report is particularly valuable for a large site because it displays a list of all files that are not part of the Web site. Deleting unused files from the Web site increases server space and streamlines your site. You can use the Link Checker to check links throughout your entire site from any Web page within the site. After you run the Link Checker, you can fix broken links in the Property inspector or directly in the Link Checker panel.

To Verify Internal Links with the Link Checker

To test the internal links in the Gallery Web site, you can change a link in the navigation bar on the Gallery Template to reference the incorrect Web page to illustrate how the Link Checker works and how to make corrections. Your results may be different from the ones shown if your page has other link errors. The following steps use the Link Checker to verify the internal links for the Gallery Template.

- If necessary, open Dreamweaver, select the Gallery site, and then expand the Gallery folder and the Templates folder.

- Double-click the Gallery Template file to display the contents of the Gallery Template.

- Select the text, services, in the navigation div region at the top of the page to display the correct link in the Link text box in the Property inspector (Figure 8–24).

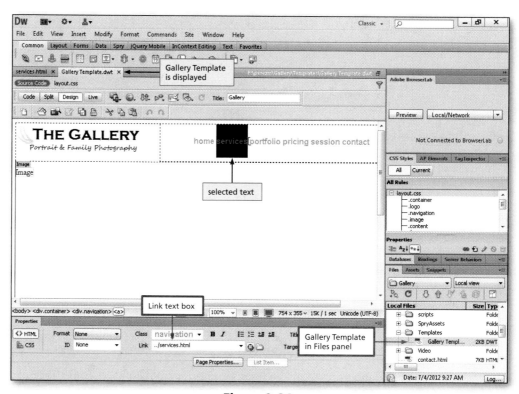

Figure 8–24

2

- Type 2 after services in the Link text box to change the link from ../services.html (the correct link) to ../`services2.html` (an incorrect link) to test the Link Checker (Figure 8–25).

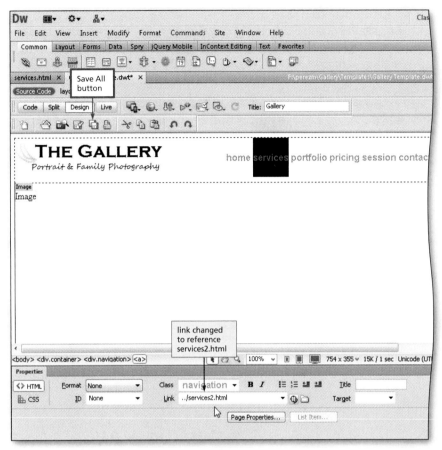

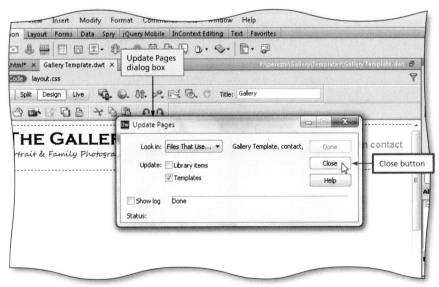

Figure 8–25

3

- Click the Save All button on the Standard toolbar to display the Update Template Files dialog box.

- Click the Update button to display the Update Pages dialog box (Figure 8–26).

 Why is the Update Pages dialog box displayed?

Recall that the Update Pages dialog box is displayed when you change the template for a site. You can select the pages to update with the same change you applied to the template, or you can click the Close button to apply the change to all the pages attached to the template.

Figure 8–26

● Click the Close button to close the Updates Pages dialog box and update all the pages attached to the Gallery Template.

● Click Site on the Application bar to display the Site menu (Figure 8–27).

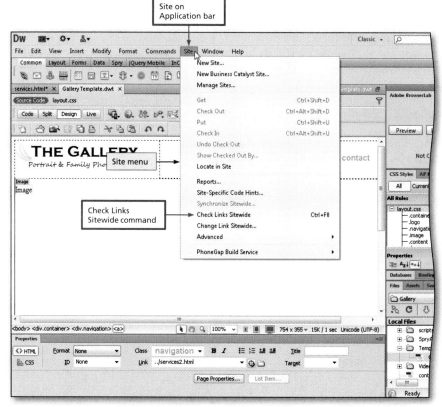

Figure 8–27

● Click Check Links Sitewide to display the Link Checker tab in the Results panel below the Property inspector (Figure 8–28).

Q&A

Why does the report state that services2.html is broken in multiple files?

The Results panel displays a report identifying the broken links in the site. When you closed the Update Pages dialog box in a previous step, you updated the incorrect link entered in the Gallery Template to all six pages within the site.

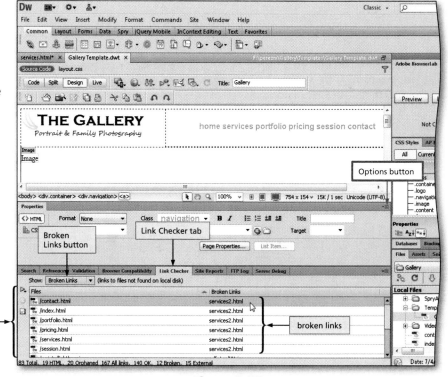

Figure 8–28

- In the Property inspector, rename the services2.html link as `services.html` to restore the correct file name for the link.

- Click the Save All button on the Standard toolbar to display the Update Template files dialog box.

- Click the Update button to apply the change to all the files attached to the template and to display the Update Pages dialog box.

- Click the Close button to close the Update Pages dialog box.

- Click the Options button on the Results panel and then click Close Tab Group to close the Results panel and return the Dreamweaver window to its Classic state (Figure 8–29).

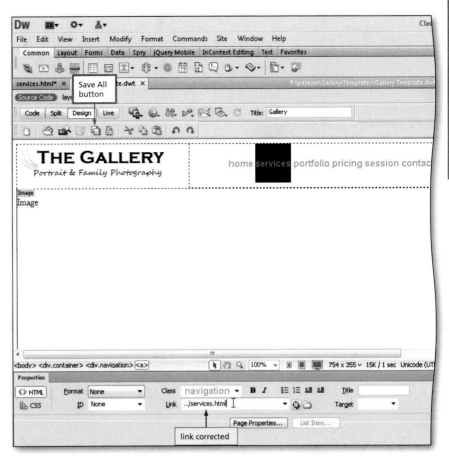

Figure 8–29

Proofreading a Web Site

In addition to checking the layout of the site, carefully proofread the site for typographical or spelling errors. Errors on a Web site detract from your image with your target audience, which might assume other Web site information such as content is inaccurate. Having too many errors can discourage visitors from returning to your site. In addition, typographical errors, grammatical inconsistencies, and spelling mistakes reduce the amount of traffic to your site by lowering the search engine ranking that is assigned to a Web site. To avoid these problems, Dreamweaver includes a spell checking feature with an English spelling dictionary.

To Check Spelling

Dreamweaver can only check the spelling in the file that is currently open in the document window. It cannot check the spelling of all of the files in a site simultaneously. The following steps check the spelling of a single page in the site. This process can be repeated on each page of the site.

- Close the Gallery Template file and, if necessary, open the services.html page.

- Click Commands on the Application bar to display the Commands menu (Figure 8–30).

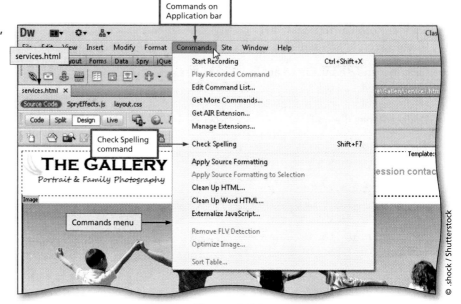

Figure 8–30

- Click Check Spelling to begin checking the spelling on the page. If necessary, check the spelling and make any necessary corrections on the Services page until the Dreamweaver dialog box shows the message, Spelling check completed (Figure 8–31).

Q&A When the Check Spelling dialog box opens, what does the 'Add to Personal' button do?

The 'Add to Personal' button adds an unrecognized word to your personal dictionary.

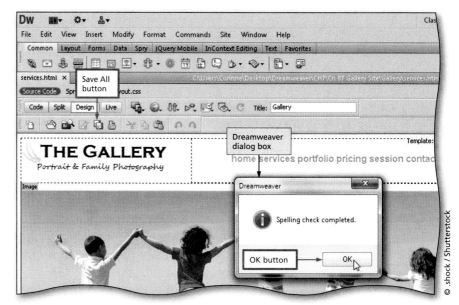

Figure 8–31

- Click the OK button to finish checking the spelling in the Services page.

- Open each HTML page within the site and check each page using the Check Spelling command. Close each page after checking the spelling in the document.

- Click the Save button on the Standard toolbar to save your work.

Other Ways

1. Press SHIFT+F7

Publishing to a Web Server

With Dreamweaver, Web designers usually define a local site and then use it to do the majority of their designing. You defined a local site in Chapter 2. In creating the projects in this book, you have added Web pages to the local Gallery site, which resides on your computer's hard drive, a network drive, SkyDrive, or a USB drive. The remote server is associated with a domain name that you can purchase at sites such as www.networksolutions.com. A domain name is an Internet address that is formed by the rules and procedures of the **Domain Name System (DNS)**. The DNS associates various pieces of information with domain names assigned to each of the participating organizations that purchased a unique name.

Determine where the site will be published.
Publish a site to a remote server. A Web site must be hosted on a Web server to be viewed on the Internet. Each page of the site is posted to a remote server using an FTP address.

Plan Ahead

To prepare a Web site and make it available for others to view, you must publish your site by uploading it to a Web server for public access. A **Web server** is an Internet- or intranet-connected computer that delivers the Web pages to online visitors. Dreamweaver includes built-in support that enables you to connect and transfer your local site to a Web server. To publish to a Web server, you must have access to a Web server. Your instructor will provide you with the location, user name, and password information for the Web server on which you will publish your site locally on your school's server or a public remote Web server.

After you establish access to a Web server, you need a remote site folder. The remote folder will reside on the Web server and contain your Web site files. Generally, the remote folder is defined by the Web server administrator or your instructor. The name of the local root folder in this example is the author's first and last names. Most likely, the name of your remote folder also will be your last name and first initial or your first and last names. You upload your local site to the remote folder on the Web server. The remote site connection information must be defined in Dreamweaver through the Site Setup dialog box. You display the Site Setup dialog box, select the Servers category, and then enter the remote site information. Dreamweaver provides the following protocols for connecting to a remote site. These methods are as follows:

- **FTP (File Transfer Protocol):** This protocol is used on the Internet for sending and receiving files. It is the most widely used method for uploading and downloading pages to and from a Web server.

- **Local/Network:** This option is used when the Web server is located on a local area network (LAN) or the intranet for a company, school, or other organization. Files on LANs generally are available for internal viewing only.

- **RDS (Remote Development Services) and WebDAV:** These protocols are systems that permit users to edit and manage files collaboratively on remote Web servers.

  Most likely you will use the FTP option to upload your Web site to a remote server.

Defining a Remote Site

You define the remote site by changing some of the settings in the Site Setup dialog box. To allow you to create a remote site using FTP, your instructor will supply you with the following information:

- **Server name:** The name of the server where your remote site will be stored
- **FTP address:** The Web address for the remote host of your Web server
- **Username:** Your user name
- **Password:** The FTP password to authenticate and access your account
- **Web URL:** The URL for your remote site

To Define a Remote Site

The following steps define a remote site for the Gallery Web site.

1

- Double-click the index.html file in the Files panel to display the home page.

- Click Site on the Application bar to display the Site menu (Figure 8–32).

Figure 8–32

© Wavebreak Media Ltd / Shutterstock

2

- Click Manage Sites on the Site menu to open the Manage Sites dialog box.

- If necessary, click Gallery to select the Gallery Web site (Figure 8–33).

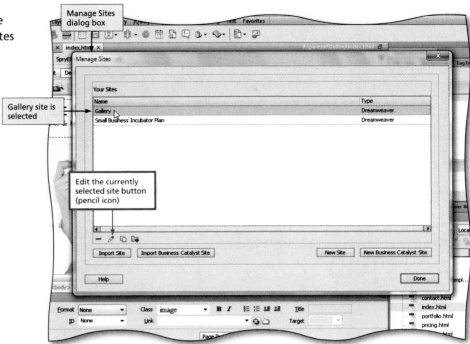

Figure 8–33

3

- Click the 'Edit the currently selected site' button (pencil icon) to open the Site Setup for Gallery dialog box (Figure 8–34).

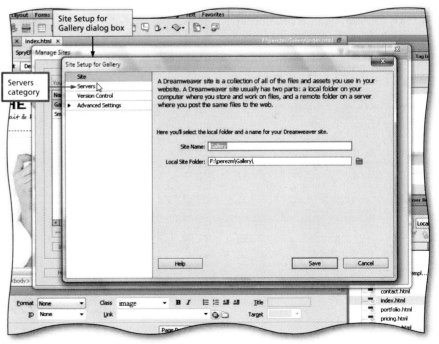

Figure 8–34

• Click Servers to display controls for entering the server settings in the Site Setup for Gallery dialog box (Figure 8–35).

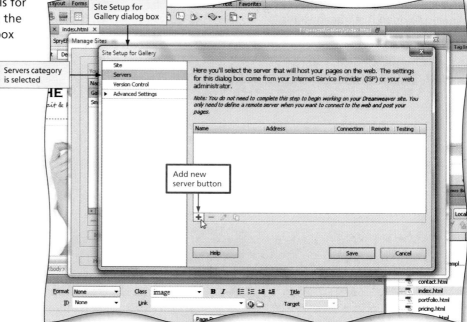

Figure 8–35

• Click the 'Add new server' button (plus button) to display the basic server information (Figure 8–36).

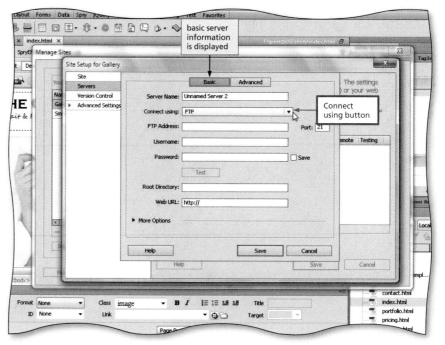

Figure 8–36

- If necessary, click the Connect using button, and then click FTP to select FTP as the protocol for connecting to a remote site.

- In the appropriate text boxes, enter the Server Name, FTP Address, Username, Password, and Web URL information provided by your instructor (Figure 8–37).

Q&A What if I am required to enter different information from that shown in Figure 8–37?

Your information will differ from that in Figure 8–37.

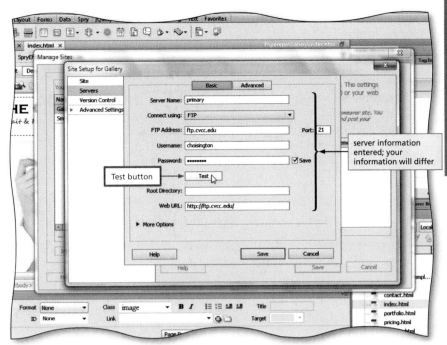

Figure 8–37

- Click the Test button to test the connection and display the Dreamweaver dialog box (Figure 8–38).

Q&A What should I do if a security dialog box is displayed?

If a Windows Security Alert dialog box is displayed, click the Allow access button.

Q&A What should I do if my connection is not successful?

If your connection is not successful, review your text box entries and make any necessary corrections. If all entries are correct, check with your instructor.

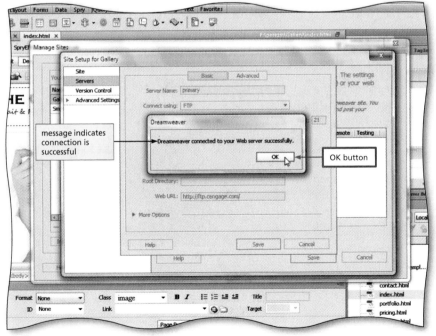

Figure 8–38

8

- Click the Save button to close the Dreamweaver dialog box and to display the server information in the Site Setup for Gallery dialog box (Figure 8–39).

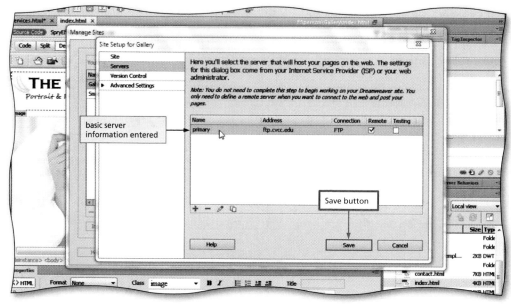

Figure 8–39

9

- Click the Save button in the Site Setup for Gallery dialog box to save the server information.

- If a Dreamweaver dialog box is displayed, click the OK button to verify changes in the settings.

- Click the Done button to close the Manage Sites dialog box and display the home page (Figure 8–40).

Q&A

What should I do if another Dreamweaver dialog box is displayed?

If another Dreamweaver dialog box is displayed, click the OK button.

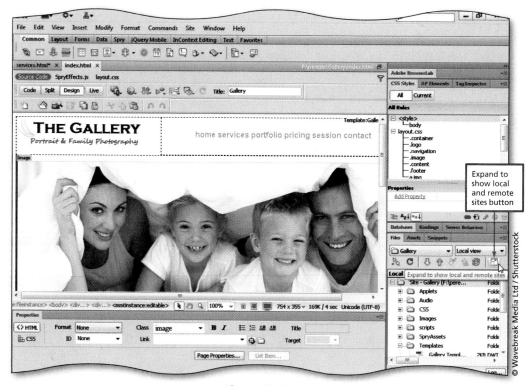

Figure 8–40

Connecting to a Remote Site

Now that you have completed the remote site information and tested your connection, you can interact with the remote server. You must create the remote site folder on the Web server for your Web site before a connection can be made. This folder, called the remote site root, generally is created automatically by the Web server administrator of the hosting company or by your instructor. The naming convention usually is determined by the hosting company.

This book uses the last name and first initial of the author for the user name of the remote site folder. Other naming conventions may be used on the Web server to which you are connecting. Your instructor will supply you with this information. If all information is correct, connecting to the remote site is easily accomplished through the Files panel.

To Connect to a Remote Site

The following steps connect to the remote server and display your remote site folder.

1

- On the Files panel toolbar, click the 'Expand to show local and remote sites' button to expand the Site pane and show both a right (Local Files) pane and a left (Remote Server) pane (Figure 8–41).

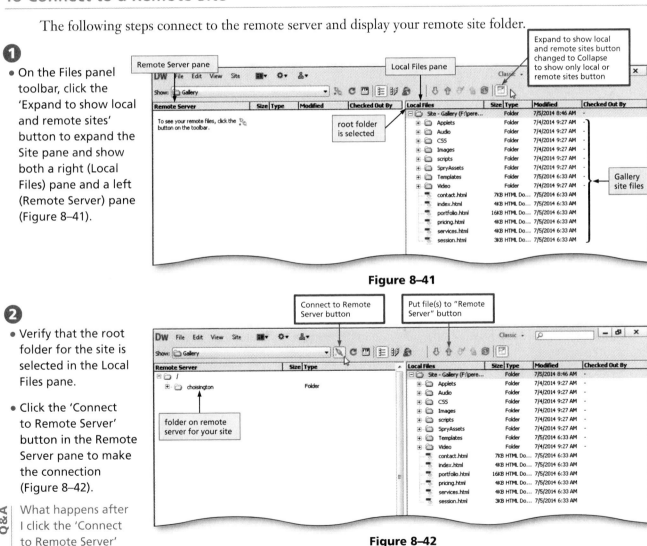

Figure 8–41

2

- Verify that the root folder for the site is selected in the Local Files pane.

- Click the 'Connect to Remote Server' button in the Remote Server pane to make the connection (Figure 8–42).

Q&A

What happens after I click the 'Connect to Remote Server' button?

Figure 8–42

The 'Connect to Remote Server' button changes to the 'Disconnect from Remote Server' button to indicate that the connection has been made, and a default Home.html folder is created in the remote root folder.

Uploading Files to a Remote Server

Uploading is the process of transferring your files from your computer to the remote server. **Downloading** is the process of transferring files from the remote server to your computer. Dreamweaver uses the term, put, for uploading and the term, get, for downloading.

To Upload Files to a Remote Server

The following steps upload your files to the remote server.

1

- If necessary, click the Gallery root folder in the Local Files panel to select the root folder.

- Click the 'Put File(s) to "Remote Server"' button on the Files panel toolbar to display a Dreamweaver dialog box (Figure 8–43).

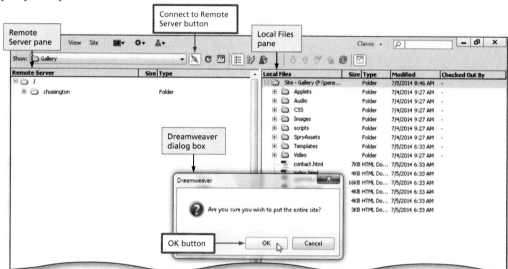

Figure 8–43

2

- Click the OK button to begin uploading the files and to display a dialog box that shows progress information (Figure 8–44).

Q&A

My files are uploaded, but they appear in a different order. Is that okay?

The files that are uploaded to the server may be displayed in a different order from that on the local site based on the server settings.

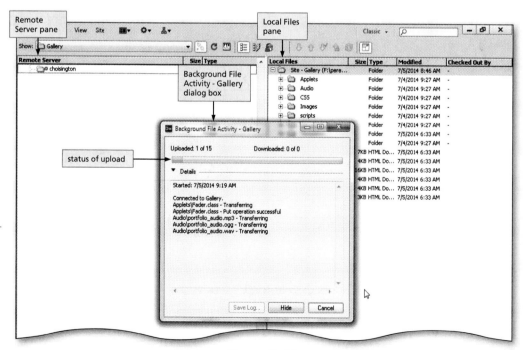

Figure 8–44

3

- When you are finished uploading the files, click the 'Collapse to show only local or remote sites' button on the toolbar to return the Files panel to its default state.

Web Site Project Management

Typically, Web developers work in teams to complete a Web project in larger companies and schools. The project team works together to develop a project plan that usually outlines the tasks and responsibilities of each team member. Roles include project manager, Web page engineer (who might use Dreamweaver to develop the site), designers for Web graphics and interactive media objects, and a site editor. The project manager is primarily responsible for defining the project scope, assigning tasks to team members, setting and tracking due dates, and allocating resources such as computer equipment, software, and people. The site editor ensures editorial quality, but also must make sure that the material on the site does not violate copyright laws. The site editor verifies that rights are secured for protected material, that material is original, or that material is free of copyright.

The project plan is organized by phase as shown in Table 8–2. Most Web site projects progress through the following phases from initial proposal of the Web site to completion in a predetermined workflow.

Table 8–2 Project Plan Phases

Project Planning Phase	Description
Site definition and planning	During this phase, the Web site project team defines objectives for the Web site and begins to collect and analyze the information it needs. When the site involves media objects, part of the planning includes identifying the technologies the site will use, such as plug-ins visitors may need. Identifying the purpose and target audience of the Web site and communicating with project stakeholders and users at this phase helps to avoid **project creep**, the gradual process of adding unplanned content or features to the Web site, resulting in a bloated Web site with poor focus.
Information design	After planning the Web site, the project team identifies the content and proposes an organization for the Web site. During this phase, the team builds prototype pages to test the design and navigation of the site. The team typically creates wireframes, which are visual guides to a Web site showing the relationship among Web pages. A storyboard is another graphic organizer that includes a series of screen shots or sketches that represent Web pages. These graphics are organized in sequence to visualize a Web site and test interactivity. A detailed storyboard includes file names, Web page titles and images, and brief content descriptions, and shows how pages are linked.
Site design	The project begins to take shape during this phase. The team defines the overall design of Web pages and the entire site. Members begin to gather the graphics and media for the site. Writers and editors begin to organize the text content. The goal is to produce the basic content so the Web pages can be created. Deliverables include style sheets, completed templates, graphics, navigation design, and text.
Site construction	Now the Web site can be built according to the design and using the content developed in previous phases. Deliverables during this phase include finished HTML files for all Web pages, images such as graphics and photos saved in appropriate folders, and other components such as style sheets and templates saved in an appropriate folder structure. When all the pages are completed and linked, the site is ready for user testing in Adobe BrowserLab and actual browsers. Users should be people outside of the project team who record errors and report on the experience of using the site. After collecting the results of user testing, the team might need to return to the site design phase or begin the site construction phase again to correct errors.
Site launch	The Web site files are published to a Web server and the URL for the site is marketed.
Post launch	The site is continually updated to keep the information and the technology current. Typically a maintenance plan is set up with the client.

Updating the Site

After publishing your Web site to a remote server, the long-term commitment to keep the site current and to continue adding additional features begins. By adding new information to the Gallery Web site, the site stays fresh and the target audience will continually return looking for recent content or photography specials. By simply changing the graphics of the site monthly, the site continues to be new and up to date. You must keep up with new Web standards and browser versions after the initial launch of your Web presence. The next chapter adds a mobile version of the site for use on smartphones to attract a mobile audience to the site.

To Quit Dreamweaver

The following steps quit Dreamweaver and return control to the operating system.

1 Click the Close button on the right side of the Application bar to close the Dreamweaver window and any open files.

2 If Dreamweaver displays a dialog box asking you to save changes, click the No button.

Chapter Summary

In this chapter, you finalized the Web site by examining copyright information and securing necessary permissions before publishing your completed site. You added meta tags to advance your Web site's standings within search engines. You learned about the Results panel and how to use Adobe BrowserLab and the Link Checker. You proofread and checked the spelling in the site. Finally, you published your site to a remote server. The following tasks are all the new Dreamweaver skills you learned in this chapter:

1. Add Keywords and a Description Meta Tag (DW 428)
2. Open Adobe BrowserLab (DW 432)
3. Create a New Browser Set in Adobe BrowserLab (DW 435)
4. Change the View in Adobe BrowserLab (DW 438)
5. Verify Internal Links with the Link Checker (DW 440)
6. Check Spelling (DW 444)
7. Define a Remote Site (DW 446)
8. Connect to a Remote Site (DW 451)
9. Upload Files to a Remote Server (DW 452)

Apply Your Knowledge

Reinforce the skills and apply the concepts you learned in this chapter.

Adding Meta Tags

Note: To complete this assignment, you will be required to use the Data Files for Students. Visit www.cengage.com/ct/studentdownload for detailed instructions or contact your instructor for information about accessing the required files.

Instructions: In this activity, you review the content of a Web page and add meta tags for keywords and description. The Web page is shown in Figure 8–45.

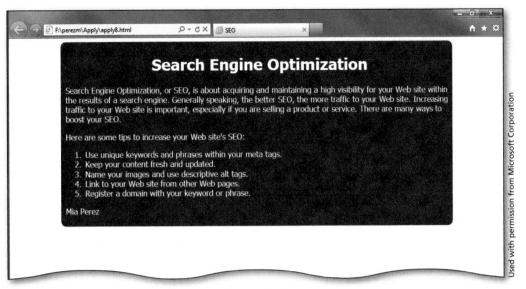

Figure 8–45

Perform the following tasks:

1. Use Windows Explorer to copy the apply8.html file from the Chapter 08\Apply folder into the *your last name and first initial*Apply folder.

2. Start Dreamweaver. Use the Sites button on the Files panel to select the Apply site. Open apply8.html and review the Web page content.

3. Insert a keywords meta tag and use `search engine optimization, SEO, Web site visibility` for the keywords content.

4. Insert a description meta tag and use `Search engine optimization tips` for the description content.

5. Replace the text, Your name here, with your first and last names.

6. Save your changes.

7. Submit the document in the format specified by your instructor.

Extend Your Knowledge

Extend the skills you learned in this chapter and experiment with new skills. You may need to use Help to complete the assignment.

Researching Copyright Law and the Internet

Instructions: As you learned in this chapter, copyright protects content creators against unauthorized use of their work. If you want to use material on your Web site that was created by someone else, you must obtain permission from the owner or use materials that are in the public domain. In this activity, you conduct research about copyright law as it relates to the Internet.

Perform the following tasks:

1. Open your Web browser and use a search engine to research copyright law and the Internet.

2. Use your word processor to answer the following questions:

   a. Name two U.S. copyright laws and briefly describe each.

   b. Identify four types of Web materials protected under copyright law.

Continued >

c. Create a list of Do's and Don'ts related to copyright law and your Web site. Include at least four Do's and four Don'ts.

d. Locate material available for use in the public domain. Identify the material (such as photos or audio recordings), the name of the site where you found them, and the URL where you obtained them.

3. Save the document with the file name Extend8_*lastname_firstname*. Submit the document in the format specified by your instructor.

Make It Right

Analyze a Web site and suggest how to improve its design.

Testing Links

Note: To complete this assignment, you will be required to use the Data Files for Students. Visit www.cengage.com/ct/studentdownload for detailed instructions or contact your instructor for information about accessing the required files.

Instructions: In this activity, you use the Link Checker to scan a Web page for broken links and then repair the broken links. The Web page and links are shown in Figure 8–46. The result of correcting the Right 2 link is shown in Figure 8–47.

Figure 8–46

Figure 8–47

Perform the following tasks:

1. Copy the right8.html file into the *your last name and first initial*\Right folder.

2. Start Dreamweaver. Use the Sites button on the Files panel to select the Right site. Open right8.html.

3. Use the Site menu to check links sitewide.

4. Review the Link Checker report in the Results panel for broken links.

5. Correct each broken link using the Property inspector. Each exercise listed should link to its respective exercise page. For example, Right 1 should be linked to right1.html.

6. Save your changes.

7. Submit the document in the format specified by your instructor.

In the Lab

Design and/or create a document using the guidelines, concepts, and skills presented in this chapter. Labs are listed in order of increasing difficulty.

Lab 1: Using Adobe BrowserLab

Problem: You have created an internal Web site for your company that features information about how to live a healthy lifestyle. Employees at your company will use this Web site as a resource for nutrition, exercise, and other health-related tips. You now need to use Adobe BrowserLab to view the site across multiple browsers.

First, open the home page for the Healthy Lifestyle Web site. Next, preview the page in Adobe BrowserLab, and then view the page using My Browser Set in Adobe BrowserLab. The home page, as displayed in Adobe BrowserLab, Internet Explorer 9.0 – Windows and Chrome 18.0 – Windows, is shown in Figure 8–48.

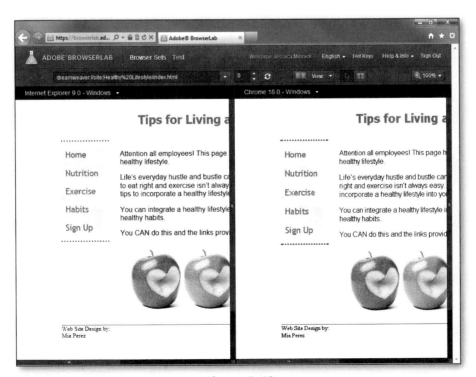

Figure 8–48

Continued >

In the Lab *continued*

Perform the following tasks:

1. Start Dreamweaver. Use the Sites button on the Files panel to select the Healthy Lifestyle site. Open index.html.

2. Use the Adobe BrowserLab panel to preview the home page.

3. Display the page in all of the listed browsers.

4. Change the view to 2-up View and then select Internet Explorer 9.0 – Windows and Chrome 18.0 – Windows as the test browsers.

5. Press the PRINT SCREEN key to take a screen shot of the 2-up View, open a graphics program such as Paint, press the CTRL+V keys to paste the image in a new file, and then save the file as a JPG with the name Lab1_*lastname_firstname*. Compare your file to Figure 8–48. Make any necessary changes, take a new screen shot, if necessary, and then save the screen shot file using the same name as the original file.

6. Close all open windows.

7. Submit the file in the format specified by your instructor.

In the Lab

Lab 2: Adding Meta Tags and Publishing to a Remote Server

Problem: You have created a Web site for Designs by Dolores, a Web site design company. The site provides information about the company and its services. Now that the Web site is complete, you need to add meta tags and publish the site to a remote server.

In this activity, you add keywords and description meta tags. Then you publish the Web site to a remote server. The code for the meta tags and the home page is shown in Figure 8–49.

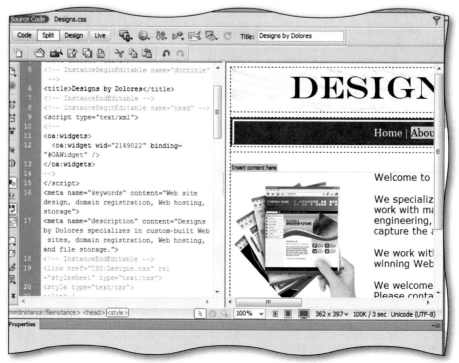

Figure 8–49

Perform the following tasks:

1. Start Dreamweaver. Use the Sites button on the Files panel to select the Designs by Dolores site. Open index.html.

2. Insert a keywords meta tag and use `Web site design, domain registration, Web hosting, storage` for the keywords content.

3. Insert a description meta tag and use `Designs by Dolores specializes in custom-built Web sites, domain registration, Web hosting, and file storage.` for the description content.

4. Use the Split button to view the code for the meta tags.

5. Use the Site menu to open the Manage Sites dialog box, and then select Designs by Dolores.

6. Open the Site Setup for Designs by Dolores dialog box and then select the Servers category.

7. Use the 'Add new server' button (plus button) to add the server information. Your instructor will provide you with the server information.

8. Save the changes and close the Manage Sites dialog box.

9. Submit the documents in the format specified by your instructor.

In the Lab

Lab 3: Adding Meta Tags and Publishing to a Remote Server

Problem: You have created a Web site for Justin's Lawn Care Service, a lawn care company. The site provides information about the company and its services. Now that the Web site is complete, you need to add meta tags and publish the site to a remote server.

In this activity, you add keywords and description meta tags. Then you publish the Web site to a remote server. The code for the meta tags and the home page is shown in Figure 8–50.

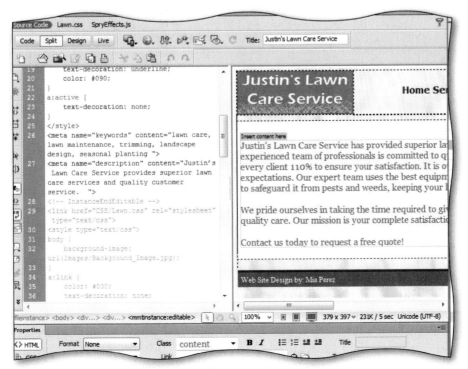

Figure 8–50

Continued >

In the Lab *continued*

Perform the following tasks:

1. Start Dreamweaver. Use the Sites button on the Files panel to select the Justin's Lawn Care Service site. Open index.html.

2. Insert a keywords meta tag and use `lawn care, lawn maintenance, trimming, landscape design, seasonal planting` for the keywords content.

3. Insert a description meta tag and use `Justin's Lawn Care Service provides superior lawn care services and quality customer service.` for the description content.

4. Use the Split button to view the code for the meta tags.

5. Use the Manage Sites dialog box to edit the Justin's Lawn Care Service site.

6. Use the 'Add new server' button to add the server information. Your instructor will provide you with the server information.

7. Save the changes and close the Manage Sites dialog box.

8. Submit the documents in the format specified by your instructor.

Cases and Places

Apply your creative thinking and problem solving skills to design and implement a solution.

1: Proofreading and Using Adobe BrowserLab for Moving Venture Tips

Personal

You have finished creating the Web site for Moving Venture Tips, a Web site dedicated to providing helpful tips and information for moving. You now need to proofread the site and then test it using Adobe BrowserLab. Proofread all pages for the Web site and correct any grammatical errors or misspelled words. Use Adobe BrowserLab to view the home page in the browsers provided in My Browser Set. Take a screen shot of the 2-up View and save the file as a JPG with the name CP1_*lastname_firstname*. Submit the file in the format specified by your instructor.

2: Adding Meta Tags and Publishing Student Campus Resources

Academic

You have finished creating the Web site for Student Campus Resources, which is dedicated to providing college students with information about campus resources. You now need to add keywords and description meta tags to the home page and then publish the site to a remote server. Using your knowledge of the Web site material, create a keywords meta tag and a description meta tag. Use the Manage Sites dialog box to edit the Web site's server information. Your instructor will provide you with the required server information. Save your changes. Submit the documents in the format specified by your instructor.

3: Adding Meta Tags and Publishing French Villa Roast Café

Professional

You have finished creating the Web site for French Villa Roast Café, a local cafe. You now need to add keywords and description meta tags to the home page and then publish the site to a remote server. Using your knowledge of the Web site material, create a keywords meta tag and a description meta tag. Use Manage Sites to edit the Web site's server information. Your instructor will provide you with the required server information. Save your changes. Submit the assignment in the format specified by your instructor.

9 Building a Mobile Web Site

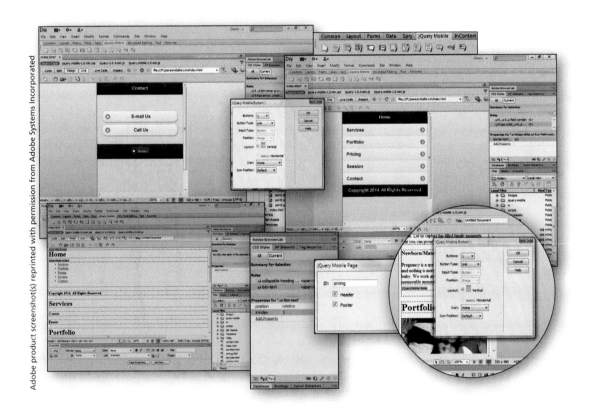

Objectives

You will have mastered the material in this chapter when you can:

- Create a mobile directory
- Identify jQuery Mobile elements
- Insert a jQuery Mobile page
- Add content including graphics to a jQuery Mobile page
- Add links to a jQuery Mobile page
- Insert a jQuery Mobile List View
- Insert a jQuery Mobile Layout Grid
- Insert a jQuery Mobile Collapsible Block
- Insert a jQuery Mobile button
- Edit a jQuery Mobile CSS rule
- Use the jQuery Mobile Swatches panel
- Review the mobile Web site in a browser

9 | Building a Mobile Web Site

Introduction

Having a standard Web site that accommodates a desktop or laptop computer is a great way to attract more business. However, users have made a paradigm shift in the way they surf the Internet. Now we live in a mobile world, where smartphones with Wi-Fi and Internet access are connecting to Web sites. As mobile Web site users view Web pages on a mobile device, they are usually busy also waiting in line, riding on public transportation, or running errands, for example. When people using mobile devices visit your Web site, their main focus is not reading the details of your main Web site, but scanning the pages to find information. Therefore, that information must be easy to access. A mobile Web presence should provide clear and concise navigation, simple graphics, and brief textual content.

In this chapter, you learn how to create a mobile Web site for the Gallery. One of the main concepts for an effective mobile Web design is simplicity. With smaller mobile processors, expensive data plans with limited bandwidth, and battery power as considerations, reducing the amount of images, text, and media helps a mobile site load quickly. Large clickable objects that link to other mobile pages are essential for touch screen users. A responsive design also ensures that your mobile site fits on a screen of any size.

Project — Mobile Web Site

The Gallery photography studio has decided to accommodate the growing mobile device market by creating a mobile version of its Web site. The Gallery's traditional Web site does not display well on mobile devices, and the studio owners are concerned about losing potential customers. Mobile sites should be more than tiny versions of the full Web site because users interact with mobile devices differently from the way they interact with desktop devices. When a customer enters the URL of the Gallery site, the page needs code to check whether the user is viewing the site on a desktop or a mobile device. The code can display the full desktop version for those using a traditional computer, and a mobile version for those using a smartphone or tablet device. The mobile optimized Gallery site presents the essential content of the full site with easy-to-use touch navigation.

The Gallery mobile Web site is shown in Figure 9–1. Professional, polished multimedia in small doses adds entertainment value to the Gallery site while showcasing additional photos in the video. The opening home page of the mobile site provides large navigational buttons for touch access to each page of the site. The information displayed on each page is brief and concise.

Adobe Dreamweaver CS6 has integrated the jQuery Mobile framework as a tool to assist in Web design for a variety of smartphones and tablets, including Google Android, Apple iOS, and Windows Phone platforms. **jQuery Mobile** is a library of JavaScript tools for designing mobile Web sites and mobile applications. Using jQuery Mobile, you can build mobile Web pages in a single HTML5 file and control the information displayed based on touch user interaction. Using jQuery Mobile tools, you can design a Web application that works on most mobile devices and adapts to the size of their screens. A **Web application** is a site that is coded using HTML and a browser-supported language such as JavaScript (the basis for the jQuery Mobile framework). Users interact with a Web application through a browser.

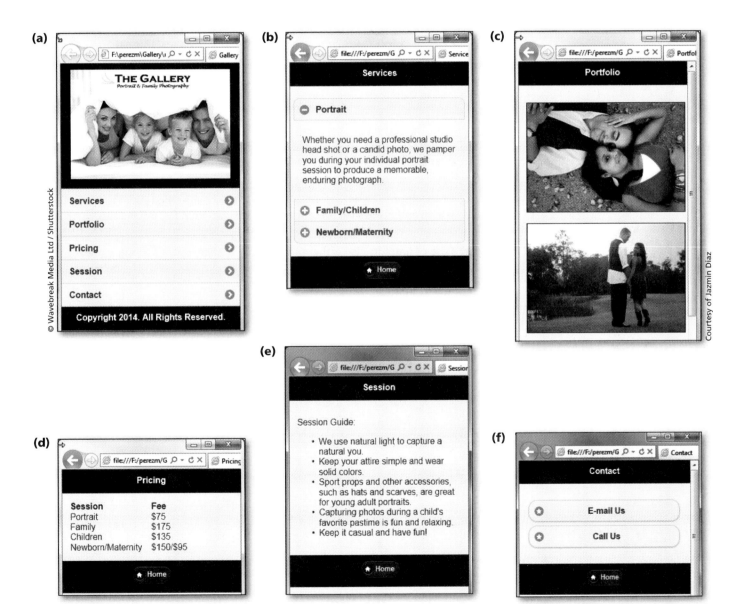

Figure 9–1 Mobile version of the Gallery site

Overview

As you read this chapter, you will learn how to create the Web page project shown in Figure 9–1 by performing these general tasks:

- Prepare the folder structure for the mobile site.
- Create a mobile page for each page in the Gallery site.
- Add content to mobile pages.
- Link mobile pages.
- Add buttons and other user interface widgets to mobile pages.
- Apply CSS rules to the mobile site.
- Preview and publish the mobile site.

Plan
Ahead

General Project Guidelines

When designing a mobile Web site, consider the purpose and audience of the site. As you create the mobile Web pages shown in Figure 9–1, you should follow these general guidelines:

1. **Design the mobile Web site.** When designing a mobile Web site, keeping it simple is the major design principle to follow. Mobile device users want to access your Web site quickly, and easily find information while they are on the go.

2. **Determine mobile Web site content.** Include the most essential information on your mobile Web site. Determine what information would be the most beneficial to a mobile device user, and make that content easily accessible from the mobile site home page.

3. **Minimize the use of graphics.** To further simplify your mobile Web site design, use graphics sparingly. Remember, the more graphics on your mobile Web site, the more time it will take for your site to load on a mobile device.

4. **Consider the mobile device display.** Because the display on a mobile device is much smaller than that of a desktop computer, accommodate the mobile device user by breaking up content using collapsible blocks and by limiting scrolling to one direction.

   More specific details about these guidelines are presented at appropriate points throughout the chapter. The chapter also identifies the actions performed and decisions made regarding these guidelines during the development of the pages within the site shown in Figure 9–1.

To Start Dreamweaver and Open the Gallery Site

Each time you start Dreamweaver, it opens to the last site displayed when you closed the program. The following steps start Dreamweaver and open the Gallery Web site.

1 Click the Start button on the Windows 7 taskbar to display the Start menu, and then type `Dreamweaver CS6` in the 'Search programs and files' box.

2 Click Adobe Dreamweaver CS6 in the list to start Dreamweaver.

3 If the Dreamweaver window is not maximized, click the Maximize button next to the Close button on the Application bar to maximize the window.

4 If the Gallery site is not displayed in the Files panel, click the Sites button on the Files panel toolbar and then click Gallery to display the files and folders in the Gallery site.

Designing for Mobile Devices

If a Web page is viewed in a mobile browser that is not specifically designed for a mobile device, you have to zoom and pan to view the site. For example, a site may have a drop-down menu used for navigation that is not displayed properly on a smaller mobile device. As you begin the design of the Gallery mobile site, consider simplifying the layout and content to increase usability. The mobile version of the site should include only the most essential information to reduce the amount of content displayed on a smartphone. Lengthy sections of text can be hard to read on a mobile device, so it is best to keep text succinct and displayed in a single column. Although the mobile Web site should look cleaner and simpler than the desktop version, the user experience should be similar. Mobile design is not about crafting an entirely new experience, but delivering a similar experience, fine-tuned for mobile devices.

Design the mobile Web site.

When designing a mobile Web site, you must consider your target audience and provide them with quick, relevant information. In general, consider the following:

- **Include a list of links on the home page.** A mobile user needs quick access to information about your core products and services. The mobile site home page should include a list of links to the most pertinent information about your business, such as products and services, pricing, location, and contact information.

- **Keep the navigation simple.** Design the navigation to be as simple as possible and to avoid horizontal scrolling. Buttons that link to the home page help your target audience return to the main list of links.

- **Use jQuery Mobile UI Widgets.** The jQuery Mobile UI Widgets provide the desired mobile look and feel, and, most important, functionality. Incorporate these widgets to improve the layout of your mobile site.

Plan Ahead

The wide variety of smartphone screen sizes makes it difficult to decide how to choose an appropriate layout size for mobile devices. One way to display the mobile site across multiple screen sizes is to create fluid layouts that change depending on the current resolution and orientation of a handheld device. Mobile devices do not all use a single standard screen size, so you should use fluid layouts to fill the mobile browser window appropriately across a range of screen sizes without having to recreate pages for different platforms.

Another significant difference between accessing a Web page on a desktop device and a handheld device is that a user clicks the mouse to navigate to pages within the site on a traditional computer. People with mobile devices use their fingertips to tap and swipe the screen and navigate a Web site. A region of the smartphone screen is touched instead of a precise location with a mouse cursor on a larger computer display. As you design the user interface, do not place touch-based controls close to one another; this will help prevent users from navigating to the incorrect page in the mobile site. Mobile Web sites should display large, clean buttons that are easy to tap instead of small text links. If you have multiple text links in the same vicinity, users may tap the wrong one.

Mobile devices typically do not support Flash and other plug-ins. Flash Web sites rely on the arrow of the mouse for rollover effects and do not support touch-based devices. In addition, Flash effects use too much battery power. To achieve long battery life when playing video, mobile devices must decode the video in hardware; decoding it in Flash software uses too much power. As you design a mobile Web site, keep your text, navigation, and effects simple.

BTW

Using Dreamweaver Starter Pages
To help you design a mobile Web site that can be displayed on multiple screen sizes, Dreamweaver provides mobile starter pages. Click File on the Application bar, click New, click Page From Sample, and then select one of three mobile starter pages.

Using jQuery Mobile Tools

The jQuery Mobile framework is a collection of tools created to simplify mobile Web site design and mobile Web application development. The jQuery Mobile framework is composed of a library of user interface (UI) widgets to incorporate within your mobile Web site. The UI widgets include pages, buttons, collapsible blocks, check boxes, buttons, icons, and more. jQuery Mobile is compatible with several popular mobile platforms, including iOS, Android, Windows Mobile, Firefox Mobile, and BlackBerry.

Adobe Dreamweaver CS6 has integrated many aspects of the jQuery Mobile framework to allow for easy mobile design. The jQuery Mobile UI widgets allow you to create multiple, well-designed mobile pages using one HTML5 page. Adobe Dreamweaver CS6 provides several jQuery Mobile UI widgets to help build and design your mobile Web site. To locate the jQuery Mobile category on the Insert bar, click the jQuery Mobile tab. The jQuery Mobile category of tools on the Insert bar is shown in Figure 9–2. Table 9–1 describes the jQuery Mobile UI widgets available through Adobe Dreamweaver CS6.

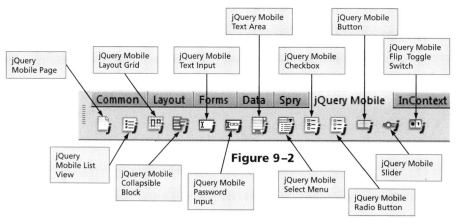

Figure 9–2

Table 9–1 jQuery Mobile UI Widgets

Command	Description
jQuery Mobile Page	Creates a new jQuery Mobile Page
jQuery Mobile List View	Creates an unordered list of linked pages within a jQuery Mobile Page
jQuery Mobile Layout Grid	Creates a specified number of columns and rows on a jQuery Mobile Page
jQuery Mobile Collapsible Block	Creates a set of three collapsible blocks within a jQuery Mobile Page
jQuery Mobile Text Input	Inserts a text box to allow for text input on a jQuery Mobile Page
jQuery Mobile Password Input	Inserts a text box to allow for text input, but does not show the characters input into the field on a jQuery Mobile Page
jQuery Mobile Text Area	Inserts a large text box to allow for multiline text inputs on a jQuery Mobile Page
jQuery Mobile Select Menu	Creates a button that, when touched, displays a list of option values available for selection on a jQuery Mobile Page
jQuery Mobile Checkbox	Creates a jQuery Mobile checkbox; represents a selection
jQuery Mobile Radio Button	Creates a jQuery Mobile radio button; represents an exclusive choice
jQuery Mobile Button	Creates a jQuery Mobile button, which can be linked to another page, an e-mail address, or a telephone number
jQuery Mobile Slider	Creates an interactive element where the user slides a handle to update a value on a jQuery Mobile Page
jQuery Mobile Flip Toggle Switch	Creates a flip switch with two options, typically an 'On' and an 'Off' switch

To Create a Mobile Directory and Copy Data Files

Before creating the mobile Web site, you must first create a mobile directory. A **mobile directory** is a folder within the Gallery site root folder that will store the mobile Web site. In this case, the folder you create is named *m*. You create a mobile directory by making a new folder within the Gallery root folder. In addition, the mobile site will showcase pictures for the Portfolio page, so you need to add the required image files to the mobile directory.

To complete the following steps, you will be required to use the Data Files for Students. Visit www .cengage.com/ct/studentdownload for detailed instructions or contact your instructor for information about accessing the required files. The following steps create a mobile directory and copy the Images folder from the Data Files for Students to the Gallery site.

1

- Use Windows Explorer to navigate to the *your last name and first initial* folder on Removable Disk F:.

- Double-click the Gallery folder to display its contents.

- Click the New folder button to create a new folder and name it m (Figure 9–3).

Q&A

Why do I need to create a folder called m?

The m folder will contain the mobile Web site and its related files. It is the mobile directory for the Gallery Web site.

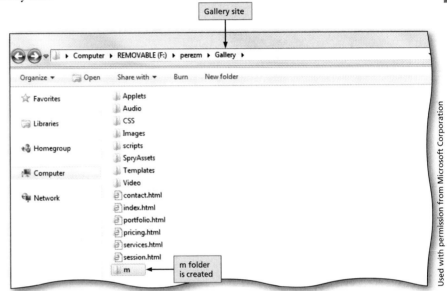

Figure 9–3

2

- Use Windows Explorer to navigate to the storage location of the Data Files for Students.

- Double-click the Chapter 09 folder to open it and then double-click the Gallery folder.

- Right-click the Images folder, click Copy on the context menu, and then navigate to the *your last name and first initial* folder on Removable Disk F: to prepare to copy the folder.

- Double-click the Gallery folder, double-click the m folder, right-click anywhere in the open window, and then click Paste on the context menu to copy the Images folder into the m folder of the Gallery site. Verify that the m folder now contains an Images folder (Figure 9–4).

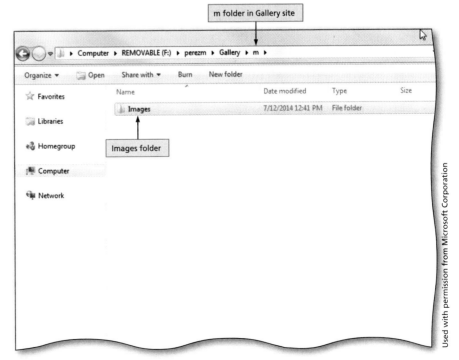

Figure 9–4

Preparing the Mobile Web Site Structure

The process of designing a new mobile Web site begins with creating and saving a new HTML5 page. You then add jQuery Mobile UI widgets to the HTML5 page. You can add several jQuery Mobile pages on just one HTML5 page. Once jQuery Mobile features have been added to the page and saved, three jQuery Mobile files are created. Dreamweaver CS6 automatically creates a jquery-mobile folder and saves the required jQuery Mobile files within this folder. The first file created is an external style sheet (CSS) for the mobile Web site. The other two files are the required jQuery and jQuery Mobile files. All three of these files are referenced in the <head> section of the HTML code. To create a mobile Web site, you use the Dreamweaver jQuery Mobile buttons on the Insert bar.

To Create and Save a Page for the Mobile Site

Before you can use jQuery Mobile to create a mobile Web site, you must create a new HTML5 document. The following steps create a new HTML5 document and save it within the mobile directory.

1

- Click the More folder on the Welcome screen to display the New Document dialog box.

- Click Blank Page in the left pane to specify you are creating a blank page.

- Click HTML in the Page Type list to select an HTML page.

- If necessary, click <none> in the Layout list to create a blank HTML page with no predefined layout.

- Click the DocType button and then click HTML 5 to set the DocType to HTML5 (Figure 9–5).

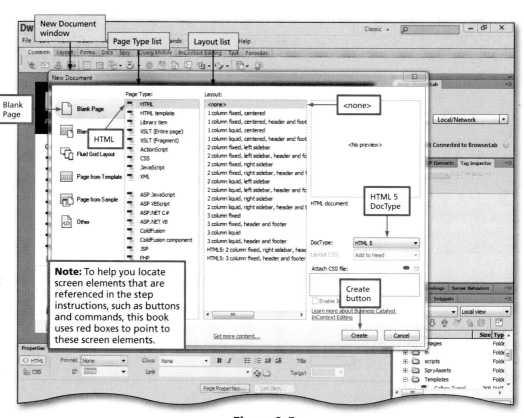

Figure 9–5

2

- Click the Create button to display the blank HTML page in the Document window.

- Click File on the Application bar and then click Save As on the File menu to display the Save As dialog box.

- If necessary, click Gallery to select the Gallery site.

- Double-click the m folder to open the mobile directory.

- Select the text in the File name text box and then type index to name the new page (Figure 9–6).

 Why is the mobile Web site page named index?

The home page for any Web site should be named index or default. This includes mobile Web sites, too.

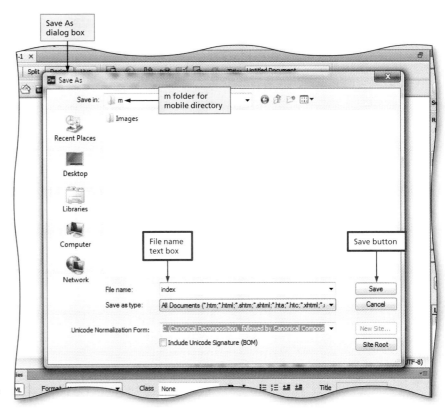

Figure 9–6

3

- Click the Save button to save index.html and display it in the document window (Figure 9–7).

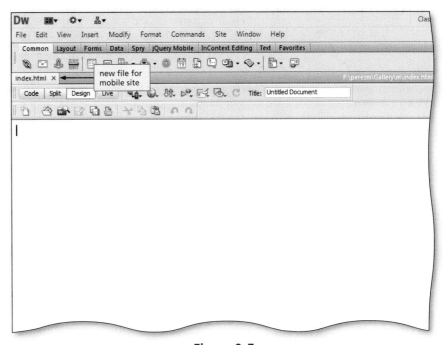

Figure 9–7

Other Ways

1. Click File on Application bar, click New
2. Press CTRL+N

Creating jQuery Mobile Pages

Recall that one way to start creating a mobile Web site is to create and save a new HTML5 page. You can then add several jQuery Mobile pages to the single HTML5 page. You do so by inserting a jQuery Mobile Page User Interface (UI) widget from the jQuery Mobile category on the Insert bar. Each jQuery Mobile page is defined as a div element on the HTML5 page. (You can display the elements in Code view or Split view.) However, when displayed in a browser, each jQuery Mobile page appears as a separate Web page in the mobile site. The div element includes a **data attribute**, which describes the content of the div and uses the form data-*attribute*, where the property name is *data* followed by a hyphen and then an attribute, such as *role*. Collectively, the word, data, and the hyphen are called a prefix.

Each div and data attribute control the structure of jQuery Mobile widgets. For example, the code <div data-role="page"> means that the div serves as a page and can contain page sections such as a header, content, and a footer. When you insert a new jQuery Mobile Page UI widget on an HTML5 page, the page consists of one primary div containing the page name. The primary div contains three additional divs for the page header, content, and footer, by default. The following is an example of the structure of a typical jQuery Mobile Page UI widget.

```
<div data-role="page">
    <div data-role="header">Header</div>
    <div data-role="content">Content</div>
    <div data-role="footer">Footer</div>
</div>
```

When you insert a new jQuery Mobile Page, you must provide a div ID to identify the page, such as home or services. The header div contains the name of the page, such as Services. The content div contains the primary information for that particular page, such as information about the Gallery's services. The footer div contains copyright information for the Home page and a Home navigation button on subsequent pages. Each time you add a new jQuery Mobile feature to the mobile site, Dreamweaver adds a new div and data attribute for the UI widget.

Plan Ahead

> **Determine mobile Web site content.**
> Because the screen on a mobile device is much smaller than the screen on a traditional desktop or laptop computer, include only the most pertinent information about your business, such as products and services, pricing, and contact information. Condense content from the original Web site to avoid the need for scrolling.

Mobile Page Content

Each new jQuery Mobile Page consists of a header area, content area, and footer area. The header area is where you provide the page title to identify the page, such as Services. The content area is where you provide the main information for the page, such as information about the Gallery's services. The footer area on the Home page contains the copyright information. The footer area on the subsequent pages will contain a navigation button that links to the Home page.

To Create the Home Page for the jQuery Mobile Site

Before creating a mobile Web site using jQuery, you should display the jQuery Mobile category of tools on the Insert bar so you can work efficiently. Next, you use the Insert bar to add a jQuery Mobile Page UI widget to the new HTML document. The following steps display the jQuery Mobile category on the Insert bar, and then insert a jQuery Mobile Page UI widget.

1

- Click the jQuery Mobile tab on the Insert bar to display the jQuery Mobile options (Figure 9–8).

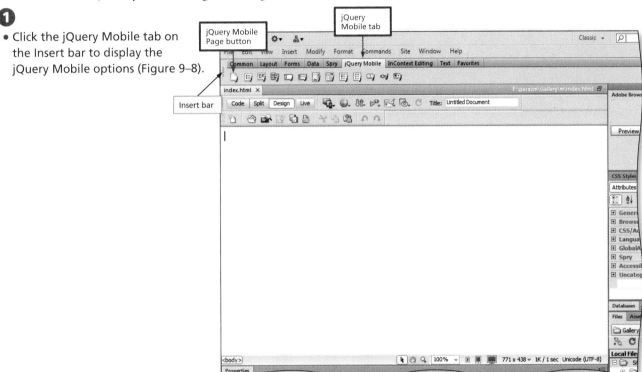

Figure 9–8

2

- Click the jQuery Mobile Page button on the Insert bar to display the jQuery Mobile Files dialog box (Figure 9–9).

Q&A

What are the files listed in the lower part of the dialog box?

Those are the required files for the style sheet and other parts of a jQuery Mobile Web site that will be created as you create the jQuery Mobile page.

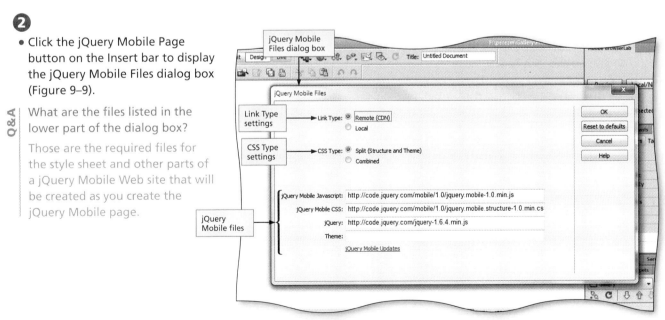

Figure 9–9

3

- If necessary, click the Local option button to select the local link type.

- Click the Combined option button to select the combined structure and theme CSS type (Figure 9–10).

 What is the local link type?

The local link type stores the required jQuery Mobile files within a jquery-mobile folder in the Gallery root folder.

 What is the combined CSS type?

The combined CSS type designates one CSS file to be used for the structure and the theme of the mobile site.

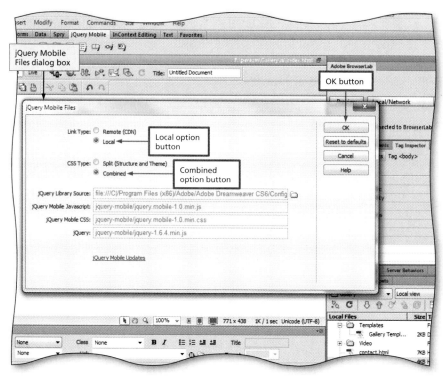

Figure 9–10

4

- Click the OK button to display the jQuery Mobile Page dialog box.

- Replace the ID placeholder text, page, with home to enter the Div ID.

- If necessary, click the Header check box and then click the Footer check box to select them (Figure 9–11).

 What is the div ID?

The div ID is the name of the mobile page.

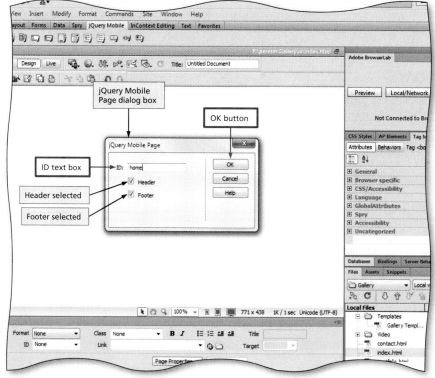

Figure 9–11

5

- Click the OK button to create the Home page for the mobile Web site.

- Click the Save button on the Standard toolbar to save the page.

- Click the OK button to copy dependent files and create the jquery-mobile folder and files (Figure 9–12).

Q&A

Why does my page show a Header, Content, and Footer area?

When a new jQuery Mobile Page is inserted, it consists of a header area, content area, and footer area by default.

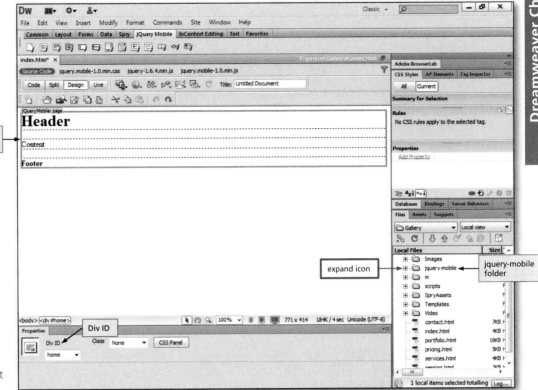

jQuery Mobile home page

expand icon

jquery-mobile folder

Div ID

Figure 9–12

6

- Replace the word, Header, with Home to name the page.

- Click the expand icon next to the jquery-mobile folder in the Files panel to display the folder's contents (Figure 9–13).

Q&A

What are the jquery-mobile files?

The jquery-mobile files are the required JavaScript and CSS files associated with the mobile site.

7

- Click the collapse icon next to the jquery-mobile folder in the Files panel to collapse the folder's contents.

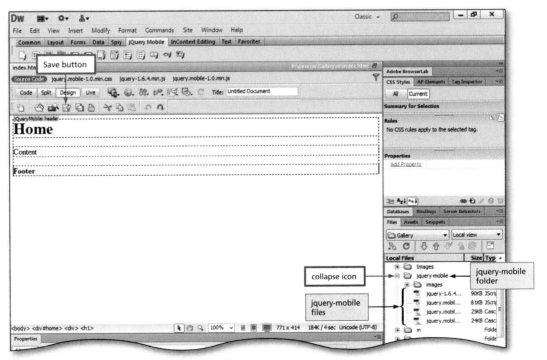

Save button

collapse icon

jquery-mobile folder

jquery-mobile files

Figure 9–13

- Click the Save button on the Standard toolbar to save your work.

Other Ways

1. Click Insert on Application bar, point to jQuery Mobile, click Page

To Create the Services Page Using jQuery Mobile

You have created the Home page for the mobile site, but now need to create the other five pages: Services, Portfolio, Pricing, Session, and Contact. Recall that you can insert multiple jQuery Mobile pages onto one HTML5 page. The following steps create the jQuery Mobile page for the Services page.

- Click below the Home page to place your insertion point to the right of the Footer section (Figure 9–14).

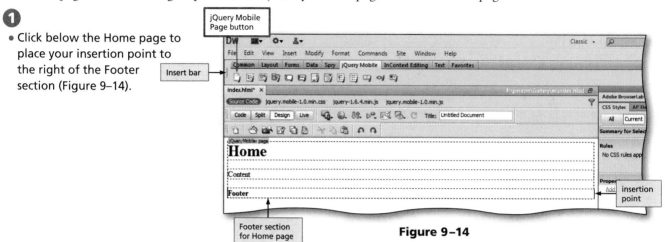

Figure 9–14

②

- Click the jQuery Mobile Page button on the Insert bar to display the jQuery Mobile Page dialog box.

- Replace the ID placeholder text, page, with `services` to enter the ID.

- If necessary, click the Header check box and then click the Footer check box to select them (Figure 9–15).

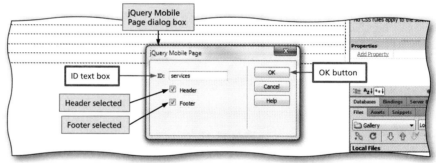

Figure 9–15

③

- Click the OK button to create the Services mobile page.

- Replace the word, Header, with `Services` to name the Services page (Figure 9–16).

Q&A

Why did I insert the Services page on the same document where the Home page is located?

The jQuery Mobile framework allows you to create all the jQuery Mobile pages you need on one HTML5 document.

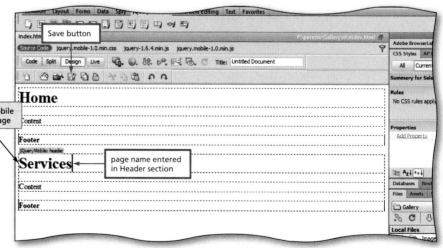

Figure 9–16

④

- Click the Save button on the Standard toolbar to save your work.

Other Ways

1. Click Insert on Application bar, point to jQuery Mobile, click Page

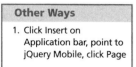

To Create the Portfolio Page Using jQuery Mobile

The following steps create the jQuery Mobile page for the Portfolio page.

 1

- Click below the Services page to place your insertion point to the right of the Footer section (Figure 9–17).

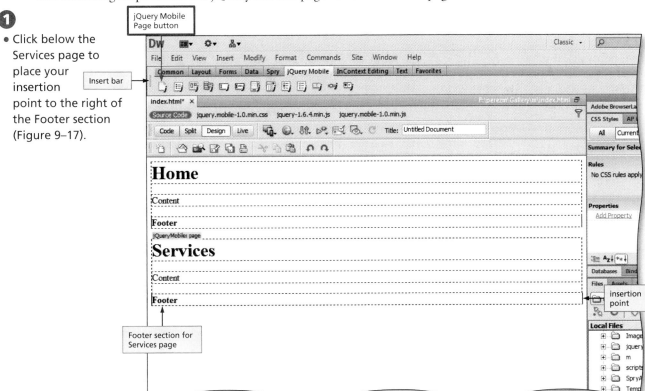

Figure 9–17

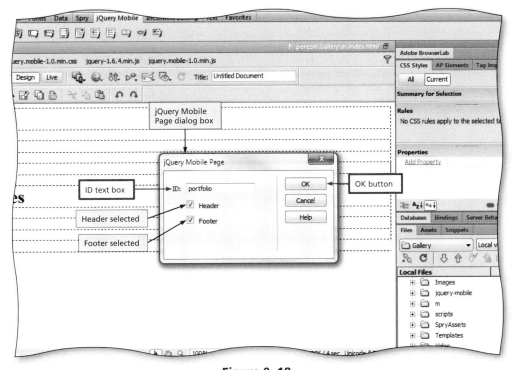

2

- Click the jQuery Mobile Page button on the Insert bar to display the jQuery Mobile Page dialog box.

- Replace the ID placeholder text, page, with `portfolio` to enter the ID.

- If necessary, click the Header check box and then click the Footer check box to select them (Figure 9–18).

Figure 9–18

3

- Click the OK button to create the Portfolio mobile page.

- If necessary, use the scroll bar to scroll down to view the Portfolio mobile page.

- Replace the word, Header, with `Portfolio` to name the Portfolio page (Figure 9–19).

4

- Click the Save button on the Standard toolbar to save your work.

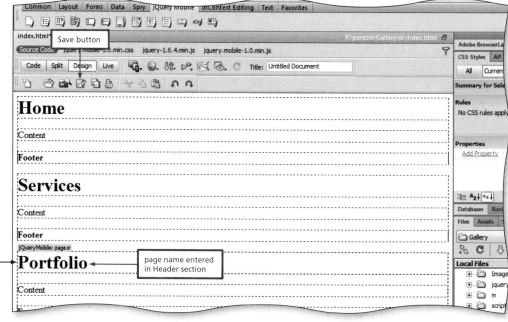

Figure 9–19

Other Ways

1. Click Insert on Application bar, point to jQuery Mobile, click Page

To Create the Pricing Page Using jQuery Mobile

The following steps create the jQuery Mobile page for the Pricing page.

1

- Click below the Portfolio page to place your insertion point to the right of the Footer section (Figure 9–20).

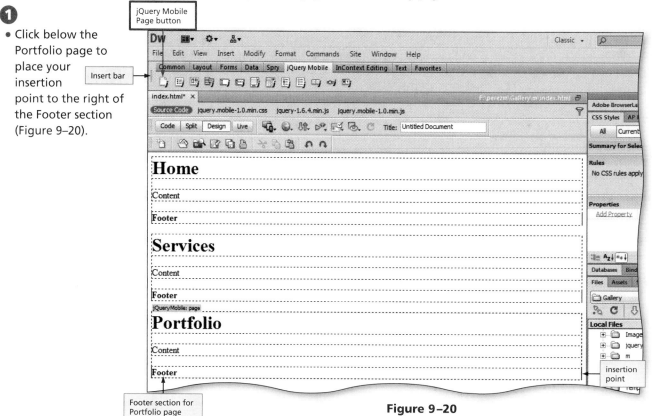

Figure 9–20

• Click the jQuery Mobile Page button on the Insert bar to display the jQuery Mobile Page dialog box.

• Replace the ID placeholder text, page, with `pricing` to enter the ID.

• If necessary, click the Header check box and then click the Footer check box to select them (Figure 9–21).

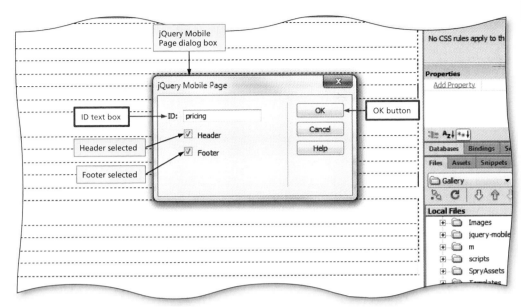

Figure 9–21

• Click the OK button to create the Pricing mobile page.

• If necessary, use the scroll bar to scroll down to view the Pricing mobile page.

• Replace the word, Header, with `Pricing` to name the Pricing page (Figure 9–22).

• Click the Save button on the Standard toolbar to save your work.

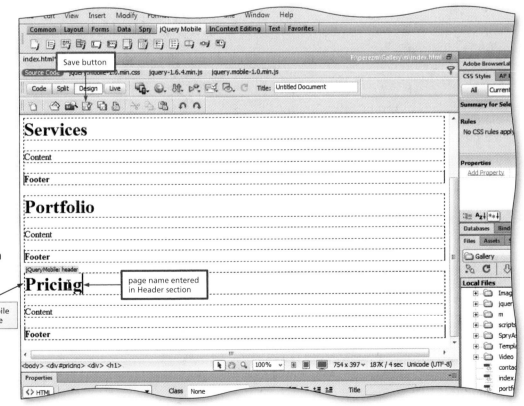

Figure 9–22

Other Ways

1. Click Insert on Application bar, point to jQuery Mobile, click Page

To Create the Session and Contact Pages Using jQuery Mobile

The following steps create the jQuery Mobile pages for the Session and Contact pages.

1

- Click below the Pricing page to place your insertion point to the right of the Footer section.

- Click the jQuery Mobile Page button on the Insert bar to display the jQuery Mobile Page dialog box.

- Replace the ID placeholder text, page, with `session` to enter the ID.

- If necessary, click the Header check box and then click the Footer check box to select them.

- Click the OK button to create the Session mobile page.

- If necessary, use the scroll bar to scroll down to view the Session mobile page.

- Replace the word, Header, with `Session` to name the Session page (Figure 9–23).

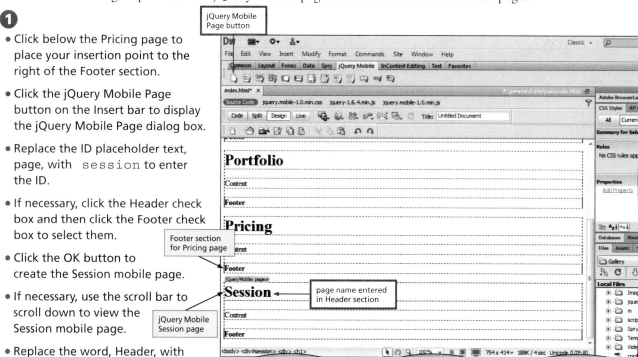

Figure 9–23

2

- Click below the Session page to place your insertion point to the right of the Footer section.

- Click the jQuery Mobile Page button on the Insert bar to display the jQuery Mobile Page dialog box.

- Replace the ID placeholder text, page, with `contact` to enter the ID.

- If necessary, click the Header check box and the Footer check box to select them.

- Click the OK button to create the Contact mobile page.

- If necessary, use the scroll bar to scroll down to view the Contact mobile page.

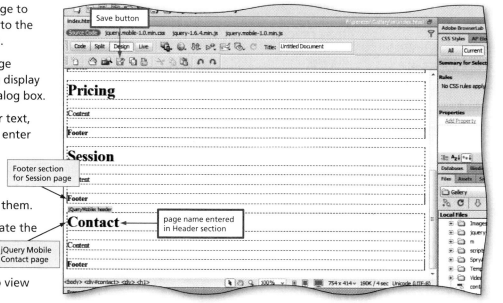

Figure 9–24

- Replace the word, Header, with `Contact` to name the Contact page (Figure 9–24).

3

- Click the Save button on the Standard toolbar to save your work.

 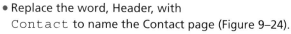

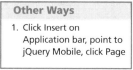

Other Ways

1. Click Insert on Application bar, point to jQuery Mobile, click Page

Adding Content to a Mobile Web Page

After you create the mobile Web pages, you need to add content to them. Content is the textual or graphical information that appears on a Web page. This includes pictures that display the Gallery's portfolio, and text describing the Gallery's services, outlining pricing information, explaining how to prepare for a photography session, and listing contact links.

After creating the additional pages, you must provide links to the pages from the Home page. Because a mobile Web page does not allow enough room for a navigation bar, these links are the only way users can navigate from one page to another on the mobile site. To create the links, you insert a jQuery Mobile List View UI widget. This UI widget displays the links as standard link text in Design view, but formats the links as bars users click when visiting the mobile site. When you insert a jQuery Mobile List View UI widget, you specify how many links you need, name each link, and then specify the linked page.

To Add Content to the Gallery Mobile Web Site

Now that you have created six jQuery Mobile pages for the Gallery Web site, you need to provide content for each page. The Home page will display the Gallery logo, a list of links to the other pages in the site, and a copyright notice. The Session page will provide a bulleted list of tips for preparing for a photography session. You can add the copyright notice and all of the session tips text directly to the jQuery Mobile Page. The following steps add content to the Home and Session mobile pages.

- Replace the word, Footer, on the Home page with `Copyright 2014. All Rights Reserved.` to provide content for the Home page footer (Figure 9–25).

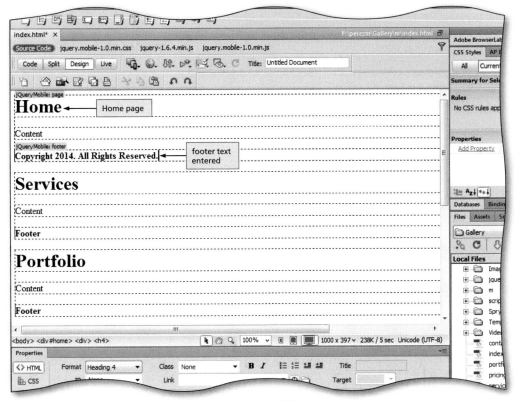

Figure 9–25

- Use the scroll bar to scroll down to the Session page.

- Replace the word, Content, with `Session Guide:` and then press the ENTER key to insert a new paragraph.

- Click the Unordered List button on the Property inspector to insert an unordered list (Figure 9–26).

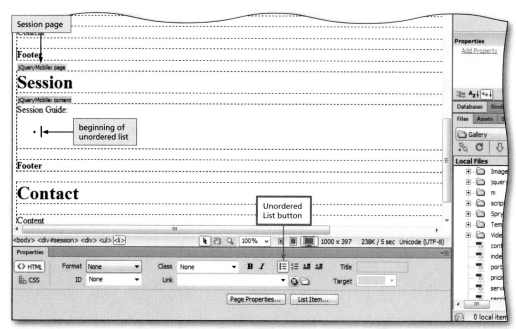

Figure 9–26

- Type `We use natural light to capture a natural you.` to provide content for the first list item, and then press the ENTER key.

- Type `Keep your attire simple and wear solid colors.` to provide content for the next list item , and then press the ENTER key.

- Type `Sport props and other accessories, such as hats and scarves, are great for young adult portraits.` to provide content for the next list item, and then press the ENTER key.

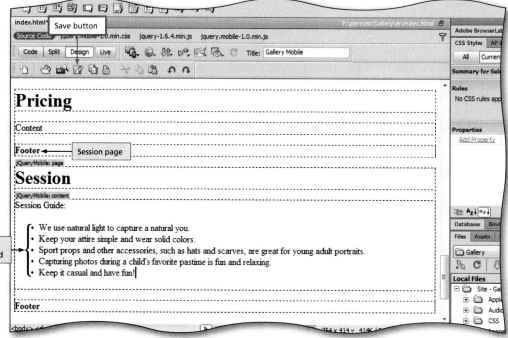

Figure 9–27

- Type `Capturing photos during a child's favorite pastime is fun and relaxing.` to provide content for the next list item, and then press the ENTER key.

- Type `Keep it casual and have fun!` to provide content for the last list item (Figure 9–27).

- Click the Save button on the Standard toolbar to save your work.

To Add a jQuery Mobile List View UI Widget to the Home Page

Next, you use a jQuery Mobile List View UI widget to create a list of page links on the Home page, with each list item linking to a page within the mobile Web site. The jQuery Mobile List View UI widget applies the required styles to create a mobile-friendly list of bars that each display specified text and a right arrow that fit the width of the mobile device. When users tap the linked bar, the specified page opens. You must insert a jQuery Mobile List View to access the other pages within the mobile Web site. The following steps add the list of links to the Home page.

1

- If necessary, use the scroll bar to scroll up to the Home page.

- Delete the word, Content, to prepare to insert the jQuery Mobile List View UI widget (Figure 9–28).

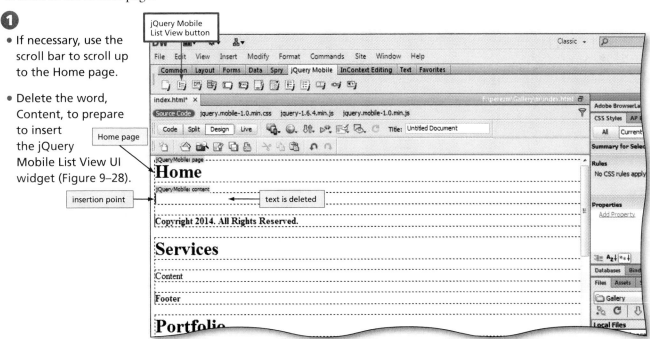

Figure 9–28

2

- Click the jQuery Mobile List View button on the Insert bar to display the jQuery Mobile List View dialog box (Figure 9–29).

Q&A What is the purpose of the jQuery Mobile List View dialog box?

You use this dialog box to designate the number of links needed for the Home page. A link is established in the list view for every page with the exception of the Home page. You can also specify additional styling options, such as icons.

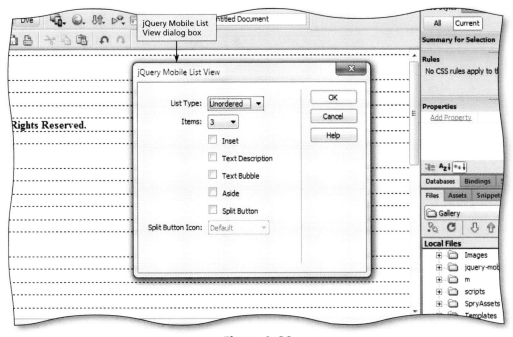

Figure 9–29

• If necessary, click the List Type button and then click Unordered to select the Unordered list type.

• Click the Items button and select 5 (Figure 9–30).

What is the purpose of the Items button?

You use this button to select the number of items in the list.

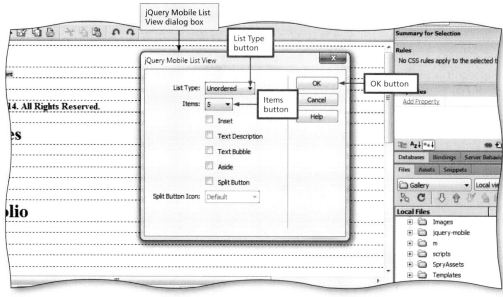

Figure 9–30

• Click the OK button to insert the jQuery List View UI widget (Figure 9–31).

• Click the Save button on the Standard toolbar to save your work.

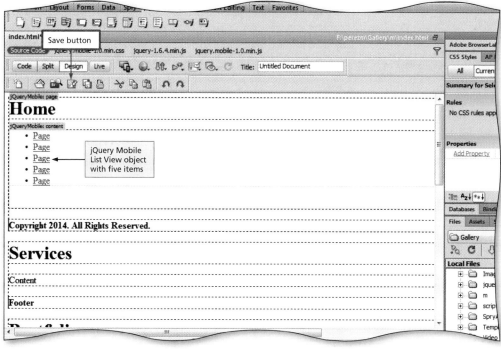

Figure 9–31

Other Ways

1. Click Insert on Application bar, point to jQuery Mobile, click List View

To Add Page Links to the jQuery Mobile List View UI Widget

After inserting a jQuery Mobile List View UI widget, you link each mobile page to an item in the jQuery Mobile List View UI widget. You name each link according to its page title, and then use the Property inspector to assign a page to each list item. The following steps assign pages to the bulleted list of links on the Home page.

1

- Replace the first list item with `Services` to name the page link (Figure 9–32).

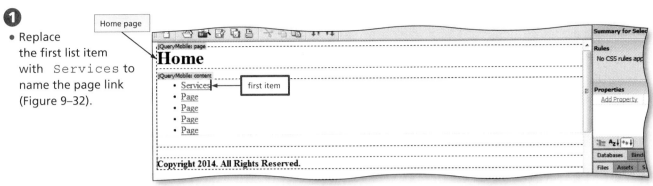

Figure 9–32

2

- Type `services` after the number sign (#) in the Link text box on the Property inspector to link to the Services page (Figure 9–33).

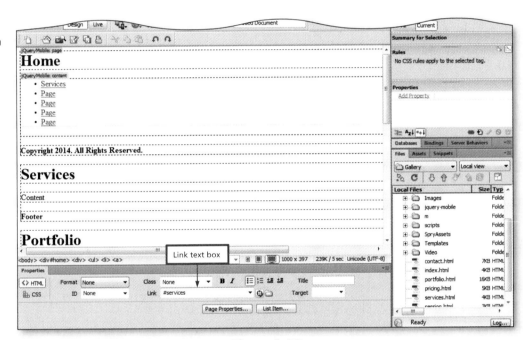

Figure 9–33

3

- Replace the second list item with `Portfolio` to name the page link.

- Type `portfolio` after the number sign (#) in the Link text box on the Property inspector to link to the Portfolio page (Figure 9–34).

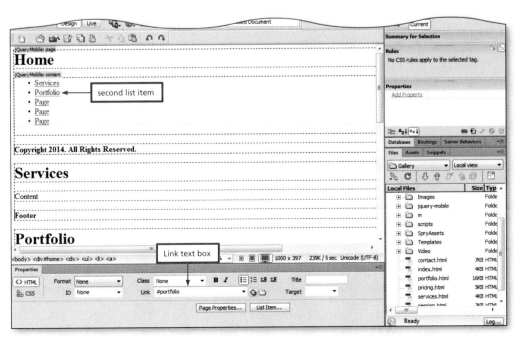

Figure 9–34

● Replace the third list item with `Pricing` to name the page link.

● Type `pricing` after the number sign (#) in the Link text box on the Property inspector to link to the Pricing page (Figure 9–35).

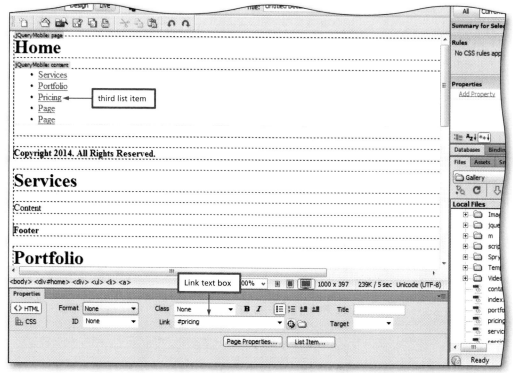

Figure 9–35

● Replace the fourth list item with `Session` to name the page link.

● Type `session` after the number sign (#) in the Link text box on the Property inspector to link to the Session page (Figure 9–36).

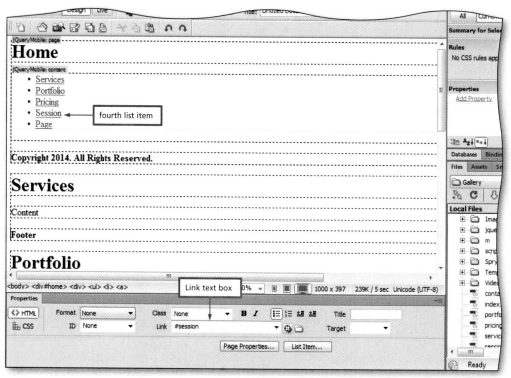

Figure 9–36

6

- Replace the last list item with `Contact` to name the page link.

- Type `contact` after the number sign (#) in the Link text box on the Property inspector to link to the Content page (Figure 9–37).

7

- Click the Save button on the Standard toolbar to save your work.

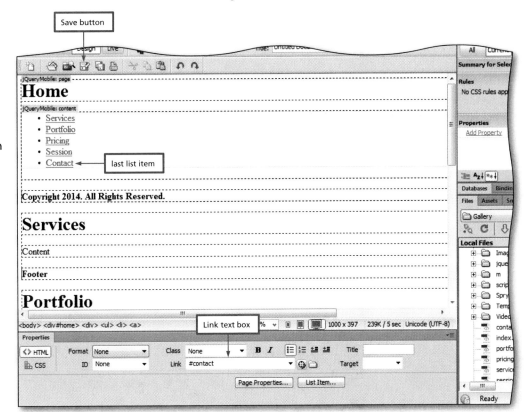

Figure 9–37

To Change the View Size and Display the Mobile Site in Live View

You have used Design view to design the mobile pages for the Gallery mobile Web site. To display the site as a user will view it, adjust the document window size to the size of a smartphone, and then display the site in Live view. The following steps change the document window view size and display the mobile site in Live view.

1

- Click the Multiscreen button on the Document toolbar to display the multiscreen view options and then click 320 x 480 Smart Phone to change the size of the document window (Figure 9–38).

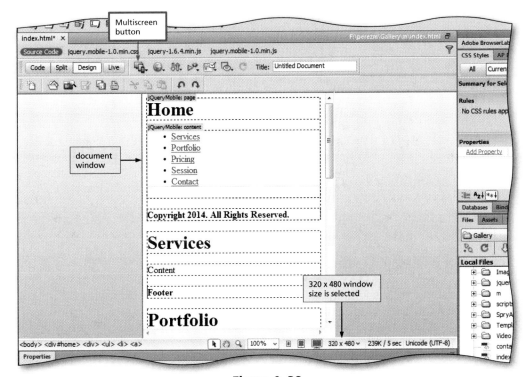

Figure 9–38

- Click the Live button on the Document toolbar to display the mobile Web site in Live view (Figure 9–39).

- Click the Live button on the Document toolbar again to exit Live view.

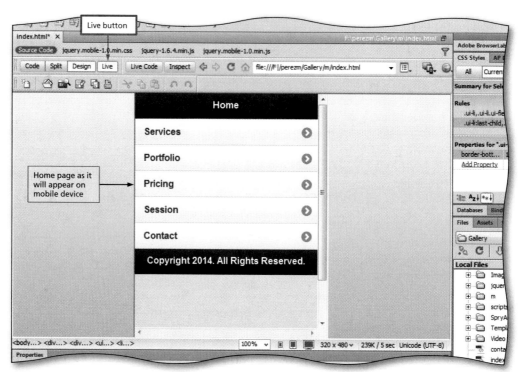

Figure 9–39

Using Graphics on a Mobile Page

While using many graphics increases the appeal of a traditional Web site, the same is not true for a mobile Web site. Graphics consume downloading time on a mobile device. If the graphics on your mobile Web site take too long to download, users are likely to leave the page, or worse — your mobile site. Branding is important, so one graphic to include is your company logo. To decrease the file size, use a lower resolution file. (The smaller the file, the less time it takes to download.) Use other graphics to display your products or services, but use fewer images than your traditional Web site. Avoid displaying videos on your Web site due to the download time.

Plan Ahead

Minimize the use of graphics.
Because it takes time to download graphics on a mobile device, keep the use of graphics to a minimum. Include a company logo and any other essential graphics as low-resolution files. Keep in mind that users will view the graphics on a small screen, so opt for clean, easy-to-recognize images instead of those with lots of detail.

To Add a Graphic to the Home Page

Instead of a page name, the Home page should display the Gallery logo. The following steps replace the Home page name with a logo graphic.

1

- Delete the text, Home, on the Home page to remove the text (Figure 9–40).

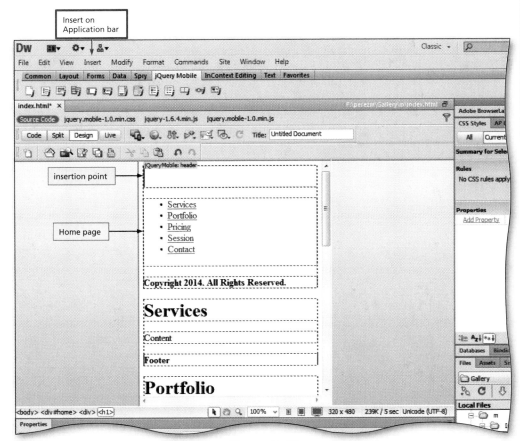

Figure 9–40

2

- Click Insert on the Application bar and then click Image to display the Select Image Source dialog box.

- In the Files panel, double-click the m folder to open it.

- Double-click the Images folder to open it and then click the m_logo file to select the file (Figure 9–41).

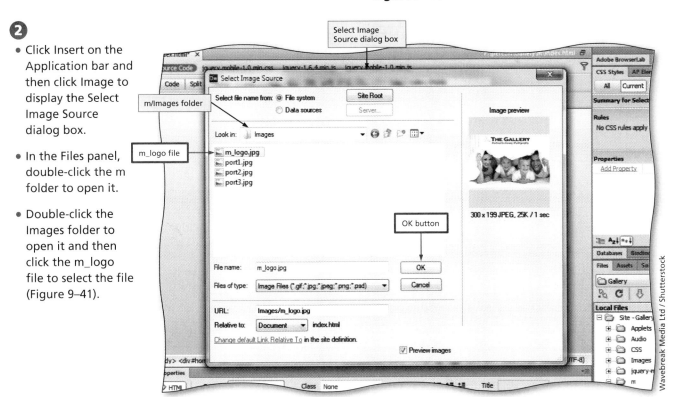

Figure 9–41

3

- Click the OK button to display the Image Tag Accessibility Attributes dialog box.

- In the Alternate text text box, type `The Gallery Mobile Logo` to enter the alternate text.

- Click the OK button to insert the m_logo file on the Home page (Figure 9–42).

4

- Click the Save button on the Standard toolbar to save your work.

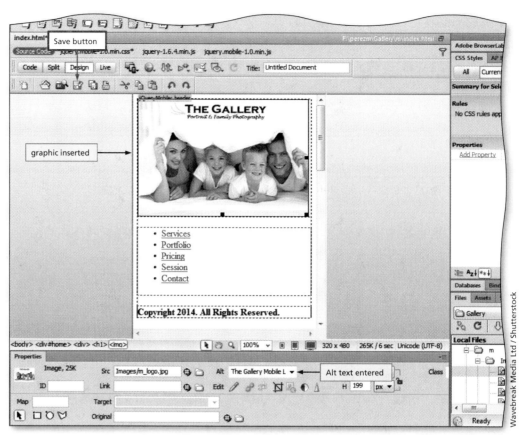

Figure 9–42

Setting Styles for Mobile Web Pages

The style and appearance of each jQuery Mobile page and additional jQuery Mobile elements are determined by an external style sheet named jquery.mobile-1.0.min.css. The default styles defined in the style sheet provide a good starting point for setting up mobile Web pages. However, you can change styles to suit your needs, such as adjusting the margin to accommodate a page layout. In fact, the default styles for the jQuery Mobile UI widgets often need to be adjusted to suit the content of a mobile site. As you create a mobile site, switch to Live view frequently to display the pages as they appear in a mobile device, and then modify CSS styles as necessary to fit your design.

To Modify jQuery Mobile CSS Rules

The following steps display the Gallery's mobile site in Live view and then modify the jquery.mobile-1.0.min CSS file to accommodate the logo graphic on the Home page.

1

- Click the Live button on the Document toolbar to display the mobile Web site in Live view (Figure 9–43).

Q&A

Why isn't the graphic displayed properly in Live view?

The placement of the graphic is determined by a CSS rule named .ui-header .ui-title, .ui-footer .ui-title in jquery.mobile-1.0.min.css, which sets the left and right margins to 90 pixels for the Header section. Those settings mean that only part of the logo graphic can be displayed. The default CSS rule for .ui-header .ui-title, .ui-footer .ui-title in jquery.mobile-1.0.min.css must be modified to show the logo.

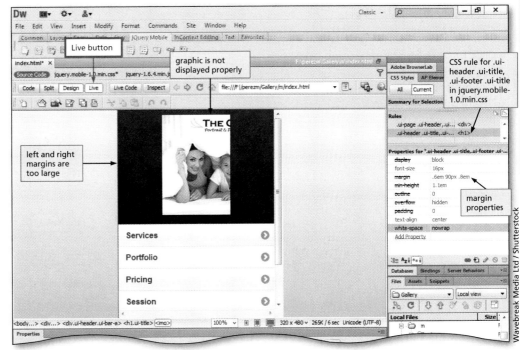

Figure 9–43

2

- Double-click the CSS Rule for .ui-header .ui-title,.ui-footer .ui-title in jquery. mobile-1.0.min.css rule on the CSS Styles panel to display the CSS Rule Definition for .ui-header .ui-title,.ui-footer .ui-title in jquery. mobile-1.0.min. css dialog box (Figure 9–44).

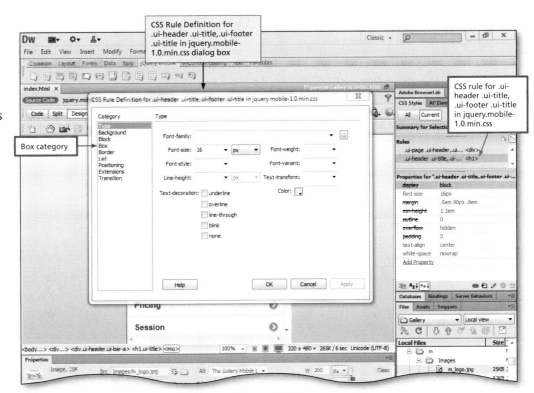

Figure 9–44

Wavebreak Media Ltd / Shutterstock

3

- In the Category list, click Box to select it.

- Click the Right text box in the Margin section, and then delete the value to use the default right margin.

- Click the Left text box in the Margin section, and then delete the value to use the default left margin (Figure 9–45).

Q&A

Why should I delete the right and left margin values to use the default margins?

By default, CSS provides 90 pixels for the right and left margins. Using these default values will provide enough room in the Header section to display the complete logo graphic.

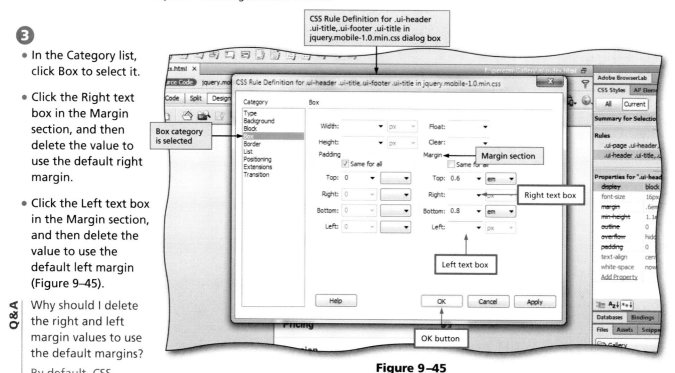

Figure 9–45

4

- Click the OK button to accept the changes and to display the page in Live view (Figure 9–46).

5

- Click the Live button on the Document toolbar to close Live view.

- Click the Save All button on the Document toolbar to save your changes.

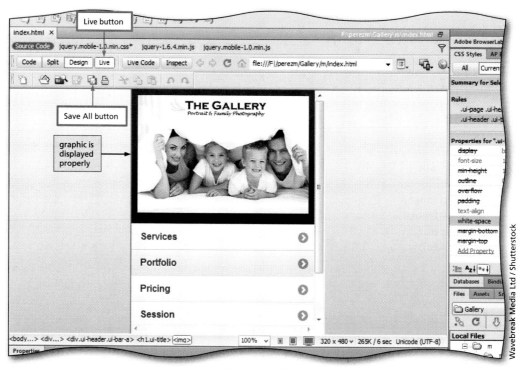

Figure 9–46

To Add Graphics to the Portfolio Page

The following steps add photos to the Portfolio mobile page.

- If necessary, use the scroll bar to scroll down to the Portfolio page.

- Delete the word, Content, to remove it.

- Click Insert on the Application bar and then click Image to display the Select Image Source dialog box (Figure 9–47).

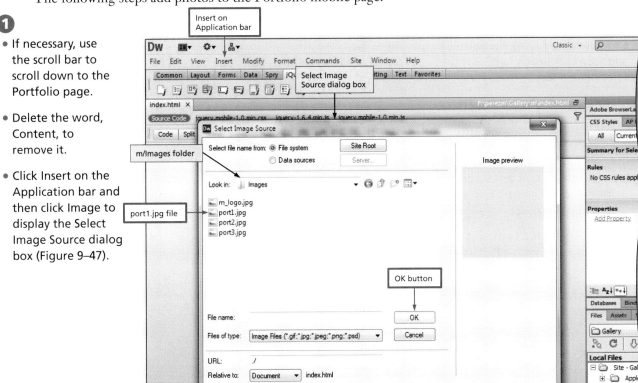

Figure 9–47

- Select the port1 file and then click the OK button to display the Image Tag Accessibility Attributes dialog box.

- In the Alternate text text box, type Portfolio image 1 to enter the Alternate text.

- Click the OK button to insert the port1 image file on the Portfolio page (Figure 9–48).

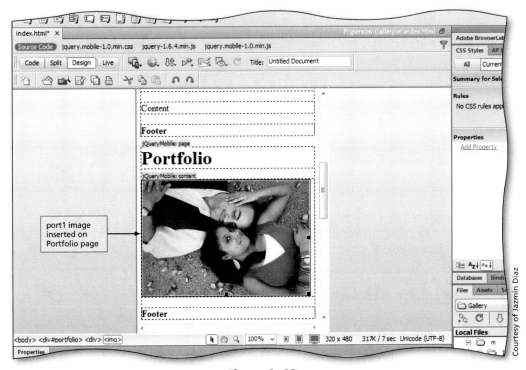

Figure 9–48

- Click to the right of the port1 image and then press the ENTER key to create a new paragraph.

- Click Insert on the Application bar and then click Image to display the Select Image Source dialog box.

- Select the port2 file and then click the OK button to display the Image Tag Accessibility Attributes dialog box.

- In the Alternate text text box, type `Portfolio image 2` to enter the Alternate text.

- Click the OK button to insert the port2 image file on the Portfolio page.

- If necessary, use the scroll bar to scroll down to display the port2 image (Figure 9–49).

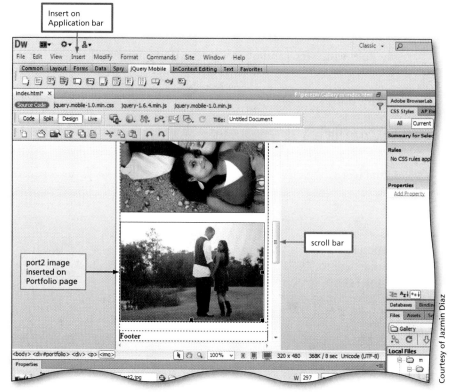

Figure 9–49

Courtesy of Jazmin Díaz

- Click to the right of the port2 image and then press the ENTER key to create a new paragraph.

- Click Insert on the Application bar and then click Image to display the Select Image Source dialog box.

- Select the port3 file and then click the OK button to display the Image Tag Accessibility Attributes dialog box.

- In the Alternate text text box, type `Portfolio image 3` to enter the Alternate text.

- Click the OK button to insert the port3 image file on the Portfolio page.

- If necessary, use the scroll bar to scroll down to display the port3 image (Figure 9–50).

Figure 9–50

Courtesy of Jazmin Díaz

- Click the Save button on the Standard toolbar to save your work.

Adding Other jQuery Mobile UI Widgets to Mobile Pages

Recall that you can incorporate several types of jQuery Mobile UI Widgets into a mobile Web site. For example, you can use the jQuery Mobile Collapsible Block UI widget to condense information into blocks, which provides the streamlined appearance typical of mobile sites. As the name suggests, users can expand each collapsible block to retrieve additional information about a topic. This widget is similar to the Spry Collapsible Panel introduced in Chapter 5, but is specifically styled for a mobile device.

Another widget is the jQuery Mobile Layout Grid UI widget, which inserts a table into your mobile site. On the Gallery mobile site, a jQuery Mobile Layout Grid UI widget organizes the pricing information on the Pricing page. You also can use jQuery Mobile Button UI widgets to integrate links within a mobile Web site. Buttons provide an easy way for users to navigate a mobile site. Buttons can be linked to pages, e-mail addresses, and even phone numbers. Each jQuery Mobile UI Widget contains appropriate styling to accommodate the mobile device user.

Consider the mobile device display.
Because the display area of a mobile device is much smaller than that of a traditional computer, condense your information and keep the navigation flowing in one direction, such as from the home page to the other pages. If certain pages are used to convey lots of information, use a collapsible block to help break up the content and decrease the need for scrolling.

To Add a jQuery Mobile Collapsible Block UI Widget to the Services Page

The Services page describes the types of services the Gallery offers: portraits, family and children photos, and newborn and maternity photos. To describe these services in the limited space of a mobile device screen, you can use a jQuery Mobile Collapsible Block UI widget, which displays each type of service on a button. Visitors can click a button to display a description of the service. The following steps add a jQuery Mobile Collapsible Block UI widget to the Services page.

1

- If necessary, use the scroll bar to scroll to view the Services page.

- Delete the word, Content, to remove it.

- Click the jQuery Mobile Collapsible Block button to insert a collapsible block set (Figure 9–51).

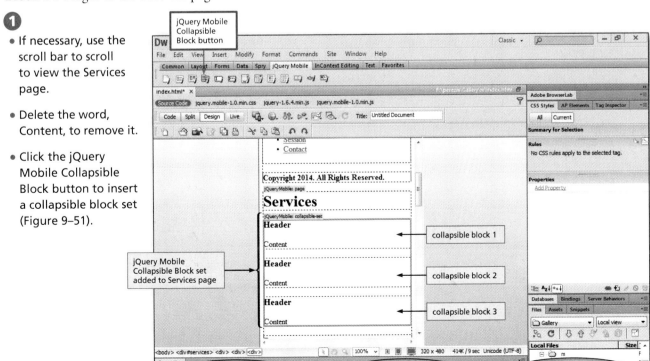

Figure 9–51

2

- Delete the text, Header, in the first collapsible block and then type `Portrait` to name the collapsible block.

- Delete the text, Content, below Portrait and type `Whether you need a professional studio head shot or a candid photo, we pamper you during your individual portrait session to produce a memorable, enduring photograph.` to provide content for the Portrait collapsible block (Figure 9–52).

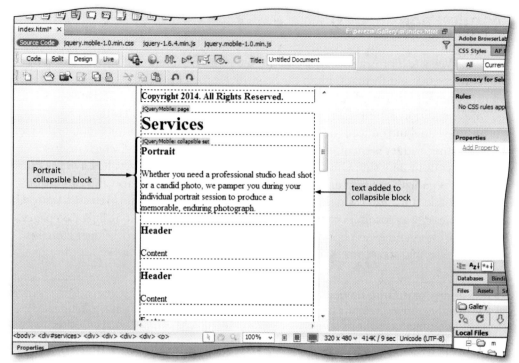

Figure 9–52

3

- Delete the text, Header, in the second collapsible block and type `Family/Children` to name the collapsible block.

- Delete the text, Content, below Family/Children and type `Family photos are an important piece for your home. Let us capture fun-filled family moments that you can proudly display in any room.` to provide content for the Family/Children collapsible block (Figure 9–53).

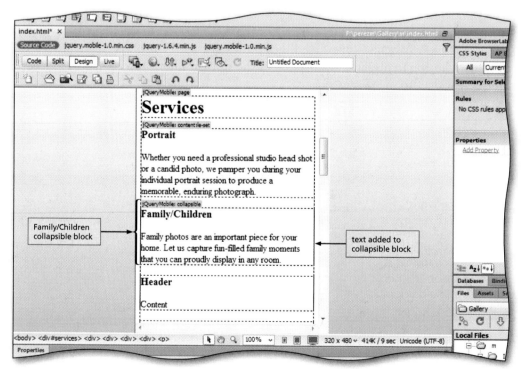

Figure 9–53

4

- Delete the text, Header, in the third collapsible block and type Newborn/ Maternity to name the collapsible block.

- Delete the text, Content, below Newborn/ Maternity and type Pregnancy is a special time for expecting parents and nothing is more precious than a newborn baby. We work around your baby to capture memorable moments in your child's life. to provide content for the Newborn/Maternity collapsible block (Figure 9–54).

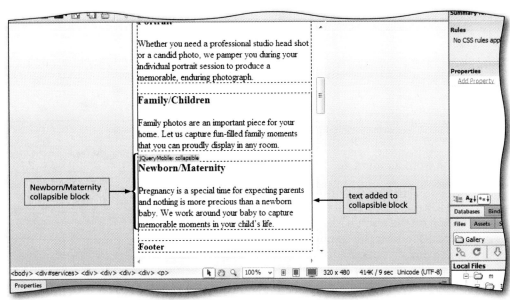

Figure 9–54

5

- Click the Live button on the Document toolbar to view the site in Live view.

- Click the Services button to view the Services page.

- Click the Portrait button to view the Portrait collapsible block.

- Click the Family/ Children button to view the Family/ Children collapsible block.

- Click the Newborn/ Maternity button to view the Newborn/Maternity collapsible block (Figure 9–55).

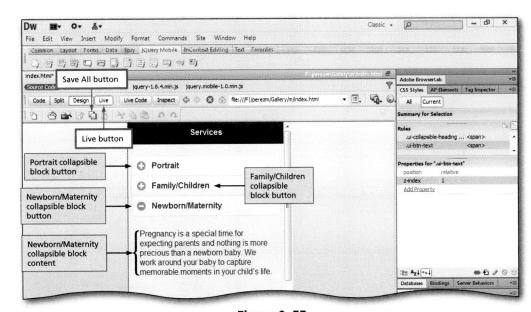

Figure 9–55

 What is the purpose of the plus and minus icons on the collapsible blocks?

The plus icon indicates that clicking the button expands the collapsible block to display the content. The minus icon indicates that clicking the button collapses the collapsible block to hide the content.

6

- Click the Live button on the Document toolbar to close Live view.

- Click the Save All button on the Standard toolbar to save your work.

Other Ways

1. Click Insert on Application bar, point to jQuery Mobile, click Collapsible Block

To Add a jQuery Mobile Layout Grid UI Widget to the Pricing Page

The Pricing page lists the prices of five types of sessions using a tabular format. To include this type of information for a range of mobile devices, you can insert a jQuery Mobile Layout Grid UI widget, and then enter the session types and prices as text. The following steps add a jQuery Mobile Layout Grid UI widget to the Pricing page.

1

- If necessary, use the scroll bar to scroll to view the Pricing page.

- Delete the word, Content, to remove it.

- Click the jQuery Mobile Layout Grid button on the Insert bar to display the jQuery Mobile Layout Grid dialog box.

- Click the Rows button and then select 5 to designate the number of rows for the layout grid.

- If necessary, click the Columns button and then select 2 to designate the number of columns for the layout grid (Figure 9–56).

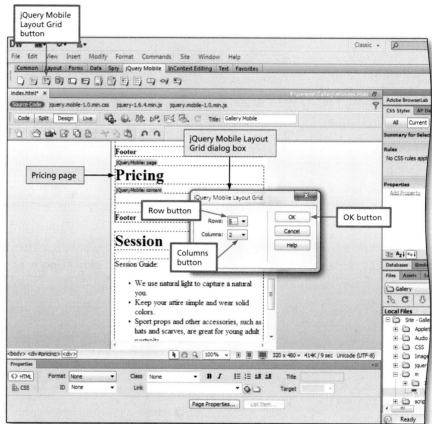

Figure 9–56

2

- Click the OK button to insert the jQuery Mobile Layout Grid UI widget on the page (Figure 9–57).

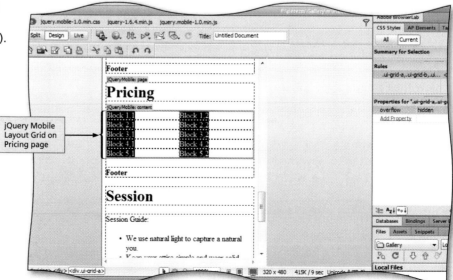

Figure 9–57

3

- Delete the text, Block 1,1, and then type `Session` to enter a title for the column heading.

- Select the word, Session, and then click the Bold button on the Property inspector to bold the text.

- Delete the text, Block 1,2, and then type `Fee` to enter a title for the column heading.

- Select the word, Fee, and then click the Bold button on the Property inspector to bold the text (Figure 9–58).

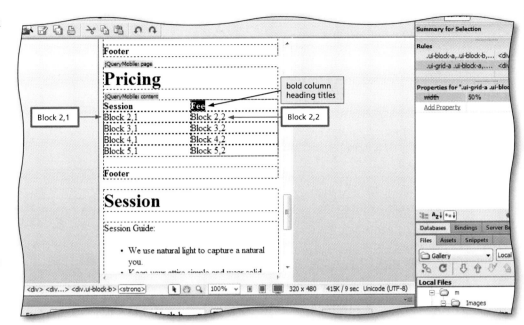

Figure 9–58

Q&A

What do the numbers in text such as Block 1,1 mean?

Block 1,1 refers to the first row and first column of the grid, Block 1,2 refers to the first row and second column of the grid, and so on.

4

- Delete the text, Block 2,1, and then type `Portrait` to enter text into the next block.

- Delete the text, Block 2,2, and then type `$75` to enter text into the next block (Figure 9–59).

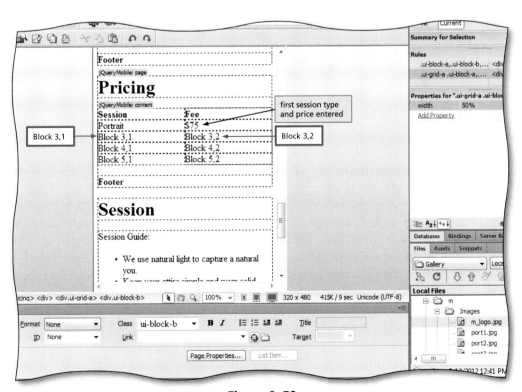

Figure 9–59

5

- Delete the text, Block 3,1, and then type `Family` to enter text into the next block.

- Delete the text, Block 3,2, and then type `$175` to enter text into the next block.

- Delete the text, Block 4,1, and then type `Children` to enter text into the next block.

- Delete the text, Block 4,2, and then type `$135` to enter text into the next block.

- Delete the text, Block 5,1, and then type `Newborn/Maternity` to enter text into the next block.

- Delete the text, Block 5,2, and then type `$150/$95` to enter text into the last block (Figure 9–60).

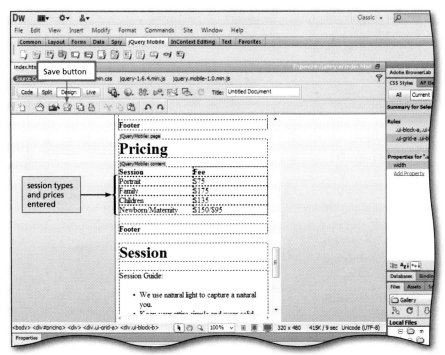

Figure 9–60

6

- Click the Save button on the Standard toolbar to save your work.

Other Ways

1. Click Insert on Application bar, point to jQuery Mobile, click Layout Grid

---

Break Point: If you wish to take a break, this is a good place to do so. To resume at a later time, start Dreamweaver and continue following the steps from this location forward.

To Add jQuery Mobile Home Buttons

A traditional Web site typically provides a navigation bar containing a button for each page in the site. Visitors click a button in the navigation bar to switch from one page to another. Mobile device screens are too small to accommodate a navigation bar. However, visitors still need a way to navigate the site. You can insert a button on each page of the site except for the home page. Visitors can click this button to return to the home page, where they can select another page in the site, if necessary. To create these buttons, use the jQuery Mobile Button UI widget. When you insert a jQuery Mobile Button UI widget, it appears as plain link text on the mobile page. In a mobile browser, this UI widget appears as a button displaying a home icon and text you specify. The following steps add jQuery Mobile buttons to the footer area of the Services, Portfolio, Pricing, Session, and Contact pages.

- If necessary, use the scroll bar to view the Services footer.

- Delete the word, Footer, to remove it.

- Click the jQuery Mobile Button button on the Insert bar to display the jQuery Mobile Button dialog box.

- Click the Icon button and then click Home to select the Home icon (Figure 9–61).

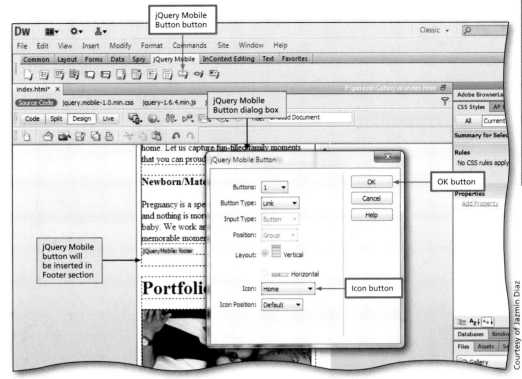

Figure 9–61

- Click the OK button to insert the jQuery Mobile Button UI widget on the page.

- In the Footer section of the Services page, type Home to change the button name.

- Type home after the number sign (#) in the Link text box on the Property inspector to link this button to the Home page (Figure 9–62).

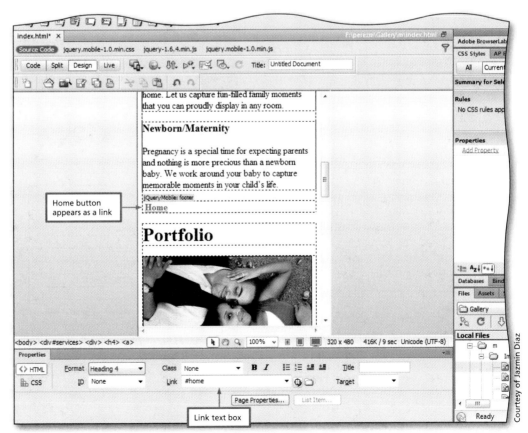

Figure 9–62

3

- Select the Home button and then press the CTRL+C keys to copy it.

- Use the scroll bar to scroll down to the Portfolio footer.

- Delete the word, Footer, to remove it.

- Press the CTRL+V keys to paste the Home button in the Portfolio footer (Figure 9–63).

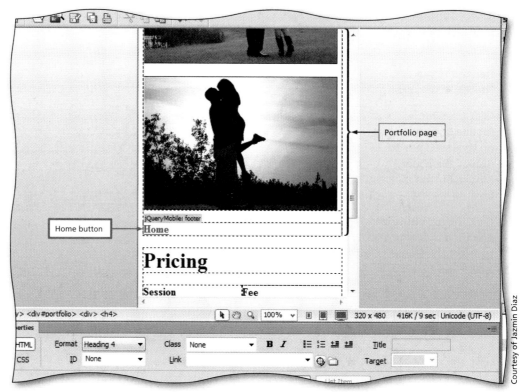

Figure 9–63

4

- Use the scroll bar to scroll down to the Pricing footer.

- Delete the word, Footer, to remove it.

- Press the CTRL+V keys to paste the Home button in the Pricing footer (Figure 9–64).

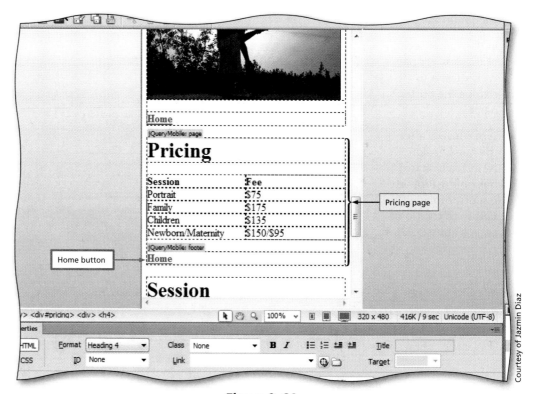

Figure 9–64

- Use the scroll bar to scroll down to the Session footer.

- Delete the word, Footer, to remove it.

- Press the CTRL+V keys to paste the Home button in the Session footer (Figure 9–65).

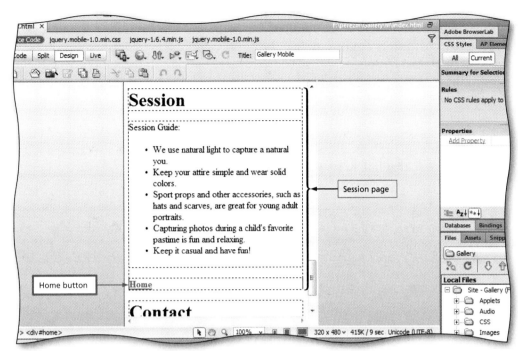

Figure 9–65

- Use the scroll bar to scroll down to the Contact footer.

- Delete the word, Footer, to remove it.

- Press the CTRL+V keys to paste the Home button in the Contact footer (Figure 9–66).

7

- Click the Save button on the Standard toolbar to save your work.

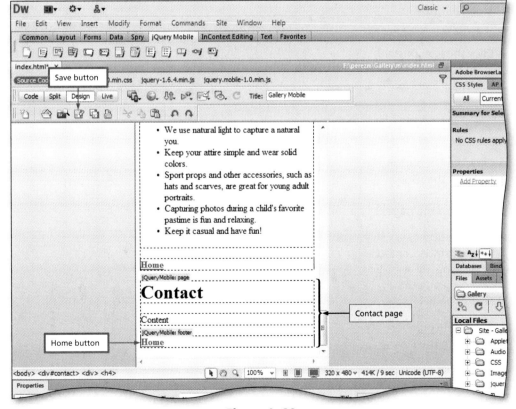

Figure 9–66

Other Ways

1. Click Insert on Application bar, point to jQuery Mobile, click Button

To Add and Link jQuery Mobile Buttons to an E-mail Address and Phone Number

Besides a Home button that links to the Home page on the Gallery's mobile site, the Contact page should contain buttons visitors can click to find the e-mail address and phone number of the Gallery studio. You can use the jQuery Mobile Button UI widget to create these buttons. In the jQuery Mobile Button dialog box, you select Link as the button type, as you did when creating the Home buttons, and then select an icon other than a home icon, such as a star icon, to appear on the button. Use the Link text box on the Property inspector to specify the e-mail address visitors can use to send an e-mail message to the Gallery studio. Because a smartphone can make telephone calls, you can enter a telephone number in the Link box for the second button. When visitors click that button, the smartphone connects to that phone number. The following steps add jQuery Mobile buttons to the Contact page that link to an e-mail address and a phone number.

- If necessary, use the scroll bar to scroll down to the Contact page.

- Delete the word, Content, to remove it.

- Click the jQuery Mobile Button button on the Insert bar to display the jQuery Mobile Button dialog box.

- Click the Icon button and then click Star to select the Star icon (Figure 9–67).

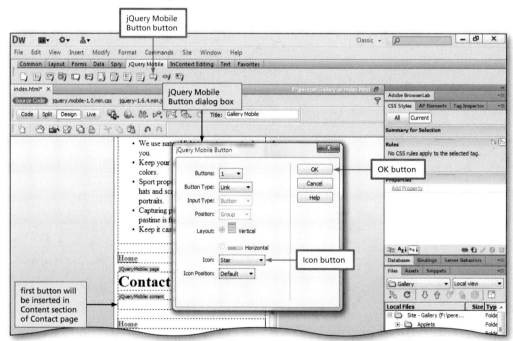

Figure 9–67

- Click the OK button to insert the jQuery Mobile Button UI widget on the page.

- In the Content section of the Contact page, type E-mail Us to change the button name.

- Delete the number sign (#) in the Link text box on the Property inspector to remove it.

- In the Link text box, type mailto: TheGalleryPFP@ thegallery.net to link this button to the Gallery's e-mail address (Figure 9–68).

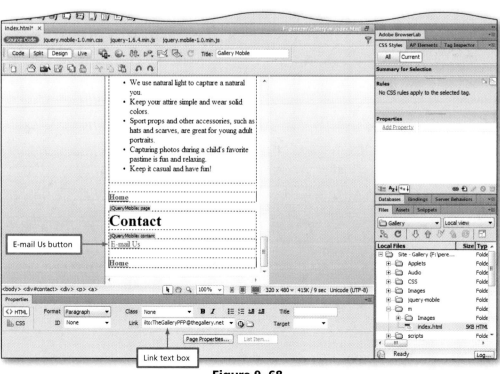

Figure 9–68

3

- Place the insertion point after the E-mail Us button on the Contact page and then press the SPACEBAR to insert a space.

- Click the jQuery Mobile Button button on the Insert bar to display the jQuery Mobile Button dialog box.

- Click the Icon button and then click Star to select the Star icon (Figure 9–69).

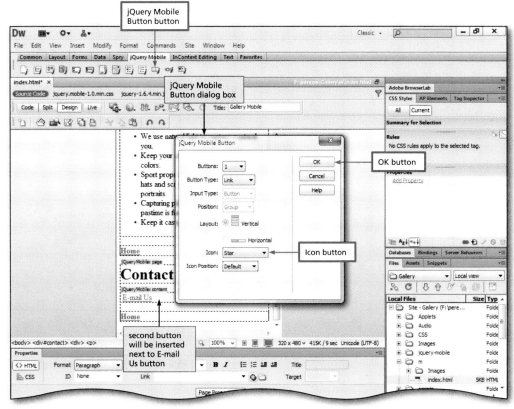

Figure 9–69

4

- Click the OK button to insert the jQuery Mobile Button on the page.

- Type `Call Us` to change the button name.

- Delete the number sign (#) in the Link text box on the Property inspector to remove it.

- In the Link text box, type `tel:6435550324` to link this button to the Gallery's phone number (Figure 9–70).

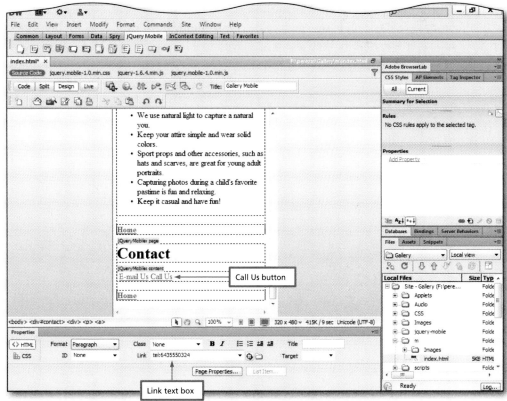

Figure 9–70

- Click the Live button on the Document toolbar to display the mobile site in Live view.

- If necessary, click the Contact link on the Home page to display the Contact page (Figure 9–71).

- Click the Live button on the Document toolbar to close Live view.

- Click the Save button on the Standard toolbar to save your work.

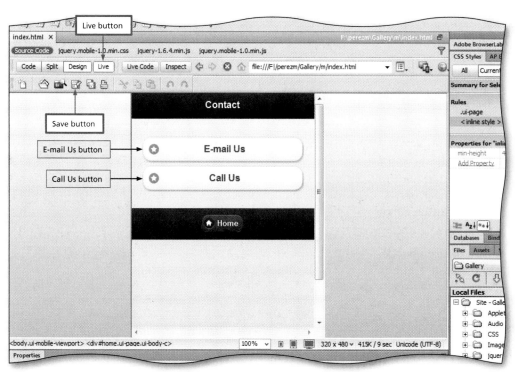

Figure 9–71

To Add a Title and View the Mobile Web Site in a Browser

To complete the mobile site, add a title on the index.html page. In a mobile site, the title appears on the browser tab when the mobile Web site is displayed in a browser. Next, you can preview the mobile site in a browser to test the site. The following steps open the Gallery mobile Web site in Internet Explorer.

- Type `Gallery Mobile` in the Title text box on the Document toolbar to name the mobile site.

- Click the Save button on the Standard toolbar to save your work (Figure 9–72).

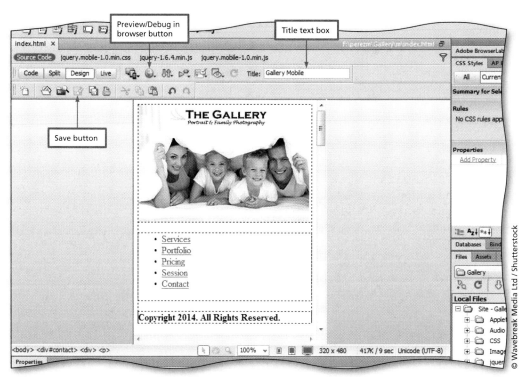

Figure 9–72

2

- Click the 'Preview/ Debug in browser' button and then select Preview in IExplore to launch the mobile site in Internet Explorer.

- If a security message appears regarding ActiveX controls, click the 'Allow blocked content' button to allow the mobile site to work properly.

- Adjust the size of the Internet Explorer window to make it smaller and similar to a smartphone screen (Figure 9–73).

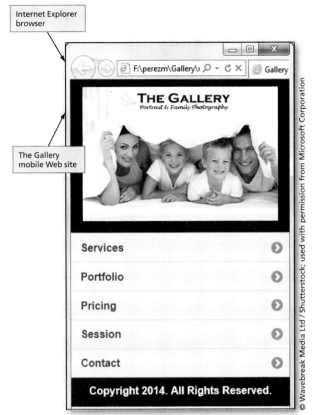

Internet Explorer browser

The Gallery mobile Web site

© Wavebreak Media Ltd / Shutterstock; used with permission from Microsoft Corporation

Figure 9–73

3

- Click each button (Services, Portfolio, Pricing, Session, and Contact) on the Home page to view the contents of each page.

- Click the Home button on each page to return to the Home page.

- Click the Close button to close Internet Explorer.

Other Ways
1. Press F12

To Use the jQuery Mobile Swatches Panel

The jQuery Mobile Swatches Panel provides five themes, Themes a-e, to further enhance the look of your mobile Web site. You can apply these themes to the page, header areas, content areas, footer areas, lists, icons, and buttons, for example. Although you will not apply a theme to the Gallery mobile Web site, the following steps display and use the jQuery Mobile Swatches panel to view themes for a mobile site.

BTW
Additional jQuery Mobile Themes
More themes are available for download at jquerymobile.com/themeroller.

1

- Use the scroll bar to scroll to the top of the page to display the Home page.

- Click the left edge of the Home page to select it and to display a blue border around the page (Figure 9–74).

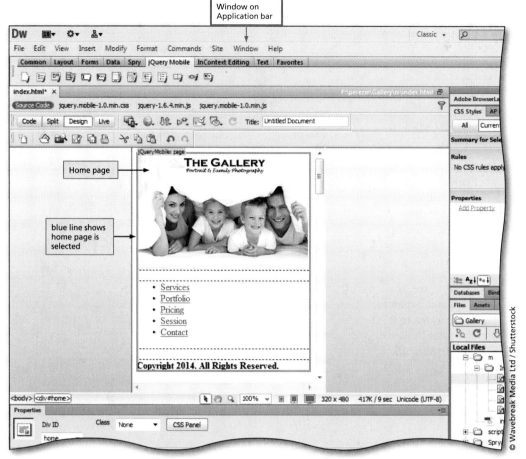

Figure 9–74

2

- Click Window on the Application bar and then click jQuery Mobile Swatches to display the jQuery Mobile Swatches panel (Figure 9–75).

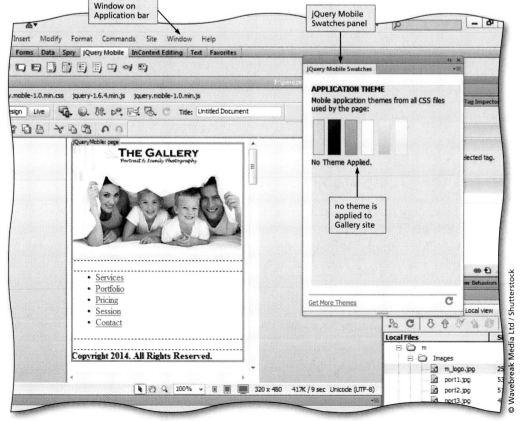

Figure 9–75

3

- Click the Live button on the Document toolbar to display the site in Live view.

- Click Theme: a (the black swatch) on the jQuery Mobile Swatches panel to display the mobile site theme color changed to black.

- Click each theme color swatch on the jQuery Mobile Swatches panel to see the site theme in different colors (Figure 9–76).

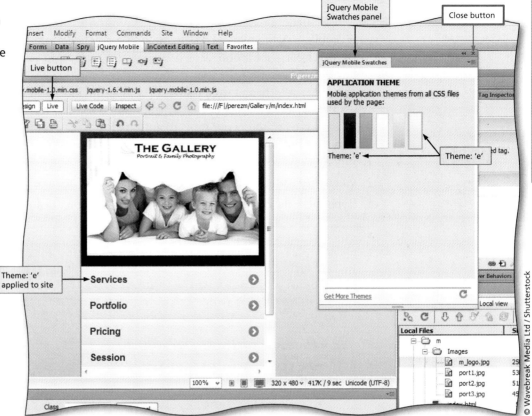

Figure 9–76

4

- Click the first theme in the jQuery Mobile Swatches panel to select No Theme Applied.

- Click the Close button for the jQuery Mobile Swatches panel to close the panel.

- Click the Live button on the Document toolbar to close Live view (Figure 9–77).

5

- Click the Save button on the Standard toolbar to save your work.

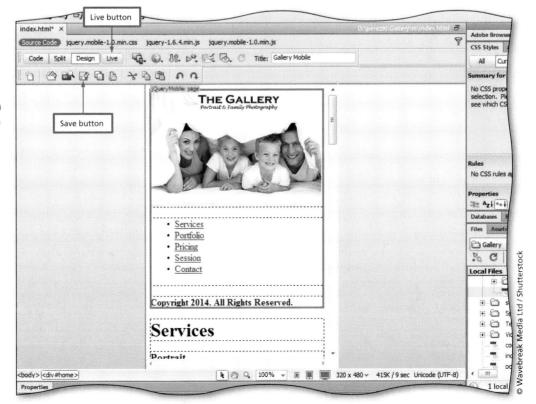

Figure 9–77

To Redirect a Mobile Device User to the Mobile Web Site

When mobile device users visit your Web site, they will not know the URL for your mobile site and may not even know that you have a mobile site. Instead of making them search for your mobile site, incorporate a script into your regular Web site that will determine whether the user is accessing the site through a mobile device, based on the screen width, and then redirect them to the mobile site if they are trying to access the site on a mobile device. If you wanted to insert scripts into the HTML code, you would use the following steps to automatically redirect mobile device users to the Gallery's mobile Web site.

1. Open the Gallery template file.
2. Click the Code button on the Document toolbar to view the code.
3. Type the following scripts within the <head> element of the HTML code.

```
<script type="text/javascript">
if (screen.width <= 699) {
document.location = "http://yourURLhere.com/your last
  name and first initial/Gallery/m/index.html";
}
</script>
<script language=javascript>
if ((navigator.userAgent.match(/iPhone/i)) ||
  (navigator.userAgent.match(/iPod/i))) {
location.replace("http://yourURLhere.com/your last
  name and first initial/Gallery/m/index.html");
}
</script>
```

4. Click the Save button and update all dependent files.

Publishing a Mobile Web Site

Use the same method you used in Chapter 8 to publish the Gallery's mobile Web site. You can publish the mobile directory folder, m, to publish the Gallery's mobile Web site. Before you publish the mobile site, add the following line of code within the <head> of the HTML code and save your changes. This meta tag adjusts the width for mobile viewing.

<meta name="viewport" content="width=device-width, initial-scale=1">

Once the mobile site has been published, view the site from a mobile device. The mobile Home page, Services page, and Contact page, after tapping the Call Us button on a mobile device, are shown in Figure 9–78.

(a) 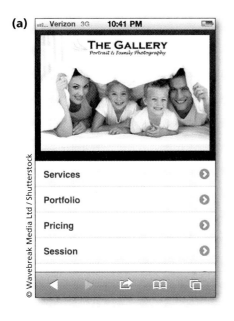   (b) 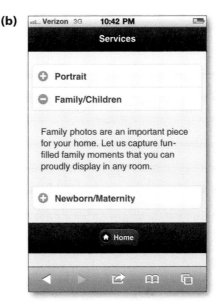   (c)

Figure 9–78

Creating PhoneGap Applications

In this chapter, you created a mobile Web site, also called a Web application for mobile devices, using jQuery that is displayed in the browser of a smartphone or tablet. In addition to creating a Web site, you can also use Dreamweaver CS6 with an additional product called **PhoneGap** (http://phonegap.com) to create mobile native apps. A **native app** is a locally installed application that you typically purchase in an app store and is designed to run on that specific type of mobile device. PhoneGap is an application container technology that allows you to create natively installed applications for mobile devices using HTML, CSS, and JavaScript. PhoneGap builds native apps for iPhone, Android, Windows Mobile, and other mobile platforms. The PhoneGap Build is an application wrapper that bridges the gap between browser and native functions. PhoneGap interacts with the native operating system so that you do not have to write native code.

To Quit Dreamweaver

The following steps quit Dreamweaver and return control to the operating system.

1 Click the Close button on the right side of the Application bar to close the Dreamweaver window.

2 If Dreamweaver displays a dialog box asking you to save changes, click the No button.

Chapter Summary

In this chapter, you learned about the jQuery Mobile framework integrated into Adobe Dreamweaver CS6 to create seamless mobile Web sites. You inserted jQuery Mobile pages and added content and graphics to them. You inserted jQuery Mobile List View UI widgets and created links to the pages within the mobile Web site. You inserted a jQuery Mobile Collapsible Block UI widget to showcase services. You inserted a jQuery Mobile Layout Grid UI widget to provide pricing information. Finally, you inserted jQuery Mobile Buttons and then created links for the buttons. The following tasks are all the new Dreamweaver skills you learned in this chapter:

1. Create a Mobile Directory and Copy Data Files (DW 467)
2. Create the Home Page for the jQuery Mobile Site (DW 471)
3. Add a jQuery Mobile List View UI widget to the Home Page (DW 481)
4. Change the View Size and Display the Mobile Site in Live View (DW 485)
5. Add a jQuery Mobile Collapsible Block UI widget to the Services Page (DW 493)
6. Add a jQuery Mobile Layout Grid UI widget to the Pricing Page (DW 496)
7. Add jQuery Mobile Home Buttons (DW 498)
8. Add and Link jQuery Mobile Buttons to an E-mail Address and Phone Number (DW 502)
9. Use the jQuery Mobile Swatches Panel (DW 505)

Apply Your Knowledge

Reinforce the skills and apply the concepts you learned in this chapter.

Creating a jQuery Mobile Web Site

Instructions: Create a mobile Web site to provide more information about jQuery Mobile. In this activity, you create three jQuery Mobile pages, add content to each page, and add a jQuery Mobile List View UI widget to the Home page and define links. First, you create the Home page. Next, you create two additional pages. Then you insert a jQuery Mobile List View UI widget and define the page links. Finally, you add content to the pages. The mobile Web site, as displayed in Live view, is shown in Figure 9–79.

(a)

(b)
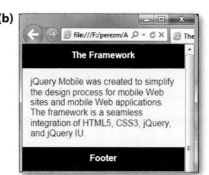

(c)

Figure 9–79

Perform the following tasks:

1. Start Dreamweaver. Use the Sites button on the Files panel to select the Apply site. Create a new, blank HTML5 document. Save it as `apply9.html`.

2. Use the jQuery Mobile Page button to insert a new mobile page. Use `home` for the ID.

3. Insert another jQuery Mobile Page and use `page2` for the ID.

4. Insert another jQuery Mobile Page and use `page3` for the ID.

5. Replace the word, Header, on the Home page with `jQuery Mobile`. Replace the word, Footer, on the Home page with your first and last names.

6. Delete the word, Content, on the Home page and insert a jQuery Mobile List View UI widget with two unordered list items.

7. Name the first link `The Framework`, and link it to `#page2`.

8. Name the second link `Device Support`, and link it to `#page3`.

9. Replace the word, Header, on page2 with `The Framework`.

10. Replace the word, Content, on page2 with `jQuery Mobile was created to simplify the design process for mobile Web sites and mobile Web applications. The framework is a seamless integration of HTML5, CSS3, jQuery, and jQuery UI.`

11. Replace the word, Header, on page3 with `Device Support`.

12. Delete the word, Content, on page3 and insert an unordered list with the following list items:

    iOS

    Android

    BlackBerry

    Bada

    Windows Phone

    Palm webOS

    Symbian

    MeeGo

13. Save your changes, copy the dependent files, and then view the mobile Web site in your browser. Allow blocked content, if necessary. Click each page link to view its contents, and then click the Back button in the browser to return to the Home page. Compare your site to Figure 9–79. Make any necessary changes, and then save your changes.

14. Submit the document in the format specified by your instructor.

Extend Your Knowledge

Extend the skills you learned in this chapter and experiment with new skills. You may need to use Help to complete the assignment.

Adding jQuery Mobile Form Elements

Note: To complete this assignment, you will be required to use the Data Files for Students. Visit www.cengage.com/ct/studentdownload for detailed instructions or contact your instructor for information about accessing the required files.

Continued >

Extend Your Knowledge *continued*

Instructions: A mobile Web page contains four jQuery Mobile pages. In this activity, you add the jQuery Mobile Text Input UI widget, jQuery Mobile Slider, and jQuery Mobile Flip Toggle Switch UI widget to a mobile site. First, you add the jQuery Mobile Text Input UI widget to the Text Input page. Next, you add the jQuery Mobile Slider to the Slider page. Finally, you add the jQuery Mobile Flip Toggle Switch UI widget to the Flip Switch page. The mobile Web site, as displayed in a browser, is shown in Figure 9–80.

Figure 9–80

Perform the following tasks:

1. Use Windows Explorer to copy the jquery-mobile folder and the extend9.html file from the Chapter 09\Extend folder into the *your last name and first initial*\Extend folder.

2. Start Dreamweaver. Use the Sites button on the Files panel to select the Extend site. Open extend9.html.

3. Delete the text, Content, on the Text Input page.

4. Use the jQuery Mobile Text Input button on the Insert bar to insert a jQuery Mobile Text Input UI widget. Replace the text, Text Input, with `Your Name`.

5. Delete the text, Content, on the Slider page.

6. Use the jQuery Mobile Slider button on the Insert bar to insert a jQuery Mobile Slider. Replace the text, Value, with `Amount`.

7. Delete the text, Content, on the Flip Switch page.

8. Use the jQuery Mobile Flip Toggle Switch button on the Insert bar to insert a jQuery Mobile Flip Toggle Switch UI widget.

9. Replace the text, Footer, on the Home page with your first and last names.

10. Save your changes and then view the mobile Web site in your browser. Click each page link to view the jQuery Mobile form elements. Press the Back button in the browser to return to the Home page.

11. Compare your site to Figure 9–80. Make any necessary changes, and then save your changes.

12. Submit the document in the format specified by your instructor.

Make It Right

Analyze a Web site and suggest how to improve its design.

Adding a jQuery List View UI Widget and Buttons

Note: To complete this assignment, you will be required to use the Data Files for Students. Visit www.cengage.com/ct/studentdownload for detailed instructions or contact your instructor for information about accessing the required files.

Instructions: Create a mobile Web site that provides information about movie rewards. In this activity, add a jQuery Mobile List View UI widget to a mobile page, name each link, link each item to its respective page, and add jQuery Mobile Buttons to link to the Home page, an e-mail address, and a telephone number. First, insert a jQuery Mobile List View UI widget. Next, name each link in the list and link each item to the appropriate page. Then insert a jQuery Mobile Button, name it, and link it to the Home page. Finally, insert jQuery Mobile Buttons to link to an e-mail address and a telephone number. The modified mobile Web site page is shown in Figure 9–81.

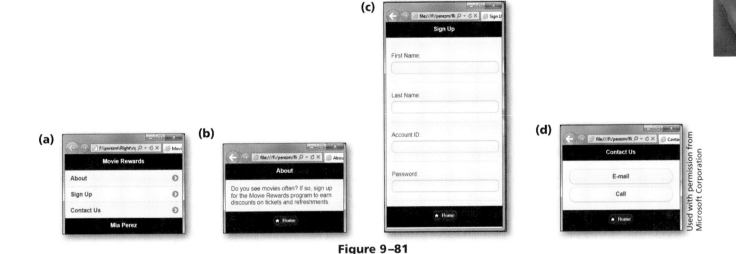

Figure 9–81

Perform the following tasks:

1. Use Windows Explorer to copy the jquery-mobile folder and the right9.html file from the Chapter 09\Right folder into the *your last name and first initial*\Right folder.

2. Start Dreamweaver. Use the Sites button on the Files panel to select the Right site. Open right9.html.

3. Delete the text, Content, on the Movie Rewards Home page. Insert a jQuery Mobile List View UI widget with three unordered list items.

4. Name the first link About, and then link it to #page2.

5. Name the second link Sign Up, and then link it to #page3.

6. Name the third link Contact Us, and then link it to #page4.

7. Delete the text, Footer, on the About page. Insert a jQuery Mobile Button with a Home icon. Name the button Home and link it to #home.

8. Copy the Home button. Delete the text, Footer, on the Sign Up page. Paste the Home button in the footer area of the Sign Up page.

Continued >

Make It Right *continued*

9. Delete the text, Footer, on the Contact Us page. Paste the Home button in the footer area of the Contact Us page.

10. Delete the text, Content, on the Contact Us page. Insert a jQuery Mobile Button. Name the button `E-mail`, and then use `mailto:movie@rewards.com` as the button link.

11. Insert another jQuery Mobile Button after the E-mail button. Name the button `Call`, and then use `tel:1235558563` as the button link.

12. Replace the text, Footer, on the Movie Rewards Home page with your first and last names.

13. Save your changes and then view the mobile Web site in your browser. Compare your site to Figure 9–81. Make any necessary changes, and then save your changes.

14. Submit the document in the format specified by your instructor.

In the Lab

Design and/or create a document using the guidelines, concepts, and skills presented in this chapter. Labs are listed in order of increasing difficulty.

Lab 1: Adding Content, Links, and jQuery Mobile UI Widgets to the Healthy Lifestyle Mobile Web Site

Note: To complete this assignment, you will be required to use the Data Files for Students. Visit www.cengage.com/ct/studentdownload for detailed instructions or contact your instructor for information about accessing the required files.

Problem: You have created an internal Web site for your company that features information about how to live a healthy lifestyle. Employees at your company will use this Web site as a resource for nutrition, exercise, and other health-related tips. You have created a mobile version of the Web site and now need to add content and link to the mobile pages.

First, add content to the mobile pages. Next, insert a jQuery Mobile Collapsible Block UI widget on the Habits page and add content. Finally, insert a jQuery Mobile List View UI widget, name the links, and then link to the appropriate pages. The mobile Web site page is shown in Figure 9–82.

(a)

(b)
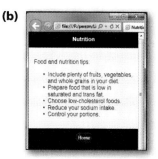

(c)
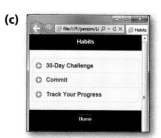

(d)

Figure 9–82

Perform the following tasks:

1. Copy the m folder and the jquery-mobile folder provided with your Data Files from the Chapter 09\Lifestyle folder into the *your last name and first initial*\ Lifestyle folder.

2. Start Dreamweaver. Use the Sites button on the Files panel toolbar to select the Healthy Lifestyle site. Open index.html in the m folder.

3. Delete the text, Content, on the Nutrition page. Type `Food and nutrition tips:` then press the ENTER key. Insert an unordered list with the following list items:

   `Include plenty of fruits, vegetables, and whole grains in your diet.`

   `Prepare food that is low in saturated and trans fat.`

   `Choose low-cholesterol foods.`

   `Reduce your sodium intake.`

   `Control your portions.`

4. Delete the text, Content, on the Habits page. Insert a jQuery Mobile Collapsible Block UI widget.

5. In the first collapsible block, replace the text, Header, with `30-Day Challenge`. Replace the text, Content, with `Challenge yourself to change the habit for the next 30 days. After 30 days, it is much easier to continue. If you fail the 30-day challenge, start again and continue until you are successful.`

6. In the second collapsible block, replace the text, Header, with `Commit`. Replace the text, Content, with `You have to commit to the change. Remind yourself of your goal daily. Place encouraging notes around your home to remind you of your goal and commitment. Update your social networking status to reflect your daily goal to help motivate you to stick to it!`

7. In the third collapsible block, replace the text, Header, with `Track Your Progress`. Replace the text, Content, with `Keep a notebook of your progress. Seeing your progress will only increase your motivation to continue.`

8. Delete the text, Content, on the Home page. Insert a jQuery Mobile List View UI widget with three unordered list items.

9. Name the first link `Nutrition`, and then link it to `#nutrition`.

10. Name the second link `Habits`, and then link it to `#habits`.

11. Name the third link `Sign Up`, and then link it to `#signup`.

12. Replace the text, Footer, on the Home page with your first and last names.

13. Save your changes and then view the mobile Web site in your browser. Compare your site to Figure 9–82. Make any necessary changes, and then save your changes.

14. Submit the document in the format specified by your instructor.

In the Lab

Lab 2: Adding Content, a jQuery Mobile Layout Grid UI Widget, and Buttons to the Designs by Dolores Mobile Web Site

Note: To complete this assignment, you will be required to use the Data Files for Students. Visit www.cengage.com/ct/studentdownload for detailed instructions or contact your instructor for information about accessing the required files.

Problem: You have created a Web site for Designs by Dolores, a Web site design company. The site provides information about the company and its services. Now you need to design a mobile version of the Web site.

First, add content to the About Us mobile page. Next, insert a jQuery Mobile Layout Grid UI widget on the Services and Pricing page, and then add text to the grid. Next, add Home buttons to the mobile pages. Finally, add buttons on the Contact Us mobile page that link to an e-mail address and a telephone number. The mobile Web site page is shown in Figure 9–83.

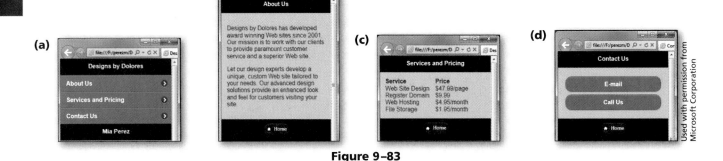

Figure 9–83

Perform the following tasks:

1. Copy the m folder and the jquery-mobile folder provided with your Data Files from the Chapter 09\Designs folder into the *your last name and first initial*\Designs folder.

2. Start Dreamweaver. Use the Sites button on the Files panel toolbar to select the Designs by Dolores site. Open index.html in the m folder.

3. Delete the text, Content, on the About Us page. Type `Designs by Dolores has developed award winning Web sites since 2001. Our mission is to work with our clients to provide paramount customer service and a superior Web site.` Press the ENTER key, and then type `Let our design experts develop a unique, custom Web site tailored to your needs. Our advanced design solutions provide an enhanced look and feel for customers visiting your site.`

4. Delete the text, Content, on the Services and Pricing page. Insert a jQuery Mobile Layout Grid UI widget with five rows and two columns. Use the information in Table 9–2 to complete the layout grid. Bold the text, Service, and, Price.

5. Delete the text, Footer, on the About Us page. Insert a jQuery Mobile Button with a Home icon. Name the button `Home` and link it to `#home`.

6. Copy the Home button. Delete the text, Footer, on the Services and Pricing page. Paste the Home button in the footer area of the Services and Pricing page.

7. Delete the text, Footer, on the Contact Us page. Paste the Home button in the footer area of the Contact Us page.

8. Delete the text, Content, on the Contact Us page. Insert a jQuery Mobile Button. Name the button, `E-mail`, and use `mail to:DBD@designsbyd.net` as the button link.

9. Insert another jQuery Mobile Button after the E-mail button. Name the button `Call`, and then use `tel:7895554810` as the button link.

10. Replace the text, Footer, on the Home page with your first and last names.

11. Save your changes and then view the mobile Web site in your browser. Compare your site to Figure 9–83. Make any necessary changes, and then save your changes.

12. Submit the document in the format specified by your instructor.

Table 9–2 jQuery Mobile Layout Grid Information

Service	Price
Web Site Design	$47.99/page
Register Domain	$9.99
Web Hosting	$4.95/month
File Storage	$1.95/month

In the Lab

Lab 3: Enhancing the Justin's Lawn Care Service Mobile Web Site

Note: To complete this assignment, you will be required to use the Data Files for Students. Visit www.cengage.com/ct/studentdownload for detailed instructions or contact your instructor for information about accessing the required files.

Problem: You have created a Web site for a lawn care company, Justin's Lawn Care Service. The site provides information about the company and its services. Now you need to design a mobile version of the Web site.

First, add the company logo to the Home page. Next, add graphics to the Landscape page and a jQuery Mobile Collapsible Block UI widget on the Services page. Also insert a jQuery Mobile List View UI widget on the Home page, name the links, and then link to the appropriate pages. Add Home buttons to the mobile pages. Finally, add buttons on the Contact Us mobile page that link to an e-mail address and a telephone number. The mobile Web site page is shown in Figure 9–84.

Perform the following tasks:

1. Copy the m folder and the jquery-mobile folder provided with your Data Files from the Chapter 09\Lawn folder into the *your last name and first initial*\Lawn folder.

2. Start Dreamweaver. Use the Sites button on the Files panel toolbar to select the Justin's Lawn Care Service site. Open index.html in the m folder.

3. Delete the text, Header, on the Home (first) page. Insert the lawn_logo graphic from the Images folder. Use `Justin's Lawn Care Service logo` for the Alternate text.

4. Delete the text, Content, on the Landscape page. Insert the landscape1 graphic from the Images folder. Use `Landscape picture 1` for the Alternate text. Click to the right of the picture and then press the ENTER key.

Continued >

STUDENT ASSIGNMENTS

(a)

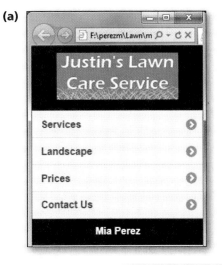

(b)

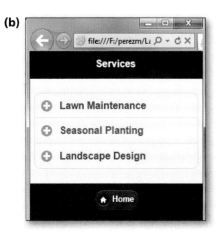

(c)

(d)

(e)

Figure 9–84

Used with permission from Microsoft Corporation

5. Insert the landscape2 graphic from the Images folder. Use `Landscape picture 2` for the Alternate text. Click to the right of the picture and then press the ENTER key. Insert the landscape3 graphic from the Images folder. Use `Landscape picture 3` for the Alternate text.

6. Delete the text, Content, on the Services page. Insert a jQuery Mobile Collapsible Block UI widget.

7. In the first collapsible block, replace the text, Header, with `Lawn Maintenance`. Replace the text, Content, with `Our standard lawn care service includes mowing the lawn and clipping around all trees, posts, and fencing. We also edge all sidewalks, patios, and driveways, and fertilize your lawn in the spring and fall.`

8. In the second collapsible block, replace the text, Header, with `Seasonal Planting`. Replace the text, Content, with `We can plant your annuals and perennials to suit your taste and property needs. We can replant your favorite types of plants up to five times a year.`

9. In the third collapsible block, replace the text, Header, with `Landscape Design`. Replace the text, Content, with `Landscaping is the first thing you notice when you look at a property. From arranging plantings to building paths and patios, we design and create outdoor living spaces for you to enjoy. We will listen to your ideas and work with you to achieve the design you want. Visit our Landscape page for ideas and then contact us to schedule an appointment.`

10. Delete the text, Content, on the Home page. Insert a jQuery Mobile List View UI widget with four unordered list items.

11. Name the first link `Services`, and then link it to `#services`.

12. Name the second link `Landscape`, and then link it to `#landscape`.

13. Name the third link `Prices`, and then link it to `#prices`.

14. Name the fourth link `Contact Us`, and then link it to `#contact`.

15. Delete the text, Footer, on the Services page. Insert a jQuery Mobile Button with a Home icon. Name the button `Home` and link it to `#home`.

16. Copy the Home button. Delete the text, Footer, on the Landscape, Prices, and Contact Us pages. Paste the Home button in the footer area of the Landscape, Prices, and Contact Us pages.

17. Delete the text, Content, on the Contact Us page. Insert a jQuery Mobile Button. Name the button `E-mail`, and then use `mailto: JustinsLCS@justin.net` as the button link.

18. Insert another jQuery Mobile Button after the E-mail button. Name the button, `Call`, and then use `tel:7775552629` as the button link.

19. Replace the text, Footer, on the Home page with your first and last names.

20. Save your changes and then view the mobile Web site in your browser. Compare your site to Figure 9–84. Make any necessary changes, and then save your changes.

21. Submit the document in the format specified by your instructor.

Cases and Places

Apply your creative thinking and problem solving skills to design and implement a solution.

Note: To complete these assignments, you will be required to use the Data Files for Students. Visit www.cengage.com/ct/studentdownload for detailed instructions or contact your instructor for information about accessing the required files.

1: Adding Content and Buttons to the Moving Venture Tips Mobile Web Site

Personal

You have created a Web site for Moving Venture Tips and now need to design a mobile Web site. Save the m and jquery-mobile folders from the Chapter 09\Move folder provided with your Data Files into the *your last name and first initial*\Move folder. You have started the basic mobile site design with four pages and are ready to add content and buttons. Add content to the Rentals and Tips pages. Add Home buttons to the Rentals, Tips, and Contact pages. Link the Home buttons to the Home page. Then add an E-mail and Phone button to the Contact page. Link the E-mail button to mailto: *your email address* and link the phone button to tel:4565558512. Add your first

Continued >

Cases and Places *continued*

and last names to the footer on the Home page. Save your changes and view the mobile site in Live view. Click each link to verify that it works. Submit the document in the format specified by your instructor.

2: Adding Links and Buttons to the Student Campus Resources Mobile Web Site

Academic

You have created the Student Campus Resources Web site and now need to create a mobile version of the site. Save the m and jquery-mobile folders from the Chapter 09\Campus folder provided with your Data Files into the *your last name and first initial*\Campus folder. You have created all of the mobile pages and now need to add a jQuery Mobile List View UI widget to the Home page. You also need to add Home buttons to the footer of the mobile pages and add an E-mail button to the Contact page. Insert a jQuery Mobile List View UI widget on the Home page and name the links Events, Activities, Committees, and Contact Us. Link each page to its respective mobile page. Add home buttons to the Events, Activities, Committees, and Contact Us pages. Link the home buttons to the Home page. Then add an E-mail button to the Contact page. Link the E-mail button to mailto:*your email address*. Add your first and last names to the footer on the Home page. Save your changes and view the mobile site in Live view. Click each link to verify that it works. Submit the document in the format specified by your instructor.

3: Adding Content, Links, a Layout Grid, and Buttons to the French Villa Roast Café Mobile Web Site

Professional

You need to create a mobile Web site for French Villa Roast Café. Save the m and jquery-mobile folders from the Chapter 09\Cafe folder provided with your Data Files into the *your last name and first initial*\Cafe folder. You have created all of the mobile pages and now need to add content to the pages. You also need to insert the jQuery Mobile List View UI widget in the Home page. You also need to add home buttons to the footer of the mobile pages and add an E-mail and Phone button to the Contact page. Add content to the About page. Insert a jQuery Layout Grid UI widget on the Menu page with five rows and four columns. Add menu contents to the layout grid and bold the column headings. Insert a jQuery Mobile List View UI widget on the Home page and name the links About, Menu, and Contact Us. Link each page to its respective mobile page. Add home buttons to the About, Menu, and Contact Us pages. Use the home icon. Link the Home buttons to the Home page. Then add an E-mail and Phone button to the Contact page. Link the E-mail button to mailto:*your email address* and link the phone button to tel:1235550103. Add your first and last names to the footer on the Home page. Save your changes and view the mobile site in Live view. Click each link to verify that it works. Submit the document in the format specified by your instructor.

Appendix A
Adobe Dreamweaver CS6 Help

Getting Help with Dreamweaver CS6

This appendix shows you how to use Dreamweaver Help. The Help system is a complete reference manual at your fingertips. You can access and use the Help system through the Help menu in Dreamweaver CS6, which connects you to up-to-date Help information online at the Adobe Web site. The Help system contains comprehensive information about all Dreamweaver features, including the following:

- A list of links to Dreamweaver help topics.
- A link to new features offered in Dreamweaver CS6.
- A link to Adobe TV, which features online tutorials.
- Search tools, which are used to locate specific topics.

The Dreamweaver Help Menu

One way to access Dreamweaver's Help features is through the Help menu and function keys. Dreamweaver's Help menu provides an easy system to access the available Help options (see Figure A–1 on the next page). Most of these commands open a Help window that displays the appropriate up-to-date Help information from the Adobe Web site. Table A–1 on the next page summarizes the commands available through the Help menu.

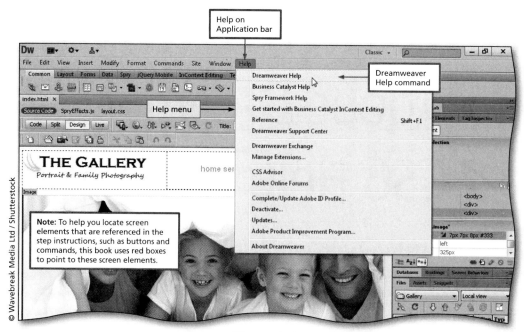

Figure A–1

Table A–1 Summary of Commands on the Help Menu

Command on Help menu	Description
Dreamweaver Help	Starts your default Web browser and displays the Dreamweaver CS6 online help system at the Adobe Web site.
Business Catalyst	Provides help for Adobe Business Catalyst, a unified hosting platform that enables Web designers to build Web sites that meet client requirements.
Spry Framework Help	Displays a complete Help document for the Spry framework for Ajax, a JavaScript library that provides the Web site developer with an option to incorporate XML data and other kinds of effects.
Get started with Business Catalyst InContext Editing	Provides information on how to make Web pages editable through any common browser so that content editors can revise Web page text while designers focus on design.
Reference	Opens the Reference panel group, which is displayed below the Document window. The Reference panel group contains the complete text from several reference manuals, including references on HTML, Cascading Style Sheets, JavaScript, and other Web-related features.
Dreamweaver Support Center	Provides access to the online Adobe Dreamweaver support center.
Dreamweaver Exchange	Links to the Adobe Exchange Web site, where you can download for free and/or purchase a variety of Dreamweaver add-on features.
Manage Extensions	Displays the Adobe Extension Manager window where you can install, enable, and disable extensions. An extension is an add-on piece of software or a plug-in that enhances Dreamweaver's capabilities. Extensions provide the Dreamweaver developer with the capability to customize how Dreamweaver looks and works.
CSS Advisor	Connects to the online Adobe CSS Advisor Web site, which provides solutions to CSS and browser compatibility issues, and encourages you to share tips, hints, and best practices for working with CSS.

Table A–1 Summary of Commands on the Help Menu (*continued*)	
Command on Help menu	**Description**
Adobe Online Forums	Accesses the Adobe Online Forums Web page. The forums provide a place for developers of all experience levels to share ideas and techniques.
Complete/Update Adobe ID Profile	Create or update your Adobe Account information.
Deactivate	Deactivates the installation of Dreamweaver CS6. If you have a single-user retail license, you can activate two computers. If you want to install Dreamweaver CS6 on a third computer, you need to deactivate it first on another computer.
Updates	Lets you check for updates to Adobe software online and then install the updates as necessary.
Adobe Product Improvement Program	Displays a dialog box that explains the Adobe Product Improvement Program and allows you to participate in the program.
About Dreamweaver	Opens a window that provides copyright information and the product license number.

Exploring the Dreamweaver CS6 Help System

The Dreamweaver Help command accesses Dreamweaver's primary Help system at the Adobe Web site and provides comprehensive information about all Dreamweaver features. You can click a topic for more information or use a search tool to look for a particular topic. The Dreamweaver CS6 Help Web site is shown in Figure A–2.

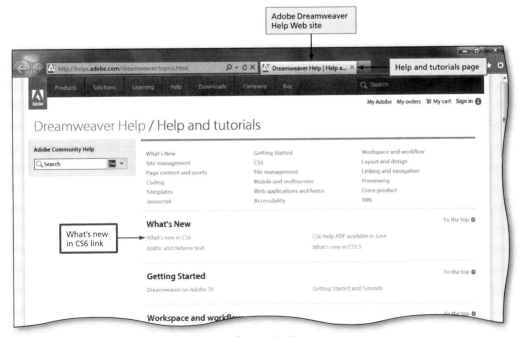

Figure A–2

Using the Help and Tutorials List

The **Help and tutorials list** is similar to a table of contents in a book and is useful for displaying Help when you know the general category of the topic in question, but not the specifics. You use the Help and tutorials list to navigate to the main topic, and then to the subtopic. When the information on the subtopic is displayed, you can read the information, click a link contained within the subtopic, or click the Previous or Next button to open the previous or next Help page in sequence. If a Comments link appears on the page, click it to view comments other users or experts have made about this topic.

To Find Help Using the Help and Tutorials List

To find help using the contents panel, you click a link to display a list of specific subtopics. You then can click a link to open a page related to that subtopic. The following steps use the Help and tutorials list to look up information about CSS3 transitions from the What's New topic on the Dreamweaver Help Web site.

- Click the link, What's new in CS6, below the What's New topic to view the list of new features in Adobe Dreamweaver CS6 (Figure A–3).

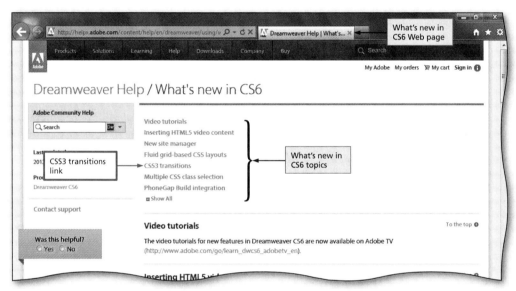

Figure A–3

- Click the topic, CSS3 transitions, to view the list of topics for CSS3 transitions (Figure A–4).

Q&A

How do I return to the main topic list?

Use your browser's Back button.

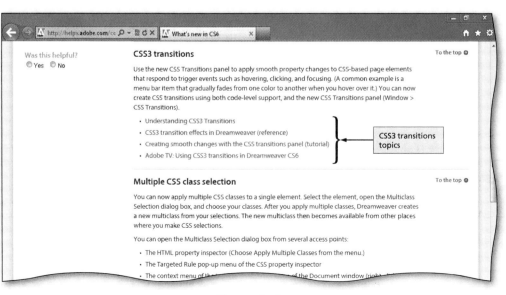

Figure A–4

Using the Search Feature

The quickest way to navigate the Dreamweaver help system is through the Adobe Community Help Search box in the left pane of the Dreamweaver Help Web window. Here you can type words, such as *CSS3 transitions* or *jquery mobile*; or you can type phrases, such as *how to insert Spry collapsible panel* or *Dreamweaver behaviors*. Adobe Community Help responds by displaying search results with a list of topics you can click.

The following are tips regarding the words or phrases you can enter to initiate a search:

1. Check the spelling of the word or phrase.

2. Keep your search specific, with fewer than seven words, to return the most accurate results.

3. If you search using a specific phrase, such as *option button*, put quotation marks around the phrase — the search returns only those topics containing all words in the phrase.

4. If a search term does not yield the desired results, try using a synonym, such as Web instead of Internet.

To Use the Adobe Community Help Search Feature

The following steps open Adobe Community Help and use the Search box to obtain useful information by entering the keywords, jquery mobile.

- If necessary, click Help on the Application bar and then click Dreamweaver Help to display the Dreamweaver Help Web site, open to the Help and tutorials page (Figure A–5).

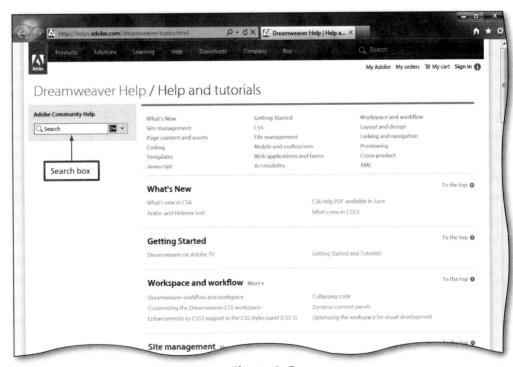

Figure A–5

2

- Click the Search box, type `jquery mobile`, and then press the ENTER key to display the search results (Figure A–6).

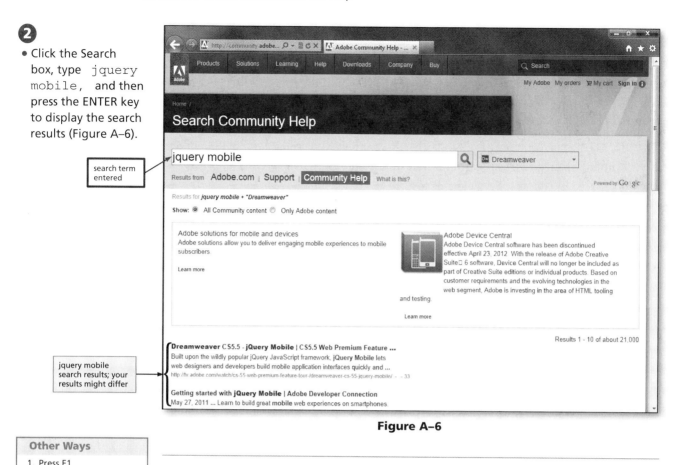

search term entered

jquery mobile search results; your results might differ

Figure A–6

Context-Sensitive Help

Using **context-sensitive help**, you can open a relevant Help topic in panels, inspectors, and most dialog boxes. To view these Help features, you click a Help button in a dialog box, choose Help on the Options pop-up menu in a panel group, or click the question mark icon in a panel or an inspector.

To Display Context-Sensitive Help on Text Using the Options Menu

Many of the panels and inspectors within Dreamweaver contain an Options Menu button in their upper-right corner. Clicking this button displays context-sensitive help. The following steps use the Options Menu button to view context-sensitive help through the Property inspector. In this example, the default Property inspector for text is displayed.

1

- Open a new document in Dreamweaver to prepare for using context-sensitive help.

- Display the Property inspector, if

Property inspector

Options Menu button

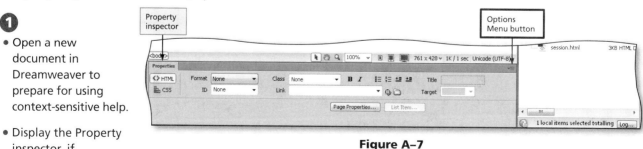

Figure A–7

necessary, to gain access to the Options Menu button (Figure A–7).

- Click the Options Menu button to display an online Help page on setting text properties in the Property inspector (Figure A–8).

Figure A–8

Using the Reference Panel

The Reference panel is another valuable Dreamweaver resource. This panel provides you with a quick reference tool for HTML tags, JavaScript objects, Cascading Style Sheets, and other Dreamweaver features.

To Use the Reference Panel

The following steps access the Reference panel, review the various options, and select and display information on the <h1> tag.

- Click Help on the Application bar, and then click Reference to open the Reference panel.

- If necessary, click the Book button, and then click O'REILLY HTML Reference to display information about HTML tags (Figure A–9).

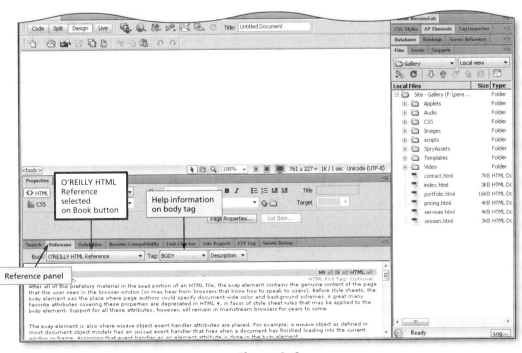

Figure A–9

2

- Click the Tag button to display the Tag menu (Figure A–10).

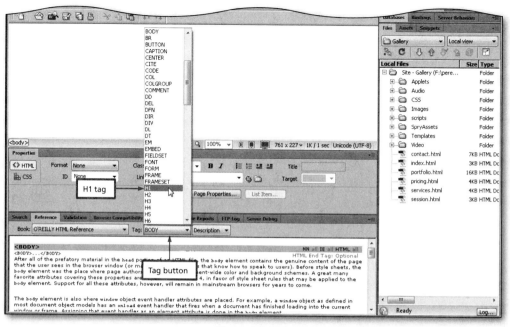

Figure A–10

3

- Click H1 to display information on the <h1> HTML tag (Figure A–11).

Help information about <h1> HTML tag

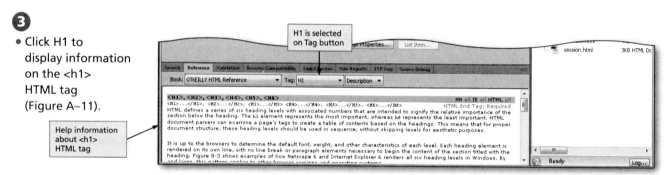

Figure A–11

4

- Click the Book button and then review the list of available reference books (Figure A–12).

5

- Click the Options Menu button on the Reference panel and then click Close Tab Group to close the panel.

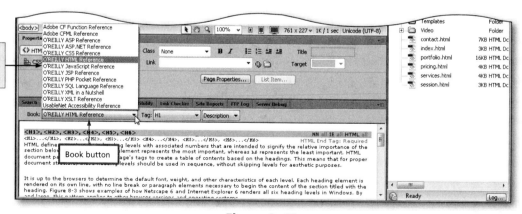

Figure A–12

Apply Your Knowledge

Reinforce the skills and apply the concepts you learned in this appendix.

Viewing the Dreamweaver Help Resources

Instructions: Start Dreamweaver. Perform the following tasks using the Dreamweaver Help command.

1. Click Help on the Application bar and then click Dreamweaver Help.
2. Click the Site management link, then click the Connect to a remote server link.
3. Read the Connect to a remote server topic, and then use a word processing program to write a short overview of what you learned.
4. Submit the document in the format specified by your instructor.

Using the Search Box

Instructions: Start Dreamweaver. Perform the following tasks using the Search box in the Dreamweaver CS6 online Help system.

1. Press the F1 key to display the Using Dreamweaver CS6 Help page.
2. Click the Search box, type **adding sound**, and then press the ENTER key.
3. Click an appropriate link in the search results to open a Help page, click a link on the Help page about embedding a sound file, and then read the Help topic.
4. Use a word processing program to write a short overview of what you learned.
5. Submit the document in the format specified by your instructor.

Using Adobe Community Help

Instructions: Start Dreamweaver. Perform the following tasks using the Search box in the Dreamweaver CS6 online Help system.

1. Press the F1 key to display the Using Dreamweaver CS6 Help page.
2. Click the Search box, type **HTML5**, and then press the ENTER key.
3. Click an appropriate link in the search results to open a Help page and read the Help topic.
4. Review the 'Designing for web publishing' article.
5. Use your word processing program to prepare a report on what you learned.
6. Submit the document in the format specified by your instructor.

Appendix B
Changing Screen Resolution

This appendix explains how to change the screen resolution in Windows 7 to the resolution used in this book.

Changing Screen Resolution

Screen resolution indicates the number of pixels (dots) that the computer uses to display the letters, numbers, graphics, and background you see on the screen. When you increase the screen resolution, Windows displays more information on the screen, but the information decreases in size. The reverse also is true: As you decrease the screen resolution, Windows displays less information on the screen, but the information increases in size.

The screen resolution usually is stated as the product of two numbers, such as 1024×768 (pronounced "ten twenty-four by seven sixty-eight"). A 1024×768 screen resolution results in a display of 1,024 distinct pixels on each of 768 lines, or about 786,432 pixels. The figures in this book were created using a screen resolution of 1024×768.

This is the screen resolution most commonly used today, although some Web designers set their computers at a much higher screen resolution, such as 2048×1536.

To Change the Screen Resolution

The following steps change the screen resolution to 1024 × 768 to match the figures in this book.

- If necessary, minimize all programs so that the Windows 7 desktop appears.

- Right-click the Windows 7 desktop to display the Windows 7 desktop shortcut menu (Figure B–1).

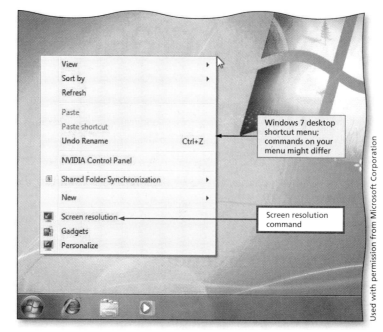

Figure B–1

- Click Screen resolution on the shortcut menu to open the Screen Resolution window (Figure B–2).

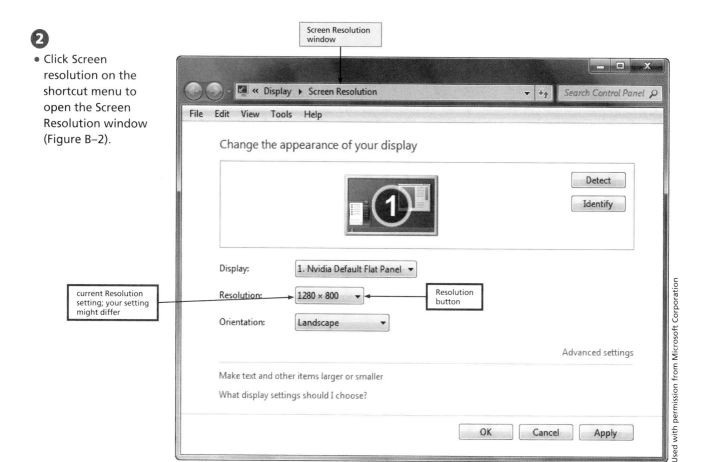

Figure B–2

3

- Click the Resolution button to display a list of resolution settings for your monitor.

- If necessary, drag the Resolution slider so that the screen resolution is set to 1024 × 768 (Figure B–3).

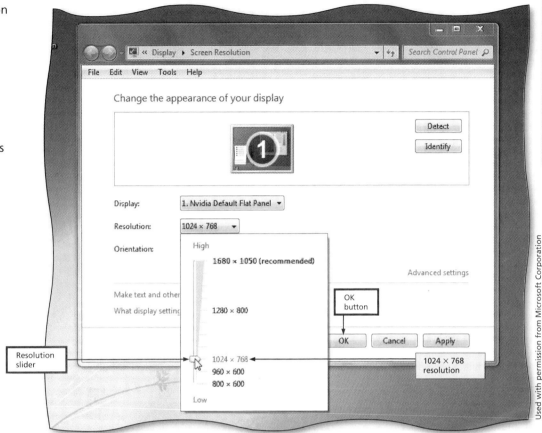

Resolution slider

OK button

1024 × 768 resolution

Used with permission from Microsoft Corporation

Figure B–3

4

- Click the OK button to set the screen resolution to 1024 × 768.

- Click the Keep changes button in the Display Settings dialog box to accept the new screen resolution (Figure B–4).

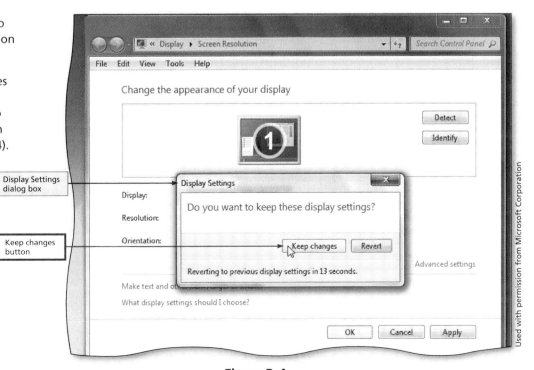

Display Settings dialog box

Keep changes button

Used with permission from Microsoft Corporation

Figure B–4

Appendix C

For Mac Users

For the Mac User of this Book

For most tasks, running Adobe Dreamweaver CS6 with the Windows 7 operating system is not different from using it with the Mac OS X Lion 10.7 operating system. For some tasks, however, you might see some differences in the appearance or location of options, or you may need to complete the tasks using different steps. This appendix demonstrates how to start Adobe Dreamweaver CS6, create an HTML file, save a file, close a file, and quit Adobe Dreamweaver CS6 using the Mac operating system.

Keyboard Differences

One difference between a Mac and a PC is in the use of modifier keys. **Modifier keys** are special keys used to modify the normal action of a key when the two are pressed in combination. Examples of modifier keys include the SHIFT, CTRL, and ALT keys on a PC, and the SHIFT, COMMAND, and OPTION keys on a Mac (Figure C–1).

(a) PC Keyboard

(b) Mac Keyboard

Figure C–1

Table C–1 compares modifier keys on a Windows PC and a Mac. For instance, if the chapter instruction is to press CTRL+S to perform a task, Mac users would press COMMAND+S. For Mac modifier keys, many menu shortcut notations use the symbols instead of the key names. In addition, Mac users with a one-button mouse can press the CTRL key and then click (or CTRL+click) to display the shortcut or context menu.

Table C–1 PC and Mac Modifier Keys		
PC	Mac	Mac Symbol
CTRL key	CMD key	⌘
ALT key	OPT key	⌥
SHIFT key	SHIFT key	⇧
Right-click	CTRL-click	

To Start Adobe Dreamweaver CS6

The following steps, which assume Mac OS X Lion 10.7 is running, start Adobe Dreamweaver CS6 based on a typical installation. You may need to ask your instructor how to start Adobe Dreamweaver CS6 for your computer.

1

- Click the Spotlight button on the Mac desktop to display the Spotlight box.

- Type Adobe Dreamweaver CS6 as the search text in the Spotlight text box and watch the search results appear (Figure C–2).

Q&A

Where does Adobe Dreamweaver CS6 appear in the search results?

If Adobe Dreamweaver CS6 is installed on the computer, it should be the Top Hit in the Spotlight search results.

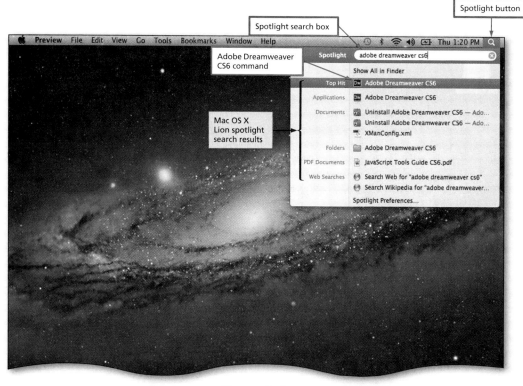

Figure C–2

2

- Click Adobe Dreamweaver CS6 to open the application.

- If the window is not maximized, click the green Zoom button on the title bar to maximize the window (Figure C–3).

Q&A

Does the PC version of Dreamweaver have a title bar?

On a PC, Dreamweaver CS6 has a title bar and a menu bar (called the Application bar). In all three applications, the clip control functions, minimize, maximize, and close, are inherited from the operating system and placed where the system user would expect to find them.

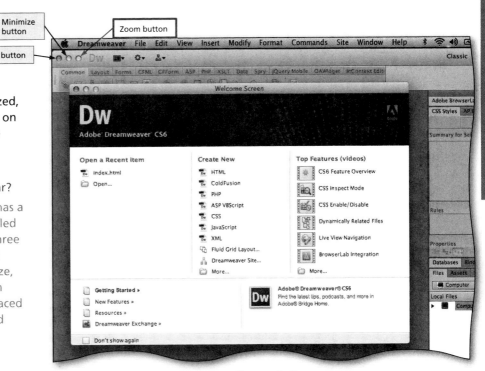

Figure C–3

Other Ways

1. Click Finder icon in Dock, navigate to Applications, double-click Adobe Dreamweaver CS6 folder, double-click

   Adobe Dreamweaver CS6 application icon

2. Click Dreamweaver icon on dock

To Create a New HTML File

The following steps, which assume Adobe Dreamweaver CS6 is open, create a new HTML file in Adobe Dreamweaver CS6.

1

- Click File on the menu bar to display the File menu (Figure C–4).

Figure C–4

2

- Click New on the File menu to display the New Document dialog box (Figure C–5).

Q&A

What does it mean when a menu command includes multiple symbols in the shortcut key notation?

Multiple symbols mean that you must hold down more than one key. For example, a notation of ⌥⌘O means to press and hold the OPTION and COMMAND keys while you press the O key. In this book, the instructions are formatted as OPT+COMMAND+O.

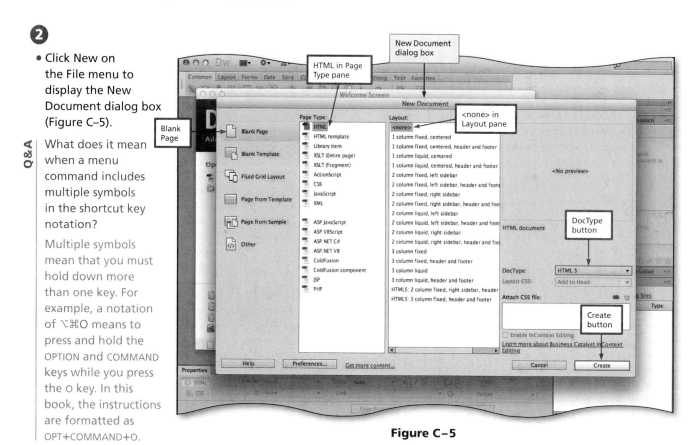

Figure C–5

3

- If necessary, click Blank Page in the left pane, click HTML in the Page Type pane, click <none> in the Layout pane, click the DocType button, and then click HTML 5 to select the settings.

- Click the Create button to create a page.

Other Ways

1. Press COMMAND+N

Save a File in Dreamweaver

The following steps, which assume a file is open, save an Adobe Dreamweaver CS6 file for the first time on the Mac operating system.

- Click File on the menu bar and then click Save As to display the Save As dialog box.

- Click the Save As text box and then type a name for the file, such as `index.html` (Figure C–6).

Q&A

Should I select a location for the file?

Dreamweaver usually saves the file in the root folder of a site. If necessary, click the Where button in the Save As dialog box to navigate to the correct location.

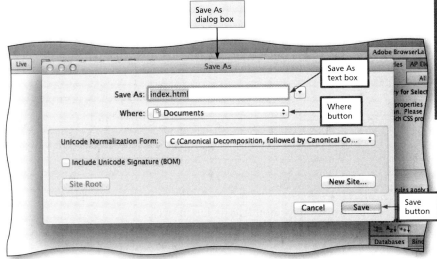

Figure C–6

- Click the Save button to save the file.

Other Ways

1. Press SHIFT+CTRL+S, type name, click Save button

Print a File in Dreamweaver

The following steps print an Adobe Dreamweaver CS6 file on the Mac operating system.

1

- Click File on the menu bar and then click Print Code to display the Print dialog box (Figure C–7).

2

- If necessary, click the Printer button and then click a printer name to select a printer.

- Click the Print button to print the document.

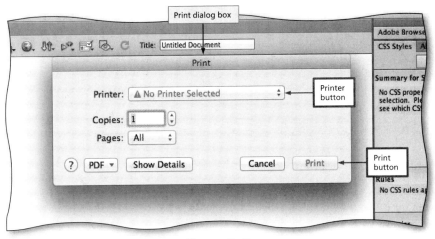

Figure C–7

Other Ways

1. Press COMMAND+P, click Printer button, click printer name, click Print button

Close a File in Dreamweaver

The following step closes an Adobe Dreamweaver CS6 file on the Mac operating system.

- Click File on the menu bar and then click Close.

Other Ways
1. Press COMMAND+W

Quit Dreamweaver

The following steps quit Adobe Dreamweaver CS6 on the Mac operating system.

- Click Dreamweaver on the menu bar to display the Dreamweaver menu (Figure C–8).

- Click Quit Dreamweaver to quit Dreamweaver.

Dreamweaver on menu bar

Dreamweaver menu

Quit Dreamweaver command

Figure C–8

Other Ways
1. Press COMMAND+Q
2. Click Dreamweaver icon on dock, click Quit

Appendix D

Project Planning Guidelines

Using Project Planning Guidelines

The process of communicating specific information to others is a learned, rational skill. Computers and software, especially Adobe Dreamweaver CS6, can help you develop ideas and present detailed information to a particular audience.

Using Adobe Dreamweaver CS6, you can create and publish Web sites. Computer hardware and Web site design software, such as Adobe Dreamweaver CS6, reduces much of the laborious work of drafting and revising projects. Some design professionals use sketch pads or storyboards, others compose directly on the computer, and others have developed unique strategies that work for their own particular thinking and design styles.

No matter what method you use to plan a project, follow specific guidelines to arrive at a final product that presents one or more Web pages clearly and effectively (Figure D–1). Use some aspects of these guidelines every time you undertake a project and others as needed in specific instances. For example, in determining content for a project, you may decide a new Web page with a different layout would communicate the idea more effectively than an existing Web page. If so, you would create this new Web page from scratch.

PROJECT PLANNING GUIDELINES

1. DETERMINE THE PROJECT'S PURPOSE
Why are you undertaking the project?

2. ANALYZE YOUR AUDIENCE
Who are the people who will use your work?

3. GATHER POSSIBLE CONTENT
What graphics exist, and in what forms?

4. DETERMINE WHAT CONTENT TO PRESENT TO YOUR AUDIENCE
What image will communicate the project's purpose to your audience in the most effective manner?

Figure D–1

Determine the Project's Purpose

Begin by clearly defining why you are undertaking this assignment. For example, you may want to create a personal Web page or modify an existing one. Or you may want to create a Web site for a business or other organization. Once you clearly understand the purpose of your task, begin to draft ideas of how best to communicate this information.

Analyze Your Audience

Learn about the people who will use, analyze, or view your work. Where are they employed? What are their educational backgrounds? What are their expectations? What questions do they have? How will they interact with your product? What kind of computer system and Internet connection will they have? Web design experts suggest drawing a mental picture of these people or finding photographs of people who fit this profile so that you can develop a project with the audience in mind.

By knowing your audience members, you can tailor a project to meet their interests and needs. You will not present them with information they already possess, and you will not omit the information they need to know.

Example: Your assignment is to raise the profile of your college's nursing program in the community. Your project should address questions such as the following: How much does the audience know about your college and the nursing curriculum? What are the admission requirements? How many of the applicants admitted complete the program? What percent of participants pass the state nursing boards?

Gather Possible Content

Rarely are you in a position to develop all the material for a project. Typically, you would begin by gathering existing information, images, and photos, or writing and designing new text and graphics based on information that may reside in spreadsheets or databases. Web design work for clients often must align with and adhere to existing marketing campaigns or other business materials. Web sites, pamphlets, magazine and newspaper articles, and books could provide insights into how others have approached your topic. Personal interviews often provide perspectives not available by any other means. Consider video and audio clips as potential sources for material that might complement or support the factual data you uncover. Make sure you have all legal rights to any photographs and other media you plan to use.

Determine What Content to Present to Your Audience

Experienced Web designers recommend identifying three or four major ideas you want an audience member to remember after viewing your Web project. It also is helpful to envision your project's endpoint, the key fact or universal theme that you want to emphasize. All project elements should lead to this ending point.

As you make content decisions, you also need to think about other factors. Presentation of the project content is an important consideration. For example, will the content of your Web pages look good when printed on paper? How will the content look when viewed on a screen for a mobile device? Determine relevant time factors, such as the length of time to develop the project, how long editors will spend reviewing your project, or the amount of time allocated for presenting your Web site designs to the customer. Your project will need to accommodate all of these constraints.

Decide whether a graphic, a photograph, or an artistic element can express or emphasize a particular concept. The right hemisphere of the brain processes images by attaching an emotion to them; so in the long run, audience members are more apt to recall themes from graphics rather than those from the text.

Finally, review your project to make sure the theme still easily is identifiable and has been emphasized successfully. Is the purpose of each page clear and presented without distraction? Does the project satisfy the requirements?

Summary

When creating a Web site project, it is beneficial to follow some basic guidelines from the outset. By taking some time at the beginning of the process to determine the project's purpose, analyze the audience, gather possible content, and determine what content to present to the audience, you can produce a project that is informative, relevant, and effective.

Index

Quick Reference Summary

Adobe Dreamweaver CS6 Quick Reference Summary

Task	Page Number	Mouse	Menu	Context Menu	Keyboard Shortcut
Absolute link, create	DW 172	Link text box on Property inspector			
Add box shadow to image	DW 409	On CSS Styles panel, click .image, click Add Property, click box arrow, click box-shadow			
Add button to form	DW 250	Button button on Forms tab of Insert bar	Insert | Form | Button		
Add Check Plugin behavior	DW 386	Click Plus button on Behaviors panel, click Check Plugin			
Add CSS transition	DW 404	On CSS Transitions panel, click plus button			
Add description to meta tag	DW 428	Click Head on Insert bar, click Description	Insert | HTML | Head Tags | Description		
Add div tag	DW 94	Insert Div Tag button on Insert bar			
Add keywords to meta tag	DW 428	Click Head on Insert bar, click Keywords	Insert | HTML | Head Tags | Keywords		
Add menu to form	DW 244	Select (List/Menu) button on Forms tab of Insert bar	Insert | Form | Select (List/Menu)		
Add Radio Group to form	DW 248	Radio Group button on Forms tab of Insert bar	Insert | Form | Radio Group		
Add Spry Collapsible Panel to page	DW 292	Spry Collapsible Panel button on Spry tab of Insert bar	Insert | Spry | Spry Collapsible Panel		
Add Spry validation to date text field	DW 284	Spry Validation Text Field button on Spry tab of Insert bar, click Type button in Property inspector, select Date	Insert | Spry | Spry Validation Text Field, Type button in Property inspector, select Date		
Add Spry validation to Radio Group	DW 287	Spry Validation Radio Group button on Spry tab of Insert bar	Insert | Spry | Spry Validation Text Field		
Add text field to form	DW 239	Text Field button on Forms tab of Insert bar	Insert | Form | Text Field		
Adobe BrowserLab, change view	DW 438	Click View, click *view*			
Adobe BrowserLab, open	DW 432	On Adobe BrowserLab panel, click Preview			
Adobe Widget Browser, display	DW 303	Extend Dreamweaver button on title bar, click Widget Browser			
Align table	DW 224	Align button in Property inspector			
Align table text	DW 214		Format | Align | alignment command	Align | alignment command	

Adobe Dreamweaver CS6 Quick Reference Summary *(continued)*

Task	Page Number	Mouse	Menu	Context Menu	Keyboard Shortcut
Alignment, horizontal, set for table cell	DW 228	Horz button in Property inspector			
Alignment, vertical, set for table cell	DW 228	Vert button in Property inspector			
Applet, insert	DW 378	Click Media button arrow on Insert bar, click Applet	Insert	Media	Applet		
Applet, set parameters	DW 381	Click Parameters button on Property inspector		Parameters	
Apply jQuery Mobile theme	DW 505		Window	jQuery Mobile Swatches, click *theme*		
Audio, add to page as plug-in	DW 338	Media button arrow on Insert bar, click Plugin	Insert menu	Media	Plugin		
Background color, set for table	DW 228	Bg button in Property inspector			
Background sound, add to page as plug-in	DW 338	Media button arrow on Insert bar, click Plugin	Insert menu	Media	Plugin		
Bold, apply to text	DW 68	Bold button on Property inspector	Format	Style	Bold	Style	CTRL+B
Box shadow, add to image	DW 409	On CSS Styles panel, click .image, click Add Property, click box arrow, click box-shadow			
Brightness and contrast, adjust for image	DW 158	Brightness and Contrast tool on Property inspector	Modify	Image	Brightness/Contrast		
Browser, select for preview	DW 64	Preview/Debug in browser on Document toolbar	File	Preview in Browser	Preview in Browser	F12
Bulleted list, create	DW 59	Unordered List button on Property inspector	Format	List	Unordered List	List	Unordered List	
Button, add to form	DW 250	Button button on Forms tab of Insert bar	Insert	Form	Button		
Center text	DW 68		Format	Align	Center	Align	Center	CTRL+ALT+SHIFT+C
Change view in Adobe BrowserLab	DW 438	Click View, click *view*			
Change view size	DW 485	Click Multiscreen, click *view size*			
Check Plugin behavior, add	DW 386	Click Plus button on Behaviors panel, click Check Plugin			
Check spelling	DW 63		Commands	Check Spelling		SHIFT+F7
Classic workspace, switch to	DW 29	Workspace switcher button, Classic	Window	Workspace Layout	Classic		
Code view, display	DW 35	Code button on Document toolbar	View	Code		
Collapse Property inspector	DW 37	Double-click Properties tab	Window	Properties	Minimize	CTRL+F3
Color, background, set for table	DW 228	Bg button in Property inspector			
Column, table, change width	DW 226	W box in Property inspector *or* Select column, drag right border			
Connect to remote site	DW 451	On Files panel toolbar, click Expand to show local and remote sites, click Connect to Remote Server			

Adobe Dreamweaver CS6 Quick Reference Summary *(continued)*

Task	Page Number	Mouse	Menu	Context Menu	Keyboard Shortcut
Contrast, adjust for image	DW 158	Brightness and Contrast tool on Property inspector	Modify | Image | Brightness/Contrast		
Create editable region	DW 146		Insert | Template Objects | Editable Region		
Create link	DW 56	Point to File button on Property inspector *or* Link text box on Property inspector	Insert | Hyperlink	Make Link	
Create page	DW 51	HTML or More folder on Welcome screen, Blank Page	File | New	New File	CTRL+N
Create page for mobile site	DW 471	Click jQuery Mobile Page button on jQuery Mobile tab of Insert bar	Insert | jQuery Mobile | Page		
Create page from template	DW 116	More folder on Welcome screen, Page from Template			
Create site	DW 46	Dreamweaver Site on Welcome screen	Site | New Site		
Create template	DW 88	More folder on Welcome screen, Blank Template	File | New		
Create unordered list	DW 59	Unordered List button on Property inspector	Format | List | Unordered List	List | Unordered List	
CSS rule, create	DW 95	New CSS Rule button on CSS Styles panel			
CSS rule, edit	DW 145	Edit Rule button on CSS Styles panel *or* CSS button on Property inspector *or* Double-click selector on CSS Styles panel	Window | CSS Styles		
CSS rule, set for new div	DW 95	Insert Div Tag button on Insert bar, enter name, click New CSS Rule button, click Rule Definition button			
CSS transition, add	DW 404	On CSS Transitions panel, click plus button			
Define remote site	DW 446	On Files panel, click Sites, click Manage Sites, click Edit the currently selected site, click Servers, click Add new server	Site | Manage Sites, Edit the currently selected site, Servers, Add new server		
Description, add to meta tag	DW 428	Click Head on Insert bar, click Description	Insert | HTML | Head Tags | Description		
Design view, display	DW 36	Design button on Document toolbar	View | Design		
Display Adobe Widget Browser	DW 303	Extend Dreamweaver button on title bar, click Widget Browser			
Display site in Live view	DW 118	Live button on Document toolbar			ALT+F11
Display Spry category on Insert bar	DW 275	Spry tab on Insert bar			
Div tag, add	DW 94	Insert Div Tag button on Insert bar			
Document, create	DW 51	HTML or More folder on Welcome screen	File | New		CTRL+N
Document, save	DW 51	Save button on Standard toolbar	File | Save *or* File | Save As	Save *or* Save As	CTRL+S CTRL+SHIFT+S
Dreamweaver, quit	DW 65	Close button	File | Exit		CTRL+Q
Dreamweaver, start	DW 29	Dreamweaver icon on desktop	Start | All Programs | Adobe Dreamweaver CS6		

Adobe Dreamweaver CS6 Quick Reference Summary *(continued)*

Task	Page Number	Mouse	Menu	Context Menu	Keyboard Shortcut
Edit CSS rule	DW 145	Edit Rule button on CSS Styles panel *or* CSS button on Property inspector *or* Double-click selector on CSS Styles panel	Window | CSS Styles		
Edit text	DW 53	Drag to select text, type text *or* Type text			
Editable region, create	DW 146		Insert | Template Objects | Editable Region		
Effect behavior, insert	DW 388	Click Plus button on Behaviors panel, point to Effects, click *effect*			
E-mail link, create	DW 174		Insert | Email Link		
E-mail text field, add Spry validation to	DW 281	Spry Validation Text Field button on Spry tab of Insert bar, click Type button in Property inspector, select E-mail	Insert | Spry | Spry Validation Text Field, Type button in Property inspector, select E-mail		
Expand Property inspector	DW 37	Properties tab	Window | Properties	Expand Panel	CTRL+F3
File, upload to remote server	DW 452	On Files panel toolbar, click Put File(s) to "Remote Server"			
Fireworks, set as primary image editor	DW 387		Edit | Preferences, File Types/Editors category		
Flash movie, insert	DW 374	Click Media button arrow on Insert bar, click SWF	Insert | Media | SWF		
Flash video, insert	DW 374	Click Media button arrow on Insert bar, click FLV	Insert | Media | FLV		
Folder, create for site files	DW 48			New Folder	CTRL+ALT+SHIFT+N
Form, insert	DW 233	Form button on Insert bar	Insert | Form | Form		
Format link as rollover text	DW 176	Page Properties button on Property inspector, Links (CSS) category			
Format, apply to paragraph	DW 56	Format button on Property inspector	Format | Paragraph Format	Paragraph Format	
Help	DW 65		Help | Dreamweaver Help		F1
Image brightness or contrast, adjust	DW 158	Brightness and Contrast tool on Property inspector	Modify | Image | Brightness/Contrast		
Image placeholder, insert	DW 182	Image button on Insert bar, Image Placeholder	Insert | Image Objects | Image Placeholder		
Image placeholder, replace with image	DW 183	Double-click placeholder *or* Browse for File button on Property inspector			
Image, align	DW 193		Format | Align	Align	
Image, insert into page	DW 157	Drag image from Assets panel or Files panel	Insert | Image		CTRL+ALT+I
Image, specify Alt text image	DW 158	Alt text box on Property inspector *or* drag image to page, enter Alt text (Image Tab Accessibility Attributes dialog box)			
Indent text	DW 68	Blockquote button on Property inspector	Format | Indent	List	CTRL+ALT+]
Insert applet	DW 378	Click Media button arrow on Insert bar, click Applet	Insert | Media | Applet		

Adobe Dreamweaver CS6 Quick Reference Summary *(continued)*

Task	Page Number	Mouse	Menu	Context Menu	Keyboard Shortcut
Insert bar, display	DW 30		Window | Insert		CTRL+F2
Insert div tag	DW 94	Insert Div Tag button on Insert bar			
Insert Effect behavior	DW 388	Click Plus button on Behaviors panel, point to Effects, click *effect*			
Insert Flash movie	DW 374	Click Media button arrow on Insert bar, click SWF	Insert | Media | SWF		
Insert Flash video	DW 374	Click Media button arrow on Insert bar, click FLV	Insert | Media | FLV		
Insert form	DW 233	Form button on Insert bar	Insert | Form | Form		
Insert image	DW 157	Drag image from Assets panel or Files panel	Insert | Image	Insert	
Insert image placeholder	DW 182	Image button arrow on Insert bar, Image Placeholder	Insert | Image Objects | Image Placeholder		
Insert jQuery Mobile Button object	DW 498	Click jQuery Mobile Button button on jQuery Mobile tab of Insert bar	Insert | jQuery Mobile | Button		
Insert jQuery Mobile Collapsible Block object	DW 493	Click jQuery Mobile Collapsible Block button on jQuery Mobile tab of Insert bar	Insert | jQuery Mobile | Collapsible Block		
Insert jQuery Mobile Layout Grid object	DW 496	Click jQuery Mobile Layout Grid button on jQuery Mobile tab of Insert bar	Insert | jQuery Mobile | Layout Grid		
Insert jQuery Mobile List View object	DW 485	Click jQuery Mobile List View button on jQuery Mobile tab of Insert bar	Insert | jQuery Mobile | List View		
Insert pop-up message	DW 394	Click Plus button on Behaviors panel, click Popup Message			
Insert Shockwave file	DW 376	Click Media button arrow on Insert bar, click Shockwave	Insert | Media | Shockwave		
Insert Smart Object	DW 386	Drag Photoshop file to insertion point			
Insert table	DW 207	Table button on Insert bar	Insert | Table		
Italic, apply to text	DW 68	Italic button on Property inspector	Format | Style	Style	CTRL+I
jQuery Mobile Button object, insert	DW 498	Click jQuery Mobile Button button on jQuery Mobile tab of Insert bar	Insert | jQuery Mobile | Button		
jQuery Mobile Collapsible Block object, insert	DW 493	Click jQuery Mobile Collapsible Block button on jQuery Mobile tab of Insert bar	Insert | jQuery Mobile | Collapsible Block		
jQuery Mobile Layout Grid object, insert	DW 496	Click jQuery Mobile Layout Grid button on jQuery Mobile tab of Insert bar	Insert | jQuery Mobile | Layout Grid		
jQuery Mobile List View object, insert	DW 481	Click jQuery Mobile List View button on jQuery Mobile tab of Insert bar	Insert | jQuery Mobile | List View		
jQuery Mobile Page object, insert	DW 471	Click jQuery Mobile Page button on jQuery Mobile tab of Insert bar	Insert | jQuery Mobile | Page		
jQuery Mobile theme, apply	DW 505		Window | jQuery Mobile Swatches, click *theme*		
Keywords, add to meta tag	DW 428	Click Head on Insert bar, click Keywords	Insert | HTML | Head Tags | Keywords		
Layout, select template	DW 50	More folder on Welcome screen, click layout	File | New		CTRL+N
Line break, insert	DW 72		Insert | HTML | Special Characters | Line Break	Insert HTML | br	SHIFT+ENTER

Adobe Dreamweaver CS6 Quick Reference Summary *(continued)*

Task	Page Number	Mouse	Menu	Context Menu	Keyboard Shortcut
Link Checker, use to verify internal links	DW 440		Site | Check Links Sitewide		
Link, create absolute	DW 172	Link text box on Property inspector			
Link, create e-mail	DW 174		Insert | Email Link		
Link, create relative	DW 170	Point to File button on Property inspector *or* Link text box on Property inspector *or* Browse for File button next to Link text box on Property inspector	Insert | Hyperlink	Make Link	
Link, format as rollover text	DW 176	Page Properties button on Property inspector, Links (CSS) category			
Links, verify with Link Checker	DW 440		Site | Check Links Sitewide		
List, create unordered	DW 59	Unordered List button on Property inspector	Format | List | Unordered List	List | Unordered List	
Live view, display	DW 118	Live button on Document toolbar			ALT+F11
Menu, add to form	DW 244	Select (List/Menu) button on Forms tab of Insert bar	Insert | Form | Select (List/Menu)		
Merge cells in table	DW 220	'Merges selected cells using spans' button in Property inspector	Modify | Table | Merge Cells	Merge Cells	
Meta tag, add description to	DW 428	Click Head on Insert bar, click Description	Insert | HTML | Head Tags | Description		
Meta tag, add keywords to	DW 428	Click Head on Insert bar, click Keywords	Insert | HTML | Head Tags | Keywords		
Open Adobe BrowserLab	DW 432	On Adobe BrowserLab panel, click Preview			
Open site	DW 159	Site button on Files panel, *site*			
Open template	DW 143	Open on Welcome screen	File | Open		CTRL+O
Page title, enter	DW 54	Title text box on Document toolbar			
Page, create	DW 51	More folder on Welcome screen	File | New	New File	CTRL+N
Page, create for mobile site	DW 471	Click jQuery Mobile Page button on jQuery Mobile tab of Insert bar	Insert | jQuery Mobile | Page		
Page, create from template	DW 116	More folder on Welcome screen, Page from Template			
Page, preview in browser	DW 64	Preview/Debug in browser on Document toolbar	File | Preview in Browser	Preview in Browser	F12
Page, save	DW 51	Save button on Standard toolbar	File | Save *or* File | Save As	Save *or* Save As	CTRL+S CTRL+SHIFT+S
Page, save as template	DW 97		File | Save As Template		
Panel, collapse or expand	DW 37	Click Collapse to Icons button or Expand Panels button		Minimize or Expand Panel	
Panel, move	DW 37	Drag panel by its tab			
Panel, open or close	DW 37	Click panel options button, click Close or click Close Tab Group	Window | Hide Panels	Close Tab Group	F4
Paragraph, center	DW 68		Format | Align | Center	Align | Center	CTRL+ALT+SHIFT+C
Parameters, set for applet	DW 381	Click Parameters button on Property inspector		Parameters	
Parameters, set for sound file	DW 341	Parameters button in Property inspector			

Adobe Dreamweaver CS6 Quick Reference Summary *(continued)*

Task	Page Number	Mouse	Menu	Context Menu	Keyboard Shortcut
Parameters, set for video file	DW 354	Parameters button in Property inspector			
Phone number text field, add Spry validation to	DW 283	Spry Validation Text Field button on Spry tab of Insert bar, click Type button in Property inspector, select Phone Number	Insert | Spry | Spry Validation Text Field, Type button in Property inspector, select Phone Number		
Placeholder, insert image	DW 182	Image button arrow on Insert bar, Image Placeholder	Insert | Image Objects | Image Placeholder		
Plug-in sound file, add to page	DW 338	Media button arrow on Insert bar, click Plugin	Insert menu | Media | Plugin		
Pop-up message, insert	DW 394	Click Plus button on Behaviors panel, click Popup Message			
Preferences, set	DW 41		Edit | Preferences		
Preview Web page	DW 64	Preview/Debug in browser on Document toolbar	File | Preview in Browser	Preview in Browser	F12
Primary image editor, set to Fireworks	DW 387		Edit | Preferences, File Types/Editors category		
Property inspector, collapse or expand	DW 37	Double-click Properties tab	Window | Properties	Minimize/ Expand Panel	CTRL+F3
Quit Dreamweaver	DW 65	Close button	File | Exit		CTRL+Q
Radio Group, add to form	DW 248	Radio Group button on Forms tab of Insert bar	Insert | Form | Radio Group		
Region, create editable	DW 146		Insert | Template Objects | Editable Region		
Relative link, create	DW 170	Point to File button on Property inspector *or* Link text box on Property inspector *or* Browse for File button next to Link text box on Property inspector	Insert | Hyperlink	Make Link	SHIFT+drag to file
Remote server, upload file to	DW 452	On Files panel toolbar, click Put File(s) to "Remote Server"			
Remote site, connect to	DW 451	On Files panel toolbar, click Expand to show local and remote sites, click Connect to Remote Server			
Remote site, define	DW 446	On Files panel, click Sites, click Manage Sites, click Edit the currently selected site, click Servers, click Add new server	Site | Manage Sites, Edit the currently selected site, Servers, Add new server		
Reset button, add to form	DW 250	Button button on Forms tab of Insert bar	Insert | Form | Button		
Rounded borders, add to Spry Collapsible Panels	DW 398	In CSS Styles panel, click .CollapsiblePanel, click Add Property, click box arrow, click border-radius, click corner radius button			
Rule definition, create	DW 95, DW 170	New CSS Rule button on CSS Styles panel			
Rule definition, edit	DW 145	Edit Rule button on CSS Styles panel *or* CSS button on Property inspector *or* Double-click selector on CSS Styles panel	Window | CSS Styles		

Adobe Dreamweaver CS6 Quick Reference Summary *(continued)*

Task	Page Number	Mouse	Menu	Context Menu	Keyboard Shortcut
Save document	DW 51	Save button on Standard toolbar	File | Save *or* File | Save As	Save *or* Save As	CTRL+S CTRL+SHIFT+S
Save page as template	DW 97		File | Save As Template		
Select List/Menu, add to form	DW 244	Select (List/Menu) button on Forms tab of Insert bar	Insert | Form | Select (List/Menu)		
Select template layout	DW 50	More folder on Welcome screen, click layout	File | New		
Shadow, add to image	DW 409	On CSS Styles panel, click .image, click Add Property, click box arrow, click box-shadow			
Shockwave file, insert	DW 376	Click Media button arrow on Insert bar, click Shockwave	Insert | Media | Shockwave		
Site, create site	DW 46	Dreamweaver Site on Welcome screen	Site | New Site		
Site, open	DW 159	Site button on Files panel, *site*			
Smart Object, insert	DW 386	Drag Photoshop file to insertion point			
Sound file, set parameters for	DW 341	Parameters button in Property inspector			
Sound, add to page as plug-in	DW 338	Media button arrow on Insert bar, click Plugin	Insert menu | Media | Plugin		
Spelling, check	DW 63		Commands | Check Spelling		SHIFT+F7
Spry category, display on Insert bar	DW 275	Spry tab on Insert bar			
Spry Collapsible Panel, add to page	DW 292	Spry Collapsible Panel button on Spry tab of Insert bar	Insert | Spry | Spry Collapsible Panel		
Spry Collapsible Panels, add rounded borders to	DW 398	On CSS Styles panel, click .CollapsiblePanel, click Add Property, click box arrow, click border-radius, click corner radius button			
Spry validation, add to date text field	DW 284	Spry Validation Text Field button on Spry tab of Insert bar, click Type button in Property inspector, select Date	Insert | Spry | Spry Validation Text Field, Type button in Property inspector, select Date		
Spry validation, add to e-mail text field	DW 281	Spry Validation Text Field button on Spry tab of Insert bar, click Type button in Property inspector, select E-mail	Insert | Spry | Spry Validation Text Field, Type button in Property inspector, select E-mail		
Spry validation, add to phone number text field	DW 283	Spry Validation Text Field button on Spry tab of Insert bar, click Type button in Property inspector, select Phone Number	Insert | Spry | Spry Validation Text Field, Type button in Property inspector, select Phone Number		
Spry validation, add to Radio Group	DW 287	Spry Validation Radio Group button on Spry tab of Insert bar	Insert | Spry | Spry Validation Text Field		
Spry validation, add to text field	DW 276	Spry Validation Text Field button on Spry tab of Insert bar	Insert | Spry | Spry Validation Text Field		
Standard toolbar, display	DW 40		View | Toolbars | Standard	Standard	
Style rule, set for new div	DW 95	Insert Div Tag button on Insert bar, enter name, click New CSS Rule button, click Rule Definition button			

Dreamweaver Quick Reference

Adobe Dreamweaver CS6 Quick Reference Summary *(continued)*

Task	Page Number	Mouse	Menu	Context Menu	Keyboard Shortcut
Submit button, add to form	DW 250	Button button on Forms tab of Insert bar	Insert | Form | Button		
Table cell, set horizontal alignment	DW 228	Horz button in Property inspector			
Table cell, set vertical alignment	DW 229	Vert button in Property inspector			
Table column, change width	DW 226	W box in Property inspector *or* Select column, drag right border			
Table text, align	DW 214		Format | Align | alignment command	Align | alignment command	
Table, align	DW 224	Align button in Property inspector			
Table, change width	DW 224	W box in Property inspector			
Table, insert	DW 207	Table button on Insert bar	Insert | Table		
Table, merge cells in	DW 220	'Merges selected cells using spans' button in Property inspector	Modify | Table | Merge Cells	Merge Cells	
Template layout, select	DW 50	More folder on Welcome screen, click layout	File | New		
Template, create	DW 88	More folder on Welcome screen, Blank Template	File | New		
Template, create from page	DW 116	More folder on Welcome screen, Page from Template			
Template, open	DW 143	Open on Welcome screen			
Text field, add Spry validation to	DW 276	Spry Validation Text Field button on Spry tab of Insert bar	Insert | Spry | Spry Validation Text Field		
Text field, add to form	DW 239	Text Field button on Forms tab of Insert bar	Insert | Form | Text Field		
Text, center	DW 68		Format | Align | Center	Align | Center	CTRL+ALT+SHIFT+C
Text, edit	DW 53	Drag to select text, type text *or* Type text			
Title, enter for page	DW 54	Title text box on Document toolbar			
Unordered list, create	DW 59	Unordered List button on Property inspector	Format | List | Unordered List	List | Unordered List	
Upload file to remote server	DW 452	On Files panel toolbar, click Put File(s) to "Remote Server"			
Video, add to page as plug-in	DW 352	Media button arrow on Insert bar, click Plugin	Insert menu | Media | Plugin		
View size, change	DW 485	Click Multiscreen, click *view size*			
Web page, preview in browser	DW 64	Preview/Debug in browser on Document toolbar	File | Preview in Browser	Preview in Browser	F12 CTRL+F12
Widget, add to page	DW 311		Insert | Widget		
Width, change for table	DW 224	W box in Property inspector			
Width, change for table column	DW 226	W box in Property inspector *or* Select column, drag right border			
Window, maximize	DW 30	Maximize button			
Workspace, reset to default settings	DW 39	Workspace Switcher button, Reset *workspace*	Window | Workspace Layout | Reset *workspace*		
Workspace, switch	DW 31	Workspace Switcher button	Window | Workspace Layout | *workspace*		